THE
Electronic Packaging
HANDBOOK

ELECTRONICS HANDBOOK SERIES

Series Editor:
Jerry C. Whitaker
Technical Press
Morgan Hill, California

PUBLISHED TITLES

AC POWER SYSTEMS HANDBOOK, SECOND EDITION
Jerry C. Whitaker

THE ELECTRONIC PACKAGING HANDBOOK
Glenn R. Blackwell

POWER VACUUM TUBES HANDBOOK, SECOND EDITION
Jerry C. Whitaker

INTERCONNECTING ELECTRONIC SYSTEMS
Jerry C. Whitaker and Gene DeSantis

FORTHCOMING TITLES

ELECTRONIC SYSTEMS MAINTENANCE HANDBOOK
Jerry C. Whitaker

FORMULAS FOR THERMAL DESIGN OF ELECTRONIC EQUIPMENT
Ralph Remsberg

THE RESOURCE HANDBOOK OF ELECTRONICS
Jerry C. Whitaker

THE Electronic Packaging HANDBOOK

Edited by
GLENN R. BLACKWELL

 **CRC PRESS**

 **IEEE PRESS**

A CRC Press Handbook Published in Cooperation with IEEE Press

TK
7870
.15
E55
2000

Library of Congress Cataloging-in-Publication Data

The electronic packaging handbook / edited by Glenn R. Blackwell.
    p.   cm. — (The electrical engineers handbook series)
    Includes bibliographical references.
    ISBN 0-8493-8591-1 (alk. paper)
    1. Electronic packaging Handbooks, manuals, etc. I. Blackwell,
Glenn R. II. Series.
TK7870.15.E44 1999
621.381′046—dc21
                                                                                                                                         99-41244
                                                                                                                                            CIP

This book contains information obtained from authentic and highly regarded sources. Reprinted material is quoted with permission, and sources are indicated. A wide variety of references are listed. Reasonable efforts have been made to publish reliable data and information, but the author and the publisher cannot assume responsibility for the validity of all materials or for the consequences of their use.

Neither this book nor any part may be reproduced or transmitted in any form or by any means, electronic or mechanical, including photocopying, microfilming, and recording, or by any information storage or retrieval system, without prior permission in writing from the publisher.

All rights reserved. Authorization to photocopy items for internal or personal use, or the personal or internal use of specific clients, may be granted by CRC Press LLC, provided that $.50 per page photocopied is paid directly to Copyright Clearance Center, 222 Rosewood Drive, Danvers, MA 01923 USA. The fee code for users of the Transactional Reporting Service is ISBN 0-8493-8591-1/00/$0.00+$.50. The fee is subject to change without notice. For organizations that have been granted a photocopy license by the CCC, a separate system of payment has been arranged.

The consent of CRC Press LLC does not extend to copying for general distribution, for promotion, for creating new works, or for resale. Specific permission must be obtained in writing from CRC Press LLC for such copying.

Direct all inquiries to CRC Press LLC, 2000 N.W. Corporate Blvd., Boca Raton, Florida 33431.

Trademark Notice: Product or corporate names may be trademarks or registered trademarks, and are used only for identification and explanation, without intent to infringe.

© 2000 by CRC Press LLC

No claim to original U.S. Government works
International Standard Book Number 0-8493-8591-1
Library of Congress Card Number 99-41244
Printed in the United States of America 1 2 3 4 5 6 7 8 9 0
Printed on acid-free paper

Preface

The *Electronic Packaging Handbook* is intended for engineers and technicians involved in the design, manufacturing, and testing of electronic assemblies. The handbook covers a range of applied technologies and concepts that are necessary to allow the user to follow reasonable steps to reach a defined goal. The user is encouraged to follow the steps of concurrent engineering, which considers aspects of design, manufacturing, and testing during the design phase of a project and/or product.

Each chapter begins with an introduction, which includes a *Where Else?* section. Because the topics considered in this handbook are interactive, this section guides the reader of a particular chapter to other sections of the handbook where similar issues are discussed.

The *Electronic Packaging Handbook* is the latest in a series of major electrical/electronics engineering handbooks from CRC Press, including several that are published jointly with the IEEE Press:

- *The Electronics Handbook,* Jerry C. Whitaker
- *The Electrical Engineering Handbook,* 2nd ed., Richard C. Dorf
- *The Circuits and Filters Handbook,* Wai-Kai Chen
- *The Control Handbook,* William S. Levine
- *The Mobile Communications Handbook,* 2nd ed., Jerry D. Gibson
- *The Transforms and Applications Handbook,* 2nd ed., Alexander D. Poularikas

This handbook covers a subset of the topics that exist in Whitaker's *The Electronics Handbook,* and as such covers the included topics in more detail than that handbook, while restricting coverage to only topics directly related to electronics packaging. Electronics packaging continues to include expanding and evolving topics and technologies, as the demands for smaller, faster, and lighter products continue without signs of abatement. These demands mean that individuals in each of the specialty areas involved in electronics packaging, such as electronic, mechanical, and thermal designers, and manufacturing and test engineers, are all interdependent on each other's knowledge. This handbook will assist each group in understanding other areas.

Organization

The two introductory chapters of this handbook are intended to provide an overview to the topics of project management and quality, and to surface mount technology generally. Following chapters then present more detailed information about topics needed to successfully design, manufacture, and test the packaging for an electronic product:

1. Fundamentals of the Design Process
2. Surface Mount Technology
3. Integrated Circuit Packages

4. Direct Chip Attach
5. Circuit Boards
6. EMC and Printed Circuit Board Design
7. Hybrid Assemblies
8. Interconnects
9. Design for Test
10. Adhesive and Its Application
11. Thermal Management
12. Testing
13. Inspection
14. Package/Enclosure
15. Electronics Package Reliability and Failure Analysis
16. Product Safety and Third-Party Certification

The last two chapters cover reliability and failure analysis issues, which are necessary to understand both failure mechanisms, and also analysis of failed products and the safety issues that must be considered for any product that is intended to be sold to corporate or public consumers.

The index is complete and was developed by the chapter authors. This index will be of great value to the reader in identifying areas of the book that cover the topics of interest.

This handbook represents a multi-year effort by the authors. It is hoped that the reader will both benefit and learn from their work.

Glenn R. Blackwell

Contributors

Bruce C. Beihoff
Rockwell Automation
Allen Bradley
Milwaukee, WI

Glenn R. Blackwell
Purdue University
W. Lafayette, IN

Constantin Bolintineanu
Digital Security Controls, Ltd.
Toronto, Ontario, Canada

Garry Grzelak
Teradyne Telecommunications
Deerfield, IL

Steli Loznen
The Standards Institution of Israel
Tel Aviv, Israel

Janet K. Lumpp
University of Kentucky
Lexington, KY

Victor Meeldijk
Diagnostic/Retrieval Systems Inc.
Oakland, NJ

Mark I. Montrose
Montrose Compliance Services, Inc.
Santa Clara, CA

Ray Prasad
Ray Prasad Consultancy, Inc.
Portland, OR

Michael C. Shaw
Design and Reliability
  Department
Rockwell Science Center
Thousand Oaks, CA

Peter M. Stipan
Rockwell Automation
Allen-Bradley
Milwaukee, WI

Contents

1. Fundamentals of the Design Process
   Glenn R. Blackwell 1.1

2. Surface Mount Technology
   Glenn R. Blackwell 2.1

3. Integrated Circuit Packages
   Victor Meeldijk 3.1

4. Direct Chip Attach
   Glenn R. Blackwell 4.1

5. Circuit Boards
   Glenn R. Blackwell 5.1

6. EMC and Printed Circuit Board Design
   Mark I. Montrose 6.1

7. Hybrid Assemblies
   Janet K. Lumpp 7.1

8. Interconnects
   Glenn R. Blackwell 8.1

9. Design for Test
   Glenn R. Blackwell 9.1

10. Adhesive and Its Application
    Ray Prasad 10.1

11. Thermal Management
    Glenn R. Blackwell 11.1

12. Testing
    Garry Grzelak and *Glenn R. Blackwell* 12.1

13 Inspection
Glenn R. Blackwell 13.1

14 Package/Enclosure
Glenn R. Blackwell 14.1

15 Electronics Package Reliability and Failure Analysis: A Micromechanics-Based Approach
Peter M. Stipan, Bruce C. Beihoff, and *Michael C. Shaw* 15.1

16 Product Safety and Third-Party Certification
Constantin Bolintineanu and *Steli Loznen* 16.1

Appendix A: Definitions A.1

Index I.1

1
Fundamentals of the Design Process

Glenn R. Blackwell
Purdue University

1.1 Handbook Introduction 1.1
1.2 Concurrent Engineering 1.1
1.3 Systems Engineering 1.7
1.4 Quality Concepts 1.22
1.5 Engineering Documentation 1.32
1.6 Design for Manufacturability 1.35
1.7 ISO9000 .... 1.37
1.8 Bids and Specifications 1.45
1.9 Reference and Standards Organizations 1.45

1.1 Handbook Introduction

This handbook is written for the practicing engineer who needs current information on electronic packaging at the circuit level. The intended audience includes engineers and technicians involved in any or all aspects of design, production, testing, and packaging of electronic products regardless of whether those products are commercial or industrial in nature. This means that circuit designers participating in concurrent engineering teams, circuit board designers and fabricators, test engineers and technicians, and others will find this handbook of value.

1.2 Concurrent Engineering*

In its simplest definition, *concurrent engineering* requires that a design team consider all appropriate issues of design, manufacturability, and testability during the design phase of a project/product. Other definitions will include repairability and marketability. Each user must define the included elements to best fit the specific needs.

1.2.1 Introduction

Concurrent engineering (CE) is a present-day method used to shorten the time to market for new or improved products. Let it be assumed that a product will, upon reaching the marketplace, be competitive in nearly every respect, such as quality and cost, for example. But the marketplace has shown that products, even though competitive, must not be late to market, because market share, and therefore

---

*Adapted from Whitaker, J., *The Electronics Engineering Handbook*, Chapter 146, "Concurrent Engineering," by Francis Long, CRC/IEEE Press, 1997.

profitability, will be adversely affected. Concurrent engineering is the technique that is most likely to result in acceptable profits for a given product.

A number of forwarding-looking companies began, in the late 1970s and early 1980s, to use what were then innovative techniques to improve their competitive position. But it was not until 1986 that a formal definition of concurrent engineering was published by the Defense Department:

> A systematic approach to the integrated, concurrent design of products and their related processes, including manufacture, and support. This approach is intended to cause the developers, from the outset, to consider all elements of the product life cycle from concept through disposal including quality, cost, schedule, and user requirements.

This definition was printed in the Institute for Defense Analyses Report R-338, 1986. The key words are seen to be *integrated, concurrent design* and *all elements of the product life cycle.* Implicit in this definition is the concept that, in addition to input from the originators of the concept, input should come from users of the product, those who install and maintain the product, those who manufacture and test the product, and the designers of the product. Such input, as appropriate, should be in every phase of the product life cycle, even the very earliest design work.

This approach is implemented by bringing specialists from manufacturing, test, procurement, field service, etc., into the earliest design considerations. It is very different from the process so long used by industry. The earlier process, now known as the *over the wall* process, was a serial or sequential process. The product concept, formulated at a high level of company management, was then turned over to a design group. The design group completed its design effort, tossed it over the wall to manufacturing, and proceeded to an entirely new and different product design. Manufacturing tossed its product to test, and so on through the chain. The unfortunate result of this sequential process was the necessity for redesign, which happened regularly.

Traditional designers too frequently have limited knowledge of a manufacturing process, especially its capabilities and limitations. This may lead to a design that cannot be made economically or in the time scheduled, or perhaps cannot be made at all. The same can be said of the specialists in the processes of test, marketing, and field service, as well as parts and material procurement. A problem in any of these areas might well require that the design be returned to the design group for redesign. The same result might come from a product that cannot be repaired. The outcome is a redesign effort required to correct the deficiencies found during later processes in the product cycle. Such redesign effort is costly in both economic and time to market terms. Another way to view these redesign efforts is that they are not *value added*. Value added is a particularly useful parameter by which to evaluate a process or practice.

This presence of redesign in the serial process can be illustrated as in Fig. 1.1, showing that even feedback from field service might be needed in a redesign effort. When the process is illustrated in this manner, the presence of redesign can be seen to be less efficient than a process in which little or no redesign is required. A common projection of the added cost of redesign is that changes made in a following process step are about 10 times more costly than correctly designing the product in the first place. If the product should be in the hands of a customer when a failure occurs, the results can be disastrous, both in direct costs to accomplish the repair and in possible lost sales due to a tarnished reputation.

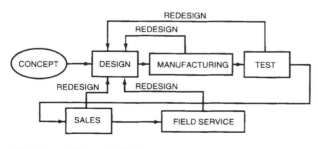

FIGURE 1.1 The serial design process.

There are two major facets of concurrent engineering that must be kept in mind at all times. The first is that a concurrent engineering process requires team effort. This is more than the customary committee. Although the team is composed of specialists from the various activities, the team members are not there

Fundamentals of the Design Process

as representatives of their organizational home. They are there to cooperate in the delivery of product to the market place by contributing their expertise in the task of eliminating the redesign loops shown in Fig. 1.1. Formation of the proper team is critical to the success of most CE endeavors. The second facet is to keep in mind is that concurrent engineering is information and communication intensive. There must be no barriers of any kind to complete and rapid communication among all parts of a process, even if they are located at geographically dispersed sites. If top management has access to and uses information relevant to the product or process, this same information must be available to all in the production chain, including the line workers.

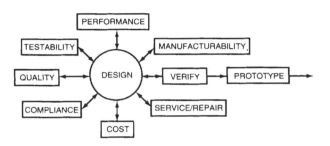

FIGURE 1.2 Concurrence of design is communication intensive.

An informed and knowledgeable workforce at all levels is essential so that they may use their efforts to the greatest advantage. The most effective method to accomplish this is to form, as early as possible in the product life cycle, a team composed of knowledgeable people from all aspects of the product life cycle. This team should be able to *anticipate and design out* most if not all possible problems before they actually occur. Figure 1.2 suggests many of the communication pathways that must be freely available to the members of the team. Others will surface as the project progresses. The inputs to the design process are sometimes called the "design for...," inserting the requirement.

The top management that assigns the team members must also be the *coaches* for the team, making certain that the team members are properly trained and then allowing the team to proceed with the project. There is no place here for the traditional *bossism* of the past. The team members must be selected to have the best combination of recognized expertise in their respective fields and the best team skills. It is not always the one most expert in a specialty who will be the best team member. Team members, however, must have the respect of all other team members, not only for their expertise but also for their interpersonal skills. Only then will the best product result. This concurrence of design, to include these and all other parts of the cycle, can measurably reduce time to market and overall investment in the product.

Preventing downstream problems has another benefit in that employee morale is very likely to he enhanced. People almost universally want to take pride in what they do and produce. Few people want to support or work hard in a process or system that they believe results in a poor product. Producing a defective or shoddy product does not give them this pride. Quite often it destroys pride in workmanship and creates a disdain for the management that asks them to employ such processes or systems. The use of concurrent engineering is a very effective technique for producing a quality product in a competitive manner. Employee morale is nearly always improved as a result.

Perhaps the most important aspect of using concurrent engineering is that the design phase of a product cycle will nearly always take more time and effort than the original design would have expended in the serial process. However, most organizations that have used concurrent engineering report that the overall time to market is measurably reduced because product redesign is greatly reduced or eliminated entirely. The time-worn phrase *time is money* takes on added meaning in this context.

Concurrent engineering can be thought of as an evolution rather than a revolution of the product cycle. As such, the principles of *total quality management (TQM)* and *continuous quality improvement (CQI)*, involving the ideas of robust design and reduction of variation, are not to be ignored. They continue to be important in process and product improvement. Nor is concurrent engineering, of itself, a type of re-engineering The concept of re-engineering in today's business generally implies a massive restructuring of the organization of a process, or even a company, probably because the rate of improvement of a process using traditional TQM and CQI is deemed to be too slow or too inefficient or both to remain competitive. Still, the implementation of concurrent engineering does require and demand a certain and often substantial change in the way a company does business.

Concurrent engineering is as much a cultural change as it is a process change. For this reason it is usually achieved with some trauma. The extent of the trauma is dependent on the willingness of people to accept change, which in turn is dependent on the commitment and sales skills of those responsible for installing the concurrent engineering culture. Although it is not usually necessary to re-engineer, that is, to restructure, an entire organization to install concurrent engineering, it is also true that it cannot be installed like an overlay on top of most existing structures. Although some structural changes may be necessary, the most important change is in attitude, in culture. Yet it must also be emphasized that there is no *one size fits all* pattern. Each organization must study itself to determine how best to install concurrent engineering. However, there are some considerations that are helpful in this study. A discussion of many fine ideas can be found in Solomon [1995]. The importance of commitment to a concurrent engineering culture from top management to line workers cannot be emphasized too strongly.

1.2.2 The Process View of Production

If production is viewed as a process, the product cycle becomes a seamless movement through the design, manufacture, test, sales, installation, and field maintenance activities. There is no competition within the organization for resources. The team has been charged with the entire product cycle such that allocation of resources is seen from a holistic view rather than from a departmental or specialty view. The needs of each activity are evident to all team members. The product and process are seen as more than the sum of the parts.

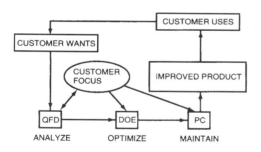

FIGURE 1.3 Functional view of the process cycle.

The usual divisions of the process cycle can he viewed in a different way. Rather than discuss the obvious activities of manufacturability and testability and the others shown in Fig. 1.2, the process can be viewed in terms of functional techniques that are used to accomplish the process cycle. Such a view might be as shown in Fig. 1.3.

In this view, it is the functions of *quality function deployment (QFD), design of experiments (DOE),* and *process control (PC)* that are emphasized rather than the design, manufacturing, test, etc. activities. It is the manner in which the processes of design, manufacturing, and other elements are accomplished that is described. In this description, QFD is equated to analysis in the sense that the customers' needs and desires must be the driver in the design of today's products. Through the use of QFD, not only is the customer input, often referred to as the *voice of the customer*, heard, it is translated into a process to produce the product. Thus, both initial product and process design are included in QFD in this view. It is important to note that the product and the process to produce the product are designed together, not just at the same time.

DOE can be used in one of two ways. First is the optimization of an existing process by removing any causes of defects and determining the best target value of the parameters. The purpose of this is to maximize the yield of a process, which frequently involves continuous quality improvement techniques. The second is the determination of a process for a new product by optimization of a proposed process before it is implemented. Today, simulation of processes is becoming increasingly important as the processes become increasingly complex. DOE, combined with simulation, is the problem solving technique, both for running processes and proposed processes.

PC is a monitoring process to ensure that the optimized process remains an optimized process. Its primary purpose is to issue an alarm signal when a process is moving away from its optimized state. Often, this makes use of statistical methods and is then *called statistical process control (SPC).* PC is not a problem solving technique, although some have tried to use it for that purpose. When PC signals a problem, then problem solving techniques, possibly involving DOE, must be implemented. The following sections will expand on each of these functional aspects of a product cycle.

Quality Function Deployment

QFD begins with a determination of the customers needs and desires. There are many ways that raw data can be gathered. Two of these are questionnaires and focus groups. Obtaining the data is a well developed field. The details of these techniques will not be discussed here as much has been written on the subject. It is important, however, that professionals are involved in the design of such data acquisition because of the many nuances that are inherent in such methods.

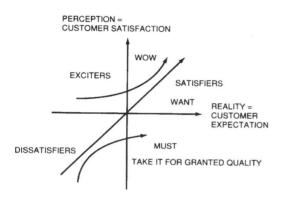

FIGURE 1.4 The Kano model.

The data obtained must be translated into language that is understood by the company and its people. It is this translation that must extract the customers' needs and wants and put them in words that the designers, manufacturers, etc., can use in their tasks. Yet, the intent of the customers' words must not be lost. This is not always an easy task, but is a vitally important one. Another facet of this is the determination of unstated but pleasing qualities of a product that might provide a marketing edge. This idea has been sketched by the Kano model, shown in Fig. 1.4.

The Kano model, developed by Noriaki Kano, a Japanese professor, describes these pleasers as the *wows*. The model also shows that some characteristics of a product are not even mentioned by customers or potential customers, yet they must be present or the product will be deemed unsatisfactory. The wows are not even thought of by the customers but give the product a competitive advantage. An often cited example is the net in the trunk of the Ford Taurus that can be used to secure loose cargo such as grocery bags.

Translating the customers' responses into usable items is usually accomplished by application of the house of quality. The *house of quality* is a matrix, or perhaps more accurately an augmented matrix. Two important attributes of the house of quality are (1) its capability for ranking the various inputs in terms of perceived importance and (2) the data in the completed house that shows much of the decision making that went into the translation of customers' wants and need into usable task descriptions. The latter attribute is often called the *archival characteristic* and is especially useful when product upgrades are designed or when new products of a similar nature are designed.

Constructing the house of quality matrix begins by identifying each horizontal row of the main or correlation matrix with a customer input, called the *customer attribute (CA)*. These CAs are entered on the left side of the matrix row. Each vertical column of the matrix is assigned an activity, called the *engineering characteristic (EC)*. The ECs are entered at the top of the columns. Together, the ECs are believed by the team to be able to produce the CAs. Because a certain amount of judgment is required in determining the ECs, such TQM techniques as brainstorming are often used by the team at this time. The team must now decide the relative importance of the ECs in realizing the CAs. Again, the techniques of TQM are used. The relative importance is indicated by assigning the box that is the intersection of an EC with a CA a numerical value, using an agreed upon rating scale, such as blank for no importance to 5 or 10 as very important. Each box in the matrix is then assigned a number. Note that an EC may affect more than one CA and that one CA may require more than one EC to be realized. An example of a main matrix with ranking information in the boxes and a summary at the bottom is shown in Fig. 1.5. In this example, the rankings of the ECs are assigned only three relative values rather than a full range of 0–9. This is frequently done to reduce the uncertainty and lost time as a result of trying to decide, for example, between a 5 or a 6 for the comparisons. Also, weighting the three levels unequally will give emphasis to the more important relationships.

Following completion of the main matrix, the augmentation portions are added. The first is usually the planning matrix that is added to the right side of the main matrix. Each new column added by the planning matrix lists items that have a relationship to one or more of the CAs but are not ECs. Such items might be assumed customer relative importance, current company status, estimated competitor's status, sales positives (wows), improvement needed, etc. Figure 1.6 shows a planning matrix with relative weights of the customer attributes for each added relationship. the range of weights for each column is arbitrary, but the relationships between columns should be such that a row total can be assigned to each CA and each row given a relative weight when compared to other CA rows.

Another item of useful information is the interaction of the ECs because some of these can be positive, reinforcing each other, whereas others can be negative; improving one can lower the effect of another. Again, the house of quality can be augmented to indicate these interactions and their relative importance. This is accomplished by adding a roof. Such a roof shown in Fig. 1.7. This is very important information to the design effort, helping to guide the designers as to where special effort might be needed in the optimization of the product.

CAs \ ECs	A	B	C	D	E	F	G	H	I	J	
1	9		3	3		1					
2		3	1			9					
3	3	3		1	1		3		9	1	
4			1					1			
5	1	1			3				1		
6			3			1				3	
7				9			3		1		
8	9				3			3			
9		3		1			1			3	
10	3		9								
TOTAL	25	10	20	14	7	11	7	4	11	7	116
%	21	9	17	12	6	10	6	3	10	6	100

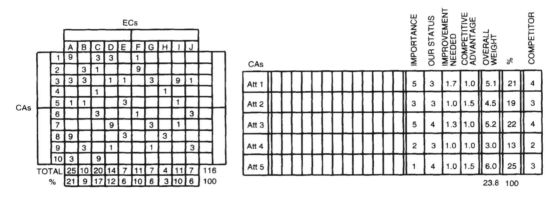

FIGURE 1.5 The house of quality main matrix.

CAs	IMPORTANCE	OUR STATUS	IMPROVEMENT NEEDED	COMPETITIVE ADVANTAGE	OVERALL WEIGHT	%	COMPETITOR
Att 1	5	3	1.7	1.0	5.1	21	4
Att 2	3	3	1.0	1.5	4.5	19	3
Att 3	5	4	1.3	1.0	5.2	22	4
Att 4	2	3	1.0	1.0	3.0	13	2
Att 5	1	4	1.0	1.5	6.0	25	3
					23.8	100	

FIGURE 1.6 CAs and the planning matrix.

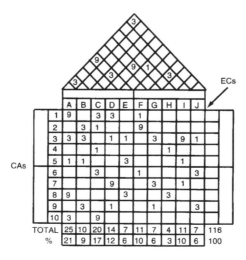

FIGURE 1.7 The house of quality with a roof.

Fundamentals of the Design Process

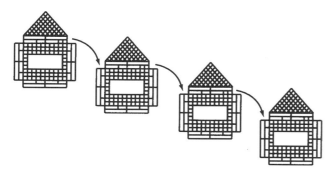

FIGURE 1.8 A series of houses of quality.

It is very likely that the ECs used in the house of quality will need to be translated into other requirements. A useful way to do this is to use the ECs from this first house as the inputs to a second house, whose output might be cost or parts to accomplish the ECs. It is not unusual to have a sequence of several houses of quality, as shown in Fig. 1.8.

The final output of a QFD utilization should be a product description and a first pass at a process description to produce the product. The product description should be traceable to the original customer inputs so that this product will be competitive in those terms. The process description should be one that will produce the product in a competitive time and cost framework. It is important to note that the QFD process, to be complete, requires input from all parts of a product cycle.

This brief look at QFD hopefully indicates the power that this tool has. Initial use of this tool will most likely be more expensive than currently used methods because of the familiarization that must take place. It should not be used for small tasks, those with fewer than 10 input statements. It should probably not be used if the number of inputs exceeds 50 because the complexity makes the relative weightings difficult to manage with a satisfactory degree of confidence. Those who have learned to use QFD do find it efficient and valuable as well as cost effective.

1.3 Systems Engineering*

1.3.1 Introduction

Modern systems engineering emerged during World War II as—due to the degree of complexity in design, development, and deployment—weapons evolved into weapon systems. The complexities of the space program made a systems engineering approach to design and problem solving even more critical, with the Department of Defense (DoD) and NASA two of the staunchest practitioners (see MIL-STD-499A). With the growth of large digital systems, the need for systems engineering has gained increased attention. Most large engineering organizations now use some level of systems engineering. The tools and techniques of this process continue to evolve to address each job better, save time, and cut costs. In these goals, systems engineering is similar to concurrent engineering, but there are many nonoverlapping concepts as well. In large part, this is due to systems engineering concentrating primarily on component functions, whereas concurrent engineering must consider both form and function of the system or product under discussion. This section will first describe systems engineering in a general sense, followed by some practical examples and implementations of the process.

1.3.2 Systems Engineering Theory and Concepts

Systems theory is applicable to the engineering of control, computing, and information processing systems. These systems are made up of component elements that are interconnected and programmed to function together and are frequently found in many industries. A *system* is defined as a set of related elements that function together as a single entity. Systems theory is a body of concepts and methods that guide the description, analysis, and design of complex systems.

---

*Adapted from Whitaker, J., *The Electronics Engineering Handbook,* Chapter 143, "Systems Engineering Concepts," by Gene DeSantis, CRC/IEEE Press, 1997.

It is important to recognize that systems engineering most commonly describes component parts of the system in terms of their functions, not their forms. As a result, *the systems engineering process does not produce the actual system itself.* Graphical models such as block diagrams, flow diagrams, timing diagrams, and the like are commonly used. Mathematical models may also be used, although systems theory says that there are *hard* and *soft* systems. Hard systems lend themselves well to mathematical descriptions, whereas soft systems are more difficult to describe mathematically. Soft systems are commonly involved with human activity, with its unpredictable behavior and nonuniformity. Soft systems introduce difficulties and uncertainties of conceptualization, description, and measurement.

Decomposition is an essential tool of systems theory and engineering. It is used by applying an organized method to complex projects or systems to break them down into simpler more manageable components. These elements are treated separately, analyzed separately, and designed separately. In the end, all of the components are combined to build the entire system. The separate analysis and recombination is similar to the circuit analysis technique of superposition.

The modeling and analytical methods used in systems engineering theoretically enable all essential effects and interactions within a system, and between a system and its surroundings, to be taken into account. Errors resulting from the idealizations and approximation involved in treating parts of the system separately are hopefully avoided.

Systems engineering uses a consistent, logical process to accomplish the system design goals. It begins with the initial description of the product or system to be designed. This involves four activities:

- Functional analysis
- Synthesis
- Evaluation and decision
- Description of system elements

To allow for improvements, the process is iterative, as shown in Fig. 1.9.

With each successive pass through the process, the description of each system/product element becomes more detailed. At each stage in the process, a decision is made to accept the results, make changes, or return to an earlier stage of the process and produce new documentation. The result of this activity is documentation that fully describes all system elements and that can be used to develop and produce the elements of the system. Again, the systems engineering process will not produce the system.

Functional Analysis

Systems engineering will include the systems engineering design process and elements of system theory (Fig. 1.10). To design a system or product, the systems, hardware, and software engineers first develop a vision of the product (initially a functional vision, then a quantitative vision) driven by customer/user specifications. It must be noted that the "customer" may be a true pay-by-money customer, an internal customer, or a hypothetical customer. An organized process to identify and validate customer needs, whether done by the marketing department or the engineering department, will minimize false starts and redesigns. System-level objectives are first defined, and analysis is carried out to identify the require-

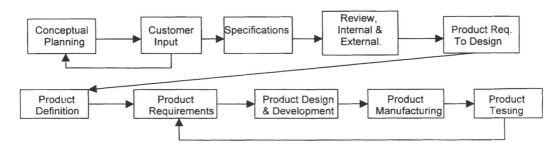

FIGURE 1.9 The systems engineering product development/documentation process.

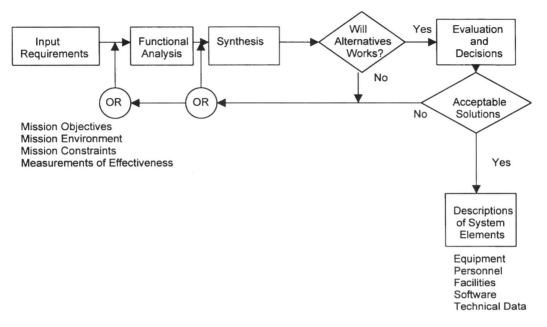

FIGURE 1.10 The systems engineering decision process.

ments and what essential functions the system must perform and why. The *functional flow block diagram* is a basic tool used to identify functional needs. It shows logical sequences and relationships of operational and support functions at the system level. Other functions, such as maintenance, testing, logistics support, and productivity, may also be required in the functional analysis. The functional requirements will be used during the synthesis phase to show the allocation of the functional performance requirements to individual system elements or groups of elements. Following evaluation and decisions, the functional requirements provide the functionally oriented data required in the description of the system elements.

Timing analysis of functional relationships is also important during the analysis phase. Determination of the specific overall schedule, as well as determination of sequential or concurrent timing of individual elements, is necessary during this phase. Time line documents or software must be setup at this phase to allow an orderly progression of development.

Synthesis

Synthesis is the process by which concepts are developed to accomplish the functional requirements of a system. Performance requirements and constraints, as defined by the functional analysis, are applied to each individual element of the system, and a design approach is proposed for meeting the requirements. Conceptual schematic arrangements of system elements are developed to meet system requirements. These documents can be used to develop a description of the system elements and can be used during the acquisition phase.

Modeling is the start of the synthesis process. It requires the determination of the quantitative specifications and performance requirements of the system. While it may seem that the model should be as detailed as possible, reality and time constraints normally dictate the simplest possible model to improve the chances of a design success. Too much detail in the model may lead to a set of specifications that are impossible to meet. The model will work best if it starts as a simple block diagram to which more detail can be added as necessary.

Systems are *dynamic* by nature. A completely static system would be of little value in the real world. Signals change over time, and components of the system determine its dynamic response to those signals. The system behavior will depend on the signal levels at any given instant of time as well as on the signals' rate of change, past values, and setpoints. Signals may be electronic signals or may include human factors such the number of users on a network or the number of degrees a steering wheel has been turned.

Optimization is making the best decision given the alternatives. Every project involves making a series of compromises and choices based on relative weighting of the merits of important aspects of the elements. Decisions may be objective or subjective, depending on the kind of (or lack of) data available.

Evaluation and Decision

Product and program costs are determined by the trade-offs between operational requirements, engineering design capabilities, costs, and other limitations. During the design and development phase, decisions must be made based on evaluation of alternatives and their relative effects on cost and performance. A documented review and evaluation must be made between the characteristics of alternative solutions. Mathematical models (see "Modeling," above) and computer simulations may be used to aid in the evaluation process.

Trade studies are used to guide the selection of alternative configurations and ensure that a logical and unbiased choice is made. They are carried out to determine the best configuration that will meet the requirements of the program. During the exploration and demonstration phases of the concepts, trade studies help define the system configuration. They are used as a detailed design analysis tool for individual system elements in the full-scale development phase. During production, trade studies are used to select alternatives when it is determined that changes need to be made. The figures that follow illustrate the trade study process.

Figure 1.11 shows that, to provide a basis for the selection criteria, the objectives of the trade study first must be defined. Functional flow diagrams and system block diagrams are used to identify trade study areas that can satisfy certain requirements. Alternative approaches to achieving the defined objectives can then be established.

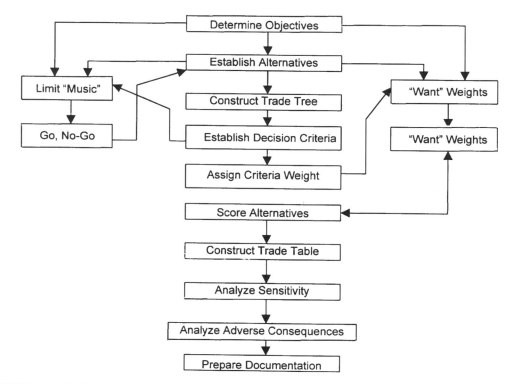

FIGURE 1.11 Trade studies process flowchart. Adapted from Defense Systems Management, *Systems Engineering Management Guide*, 1983.

Fundamentals of the Design Process

As shown in Fig. 1.12, a trade study tree can be constructed to show the relationship and dependencies at each level of the selection process. The trade tree shown allows several trade study areas to be identified as possible candidates for accomplishing a given function. The trade tree shows relationships and the path through selected candidate trade areas at each level to arrive at a solution.

Several alternatives may be candidates for solutions in a given area. The selected candidates are then submitted to a systematic evaluation process intended to weed out unacceptable candidates. Criteria are determined that are intended to reflect the desirable characteristics. Undesirable characteristics may also be included to aid in the evaluation process. Weights are assigned to each criteria to reflect its value or impact on the selection process. This process is subjective and should take into account costs, schedules, and technical constraints that may limit the alternatives. The criteria on the candidates are then collected and tabulated on a decision analysis worksheet, as shown in Table 1.1.

TABLE 1.1 Decision Analysis Worksheet Example (Adapted from Defense Systems Management, 1983, *Systems Engineering Management Guide*, Contract no. MDA 903-82-C-0339, Defense Systems Management College, Ft. Belvior, Virginia.)

Alternatives		Candidate 1			Candidate 2			Candidate 3		
Wanted	WT		SC	WT SC		SC	WT SC		SC	WT SC
Video bandwidth, MHz	10	5.6	10	100	6.0	10	100	5.0	9	90
Signal-to-noise ratio, dB	10	60	8	80	54	6	60	62	10	100
10-bit quantizing	10	yes	1	10	yes	1	10	yes	1	10
Max. program length, h	10	2	2	20	3	3	30	1.5	1.5	15
Read before write correction avail.	5	yes	1	5	yes	1	5	no	0	0
Capable of 16:9 aspect ratio	5	yes	1	5	no	0	0	yes	1	5
Employs compression	10	no	0	0	yes	1	10	yes	1	10
SDIF (series digital interface) built in	−5	yes	1	−5	no	0	0	yes	1	−5
Current installed base	10	yes	1	10	yes	1	10	yes	1	10
	8	medium	2	16	low	1	8	low	1	8
Total weighted score				241			234			243

The attributes and limitations are listed in the first column, and the data for each candidate is listed in adjacent columns to the right. The performance data may be available from vendor specification sheets or may require testing and analysis. Each attribute is given a relative score from 1 to 10 based on its

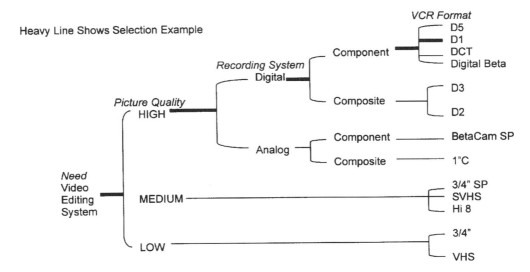

FIGURE 1.12 An example trade study tree.

comparative performance relative to the other candidates. Utility function graphs of Fig. 1.13 can be used to assign logical scores for each attribute.

The utility curve represents the advantage rating for a particular value of an attribute. A graph is made of ratings on the y-axis vs. attribute value on the x-axis. Specific scores can then be applied that correspond to particular performance values. The shape of the curve may take into account requirements, limitations, and any other factor that will influence its value regarding the particular criteria being evaluated. The limits to which the curves should be extended should run from the minimum value, below which no further benefit will accrue, to the maximum value, above which no further benefit will accrue.

The scores from the curves are filled in on the decision analysis work sheet and multiplied by the weights to calculate the weighted score. The total of the weighted scores for each candidate then determines their overall ranking. Generally, if a difference in scores is at least 10%, that difference is considered meaningful.

Further analysis can be applied in terms of evaluating the sensitivity of the decision to changes in the value of attributes, weights, subjective estimates, and cost. Scores should be checked to see if changes in weights or scores would reverse the choice. The decision should be evaluated to determine how sensitive it is if there are changes in system requirements and/or technical capabilities.

A *trade table* can be prepared to summarize the selection results (see Table 1.2). Pertinent criteria are listed for each alternative solution. The alternatives may be described in a quantitative manner, such as high, medium, or low. Finally, the results of the trade study are documented in the form of a report, which discusses the reasons for the selections and may include the trade tree and the trade table.

There must also be a formal system for controlling changes throughout the systems engineering process, much like engineering change orders in the product engineering process. This prevents changes from being made without proper review and approval by all concerned parties and to keep all parties informed. Change control helps control project costs, can help eliminate redundant documents, and ensures that all documentation is kept up to date.

Description of System Elements

Five categories of interacting system elements can be defined, although they may not all apply to a given project.

1. equipment (hardware)
2. software
3. facilities
4. personnel
5. procedural data

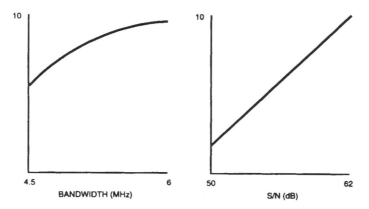

FIGURE 1.13 Attribute utility trade curve example.

Fundamentals of the Design Process 1-13

TABLE 1.2 Trade Table Example

Criteria	Cool room only; only normal convection cooling within enclosures.	Forced cold air ventilation through rack then directly into return.	Forced cold air ventilation through rack, exhausted into the room, then returned through the normal plenum.
Cost	Lowest Conventional central air conditioning system used.	High Dedicated ducting required, separate system required to cool room.	Moderate Dedicated ducting required for input air.

Performance
  Equipment temp.
  Room temp.
Control
	Poor
	80–120° F+
	65–70° F typical as set
		Very good
		55–70° F typical
		65–70° F typical as set
			Very good
			55–70° F typical
			65–70° F typical as set

Control of equipment temperature	Poor Hot spots will occur within enclosures.	Very good	Very good When the thermostat is set to provide a comfortable room temperature, the enclosure will be cool inside.
Control of room temperature	Good Hot spots may still exist near power-hungry equipment.	Good	Good If the enclosure exhaust air is comfortable for operators, the internal equipment must be cool.
Operator comfort	Good	Good Separate room ventilation system required can be set for comfort.	Good When the thermostat is set to provide a comfortable temperature, the enclosure will be cool inside.

Performance, design, and test requirements must be specified and documented for equipment, components, and computer software elements of the system. It may be necessary to specify environmental and interface design requirements, which are necessary for proper functioning of system elements within the project. The documentation produced by the systems engineering process controls the evolutionary development of the system. Figure 1.14 illustrates the special purpose documentation used by one organization in each step of the systems engineering process.

The requirements are formalized in written specifications. In any organization, there should be clear standards for producing specifications. This can help reduce the variability of technical content and improve product quality. It is important to remember that the goal of the systems engineering process is to produce functional performance specifications, *not* to define the actual design or make the specifications unnecessarily rigid and drive up costs. The end result of this process results in documentation which defines the system to the extent necessary to allow design, development, and testing of the system. It also ensures that the design requirements reflect the functional performance requirements, that all functional performance requirements are satisfied by the combined system elements, and that such requirements are optimized with respect to system performance requirements and constraints.

1.3.3 Example Phases of a Typical System Design Project

The design of a complex video production facility is used to illustrate systems engineering concepts in a medium-size project.

Design Development

Systems design is carried out in a series of steps that lead to an operational system. Appropriate research and preliminary design work is completed in the first phase of the project, the design development phase.

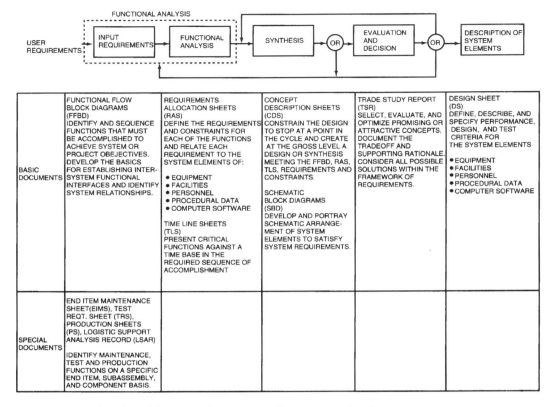

FIGURE 1.14 Basic and special purpose documentation for systems engineering.

It is the intent of this phase to fully define all functional requirements of the project and to identify any constraints. Based on initial concepts and information, the design requirements are modified until all concerned parties are satisfied and approval is given for the final design work to proceed. Questions that should be addressed during the design development phase include:

- What are the functional requirements of the facility?
- What are the physical requirements of this facility?
- What are the performance requirements of this facility?
- What are the constraints limiting design decisions?
- Will any existing equipment be used?
- Is the existing equipment acceptable to the new specifications?
- Will this be a new facility or a renovation?
- Will this be a retrofit or upgrade to an existing system?
- Will this be a stand-alone system?
- How many personnel are expected to be able to operate the facility?
- Are any future needs or expansions defined at this time?
- What future technical upgrades and/or expansions can be anticipated and defined?

Working closely with the customer's representatives, the equipment and functional requirements of each of the major technical areas of the facility are identified. The engineer must identify, define, and meet the needs of the customer within the projected budget. If the budget does not allow for meeting all the customer's needs, the time to have this discussion is as early in the project development as possible.

It is extremely important at this phase that the systems engineer adhere closely to the customer's wishes and needs. If the wishes and needs conflict with each other, agreement must be reached with the customer on the priorities before the project proceeds. Additionally, the systems engineer must not overengineer the project. This can be a costly mistake and easy to make, since there is always a desire to do "a little more" for the customer.

If the project is a renovation, the system engineer must first conduct a site visit to analyze the condition and specifications of the existing equipment, and determine the layout of existing equipment and of the facility. Any equipment whose condition will prevent it from being reused is noted.

Once the equipment list is proposed and approved by the customer, preliminary system plans are drawn up for review and further development. For a renovation project, architectural drawings of the facility should be available and can be used as a starting point for laying out an equipment floor plan. For a new facility, the systems engineer and the architect must work closely together. In either case, the floor plan is used as a starting point for laying out an equipment floor plan. Included in floor plan considerations are the following:

- Space for existing equipment
- Space for new equipment
- Space for future equipment
- Clearance for movement/replacement of equipment
- Clearance for operational needs
- Clearance for maintenance needs

Major equipment documentation for this video facility includes, but is not limited to

- Technical system functional block interconnect diagrams
- Technical system specifications for both stock and custom items
- Equipment prices
- Rack and console floor plans and elevations
- Equipment floor plans and elevations

As in any systems and/or design work, accurate records, logs, and notes must be kept as the project progresses. Ideas and concepts exchanged between the customer's representatives and the engineers must be recorded, since the bulk of the creative work will occur during the design development phase. The physical layout, which is the look and feel of the finished project, and the functionality of the facility will all have been decided and agreed upon during this phase. If the design concepts appear feasible, and the cost appears to be within the anticipated budget, management can authorize work to proceed on the final detailed design.

Electronic System Design

For a technical facility such as this, the performance standards and specifications must be established in the first phase. This will determine the performance level of equipment that will be acceptable for use in the system and affect the size of the budget. Signal quality, stability, reliability, distortion, and accuracy are examples of the kinds of parameters that will have to be specified. Access and processor speeds are important parameters when dealing with computer-driven products. It is the systems engineer's job to select equipment that conforms to the specifications.

From the decisions made during the first phase, decisions will now be made with regard to what functions each piece of equipment must have to fulfill the overall functional requirements. It is also crucial to work with the operations staff of the facility to understand what controls they expect to have available for their work. If they have already selected equipment which they feel is appropriate, much of the engineer's work in this area may already be done. Questions for the operations staff include the following:

- What manual functions must be available to the operators?
- What automatic functions must be available?
- What level of automation is necessary?
- What functions are not necessary? (There is no sense in overengineering or overbuying.)
- How accessible should the controls be?
- What maintenance operations are the operators expected to perform?

Care should be exercised to make sure that seemingly simple requests by the operators do not result in serious complexity increases. Compromises may be necessary to ensure that the required functions and the budget are both met.

When existing equipment is going to be used, it will be necessary to make an inventory list, then determine which of the existing equipment will allow the project to meet the new specifications. Equipment which will not meet the specifications must be replaced. After this information is finalized, and the above questions have been answered, the systems engineer develops a summary of equipment needs for both current and future acquisitions. Again, the systems engineer must define these needs within the available budget. In addition to the responses to the questions previously discussed with the operations staff, the engineer must also address these issues:

- Budget
- Space available
- Performance requirements
- Ease of operation
- Flexibility of use
- Functions and features
- Past performance history (for existing equipment)
- Manufacturers'/vendors' support

Consideration of these issues should complete this phase of the systems engineers job, and allows him/her to move on to the detailed design phase.

Detailed design of a project such as this requires that the systems engineer prepare complete detailed documentation and specifications necessary for the purchase, fabrication, and installation of all major and minor components of the technical systems. Drawings must show the final configuration and the relationship, including interconnection, of each component to other elements of the system as well as how they will interface with other building services such as air conditioning and electrical power. This documentation must communicate the design requirements to purchasing and to the other design professionals, including the construction and installation contractors.

This phase requires that the engineer develop final, detailed flow diagrams, and schematics. Complete cabling information for each type of signal is required, and from this a cable schedule will be developed. Cable paths are measured, and timing calculations performed. Timed cable lengths for video and other services are entered onto the cable schedule.

The flow diagram is a schematic drawing used to show the interconnections between all equipment that will be installed. It is more detailed than the block diagram, and shows every wire and cable. An example is shown in Fig. 1.15.

If the project uses existing equipment and/or space, the starting point for preparing a flow diagram is the original diagram, if one exists. New equipment can be shown in place of old equipment being replaced, and wiring changes can be added as necessary. If the facility is new, the block diagram is the starting point for the flow diagram. Details are added to show all of the equipment and their interconnections and to show any details necessary to describe the installation and wiring completely. These details will include all separable interconnects for both racks and individual equipment. The separable interconnects are important, since equipment frequently gets moved around in a facility such as this,

Fundamentals of the Design Process   1-17

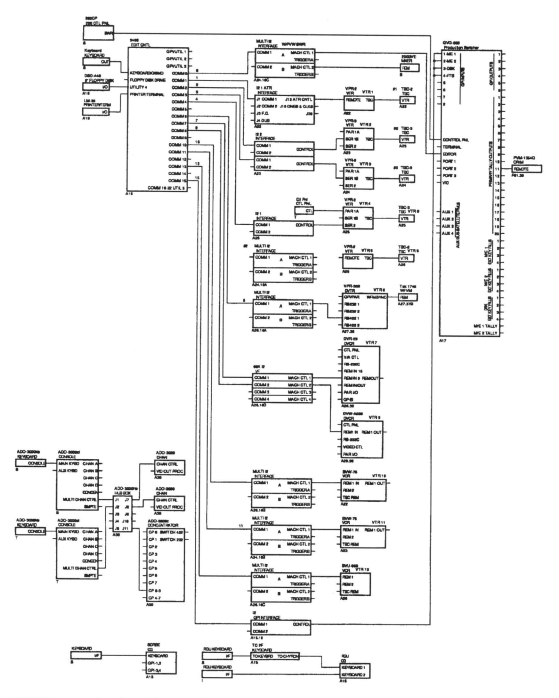

FIGURE 1.15 Basic and special purpose documentation for systems engineering.

and it is also important that labels be defined in the flow diagram—and required of the installation contractor. Any color codes are also defined on the flow diagrams, based on the customer's wishes or industry color code standards that apply.

The systems engineer will also provide layouts of cable runs and connections to the architect. This information must be included on the architect's drawings, along with wire ways, conduits, duct, trenches, flooring access, and overhead wire ways. These drawings will also include dimensioned floor plans and elevations to show the placement of equipment; lighting; and heating, ventilating, and air conditioning (HVAC) ducting, as well as the quantity and type of acoustical treatments.

Equipment and personnel heat loads must be calculated and submitted to the HVAC consultant. This consultant will also need to know the location of major heat-producing equipment, so the ac equipment can be designed to prevent hot spots within the facility. Additionally, electrical loads must be calculated and submitted to the electrical contractor, as well as layout requirements for outlets and drops.

Customer support is an important part of the systems engineer's job. The engineer can aid the customer and the project by developing appropriate purchasing specifications, setting up a move schedule (if existing equipment and/or facilities must be kept on-line during the construction and movement to the new/remodeled facility), and testing all new equipment prior to the its intended turn-on date. The engineer must also be certain that all necessary documentation from suppliers is in fact present and filed in a logical place.

1.3.4 Large Project Techniques

Budget Requirements Analysis and Project Proposal

The need for a project may originate with customers, management, operations, staff, technicians, or engineers. Some sort of logical reasoning or a specific production requirement will be needed to justify the cost. The overall cost is rarely apparent on the initial consideration of a large project. A true cost estimate will require consideration of all the elements mentioned in the video facility example, plus items such as major physical facility costs, additional personnel necessary to run the new facility/equipment, and maintenance costs associated with both the facility and equipment. A capital project budget request containing a detailed breakdown of all elements can provide the information needed by management to determine the return on investment (ROI) and make an informed decision on whether to proceed.

A capital project budget request will normally contain at least the following information:

- Name of the project
- Assigned number of the project
- Initiating person and/or department
- Project description (an overview of what the project will accomplish)
- Project justification (This may include items such as productivity increase expected, overall production increase expected, cost savings expected, maintenance/reliability issues with current equipment, etc.)
- Initiation date
- Anticipated completion date
- Time action plan, Gantt chart, etc., for the project
- Results of a feasibility study, if conducted
- Material and equipment cost estimate
- Labor cost estimate
- Miscellaneous cost estimate (This would include consultants' fees, impact of existing work [e.g. interruption of an assembly line].)
- Total project cost estimate
- Payment schedule (an estimate of payouts required during the course of the project and their timing)
- Return on investment
- Proposal preparers' name and date prepared
- Place for required approvals

The feasibility study must include the impact of any new technology, including an appropriate learning curve before that technology contributes to a positive cash flow. The time plan must include the impacts of new technology on each involved department as well as a signoff by each department head that the manpower required for the project will be available. The most common time tools are the Gantt chart, the critical path method (CPM), and the project evaluation and review (PERT) chart. Computerized versions of all these tools are available. This will allow tracking and control of the project and well as generation of periodic project status reports.

Project Management

The Defense Systems Management College[1] defines systems engineering as follows:

> Systems engineering is the management function which controls the total system development effort for the purpose of achieving and optimum balance of all system elements. It is a process which transforms an operational need into a description of system parameters and integrates those parameters to optimize the overall system effectiveness.

Systems engineering is both a technical process and a management process. Both processes must be applied throughout a program if it is to be successful. The persons who plan and carry out a project constitute the project team. The makeup of a project team will vary depending on the size of the company and the complexity of the project. It is up to management to provide the necessary human resources to complete the project.

The *executive manager* is the person who can authorize that a project be undertaken but is not the person who will shepherd the project through to completion. This person can allocate funds and delegate authority to others to accomplish the task. Motivation and commitment is toward the goals of the organization. The ultimate responsibility for a project's success is in the hands of the executive manager. This person's job is to get tasks completed through other people by assigning group responsibilities, coordinating activities between groups, and resolving group conflicts. The executive manager establishes policy, provides broad guidelines, approves the master plan, resolves conflicts, and ensures project compliance with commitments. Executive management delegates the project management functions and assigns authority to qualified professionals, allocates a capital budget for the project, supports the project team, and establishes and maintains a healthy relationship with project team members.

Management has the responsibility to provide clear information and goals, up front, based on the needs and initial research. Before initiating a project, the executive manager should be familiar with daily operation of the facility and analyze how the company works, how jobs are done by the staff, and what tools are needed to accomplish the work. For proper consideration of a project proposal, the executive manager may chose to bring in expert project management and engineering assistance, and accounting/controller expertise.

The *project manager* will be assigned at the time of the initiation of the project and is expected to accomplish large complex projects in the shortest possible time, within the anticipated cost, and with the required performance and reliability. The project manager must be a competent systems engineer, accountant, and personnel manager. As systems engineer, this individual must have an understanding of analysis, simulation, modeling, and reliability and testing techniques. There must be an awareness of state-of-the-art technologies and their limitations. As accountant, there must be awareness of the financial implications of planned decisions and knowledge of how to control them. As manager, the planning and control of schedules is an important part of controlling the costs of a project and completing it on time. Also, the manager must have the skills necessary to communicate clearly and convincingly with subordinates and superiors to make them aware of problems and their solutions. The manager must also be able to solve interdepartmental squabbles, placing full responsibility on all concerned to accomplish their assigned missions.

The project manager must have the ability and control to use whatever resources are necessary to accomplish the goals in the most efficient manner. The manager and staff provide and/or approve the project schedule, budget, and personnel needs. As the leader, the project manager will perform many tasks.

- Assemble the project organization
- Develop the project plan
- Publish the project plan
- Secure the necessary commitments from top management to make the project a success
- Set measurable and attainable project objectives
- Set attainable project performance standards
- Determine which time scheduling tools (PERT, CPM, Gantt, etc.) are appropriate for the project
- Using the scheduling tools, develop and coordinate the project plan, including the budget, resources, and schedule
- Develop the project schedule
- Develop the project budget
- Manage the budget
- Work with accounting to establish accounting practices that help, not hinder, successful completion of the project.
- Recruit appropriate personnel for the project, who will work together constructively to ensure success of the project
- Select subcontractors
- Assign work, responsibility, and authority so that team members can make maximum use of their abilities
- Estimate, allocate, coordinate, and control project resources
- Deal with specifications and resource needs that are unrealistic
- Decide on the appropriate level of administrative and computer support
- Train project members on how to fulfill their duties and responsibilities
- Supervise project members, giving them day-to-day instructions, guidance, and discipline as required to fulfill their duties and responsibilities
- Design and implement reporting and briefing information systems or documents that respond to project needs
- Require formal and informal reports that will measure the status of the project
- Maintain control of the project
- Be certain to complement and reward members of the project team when exceptional work is being done
- Be ready to correct the reasons for unsatisfactory results
- Be certain the team members believe the project manager understands their interests as well as the interests of the project

By fostering a good relationship with associates, the project manager will have less difficulty communicating with them. The fastest, most effective communications takes place among people when needs are understood and agreed to by all.

The term *systems engineer* means different things to different people. The systems engineer is distinguished from the engineering specialist, who is concerned with only one specific engineering discipline, in that the systems engineer must be a generalist who is able to adapt to the many different requirements of a system. However, the systems engineer is expected to be an expert in at least one of the engineering specialties, relieving budget resources in that area.

The systems engineer uses management techniques to develop overall functional specifications for a project, while the engineering specialist will use those specifications to do design work to implement the specifications. The systems engineer will prepare necessary documentation for consultants, contractors, and technicians, who will design, build, and install the systems. A competent systems engineer will help in making cost-effective decisions and will be familiar enough with the included engineering disciplines to determine that equipment, construction, and installation work is being done correctly.

The systems engineer performs studies that compare trade-offs so that all decisions are based on the best information available. This individual works during the construction and installation phases to answer questions (or find the most appropriate person to answer questions) and to resolve problems that may arise.

Other project team members include

- *Architect,* responsible for the design of any structure
- *Engineering specialists,* if these areas are not handled by the systems engineer
    - *Electrical engineer,* responsible for power system design
    - *Electronics engineer,* responsible for computer systems, telecommunications, and related fields
    - *Mechanical engineer,* responsible for HVAC, plumbing, and structural
    - *Structural engineer,* responsible for concrete and steel structures
- *Construction contractors,* responsible for executing the plans developed by the architects and mechanical and structural engineers
- *Other outside contractors,* responsible for certain customized work and/or items that cannot be developed by team members already mentioned

For the systems engineer and all others on the project, control of any single phase of the project must be given to the member of the team who has the most to gain by successful completion of that phase of the project, and the most to lose if that phase of the project is not successfully completed.

Time Control of the Project

The time tool chosen and the approved budget will allow the project to remain under reasonable control. After these two items are developed, and money is allocated to the project, any changes may increase or decrease the overall cost of the project. In addition, it is mandatory that all involved personnel understand the need for and use engineering change orders (ECOs) for any and all changes to the project. There must be a method for ECOs to be generated, approved, and recorded. Additionally, there must be a method for all personnel to be able to immediately determine whether they are working from the latest version of any document.

The ECO must include

- The project name and number
- Date of the proposal for the change
- Preparer's name
- A brief description of the change
- The rationale for the change
- The total cost or savings of the change, including specific material and labor costs
- Impact on the schedule

It is appropriate that there be at least two levels of ECO approval. If an ECO is to be totally funded within one department and will not impact the schedule or engineering plans of any other department, approval may be given by the department head, with copies of the approved ECO distributed to any and all departments that need to know of the change. An ECO that affects multiple departments must be approved at the systems engineering level with no approval given until all affected departments have been consulted. And again, copies of the approved ECO must be distributed to any and all departments that are affected by the change, as well as to accounting.

1.3.5 Defining Terms for Systems Engineering

Abstraction. Although dealing with concrete systems, abstraction is an important feature of systems models. Components are described in terms of their function rather than in terms of their form. Graphical models such as block diagram, flow diagrams, and timing diagrams are commonly used. Mathematical models may also be used. Systems theory shows that, when modeled in formal language, apparently diverse kinds of systems show significant and useful similarities of structure and function. Similar interconnection structures occur in different types of systems. Equations that describe the behavior of electrical, thermal, fluid, and mechanical systems are essentially identical in form.

Decomposition. This refers to treating a large complex system by breaking it down into simpler, more manageable component elements. These elements are then reassembled to result in the large system.

Dynamics. These are systems that change with time and require a dynamic response. The system behavior depends on the signals at a given instant as well as the rates of change of the signals and their past values.

Emergent properties. These properties result from the interaction of systems components rather than being properties unique to the components themselves.

Hard and soft systems. In hard systems, the components and their interactions can be described mathematically. Soft systems cannot be easily or completely described mathematically. Soft systems are mostly human activities, which implies unpredictable and nonuniform behavior.

Isomorphism. This refers to similarity in elements of different kinds. Similarity of structure and function in elements implies isomorphism of behavior of a system. Different systems that nonetheless exhibit similar dynamic behavior, such as response to a stimulus, are isomorphic.

Modeling. Modeling requires the determination of the quantitative features that describe the operation of the system. The model is always a compromise, as with most real systems it is not possible to completely describe it, nor is it desirable in most cases.

Optimization. This is the process of making an element of the system as effective or functional as possible. Normally done by examining the alternatives and selecting the best, in terms of function and cost-effectiveness.

Synthesis. This is the process by which concepts are developed to accomplish the functional requirements of the system. Performance requirements and constraints, as defined by the functional analysis, are applied to each individual element of the system, and a design approach is proposed for meeting the requirements.

1.4 Quality Concepts

The ultimate goal of a quality control program would be to have the design and assembly processes under such excellent control that no testing would be necessary to have a reliable product. Reality prevents this, but a total quality program will result in the following:

- Cost reduction
- Improved product reliability
- Reduction of rework and repair

Quality is an elusive issue—not only making a *quality* product, but just defining the term itself. Dobyns, et al., in *Quality or Else,* after interviewing quality professionals, concluded that "…no two people we've talked to agree…on how to define quality." For our purposes, a *quality* product will be defined as one that meets its specifications during the manufacturing and testing phases prior to shipment. This is different from *reliability,* which can be defined as a product meeting its specifications during its expected lifetime.

The type of quality program chosen is less important than making sure that all employees, from the CEO to designers to line workers to support personnel and suppliers, believe in the program and its potential for positive results if participants perform their jobs properly. For virtually every quality-implementation technique, there are both followers and detractors. For instance, the Taguchi method is widely accepted. However, not all practicing engineers believe Taguchi is appropriate at the product level.[5,9]

All operating areas and personnel in the process must be included in the quality program. For example, in a surface mount technology (SMT) design, representatives from these areas should be involved:

- Circuit design
- Substrate design
- Substrate manufacturing and/or acquisition
- Parts acquisition and testing
- Solder paste selection, acquisition and testing
- Solder paste deposition (printing or syringe dispense)
- SMD placement
- Placement equipment acquisition and use
- Reflow oven acquisition and soldering
- Cleaning equipment acquisition and use
- Test system acquisition and use
- Documentation

Note that, as indicated by the inclusion of the term *acquisition* in the list, vendors are very important to the overall quality process. They must be included in decisions relating to their products and must believe that their input matters. It also important to keep the entire process under control and have enough information to detect when control of the process is declining. Defects must also be analyzed to allow assignment of a cause for each defect. Without determination of the cause of each defect, there is no way to improve the process to minimize the probability of that defect occurring again. The types of inspections to perform during the process will have to be determined by the people who best know each of the steps in the process.

One of the best indicators of in-process quality in an electronic assembly is the quality of each soldered joint. Regardless of the quality of the design, or any other single portion of the process, if high-quality reliable solder joints are not formed, the final product is not reliable. It is at this point that PPM levels take on their finest meaning. For a medium-size substrate (nominal 6 × 8 in), with a medium density of components, a typical mix of active and passive parts on the top side and only passive and three- or four-terminal active parts on bottom side, there may be in excess of 1000 solder joints/board. If solder joints are manufactured at the 3 sigma level (99.73% good joints, or 0.27% defect rate, or 2700 defects/million joints) there will be 2.7 defects per board! At the 6 sigma level of 3.4 PPM, there will be a defect on 1 board out of every 294 produced. If your anticipated production level is 1000 units/day, you will have 3.4 rejects based solely on solder joint problems, not counting other sources of defects.

Using solder joints rather than parts as the indicator of overall quality also indicates the severity of the problem. If a placement machine places a two-lead 0805 resistor incorrectly, two solder joints are bad. If the same placement machine places a 208-lead PQFP incorrectly, 208 solder joints may be bad. But in each case, only one part is faulty. Is the incorrect placement of the PQFP 104 times as bad as resistor placement? Yes, because it not only results in 104 times as many bad solder joints but most likely also results in far more performance problems in the completed circuit.

Examples of design and process variables which affect the quality of the solder joint include

- Design of the substrate lands
- Accuracy of substrate production and fiducial locations

- Initial incoming quality of the solder paste
- Continuing inspection of solder paste during the duration of production
- Initial inspection of part quality for performance and/or adherence to specs
- Initial inspection of the quality of part leads for solderability and coplanarity
- Handling of parts without damage to any leads
- For stencil deposition, the quality of the solder paste stencil—proper opening shape, proper opening polish, and proper speeds and angles of the squeegee(s)
- For syringe deposition, proper pressure and x-y-z motions
- Proper volume of solder paste dispensed
- Accuracy of placement machine x-y-z motions and downward pressure
- Correctness of the reflow profile

Determination of which quality program, or combination of programs, is most appropriate for the reader's process is beyond the scope of this book. SPC, Taguchi, quality function deployment (QFD), design of experiments (DOE), process capability, and other quality programs and techniques should be considered. Remember that the key is to not just find the faults, it is to assign a cause and improve the process. Emphasizing the process, rather than separate design, manufacturing, and test issues, will typically lead to use of techniques such as QFD, DOE, and process control (PC).

1.4.1 Design of Experiments

While QFD (Section 1.2) is used in the development of a product and a corresponding process, design of experiments (DOE) is an organized procedure for identifying those parts of a process that are causing less than satisfactory product and then optimizing the process. This process might already be used in production, or it might be one that is proposed for a new product. A complex process such as an SMT assembly line cannot be studied effectively by using the simple technique of varying one parameter while holding all others steady. Such a process ignores the interaction between parameters, a condition that normally prevails in the real world. If all interactions as well as primary parameters are to be tested, the number of experiments rapidly becomes large enough to be out of the question even for only a few variables. DOE has been developed to help reduce the number of experiments required to uncover a problem parameter. DOE relies on statistics, particularly factorial analysis, to determine the relative importance of relationships between parameters. Initially, the implementation of DOE was the purview of statisticians, which meant that it was outside the realm of most engineers and line workers.

Factorial analysis is the study of the chosen parameters and all their possible interactions. It is neither time efficient nor easy to calculate if more than four parameters are involved. To allow more use of factorial analysis, fractional factorial analysis was developed, using only identified primary parameters and selected interactions. Fractional factorials can be used effectively only with the correct selection of primary parameters and their interactions. Use of brainstorming and similar techniques from total quality management (TQM) can help, but it cannot guarantee that correct selections were made.

Taguchi introduced a technique called *orthogonal arrays*[11] in an attempt to simplify the selection of parameters for fractional factorial analyses. The technique is not simple to use and requires in-depth study. There are no guidelines for the selection of the possible interactions, a shortcoming also present in the original fractional factorial technique. If the quality problems are due to one of the main parameters, this is less of an issue, but it still begs the question of selecting interactions. As with many quality techniques, one must make one's own analysis of what works and what doesn't. In an interchange with Robert Pease of National Semiconductor, regarding his analysis of a voltage regulator circuit, Taguchi said, "We are not interested in any actual results, because quality engineering deals with only optimization."

A lesser-known but simpler DOE system was developed by Dorian Shanin. Based on sound statistical techniques, Shanin's techniques use much simpler calculations, typically with a knowledge of mean, median, and standard deviation. Questions may still arise that require consultation with a statistician, but Shanin's techniques are designed to be used by engineers and operators and to identify when the techniques will not be able to adequately work.

Shanin's general procedure is to reduce a large number of possible causes of a problem to four or fewer, then use the full factorial method of analysis to identify the most likely cause or causes. The underlying principle is that most real-world problems can have their causes reduced to four or fewer primary causes plus their interactions. With four or fewer causes, the full factorial analysis is very appropriate. Once the causes have been identified, the process can be improved to produce the best possible product.

As shown in Fig. 1.16, seven DOE procedures make up Shanin's system. First, variables are eliminated that are not a cause. The multivari charts are used in determining what type or family a variation belongs to and eliminating causes that are not in this family. Other first-level procedures include *components search* and *paired comparisons,* which are mutually exclusive as techniques, but either of which can be used in conjunction with multivari charts. Components search requires disassembly and reassembly of the product a number of times to rule out assembly problems, as opposed to component problems. Paired comparisons are used when the product or part cannot be disassembled and must be studied as a unit. *B vs. C* is used as a final validation of the previous techniques. Scatter plots are used primarily for relating the tolerance values of identified input variables to quality requirements presented by the customer. Shainen's techniques will now be examined in some detail.

Multivari Chart

The multivari chart is used to classify the family into which the red X or pink Xs fall. A "red X" is most certainly a cause of variation, and a "pink X" has a high probability of being a cause of variation. A parameter that is indicative of the problem, and can be measured, is chosen for study. Sets of samples are then taken and the variation noted. The categories used to distinguish the parameter output variation are: (1) variation within sample sets (cyclical variation) is larger than variation within samples or variation over time, (2) time variation (temporal variation) between sample sets is larger than variation within sample sets or variation of the samples, and (3) variations within samples (positional variation) are larger than variation of sample sets over time or variation within the sample sets. These are shown in Fig. 1.17.

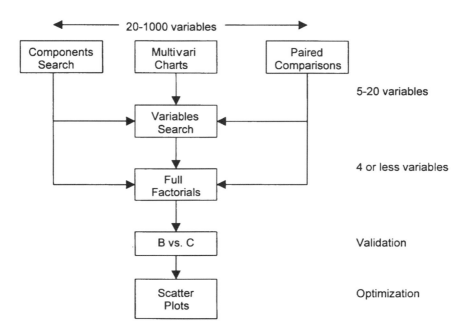

FIGURE 1.16 Shanin's seven DOE procedures.

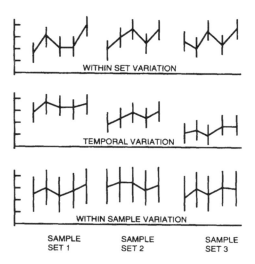

FIGURE 1.17 Multivari types of variation.

To illustrate, assume a process has been producing defective product at a known historical rate, that is, at an average rate of X ppm, for the past weeks or months. Begin the study by collecting, *consecutively,* three to five products from the process. At a later time, after a number of units have been produced in the interim, collect three to five products again. Repeat this again and again, as often as necessary; three to five times is frequently sufficient to capture at least 80% of the historical defect rate in the samples, that is, these samples should include defects at least at 80% of the historical rate, X, that the process has produced defects in historical samples. This is an important rule to observe to provide statistical validity to the samples collected.

In the language of statistics, this is a *stratified* experiment; that is, the samples are grouped according to some criterion. In this case, the criterion is consecutive production of the units. This is not, therefore, a random selection of samples as is required in many statistical methods. It also is not a control chart, even though the plot may resemble one. It is a snapshot of the process taken at the time of the sampling. Incidentally, the multivari chart is not a new procedure, dating from the 1950s, but it has been incorporated into this system by Shainen.

The purpose of the multivari chart is to discover the family of the red X, although on rare occasions the red X itself may become evident. The anticipated result of a successful multivari experiment is a set, the family, of possible causes that includes the red X, the pink Xs, or the pale pink Xs. The family will normally include possible causes numbering from a few up to about 20. Further experiments will be necessary to determine the red X or the pink Xs from this set.

The example displays in Fig. 1.18 show the variations of four 98-Ω nominal valued resistors screen printed on a single ceramic substrate. They will later be separated from each other by scribing. Data recorded over a two shift period, from units from more than 2800 substrate printings, are shown in the chart. A graph of this data is also shown, with the range of data for each substrate indicated and the average of the four resistors for each substrate as well as the overall average for the three substrates sampled at that time.

Note:

1. Unit-to-unit variation within a group is the largest.
2. Time variation is also serious as the time averages increase sharply.

With this information about the family of the cause(s) of defects, additional experiments to find the cause(s) of the variation can be designed according to the methodology to be described later.

Components Search

Components search is used only when a product can be disassembled and then reassembled. It is a part-swapping procedure that is familiar to many who have done field repair. The first step is to select a performance parameter by which good and bad units can be identified. A good unit is chosen at random, measured, then disassembled and reassembled two times, measuring the performance parameter each time. This establishes a range of variability of the performance parameter that is related to the assembly operation for good units. Repeat this for a randomly selected bad unit, once again establishing the range of variability of the performance parameter for assembly of bad units. The good unit must remain a good unit after disassembly and reassembly, just as the bad unit must remain a bad unit after disassembly and reassembly. If this is not the case, then the parameter chosen as performance indicator needs to be reviewed.

Fundamentals of the Design Process

(a)

	9:00 A.M.			1:30 P.M.			6:00 P.M.			11:00 P.M.		
PRINT	200	201	202	900	901	902	1800	1801	1802	2800	2801	2802
R1	92	90	91	98	100	92	97	98	96	102	95	98
R2	94	93	92	96	93	95	102	100	100	98	101	102
R3	95	98	97	92	95	98	95	96	98	96	102	101
R4	93	96	95	94	98	102	102	98	96	99	98	100
GRP MEAN	93.5	94.3	93.8	95.0	96.5	96.8	99.0	98.0	97.5	98.8	99.0	100.3
SET MEAN	93.8			96.1			98.2			99.2		

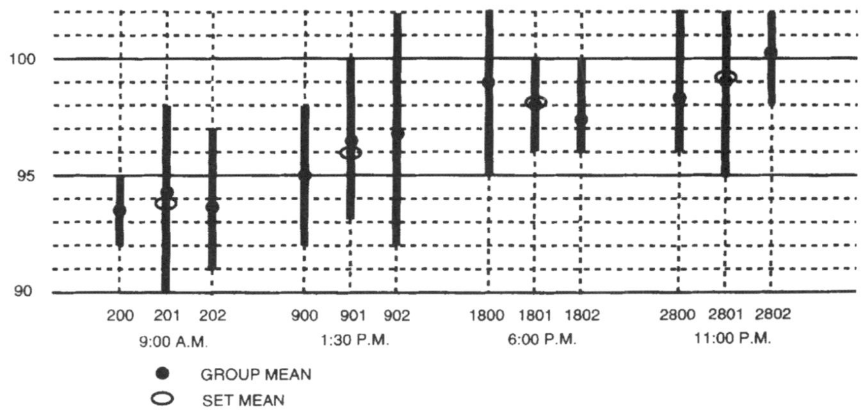

(b) [graph of data with GROUP MEAN ● and SET MEAN ○]

FIGURE 1.18 Example resistor value variation (a) table of data and (b) graph of data.

Because there are only three data points for each type of unit, the statistics of small samples is useful. The first requirement here is that the three performance parameter measurements for the good unit must all yield values that are more acceptable than the three for the bad unit. If this is so, then there is only 1 chance in 20 that this ranking of measurements could happen by accident, giving a 95% confidence in this comparison. The second requirement is that the minimum separation between the medians of variability of the good unit and the bad unit exceeds a minimum. This is illustrated in Fig. 1.19, showing the three data points for the good and bad units. The value of 1.25 for the ratio D/d is based on the classical F Table at the 0.05 level. This means that the results of further tests conducted by swapping parts have at least a 95% level of confidence in the results.

The next step is to identify the parts to be swapped and begin doing so, keeping a chart of the results, such as that in Fig. 1.20. In this plot, the left three data points are those of the original good and bad units, plus their disassembly and reassembly two times. The remaining data points represent those measurements for swapping one part for parts labeled A, etc.

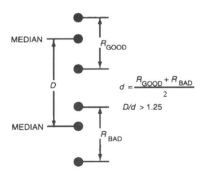

FIGURE 1.19 Test of acceptability of data for component search.

Three results are possible: (1) no change, indicating the part is not at fault, (2) change in one of the units outside its limits but the other unit remains within its limits; or (3) the units flip-flop, the good unit becoming a bad unit and vice versa. A complete flip-flop indicates a part that is seriously at fault—call it a red X. Parts with measurements that are outside the limits but do not cause a complete reversal of good and bad are deemed pink Xs, worthy of further study. A pink X is a partial cause, so that one or more additional pink Xs should be found. Finally, if pink Xs are found, they should be bundled together, that is, all the parts with a pink X result should be swapped as a block

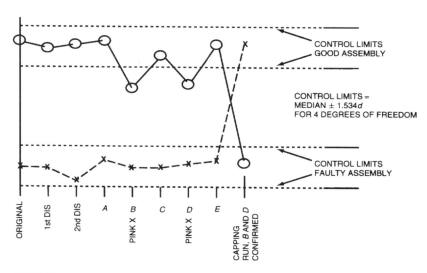

FIGURE 1.20 Components search.

between units. This is called a *capping run* and is illustrated as well. A capping run should result in a complete reversal, as shown, indicating that there are no other causes. Less than a full reversal indicates that other, unidentified, causes exist or that the performance measure is not the best that could have been chosen.

If a single cause, a red X has been found, and remedial action can be initiated. If two to four pink Xs are found, a full factorial analysis, to be described later, should be done to determine the relative importance of the revealed causes and their interactions. Once the importance has been determined, then allocation of resources can be guided by the relative importance of the causes.

Paired Comparisons

If the product cannot be disassembled and reassembled, the technique to use is paired comparisons. The concept is to select pairs of good and bad units and compare them, using whatever visual, mechanical, electrical, chemical, etc., comparisons are possible, recording whatever differences are noticed. Do this for several pairs, continuing until a pattern of differences becomes evident. In many cases, a half-dozen paired comparisons is enough to detect repeatable differences. The units chosen for this test should be selected at random to establish statistical confidence in the results. If the number of differences detected is more than four, then use of *variables search* is indicated. For four or fewer, a full factorial analysis can be done.

Variables Search

Variables search is best applied when there are 5 or more variables with a practical limit of about 20. It is a binary process. It begins by determining a performance parameter and defining a *best* and a *worst* result. Then a ranking of the variables as possible causes is done, followed by assigning for each variable two levels—call them best and worst or good and bad, or some other distinguishing pair. For all variables simultaneously at the best level, the expected result is the best for the performance parameter chosen, similarly for the worst levels. Run two experiments, one with all variables at their best levels and one with all variables at their worst levels. Do this two more times, randomizing the order of best and worst combinations. Use this set of data in the same manner as that for components search using the same requirements and the same limits formula.

If the results meet the best and worst performance, proceed to the next step. If the results do not meet these requirements, interchange the best and worst levels of one parameter at a time until the requirements are met or until all pair reversals are used. If the requirements are still not met, an important factor has

Fundamentals of the Design Process

been left out of the original set, and additional factors must be added until all important requirements are met.

When the requirements are met, then proceed to run pairs of experiments, choosing first the most likely cause and exchanging it between the two groupings. Let the variables be designated as A, B, etc., and use subscripts B and W to indicate the best and worst levels. Let R designate the remainder of the variables. If A is deemed the most likely cause, then this pair of experiments would use $A_W R_B$ and $A_B R_W$, where R is all remaining variables B, C, etc. Observe whether the results fall within the limits, outside the limits but not reversal, or complete reversal, as before. Use a capping run if necessary. If the red Xis found, proceed to remedial efforts. If up to four variables are found, proceed to a full factorial analysis.

Full Factorial Analysis

After the number of possible causes, variables, has been reduced to four or fewer but more than one, a full factorial analysis is used to determine the relative importance of these variables and their interactions. Once again, the purpose of DOE is to direct the allocation of resources in the effort to improve a product and a process. One use of the results is to open tolerances on the lesser important variables if there is economic advantage in doing so.

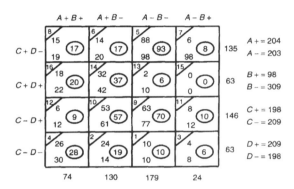

FIGURE 1.21 Full factorial chart.

The simplest 2-factor factorial analysis is to use 2 levels for each factor, requiring that 16 experiments be performed in random order. Actually, for reasons of statistical validity, it is better to perform each experiment a second time, again in a different random order, requiring a total of 32 experiments. If there are fewer than four factors, then correspondingly fewer experiments would need to be performed. The data from these experiments are used to generate two charts—a full factorial chart and an *analysis of variance* (ANOVA) chart. Examples of these two are shown in Figs. 1.21 and 1.22, where the factors are A, B, C, and D with the two levels denoted by + and −. The numbers in the circles represent the average or mean of the data for the two performances of that particular combination of variables. These numbers are then the data for the input column of the ANOVA chart. The numbers in the upper left-hand corner are the cell or box number corresponding to the cell number in the left-hand column of the ANOVA chart. In the ANOVA chart, the + and − signs in the boxes indicate

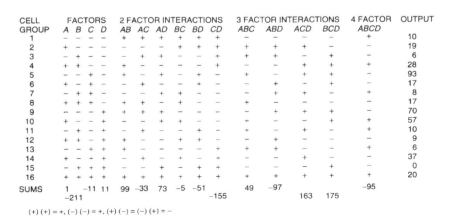

FIGURE 1.22 ANOVA table.

whether the output is to added to or subtracted from the other outputs in that column, with the sum given at the bottom of that column. A column sum with small net, plus or minus, compared to other columns is deemed to be of little importance. The columns with large nets, plus or minus, are deemed the ones that require attention. These two charts contain the data necessary to make a determination of resource allocation.

B vs. C Method

At this point it might be desirable to validate these findings by an independent means. The *B (better)* vs. *C (current)* method is useful for this purpose. There are two parts to this validation: (1) rank a series of samples to see if *B* is better than *C* and (2) determine the degree of risk of assuming that the results are valid. For example, if there are two *B* and two *C*, then there is only one ranking of these four parameters in which the two *B* outrank the two *C*. Therefore, there is only a one in six probability that this ranking occurred by chance. There is a 16.7% risk in assuming that this ranking actually occurred when it should not have. If there are three *B* and three *C*, then requiring that the three *B* outrank the three *C* has only a 1 in 20 probability of happening by chance, a 5% risk. These risk numbers are simply the calculation of the number of combinations of the inputs that can result in the required ranking vs. the total number of combinations that exist. This risk is called the α risk. A definition of an α risk is the risk of assuming improvement when none exists. This is also referred to as a type I error risk. There is also a β risk that is the risk of assuming no improvement when improvement actually does exist, referred to as a type II error risk. It is worthy of note that decreasing one type of risk increases the other for a given sample size. Increasing the sample size may permit decreasing both. It is also true that increasing the sample size may allow some overlap in the *B* vs. *C* ranking, that is, some *C* may be better than some *B* in a larger sample size. Please refer to the references for further discussion.

Realistic Tolerances Parallelogram Plots

The final step in this set of DOE procedures is the optimization of the variables of the process. The tool for this is the realistic tolerances parallelogram plot, often called the scatter plot. The purpose is to establish the variables at their optimum target values. Although there are a number of other techniques for doing this, the use of scatter plots is a simpler process than most, generally with equally acceptable results.

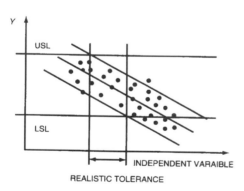

FIGURE 1.23 Scatter plot (realistic tolerance parallelogram plot).

The procedure begins with acquiring 30 output data points by varying the variable over a range of values that is assumed to include the optimum value and plotting the output for these 30 data points vs. the variable under study. An ellipse can be drawn around the data plot so as to identify a major axis. Two lines parallel to the major axis of the ellipse are then drawn on either side of the ellipse to include all but one or one and one-half of the data points. Assuming that specification limits exist for the output, these are drawn on the plot. Then vertical lines are drawn to intersect these specification limit lines at the same point that the parallelogram lines intersect the specification limits, as shown in Fig. 1.23. The intersection of these vertical lines with the variable axis determines the realistic tolerance for the variable.

Additional information can be found in this plot. Small vertical scatter of the data indicates that indeed this is an important variable, that most of the variation of the output is caused by this variable. Conversely, large vertical scatter indicates that other variables are largely responsible for variation of the output. Also, little or no slope to the major axis of the ellipse indicates little or no importance of this variable in influencing the output. The actual slope is not important, as it depends on the scale of the plot. Scatter plots can be made for each of the variables to determine their optimum, target value.

Fundamentals of the Design Process

The techniques presented here are intended to be easy to implement with pencil and paper. As such, they may not be the best, although they are very likely the most economical. The use of recently developed software programs is gaining acceptance, and some of these programs are able to do very sophisticated data manipulation and plotting. As mentioned previously, the advice or direction of a professional statistician is always to be considered, especially for very complex problems.

Process Control

Once the process has been optimized and is in operation, the task becomes that of maintaining this optimized condition. Here, Shainen makes use of two tools, *positrol* and *precontrol*. Again, these are easy to use and understand and provide, in many instances, better results than previously used tools such as control charts from TQM.

Positrol

In simple terms, positrol (short for positive control) is a plan, with appropriate documentation recorded in a log, that identifies *who* is to make *what* measurements, *how* these measurements are to be made, and *when* and *where* they are to be measured. It establishes the responsibility for and the program of measurements that are to be a part of the process control plan. A simple log, although sometimes with detailed entries, is kept so that information about the process is available at any time to operators and managers. Those responsible for keeping the log must be given a short training period on how to keep the log, with emphasis on the importance of making entries promptly and accurately.

An example of a positrol log entry for a surface mount soldering process would be

What	Who	How	Where	When
Copper patterns	Etcher	Visual	Etcher	After etch

Such a positrol log contains complex steps, each of which might well be kept in its own log. For example, the reflow soldering operation could have a number of what steps, such as belt speed; furnace zone temperatures for preheat; soldering zone; cool down zone; atmosphere chemistry control; atmosphere flow speed; visual inspection for missing, misaligned, or tombstoned parts; solder bridges; and solder opens.

A positrol log should contain whatever steps or specifications *(whats)* in a process are to be monitored. As such, it should be the best log or record of the process that the team, in consultation with the operators, can devise. It should be clear that no important steps are to be omitted. The log provides documentation that could be invaluable should the process develop problems. However, it is not the purpose of the log to identify people to blame, but rather to know the people who might have information to offer in finding a solution to the problem.

Precontrol

The second procedure in process control is precontrol, developed by Frank Satterthwaite and described in the 1950s. One of the problems with control charts of the TQM type is that they are slow to indicate a process that is moving to an out of control state. In part this is because of the way the limits on the variability of a process are determined. Satterthwaite suggested that it would be better if the specification limits were used rather than the traditional control limits. He then divided the range between the *upper specification limit (USL)* and the *lower specification limit (LSL)* into four equal-sized regions. The two in the middle on either side of the center of the region between the limits were called the *green zones*, indicating a satisfactory process. The two regions bordering the green zones on one side and the USL and LSL on the other were called the *yellow zones*. Zones outside the limits were called the *red zones*. For an existing process being studied, simple rules for an operator to follow are:

1. If two consecutive samples fall in the green zones, the process may continue.
2. If one sample falls in a green zone and one in a yellow zone, the process is still OK.
3. If both are in the same yellow zone, the process needs adjustment but does not need to be stopped.
4. If the two are in different yellow zones, the process is going out of control and must be stopped.
5. Even one in red zones the process must be stopped.

A stopped process must then be brought back to control.

A new process must be in control before it is brought into production. Here, the procedure is to select five samples. If all five are in green zones, the process may be implemented. If even one is not in the green zones, the process is not yet ready for production and must be further studied.

Another important aspect of precontrol is that, for a process that is in control, the time between samples can be increased the longer it remains in control. The rule is that the time between samples is the time between the prior two consecutive stops divided by six. If it is determined that a more conservative sampling is in order, then divide by a larger number.

Experience has shown that the α risk of precontrol is less than 2%, that is, the risk of stopping a good process. The β risk, not stopping a bad process, is less than 1.5%. These are more than acceptable risks for most real processes.

There have been numerous books and articles written on quality and on quality within the electronic design, manufacturing, and test arenas. The references at the end of this chapter highlight some of them. This handbook will not attempt to duplicate the information available in those and other references.

1.5 Engineering Documentation

1.5.1 Introduction

Little in the technical professions is more important, exacting, or demanding than concise documentation of electronics, electronic products, physical plants, systems, and equipment. Yet this essential task, involving as it does both right- and left-brain activities, a combination of science and art, is all too often characterized as an adjunct skill best left to writers and other specialized talents. The predictable result is poor documentation or, worse, incomplete or incorrect documentation. We underestimate the need for documentation because we underestimate the need to change our systems as time and technology advance.

Neglecting the task of documentation will result, over time, in a product or technical facility where it is more economical and efficient to gut the existing design, assembly, and/or wiring and start over than to attempt to regain control of the documentation. Retroactive documentation is physically difficult and emotionally challenging, and it seldom generates the level of commitment required to be entirely successful or accurate. Inadequate documentation is a major contributor to the high cost of systems maintenance and for the resulting widespread distaste for documentation work; in that sense, bad documentation begets worse documentation.

Yet, documentation is a management function every bit as much as project design, budgeting, planning, and quality control. Documentation is often the difference between an efficient and reliable operation and a misadventure. If the designer does not feel qualified to attempt documentation of a project, that engineer must at the very least oversee and approve of the documentation developed by others.

The amount of time required for documentation can vary from 10 to 50% of the time actually required for the physical project. Because this is often viewed as unseen work, few owners or managers understand its value, and many engineers and technicians simply disdain paperwork; documentation often receives a low priority. In extreme cases, the technical staff may even see "keeping it in my head" as a form of job security, although that threat is often not recognized by today's bottom-line oriented managers. One of the strongest arguments in favor of proper emphasis on documentation is for customer satisfaction. This is true whether a product will be self-serviced by the customer or serviced by the manufacturer's own personnel, and is also true when the "product" is a project that must operate reliably at the customer's site and face periodic maintenance and/or upgrades. All of these situations call strongly for good documentation.

A well-documented project pays dividends in a number of areas.

- Good documentation encourages efficient use of the product or project by providing clear explanation of its purpose and design. Many products or projects are rebuilt, replaced, or retired because of supposed obsolescence, when in fact the designers and builders anticipated future requirements and prepared the product or system to accept them. All this is lost without adequate documentation.

- Good documentation encourages maximum utilization of the product or project by providing a clear explanation of its parts, operating details, maintenance needs, and assembly/construction details. Future modifications can be made with full knowledge of the limits that must be respected as the system is used throughout its operational life.
- Good documentation permits a longer effective operational life for the product or project as changes in technology or task may require periodic updating. A product or facility project that is poorly documented has little or no chance of being expanded to incorporate future changes in technology, simply because the task of writing documentation for an existing system is considered more onerous than reconstruction of the entire work. Or perhaps because there "isn't enough time to do it right" but, as always, we discover there is always enough time to do it over again.
- Good documentation provides maintainability for a product or project without requiring the excessive time and reinvention of the wheel that a poorly documented project requires.
- Good documentation allows any skilled personnel to work on the project, not just someone with personal knowledge of the product or system.

Conventional wisdom asserts that engineering talent and writing talent are not often present in the same individual, providing a lack of incentive for engineers to attempt to provide proper documentation. However, it must be remembered that the scientific method (also the engineering method) requires that experimenters and builders keep careful documentation; yet, outside the laboratory, those engaged in what they see as nonresearch projects forget that the same principles should apply.

Ideally, documentation should begin at the moment of the beginning of a new product. The product log book provides the kernel for documentation by providing information about the rationale for all the design and construction actions taken. In addition, the log book will provide information about rework, repairs, and periodic maintenance the prototype product originally required.

1.5.2 Computers and Types of Documentation

Since creation of documentation involves constant updating and often requires that the same event be recorded and available in two or more locations, it only makes sense that the documentation be produced on a PC or workstation computer. Specific programs that may be used during the creation of documentation include word processors, data bases, spreadsheets, and CAD drawings and data. Ideally, all programs will allow data transfers between programs and permit a single entry to record in several documents. A simple way to allow engineering change orders to be recorded in all applicable documents with one entry is also desirable. It is imperative that all files are backed up on a regular basis and that the copies are stored in a location separate from the main files. In this way, a catastrophe that affects the main files will not affect the backup copies.

Any of the types of documentation need to be written such that a technically skilled person who is not familiar with the specific product can use and follow the documentation to a successful end. Any documentation should be read and understood by a person not involved with the product or project before it is released for use.

There are a number of types of documentation, all or some of which may be necessary, depending on the nature of the project.

- Self-documentation
- Design documentation
- Manufacturing/assembly documentation
- Installation documentation
- Maintenance documentation

- Software documentation
- User/operator documentation

Self-documentation is used for situations that rely on a set of standard practices that are repeated for almost any situation. For example, telephone installations are based on simple circuits that are repeated in an organized and universal manner. Any telephone system that follows the rules is easy to understand and repair or modify, no matter where or how large. For this reason, telephone systems are largely self-documenting, and specific drawings are not necessary for each, e.g., house with a phone system installed.

For self-documentation, like a phone system, the organization, color codes, terminology, and layout must be recorded in minute detail. Once a telephone technician is familiar with the rules of telephone installations, drawings and documentation specific to an individual installation are not necessary for routine installation, expansion, or repair. The same is true of systems such as video and audio recording equipment and small and medium-size computer systems. Likewise, building electrical systems follow a set of codes that are universally applied. The drawings associated with a building wiring installation are largely involved with where items will be located, and not with what wiring procedures will be followed or what type of termination is appropriate.

The key to self-documentation is a consistent set of rules that are either obvious or clearly chronicled such that all engineers and technicians involved are familiar with and rigidly adhere to those rules. If the rules are not all-encompassing, self-documentation is not appropriate. The best rules are those that have some intuitive value, such as using the red cable in a stereo system for the right channel. Both *red* and *right* start with *R*. It must also be noted that all the rules and conventions for a self-documented operation must exist such that they are available to anyone needing them.

Design documentation involves compete information regarding how design decisions were made, the results and justification for each final decision, and inputs from other members of the design team, including manufacturing, test, and installation professionals. It must also include information from suppliers, as appropriate to the product or project. For example, use of a specialized IC, not currently in use by the designer's company, must include copies of the IC specification sheets as part of the design documentation, not just a copy buried somewhere on an engineer's desk. Purchase of a specialized transducer for use in a control system likewise must have spec sheets included not only in the design documentation but also in the manufacturing and installation documentation.

Manufacturing and assembly documentation involves both assembly information for a product that is to be built repeatedly and information for a custom product/project that may only be assembled once. It includes part qualifications, specialized part information, assembly drawings, test procedures, and other information necessary for complete assembly. It should also include traceable source information (who are we buying this part from?) and second-source information for the parts used in the assembly. Manufacturing documentation must also include information about the equipment setup used in the manufacturing and testing of the assembly.

Installation documentation involves information for assembly/installation of a large field system that may only be installed one time. This may be as simple as the instructions for installation of a new float for a tank-level transducer to the complex instructions necessary for the installation of a complete control system for a complex manufacturing line. If it is reasonably expected that the product will be integrated into a system incorporating other products that may or may not be from the same manufacturer, any necessary integration information must be included. Troubleshooting information should be included as appropriate.

Maintenance documentation involves documentation for anyone who will be involved with the maintenance of the product or system. It must be written at a level appropriate for the reader. In the case of a complex product or system, this may reasonably be someone who can be expected to have been through the manufacturer's training of the product or system. In other cases, it can be expected that a maintenance electrician without specific training will be the user of the manual. Necessary drawings, calibration information, and preventive maintenance information must be included, or other manuals provided that complete a total documentation package for the user. The maintenance manual must also include a listing

Fundamentals of the Design Process

of any spare parts the manufacturer believes are necessary on a one-time or a repetitive basis for the proper operation and maintenance of the product throughout its expected lifetime. It is crucial that the location/owners of all copies of maintenance manuals be recorded, so that any necessary changes can be made to all affected manuals.

Maintenance manuals must also include pages for the maintenance personnel to record dates and repairs, upgrades, and preventative maintenance. It may also be appropriate for the manual to suggest that a brief, dated note be made with a permanent marker inside the cover of the product or system enclosure anytime a repair or upgrade is made to that product or system. It is also helpful to have a notice on the enclosure which specifies the location of pertinent manuals for the benefit of new personnel.

Software documentation may or may not include complete software listings. Depending again on the expected reader, software documentation may only include information allowing a user to reach one more level of software changes than the level available to the operator. On the other hand, complete software listings may be requested even by users who have no intention of modifying the software themselves. They may use it to allow others to interface with the given software, or they may want it as a hedge in the event the provider's company goes out of business to prevent having an orphan system without enough documentation for upgrades and necessary changes. Like maintenance manuals, software documentation must have its location/owners recorded so that any necessary changes can be made to all affected manuals.

User/operator documentation is perhaps the most common form of documentation. If the operator of a piece of equipment is not happy with it or has problems operating it due to lack of information and understanding, the report will be that the equipment "doesn't work right." Good documentation is important to prevent this from happening. The operator not only needs to know which buttons to push but also may need to know why a button has a particular function and what "magic" combination of buttons will allow the operator to perform diagnostic procedures. Like maintenance and software manuals, operator documentation must be registered so that any necessary changes can be made to all affected manuals.

Purchasers of any product or system have the right to specify any and all documentation needed in the bid specs for that product or system. Suppliers who choose not to include, or refuse to include, documentation as part of the bid package are providing an indication of their attitude and type of service that can be expected after the sale. Caveat emptor.

1.6 Design for Manufacturability

The principles of design for manufacturing (or manufacturability) are not new concepts, and in their simplest form they can be seen in assembling Legos, as follows:

- Limited parts types
- Standard component sizes
- No separate fasteners required
- No assembly tools required
- Minimized assembly time and operator skills

Electronic assembly is not as simple as Legos and, as discussed in Section 1.2, "Concurrent Engineering," DFM for electronics must be integrated with design for test (DFT). DFM must include quality techniques such as QFD and DOE as well, and the reader must decide which of the many techniques introduced in this chapter will be of the most value to a particular situation. This section will introduce the reader to Suh's DFM technique, the *axiomatic theory of design*.

Axiomatic Theory of Design

The axiomatic theory of design (ATD) is a general structured approach to implementing a product's design from a set of functional requirements. It is a mathematical approach that differentiates the attributes of successful products. The approach develops the functional requirements of the product/assembly, which can then be mapped into design parameters through a design matrix and then into manufacturing process variables. The functional requirements and the design parameters are hierarchical and should decompose into subrequirements and subparameters.

The design function is bounded by the following two constraints:

- Input constraints, which originate from the original functional specifications of the product
- Systems constraints, which originate from use-environment issues

Using functional requirements is similar to using quality function deployment (QFD). Typically, the constraints in ATD do not include customer issues, whereas QFD includes those issues and does not address the specifics of design and manufacturing. QFD is discussed briefly in Section 1.4, "Quality Concepts."

The constraints drive the design parameters to form a boundary inside which the implementation of the design must rest. For example, if operation requires deployment of the product in an automotive environment, then one must design for 12 V nominal battery power. Designing for, e.g., 5 V power would be designing outside the defined border.

Generally, it is assumed the design will result in a series of functional modules, whether they are packaged separately or not. A complete audio system may be seen as having separate CD, tape, amplifier, and video modules, whereas a CD player may be seen as including laser, microprocessor, and amplifier modules.

The two axioms of design are

1. The *independence axiom,* which requires the independence of functional requirements. This axiom is best defined as having each functional block/module stand alone, not requiring individualized tuning of the module to its input and output modules.
2. The *information axiom,* which minimizes the information content of the design. This axiom is intended to minimize both the initial specifications as well as minimizing the manufacturability issues necessary for the product.

Remember that every part in every design has a "range" in its specifications. For example, a 10% tolerance 10 kΩ resistor may fall anywhere in the range of 10k − 10% to 10k + 10% (9k to 11k). An op-amp has a range to its frequency response and its CMRR, usually specified at their minimum limits.

Design Guidelines

1. *Decouple designs.*   Each module should stand alone, and be able to be manufactured, tested, and assembled without depending on the individual parts characteristics of another module, as long as that module performs within its specifications.
    - No final assembly adjustments should be required after modules are assembled. If each module meets its specs, the final assembled device should function within the overall specifications without further adjustments to any of the individual modules.
    - Provide self-diagnosis capability as appropriate for each module. Microprocessor/controller-based modules can have this ability, while other modules may incorporate diagnoses as simple as battery level indicators and nonfunctional light indicators.
2. *Minimize functional requirements.*   If a requirement is not necessary for a product to meet its overall functional requirements, eliminate it. This, for example, may be a performance requirement, an environmental requirement, or an internal test requirement.
    - Modular design assumes that a module is discrete and self-contained. This means that the overall functional specs will be broken into appropriate module specs such that each module assembly, within the design and manufacturing specs, will meet its own functional specs. Inputs to the module are specified, along with their ranges, based on the output of the previous module. Output specs are developed based on the needs of the next module.

Fundamentals of the Design Process

- Build on a suitable base. This means considering the form and fit of the module as required by the overall system. Consider the final orientation of the module if it may affect performance/adjustment. Depending on how the module will be assembled into the final assembly, fasteners should be eliminated or minimized. The module should be designed for whatever automated assembly processes are available.
3. *Integrate physical parts.*
    - Follow DFM guidelines, e.g., those in Prasad.[6]
    - Minimize excess parts both in part count (Is it necessary to have a decoupling capacitor at every IC?) and in types of parts (Is it necessary to have an LF351 op-amp when the rest of the op-amp applications use an LF353?). Doing this requires analyzing the value of each part:
        - Identify necessary functions of each part
        - Find the most economical way to achieve the functions

Remember that the final cost of the module and assembly is directly related to the number and cost of the parts. Furthermore, each additional part impacts the module/assembly reliability and quality.

4. *Standardize parts.*
    - Stock issues impact manufacturing and rework/repair as well as cost. The fewer part types used in an assembly, the easier both of these functions become.
    - Standardization issues include not only part values/types but also tolerances and temperature ranges.

1.7 ISO 9000[*]

1.7.1 Introduction and Definitions

Developing a product for international exposure will almost certainly require knowledge of and adherence to ISO 9000 standards. This section is intended to make the design team knowledgeable about the ISO 9000 series of standards and the regulatory environment in which an ISO 9000-registered product must exist. It does not attempt to repeat the information in the standards themselves.

ISO 9000 and related documents make up a set of standards developed and promulgated by the International Organization for Standardization (ISO) in Geneva, Switzerland. Representing 91 countries, the purpose of the ISO is to promote the worldwide standardization of manufacturing practices with the intent of facilitating the international exchange of goods and services. In 1987, the ISO released the first publication of the ISO 9000 series, which was and continues to be composed of five international standards. These standards are designed to (1) guide the development of an organization's internal quality management programs and (2) help an organization ensure the quality of its externally purchased goods and services. To this end, the ISO 9000 standards apply to both suppliers and purchasers. They pertain not only to the manufacturing and selling of products and services but to the buying of them as well.

The rationale behind the design of the ISO 9000 standards is as follows. Most organizations—industrial, governmental, or commercial—produce a product or service intended to satisfy a user's needs or requirements. Such requirements are often incorporated in specifications. Technical specifications, however, may not in themselves guarantee that a customer's requirements will be consistently met if there happen to be any deficiencies in the specification or in the organizational system to design and produce the product or service. Consequently, this has led to the development of quality system standards and guidelines that complement relevant product or service requirements given in the technical specification.

[*]Adapted from Whitaker, J, *The Electronics Engineering Handbook,* Chap. 148, "ISO 9000," by Cynthia Tomovic.

The ISO series of standards (ISO 9000 through ISO 9004) embodies a rationalization of the many and various national approaches in this sphere.

If a purchaser buys a product or service from an organization that is ISO 9000 certified, the purchaser will know that the quality of the product or service meets a defined series of standards that should be consistent, because the documentation of the processes involved in the generation of the product or service *have been verified by an outside third party* (auditor and/or registrar).

As defined by the ISO, the five standards are documents that pertain to quality management standards. Individually, they are

1. *ISO 9000: Quality Management Assurance Standards—Guide for Selection and Use.* This standard is to be used as a guideline to facilitate decisions with respect to selection and use of the other standards in the ISO 9000 series.
2. *ISO 9001: Quality Systems—Model for Quality Assurance in Design/Development, Production, Installation, and Services.* This is the most comprehensive ISO standard, used when conformance to specified requirements are to be assured by the supplier during the several stages of design, development, production, installation, and service.
3. *ISO 9002: Quality Systems—Model for Quality Assurance in Production and Installation.* This standard is to be used when conformance to specified requirements are to be assured by the supplier during production and installation.
4. *ISO 9003: Quality Systems—Model for Quality Assurance in Final Inspection and Test.* This standard is to be used when conformance to specified requirements are to be assured by the supplier solely at final inspection and test.
5. *ISO 9004: Quality Management and Quality System Elements.* This standard is used as a model to develop and implement a quality management system. Basic elements of a quality management system are described. There is a heavy emphasis on meeting customer needs.

From these definitions, the ISO states that only ISO 9001, 9002, and 9003 are contractual in nature and may be required in purchasing agreements. ISO 9000 and 9004 are guidelines, with ISO 9000 serving as an index to the entire ISO 9000 series and ISO 9004 serving as a framework for developing quality and auditing systems.

1.7.2 Implementation

In 1987, the United States adopted the ISO 9000 series as the American National Standards Institute/American Society for Quality Control (ANSI/ASQC) Q9000 series. These standards are functionally equivalent to the European standards. For certain goods known as registered products, which must meet specific product directives and requirements before they can be sold in the European market, ISO 9000 certification forms only a portion of the export requirements. As an example, an electrical device intended for the European market may be expected to meet ISO 9000 requirements and additionally may be required to meet the electrotechnical standards of the International Electrotechnical Commission (IEC).

In the U.S., quality standards continue to develop. In the automotive arena, Chrysler (now Daimler-Chrysler), Ford, and General Motors developed QS9000 and QS13000, which go beyond the requirements of the ISO series in areas the developers feel are important to their specific businesses.

1.7.3 Registration and Auditing Process

The following is an introduction to the ISO series registration and auditing process. As will be seen, choosing the correct ISO standard for certification is as important as completing the registration requirements for that standard.

Registration, Europe

Initial embracing of the ISO series was strongest in Europe. In conjunction with the development of ISO 9000 certification for products and services sold in the European markets, a cottage industry of consultants, auditors, and registrars has developed. To control this industry, a number of Memorandums of Understanding (MOUs) were agreed upon and signed between many European nations. The European Accreditation of Certification (EAC) was signed by 13 European countries so that a single European system for recognizing certification and registration bodies could be developed. Likewise, the European Organization for Testing and Certification (EOTC) was signed between member countries of the European community and the European Free Trade Association to promote mutual recognition of test results, certification procedures, and quality system assessments and registrations in the nonregulated product groups.

A number of such MOUs have been signed between countries, including:

- European Organization for Quality (EOQ), to improve the quality and reliability of goods and services through publications and training
- European Committee for Quality Systems Certification (EQS), to promote the blending of rules and procedures used for quality assessment and registration among member nations.
- European Network for Quality System Assessment and Certification (EQNET), to establish close cooperation leading to mutual recognition of registration certificates.

In addition to these MOUs, several bilateral association agreements have been signed with the European community and other European countries, including former Soviet-bloc states. In reality, most accrediting bodies in Europe continue to be linked country-specific boards. Therefore, the manufacturer intending to participate in international commerce must investigate, or have investigated for it, the applicable standards in any and all countries in which the manufacturer expects to do business.

Registration, U.S.A.

As in Europe, the development of ISO 9000 certification as a product and service requirement for European export has promoted the development of ISO 9000-series consultants, auditors, and registrars, not all of whom merit identical esteem. In an attempt to control the quality of this quality-based industry, the European community and the U.S. Department of Commerce designated that, for regulated products, the National Institute of Standards and Technology (NIST, formerly the NBS) would serve as the regulatory agency responsible for conducting conformity assessment activities, which would ensure the competence of U.S.-based testing, certification, and quality system registration bodies. The program developed by the NIST is called the *National Voluntary Conformity Assessment System Evaluation (NVCASE)*.

For nonregulated products, the Registrar Accreditation Board (RAB), a joint venture between the American Society for Quality Control (ASQC) and the American National Standards Institute (ANSI), was designated as the agency responsible for certifying registrars and their auditors and for evaluating auditor training. In addition to the RAB, registrars in the U.S.A. may also be certified by the Dutch Council for Accreditation (RvC) and the Standards Council of Canada (SCC). Some U.S. registrars, on the other hand, are registrars in parent countries in Europe and are certified in their home countries.

Choosing the right ISO 9000 registrar is no easy matter, whether European based, or U.S. based. Since the RAB in the U.S. is not a governmental agency, there is little to prevent anyone from claiming to be an ISO 9000 consultant, auditor, or registrar. For that reason, applicants for registration in the U.S. may be wise to employ auditing bodies that have linked themselves with European registrars or to ask the NIST (for regulated products) or the RAB (for nonregulated products) for a list of accepted U.S. accredited agencies. In any event, the following questions have been suggested as a minimum to ask when choosing a registrar:[7b]

- Does the registrar's philosophy correspond with that of the applicant?
- Is the registrar accredited and by whom?
- Does the registrar have experience in the applicant's specific industry?
- Does the registrar have ISO 900 certification itself? In general, it should. Ask to see its quality manual.

- Will the registrar supply references from companies it has audited and then registered to ISO 9000?
- Is the registrar registered in the marketplace into which the applicant wants to sell?

Remember that auditors within the same registering body can differ. If you are confident of the registering body but not with the auditor, ask that a different auditor be assigned to your organization.

U.S. Regulatory Agencies

Regulated Products
    National Center for Standards and Certification Information
    National Institute for Standards and Technology
    TRF Bldg. A 163
    Gaithersburg, MD 20899
    301 975-4040

Nonregulated Products
    Registrar Accreditation Board
    611 East Wisconsin Ave.
    P.O. Box 3005
    Milwaukee, WI 53202
    414 272-8575

Auditing Process

The auditing process[4] will begin with a preliminary discussion of the assessment process among the parties involved, the auditing body, and the applicant organization pursuing certification. If both sides agree to continue, the auditing body should conduct a preliminary survey, and the organization should file an application. Next, dates for conducting a preaudit visit should be set as well as for subsequent on-site audits. Estimates should be prepared of the time and money required for the registration process.

Depending on which ISO document the applicant has as a certification goal, different areas of the organization will be involved. ISO 9003 will, e.g., require primary involvement with the inspection and test areas. If it is possible, or planned, that other documents will become goals at a later date, this is the time to get representatives from those areas involved as "trainees" to the certification process. If it is intended to later pursue ISO 9002 certification, representatives from production and installation should participate as silent members of the 9003 certification team. When 9002 certification is pursued later, these members will understand what will be required of their areas.

During the preaudit visit, the auditing body should explain how the assessment will be conducted. After an on-site audit is conducted, the applicant organization should be provided with a detailed summary of the audit outcome, including areas requiring attention and corrective action.

1.7.4 Implementation: Process Flow

The first step to implementing ISO 9000 is to recognize the need for, and to develop the desire for, a continuous quality improvement program throughout the entire organization. Second, organizations that previously employed closed-system/open-loop quality practices must forego those practices in favor of practices more appropriate to an open-system/closed-loop approach. Some organizations, for example, depend on a technical specialist to develop and update quality documents with little or no input from operations personnel who must believe in and implement any quality program. How many walls have SPC charts which are rarely, if ever, looked at from the standpoint of making real-time process corrections to reduce process errors? These companies must change their quality practices and deploy a system of practices that constantly solicits employee input on matters of quality improvement, and that documents such suggestions if they are implemented. Documentation must move from being a static, one-time procedure to becoming a dynamic, quality improvement process whose benefits are realized as a function of improving organizational communication. Believe in and use the information on the SPC charts.

Fundamentals of the Design Process

Clearly this type of process better supports day-to-day operations, which should lead to greater profitability.

An organization interested in becoming ISO 9000 certified needs to:[7a]

- Conduct a self-assessment relative to the appropriate standard document.
- Analyze the results of the self-assessment and identify problem areas.
- Develop and implement solutions to the problems identified in the self-assessment.
- Create a detailed quality process manual after solutions have been implemented. This manual must be submitted during the registration process.
- Hire a registered, independent third-party registrar who will determine whether the organization qualifies for certification. If an organization passes, the auditor will register the organization with the ISO and schedule subsequent audits every two years, which are required for the organization to maintain its certified status.

Based on published ISO 9000 implementation guides, the flowchart shown in Fig. 1.24 illustrates an overview of an ISO 9001 implementation scheme. Detailed flowcharts for each program are available in Ref. 10.

First Stage

Identify the major elements of the standard for which you are seeking registration. In addition, assign a champion to each element (a person responsible for each element) along with appropriate due dates.

Second Stage

Develop and implement the following three primary programs that permit a quality system to operate:

1. document control
2. corrective action
3. internal quality auditing

The *document control program* describes the processes, procedures, and requirements of the business operation. Steps and activities in this program should include (1) defining the process, (2) developing a procedure for each task identified in the process, (3) establishing requirements for performance, and (4) establishing a method for measuring actual performance against the requirements.

The *corrective action program* describes the manner in which corrective action is to be conducted in a business operation. Steps and activities in this program should include (1) writing a corrective action request when a problem is identified, (2) submitting the corrective action request to the corrective action coordinator who logs the request, (3) returning the request to the coordinator after the corrective action is completed, for updating the log, (4) establishing requirements for performance, and (5) establishing a method for measuring actual performance against the requirements.

The *internal quality auditing program* describes the manner in which internal quality auditing is to be conducted in a business operation. Stops and activities in this program should include (1) planning and scheduling the audit, (2) developing an audit checklist based on the functions of the audit, (3) preparing a written audit report that describes the observations of the audit, (4) establishing requirements for performance, and (5) establishing a method for measuring actual performance against the requirements.

Third Stage

Develop and implement the following programs: contract review and purchasing.

The *contract review program* describes the manner in which a contract review is to be conducted in a business operation. Steps and activities in the program should include (1) developing a process whereby customer orders are received, (2) developing a process for verifying customer information and needs, (3) fulfilling and verifying whether customer needs have been met, (4) establishing requirements for performance, and (5) establishing a method for measuring actual performance against requirements.

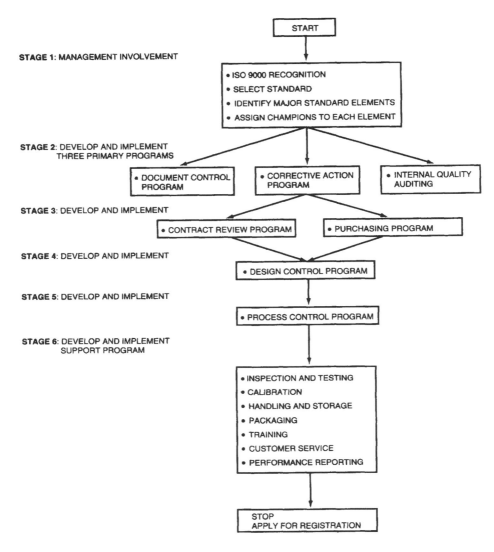

FIGURE 1.24 ISO 9001 program development process flowchart.

The *purchasing program* describes the manner in which purchasing is to be conducted in a business operation. Steps and activities in this program should include (1) identifying supplier evaluation requirements, (2) developing a purchase order process review procedure, (3) creating a method for identifying material requirements, (4) establishing requirements for performance, and (5) establishing a method for measuring actual performance against the requirements.

Fourth Stage

Develop and implement the design control program.

The *design control program* describes the manner in which to control the design process. Steps and Activities in this program should include (1) providing engineers with design input from the sales/marketing departments at the start of any project, (2) establishing a design plan that would include appropriate identification, approval signatures, designated design activities, identification of responsible persons, and tracking of departmental interfaces, (3) establishing requirements for performance, and (4) establishing a method for measuring actual performance against the requirements.

Fifth Stage

Develop and implement the process control program.

The *process control program* describes the manner in which to implement a manner of controlling the process. The steps and activities in this program should include (1) planning and scheduling production, (2) developing a bill of material based on the product to be produced, (3) developing product requirements, (4) establishing requirements for performance, and (5) establishing a method for measuring actual performance against the requirements.

There is a sixth stage, in which support programs are developed and implemented. These programs include inspection and testing, calibration, handling and storage, packaging, training, service, and performance reporting.

It is important to note that, in an organization that already embraces design for manufacturability (DFM) and/or design for testability (DFT), along with other modern process integration techniques and process quality techniques such as total quality management (TQM), many of the activities needed for ISO 9000 implementation already exist. Implementation should not require dismantling existing programs; rather, it should involve blending existing activities with additional ones required by the ISO documentation.

Many of the process and quality techniques embrace some form of empowerment of individuals. With the proper training and education in the organizational and documentation issues related to the ISO series, this additional knowledge aids the empowerment by allowing increasingly informed decisions to be made that will continue to have a positive effect on the success of the organization.

1.7.5 Benefits vs. Costs and Associated Problems

Based on a limited review of the literature, the following benefits,[8] and costs and associated problems[12] with ISO 9000 have been identified.

Benefits of ISO 9000

The transformation of regional trade partnerships into global exchange networks has spurred the need for the standardization of products and services worldwide. Although the original intent of the ISO 9000 series was to provide guidelines for trade within the European Community, the series has rapidly become the world's quality process standards. Originally, it was thought that the ISO 9000 series would be of interest only to large manufacturing organizations with international markets. However, it has become apparent that medium- and smaller-sized organizations with a limited domestic market base are interested as well. Much of this is driven by major manufacturers. As they become ISO 9000 certified, they expect their suppliers to do the same, since they must certify their parts acquisition process as well as their final product. In many markets, certification has become a de facto market requirement, and certified suppliers frequently beat out their noncertified competitors for contracts.

In addition to gaining a competitive edge, other reasons for small- and medium-sized organizations to consider certification include

- Employee involvement in the audit preparation process fosters team spirit and a sense of communal responsibility.
- Audits may reveal that critical functions are not being performed well or at all.
- ISO 9001 and 9002 provide the foundation for developing a disciplined quality system.
- Outside auditors may raise issues that an inside observer do not see because the inside observer is too close to the business.

Last, in many organizations, after the initial work to bring documentation and quality process up to acceptable standards, significant cost savings can accrue. The British Standards Institute estimates that certified organizations reduce their operating costs by 10% on average.[2]

Costs and Associated Problems

A common complaint among ISO seekers is that the amount of time required to keep up with the paperwork involved in developing a comprehensive, organizationwide quality system robs middle managers of the time necessary to accomplish other important job-related tasks. Thus, the management of time becomes a problem. Again, among middle managers, the ISO 9000 series is frequently perceived as another management gimmick that will fade in time. Thus, obtaining the cooperation of middle management becomes a problem.

Relative to time and money, ISO 9000 certification typically takes 12 to 18 months, and it is not cheap. For a medium-sized organization, expect a minimum of 15,000 to 18,000 man-hours of internal staff time to be required. Expect to spend $30,000 to $40,000 for outside consultants the first year. Also, expect to spend $20,000 plus travel costs for the external ISO 9000 auditing body you hire to conduct the preassessment and final audits. Clearly, organizations should expect to spent a considerable amount of both time and money resources. A lack of willingness to designate resources has been cited as one of the primary reasons organizations either failed or did not achieve full benefits from the ISO audits.[4]

1.7.6 Additional Issues

The following potpourri of issues may relate to your organization's consideration of the ISO 9000 process and ramifications.

An organization can seek certification for activities related to a single product line. This is true even if other products are produced at the same facility.

An organization's quality system is not frozen in time with ISO 9000 certification. Changes to the quality system can be made as long as they are documented and the changes are acceptable to future ISO 9000 surveillance audits.

The value of ISO certification must be examined by each organization. A cost-benefit analysis must be conducted to determine whether the expenses associated with ISO 9000-series certification is cost effective.

In addition to the other issues presented in this section, consider these issues in evaluation of the worth of ISO 9000 certification:

- What percentage of your current customers are requesting certification?
- What percentage of your total business is with the customers who are requesting certification?
- Can you afford to lose the customers who are requesting certification?
- Are most of your customers certified, giving them a competitive edge over you?
- Do you stand to gain a competitive edge by being one of the first in your industry to obtain certification?
- What percentage of your business is currently conducted in the European community? Do you expect that business to increase or decrease?

1.7.7 ISO 9000 Summary

In summary, the ISO 9000 series does not refer to end-state products or services but to the system that produces them. Although the ISO does not mandate a specific approach to certification, it does mandate that organizations "say what they do and do what they say." The purpose of the ISO is to promote the standardization of manufacturing practices with the intent of facilitating the international exchange of goods and services. The ISO 9000 series was designed to aid in the development of internal quality programs and to give purchasers the confidence that certified organizations consistently deliver the quality buyers expect. Although applying to all products and services, the ISO 9000 series is intended to complement industry-specific product and service standards.

Clearly, obtaining ISO 9000 certification, or the even more rigorous QS 9000 and QS 13000 certifications, is no easy task. The decision to pursue ISO certification and the resultant activities have resulted in decreased costs and increased quality for organizations such as IBM, Apple, Motorola, Hewlett-Packard, Solectron, and others. However this decision is neither easy nor without risk. IBM's Baldridge-award winning Rochester, MN, site failed its first ISO audit.

Defining Terms

- *De facto market requirements.* The baseline expectations of the marketplace.
- *Nonregulated products.* Products presumed to cause no bodily harm.
- *Regulated products.* Products known to potentially cause a fatality or result in bodily harm if used inappropriately.

1.8 Bids and Specifications

Most companies have general specifications that must be followed when bidding, whether the bid request is for the purchase of a single item or for a system. These specifications must be followed. In addition, any bidding done as part of a government contract must follow a myriad of rules with which the company's purchasing department can assist. No attempt will be made here to address the various governmental and military purchasing requirements.

Specifications for bidding must be written with care. If they are written to a specific component, product, instrument, or device, the writer will have problems justifying another device that may be a better performer but doesn't meet the written specifications. One technique used is to write bid specifications such that no known vendor can meet any of them. This prevents accusations that the writer had a particular vendor in mind and allows any and all of the bids to be thrown out if that seems appropriate. It also allows justification of any acceptable unit. This technique must result in a set of bid specifications for which each item is in the best interests of the company. Any hint of conflict of interest or favoritism leaves the writer open to reprimand, lawsuits, or dismissal.

1.9 Reference and Standards Organizations

See also the abbreviations list at the end of this section.

American National Standards Institute (ANSI)
11 West 42nd Street
New York, NY 10036
(212) 642-4900 fax: (212) 398-0023
http://www.ansi.org

American Society for Testing and Materials (ASTM)
100 Barr Harbor Drive
West Conshohocken, PA 19428-2959
(610) 832-9585 fax: (610) 832-9555
http://www.astm.org

Canadian Standards Association (CSA)
178 Rexdale Boulevard
Etobicoke (Toronto), ON M9W 1R3
(416) 747-4000 fax: (416) 747-4149
http://www.csa.ca

Department of Defense Standardization Documents
Order Desk, Building 4D
700 Robbins Ave.
Philadelphia, PA 19111-5094
(215) 697-2667

Electronic Industries Alliance (EIA)
2500 Wilson Blvd.
Arlington, VA 22201-3834
(703) 907-7500 fax: (703)907-7501
http://www.eia.org

Federal Communications Commission (FCC)
445 12th St. S.W.,
Washington DC 20554
(202) 418-0200
http://www.fcc.gov

Institute for Interconnecting and Packaging Electronic Circuits (IPC)
2215 Sanders Rd.
Northbrook, IL 60062-6135
(847) 509-9700 fax (847) 509-9798
http://www.ipc.org/index.html

International Electrotechnical Commission (IEC)
Rue de Varembé
1121 Genève 20, Switzerland
IEC documents are available in the USA from ANSI.

International Organization for Standardization (ISO)
1, Rue de Varembé
Case Postale 56
CH-1211 Genève 20
Switzerland
Telephone +41 22 749 01 11 Telefax +41 22 733 34 30
http://www.iso.ch
The ISO Member for the USA is ANSI, and ISO documents are available from ANSI.

Joint Electron Device Engineering Council (JEDEC)
Electronic Industries Alliance
2500 Wilson Boulevard
Arlington, VA 22201-3834
(703) 907-7500 fax: (703) 907-7501
http://www.jedec.org

National Fire Protection Association (NFPA)
1 Batterymarch Park
PO Box 9101
Quincy, MA 02269-9101
(617) 770-3000 fax: (617) 770-0700
http://www.nfpa.org

Surface Mount Technology Association (SMTA)
5200 Willson Road, Suite 215
Edina, MN 55424
(612) 920-7682 fax: (612) 926-1819
http://www.smta.org

Underwriters Laboratories
333 Pfingsten Rd.
Northbrook, IL 60062

(800) 595-9844 fax: (847) 509-6219
http://www.ul.com

Abbreviations for Standards Organizations

ASTM	American Society for Testing and Materials
CSA	Canadian Standards Association
DODISS	Department of Defense Index of Specifications and Standards
IEC	International Electrotechnical Commission
EIA	Electronic Industries Alliance
EIAJ	Electronic Industries Alliance of Japan
FCC	Federal Communications Commission
IMAPS	International Microelectronics and Packaging Society
IPC	Institute for Interconnecting and Packaging Electronic Circuits
ISHM	now IMAPS
ISO	International Standards Organization
JEDEC	Joint Electronic Devices Engineering Council of the EIA
NEMA	National Electrical Manufacturers Association
NFPA	National Fire Protection Association
UL	Underwriters Laboratories

References

1. Hoban, FT, Lawbaugh, WM, *Readings in Systems Management,* NASA Science and Technical Information Program, Washington, 1993.
2. Hayes, HM, "ISO 9000: The New Strategic Consideration," *Business Horizons,* vol. 337, Oct. 1994, 52–60.
3. *ISO 9000: International Standards for Quality,* International Organization for Standardization, Geneva, Switzerland, 1991.
4. Jackson, S, "What you Should Know About ISO 9000," *Training,* vol. 29, May 1993, 48–52.
5. (a) Pease, R, "What's All This Taguchi Stuff?" *Electronics Design,* June 25, 1992, 95ff.
   (b) Pease R, "What's All This Taguchi Stuff, Anyhow (part II)?" *Electronics Design,* June 10, 1993, 85ff.
6. Prasad, RP, *Surface Mount Technology Principles and Practice,* 2/e, 7.5–7.8. Van Nostrand Reinhold, New York, 1997.
7. (a) Russell, JF, "The Stampede to ISO 9000." *Electronics Business Buyer,* vol. 19, Oct. 1993, 101–110.
   (b) Russell, JF, "Why the Right ISO 9000 Registrar Counts," *Electronics Business Buyer,* vol. 19, Oct. 1993, 133–134.
8. Schroeder, WL, "Quality Control in the Marketplace." *Business Mexico,* vol. 3, May, 1993, 44–46.
9. Smith, J, Oliver, M, "Statistics: The Great Quality Gamble," *Machine Design,* October 8, 1992.
10. (a) Stewart, JR, Mauch, P., and Straka, F, *The 90-Day ISO Manual: The Basics.* St. Lucie Press, Delray Beach, FL, 1994, 2–14
    (b) Stewart, JR, Mauch, P., and Straka, F, *The 90-Day ISO Manual: Implementation Guide.* St. Lucie Press, Delray Beach, FL, 1994.
11. Suh, 1993.
12. Zuckerman, A., "Second Thoughts about ISO 9000.", vol. 31, Oct. 1994, 51–52.

2
Surface Mount Technology

Glenn R. Blackwell
Purdue University

2.1 Introduction 2.1
2.2 SMT Overview 2.1
2.3 Surface Mount Device Definitions 2.6
2.4 Substrate Design Guidelines 2.8
2.5 Thermal Design Considerations 2.10
2.6 Adhesives 2.12
2.7 Solder Joint Formation 2.25
2.8 Parts 2.32
2.9 Reflow Soldering 2.36
2.10 Cleaning 2.39
2.11 Prototype Systems 2.40

2.1 Introduction

This chapter on surface mount technology (SMT) is to familiarize the reader with the process steps in a successful SMT design. It assumes basic knowledge of electronic manufacturing. Being successful with the implementation of SMT means the engineers involved must commit to the principles of concurrent engineering. It also means that a continuing commitment to a quality technique is necessary, whether that is Taguchi, TQM, SPC, DOE, another technique, or a combination of several quality techniques.

Related information is available in the following chapters of this book:

- Concurrent engineering, quality—Chapter 1
- IC packaging—Chapter 3
- Circuit boards—Chapter 5
- Design for test—Chapter 9
- Adhesives—Chapter 10
- Thermal management—Chapter 11
- Inspection—Chapter 13

2.2 SMT Overview

Surface mount technology is a collection of scientific and engineering methods needed to design, build, and test products made with electronic components that mount to the surface of the printed circuit board without holes for leads.[1] This definition notes the breadth of topics necessary to understand SMT and also clearly says that the successful implementation of SMT will require the use of concurrent engineering.[2] Concurrent engineering, as discussed in Chapter 1, means that a team of design, manufacturing, test, and marketing people will concern themselves with board layout, parts and part placement issues, soldering, cleaning, test, rework, and packaging—before any product is made. Concurrent engineering is discussed in Chapter 1. The careful control of all these issues improves both yield and reliability of the final product. In fact, SMT cannot be reasonably implemented without the use of concurrent engineering and/or the principles contained in *design for manufacturability (DFM)* and *design for testability (DFT)*, and therefore any facility that has not embraced these principles should do so if implementation of SMT is its goal. DFM and DFT are also discussed in Chapter 1, while DFT is discussed in detail in Chapters 4 and 16.

Note that, while many types of substrate are used in SMT design and production, including FR-4, ceramic, metal, and flexible substrates, this chapter will use the generic term *board* to refer to any surface upon which parts will be placed for a production assembly.

Considerations in the Implementation of SMT

The main reasons to consider implementation of SMT include

- reduction in circuit board size
- reduction in circuit board weight
- reduction in number of layers in the circuit board
- reduction in trace lengths on the circuit board, with correspondingly shorter signal transit times and potentially higher-speed operation
- reduction in board assembly cost through automation

However, not all of these reductions may occur in any given product redesign from through-hole technology (THT) to SMT. Obviously, many current products, such as digital watches, laptop computers, and camcorders, would not be possible without the size and cost advantages of SMT.

Important in all electronic products are both quality and reliability.

- Quality = the ability of the product to function to its specifications at the conclusion of the assembly process.
- Reliability = the ability of the product to function to its specifications during its designed lifetime.

Most companies that have not converted to SMT are considering doing so. All, of course, is not golden in SMT Land. During the assembly of a through-hole board, either the component leads go through the holes or they don't, and the component placement machines typically can detect the difference in force involved and yell for help. During SMT board assembly, the placement machine does not have such direct feedback, and accuracy of final soldered placement becomes a stochastic (probability-based) process, dependent on such items as component pad design, accuracy of the PCB artwork and fabrication (which affects the accuracy of trace location), accuracy of solder paste deposition location and deposition volume, accuracy of adhesive deposition location and volume if adhesive is used, accuracy of placement machine vision system(s), variations in component sizes from the assumed sizes, and thermal issues in the solder reflow process. In THT test, there is a through-hole at every potential test point, making it easy to align a bed-of-nails tester. In SMT designs, there are not holes corresponding to every device lead. The design team must consider form, fit and function, time-to-market, existing capabilities, testing, rework capabilities, and the cost and time to characterize a new process when deciding on a change of technologies. See chapters 4 and 16.

2.2.1 SMT Design, Assembly, and Test Overview

The IPC has defined three general end-product classes of electronic products.

- Class 1: General consumer products

Surface Mount Technology

- Class 2: Dedicated service electronic products—including communications, business, instrumentation, and military products, where high performance and extended life are required, and where uninterrupted service is desired but not critical.
- Class 3: High reliability electronic products—commercial and military products where equipment downtime cannot be tolerated.

All three performance classes have the same needs with regard to necessary design and process functions.

- Circuit design (not covered in this handbook)
- Substrate [typically, printed circuit board (PCB) design (Chapter 5)]
- Thermal design considerations (Chapter 11)
- Bare PCB fabrication and test (not covered in this chapter)
- Application of adhesive, if necessary (Chapter 10)
- Application of solder paste
- Placement of components in solder paste
- Reflowing of solder paste
- Cleaning, if necessary
- Testing of populated PCB (Chapters 9 and 12)

Once the circuit design is complete, substrate design and fabrication, most commonly of a printed circuit board (PCB), enter the process. Generally, PCB assemblies are classified into types and classes as described in IPC's "Guidelines for Printed Board Component Mounting," IPC-CM-770. It is unfortunate that the IPC chose to use the term *class* for both end-product classification and for this definition. The reader of this and other documents should be careful to understand which *class* is being referenced. The *types* are as follows:

- Type 1: components (SMT and/or THT) mounted on only one side of the board
- Type 2: components (SMT and/or THT) mounted on both sides of the board

The types are further subdivided by the types of components mounted on the board.

- A: through-hole components only
- B: surface mounted components only
- C: simple through-hole and surface mount components mixed
- X: through-hole and/or complex surface mount components, including fine pitch and BGAs
- Y: through-hole and/or complex surface mount components, including ultrafine pitch and chip scale packages (CSPs)
- Z: through-hole and/or complex surface mount, including ultrafine pitch, chip on board (COB), flip chip, and tape automated bonding (TAB)

The most common type combinations, and the appropriate soldering technique(s) for each, are

- 1A: THT on one side, all components inserted from top side
  Wave soldering
- 1B: SMD on one side, all components placed on top side
  Reflow soldering
- 1C: THT and SMD on top side only
  Reflow for SMDs and wave soldering for THTs
- 2B: SMD on top and bottom
  Reflow soldering
- 2C/a: THT on top side, SMD on bottom side

Wave soldering for both THTs and bottom-side SMDs
- 2C/a: THT (if present) on top side, SMD on top and bottom
  Reflow and wave soldering (if THTs are present)
- 1X: THT (if present) on top side, complex SMD on top
  Reflow and wave soldering (if THTs are present)
- 2X/a: THT (if present) on top side, SMD/fine pitch/BGA on top and bottom
  Reflow and wave soldering (if THTs are present)
- 2Y, 2Z: THT (if present) on top side, SMD/ultrafine pitch/COB/flip chip/TAB on top and bottom
  Reflow and wave soldering (if THTs are present)

Note in the above listing that the "/a" refers to the possible need to deposit adhesive prior to placing the bottom-side SMDs. If THTs are present, bottom-side SMDs will be placed in adhesive, and both the bottom-side SMDs and the protruding THT leads will be soldered by passing the assembly through a dual-wave soldering oven. If THTs are not present, the bottom-side SMDs may or may not be placed in adhesive. The surface tension of molten solder is sufficient to hold bottom-side components in place during top-side reflow. These concepts are discussed later.

A Type 1B (top side SMT) bare board will first have solder paste applied to the component pads on the board. Once solder paste has been deposited, active and passive parts are placed in the paste. For prototype and low-volume lines, this can be done with manually guided x-y tables using vacuum needles to hold the components, whereas, in medium- and high-volume lines, automated placement equipment is used. This equipment will pick parts from reels, tubes, or trays and then place the components at the appropriate pad locations on the board, hence the term *pick-and-place* equipment.

After all parts are placed in the solder paste, the entire assembly enters a reflow oven to raise the temperature of the assembly high enough to reflow the solder paste and create acceptable solder joints at the component lead/pad transitions. Reflow ovens most commonly use convection and IR heat sources to heat the assembly above the point of solder liquidus, which for 63/37 tin-lead eutectic solder is 183° C. Due to the much higher thermal conductivity of the solder paste compared to the IC body, reflow soldering temperatures are reached at the leads/pads before the IC chip itself reaches damaging temperatures. For a Type 2B (top and bottom SMT), the board is inverted and the process repeated.

If mixed-technology Type 2C (SMD only on bottom) is being produced, the board will be inverted, an adhesive will be dispensed at the centroid of each SMD, parts will be placed, the adhesive will be cured, the assembly will be re-righted, through-hole components will be mounted, and the circuit assembly will then be wave soldered, which will create acceptable solder joints for both the through-hole components and bottom-side SMDs. It must be noted that successful wave soldering of SMDs requires a dual-wave machine with one turbulent wave and one laminar wave.

For any of the type assemblies that have THT on top side and SMDs (including SMT, fine pitch, ultrafine pitch, BGA, flip chip, etc.) top and bottom, the board will first be inverted, adhesive dispensed, SMDs placed on the bottom-side of the board, the adhesive cured, the board re-righted, through-hole components placed, and the entire assembly wave soldered. It is imperative to note that only passive components and small active SMDs can be successfully bottom-side wave soldered without considerable experience on the part of the design team and the board assembly facility. It must again be noted that successful wave soldering of SMDs requires a dual-wave machine with one turbulent wave and one laminar wave. The board will then be turned upright, solder paste deposited, the top-side SMDs placed, and the assembly reflow soldered.

It is common for a manufacturer of through-hole boards to convert first to a Type 2C (SMD bottom side only) substrate design before going to an all-SMD Type I design. Since this type of board requires only wave soldering, it allows amortization of through-hole insertion and wave soldering. Many factors contribute to the reality that most boards are mixed-technology boards. While most components are available in SMT packages, through-hole connectors may still be commonly used for the additional strength that the through-hole soldering process provides, and high-power devices such as three-terminal regulators are still commonly through-hole due to off-board heat-sinking demands. Both of these issues are actively being addressed by manufacturers and solutions exist that allow all-SMT boards with connectors and power devices.[3]

Again, it is imperative that all members of the design, build, and test teams be involved from the design stage. Today's complex board designs mean that it is entirely possible to exceed the ability to adequately test a board if test is not designed in, or to robustly manufacture a the board if in-line inspections and handling are not adequately considered. Robustness of both test and manufacturing are only assured with full involvement of all parties to overall board design and production.

There is an older definition of *types* of boards that the reader will still commonly find in books and articles, including some up through 1997. For this reason, those three types will be briefly defined here, along with their soldering techniques. The reader is cautioned to be sure which definition of board types is being referred to in other publications. In these board definitions, no distinction is made among the various types of SMDs. That is, SMD could refer to standard SMT, fine-pitch, ultrafine-pitch, BGAs, etc. This older definition was conceived prior to the use of the various chip-on-board and flip chip technologies and does not consider them as special cases.

- Type 1: an all-SMT board, which could be single- or double-sided. Reflow soldering is used and is one-pass for a single-sided board. For a double-sided board, several common techniques are used:
  – Invert board, deposit adhesive at centroid of parts. Deposit solder paste. Place bottom side in eutectic paste, cure adhesive, then reflow. Invert board, place top-side parts in eutectic paste, reflow again.
  – Invert board, place bottom-side parts in eutectic paste, reflow. Invert board, place top-side parts in eutectic paste, reflow again. Rely on the surface tension of the molten solder paste to keep bottom-side parts in place.
  – Invert board, place bottom-side parts in high-temperature paste. Reflow at appropriate high temperature. Invert board, place top-side components in eutectic paste, reflow again. The bottom-side paste will not melt in eutectic reflow temperatures.
- Type 2: a mixed-technology board, composed of THT components and SMD components on the top side. If there are SMD components on the bottom side, a typical process flow will be:
  – Invert board, place bottom-side SMDs in adhesive. Cure adhesive.
  – Invert board, place top-side SMDs. Reflow board.
  – Insert THT parts into top side.
  – Wave solder both THT parts and bottom-side SMDs.

  Typically, the bottom-side SMDs will only consist of chip devices and SO transistors. Some publications will also call Type 2 a Type 2A (or Type IIA) board.
- Type 3: a mixed technology board, with SMDs only on the bottom side, and typically only chip parts and SO transistors. The typical process flow would be:
  – Invert board, place bottom-side SMDs in adhesive. Cure adhesive.
  – Insert THT parts into top side.
  – Wave solder both THT parts and bottom-side SMDs.

  Due to this simplified process flow, and the need for only wave soldering, Type 3 boards are an obvious first step for any board assembler moving from THT to mixed-technology boards. Some publications will also call Type 3 a Type 2B (or Type IIB) board (Fig. 2.1).

It cannot be overemphasized that the speed with which packaging issues are moving requires anyone involved in SMT board or assembly issues to stay current and continue to learn about the processes. If that's you, please subscribe to one or more of the industry-oriented journals noted in the "Journal References" at the end of this section, obtain any IC industry references you can, and attend the various

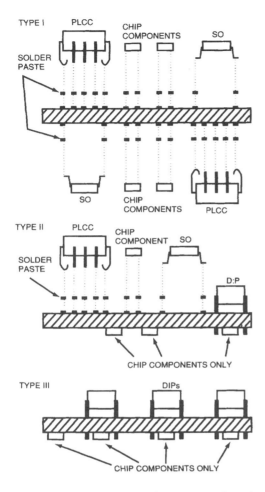

FIGURE 2.1 Type I, II, and III SMT circuit boards. *Source:* Intel. 1994. *Packaging Handbook.* Intel Corp., Santa Clara, CA. Reproduced with permission.

conferences on electronics design, manufacturing, and test. Conferences such as the National Electronics Production and Productivity Conference (NEPCON),[4] Surface Mount International (SMI),[5] as well at those sponsored by SMTA and the IPC, are invaluable sources of information for both the beginner and the experienced SMT engineer.

2.3 Surface Mount Device (SMD) Definitions*

As in many other areas, there are still both English (inch) and metric-based packages. Many English-dimensioned packages have designations based on mils (1/1000 in), while metric-dimensioned package designations are based on millimeters (mm). The industry term for the inch or millimeter dimension base is "controlling dimension." The confusion this can cause is not likely to go away anytime soon, and users of SMDs must become familiar with the various designations. For example, chip resistors and capacitors commonly come in a rectangular package, which may be called an 0805 package in English dimensions. This translates to a package that is 80 mils (0.080 in) long and 50 mils (0.050 in) wide. Height of this package is not an issue, since height does not affect land design, amount of solder dispensed, or placement. The almost identical metric package is designated a 2012 package, which is 2.0 mm long and 1.2 mm wide, equivalent to 79.2 mils long by 47.52 mils wide. While these differences are small and, in most cases, not significant, the user must still be able to correctly interpret designations

---

*See also Chapter 3 for more detailed descriptions of devices.

such as "0805" and "2012." A "1610" package, e.g., is inch, not metric, and is 160 mils long by 100 mils wide (see Fig. 2.2).

In multi-lead packages, a much larger issue faces the user. The pitch of packages such as QFPs may be metric or inch-based. If a pitch-based conversion is made, the cumulative error from one corner to another may be significant and may result in lead-to-pad errors of as much as one-half the land width. Typical CAD packages will make a conversion, but the user must know the conversion accuracy of the CAD package used. For example, consider a 100-pin QFP whose controlling dimension is millimeters with a pitch of 0.65 mm. With a common conversion that 1 mm = 39.37 mils, this would be an equivalent pitch of 25.59 mils. Over 25 leads, the total center-to-center dimension for the lands would be 0.65 mm × 24 = 15.6 mm. The 15.6 mm dimension would convert exactly to 614.17 mils, or 0.61417 in. If the ECAD (electronic CAD) program being used to lay out the board only converted to the nearest mil, a pitch conversion of 26 mils would be used, and over 25 leads the total center-to-center land dimension would be 24 × 26 = 624 mils. This would be an error of 10 mils, or almost the pad width over the width of the package (see Fig. 2.3).

Pads placed at the 0.624-in conversion would be virtually off the pads by the "last" lead on any side. The conversion accuracy of any CAD package must be determined if both mil and mm controlling dimensions exist among the components to be used.

SMD ICs come in a wide variety of packages, from 8-pin *small outline* packages (SOLs) to 1000+ connection packages in a variety of sizes and lead configurations, as shown in Figure 2.4. The most common commercial packages currently include *plastic leaded chip carriers (PLCCs), small outline packages (SOs), quad flat packs (QFPs)*, and *plastic quad flat packs (PQFPs)*, also know as *bumpered quad flat packs (BQFPs)*. Add in *tape automated bonding (TAB), ball grid array (BGA)*, and other newer technologies, and the IC possibilities become overwhelming. The reader is referred to the Chapter 3 for package details, and to the standards of the Institute for Interconnecting and Packaging Electronic Circuits (IPC) to find information on the latest packages.

FIGURE 2.2 Example of passive component sizes, top view (not to scale).[19]

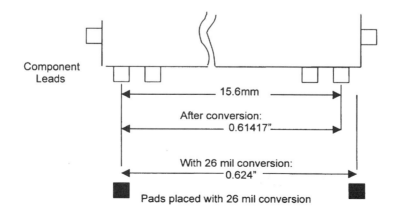

FIGURE 2.3 Dimension conversion error in 25 leads.

20-lead PLCC, 0.050-in pitch

20-lead SOW, 0.050-in pitch

132-lead BQFP, 0.025-in pitch

FIGURE 2.4 Examples of SMT plastic packages.

The examples shown above are from the author's lab, and they are examples of standard leaded SMT IC packages. The PLCC uses J-leads, whereas the SOW and BQFP packages use gull-wing leads. The IC manufacturer's data books will have packaging information for the products, and most of those data books are now available on the World Wide Web (WWW). Some WWW references are provided at the end of this chapter.

For process control, design teams must consider the minimum and maximum package size variations allowed by their part suppliers, the moisture content of parts as received, and the relative robustness of each lead type. Incoming inspection should consist of both electrical and mechanical tests. Whether these are spot checks, lot checks, or no checks will depend on the relationship with the vendor.

2.4 Substrate Design Guidelines

As noted in the "Reasons to Implement SMT" section in Section 2.2, substrate (typically PCB) design has an effect not only on board/component layout but also on the actual manufacturing process. Incorrect land design or layout can negatively affect the placement process, the solder process, the test process, or any combination of the three. Substrate design must take into account the mix of surface mount devices (SMDs) and through-hole technology (THT) devices that are available for use in manufacturing, and which are being considered during circuit design.

The considerations that will be noted here are intended to guide an engineer through the process, allowing access to more detailed information as necessary. General references are noted at the end of this chapter, and specific references will be noted as applicable. Although these guidelines are noted as *steps*, they are not necessarily in an absolute order and may require several iterations back and forth among the steps to result in a final, satisfactory process and product. Again, substrate design and the use of ECAD packages is covered in detail in Chapters 5 and 6.

Surface Mount Technology

After the circuit design (schematic capture) and analysis, step 1 in the process is to determine whether all SMDs will be used in the final design or whether a mix of SMDs and THT parts will be used. This is a decision that will be governed by some or all of the following considerations:

- Current parts stock
- Existence of current through-hole placement and/or wave solder equipment
- Amortization of current TH placement and solder equipment
- Existence of reflow soldering equipment, or cost of new reflow soldering equipment
- Desired size of the final product
- Panelization of smaller boards
- Thermal issues related to high power circuit sections on the board

It may be desirable to segment the board into areas based on function—RF, low power, high power, etc., using all SMDs where appropriate, and mixed-technology components as needed. Power and connector portions of the circuit may point to the use of through-hole components, if appropriate SMT connectors are not available (see also Chapter 4). Using one solder technique (reflow or wave) simplifies processing and may outweigh other considerations.

Step 2 in the SMT process is to define all the lands of the SMDs under consideration for use in the design. The land is the copper pattern on the circuit board upon which the SMD will be placed. Land examples are shown in Figs. 5a and 5b, and land recommendations are available from IC manufacturers in the appropriate data books. They are also available in various ECAD package used for the design process, or in several references that include an overview of the SMT process.[7,8] A footprint definition will include the land and will also include the pattern of the solder resist surrounding the copper land. Footprint definition sizing will vary depending on whether reflow or wave solder process is used. Wave solder footprints will require recognition of the direction of travel of the board through the wave, to minimize solder shadowing in the final fillet, as well as requirements for solder thieves. The copper land must allow for the formation of an appropriate, inspectable solder fillet. These considerations are covered in more detail in Chapter 7.

If done as part of the EDA process (electronic design automation, using appropriate electronic CAD software), the software will automatically assign copper directions to each component footprint as well as appropriate coordinates and dimensions. These may need adjustment based on considerations related to wave soldering, test points, RF and/or power issues, and board production limitations. Allowing the software to select 5-mil traces when the board production facility to be used can only reliably do 10-mil traces would be inappropriate. Likewise, the solder resist patterns must be governed by the production capabilities of the board manufacturer.

Figures 2.5a and 2.5b show two different applications of resist. In the 50-mil pitch SO pattern, the resist is closely patterned around each solder pad. In the 25-mil QFP pattern, the entire set of solder pads

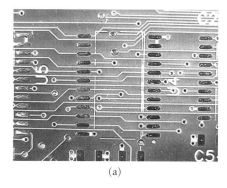
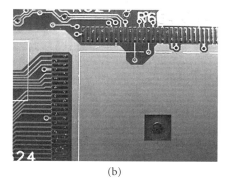

(a)   (b)

FIGURE 2.5 (a) SO24 footprint and (b) QFP132 footprint, land and resist.

is surrounded by one resist pattern. This type of decision must be made in consultation with both the resist supplier and the board fabricator. Note the local fiducial in the middle of the QFP pattern. This aids the placement vision system in determining an accurate location for the QFP pattern, and it is commonly used with 25-mil pitch and smaller lead/pad patterns.

Final land and trace decisions will

- Allow for optimal solder fillet formation
- Minimize necessary trace and footprint area
- Consider circuit requirements such as RF operation and high-current design
- Allow for appropriate thermal conduction
- Allow for adequate test points
- Minimize board area, if appropriate
- Set minimum interpart clearances for placement and test equipment to safely access the board (Fig. 2.6)
- Allow adequate distance between components for post-reflow operator inspections
- Allow room for adhesive dots on wave-soldered boards
- Minimize solder bridging
- Decisions that will provide optimal footprints include a number of mathematical issues, including:
- Component dimension tolerances
- Board production capabilities, both artwork and physical tolerances across the board relative to a 0-0 fiducial
- How much artwork/board shrink or stretch is allowable
- Solder deposition volume consistencies with respect to fillet sizes
- Placement machine accuracy
- Test probe location controls and bed-of-nails grid pitch

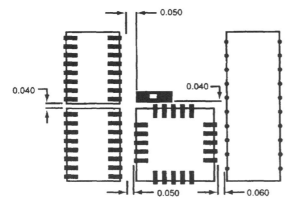

FIGURE 2.6 Minimum land-to-land clearance examples. *Source:* Intel, 1994, *Packaging Handbook,* Intel Corp., Santa Clara, CA. Reprinted with permission.

Design teams should restrict wave-solder-side SMDs to passive components and transistors. While small SMT ICs can be successfully wave soldered, this is inappropriate for an initial SMT design, and is not recommended by some IC manufacturers. Before subjecting any SOIC or PLCC to wave soldering, the IC manufacturer's recommendation must be determined.

These decisions may require a statistical computer program, if available to the design team. The stochastic nature of the overall process suggests a statistical programmer will be of value.

2.5 Thermal Design Considerations*

Thermal management issues remain major concerns in the successful design of an SMT board and product. Consideration must be taken of the variables affecting both board temperature and junction

---

*See also Chapter 11.

temperature of the IC. The reader is referred to Bar-Cohen (see Recommended Readings) for a more detailed treatment on thermal issues affecting ICs and PCB design.

The design team must understand the basic heat transfer characteristics of affected SMT IC packages.[9] Since the silicon chip of an SMD is equivalent to the chip in an identical-function DIP package, the smaller SMD package means the internal lead frame metal has a smaller mass than the lead frame in a DIP package. This lesser ability to conduct heat away from the chip is somewhat offset by the lead frame of many SMDs being constructed of copper, which has a lower thermal resistance than the Kovar and Alloy 42 materials commonly used for DIP packages. However, with less metal and shorter lead lengths to transfer heat to ambient air, more heat is typically transferred to the circuit board itself. Several board thermal analysis software packages are available (e.g., Flotherm[10]) and are highly recommended for boards which are expected to develop high thermal gradients.

Since all electronics components generate heat in use, and elevated temperatures negatively affect the reliability and failure rate of semiconductors, it is important that heat generated by SMDs be removed as efficiently as possible. The design team needs to have expertise with the variables related to thermal transfer:

- Junction temperature, T_j
- Thermal resistances, Θ_{jc}, Θ_{ca}, Θ_{cs}, Θ_{sa}
- Temperature sensitive parameter (TSP) method of determining Θs
- Power dissipation, P_D
- Thermal characteristics of substrate material

SMT packages have been developed to maximize heat transfer to the substrate. These include PLCCs with integral heat spreaders, the SOT-89 power transistor package, the DPAK power transistor package and many others. Analog ICs are also available in power packages. Note that all of these devices are designed primarily for processing with the solder paste process, and some specifically recommend against their use with wave-solder applications. Heat sinks and heat pipes should also be considered for high-power ICs.

In the conduction process, heat is transferred from one element to another by direct physical contact between the elements. Ideally, the material to which heat is being transferred should not be adversely affected by the transfer. As an example, the glass transition temperature T_g of FR-4 is 125° C. Heat transferred to the board has little or no detrimental affect as long as the board temperature stays at least 50° C below T_g. Good heat sink material exhibits high thermal conductivity, which is not a characteristic of fiberglass. Therefore, the traces must be depended on to provide the thermal transfer path.[11] Conductive heat transfer is also used in the transfer of heat from IC packages to heat sinks, which also requires use of thermal grease to fill all air gaps between the package and the "flat" surface of the sink.

The previous discussion of lead properties of course does not apply to leadless devices such as *leadless ceramic chip carriers (LCCCs)*. Design teams using these and similar packages must understand the better heat transfer properties of the alumina used in ceramic packages and must match TCEs between the LCCC and the substrate, since there are no leads to bend and absorb mismatches of expansion.

Since the heat transfer properties of the system depend on substrate material properties, it is necessary to understand several of the characteristics of the most common substrate material, FR-4 fiberglass. The glass transition temperature has already been noted, and board designers must also understand that multilayer FR-4 boards do not expand identically in the x, y, and z directions as temperature increases. Plated-through holes will constrain z-axis expansion in their immediate board areas, whereas non-through-hole areas will expand further in the z-axis, particularly as the temperature approaches and exceeds Tg.[12] This unequal expansion can cause delamination of layers and plating fracture.

If the design team knows that there will be a need for higher abilities to dissipate heat and/or needs for higher glass transition temperatures and lower coefficients of thermal expansion (TCE) than FR-4 possesses, many other materials are available, examples of which are shown below.

Substrate material	T_g, glass transition temperature	TCE, thermal coefficient of x-y expansion	Thermal conductivity	Moisture absorption
Units	°C	PPM/°C	W/M°C	%
FR-4 epoxy glass	125	13–18	0.16	0.10
Polyamide glass	250	12–16	0.35	0.35
Copper-clad Invar	Depends on resin	5–7	160XY, 15–20Z	NA
Poly aramid fiber	250	3–8	0.15	1.65
Alumina/ceramic	NA	5–7	20–45	NA

Note in the above table that copper-clad Invar has both variable T_g and variable thermal conductivity, depending on the volume mix of copper and Invar in the substrate. Copper has a high TCE, and Invar has a low TCE, so the TCE increases with the thickness of the copper layers. In addition to heat transfer considerations, board material decisions must also be based on the expected vibration, stress, and humidity in the application.

Convective heat transfer involves transfer due to the motion of molecules, typically airflow over a heat sink, and depends on the relative temperatures of the two media involved. It also depends on the velocity of air flow over the boundary layer of the heat sink. Convective heat transfer is primarily effected when forced air flow is provided across a substrate, and when convection effects are maximized through the use of heat sinks. The rules with which designers are familiar when designing THT heat-sink device designs also apply to SMT design.

The design team must consider whether passive conduction and convection will be adequate to cool a populated substrate or whether forced-air cooling or liquid cooling will be needed. Passive conductive cooling is enhanced with thermal layers in the substrate, such as the previously mentioned copper/Invar. There will also be designs that will rely on the traditional through-hole device with heat sink to maximize heat transfer. An example of this would be the typical three-terminal voltage regulator mounted on a heat sink or directly to a metal chassis for heat conduction, for which standard calculations apply.[13]

Many specific examples of heat transfer may need to be considered in board design, and of course most examples involve both conductive and convective transfer. For example, the air gap between the bottom of a standard SMD and the board effects the thermal resistance from the case to ambient, Θ_{ca}. A wider gap will result in a higher resistance due to poorer convective transfer, whereas filling the gap with a thermally-conductive epoxy will lower the resistance by increasing conductive heat transfer. Thermal-modeling software is the best way to deal with these types of issues, due to the need for rigorous application of computational fluid dynamics (CFD)[14]

2.6 Adhesives

Adhesives in electronics manufacturing have a number of potential uses. Although most of them involve an attachment function, there are other primary functions for adhesives in addition to attachment.

- Attachment of components to wave-solder side of the circuit board prior to soldering
- Thermal bonding of component bodies with a substrate, to allow maximum heat transfer out of a chip
- Dielectric insulation to reduce crosstalk in RF circuits
- Electrical connections, e.g between component leads and board pads

In the surface mount assembly process, any SMDs mounted on the bottom side of the board and subjected to the wave solder process will always require adhesive to mount the SMDs for passage through the solder wave. This is apparent when one envisions components on the bottom side of the substrate with no through-hole leads to hold them in place. Adhesives will stay in place after the soldering process, and throughout the life of the substrate and the product, since there is no convenient means for adhesive removal once the solder process is complete. This means that the adhesive used must have a number of

Surface Mount Technology 2-13

both physical and chemical characteristics, and these should be considered during the three phases of adhesive use in SMT production.

1. Preapplication properties relating to storage and dispensing issues
2. Curing properties relating to time and temperature needed for cure, and mechanical stability during the curing process
3. Post-curing properties relating to final strength, mechanical stability, and reworkability

2.6.1 Adhesive Characteristics

Physical characteristics to be considered for attachment adhesives are as follows:

- Electrical nonconductivity (conductive adhesives are discussed in Section 2.6.6)
- Coefficient of thermal expansion (CTE) similar to the substrate and components, to minimize thermal stresses
- Stable in both storage (shelf life) and after application, prior to curing
- Stable physical drop shape—retains drop height and fills z-axis distance between the board and the bottom of the component; thixotropic with no adhesive migration
- Green strength (precure tackiness) sufficient to hold parts in place
- Noncorrosive to substrate and component materials
- Chemically inert to flux, solder, and cleaning materials used in the process
- Curable as appropriate to the process: UV, oven, or air-cure
- Once cured, unaffected by temperatures in the solder process
- Post-cure bond strength
- Adhesive colored, for easy identification of its presence by operators
- Minimum absorption of water during high humidity or wash cycles, to minimize the impact of the adhesive on surface insulation resistance (SIR)

Process considerations to be considered are as follows:

- Application method to be used: pin-transfer, stencil print, syringe dispense
- Pot-life (open time) of the adhesive if pin-transfer or stencil print techniques will be used
- One-part or two-part epoxy
- Curing time and curing method: UV, oven, or air-cure
- Tacky enough to hold the component in place after placement
- Post-cure strength adequate for all expected handling processes
- Post-cure temperature resistance to wave solder temperatures
- Repair procedures: are parts placed in the adhesive removable without damage to the part, if part analysis is expected, and without damage to the substrate/board under any conditions (As discussed below, cohesive failure is preferred to adhesive failure.)

Environmental characteristics to be considered are as follows:

- Flammability
- Toxicity
- Odor
- Volatility

One-part adhesives are easier to work with than two-part adhesives, since an additional process step is not required. The user must verify that the adhesive has sufficient shelf life and pot life for the perceived process requirements. Both epoxy and acrylic adhesives are available as one-part or two-part systems, with the one-part systems cured thermally. Generally, epoxy adhesives are cured by oven heating, while acrylics may be formulated to be cured by long-wave UV light or heat. Three of the most common adhesive types are

- *Elastomeric adhesives.* A type of thermoplastic adhesives, elastomerics are, as their name implies, elastic in nature. Examples are silicone and neoprene.
- *Thermosetting adhesives.* These are cured by their chemical reaction to form a cross-linked polymer. These adhesives have good strength, and are considered structural adhesives. They cannot readily be removed. Examples are acrylic, epoxy, phenolic, and polyester.
- *Thermoplastic adhesives.* Useful in low-strength applications, thermoplastics do not undergo a chemical change during cure and therefore can be resoftened with an appropriate solvent. They will recure after softening. An example is EVA copolymer.

Of the above-noted types, one-part heat-cured epoxy is the most commonly used in part-bonding applications. An appropriate decision sequence for adhesive is

1. Select the adhesive-application process, and adhesive-cure process, best suited to the board and board materials being used.
2. Select the application machine capable of this process.
3. Select the type of adhesive to be applied by this machine. Frequently, the application-machine vendor will have designed the machine with a particular type of adhesive in mind, and the vendor's recommendations should be followed.
4. Select the cure machine (if needed) for the cure process required by the specific adhesive chosen.

Failure Characteristics during Rework

It is important to understand the failure characteristics of the adhesive that is chosen. While it is unlikely that the adhesive will be driven to failure under any reasonable assembly procedures, it must be driven to failure during rework or repair since the part must be removed. The failure characteristics of the adhesive in those operations is its reworkability.

During rework, the adhesive must reach its glass transition temperature, T_g, at a temperature lower than the melting point of solder. At T_g, the adhesive softens and, using one of several procedures, the rework/repair operator will be able to remove the part. For many adhesives used in electronics, T_g will be in the range of 75 to 100° C. During the rework process, the part will commonly be twisted 90° after the solder has melted. This will shear the adhesive under the part. The location of the shear failure can be important to the user. The adhesive will either have a shear failure within itself, a *cohesive* failure (Fig. 2.7), or it will have a shear failure at its interface with the substrate or the part body, and *adhesive* failure (Fig. 2.8). The cohesive failure indicates that, at the rework temperatures, the weakest link is the adhesive itself.

Since one of the objects of rework is to minimize any damage to the substrate or, in some cases, the component, the preferred mode of failure is the cohesive failure.

The primary reason for preferring a cohesive failure is that an adhesive failure brings with it the risk that the adhesive could lift the solder mask or pad/traces during rework.

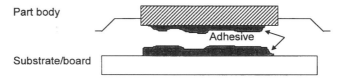

FIGURE 2.7 Cohesive failure of an adhesive.[20]

Surface Mount Technology

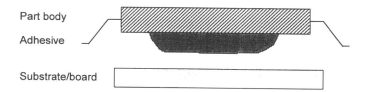

FIGURE 2.8 Adhesive failure of an adhesive.

2.6.2 Adhesive Application Techniques

Adhesive can be applied by screening techniques similar to solder paste screen application, by pin transfer techniques, and by syringe deposition. Screen and pin-transfer techniques are suitable for high-volume production lines with few product changes over time. Syringe deposition, which uses an x-y table riding over the board with a volumetric pump and syringe tip, is more suitable for lines with a varying product mix, prototype lines, and low-volume lines where the open containers of adhesive necessary in pin-transfer and screen techniques are avoided. Newer syringe systems are capable of handling high-volume lines. See Figure 2.9 for methods of adhesive deposition.

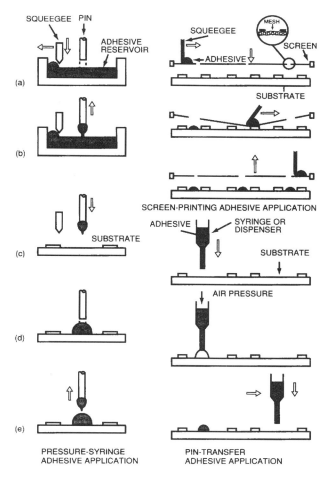

FIGURE 2.9 Methods of adhesive deposition. *Source:* Phillips, *Surface Mount Process and Application Notes,* Philips Semiconductor Corp., 1991. Reprinted with permission.

Regardless of the application method used, different components will require different deposition patterns. The pattern will depend both on the component type and size and on the cure method. If UV cure will be used, some part of every dot deposited must be "visible" to the UV source. Heat-cure adhesives do not depend on this, and the entire deposition may be under the component (Fig. 2.10). One important distinction between heat cure and UV cure is that heat cure takes minutes to accomplish, while UV cure happens in seconds, if all under-part adhesive applications (dots, stripes, etc.) have at least some direct exposure to the UV source.

Note that UV cure adhesives are not usable with ICs that have leads on all four sides. There is not a location where adhesive deposits can be placed, be "visible" to the UV lights, and not have a risk of bleeding onto component pads/leads. Also note that some manufacturers, e.g., Intel, specifically do not recommend that any of their ICs be run through the wave-solder process, which would be the intent of gluing them on.

Pin-Transfer

The pin-transfer technique is simplest method for deposition of high volumes of adhesive drops. Deposition requires a custom-made pin array panel for each board. The positions of the pins on their backplane must exactly match the desired locations of the adhesive drops on the board. Dot size is dependent on

- Diameter of each pin
- Shape of each pin
- Depth to which the pin is dipped into the adhesive reservoir
- Viscosity of the adhesive

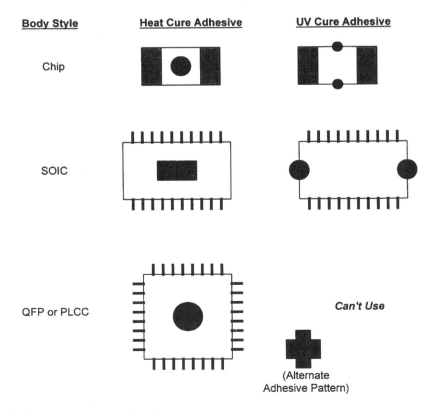

FIGURE 2.10 Examples of adhesive deposits.

- Clearance of the pin at its closest point to the board (pin must not touch board)
- Wait time after pin lowering

See Fig. 2.9 for an example of the pin-transfer technique. Since the dot size does depend on the clearance between the pin and the board, if one assumes that all pins will be in the same plane, dot size will vary if a board is warped. The pin-transfer technique may be used to place bottom-side adhesive drops after THT components have been placed and clenched from the top side.

Stencil Printing

The stencil-printing technique cannot be used after THT parts have been placed on top side. Like stencil-printing of solder paste, this method requires a separate stencil for each board design and deposits adhesive at all adhesive locations with the passage of a squeegee over the stencil. It requires a flat board, since any variation in clearance between the stencil and the board will result in variation of dot sizes. The adhesive manufacturer, the printer manufacturer, and the squeegee manufacturer all need to be consulted to be certain that all items form a compatible process. Dot size is dependent on

- Board warp
- Squeegee characteristics
- Stencil opening
- Adhesive viscosity

See Fig. 2.9 for an example of stencil-printing adhesives. Like printing of solder paste, the adhesive manufacturer will recommend the best stencil print method. Typically, the main concern is whether printing should be done with snap-off or without.

Syringe Deposition

Syringe dispensing is becoming very popular. In the United Kingdom, about 95% of adhesive deposition is done by dispensing. Most adhesive-deposition applications will require a constant-volume syringe system, as opposed to a time-pressure syringe system which results in a less consistent dot size. For small-chip (e.g., 0603 inch chips) especially, constant-volume systems are required. This method is nominally slower than the pin-transfer or stencil-print techniques, since it deposits one dot at a time. However, current technologies have resulted in fast tack times, so this has become less of an issue.

For low-volume, high-mix lines, or prototype systems, syringe deposition has the major advantage that changing location and/or volume of the drops requires no hardware changes—only a software change. All syringe systems allow storage of a number of dispense files, so all the operator has to do is select the correct file for a board change. Like solder dispensing, cleanliness of the needle and system is an absolute necessity. The systems allow direct loading of CAD data, or "teaching" of a dot layout by the operator's use of a camera-based vision system.

If thermal transfer between wave-soldered components and the substrate is a concern, the design team should consider thermally conductive adhesives. These adhesives may also be used with non-wave-soldered components to facilitate maximum heat transfer to the substrate. See Fig. 2.9 for an example of syringe deposition of adhesives.

2.6.3 Dot Characteristics

Regardless of the type of assembly, the type of adhesive used, or the curing technique used, adhesive volume and height must be carefully controlled. Slump of adhesive after application is undesirable, since the adhesive must stay high enough to solidly contact the bottom of the component. Both slump and volume must be predictable so that the dot does not spread and contaminate any pad or termination associated with the component (Fig. 2.11).

If adhesive dot height = X, substrate metal height = Y, and SMD termination thickness = Z, then $X > Y + Z$, allowing for all combinations of potential errors, e.g.:

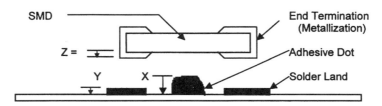

FIGURE 2.11 Relation of adhesive dot, substrate, and component.[19]

- End termination minimum and maximum thickness
- Adhesive dot minimum and maximum height
- Substrate metal minimum and maximum height

Typically, end termination thickness variations are available from the part manufacturer. Nominal termination thickness for 1206 devices, e.g., is 10–50 μm. Solder land/trace thickness variations may be as much as 30–100 μm and are a result of the board manufacturing process. They will vary not only on the type of board metallization (standard etch vs. plated-through hole) but also on the variations within each type. For adequate dot height, which will allow for the necessary dot compression by the part, X should be between 1.5× and 2.5× the total Y + Z, or just Z when dummy lands are used. If adhesive dots are placed on masked areas of the board, mask thickness must also be considered.

For leaded devices such as SOT-23 transistors, which must be glued down, the measurement Z will correspond to lead standoff, nominally 200 μm. Here, standoff variations can be considerable and must be determined from the manufacturer.

A common variation on the above design is to place "dummy" copper lands, or route actual traces, under the center of the part. Since these lands or traces are etched and plated at the same time as the actual solder pads, the variation in metal height Y is eliminated as an issue. However, if the traces are not as wide as the diameter of the dot, this does not completely eliminate the height Y issue. When adhesive dots are placed on the dummy or real pads, X > Z is the primary concern. For the higher leaded parts, solder mask may also need to be placed over the dummy lands to further reduce the necessary dot thickness. The adhesive manufacturer can suggest maximum dot height that the adhesive will maintain after deposition without slumping or flowing (see Fig. 2.12).

One dot of adhesive is typical for smaller parts such as chip resistors and capacitors, MELF packages, and SOT transistors. Larger parts such as SOICs will need two dots, although some manufacturers, such as Intel, recommend that their ICs not be wave soldered.

Adhesive dispensing quality issues are addressed by considerations of

- Type of adhesive to be used
- Process-area ambient temperature and humidity
- Incoming quality control
- No voids in cured adhesive to prevent trapping of flux, dirt, etc.
- Volume control

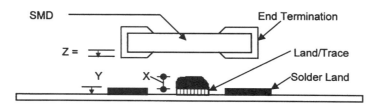

FIGURE 2.12 Adhesive dot-height criteria with lands/traces under the part. *Source*: Cox, R.N., *Reflow Technology Handbook*, Research Inc., Minneapolis, MN, 1992. Reprinted with permission.

- Location control
- Consideration of all combinations of termination, dot, and substrate height/thickness'

Prasad (see Suggested Readings at the end of this chapter) provides an excellent in-depth discussion of adhesives in SMT production.

2.6.4 Adhesive Process Issues

Problems that can occur during the use of adhesives include

- Insufficient adhesive
- Excess adhesive
- Stringing
- Clogging
- Incomplete cure
- Missing components

Insufficient adhesive causes depend on the deposition technique. If using time-pressure dispense, causes include

- Clogged nozzle tip
- Worn or bent tip
- Insufficient dispense height above board
- Viscosity too high

If using pin-transfer deposition, causes include

- Reservoir temperature too low
- Viscosity too low, adhesive drips off before deposition
- Pin size too small
- Pin shape inappropriate
- Pins dirty
- Pin immersion depth in adhesive reservoir wrong
- Pin withdrawal speed from reservoir incorrect
- Pin withdrawal speed from board too high
- Too much clearance between pins and board

Other causes may include

- Warped or incorrectly held board

Excess adhesive causes depend on the deposition technique. If using time-pressure dispense, causes include

- Pressure too high
- Low viscosity
- Dispense tip too large

If using pin-transfer deposition, causes include

- Pin size too large
- Pin depth in reservoir too deep
- Pin too close to board
- Pins with adhesive accumulation, need to be cleaned

- Viscosity incorrect
- Temperature of reservoir too high

Stringing causes include

- Dispense pressure too high
- Dispense time too long
- Nozzle or pins too far from substrate
- Viscosity of adhesive too high or too low
- Air in adhesive
- Pin withdrawal too fast

Clogging causes include

- Tip too small
- Moisture absorption by epoxy
- Too long between dispense actions (down time)
- Adhesive reacts with tip material

Incomplete cure, leaving adhesive sticky, may be caused by

- Insufficient cure oven temperature
- Insufficient cure oven time
- Thermal inequalities across board, leaving cool spots
- Wrong mixture of two-part epoxy

Missing components may be caused by

- Incomplete cure
- Adhesive skinning over if too much time elapses between dispense and placement
- Dot height too low for component (see above dot criteria)
- Poor placement, not enough down force in pipette
- Component mass too high for amount of adhesive (Increase number of dots used.)
- Wrong mixture of two-part epoxy

2.6.5 Adhesives for Thermal Conduction

As with other types of adhesives, there are a variety of thermally conductive adhesives on the market. Selection of the proper thermally conductive adhesive requires consideration of the following items:

- One- or two-part adhesive
- Open time
- Cure type, heat, UV, or activator
- Cure time, to fix and to full strength
- Thermal conductivity, Θ_{JC} in watts/mil°C
- Dielectric strength, in volts/mil
- Elongation
- Coefficient of thermal expansion, CTE
- Repairable? Can the part be removed later? T_g
- Color
- Shelf life

One or Two-Part Adhesive

If two-part adhesives are defined as those that require mixing prior to application, virtually all thermally conductive adhesives are one-part as of this writing, although a number of post-assembly potting products are two-part. The logistics of mixing and handling two-part systems have encouraged all adhesive manufacturers to develop one-part products for most applications. There are a number of "one-part" adhesives that do require the application of an activator to one of the two surfaces being joined.

A true one-part adhesive would be applied to the all locations on the board, all parts placed into the adhesive, and then the adhesive cured using either UV or heat. The one-part adhesives thereby have the advantage of only one chemical to be applied, but they require a second process step—either UV or heat—for fix and curing.

An example of the use of an activator adhesive would be in the adhesion of a power transistor to a board. The activator may be applied to the bottom/tab of the part, and the adhesive is dispensed onto the board. When the part, with the activator, is placed into the adhesive on the board, the adhesive will begin to fix. In this case, no adhesive process steps are required after the placement.

Open Time

The definition of open time for adhesives is based on the type of adhesive. For true one-part adhesives, open time is the maximum time allowed between the deposition of the adhesive and the placement of the part. For an activator-type adhesive, open time is the maximum time between deposition/application of the adhesive and the activator and the placement of the part in a manner that joins the adhesive and the activator.

Cure Type: Heat, UV, or Activator

The cure type determines if existing process equipment can be used and if a separate cure step is required. Heat cure requires an oven as the process step immediately following the placement of the part in adhesive, although in some applications a reflow oven can accomplish both adhesive cure and reflow. However, if the reader is considering a "one-step" cure and reflow operation, the information in Heisler and Kyriacopoulos (1998) is valuable. One-step operations may present problems if the adhesive has a large amount of elongation/expansion and cures before reflow.

UV cure can be done in a UV oven, or with UV lights either in an enclosure, or with a spot UV light during, e.g., rework or repair.

Activator cure, as discussed above, requires two process steps prior to placement; one to apply the activator, and one to apply the adhesive itself

Cure Time to Fix and to Full Strength

Normally, two cure times are given. One would be the time in minutes, at a given temperature, for the adhesive to set up well enough to hold the part in place, know as the *fix* time. The other would be the time to full cure under the conditions given. Some heat-cured silicone adhesives reach fix and full cure at virtually the same time.

Thermal Resistance, Θ_{JC}, in °C/Watt

This is a property of prime interest in the removal of heat from the component. It must be considered in the overall thermal calculations (see Chapter 11) to determine the amount of potential heat transfer from the chip/part to the board, thermal spreader, or heat sink. Alternatively the calculations will determine the maximum operating temperature of the part based on the power to be dissipated.

Dielectric Strength, in Volts/Mil

When used in a situation that also requires the adhesive to have the property of electrical insulation, the dielectric strength defines that property. This is important if there are electrical traces on the board under the part being bonded to the board.

In applications that do not involve insulative properties, dielectric strength is not an issue. If the thermal adhesive will be used, e.g., to bond a free-standing (not connected to any other physical device or to a portion of the enclosure) heat sink, dielectric strength should not be an issue.

Elongation

Elongation refers to the propensity of the cured adhesive to elongate when subjected to a steady force. While normally not a problem in part-bonding applications, elongation may be important in applications where the adhesive is subjected to a steady force, such may occur as in flexible circuit applications.

Coefficient of Thermal Expansion, CTE

If the thermal adhesive will be used to bond an IC to a free-standing heat sink, CTE is of little concern. If, on the other hand, it will be used to bond the IC to any type of constrained heat sink, such as the substrate or a portion of the enclosure, then CTE must be considered. The user must consider the CTE of the IC, the substrate/enclosure material, and the adhesive. A mismatch of CTEs can result in the possibility of adhesive or cohesive failure of the adhesive, damage to traces on the substrate if bonding is done to the substrate, and/or lead damage to the IC if stresses are placed on the solder joints as a result of CTE mismatch.

Repairable? Can the Part Be Removed Later? T_g

During the design of any assembly which will use adhesives, the design team must include consideration of whether the assembly is to be considered reworkable and/or repairable. If not, then these issues do not enter into the consideration of adhesive properties. However, if the assembly is being designed to include the possibility of rework and/or repair, then personnel who will be involved in these activities should be involved in the adhesive considerations.

Some adhesives are removable with solvents. Some can be heated above their glass temperature, Tg, at which point they soften enough to allow the part to be rotated 90°, which will break the adhesive bond and allow the part to be removed. See also the discussion on repairability in Section 2.6.1.

Regardless of the method of removing the part, there must also be a method for removing all adhesive residue after the part is removed and for applying the appropriate quantity of new adhesive with the new part. If the adhesive is UV or heat cured, there must be a method of exposing the assembly to UV or heat. UV can be applied locally with a fine UV light "pen." For heat cure, the user must consider whether local application will be used, with its attendant concerns of thermal shock to small areas of the board, or whether oven cure will be used, with its attendant concerns of thermal cycling of the entire assembly. If the entire assembly will be reheated, there may be the risks of

- Weakening previously applied adhesive
- Deteriorating the protection of previously applied OSPs
- Subjecting plastic ICs, which may have absorbed moisture, to high temperatures

Color

Color may be important for three reasons.

- It assists in operator inspection of adhesive deposition.
- It can alert a rework or repair operator of the type of adhesive.
- It can make the adhesive visible to an automated optical inspection (AOI) system.
- It can indicate that a component has been reworked/replaced, by using a color adhesive different from the color used for initial assembly.

While adhesives are not available in a wide array of colors, the above considerations are important. Operators would have difficulty with reliability inspection for the presence of adhesive if the adhesive were clear or neutral/translucent in appearance. Assuming that the available colors allows selecting different colors for different types of adhesive, e.g., acrylic one-part with UV cure vs. silicone one-part with heat cure, the rework operators can more easily distinguish what removal procedures to use on a part. Certain AOI systems are better able to distinguish adhesives if the color of the adhesive falls within a certain wavelength range.

Adhesive manufacturers recognize these issues and work with users and manufacturers of AOI systems to allow for the various colors required. As in the consideration of many of the other characteristics of adhesives, it is best to work with the users and AOI vendors to determine the best choice of available colors.

Shelf Life

Adhesive shelf life typically varies from 3 to 12 months. Part of the consideration of adhesives includes appropriate storage conditions and reorder policies for the users.

2.6.6 Electrically Conductive Adhesives

The initial interest in conductive adhesives was sparked by concerns about environmental legislation that could limit the use of lead in electronic manufacturing, and by an interest in a conductive joining process that did not require flux and cleaning. Work has also been done to consider their potential advantages, such as low-temperature assembly and better resistance to thermal cycle cracking than their metallic counterparts. Conductive adhesives are now used in electronics manufacturing where their dual properties of adhesion and electrical conduction are of value.

There are three types of conductive adhesives: thermoplastic composite conductive adhesives, thermosetting composite conductive adhesives, and z-axis conductive adhesives. The composite adhesives consist of a polymer base, which may be an epoxy, a silicone, a polyurethane, or a cyanoacrylate. The conductive filler may be gold, silver, nickel, copper, or graphite. Sizes of the conductive materials range from 100 μm (micron) spheres, to submicron flakes. The thermoplastic base is applied hot and undergoes no chemical change during its setting; therefore, it can be repeatedly heated, such as for rework, without losing any of its properties. The thermoplastic has the disadvantage that its tensile strength decreases with heat, and it may have considerable outgassing. Conversely, the thermosetting base goes through an irreversible chemical change as it sets and therefore cannot be reworked without complete removal. It is, however, more temperature stable than thermoplastic bases.

There is considerable interest in the z-axis conductive adhesives, which do not conduct electricity in the x- or y-axis but do conduct electricity in the z-axis when properly compressed. Generally, z-axis adhesives, also known as *anisotropic adhesives*, use a standard adhesive as a base, with a filler of conductive particles. These conductive particles are in relatively low concentration, giving the mechanical property of no x-y conduction, but when the part is placed on the adhesive, compressing it in the z-axis, the particles are moved into conduction with each other, electrically bridging between the lead and land. The adhesive is then cured.

One disadvantage to z-axis adhesives is that, like all adhesives, they have a lower thermal conductivity than metallic solder. This is an issue when high-power components are used and heat removal from the component through the leads to the board is necessary. Other disadvantages include a lower tensile strength, a lack of a joint filet for inspection purposes, and the tendency to slump before the curing process. Their higher cost, approximately 10× the cost of tin-lead solder, also needs to be overcome before any high-volume use of these adhesives is forthcoming.

Like the use of thermosetting adhesives, rework is of z-axis adhesives is difficult. The residual adhesive is not easy to remove, and manual application of the adhesive is difficult. If the leads of the replaced part are not coplanar with the pads, fewer particles will make contact at the end with the wider gap. While this is also true with initial automatic placement, it is more of a concern with rework, since it is assumed that part placement will be manual during rework. Additionally, the application of appropriate and even pressure is necessary for proper use of z-axis adhesives, and this too is difficult with manual rework.

The quality and reliability of connections made with z-axis adhesives are not considered to be as high as with a soldered joint. More data are needed on long-term tests under a variety of environmental conditions before they can be considered for high-reliability applications. Their resistance varies with time and temperature and, therefore, care must be taken in applications where the overall resistance of the joint and the leads may matter, such as with voltage regulators or with power MOSFETs, in which low output lead + joint impedance is necessary. Since most z-axis adhesives are silver-loaded, silver leaching may occur, as it does in silver-bearing capacitor terminations if they are not stabilized with a nickel coating. Prevention of this migration with silver-loaded adhesives is accomplished by conformally coating the joints. This of course adds to the expense and to the rework problem.

Iwasa[15] provides an overview of silver-bearing conductive adhesives. He states the following criteria for these adhesives:

- Single component adhesive
- Low temperature heat curing
- No bleeding when printed with an stencil opening of 4×12 mil (0.1×0.3 mm)
- Repair/removable with hot air, 30 sec at 200 to 250° C
- Tackiness and green strength equivalent to solder paste
- Final strength equal to tin-lead solder
- Contact resistance less than 100 mΩ per connection
- Must pass MIL-STD tests for contact reliability

He shows the results of tests with six different silver formulations as well as with silver-plated copper particles. The results showed that the silver-bearing adhesive maintained a contact resistance similar to tin-lead solder over 100 thermal cycles which, per MIL-STD 202F, Method 107G, cycled the boards from –65 to +125° C in approximately 1-hr cycles. Over this same test, the copper-based adhesive showed a rise in contact resistance after 20 cycles. Long-term exposure to 85° C showed a greater rise in tin-lead contact resistance than in the adhesive contact resistance after 100 hr of continuous exposure.

More development work is occurring with conductive adhesives, and it is reasonable to expect that their market penetration will continue to increase. Users must now consider what process changes will be needed to allow acceptance of these adhesives.

2.6.7 Adhesives for Other Purposes

Certain adhesives have dielectric properties, which can reduce crosstalk. The adhesive will reduce the parasitic capacitance between traces and leads. Filling the area between leads of, e.g., a gull wing package will reduce crosstalk. Likewise, filling the areas between contact solder joints in a board-mounted connector will reduce crosstalk in the connector. However, any use of adhesives in this manner must consider the difficulty in reworking solder joints that have been covered in an adhesive.

Thermal conductive adhesives are used to enhance the thermal conductivity between a component body and the substrate or a heat sink and to allow assembly of a component and a heat sink without mechanical fasteners. Thermal conductive adhesives are discussed further in Chapter 10.

One of the newest uses of adhesives is in the underfill process used with flip chips. Explained in more detail in Chapter 4, flip chips are silicon die mounted with their bond pads down, facing the board. The differences in CTE between the typical FR-4 board and the silicon flip chip initially lead to cracking of both solder joints and die. To compensate for the differences, an adhesive underfill is used. The underfill may be applied before the flip chip is placed and cured during the reflow solder process step, or it may be applied after the chip has been soldered. If applied afterward, the underfill is dispensed on one or more sides of the die with a syringe dispenser, after which it flows under the chip by capillary action. It is normally not dispensed on all four sides to eliminate the risk of an air bubble being trapped under the die. After the capillary action has largely taken place and the bulk of the dispensed underfill has wicked under the chip, another application may take place to form a fillet around the periphery of the chip.

This fillet not only provides additional stabilization, it also acts as a reservoir if more underfill wicks between the chip and the board than originally calculated. This can happen, e.g., if the bump size on a particular flip chip is smaller than the average bump size used in the original underfill volume calculation (Fig. 2.13). Alternatively, the underfill and fillet volumes may be dispensed at the same time.

As can be seen in the figure, the volume of underfill must be calculated by finding the volume of space under the die ($L_D \times W_D \times C_D$), subtracting V_B the volume of all the bumps under the chip, and adding the total expected fillet volume VF.

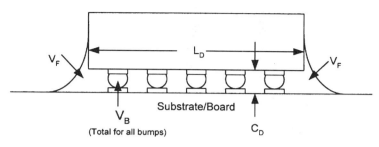

FIGURE 2.13 Example of flip chip with underfill.[19]

The underfills contain thermoset polymers and silica fillers that are used to lower the underfills CTE. They may also contain additives which will reduce their resistance to flow, allowing easier wicking under the flip chip.

Another use of adhesive is gluing and encapsulating *chip on board (COB)* products. Common uses of encapsulant on one-chip products are watches and calculators. In these applications, the die is glued onto the board, and wire standard bonding techniques are used to connect the die to the board. The die, with their associated wire bonds, are then encapsulated to both mechanically and environmentally protect them. The application of the encapsulant is critical, since it is possible to move the bonding wires, shorting them and making the product worthless, since it is not possible to rework encapsulated COB boards.

2.7 Solder Joint Formation

Solder joint formation is the culmination of the entire process. Regardless of the quality of the design, or any other single portion of the process, if high-quality reliable solder joints are not formed, the final product is not reliable. It is at this point that PPM levels take on their finest meaning. For a medium-size substrate (nominal 6 × 8 in), with a medium density of components, a typical mix of active and passive parts on the top side and only passive and three- or four-terminal active parts on bottom side, there may be in excess of 1000 solder joints/board. If solder joints are manufactured at the 3 sigma level (99.73% good joints, or 0.27% defect rate, or 2700 defects/1 million joints), *there will be 2.7 defects per board!* At the 6 sigma level, of 3.4 PPM, there will be a defect on 1 out of every 294 boards produced. If your anticipated production level is 1000 units/day, you will have 3.4 rejects based solely on solder joint problems, not counting other sources of defects.

2.7.1 Solderability

The solderability of components and board lands/traces can be defined by

- *Wettability.* The nature of component terminations and board metallizations must be such that the surface is wetted with molten solder within the specified time available for soldering, without subsequent dewetting.
- *Metallization dissolution.* The component and board metallizations must be able to withstand soldering times and temperatures without dissolving or leaching.
- *Thermal demand.* The mass of the traces, leads, and thermal aspects of packages must allow the joint areas to heat to the necessary soldering temperature without adversely affecting the component or board materials.

Wettability of a metal surface is its ability to promote the formation of an alloy at the interface of the base material with the solder to ensure a strong, low-resistance joint.

Solderability tests are typically defined for component leads by dip tests. The components leads are dipped vertically into 235° C molten solder, held steady for two seconds, then inspected under 10× to

20× magnification. Good solderability requires that 95% of the solderable surfaces be wetted by the molten solder. In Fig. 2.14, examples are shown of dipping SMT components into a solder pot, typically by holding them with tweezers. After the two-second immersion, they are removed from the pot, allowed to cool, then examined.

Examination of the terminations on leadless parts such as chip components should show a bright solder coating with no more than 5% of the area covered with scattered imperfections such as pinholes, dewetted areas, and non-wetted areas. The inspection should also look for any leaching or dissolution of terminations.

Examination of terminations on leaded parts will depend on the area of the lead being inspected. As shown in Fig. 2.15, different areas of the leads have different requirements.

The areas labeled A are defined as the side faces, underside, and outside bend of the foot, to a height above the base equal to the lead thickness. This area must be covered with a bright, smooth solder coating with no more than 5% imperfections, and the imperfections must not be concentrated in any one location. The area labeled B is defined as the top of the foot and angled portion of the lead. The surface must all be visibly wetted by solder, but there may be more imperfections in this area than in A. The area labeled C is the underside of the lead (except that underside defined for area A) and the cut end of the lead. For area C no solderability requirements are defined.

Board solderability can be preserved in several ways. The most common is solder coating the base metal. This coating is typically subjected to the *hot-air solder leveling (HASL)* process. The HASL process results in coatings that are not as planar as required by many fine- and ultrafine-pitch and BGA components. For the stricter requirements of these components, the most common solderability protection

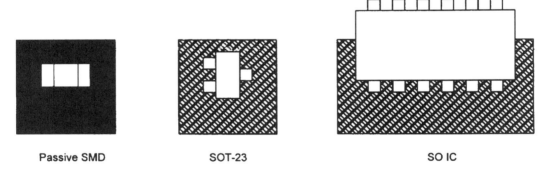

FIGURE 2.14 Dip tests for SMT components.[20]

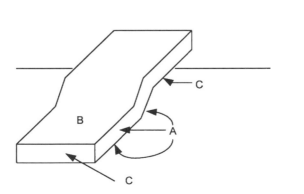

FIGURE 2.15 Solderability inspection of gull-wing leads.

technique is the use of *organic solderability preservatives (OSPs)*. For OSP applications, the board is etched in the typical manner and cleaned, then the OSP is applied directly over the bare copper. Considerations in the use of OSPs include

- Shelf life of the coated board.
- Boards that will undergo multiple soldering phases (e.g., a Type 2C SMT board that will have bottom-side placement and reflow followed by inversion of the board and top-side placement and reflow) must use OSPs that will survive through several solder cycles.
- Copper protection after soldering.

2.7.2 Flux

Flux is mixed with the metallic solder materials to create one of the forms of "solder" used in electronic soldering, typically paste. Flux

- Removes/reduces surface oxidation
- Prevents reoxidation at the elevated soldering temperatures
- Assists in heat transfer
- Improves wettability of the joint surfaces.

It is important to note that while flux has some cleaning ability, it will not make up for poor cleanliness of the board. A bare board (the finished board without any components) must be stored and handled in a manner that will minimize the formation of oxides, the accumulation of any airborne dirt, and any handling with bare hands with the resultant accumulation of oils. Flux is applied as a separate function at the beginning of the wave solder process and as part of the solder paste in the reflow solder process.

There are three primary types of fluxes in use: rosin mildly activated (RMA), water-soluble, and no-clean. Each type of flux has its strengths and weaknesses:

RMA flux is composed of colophony in a solvent, but with the addition of activators, either in the form of di-basic organic acids or organic salts. Chlorine is added as a solvent, and the activator-to-solvent ratio determines the activity and therefore the corrosivity. RMA fluxes typically have their activity and corrosivity functions primarily occur during the soldering process. Consequently, they have very little activity remaining after that process and may be left on the board. They do, however, form a yellowish, tacky residue that may present the following problems:

- Difficulty in automatic test system probe penetration
- Moisture accumulation
- Dirt accumulation

For these reasons, RMA flux is usually cleaned in commercial processes. The major disadvantage to RMA shows up in the cleaning process. The three most common cleaning agents are CFCs, which are no longer legal; alcohol, which is very flammable; and water with saponifiers.

Water-soluble fluxes are typically more active that RMA fluxes and are used when high flux activity is needed. Their post-solder residues are more corrosive than those of RMA fluxes, and therefore cleaning is mandatory when water-soluble flux is used. The term *water-soluble* means that the cleaning can be done using deionized water without any solvents. Many water-soluble fluxes do not contain water, since water tends to boil off and splatter during the reflow heating process.

No-clean fluxes are specialized RMA-based fluxes designed to leave very little noticeable residue on the board. They tend to have less activity than either RMA or water-soluble fluxes, and therefore it is imperative that board cleanliness be scrupulous before the placement process. The main advantage of no-clean flux is, of course, that it does not need to be cleaned, and this gives two positive results to the user. One is that it eliminates the requirement of cleaning under low-clearance parts (which is difficult), and the other is that the residue is self-encapsulating so that it will not accumulate moisture or dirt. This self-encapsulation also ends any activity on the part of the flux residues.

Some users feel that even the small amount of typically grayish residue is undesirable and will clean the boards. This process cannot be taken lightly, with a "we'll get what we can" attitude. A partially-cleaned no-clean-flux board will result in some of the flux residue being unencapsulated, resulting in continuing activity on the board, with the possible result of corrosion occurring. If a no-clean board is cleaned, it must be as thorough as the cleaning of an RMA or water-soluble board.

If the user is changing from a solvent-based RMA cleaning to water-based cleaning, one issue to be faced is that the surface tension of water is much higher than the surface tension of solvents. This means that it is more difficult for the water to break up into small enough particles to easily get into and clean under low-clearance parts, especially fine-pitch parts. Part of the drive to no-clean fluxes is the difficulty in cleaning under many newer part styles.

Generally, the same fluxes are used in wave soldering as are used in solder paste and preforms. However, the activity is higher in the pastes, since the spherical solder balls in the paste tend to have a relatively large surface area with resulting high levels of oxidation. These additional liquid activators make it particularly important that the thermal profile used during the reflow process allow sufficient time during the preheat phase for the liquid portion of the flux to dry out and prevent sudden heating with resultant splattering during the reflow phase.

Flux Application

In the wave-solder process, flux is applied prior to the preheat phase, prior to immersion of the bottom of the board in the solder wave. This allows the flux to dry during the preheat phase. The three most common methods of applying flux are by wave, spray, and foam.

Wave fluxing is done by an impeller in the flux tank generating a standing wave of flux, much like the solder wave is generated. This standing wave (frequently a double wave) creates a washing action on the bottom of the board and also results in application of the flux for deoxidation. The wave is commonly followed by a soft brush to wipe off excess flux. Wave fluxing can be used with virtually any flux and will work well with bottom-mounted SMDs. As with the solder wave itself, flux wave height must be closely controlled to prevent flooding the top of the board.

Spray fluxing, as the name implies, uses a system to generate a linear spray of flux that will impinge evenly on the bottom of the board. There are several techniques used to generate the spray. Spray fluxing can be used with most fluxes.

Foam fluxing involves forcing a clean gas through an aerator inside a flux tank. The small bubbles produced are then forced through a nozzle above the surface of the flux liquid. This nozzle guides the foam to the level of the bottom of the board, and a layer of flux is applied to the board. As the bubbles burst, they create a micro-cleaning action on the board. The height of the foam peak is not critical, and the micro-cleaning action makes foam fluxing a good technique for boards that have bottom-mounted SMT ICs. A disadvantage to foam fluxing is the losses that occur due to evaporation.

All the fluxing methods must have sufficient board contact and pressure to not only clean and deoxidize the bottom of the board but also any through-holes in the board. Activity is affected by

- Flux composition
- Flux density (controlled by flux monitoring systems)
- Board conveyor speed
- Flux pressure on the bottom of the board

All pre-applied fluxes must be dried during the preheat phase. This occurs during the preheat phase, at temperatures between 80 and 110° C. Specific preheat temperatures depend on which flux is used and should be determined from the flux/paste supplier. The drying accomplishes several purposes. Drying of the flux increases the density of the remaining flux, which increases and speeds up its activity. Drying of the flux is also necessary to evaporate most of the solvents in the flux. In reflow ovens, if the solvents are not minimized during preheat, the sudden expansion of the liquids during the reflow bump will expel portions of the metals in the paste, resulting in unwanted solder balls on non-wettable surfaces of

Surface Mount Technology

the board. Preheat time is affected by characteristics of the flux and should be recommended by the flux supplier.

The preheat phase also serves to bring the components and the board up to temperatures that will minimize risks of cracking the components or the board and also minimize warping of the board. Preheating can be done with quartz tubes (IR heating), quartz lamps, forced air, or a radiation panel. Some wave solder systems use a combination of these techniques.

2.7.3 Solder Paste

Solder paste may be deposited by syringe or by screen or stencil printing techniques. Stencil techniques are best for high-volume/speed production, although they do require a specific stencil for each board design. Syringe and screen techniques may be used for high-volume lines and are also suited to mixed-product lines where only small volumes of a given board design are to have solder paste deposited. Syringe deposition is the only solder paste technique that can be used on boards that already have some components mounted. It is also well suited for low-volume high-mix lines, prototype lines, and for any use requires only software changes to develop a different deposition pattern.

The most common production technique for solder deposition is stencil printing. The stencil openings will align with the component terminations and the land patterns designed for the substrate. The opening area and the stencil thickness combine to determine the volume of each solder paste deposit. The stencil thickness is usually a constant across the print surface (*stepped* stencils are possible and are discussed later), and IPC-recommended stencil thicknesses are as follows:

Stencil thickness	Components
8–20 mils	chip components only
8 mils	leaded components with >30 mil pitch
6 mils	leaded components with 20–25 mil pitch
<6 mils	leaded components with <20 mil pitch

Burr[16] reports the results of a study showing that electroformed and laser-cut/nickel-plated stencils provide more consistent print volumes than chemically etched, bimetal, or laser-cut (without plating) stencils. He also reports that square aperture openings with 0.010-in radius corners provide better solder paste release than round openings, with metal squeegees.

Solder joint defects have many possible origins.

- Poor or inconsistent solder paste quality
- Inappropriate solder pad design/shape/size/trace connections
- Substrate artwork or production problems, e.g., mismatch of copper and mask, warped substrate
- Solder paste deposition problems, e.g., wrong volume or location
- Component lead problems, e.g., poor coplanarity or poor tinning of leads
- Placement errors, e.g., part rotation or x-y offsets
- Reflow profile, e.g., preheat ramp too fast or too slow, wrong temperatures created on substrate
- Board handling problems, e.g., boards getting jostled prior to reflow

Once again, a complete discussion of all of the potential problems that can affect solder joint formation is beyond the scope of this chapter. Many references are available that address the issues. An excellent overview of solder joint formation theory is found in Klein-Wassink (see Suggested Readings).

While commonly used solder paste for both THT and SMT production contains 63-37 eutectic tin-lead solder, other metal formulations are available, including 96-4 tin-silver (a.k.a. silver solder). The fluxes available include RMA, water-soluble, and no-clean. The correct decision rests as much on the choice of flux as it does on the proper metal mixture. A solder paste supplier can best advise on solder pastes for specific needs. Many studies are in process to determine a no-lead replacement for lead-based solder in commercial electronic assemblies. The design should investigate the current status of these studies as well as the status of no-lead legislation as part of the decision-making process.

To better understand solder joint formation, one must understand the make-up of solder paste used for SMT soldering. The solder paste consists of microscopic balls of solder, most commonly tin-lead with the accompanying oxide film, flux, and activator and thickener solvents as shown in Fig. 2.16.

The fluxes are an integral part of the solder paste and are discussed further in Section 2.11. RMA, water soluble, and no-clean flux/pastes are available. An issue directly related to fluxes, cleaning, and fine-pitch components (25 mil pitch and less) is reflowing in an inert environment. Inert gas blanketing the oven markedly reduces the development of oxides in the elevated temperatures present. Oxide reduction needs are greater with the smaller metal balls in paste designed for fine-pitch parts since there is more surface area on which oxides can form. No-clean fluxes are not as active as other fluxes and therefore have a lesser ability to reduce the oxides formed on both the paste metal and substrate metallizations. Inerting the oven tends to solve these problems; however, it brings with it control issues that must be considered.

Regardless of the supplier, frequent solder paste tests are advisable, especially if the solder is stored for prolonged periods before use. At a minimum, viscosity, percent metal, and solder sphere formation should be tested.[17] Solder sphere formation is particularly important, since acceptable particle sizes will vary depending on the pitch of the smallest-pitch part to be used, and the consistency of solder sphere formation will affect the quality of the final solder joint. Round solder spheres have the smallest surface area for a given volume and therefore will have the least amount of oxide formation. Uneven distribution of sphere sizes within a given paste can lead to uneven heating during the reflow process, with the result that the unwanted solder balls will be expelled from the overall paste mass at a given pad/lead site. Fine-pitch paste has smaller ball sizes and consequently more surface area on which oxides can form.

It should be noted at this point that there are three distinctly different "solder balls" referred to in this chapter and in publications discussing SMT. The solder sphere test refers to the ability of a volume of solder to form a ball shape due to its inherent surface tension when reflowed (melted) on a non-wettable surface. This ball formation is dependent on minimum oxides on the microscopic metal balls that make up the paste—the second type of "solder ball." It is also dependent on the ability of the flux to reduce the oxides that are present as well the ramp-up of temperature during the preheat and drying phases of the reflow oven profile. Too steep a time/temperature slope can cause rapid escape of entrapped volatile solvents, resulting in expulsion of small amounts of metal that will form undesirable "solder balls" of the third type—small metal balls scattered around the solder joint(s) on the substrate itself rather than on the tinned metal of the joint. This third type of ball can also be formed by excess solder paste on the pad and by misdeposition on non-wettable areas of the substrate.

The reader is referred to Lau (Suggested Readings) for discussions of finite element modeling of solder joints and detailed analytical studies of most aspects of basic joints and of joint failures. Various articles

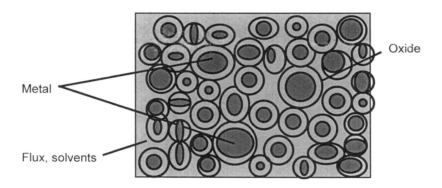

FIGURE 2.16 Make-up of SMT solder paste.

Surface Mount Technology

by Werner Engelmaier and Associates also address many solder joint reliability issues and their analytical analysis. These and other sources discuss in detail the quality issues that effect solder paste, such as:

- Viscosity and its measurement
- Printability
- Open time
- Slump
- Metal content
- Particle/ball size in mesh
- Particle/ball size consistency
- Wetting
- Storage conditions

Note that, with regard to viscosity measurements, some paste manufacturers will prefer the spindle technique and some the spiral technique. To properly compare the paste manufacturer's readings with your tests, the same technique must be used.

2.7.4 Paste-In-Hole Technology

Paste-in-hole (PIH) technology is a relatively recent development that allows the use of THT parts with the reflow soldering technology. Simple put, paste is deposited in the holes, a THT part is mounted into the paste/holes and, at the conclusion of the placement process, the board is reflowed. The two major concerns in implementing PIH are use of the appropriate volume of solder paste and that the THT part body must be able to withstand reflow temperatures.

Solder paste volume calculations are straightforward:

$$\text{Paste volume} = [(\text{volume of hole}) - (\text{volume of pin}) + (\text{volume of solder fillet})] \times 2$$

where the $\times 2$ results from the 50% reduction in solder paste volume after reflow.

Volume of solder fillets is normally estimated using straight-line approximations of a fillet, resulting in calculation of the volume of a truncated cone. An example is shown below for a DIP socket using the PIH technique.

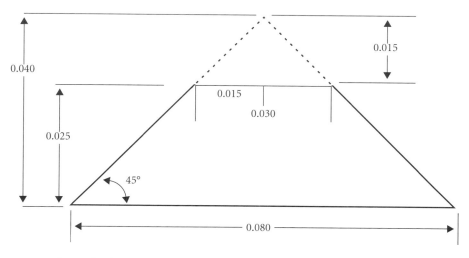

Measurement determinations:

    Board thickness, $T = 0.0625$ in
    Radius of home, $R = 0.015$ in

Width of pin, $W = 0.020$ in
Diameter of pad, $D = 0.080$ in, radius $R = 0.040$ in
Height of fillet, $H = 0.050$ in
Angle of fillet approximation $= 45°$

Volume of hole $= V_H = \pi R^2 T = \pi \times 0.015^2 \times 0.0625$ in $= 44.18$ E^{-6} in^3
Volume of pin through board $= V_P = W^2 T = 0.020^2 \times 0.0625$ in $= 25^{-6}$ in^3
Volume of solder fillets = volume of overall cone − (volume of truncated section + volume of DIP pin inside cone), where the volume of a cone is

$$= \frac{\pi \times R^2 \times H}{2 V_F}$$

$$= V_c - (V_T + V_D) = \frac{\pi \times 0.040^2 \times 0.040}{2} - \frac{\pi \times 0.015^2 \times 0.015}{2} + (0.020^2 \times 0.025)$$

$$= 85.2 \text{ E}^{-6} \text{ in}^3 \text{ of solder paste in the fillet}$$

Therefore,

paste volume = [(volume of hole) − (volume of pin) + (volume of solder fillet)] × 2
$= (44.18$ E^{-6} in$^3 - 25$ E^{-6} in$^3 + 85.2$ E^{-6} in$^3) \times 2 = 208.76$ in^3

This calculation uses inches, since the controlling dimension for DIP sockets is inches. This would be the amount of solder paste to be deposited, whether through the same volume hold in a stencil or by a syringe dispensing unit.

From this calculation, 209 in^3 of solder paste would be deposited at each hole. Final deposited volume would be adjusted if necessary from reflow tests of the final board design.

2.8 Parts

2.8.1 Part Inspection

Briefly, all parts must be inspected prior to use. Functional part testing should be performed on the same basis as for through-hole devices. Each manufacturer of electronic assemblies is familiar with the various processes used on through-hole parts, and similar processes must be in place on SMDs. Problems with solderability of leads and lead planarity are two items that can lead to the largest number of defects in the finished product. Solderability is even more important with SMDs than with through-hole parts, since all electrical and mechanical strength rests within the solder joint, there being no through-hole with lead to add mechanical strength.

Lead coplanarity is defined as follows. If a multi-lead part (e.g., an IC), is placed on a planar surface, lack of ideal coplanarity exists if the lowest solderable part of any lead does not touch that surface. Coplanarity requirements vary depending on the pitch of the component leads and their shape, but out-of-plane measurements generally should not exceed 4 mils (0.004 in) for 50-mil pitch devices, and 2 mils for 25-mil pitch devices. In Fig. 2.17, the distance that lead "C" is off the plane of the other leads would be the measurement of co-planarity for these four leads.

2.8.2 Moisture Sensitivity of Plastic Packages

All SMDs undergo thermal shocking during either reflow or wave soldering processes. In THT, the board is between the parts and the solder wave, protecting the package from the high temperature gradients experienced by SMDs. A THT part undergoing wave soldering may have a body temperature rise to 40-50°, while an SMD will have a body-temperature rise to over 200° C. Therefore, all plastic-packaged

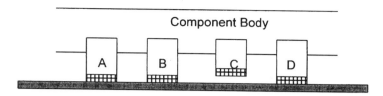

FIGURE 2.17 Coplanarity.

SMDs must be controlled for moisture content. Additionally, any THT components subjected to reflow soldering must also be controlled, since the risk is due to the combination of moisture absorption and reflow soldering, and THT IC packages use the same encapsulants as SMDs.

In addition to problems during the reflow process, moisture absorption can also result in problems during rework and repair. Repair is particularly hazardous, since moisture absorption has most likely occurred during the life cycle of the product, and repair tools and procedures typically result in higher thermal heating rates (°C/sec) than the initial reflow soldering.

The epoxy encapsulants used with all ICs are hydroscopic and absorb moisture from their surrounding environments. Additionally, thermal cycling stresses cause the metal lead frame to expand and contract at a rate different from that of the encapsulant, leading to the risk of stress fractures at the metal/epoxy interface. Moisture can then be absorbed through these fractures. Any moisture absorbed into the package, regardless of its route, will then expand rapidly during reflow or wave soldering, resulting in swelling of the package and the risk of delamination of the package from the die, bond-wire, and/or lead frame, as well as the risk of cracking of the package and bond-wire breakage. Continued exposure to moisture, flux, and airborne contaminants leads to the risk of circuit failure due to resulting corrosion.

The fluxes used in solder paste are corrosive to IC bond pad metallization if they are allowed to enter the package due to cracking or delamination. This is particularly true for fluxes with higher activity, such as water-soluble flux.

The effect of moisture absorption and the resulting risk for damage is a function of a number of package, package design, and material property issues. These include

- Die size
- Encapsulant moisture absorption properties
- Body size/encapsulant thickness
- Adhesive strength between the encapsulant, die, and lead frame
- Coefficient of thermal expansion (CTE) differences between the encapsulant, die, and lead frame

The body crack resulting from the moisture expansion results in an audible "pop," so this phenomenon is commonly known as *popcorning*. IC manufacturers evaluate their products for susceptibility to moisture absorption. Their requirements for storage and handling prior to placement and reflow must be followed. Moisture-sensitive parts are shipped from the manufacturer in dry packs that include desiccant and moisture-resistant, ESD-safe bags. The bags are to be opened only when the parts are ready for use. The manufacturer will also specify recommended shelf life, storage time and temperature conditions, and baking conditions if necessary. The user is responsible for determining if any of the storage and handling requirements have not been followed and if remedial procedures such as baking are necessary.

If IC packages have been found to exceed the manufacturer's requirements for storage prior to use, the packages must be baked to drive out the absorbed moisture. It must be remembered that SMT ICs typically have tin-lead plated copper lead frames, with the resulting intermetallic layer formation at the tin-lead-copper interface. High temperatures result in growth of the intermetallic layer, which is not a desirable phenomenon. Additionally, oxidation rates of the lead-plating materials is higher at elevated temperatures, resulting in decreased solderability. Therefore, baking must be limited to only the required time and temperature. It is far better to store and handle packages so that baking is not necessary.

IPC-SM-786A defines several "levels" of moisture sensitivity. IC shipping packages will include labelling indicating the level and the floor life of the ICs once the shipping package is opened. Floor life is the time after opening of the shipping package for which baking is not necessary. If the floor life is exceeded, baking is necessary prior to placement and reflow. These levels are as follows:

Level	Description	Floor life	T/RH	Time
1	Non-moisture sensitive	Unlimited at 30° C/85% RH	85° C/85% RH	168 hr
2	Limited moisture sensitivity	1 year at 30° C/60% RH	85° C/60% RH	168
3	Moisture sensitive	1 week at 30° C/60% RH	30° C/60% RH	168 hr + MET
4	Highly moisture sensitive	72 hr at 30° C/60% RH	30° C/60% RH	72 hr + MET
5	Extremely moisture sensitive	48 hr at 30° C/60% RH	30° C/60% RH	48 hr + MET
6	Bake before use	6 hr at 30° C/60% RH	Bake before use	Bake before use

MET = Manufacturer's exposure time, the compensation factor that accounts for the time after bake that the component manufacturer requires to process the components prior to initial bag seal, including a factor for distribution handling.

The most valuable information the above table is the floor life, assuming the temperature and relative humidity listed are not exceeded. Once the shipping container has been opened, the components must be placed and reflowed during the allowed time. If only a portion of the contents are to be used, the remaining components must be resealed in the original container, with fresh desiccant, as soon as possible. The container markings must be changed to indicate that the floor life remaining is the original floor life minus the amount of time the components were exposed to ambient conditions. Alternatively, most manufacturers allow long exposure times if the ambient RH is <20%.

Baking can typically follow either a high-temperature or a low-temperature profile. The high-temperature profile depends on package thickness. For packages 2 mm thick or greater, the bake is at 125° C for 24 hr. For packages <2 mm thick, 125° C for 6 hr is sufficient. These high temperatures contribute both to growth of the intermetallic layer and to increased oxidation rates, and they should be limited to one bake cycle. At this high temperature, the components must be removed from their original shipping container, with the risk of lead damage, and placed in a metal or high-temperature plastic container.

The low-temperature profile allows the components to remain in their original container. This bake is at 40° C for 192 hr (8 days) at <5% RH during the bake. Although this bake temperature is not detrimental to the IC die, it will ultimately result in increased oxidation, increased intermetallic formation, and reduced antistatic properties of the tubes or tape and reel, so long-term storage at this temperature is not recommended. After baking for one of the above profiles, the components will once again have the floor life listed for their moisture classification level.

2.8.3 Part Placement

Proper part placement not only places the parts within an acceptable window relative to the solder pad pattern on the substrate, the placement machine also will apply enough downward pressure on the part to force it halfway into the solder paste as shown in Fig. 2.18. This assures both that the part will sit still when the board is moved and that coplanarity offsets within limits will still result in an acceptable solder joint. The effects of coplanarity can be done mathematically by considering

- The thickness of the solder paste deposit
- Maximum coplanarity offsets among leads
- Lead penetration in paste

FIGURE 2.18 Part placed into solder paste with a passive part.

Surface Mount Technology

If T is the thickness of paste, C is maximum allowable coplanarity, and P is penetration in paste (as a percentage of overall average paste thickness), then (Fig. 2.19):

$$P = \frac{C}{T} \times 100\%$$

Part placement may be done manually for prototype or low-volume operations, although this author suggests the use of guided x-y tables with vacuum part pickup for even the smallest operation. Manual placement of SMDs does not lend itself to repeatable work. For medium- and high-volume work, a multitude of machines are available. See Fig. 2.20 for the four general categories of automated placement equipment.

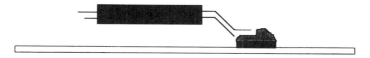

FIGURE 2.19 Part placed into solder paste with a active part.

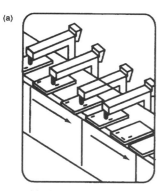

(a) Moving board/fixed head
Each head places one component
1.8 to 4.5 seconds/board

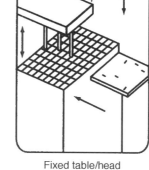

(b) Fixed table/head
All components placed simultaneously
7 to 10 seconds/board

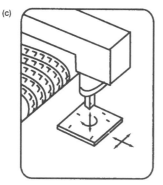

(c) X-Y movement of table head
Components placed in succession individually
0.3 to 1.8 seconds/component

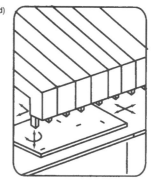

(d) X-Y table/fixed head
Sequential/simultaneous firing of heads
0.2 seconds/component

FIGURE 2.20 Four major categories of placement equipment.[19]

A good source for manufacturer's information on placement machines and most other equipment used in the various SMT production and testing phases is the annual *Directory of Suppliers to the Electronics Manufacturing Industry,* published by *Electronic Packaging and Production.*[18] Among the elements to consider in the selection of placement equipment, whether fully automated or x-y vacuum assist tables, are

- Volume of parts to be placed/hour
- Conveyorized links to existing equipment
- Packaging of components to be handled; tubes, reels, trays, bulk, etc.
- Ability to download placement information from CAD/CAM systems
- Ability to modify placement patterns by the operator
- Vision capability needed for board fiducials and/or fine-pitch parts

Checks of new placement equipment, or in-place equipment when problems occur, should include

- x-accuracy
- y-accuracy
- z-accuracy
- Placement pressure
- Both upward and downward viewing vision system checks

It must be emphasized that placement accuracy checks cannot be made using standard circuit boards. Special glass plates with engraved measurement patterns and corresponding glass parts must be used, since standard circuit boards and parts vary too widely to allow accurate x, y, and Θ measurements.

Placement machines use a combination of mechanical precision and/or vision-assisted targeting to provide the required placement accuracy. Component placement can be described in terms of three targeting techniques that can be called *ballistic, tracer,* and *guidance* targeting. A ballistic targeting system has a fixed machine origin (0,0) from which all other coordinates and coordinate systems (camera, placement head, etc.) are calculated. The machine must obtain (from CAD data or from a "teaching technique") the x-y coordinates of the location of the centroid of each part to be placed. The boards enter the machine and are located as accurately as possible. Likewise, the component placement nozzle/head uses one of several techniques to accurately locate the centroid of the part itself. These techniques may aid in x-y and Θ (rotation) correction of the part while the machine is handling it. Ballistic machines are very fast and require precision and repeatability in their handling systems. They may also incorporate basic vision systems to assist in the determination of both board and component locations. These vision systems employ fairly straightforward software algorithms. This means that, while basic board locating is done with vision assistance, there are still cumulative errors from the corner of the board that is closest to the machine's 0,0 origin.

Tracer targeting uses more sophisticated vision systems and more sophisticated algorithms to allow correction for both machine wear and for circuit board *shrink* and *stretch.* These terms mean that the circuit board may be larger or smaller than the actual CAD data. Using multiple fiducials around the board, the vision system allows the placement machine's software-controlled x-y system to correct for board size variations, within limits. Typically, correction will not be made for errors in excess of 10 mils (0.25 mm). As the vision system determines the exact board and artwork size, it will adjust the placement algorithms so that the placement system continues to place components correctly. This system effectively calculates the center of the board, and any errors are referred from that location. Once the board center is located, the stored CAD data is used to determine the location of each component, after adjusting each location for measured shrink or stretch. Since the vision system camera is typically riding on the placement system, tracer targeting systems fiducial determination will correct to some degree for x-y system wear, as long as good repeatability still exists.

Guidance targeting systems use even more sophisticated vision systems and algorithms to find the placement origin for each component. Their software allows them to determine each component's intended location on the board from the board land patterns. Each component passes over an upward-looking camera, which allows accurate calculation of each component's exact location compared to the placement system's x-y coordinates. Matching the board location and the component location allows for very accurate placement.

While tracer and guidance-targeting systems were once slower than ballistic systems due to the time overhead of the vision system algorithms, advancements in vision system computers have overcome this limitation. They are both considerably more accurate than ballistic systems and are required for fine and ultrafine-pitch parts, including CSPs and COBs.

2.9 Reflow Soldering*

Once SMDs have been placed in solder paste, the assembly will be reflow soldered. This can be done in either batch-type ovens or conveyorized continuous-process ovens. The choice depends primarily on the board throughput/hour required. While early ovens were of the vapor phase type, most ovens today use infrared (IR) heating, convection heating, or a combination of the two.

Vapor phase ovens may be batch type, where a board is lowered into the oven, or conveyorized. They operate by boiling an inert liquid compound in a vat beneath the reflow location of the board. *Inert* in this case means that it will not interact with the board or components on it, not that the fluid is inert in the true chemical sense. In fact, it is often poisonous. The liquid is chosen to have a boiling point appropriate for the reflow temperatures, approximately 215° C. The oven will include a preheat zone. As the board moves from the preheat zone to the vapor zone, the board is at a lower temperature than the vapor. The vapor will condense onto the board, components, and solder paste. As it does, the release of the latent heat of vaporization will heat all parts to the temperature of the vapor. The solder reflows and solder joints are formed. The vapor phase process is easy to control but suffers from several limitations. The vapor only allows one temperature to be present in the reflow zone. The liquid is very expensive and must be controlled by condensing it out as it tries to rise above the reflow zone. Since the fluid will not support life, it must not be allowed to enter the work area. A main issue in the purchase of a vapor phase oven is the manufacturer's technique for vapor control.

In IR ovens, the heat source may be IR panels, quartz lights, or other IR sources. The differing rate of heat absorption of the paste, parts, glue, etc., as a function of color and materials, should considered. For example, the metallized termination of chip resistors and capacitors are shiny and therefore good reflectors of IR energy, while the black epoxy encapsulation on ICs is an excellent IR absorber. While IR ovens can be successfully profiled, they do not allow the profile to stay constant when board or component changes are made. These factors limit the ability of IR ovens to produce consistent quality.

Convection ovens tend to be more forgiving with variations in color and thermal masses on the substrates. Conveyorized convection ovens are available with 3 to 10 zones of heating, allowing for close control of the profile. Batch (one-zone) ovens are discussed in Section 2.11, "Prototype Systems." All ovens are zoned to allow a thermal profile necessary for successful SMD soldering. Ovens with a higher number of zones can be more closely controlled but are also more expensive and need more floor space than ovens with fewer zones. An example of an oven profile is shown in Fig. 2.21, and the phases of reflow soldering that are reflected in that example include the following:

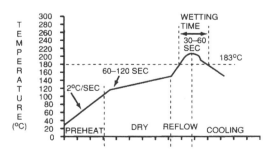

FIGURE 2.21 Typical thermal profile for SMT reflow soldering (types I or II assemblies). *Source:* Cox, R.N., *Reflow Technology Handbook,* Research Inc., Minneapolis, MN, 1992. Reprinted with permission.

---

*Wave soldering is not discussed in this handbook.

- *Preheat.* The substrate, components, and solder paste preheat.
- *Dry.*
  - Solvents evaporate from the solder paste.
  - Flux activates, reduces oxides, and evaporates.
  - Both low- and high-mass components have enough soak time to reach temperature equilibrium.
- *Reflow.* The solder paste temperature exceeds the liquidus point and reflows, wetting both the component leads and the board pads. Surface tension effects occur, minimizing wetted volume.
- *Cooling.* The solder paste cools below the liquidus point, forming acceptable (shiny and appropriate volume) solder joints.

Preheat and dry times are particularly important in the prevention of solder balls. If the reflow bump occurs before the solvents in the paste have had time to evaporate, they will boil during the reflow bump and eject small quantities of paste, forming solder balls away from the lands. On the other hand, if the drying time is too long, the paste will overdry, the deoxidation characteristics of the flux will deteriorate, and the result will be the re-formation of oxides. This will degrade the solderability of the land, and solder balls will form due to poor solderability. Solder balls may also form if solder deposition has resulted in paste being applied on non-wettable surfaces, or if the placement process has resulted in paste being pushed outside the land profile onto, e.g., the solder mask.

The setting of the reflow profile is not trivial. It will vary on whether the flux is RMA, water soluble, or no-clean, and it will vary depending on both the mix of low- and high-thermal mass components and on how those components are laid out on the board. The profile should exceed the liquidus temperature of the solder paste by 20 to 25° C. While final setting of the profile will depend on actual quality of the solder joints formed in the oven, initial profile setting should rely heavily on information from the solder paste vendor as well as the oven manufacturer.

The profile shown is the profile to be developed on the substrate, and the actual control settings in various stages of the oven itself may be considerably different, depending on the thermal inertia of the product in the oven and the heating characteristics of the particular oven being used. The final profile should be determined not by the oven settings but by instrumenting actual circuit boards with thermocouples and determining that the profiles at various locations on the circuit board meet the specifications necessary for good soldering. When determining board profiles, it is important to have thermocouples near the largest (highest thermal mass) components and at the location of the highest density of components, as well as near small and less-dense areas (Fig. 2.22).

Defects as a result of a poor profile may include

- Component thermal shock
- Solder splatter
- Formation of solder balls
- Dewetted solder
- Cold or dull solder joints

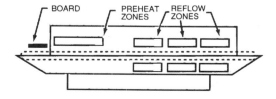

FIGURE 2.22 Conveyorized reflow oven showing the zones that create the profile. *Source:* Intel, *Packaging Handbook*, Intel Corp., Santa Clara, CA, 1994. Reprinted with permission.

Surface Mount Technology

It should be noted that many other problems may contribute to defective solder joint formation. One example would be placement misalignment, which contributes to the formation of solder bridges, as shown in Figure 2.23.

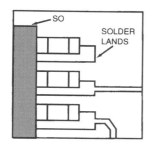

FIGURE 2.23 Solder bridge risk due to misalignment. *Source:* Philips, *Surface Mount Process and Application Notes*, Philips Semiconductor Corp., 1991. Reprinted with permission.

Other problems which may contribute to defective solder joints include poor solder mask adhesion, and unequal solder land areas at opposite ends of passive parts, which creates unequal moments as the paste liquefies and develops surface tension. This is discussed in substrate design criteria in Chapter 7. Incorrect solder paste volumes, whether too much or too little, will create defects, as will board shake in placement machines and coplanarity problems in IC components. Many of these problems should be covered and compensated for during the design process and the qualification of SMT production equipment.

Post-Reflow Inspection

Final analysis of the process is performed based on the quality of the solder joints formed in the reflow process. Whatever criteria may have been followed during the overall process, solder joint quality is the final determining factor of the correctness of the various process steps. As noted earlier, the quality level of solder joint production is a major factor in successful board assembly. A primary criteria is the indication of wetting at the junction of the reflowed solder and the part termination. This same criteria shown in Fig. 2.24 applies to both through-hole and SMDs, with only the inspection location being different.

Note that criteria shown in Fig. 2.24 are for any solderable surface, whether component or board, SMT, or THT. Some lead surfaces are defined as not solderable; e.g., the cut and not-tinned end of an SO or QFP lead is not considered to be solderable. Part manufacturers will define whether a given surface is designed to be solderable.

Presentation of criteria for all the various SMD package types and all the possible solder joint problems is beyond the scope of this chapter. The reader is directed to Recommended Readings C, D, E, G, and I for an in-depth discussion of these issues.

Solder balls are among the most common problems noted during visual inspections of soldered boards. As mentioned in the "Reflow Soldering" section, if the reflow bump occurred before the solvents in the paste have had time to evaporate, they will boil during the reflow bump and eject small quantities of paste, forming solder balls away from the lands. On the other hand, if the drying time was too long, the paste overdried, the deoxidation characteristics of the flux deteriorated, and the result was the reformation of oxides. This degraded the solderability of the land, and solder balls formed due to the resulting poor solderability. Solder balls may also form if solder deposition has resulted in paste being applied on non-wettable surfaces or if the placement process has resulted in paste being pushed outside the land profile onto, e.g., the solder mask.

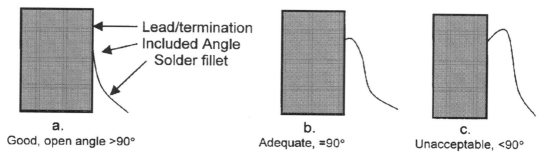

FIGURE 2.24 Solder joint inspection criteria.

Solder balls may or may not survive the cleaning process. However, inspection prior to cleaning is required to determine if solder ball formation has occurred. It is not appropriate to depend on the cleaning process to remove solder balls, since that is a very unreliable result of the cleaning process, especially under low-profile parts and BGAs. Process control ideally results in no solder ball formation. Exact determination of the cause of solder ball formation is very difficult, but it must be addressed if a quality product is expected.

2.10 Cleaning

Cleaning, like all other parts of the process, should be considered during the design phase. Cleaning requirements are determined largely by the flux in the solder paste and should be determined before production is ever started. Design issues may affect the choice of flux, which determines cleaning. For example, low-clearance (close to the substrate) parts are difficult to clean under and may suggest the use of no-clean flux.

Rosin mildly activated (RMA) is the old standard, and if cleaning is needed with RMA, either solvent-based cleaners or water with saponifiers must be used. (Saponifiers are alkaline materials that react with the rosin so that it becomes water soluble.) RMA tends to be nonactive at "room" temperatures and may not need to be cleaned on commercial products designed for indoor use. The major limitation on RMA at this point is the need for chemicals in the cleaning process.

Water-soluble fluxes are designed to be cleaned with pure water. They remain active at room temperatures and therefore must be cleaned when used. Beyond their activity, the other disadvantage to water-soluble fluxes is that their higher surface tension relative to solvent-based cleaners means they have more difficulty cleaning under low-clearance components.

No-clean fluxes are designed to not be cleaned, and this means they are of low activity (they don't reduce oxides as well as other types of flux), and when used they should not be cleaned. No-clean fluxes are designed so that, after reflow, they microscopically encapsulate themselves, sealing in any active components. If the substrate is subsequently cleaned, the encapsulants may be destroyed, leaving the possibility of active flux components remaining on the board.

2.11 Prototype Systems

Systems for all aspects of SMT assembly are available to support low-volume/prototype needs. These systems will typically have manual solder-paste deposition and part placement systems, with these functions being assisted for the user. Syringe solder paste deposition may be as simple as a manual medical-type syringe dispenser that must be guided and squeezed freehand. More sophisticated systems will have the syringe mounted on an x-y arm to carry the weight of the syringe and will apply air pressure to the top of the syringe with a foot-pedal control, freeing the operator's arm to guide the syringe to the proper location on the substrate and perform the negative z-axis maneuver that will bring the syringe tip into the proper location and height above the substrate. Dispensing is then accomplished by a timed air pressure burst applied to the top of the syringe under foot-pedal control. Paste volume is likewise determined by trial and error with the time/pressure relation and depends on the type and manufacturer of paste being dispensed.

Part placement likewise may be as simple as the use of tweezers, progressing to hand-held vacuum probes to allow easier handling of the components. As mentioned in the "Part Placement" section, x-y arm/tables are available that have vacuum-pick nozzles to allow the operator to pick a part from a tray, reel, or stick, and move the part over the correct location on the substrate (Fig. 2.25). The part is then moved down into the solder paste, the vacuum is turned off manually or automatically, and the nozzle is raised away from the substrate.

Soldering of prototype/low-volume boards may be done by contact soldering of each component, by a manually-guided hot-air tool, or in a small batch or conveyorized oven. Each step up in soldering sophistication is of course accompanied by a increase in the investment required.

Surface Mount Technology 2-41

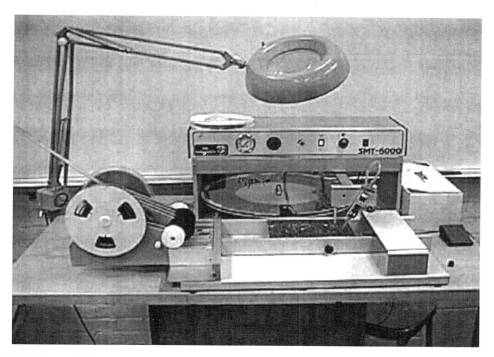

FIGURE 2.25 x-y table for SMT paste and part placement.

For manufacturers with large prototype requirements, it is possible to set up an entire line that would involve virtually no hardware changes from one board to another. This involves the following:

- ECAD design and analysis produces Gerber files.
- CNC circuit board mill takes Gerber files and mills out two-sided boards.
- The software translation package generates solder-pad centroid information.
- The syringe solder paste deposition system takes translated Gerber file and dispenses the appropriate amount at each pad centroid.
- The software translation package generates part centroid information.
- Part placement equipment places parts based on translated part centroid information.
- The assembly is reflow soldered.

The only manual process in the above system is adjustment of the reflow profile based on the results of soldering an assembly. The last step in the process would be to test the finished prototype board. This system could also be used for very small-volume production runs, and all components as described are available. With a change from milled boards to etched boards, the system can be used as a flexible assembly system.

Terms and Definitions

Coefficient of thermal expansion (CTE, a.k.a. TCE).   A measure of the ratio between the measure of a material and its expansion as temperature increases. It may be different in x, y, and z axes. CTE is expressed in PPM/°C. It is a measure that allow comparison of materials to be joined.

Coplanarity.   A simplified definition of planarity, which is difficult to measure. Noncoplanarity is the distance between the highest and lowest leads and is easily measured by placing the IC on a flat surface such as a glass plate. The lowest leads will then rest on the plate, and the measured difference to the lead highest above the plate is the measurement of coplanarity, or more correctly the lack of coplanarity.

Glass transition temperature (T_g).   Below T_g, a polymer substance, such as fiberglass, is relatively linear in its expansion/contraction due to temperature changes. Above T_g, the expansion rate increases dramatically and becomes nonlinear. The polymer will also lose its stability; i.e., an FR-4 board "droops" above T_g.

Gull Wing.   An SMD lead shape as shown in Fig. 2.4 for SOPs and QFPs. It is so called because it looks like a gull's wing in flight.

J-Lead.   An SMD lead shape as shown below in the PLCC definition; so called because is in the shape of the capital letter J.

Land.   A metallized area intended for the placement of one termination of a component. Lands may be tinned with solder, or be bare copper in the case of SMOBC circuit board fabrication.

Plastic leaded chip carrier (PLCC).   Shown below, it is a common SMT IC package and is the only package that has the leads bent back under the IC itself.

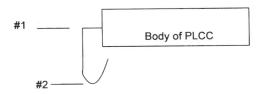

Planarity.   Lying in the same plane. A plane is defined by the exit of the leads from the body of the IC (#1 above). A second plane is defined at the average of the lowest point all leads are below the first plane (#2 above). Nonplanarity is the maximum variation in mils or mm of any lead of an SMD from the lowest point plane.

Quad flat pack.   Any flat pack IC package that has leads on all four sides.

Through hole.   Also, a plated-through hole (PTH). A via that extends completely through a substrate and is solder plated.

SMOBC.   An abbreviation for solder mask on bare copper, a circuit board construction technique that does not tin the copper traces with solder prior to the placement of the solder mask on the board.

Through-hole technology.   The technology of using leaded components that require holes through the substrate for their mounting (insertion) and soldering.

References

1. Higgins, C., Signetics Corp., presentation, 11/91.
2. See general references B and K below for further information about concurrent engineering.
3. Holmes, JG, "Surface Mount Solution for Power Devices," *Surface Mount Technology,* vol. 7, no. 9, 9/93, 18–20.
4. NEPCON, rep. by Reed Exhibition Co., Norwalk, CT.
5. Surface Mount International, Surface Mount Technology Association, 5200 Wilson Rd., Suite 100, Minneapolis, MN 55424.
6. Institute for Interconnecting and Packaging Electronic Circuits, 7380 N. Lincoln Ave., Lincolnwood, IL 60646-1705, 708 677-2850.
7. Hollomon, JK, Jr., Surface Mount Technology for PC Board Design, Chap. 2, Prompt Publ., Indianapolis, 1995.
8. Capillo, C, Surface Mount Technology, Materials, Processes and Equipment, Chap. 3, McGraw Hill, New York, 1990.
9. ____, "Conduction Heat Transfer Measurements for an Array of Surface Mounted Heated Components," Am. Soc. of Mech. Engr., Heat Transfer Div., *Proceedings of the 1993 ASME Annual Meeting,* vol. 263, 69–78, 1993.

10. Flotherm, "Advanced Thermal Analysis of Packaged Electronic Systems," Flomerics, Inc., 57 E. Main St., Suite 201, Westborough, MA 01581, 1995.
11. Choi, CY, Kim, SJ, Ortega, A, "Effects of Substrate Conductivity on Convective Cooling of Electronic Components," *Journal of Electronic Packaging,* vol. 116 no. 3, September 1994, 198–205.
12. Lee, LC, et al., "Micromechanics of Multilayer Printed Circuit Board," IBM Journal of Research and Development, vol. 28, no. 6, 11–84.
13. *Linear/Interface IC Device Databook,* vol. 1, Section 3, Addendum. Motorola, Inc., 1993.
14. Lee TY, "Application of a CFD Tool for System-Level Thermal Simulation," *IEEE Transactions on Components, Packaging, and Manufacturing Technology,* Part A, vol. 17, no. 4, Dec. 1994, 564–571.
15. Iwasa Y, "Conductive adhesives for surface mount devices," *Electronic Packaging and Production,* vol. 37, no. 15, November 1997.
16. Burr, DC, "Solder paste printing guidelines for BGA and CSP assemblies," SMT Magazine, vol. 13, no. 1, January 1999.
17. Capillo, C, *Surface Mount Technology Materials, Processes and Equipment,* Chapters 7 and 8, McGraw Hill, New York, 1990.
18. *Electronic Packaging and Production,* Cahners Publishing Co., 8773 South Ridgeline Blvd., Highlands Ranch, CO 80126-2329.
19. *Packaging Handbook,* Intel Corp., Santa Clara, 1994.
20. *Source:* Phillips, *Surface Mount Process and Application Notes,* Philips Semiconductor Corp., 1991. Reprinted with permission.

Recommended Readings on Surface Mount Technology

A. Capillo, C, *Surface Mount Technology Materials, Processes, and Equipment.* McGraw-Hill, New York, 1991.
B. Classon, F, *Surface Mount Technology for Concurrent Engineering and Manufacturing.* McGraw-Hill, New York, 1993.
C. Hollomon, JK, Jr, *Surface Mount Technology for PC Board Design.* Prompt Publishing, Indianapolis, 1995.
D. Hwang, JS, *Solder Paste in Electronics Packaging.* Van Nostrand Reinhold, New York, 1989.
E. Lau, J, ed., *Solder Joint Reliability.* Van Nostrand Reinhold, New York, 1991.
F. Lea, C, *A Scientific Guide to SMT.* Electrochemical Publishing Co. Ltd., 1988.
G. Klein-Wassink, RJ, *Soldering in Electronics.* Electrochemical Publishing Co Ltd, 1989.
H. Marcoux, PP, *Fine Pitch Surface Mount Technology.* Van Nostrand Reinhold, New York, 1992.
I. Prasad, RP, *Surface Mount Technology Principles and Practice.* 2nd Edition. Van Nostrand Reinhold, New York, 1997
J. Rowland, R, *Applied Surface Mount Assembly.* Van Nostrand Reinhold, New York, 1993
K. Shina, SG, *Concurrent Engineering and Design for Manufacture of Electronic Products.* Van Nostrand Reinhold, New York, 1991.
L. *Packaging.* Intel Corp., Santa Clara, 1994
M. *Signetics Surface Mount Process and Application Notes.* Philips Semiconductor, Sunnyvale, 1991.
N. Bar-Cohen, A, Kraus, AD, *Advances in Thermal Modelling of Electronic Components and Systems.* ASME Press, New York, 1988.

Other Sources

Specific journal references are available on any aspect of SMT. A search of the COMPENDEX Engineering Index 1987-present will show over 1500 references specifically to SMT topics.

Education/Training

A partial list of organizations which specialize in education and training directly related to issues in surface mount technology includes

- Electronic Manufacturing Productivity Facility (a joint operation of the US Navy-Naval Avionics Center and Purdue University in Indianapolis). 714 North Senate Ave., Indianapolis, IN 46202-3112. 317 226-5607
- SMT Plus, Inc. 5403-F Scotts Valley Drive, Scotts Valley, CA 95066; 408 438-6116
- Surface Mount Technology Association (SMTA), 5200 Wilson Rd., Ste. 100, Edina, MN 55424-1338. 612 920-7682.

Conferences Directly Related to SMT

- Surface Mount International (SMI). Sponsored by SMTA-see above.
- National Electronics Packaging and Production Conference (NEPCON). Coordinated by Reed Exhibition Co., P.O. Box 5060, Des Plaines, IL 60017-5060. 708-299-9311.

3
Integrated Circuit Packages*

Victor Meeldijk
Diagnostic/Retrieval Systems Inc.

3.1 Introduction 3.1
3.2 Surface Mount Packages 3.5
3.3 Chip-Scale Packaging 3.13
3.4 Bare Die 3.14
3.5 Through-Hole Packages 3.15
3.6 Module Assemblies 3.16

3.1 Introduction

In 1958, the first integrated circuit (IC) invented had one transistor on it; today the Motorola PowerPC 603 has more than 1.6×10^6 transistors, the Intel Pentium has roughly 3.1×10^6 transistors, the DEC Alpha microprocessor version has 9.3×10^6 transistors, and programmable parts are available with 10,000 gate arrays. As semiconductor devices become more complex, the interconnections from the die to the circuit hardware keep evolving. Devices with high clock rates and high power dissipation, or with multiple die, are leading to various new packages.

The use of bare die in *chip on board (COB)* and *multichip module (MCM)* applications is increasing, with designers looking for end item size and weight reduction (automotive applications have bond wires going directly from the die to a connector). The use of bare die eliminates the timing delays (caused by stray inductance and capacitance) associated with the leadframes and device input/outputs (I/Os). For burst mode static RAMs, 0.5 to 2 ns or a 20% improvement in access time is achieved with bare die product. Using bare die is not without problems, including testing issues and cost. At this time, it is more costly for vendors to handle and ship bare die than packaged devices. Using bare parts makes it important to use *known good die (KGD);* otherwise, the final assembly has to be scrapped as the device cannot be removed. Industry standards on bare die testing are still evolving. Small portable devices can also use flip chip bonding to a circuit. Current technology bonding machines can bond IC chips with 1×10^6 bumps of 1-μm size, with a 30-μm pitch. IC packaging can be divided into the following categories:

- Surface mount packages (plastic or ceramic)
- Chip-scale packaging
- Bare die
- Through-hole packages
- Module assemblies

---

*Portions of this chapter were adapted from Meeldijk, V., 1995. *Electronic Components: Selection and Application Guidelines,* Chap. 12, Wiley-Interscience, New York, with permission.

In addition to bare die and surface mount techniques, there are still the older device packages for through-hole applications, where the lead of the package goes through the printed wiring board (see Figs. 3.1 and 3.2). Modules are also available that are assemblies of either packaged parts or die.

Although there are standards for some IC packages, such as the dual-in-line packages and the transistor outline (TO) registered Joint Electron Device Engineering Council (JEDEC) packages, many IC packages

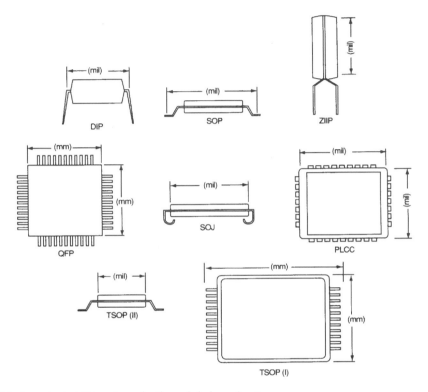

FIGURE 3.1 Some typical IC packages for through-hole applications.

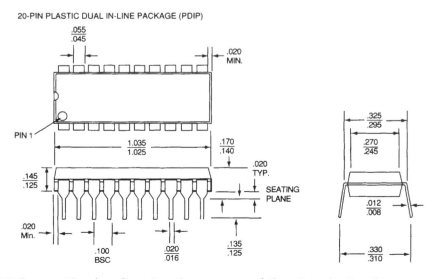

FIGURE 3.2 Common IC package dimensions. *Source:* courtesy of Altera Corp., San Jose, CA.

Integrated Circuit Packages

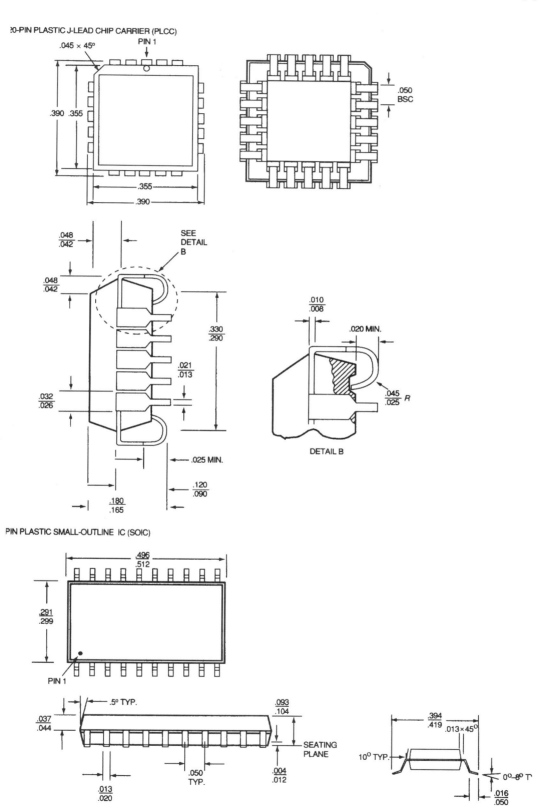

FIGURE 3.2 (continued) Common IC package dimensions. *Source:* courtesy of Altera Corp., San Jose, CA.

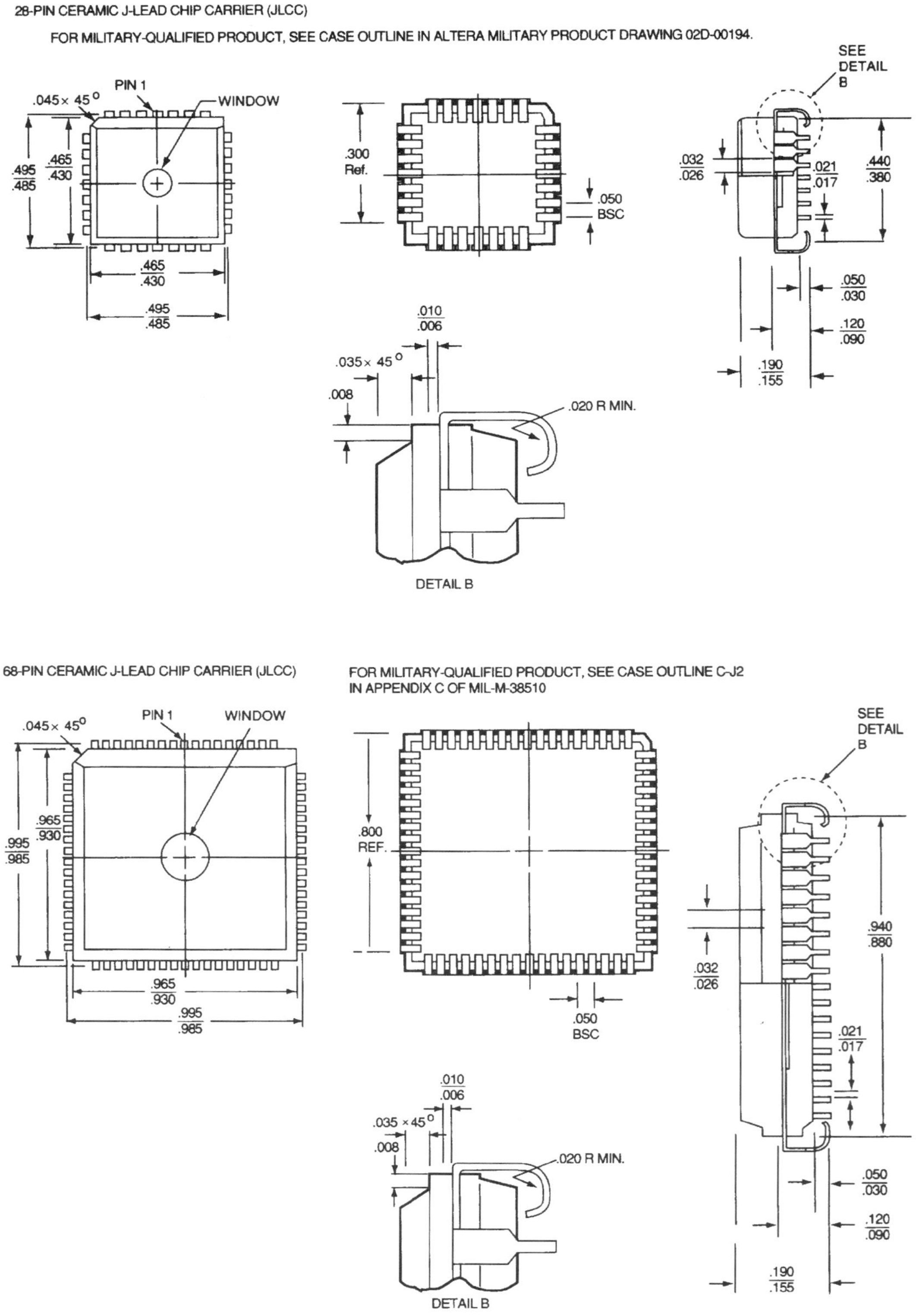

FIGURE 3.2 (continued) Common IC package dimensions. *Source:* courtesy of Altera Corp., San Jose, CA.

Integrated Circuit Packages   3-5

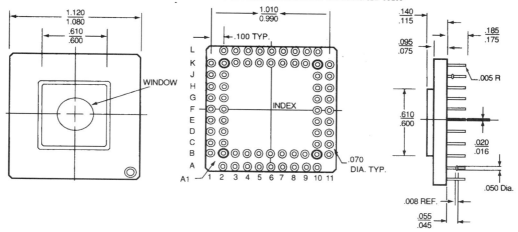

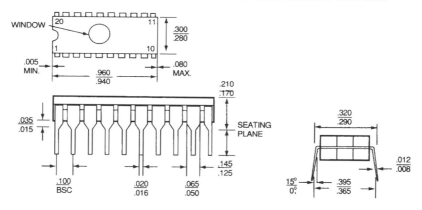

FIGURE 3.2 (continued) Common IC package dimensions. *Source:* courtesy of Altera Corp., San Jose, CA.

styles are (initially) unique to the vendor that developed them (such as in 1990, when the Intel developed the molded plastic quad flat package).

All JEDEC (a subdivision of the Electronic Industries Alliance [EIA]) semiconductor outlines are included in their Publication 95 and terms are defined in their Publication 30, *Descriptive Designation System for Semiconductor Device Packages* (note: EIAJ is the Japanese EIA).

3.2 Surface Mount Packages

Plastic surface mount packages result in a device that is light, small, able to withstand physical shock and g forces, and inexpensive due to a one-step manufacturing process. The plastic is molded around the lead frame of the device. Work is still continuing on developing coatings for the die, such as polyimide, that may resolve hermeticity problems and coefficient of thermal expansion mismatches between the die and the plastic package.

Plastic parts shipped in sealed bags with desiccant are designed for 12-month storage and should be opened only when the parts are to be used. Parts stored for longer than this time, especially *plastic quad flatpack (PQFP)* packaged devices, should be baked to remove moisture that has entered the package. Plastic packages are hydroscopic and absorb moisture to a level dependent on the storage environment.

This moisture can vaporize during rapid heating, such as in a solder reflow process, and these stresses can cause package cracking (known as the *popcorn effect*). Subsequent high-temperature and moisture exposures can allow contaminants to enter the IC and cause failure at a later time due to corrosion. Consult the device manufacturer for details on proper device handling.

Hermetic packages, such as the older flatpack design or *ceramic leadless chip carriers (CLCCs)*, are used in harsh applications (such as military and space applications) where water vapor and contaminants can shorten the life of the device. Thus, they are used in mission-critical communication and navigation and avionics systems. The reliability of CLCC package connections can be further improved by the attachment of leads to the connection pads to eliminate differences in the thermal coefficient of expansion between the parts and the circuit board material.

Metal packages, with glass seals, provide the highest level of hermetic sealing followed by glasses and ceramics (Fig. 3.3). These parts have a higher temperature range than plastic encapsulated parts (typically, from –55° to +125° C vs. 0° to 70° C, or 85° C, for plastic encapsulated devices). Although aluminum oxide, the most commonly used ceramic material, has a thermal conductivity that is an order of magnitude less than a plastic packaged device, in a ceramic sealed device the circuit die does not come into contact with the ceramic packaging material. Thus, temperature cycling will not effect the die, and parts

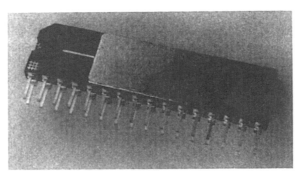

FIGURE 3.3 Typical hermetic IC packages (from left going clockwise): PGA, flatpack, VIL, LCC, and DIP. The bottom photo shows the VIL from a different angle so both rows of pins are visible.

Integrated Circuit Packages

can withstand thousands of temperature cycles without damage (1000 temperature cycles would simulate 20 years use on a commercial airliner). Large die plastic encapsulated parts can fail after only 250 temperature cycles. As mentioned earlier, die coatings being developed may resolve this problem. One of the main drawbacks in specifying ceramic packaged devices is the more complicated (than plastic encapsulated parts) manufacturing process. These devices are made up of three ceramic layers, which result in a higher cost device. Availability is limited, and lead times are often longer than standard plastic devices (this excludes parts that are relatively new and may only be available in a ceramic package, such as a *pin grid array [PGA]*).

Surface mount packages (and related packaging terms) include the following.

FIGURE 3.4 A 256-pin perimeter BGA (no connections are under the die). *Source:* courtesy of AMKOR/ANAM-AMKOR Electronics, Chandler, AZ.

First is ball grid array (BGA) (Figs. 3.4–3.6), a packaging technique developed by IBM (also known as an *overmolded pad-array carrier [OMPAC]*, a plastic version developed by Motorola and shown in Fig. 3.7). Leads, or pads, are replaced by solder balls that replace high pin count *quad flatpacks (QFPs)*. Several hundred arrayed leads can be accommodated in areas that a few hundred peripherally arranged leads would occupy.

The package transfers heat more efficiently than a QFP. To further assist heat dissipation, there is a metal-lid BGA that uses metal ribbons and thermal silicon to transfer heat to the package's metal lid (pioneered by Samsung Electronics). In late 1994, Olin International Technologies announced an aluminum (metal) BGA package using aluminum for both the substrate and IC housing. Thin-film traces as fine as 1 mil can be screened directly on the aluminum. Lead inductance and capacitance are reduced by as much as 50%. The case acts as a heatsink, and popcorning is eliminated, as there is only one layer.

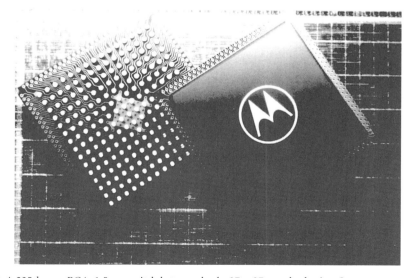

FIGURE 3.5 A 225-bump BGA, 1.5-mm pitch between leads, 27 × 27-mm body size. *Source:* courtesy of Motorola, Inc., Phoenix, AZ, © Motorola, with permission.

FIGURE 3.6 A 127-mm staggered pitch, 313-bump BGA with distributed vias, 35-mm body size. *Source:* courtesy of Motorola, Inc., Phoenix, AZ, © Motorola, with permission.

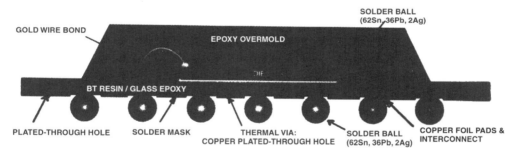

FIGURE 3.7 A plastic (OMPAC) BGA. *Source:* courtesy of Motorola, Inc., Phoenix, AZ, © Motorola, with permission.

Disadvantages of the BGA package are that solder connections cannot be visually inspected, and removed parts cannot be reused (the solder balls are melted). Problems generally occur only from defective components, bad boards, setup errors, or worn/damaged machines in the assembly line. BGA package disadvantages according to users (when comparing BGAs to quad flatpacks, tape automated bonding, and PGAs) are that the package has low rated switching performance, heat sinkability, lead compliance, thermal cycling, and hermeticity.

Other BGA versions include the perimeter lead BGA, which does not have any solder balls under the chip, and the SuperBGA (Amkor Electronics), which has a height of 1.1-1.5 mm vs. 2.2–2.5 mm for standard BGA package.

Other surface mount packages (and related packaging terms) follow:

- Ceramic ball grid array (CBGA).
- Ceramic column grid array (CCGA), invented by IBM, where solder columns replace the solder balls.
- Cerpack, a flatpack composed of a ceramic base and lid. The leadframe is sealed by a glass frit.
- Cerquad, a ceramic equivalent of plastic leaded chip carriers consisting of a glass sealed ceramic package with J leads and ultraviolet window capability.

Integrated Circuit Packages

- Chip carrier (CC), a rectangular, or square, package with I/O connections on all four sides.
- Ceramic leaded chip carrier (CLDCC).
- Ceramic leadless chip carrier [CLCC (or CLLCC)], see LCCC.
- Ceramic pin grid array (CPGA).
- Ceramic quad flat package with J leads (CQFJ).
- Ceramic quad flatpack (CQFP), an aluminum ceramic integrated circuit package with four sets of leads extending from the sides and parallel to the base of the IC.
- Ceramic small outline with J leads (CSOJ).
- Chip-scale grid array (CSGA), like a BGA, but with a smaller solder ball pitch (0.5–0.65 vs. 1.3–2.5 mm). This is used for devices with 200 or more leads.
- Dual small outline package (DSOP).
- EDQUAD, a trademark of ASAT Inc. for a plastic package that incorporates a heat sink; available as thin QFP (TQFP) and QFP (see Fig. 3.8).
- Flatpack (or quad flatpack), one of the oldest surface mount packages, used mainly on military programs. Typically, flatpacks have 14-50 leads on both sides of the body on 0.050-in centers.
- Flat SIP, a single in-line package except the leads have a 90° bend.
- Flatpack (FP).
- Gull winged leadless chip carrier (GCC). Although easier to inspect, it is weaker than the J lead.
- Gull wing, leads which exit the body and bend downward, resembling a seagull in flight. Gull wings are typically used on small outline (SO) package, but are very fragile, easily bent, and difficult to socket for testing or burn-in. J leads solve these problems.
- Hermetic chip carrier (HCC).
- HD-PQFP, originated by Intel for a high density PQFP package with more than 196 leads with a 0.4-mm pitch.
- Heat spreader QFP package (HQFP), which has a copper heat spreader and heat sink.

FIGURE 3.8 EDQUAD package. *Source:* courtesy of ASAT, Inc., Palo Alto, CA.

- Hermetic small outline transistor (H-SOT) packaged device.
- I lead, IC leads that are formed perpendicular to the printed circuit board, making a butt solder joint.
- J leads, leads that are rolled under the body of the package in the shape of the letter J. They are typically used on plastic chip carrier packages.
- J bend leads, leaded chip carrier (JCC).
- J leaded ceramic chip carrier (JLCC).
- Leadless chip carrier (LCC), a chip package with I/O pads on the perimeter of the package. Cavity (lid) up packages, with the backside of the die facing the substrate, have heat dissipated through the substrate. These JEDEC B and C ceramic packages are not suitable for air-cooled systems or for the attachment of heat sinks. Cavity (lid) down parts, (JEDEC A and D) with the die facing away from the substrate, are suitable for air-cooled systems. Rectangular, ceramic leadless E and F are intended for memory devices, with direct attachment in the lid-up position.

Computing Devices International devised patented C and S leads that can be attached, as shown in Fig. 3.9.

- Leadless ceramic chip carrier (LCCC), or ceramic leadless chip carrier. JEDEC registered type A must be socketed when on a printed circuit board or a ceramic substrate board, and type B must he soldered. The LCC minipack must be soldered into printed circuit boards.
- Leaded ceramic chip carrier (LDCC), leaded ceramic chip carrier packages include JEDEC types A, 13, C, and D. t.eaded type B parts are direct soldered to a substrate. Leaded type A parts can be socketed or direct soldered, and include sub-categories: leaded ceramic, premolded plastic, and postmolded plastic (which are not designated as LDCC devices).
- Land grid array (LGA), an Intel package used for parts such as the 80386L microprocessor. The package is similar to a PGA except that it has gold-plated pads (called *landing pads*) vs. pins.
- Leadless inverted device (LID), a shaped metallized ceramic form used as an intermediate carrier for the semiconductor chip (die). It is especially adapted for attachment to conductor lands of a thick film or thin film network by reflow solder bonding.
- Little foot (a trademark of Siliconix), a tiny small outline integrated circuit (SOIC) package.
- Leadless ceramic chip carrier (LLCC).
- Mini-BGA (mBGA), developed by Sandia National Laboratories. Layers of polyimide and metal are applied to the die to redistribute original die pads from a peripheral arrangement into an area array configuration.
- Miniflat, a flat package (MFP) approximately (0.102–0.113 in (2.6–85 mm) high that may have leads on two or four sides of the package.
- Multilayer molded (MM) package, an Intel PQFP package with improved high-speed operation due to separate power and ground planes, which reduces device capacitance.
- MM-PQFP, isolated ground and power planes are incorporated within the molded body of the PQFP.
- Metal quad flatpack (MQFP), sometimes used to refer to a plastic metric quad flatpack.
- MQuad, designed by the Olin Corporation, a high thermal dissipation aluminum package available in plastic leaded chip carrier (PLCC) and QFP packages (with 28–300 leads and clock speeds of 150 M Hz) (see Fig. 3.10).
- OMPAC, see BGA.
- Pad array carrier (PAD), a surface mount equivalent of the PGA package. This package extends surface mount silicon efficiency from 15 to 40%. A disadvantage of this package is that the solder

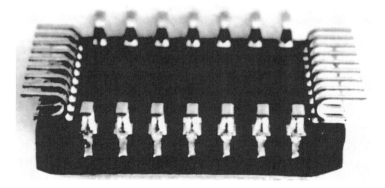

FIGURE 3.9 An LCC package with C leads attached to compensate for the thermal coefficient of expansion mismatch between the part and the circuit board, thus lowering solder joint stresses. *Source:* courtesy of Computing Devices International, Bloomington, MN.

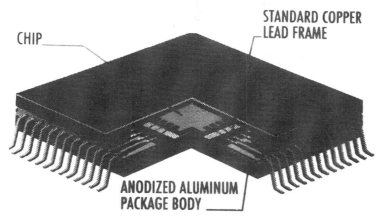

FIGURE 3.10 An MQuad (trademark Olin Corp.) package. *Source:* courtesy of Swire Technologies, San Jose, CA.

joints cannot be visually inspected, although X-ray inspection can he used to verify the integrity of the blind solder connections (and to check for solder balls and bridging). Individual solder joints, however, cannot be repaired.

- Plastic (encapsulated) ball grid array (PBGA), see OMPAC.
- Plastic leaded chip carrier [(PLCC) also known as PLDCC or a quad pack by some manufacturers, up to 100 pin packages, commonly 0.050-in pin spacing. Depending on the number of pins, pin 1 is in different locations but is confirmed by a dot in the molding, and the progression of the pin count continues counterclockwise. The leads are J shaped and protrude from the package. These parts can be either surface mounted or socketed. (This package was originally known as a post-molded type A leaded device.)
- Plastic quad flatpack (PQFP), this JEDEC approved package is used for devices that have 44–256 1/Os, 00.0025-in lead spacing (EIA-JEDEC packages), 0.0256- and 00.0316-in spacing (EIAJ), and 00.0135-in pin spacing. This package has gull wing leads on all four sides and is characterized by bumpers on the corners. The body is slightly thicker than the QFP. PQFP packages are susceptible to moisture-induced cracking in applications requiring reflow soldering.

FIGURE 3.11 A QFP package. *Source:* courtesy of ASAT, Inc., Palo Alto, CA.

- Plastic surface mount component (PSMC).
- Quad flat J (QFJ,) leaded package.
- Quad flatpack (QFP), a flatpack with leads on all four sides. The typical JEDEC approved pinouts are 100, 132, and 196, with lead pitch from 0.040 to 0.016 in. The EIAJ types are slightly thinner than the JEDEC equivalents (Fig. 3.11).
- Quarter quad flatpack (QQFT), see SQFT.
- Quad surface mount (QSM).
- Quarter-size small outline package (QSOP).
- Quad pack, see PLCC.

- Quad very small outline package (QVSOP), introduced by Quality Semiconductor. This package combines the body size of a half-width, 14-pin SOIC, with the 25-mil lead pitch of a PQFP (or a 150 × 390 mil, 48-pin package). The package has designation JEDEC MO-154.
- Repatterned die, see BGA.
- Square (body) chip carrier (SCC).
- Slam pack, a square ceramic package that looks like an LCC and is always used with a socket.
- Small outline (SO) (with versions such as SO wide, SO narrow, or SO large), this design originated with the Swiss watch industry in the 1960s (and was reportedly nicknamed SO for Swiss Outline). It was used in the modern electronics industry by N.V. Philips (formerly known as Signetics in the U.S.) in 1971. This package is also known as a *mini-flat* (which has slightly different dimensions from the JEDEC parts) by the Japanese.
- Small outline IC (SOIC), 8–28 pin packages, commonly 0.050-in pin spacing, with gull wing leads. This package style is about 50–70% of the size of a standard dual in-line package (DIP) part (30% as thick).
- Small outline package with J leads on two sides (SOJ). The leads are bent back around the chip carrier.
- Small outline large (SOL), generally refers to a package that is 0.300 mils wide vs. 150 mils wide (for an SO package). It is a larger version of the small shrink outline package (SSOP).
- Small outline package (SOP), a package with two rows of narrowly spaced gull-wing leads (same as SOIC).
- Small outline transistor (SOT), a plastic leaded package originally for diodes and transistors hut also used for some ICs (such as Hall effect sensors).
- Small outline wide (SOW), see SOI.
- Shrink quad flatpack (or flat package) (SQFP), typically a 64-pin small quad flat package (1/4 height of QFT) with a lead pitch of 0.016 in or less. Also known as quarter quad flatpack (QQFT).
- Square surface mounting (SSM).
- Shrink small outline package (SSOP).
- Shrink small outline IC (SSOIC), a plastic package with gull-wing leads on two sides, with a lead pitch equal to or less than 0.025 in.
- Tape ball grid array (TBGA).
- Thin QFP (TQFP). Typical sizes are 10-and 1.4-mm body thickness, in sizes ranging from 10 × 10 mm to 20 × 20 mm and lead count from 64 to 144 leads.
- Thin small outline package (TSOP). The type-1 plastic surface mount parts have 0.5-mm gull-wing leads on the edges of the parts (the shorter dimension) rather than along the sides (longer dimension). The type-2 parts have the leads on the longer dimension. These packages are generally used for memory ICs.
- Thin shrink (or sometimes called *scaled*) small outline package (TSSOP), half the height of a standard SOIC.
- Vertical mount package (VPAK), a package conceptually like a zig-zag package (ZIP) except instead of through-hole leads it has surface-mount leads (L-shaped leads). This package was introduced by Texas Instruments in late 1991 for 4- and 16-Mb direct random access memories (DRAMs).
- Very small (and thin) quad flat package (VQFP). Lead pitch is 0.5 mm, with 32, 48, 64, 80, 100, 128, and 208 pins pet package.
- Very small outline (VSO), usually used to denote 25-mil pitch packages with gull-wing leads.
- Very small outline package (VSOP), with 25-mil spaced leads at the ends of the package (also TSOP and SSOP and SSOIC).

Integrated Circuit Packages

- Very small peripheral array (VSPA™), a high pin count package, smaller than an equivalent QEP, developed by Archistrat Technologies (Fig. 3.12). The IC die is mounted to a metal plate that is bonded to a high temperature plastic (Vectra) carrier. The sides of the plastic carrier have rows of holes to accept 4 rows of pins. The pins are connected to the die via wire bonds.

3.3 Chip-Scale Packaging

This form of packaging technique is designed to have the size and performance of bare die parts but with the handling and testability of packaged devices. The package size is no more than 1.2 times the original chip with various techniques to connect the part to the circuit. Flip chips have enlarged solder pads (bumps or balls). Flip chip packages include the controlled collapse chip connection (C4), developed by IBM, and the direct chip attachment (DCA), developed by Motorola. In 1964, IBM invented flip chip interconnection technology to enable production-level joining of discrete transistors. General designations are as follows:

- Bumped chip, see flip chip.
- Flip chip, a semiconductor package where the I/O terminations are in the form of bumps on one side of the package (also called *bumped chip*). After the surface of the chip has been passivated, or treated, it is flipped over and attached to a matching substrate.
- MiniBGA, uses a predetermined grid array for the solder bumps and is similar to the flip chip.
- Repatterned die, see MiniBGA.
- Microsurface mount packages (MSMT), a deposited horizontal metallization, instead of bond wires, goes from the bond area to a connection post. In 1994, Micro MST, Inc. patented this technology where semiconductors are mass packaged in the wafer state and can be tested as individual devices. This package is approximately the same size as the semiconductor die and is used for devices with less than 200 leads. It generally has lower parasitic inductance than BGA or QFP packages.
- Slightly larger than IC carrier (SLICC) package.
- MicroBGA (μBGA), also known as *chip-scale packaging* (Mitsubishi); slightly larger than IC or SLICC packaging (Motorola, Inc.). A package designed by Tessera, μBGA consists of a flex circuit

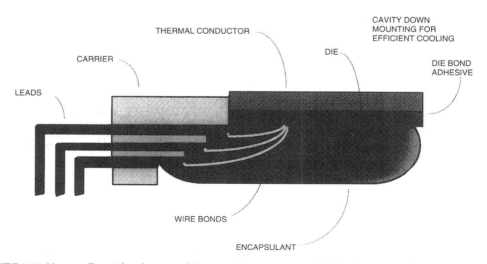

FIGURE 3.12 Very small peripheral array architecture. *Source:* courtesy of ARCHISTRAT Technologies, Division of the PANDA Project, Boca Raton, FL.

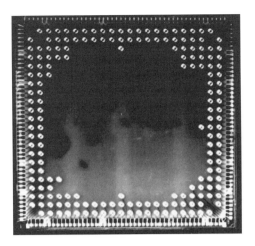

FIGURE 3.13 A μBGA package. *Source:* courtesy of Tessera, San Jose, CA.

with gold traces bonded to the die pads. The die pads fan in to an array of metal bumps used for the second-level assembly. An elastomeric adhesive is used to attach the flexible circuit to the chip. The package compensates for thermal mismatches between the die and the substrate, and the μBGA can be clamped into a socket for tests before attachment in the final assembly (see Fig. 3.13).

3.4 Bare Die

Bare, or unpackaged, parts offer the smallest size, with no signal delays associated with the device package. Current issues related to unpackaged devices are packaging, handling, and testing. The most common bare die and tape packages (bare die mounted on tape) include the following:

- C4PBGA, an IC package developed by Motorola that attaches the IC die to a plastic substrate using the C4 process. It uses a multilayer substrate, unlike the chip-scale or SLICC package.
- Chip-scale package (CSP), a Mitsubishi package where the IC is surrounded by a protective covering through which external electrode bumps on the bottom provide electrical contacts. The package, only slightly larger than the chip it houses, has a height about 0.4 mm.
- Chip on board (COB), the die is mounted directly on the printed circuit substrate (or board). See also tape automated bonding (TAB) and tape carrier packages (TCB).
- Chip on flex (COF), a variation of COB but, instead of bonding the bare die to a substrate, it is bonded to a piece of 0.15 in-thick polyimide film (such as FR4) that has a top layer of gold-plated copper. The traces provide bonding areas for wires from the die and a lead pitch similar to QFP packages. It has approximately the same weight and height as COB, but failed parts can be removed. It, however, uses about the same board real estate as QFP packages and is more expensive than COB because of the polyimide substrate.
- Demountable tape automated bonding (DTAB), developed by Hewlett-Packard. Mechanical screw and plates align and hold the IC to the board (pressure between the tape and the board provides the contact). The IC can be replaced by removing the screws (vs. a standard TAB package, which is soldered in place).
- Sealed chips on tape (SCOT), chips mounted on tape supported leads and sealed (usually with a blob of plastic).
- Tape automated bonding (TAB), see TCP.
- Tape carrier packages (TCP), formerly called TAB packages. The chip is mounted to a dielectric film, which has copper foil connection patterns on it. The chip is sealed with a resin compound. This device assembly is mounted directly to a circuit without a plastic or ceramic package. Some tape carrier packages resemble TSOP packages (with gull-wing leads) but are thinner because the bottom of the IC die is exposed. The TCP design is about one-half the volume and one-third the weight of an equivalent pin count TSOP.
- Tape carrier ring (TCR), or a guard ring package, similar to the TCP package but includes a plastic ring to support the outer rings during test, burn-in, and shipment.
- Ultra-high volume density (UHVD), a process developed by General Electric to interconnect bare ICs on prefabricated laminated polyimide film.

3.5 Through-Hole Packages

These packages, as the name implies, mount in holes (usually plated through with deposited metal) on the printed circuit board. These device packages include the following:

- Batwing, a package (sometimes a DIP type, but can be surface mounted) with two side tabs used for heat dissipation.
- Ceramic dual in-line package (CERDIP). The DIP package was developed by Fairchild Semiconductor and Texas Instruments followed with a metal topped ceramic package that resolved problems with the early ceramic packaged parts.
- Dual in-line (DIL).
- Dual in-line package (DIP), a component with two straight parallel rows of pins or lead wires. The number of leads maybe from 8–68 pins (although more than 75% of DIP devices have 14–16 pins), 0.100-in pin spacing with width anywhere from 0.300-mil centers to 0.900-mil centers. The skinny (or shrink) DIP (SDIP) has 0.300-mil centers (spacing between the rows) vs. 0.600-mil centers. The SDIP usually has 24–28 pins. DIP parts may be ceramic (pins go through a glass frit seal), sidebrazed (pins are brazed onto metal pads on the side of the package), or plastic (where the die is molded into a plastic package). There is also a shrink DIP, another small package through-hole part.
- Plastic DIP (PDIP).
- Pin grid array (PGA), a plastic or ceramic square package with pins covering the entire bottom surface of the package. Lead pitch is either 0.1 or 0.05 in perpendicular to the plane of the package. Packages have various pin counts (68 or more). The chip die can be placed opposite the pins (cavity up) or nested in the grid array (cavity array).
- Pinned uncommitted memory array (PUMA II), a PGA package application specific integrated circuit (ASIC) memory array with four 32-pad LCC sites on top of a 66 PGA. Each of the four sites can be individually accessed via a chip select signal thus allowing a user definable configuration (i.e., ×8, ×16, ×24, ×32). It provides for an ASIC memory array without tooling. The substrate is a multilayered cofired alumina substrate with three rows of 11 pins. There is a channel between the pins so that the part can be used with a heat sink rail (or ladder). Onboard decoupling capacitors are mounted in a recess in this channel. This type of device is available from various companies, including Mosaic Semiconductor (1.12 × 1.12 in square), Cypress Semiconductor (their 66 pin PGA module, the HG01, is 1.09 × 1.09 max), and Dense-Pac Microsystems (Veraspac or VPAC family, 1.09 × 1.09 max.).
- Plastic pin grid array (PPGA).
- Quad in-line package (QIP or QUIL).
- Quad in-line package (QUIP)-similar to a DIP except the QUIP has a dual row of pins along the package edge. Row to row spacing is 0.100 in, with adjacent rows aligned directly across from each other.
- Skinny (or shrink) DIP (SDIP), see DIP.
- Shrink DIP, a dual in-line package with 24–64 pins with 0.070-in lead spacing.
- Single in-line (SIL).
- Single in-line module (SIM), electrical connections are made to a row of conductors along one side.
- Single in-line package (SIP), a vertically mounted module with a single row of pins along one edge for through-hole mounting. The pins are 0.100 in apart. SIPs may also include heatsinks.
- Skinny DIP (SK-DIP), see SDIP.
- Transistor outline-XX (TO-XX), refers to a package style registered with JEDEC.
- Vertical in-line package (VIL).

- Zig-zag in-line package (ZIP). This may be either a DIP package that has all the leads on one edge in a staggered zig-zag pattern or a SIP package. Lead spacing is 0.050 in from pin to pin. In modules, the leads are on both sides in a staggered zig-zag pattern. Lead spacing is 0.100 in between pins on the same side (or 0.050 in from pin to pin).

3.6 Module Assemblies

We use this terminology to refer to packaging schemes that take either packaged parts or bare die and use them to make an assembly. This maybe by either mounting the parts to a substrate or printed circuit board or by stacking parts or die to create dense memory modules. This packaging scheme maybe surface mount or through-hole. The circuit board assemblies may be through-hole (having pins) or designed for socket mounting (with conductive traces). Variations include the following:

- Dual-in-line memory module (DIMM), pioneered by Hitachi, has memory chips mounted on both sides of a PC board (with 168 pins, 84 pins on each side). The DIMM has keying features for 5-V, 3.3-V, and 2.5-V inputs and for indicating whether it is made with asynchronous DRAMs or flash, static random access memory (SRAM) or DRAM ICs.
- Flexible-rigid-assembly memory module (FRAMM), a memory packaging scheme by Memory X. FRAMM modules use a combination of frigid and flexible PC board assemblies, with the flexible board interconnecting to rigid PC boards. The modules have standard JEDEC 30- and 72-pin SIMM outputs. TSOP DRAMS are mounted on both sides of the rigid PC boards.
- Full stack technology, packing 20–100 dice horizontally in a loaf-of-bread configuration. This manufacturing method was developed by Irvine Sensors Corporation (see also *short stack*).
- HDIP module, a hermetic DIP module that has hermetically sealed components mounted on the top and bottom of a ceramic substrate. This package style is generally used in anticipation of a monolithic part (such as a memory device) that will be available at a later time that will fit the same footprint as the module.
- Hermetic vertical DIP (HVDIP) module, a vertically mounted ceramic module with pins along both edges (through-hole mounting). Components used in this module are hermetically sealed. Pins on opposite sides of the module are aligned and are on 0.100-in spacing.
- Leaded multichip module (LMM). An LMMC is an LMM connector.
- Multichip module (MCM), a circuit package with SMT (surface mount) IC chips mounted and interconnected via a substrate similar to a multilayer PC board. MCM-L has a substrate that can be FR-4 dielectric but is often polyimide based laminations (first standardized by JEDEC JC-11 committee). Other dielectric substrates are ceramic (MCM-C), which can have passive devices (resistors, capacitors, and inductors) built into the substrate; thin film (i.e., silicon and ceramic construction), with deposited conductors (which can have capacitors built into the substrate) (MCM-D); laminated film (MCM-LF) made up of layers of modified polyimide film; or multichip on a flexible circuit (MCM-F). In 1994, some manufacturers started manufacturing substrates with MCM-D thin-film deposits onto MCM-L laminate substrates in a technique pioneered by IBM. MCM devices inherently offer higher speed and performance at lower cost than conventional devices. Manufacturing issues associated with MCM design include availability of unpackaged die, how to test/inspect the die, and test/rework of the MCM module. MCM modules for military uses fall under MIL-PRF-38534 (see *memory cube*). There is also an MCM-V, with V standing for a vertical, which is a three-dimensional module. This design is also known as Trimrod, by the developer Thomson-CSF and the European Commission (see *memory cube*).
- Memory cube, a three-dimensional module consisting of stacked memory devices such as DRAMs or SRAMs (see *Ribcage* and *Uniframe Stakpak Module*). This design was developed by RTB Technology and was licensed to various other semiconductor companies. The concept is similar to the

Integrated Circuit Packages

stacked chip design, by Irvine Sensors, but the cubes are manufactured with standard parts vs. IC die that have all I/Os on one edge (which are stacked and electrically connected together).

- Ribcage, a trademark of Staktek Corporation for a three-dimensional memory module design. Also see *Uniframe Stakpack Module*.
- Short stack, a method of stacking semiconductor dice vertically, where memory ICs (such as 4 SRAMs) are assembled into a three-dimensional thin-film monolithic package with the same footprint as a single SRAM. A variation on the full stack technology, also developed by Irvine Sensors Corporation, short stacks hold up to 10 ICs. The stacks maybe unpackaged for use in hybrid, multichip, and chip-on-board applications. Interconnect techniques include wire-bond, tab, and flip chip methods. Another stacked memory method, which uses special silicon wafer segments and vertical interconnects between the wafers in the interior of the silicon segments, was developed and patented by Wafer Drive Corporation. This method was developed for PCMCIA card applications.
- Single in-line memory module (SIMM), an assembly containing memory chips. The bottom edge of the SIMM, which is part of the substrate material, acts as an edge card connector SIMM modules are designed to be used with sockets that may hold the SIMM upright or at an angle, which reduces the height of the module on the circuit board. Typical SIMM parts are 4 × 9 (4 meg memory by 9), 1 × 9, 1 × 8, 256 × 9, and 256 × 8. Also 9-b data width SIMM modules are produced tinder license to Wang Laboratories, which developed the SIMM module and socket in the early 1980s as an inexpensive memory expansion for a small workstation.
- Stackable leadless chip carrier (SLCC), developed by Dense-Pac Microsystems, this multidimensional module consists of stacked chip carriers. Stacking is accomplished by aligning the packages together and tin dipping each of the four sides. SLCC can achieve a density of 40:1 over conventional packages.
- Uniframe Stakpak Module, a trade name for a three-dimensional stack of TSOP memory ICs by Staktek Corporation (Fig. 3.14).
- VDIP module, vertically mounted modules with plastic encapsulated components and epoxy encapsulated chips on them. VDIP modules have 0.100-in spaced pins along both sides of the substrate, with the pins on the alternate sides aligned.

FIGURE 3.14 An 8-Mb DRAM Uniframe STAKPAK™ memory module. *Source:* courtesy of STAKTEK, Austin, TX).

- Other Definitions:
    - Plastic encapsulated microcircuit (PEM).
    - Premolded plastic package (PMP).

References

Bindra, A. Nov.28, 1994. Very small package hopes for big IC impact; May 16, 1994. Sandia shrinks size of BGA packages. *Electronic Engineering Times*.

Costlow, T. Jan.16, 1995. MCM substrates mixed; Oct.10, 1994: BGAs honed to meet diverse needs; Sept.19, 1994. Chip-scale hits the fast track; Sept. 12, 1994. Olin mines metal for ball-grid arrays; Aug. 12,1994. Mitsubishi scales down IC package; July 11, 1994: Amkor trims BGA height; May 9, 1994: MCM design details remain perplexing; April 18, 1994: MCM standard created. *Electronic Engineering Times*.

Crum, S. Sept.1994. Minimal IC packages to be available in high volume. *Electronic Packaging and Production*, News column.

Derman, G. Feb.28, 1994. Interconnects and packaging. *Electronic Engineering Times*.

Hutchins, C.L. Nov.1994. *Understanding grid array packages. SMT (Surface Mount Technology)*. Hutchins and Associates, Raleigh, NC.

Karnezos, M. Feb.28, [994. DTAB mounts speed and power threat. *Electronic Engineering Times*.

Lineback, J.R. Nov. 1994. Chip makers try to add quality while removing the package. *Electronic Business Buyer*.

Locke, D. Nov.28, 1994. Plastic packs pare mil IC pricing. *Electronic Engineering Times*. American Microsystems.

Meeldijk, V. 1995. *Electronic Components: Selection and Application Guidelines*. Wiley-Interscience, New York.

Murray, J. June 1 1, 1994. MCMs pose many production problems. *Electronic Engineering Times*.

O'Brien, K., managing ed. May/June 1994. *Advanced Packaging*. µBGA offers flip chip alternative; copper heat sink QEP, dual use sought in Denver News column.

Reuning, K. Aug. 1994. Squeezing the most from BGAs. *U.S. Tech*. deHaart Inc.

Woolnough, R. April 25, 1994. Thomson 3-D modules ready for prime time. *Electronic Engineering Times*.

Further Information

The following sources can he referenced for additional data on new IC packaging and related packaging issues (i.e., converting IC packages, lead finish considerations, packaging and failure modes):

Computer Aided Life Cycle Engineering (CALCE), Electronic Packaging Research Center, University of Maryland, College Park, MD, 301-405-5323. CALCE is involved in research and software development associated with the design of high reliability advanced electronic packages.

Electronic components: Selection and Application Guidelines, by Victor Meeldijk, John Wiley and Sons Interscience Division, ©1995. The integrated circuits chapter discusses IC packages, lead material, and failure mechanisms related to packaging.

High Density Packaging Users Group (HDP User Group), a nonprofit organization, formed in 1990, serving users and suppliers. Scottsdale, AZ, 602-951-1963; Alvsjo, Sweden, 46 8 869868.

IEEE Components, Packaging, and Manufacturing Technology Society, New York. Periodicals, Piscataway, NJ 800-678-4333, 908-981 -0060.

International Electronic Packaging Society (IEPS), Wheaton, IL, 708-260-1044.

International Society for Hybrid Microelectronics (also known as the Microelectronics Society) (ISHM), Reston, VA, 800-535-4746, 703-758-1060. This technical society was founded in 1967 and covers various microelectronic issues such as MCMs and advanced packaging.

The Institute for Interconnecting and Packaging Electronic Circuits (IPC), Northbrook, IL., 847-509-9700. This organization issues standards on packaging and printed circuitry.

Microelectronics and Computer Technology Corporation, Austin TX, along with Lehigh University worked toward replacing hermetic packages with more effective lightweight protective coatings. This Reliability without Hermeticity (RwoH) research contract was awarded by the U.S. Air Force Wright Laboratory. The RwoH is an industry working group, initiated in 1988. For more information call 512-338-3740, David Clegg.

Semiconductor Technology Center, Neffs, PA, 610-799-0419 This organization does market research on manufacturing design, holds training seminars, and is involved in the Flip Chip/Tab and Flex Circuit symposiums held each year

The Surface Mount Technology Association, Triangle Park, NC, 612-920-7682.

The following publications, which offer free subscriptions to qualified readers, cover packaging issues:

- *Advanced Packaging,* IHS Publishing Group, Box 159, 17730 West Paterson Road, Libertyville, IL 60048-0159, 708-362-8711, Fax: 708-362-3484.

- *Electronic Engineering Times,* a CMP Publication, 600 Community Drive, Manhasset, N.Y. 11030, 516-562-5000, 516-562-5882 (subscription services). Subscription inquiries: 516-733-6800, Fax: 516-562-5409. This publisher also issues *Surface Mount Technology.*

- *Electronic Packaging and Production,* a Cahners Publication, Division Reed Publishing, 275 Washington Street, Newton, MA 02158-1630. Circulation/subscriptions, Cahners Publishing Co., Box 7541, Highlands Ranch, CO 80163-9341, 303-470-4700.

4
Direct Chip Attach

Glenn R. Blackwell
Purdue University

4.1 Introduction 4.1
4.2 Overview of Die Assemblies 4.2
4.3 Known Good Die 4.2
4.4 Chip on Board 4.3
4.5 Flip Chips 4.4
4.6 Chip-Scale Packages 4.25

4.1 Introduction

Bare IC die have been used since the 1970s in hybrid assemblies. The standard assembly technique of semiconductor die, whether in a standard package or in a hybrid such as a multichip module (MCM), is to glue the die onto the leadframe or substrate with its contact bond pads facing up. These bond pads are then connected to the leadframe or substrate with a wire bonding technique as shown in Fig. 4.1.

This technique, when used on standard circuit board materials, is known as *chip on board (COB)*. The die is considered right-side up, or "face" up, when its contact pads face up from the substrate.

In the 1980s and 1990s, a number of IC assembly technologie-s were developed to utilize either bare die or very small packages that approximated the size of bare die. *Flip chips* use the same bare die as COB, but the bond pads are placed toward the board, or face down. Also placed face down are *chip-scale packages (CSPs)*, which are die enclosed in a package that is no more than 120% of the bare die x-y dimensions. All of these die-size components are responses to the increasing demands for higher density, higher-speed performance, and increased functionality.

The traditional IC package provides protection from high humidity and corrosive/contaminated environments. The package also protects the die from handling machines and humans and allows some stress relief from environmental thermal excursions. It also facilitates mounting the component in a test fixture with little risk of damage. Direct chip attach techniques must address these issues.

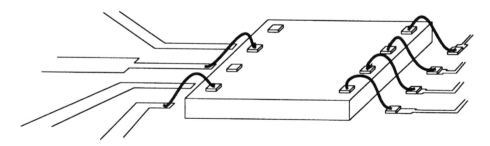

Die-to-leadframe (only 2 wires shown for clarity)   Die-to-bond pads, a la hybrid circuits

FIGURE 4.1 Wire bonding examples.

Where Else?

Also see the information in Chapter 3, "IC Packaging Technologies"; Chapter 7, "Hybrid Assemblies"; and Chapter 15, "Reliability."

4.2 Overview of Die Assemblies

Using bare die, in either COB or flip chip techniques, or using CSPs will not provide the same level of protection for the die or ease of component handling that a packaged component does, although the intent of CSPs is to approach those characteristics. This chapter will discuss chip-on-board (COB), flip chip, and CSP. COB was described in Section 4.1 as a technique that uses the same assembly techniques as standard die techniques, except that the die is placed on circuit board material instead of an IC leadframe or a hybrid/ceramic substrate. Flip chip assembly (FCA) mounts the die "upside down" with its bond pads facing the substrate. The die first has its bond pads "bumped" to create a solderable termination on the bond pads. This is described in Section 4.5.2. CSPs use a variety of techniques to encapsulate the die in an environmentally-protecting material while not increasing the size of the package beyond 120% of the x-y-z dimensions of the die. Like flip chips, CSPs use some type of bumped pads. A major issue affecting all three of these technologies is the mismatch of the coefficient of thermal expansion (CTE) between the component and any substrate other than ceramic.

4.3 Known Good Die

Any operation that uses bare IC die is faced with the issue of *known good die (KGD)*. IC manufacturers test bare die for certain performance parameters but do not do full tests until the die is packaged in its final form. Assemblers considering any of the assembly options discussed in this chapter will need to determine the "goodness" of the bare die they are using. While some IC manufacturers have put KGD testing into place, not all have.

One approach to the use of bare die in an assembly is for the user to put into place all the necessary new equipment for not only die mounting and wire bonding, but also for full test and burn-in of the bare die. For these purposes, die probe and test carriers are available, and burn-in carrier and socket assemblies have been developed. These allow temporary connection with the die without damaging the die pad area for any subsequent assembly operations.

The Electronic Industries Alliance (EIA) through its Joint Electron Devices Engineering Council (JEDEC) has put forth a procurement standard, EIA/JEDEC Standard #49, "Procurement Standard for Known Good Die (KGD)." The intent of this standard is to make suppliers of die aware of "the high levels of as-delivered performance, quality, and long term reliability expected of this product type." The die procured may be used in multichip modules (MCMs) or hybrid assemblies, or in board-level assemblies such as chip on board or flip chip components. The standard applies to both military and commercial uses of KGD. It notes that KGD users must recognize that the levels of quality and reliability users have come to expect from packaged devices cannot be assured with bare die, although it does also note that KGD are "intended" to have equivalent or better performance than equivalent packaged parts. It also notes that close cooperation between supplier and user of KGD is a necessity. The standard notes that the user accepts the responsibility for final hermetic sealing and/or plastic encapsulation.

It is reasonable for the user of bare die to work with the supplier to agree on the following:

- Design data, e.g., Spice models, power dissipation and thermal resistance, die dimensions, bond pad locations, surface finish, storage and operating thermal limits, and final use of the die including supply and other voltage potentials.
- Quality expectations, e.g., ISO qualification(s), military certifications, percent fault coverage of final testing, infant mortality expectations, stress tests, long-term survival rates, and supplier's quality control techniques.

Direct Chip Attach

- Test data, e.g., die-bottom potential, any built-in testability provisions (e.g., IEEE 1149.1 compatibility), and electrical test specifications and limits.
- Handling provisions, e.g., supplier's handling, packaging, and shipping techniques; user's storage, handling, and placement and wire-bonding techniques; ESD protection by both supplier and user; and marking and tracing techniques.
- Records, e.g., reliability test results, burn-in test results, and supplier's engineering changes.

4.4 Chip on Board

Chip on board (COB) refers to the placement of bare IC die onto the substrate, with bond pads up. Since most IC die are available in bare die configuration, this assembly technique is usable by any manufacturer that is willing to make the investment in bonding and test equipment. This technique then substitutes unpackaged semiconductor chips for packaged chips, with all the KGD issues described in Section 4.3.

In COB, the die is placed into epoxy on the substrate, and the bond pads are then connected to the substrate using standard IC wire bonding techniques. Once the bare die are bonded, the die and its associated wires are covered with epoxy for both mechanical and environmental protection. Die bond-pad pitch is commonly 0.25mm (0.010 in), with a corresponding requirement for very accurate placement. COB is attractive when space is extremely limited, e.g., on PCMCIA cards, and when maximum interconnect density is required. It is also attractive for low-cost high-volume consumer products, and is used extensively, e.g., in digital watches and calculators. The die used for COB are commonly know as a *back-bonded* semiconductor chip, since the back of the chip is bonded to the substrate. This also results in very good heat transfer to the substrate.

Besides the placement accuracy required, the main drawbacks to COB are having a wire-bonding machine in the assembly line, dealing with very fragile packages, and using a package for which pre-assembly testing is not possible without specialized die-test systems. If the reader is considering the assembly of COB, the KGD issue must be discussed with all potential die suppliers prior to embarking on this assembly technique.

Wire bonding requires special considerations on the part of the assembler. These considerations include:

- A narrow range of height differences between the bond pads and substrate lands is required.
- A level substrate is required.
- Each substrate land should be at least 0.5mm (0.020 in) from the bond pad.
- The area of the substrate land should be at least 0.25 mm (0.010 in) by 0.75 mm (0.030 in).
- Substrate land pitch should be at least 0.25 mm (0.010 in), matching the bond pad pitch.
- Solder mask should be kept at least 1.25 mm (0.050 in) from the edge of the substrate land.

IPC-CM-770 indicates the following wire bonding feature limits (see Fig. 4.2):

Bonding Process	A (min)	B (min)	C (max)	D (min)	D (max)	E (min)
Gold wire: thermocompression	0.01 mm (0.004 in)	0.01 mm (0.004 in)	0.75 mm (0.030 in)	0.63 mm (0.025 in)	2.5 mm (0.100 in)	0.25 mm (0.010 in)
Gold wire: thermosonic	0.01 mm (0.004 in)	0.01 mm (0.004 in)	0.4 mm (0.016 in)	1.0 mm (0.040 in)	2.5 mm (0.100 in)	0.25 mm (0.010 in)
Aluminum wire: ultrasonic	0.05 mm (0.020 in)	0.01 mm (0.004 in)	0.75 mm (0.030 in)	0.63 mm (0.025 in)	2.5 mm (0.100 in)	0.25 mm (0.010 in)

After assembly and wire-bonding, the chip and its associated gold wires are covered with epoxy for protection. This is a common assembly technique for small, special-purpose products which employ one custom IC, such as calculators and digital watches/clocks. This epoxy covering, or "glob-top," is typically a two-component liquid epoxy with a glass transition temperature of 165–180° C.

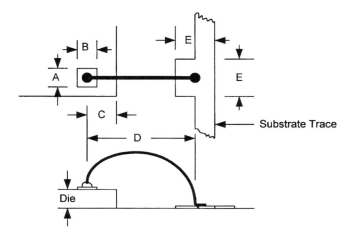

FIGURE 4.2 Wire bond layout dimensions (adapted from IPC CM-770).

Handling and storage of bare IC die must conform to the same standards of ESD and moisture protection that are used for packaged ICs, discussed at the end of this chapter.

4.5 Flip Chips

The first subsection on flip chips is an overview. Given the importance of bumping die in flip chip manufacturing, the second subsection is devoted to that topic. Use of flip chips in the assembly process is *flip chip assembly (FCA)*.

4.5.1 Flip Chip Overview

Flip chips are bare IC die placed on the board with bond pads down. Because of the reduction in "package" size (there is none), and subsequent reduction in electrical path length and associated parasitic capacitance and inductance, it is particularly suited for high-density, high-performance applications. High-performance/high-speed devices depend on minimization of the parasitic inductances and capacitances in their circuits. These parasites show up in the output and transmission circuits and degrade both overall frequency capabilities and edge-speed in digital circuits.

A choice that must be made when modeling a circuit is whether to use a lumped-circuit-element model or a transmission-line model. A general guideline is to consider the rise time, t_r, of the output/signal circuit and the relative dielectric constant ε_r and length l of the circuit path. For FR-4, $\varepsilon_r = 4$, and the speed of light, $c = 3 \times 10^8$ m/sec $= 3 \times 10^{10}$ cm/sec. If the rise time is less than 2 times the propagation delay in the trace and dielectric to be used, the model must use transmission-line characteristics.

$$2t_p l = 2 \times \frac{\sqrt{\varepsilon_r}}{c} \times l = 2 \times \frac{\sqrt{4}}{3 \times 10^{10}} \times 10 \text{ cm} = 1.3 \text{ ns}$$

This tells us that for a 10 cm (3.94 in) length of copper trace on FR-4, the critical rise time is 1.3 ns. A rise time faster than this will require the use of transmission line models.

Assuming that a lumped model can be used, a typical equivalent circuit for this type of circuit is as shown in Fig. 4.3. As shown earlier, the propagation time for the circuit depends directly on the parasitics.

$$t_{\text{propagation}} = \frac{l}{c/\sqrt{\varepsilon_r}}$$

Direct Chip Attach

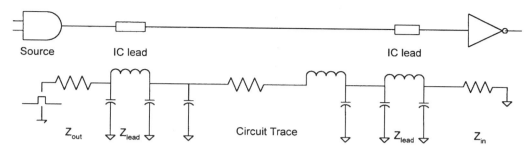

FIGURE 4.3 Lumped circuit model of an interconnect trace.

The table below shows examples of the parasitics that exist in a variety of packages. Flip chip advantages are obvious.

TABLE 4.1 Parasitics in Various Electronic Packages

Package type	Parasitic capacitance, pF	Parasitic Inductance, nH
Flip chip	0.1	0.01
Chip on board/wire bond	0.5	1–2
Pin grid array	1	2
Quad flat pack	1	1–6
Through-hole DIP	3	8–20

The EIA and IPC have jointly issued J-STD-026, "Semiconductor Design Standard for Flip Chip Applications," and J-STD-028, "Performance Standard for Flip/Chip Scale Bumps." The -026 standard defines wafer- and die-level concerns such as design rules for wafer/die and bump technologies, concentrating on the specifics for the development of evaporation bumps and solder paste deposition bumps. The -028 standard classifies bumps as follows:

- *Meltable solder bumps,* full solder balls that are expected to partially collapse upon reflow soldering.
- *Partially meltable bumps,* copper studs tipped with meltable solders.
- *Nonmeltable bumps,* such as copper or gold bumps or direct chip attach with no bump, all designed to create a nonmelting standoff that can be soldered or adhesively bonded to the next level of packaging.
- *Polymeric/conductive adhesive bumps.*

These techniques for bumping will be discussed in detail in Section 4.5.2.

The lands on the substrate must match the bond pad geometry on the chip. As in COB, die bond-pad pitch is commonly 0.25 mm (0.010 in), with a corresponding requirement for very accurate placement. The die used for flip chip assembly are face-bonded chips. Rather than the "bare" passivated bond pads used on back-bonded chips (with wire-bond connections), face-bonded chips must have prepared bond pads suitable for direct placement and soldering to the substrate lands. The technique for pad preparation is known as *bumping,* since the end result is that each die bond pad has a solderable *bump* on it.

Pad bumping is typically done with solder at each bond pad while the die are still in wafer form, requiring special processing by the die manufacturer. After individual dice are sliced from the wafer, the chip will be placed on the board lands and reflowed in a conventional reflow oven. Land bumping is done on the substrate, not on the bond pads, so most available die can be assembled using this technique. For bumped die, the flip chip process begins with bumping the contact pads of the die. As noted previously, bumping can be done in several ways, but in the most typical process results in tin-lead solder "bumps" being attached to each contact pad as shown in Fig. 4.4.

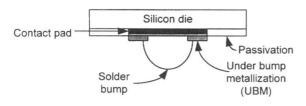

FIGURE 4.4 Typical solder bump.

This type of bump will result in the *controlled collapse chip connect (C4)* technique. Assembly of the flip chip to a substrate is by placing it face down in either solder paste or flux, then reflow soldering it like other surface mount devices. Alternatively, a gold stud bumped flip chip may be thermocompression bonded to the substrate. This direct bonding to the substrate means the flip chip has real estate reductions compared to standard hybrid wire bonded die in much the same way that a ball grid array (BGA) has real estate reductions compared to a plastic leaded chip carrier (PLCC) or a quad flat pack (QFP). The real estate reduction is what gives flip chips their advantage in high-density circuits.

The differences in flip chips are typically in their surface finish (passivation) and their bump and underbump metallurgies. The primary passivation types are nitride and polyimide, while the bump metallurgies consist of solder, polymeric, or gold stud.

Along with its advantages, flip chip, of course, has its disadvantages. Unprotected flip chips are prone to thermally induced cracking due to the CTE mismatch of 2.5ppm/°C for silicon and 16ppm/°C for an FR-4 circuit board. This mismatch can cause cracked solder connections and even damage the device itself. The effects of this thermal mismatch can be lessened considerably with the use of an underfill, which is an adhesive designed to wick into the space between the bottom of the die and the top of the board. The underfill will tend to bond the die and the board together, equalize the CTEs, and prevent the two structures from pulling apart. In its *1996 Roadmap*, the National Electronics Manufacturing Initiative (NEMI) identified the lack of a robust underfill technology as hindering the implementation of flip chip. NEMI then launched an underfill project that has been instrumental in developing more robust procedures. There are two underfill processes. One dispenses a combination flux/adhesive material prior to placement of the chip. The chip is then placed, and it cures during the reflow process step. An example of this encapsulant is Kester Solder's SE-CURE™. The other underfill process occurs after the reflow process and dispenses liquid encapsulant at elevated temperatures along one or more edges of the die. Capillary action will then draw the liquid into the gap between the die and the substrate, surrounding the interconnect bumps. This capillary action typically takes 10 to 120 seconds. A number of manufacturers can supply wicking underfill, including Dexter Electronic Materials, Thermoset Plastics, and Alpha Metals.

The two major issues with wicking underfill are the time it takes to wick under the component and the time it takes to cure. For either type of underfill, one of the most critical calculations is the amount of underfill to be dispensed. Too much underfill results in flow out onto the board in the area surrounding the chip, while insufficient underfill results in incomplete coverage of the solder bumps, which can cause the die to crack and possibly collapse. Calculation of the underfill volume includes the following:

- Volume of the total area between the die and the board, $V_A = L_D \times W_D \times C_D$, where L_D = length of the die, W_D = width of the die, C_D = clearance between the die and the board
- Volume occupied by the solder bumps, V_B
- Volume of the fillet around the die, V_F

The total volume to be dispensed, V_D, is then $V_D = V_A - V_B + V_F$ (Fig. 4.5).

The fillet serves as a "fudge factor" in the volume dispensed. Since the volume of the individual bumps will vary, and since the exact volume occupied by the pads on the board and their associated solder mask imperfections cannot be calculated for each individual board, it is impossible to calculate the exact amount necessary for underfill. The fillet will act as a reservoir to allow for more underfill when the volume occupied by the bumps is less than calculated. It will also take up the extra dispensed underfill volume when the bump volume is greater than that calculated. This means the fillet will not always be at the same level on the sides of each chip, but as long as there is some fillet remaining and the fillet does

Direct Chip Attach

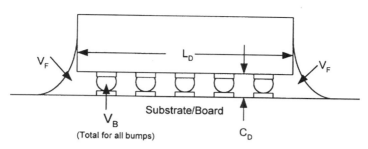

FIGURE 4.5 Typical solder bump.

not extend over the top of the die, it is an acceptable fillet. One volume calculation should be made assuming lowest acceptable bump volume, and a second volume calculation should be made with the highest acceptable bump volume. These two calculations will serve as the upper and lower specification limits on the volume to be dispensed. One final factor that must be determined from the supplier is if the underfill is expected to shrink or expand as it cures. Any expected shrinkage or expansion must be included in the final calculation.

The Hele-Shaw model, as described by Han and Wang, has been developed to realistically model the flow of the heated encapsulant. The dispensed encapsulant cools as it flows. Gordon, Schmidt, and Selvam compare the results of one-side encapsulant application vs. two-sided application. Their results show that, not surprisingly, application of encapsulant to two sides of the die results in greater coverage of the area between the die and the board than does application on only one side. Application of the encapsulant can be done more than two sides; however, this brings with it the risk of trapping an air bubble under the chip, which will once again bring with it the risk of cracking the die.

4.5.2 Wafer Bumping Technologies: A Comparative Analysis of Solder Deposition Processes and Assembly Considerations*

Abstract

Flip chip technology is experiencing a new period of explosive growth. This growth is driven by the need for low profiles and high performance for high volume, non-traditional consumer applications. An enabling infrastructure for flip chip packaging technology is being shaped and includes wafer bumping services, equipment manufacturers, and material and substrate suppliers. In addition, high volume die packaging and contract assembly houses are becoming well-educated in the handling of flip chip devices. The emergence of a global infrastructure will speed the adaptation of flip chip into a mainstream packaging technology. This paper explores the principal solder bumping technologies offered through the merchant market. Assembly considerations will also be discussed.

Key words: flip chip, wafer bumping, solder bumping, conductive adhesive, FCA, ACF, DCA, CSP, IC Packaging

Introduction

The goal of modern semiconductor design is to achieve shorter electron pathways for increased speed, power, and total device functionality. Ongoing developments in silicon and gallium arsenide wafer-scale technology continue to address these goals while maintaining or decreasing device size. Just as advancements in interconnection technology are emphasized at the wafer level, it has become more apparent that this same understanding must extend beyond wafer-level design and manufacture to the system-level interface.

---

*With permission, reprinted from Patterson, Deborah S, Elenius, Peter, Leal, James A., "Wafer Bumping Technologies—A Comparative Analysis of Solder Deposition Processes and Assembly Considerations." INTERPack '97, June 15–19, 1997, *EEP* Vol. 19-1, Advances in Electronic Packaging—1997, pp. 337–351, ASME, New York, 1997.

Historically, packaging technologies have taken a back seat to IC design technology for the simple reason that device performance was not substantially compromised by its packaging or board attachment techniques. In addition, high-speed automatic wire bonders met the cost targets for IC package assembly.

Today, many sophisticated electronic devices cannot be satisfactorily packaged with wire bonding technology. Mobile applications cannot afford the size or weight of a standard wire bonded IC package, and high-performance applications cannot afford the penalty in signal propagation delay and power distribution constraints.

Flip Chip Benefits

Chip-scale packaging (CSP) and other direct chip attachment (DCA) technologies offer a viable solution for the needs of today's IC suppliers and OEMs (other equipment manufacturers). Of the various CSP technologies, flip chip attachment (FCA) is the most advanced form of IC level interconnect available. A flip chip packaging approach offers the following advantages:

Area Array Packaging
By allowing the IC designer to utilize the entire device area for I/O, power, and ground pads, the IC size can shrink, yielding more chips per wafer. Area array bumps optimize signal propagation as well as power and ground distribution.

Electrical Performance
For high-performance requirements, the electrical characteristics of a solder bump vs. a wire bond packaging approach are critical. The inductance of a solder bump is less than 10% of a wire bond. This is especially significant in high-speed and high-frequency applications where the incorrect choice of packaging can seriously degrade signal integrity. For example, Table 4.2 illustrates best and worst case signal propagation characteristics for ICs packaged in cavity down wire-bonded pin grid arrays (PGAs) vs. flip chip on ball grid array (BGA) packages.

TABLE 4.2 Signal Effects of Two Package Approaches[1]

	Worst case		Best case	
	W/B w/PGA	F/C w/BGA	W/B w/PGA	F/C w/BGA
Inductance	19.6 nH	7.9 nH	5.6 nH	0.3 nH
Capacitance	15.9 pF	6.2 p	9.1 pF	2.5 pF
Resistance	21.0 Ω	2.1 Ω	20.2 Ω	1.7 Ω
Propagation delay	946 ps	243 ps	508 ps	51 ps

Low-Cost Assembly
Flip chip is proving itself to be a low-cost, high-volume assembly technique. Typical bumping and underfill costs result in a total cost of 0.1 to 0.3 cents per I/O.[2] As I/O densities and wafer sizes continue to increase, flip chips will exhibit even greater cost advantages. Solder bumps provide assemblers with the option of mounting the IC onto the circuit board using the same techniques that are employed for other surface mountable components. By providing a 63Sn/Pb eutectic solder material on the IC itself, assembly costs can be further reduced by eliminating the fine pitch eutectic solder that would otherwise be deposited onto the circuit board. Depositing fine pitch solder patterns on the substrate can increase the cost of the board by 15 to 20%.[3]

Reliability
Flip chip attachment has been practiced for approximately 30 years at both IBM and Delco Electronics. IBM first introduced its *controlled collapse chip connection (C4)* technology for the System/360 to eliminate the expense, unreliability, and low productivity of manual wire bonding.[4,5] Similarly, Delco Electronics introduced its own controlled collapse solder bump technology, termed *flex-on-cap (FOC)*, for under-the-hood automotive electronics to attain higher reliability and achieve lower-cost assembly. Delco is the

largest U.S. consumer of flip chip devices, internally bumping more than 300,000 devices per day. Both Delco and IBM report that there has never been a single field failure attributed to the flip chip solder joint.[6] Flip chip interconnection failures, when seen, are typically due to faulty assembly such as not having the solder bump properly wet to the board.

The long-term use of flip chip devices in mainframe computers and automotive electronics has done much toward demonstrating the robustness and reliability of the process. Furthermore, the use of flip chip technology in the cost sensitive automotive environment provides reassurance to potential users that the cost impacts of implementing flip chip are favorable.

The Solder Joint

It is necessary to understand the characteristics that make up the structure of the solder bump, since this structure is the key to long-term reliability and shorter-term assembly considerations. Not all bumps are alike, and the choice of bump material and construction can affect overall reliability as well as assembly considerations. This section briefly discusses the five basic solder bump technologies and their differences in terms of reliability, cost, and assembly. Before describing each solder deposition process, a brief outline of desired solder bump characteristics is presented.

Desired Bump Characteristics

Under-Bump Metallurgy. A high-reliability solder bump interconnect is composed of two features: the under-bump metallurgy (UBM), also known as ball-limiting metallurgy (BLM), and the solder ball itself. A UBM should provide the following features or capabilities:

- Good adhesion to the wafer passivation
- Good adhesion to the IC final metal pad
- Protection of the IC final metal from the environment
- Low resistance between IC final metal and solder bump
- An effective solder diffusion barrier
- A solder wettable metal of appropriate thickness
- Ability to be used on probed wafers

Solder Ball. The ideal solder joint will provide a controlled collapse of the bump upon assembly. A collapsible bump increases the assembly process window by accommodating less than planar boards and by being able to self-align on the circuit board pad even if it has not been completely centered.

The predictive collapse of the solder bump is important to determine pitch limitations and the suitable bump standoff for best reliability. Once the use of a particular solder bumping approach is determined, the design engineer can work within the boundary conditions of the chosen process.

Solder joint fatigue is the principle failure mechanism of flip chip devices. Figure 4.6 illustrates a first-order approximation of the fatigue life of a solder joint.[7] As shown, the solder joint fatigue life is directly proportional to the square of the solder joint height (h), and inversely proportional to the square of the distance from the neutral point to the center of a solder bump (L_e), the coefficient of thermal expansion ($\Delta\alpha$), and the temperature change that the device experiences (ΔT).

This means that the following features will increase reliability:

- Higher solder joint standoff
- Smaller die
- Closer CTE between the device and substrate
- Smaller temperature range of operation

Solder bumping technologies undergo continuous evolution to increase reliability. For example, Delco Electronics has refined its bump technology over a span of 28 years. Perhaps the most significant improvement to its bump structure was the elimination of the "minibump," a feature that is part of many UBM structures. The function of the minibump is to provide for or increase the solder wettable area as

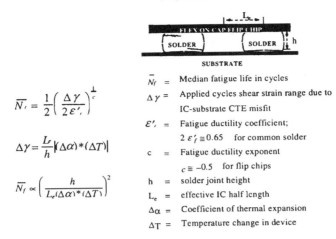

FIGURE 4.6 First-order approximation of Coffin-Manson low cycle fatigue model.

well as to provide for a solder diffusion barrier. However, the minibump structure impacts the solder fatigue life and has been shown to be less reliable than other processes.[8]

Flip Chip Solder Bump Deposition Processes

Five solder bump deposition processes will be reviewed in terms of their respective manufacturing processes. The deposition techniques are as follows:

1. Evaporated solder bump formation
2. Electroplated solder bump formation
3. Printed solder bump formation
4. Solder ball bumping (or *stud bump bonding*)
5. Electroless nickel under-bump metallurgy paired with either printed solder bumps or conductive adhesives

The benefits and drawbacks of each approach will be presented in terms of the following considerations:

- Costs of implementation
- Bump structure—under-bump metallurgy
- Bump structure—solder ball
- Alternative solder alloys
- Environmental considerations
- Assembly with eutectic (63Sn/Pb) solder
- Compatibility with probed wafers
- Option to preferentially lay down UBM without solder
- Production history

Evaporated Solder Bumping Technology
The formation of the UBM and solder bump by evaporation typically involves the following process flow as practiced by IBM and licensees of their C4 process. A schematic representation of an evaporated UBM and solder bump is shown in Fig. 4.7. The steps, shown in Fig. 4.8, are described below.

(a) In-situ sputter clean. An in-situ sputter clean of the wafer is performed to remove oxides or photoresist prior to metal deposition. The cleaning also serves to roughen the wafer passivation and surface of the bond pad to promote better adhesion of the UBM.

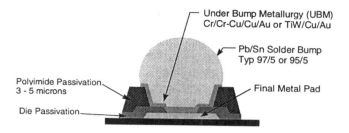

FIGURE 4.7 Cross section of an evaporated UBM and solder bump.

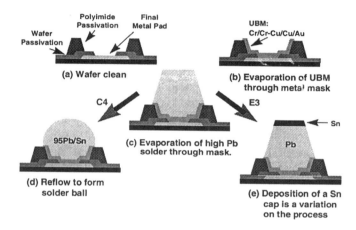

FIGURE 4.8 Evaporative solder bumping process.

(b) Metal mask. A metal mask (typically molybdenum) is used to pattern the wafer for UBM and bump deposition. The metal mask assembly typically consists of a backing plate, a spring, the metal mask, and a clamp ring. The wafer is clamped between the backing plate and molybdenum mask. The assembly must be manually aligned. Alignment tolerances to within 25 mm can be achieved with this procedure.[9] However, this configuration limits via size to 50 mm for a 100–125 mm diameter bump, impacting current carrying capacity.

(c) UBM evaporation. The sequential evaporation of a chromium layer, a phased chromium/copper layer, a copper layer, and a Au layer are deposited to form a thin film UBM. This is illustrated in Fig. 4.8b.

(d) Solder evaporation. A high-lead solder is then evaporated on top of the UBM to form a thick deposit of 97Pb/Sn or 95Pb/Sn. The deposited solder will typically produce a 100–125 mm high bump. The height of the bump is determined by the volume of the evaporated material that is deposited. This is a function of the distance between the metal mask and the wafer as well as the size of the mask opening. The deposited solder is conical in shape due to the way that the solder is formed in the openings of the solder mask. Figure 4.8c illustrates the cross section of the solder and UBM.

Note the extra layer of tin at the top of the lead bump shown in Fig. 4.8e. This process was introduced by Motorola and is called *evaporated, extended eutectic,* abbreviated as "E3." This tin "cap" allows the device to be "attached to organic boards without the need for intermediate eutectic deposits on the board."[10] This is done because the high lead bump reflows at a temperature above 300° C and is not suitable for organic substrates, which cannot tolerate such high reflow temperatures. The tin layer allows the assembler to heat the structure well below the melting point of the 95Pb/Sn solder. The goal of this procedure is to form a Pb/Sn eutectic at the tip of the solder ball, allowing the device to be placed on the board without incurring the added costs of applying eutectic solder onto the board itself.

(e) Solder balling. In the C4 process, the solder can be reflowed into a sphere. This is illustrated in Fig. 4.8d.

Benefits and Drawbacks of Evaporative Bumping

Cost considerations. The process is expensive in terms of licensing fees, capital expenditures for equipment, and materials. This equipment cannot be scaled up economically past 8 inches. Throughput is limited with the process, averaging between 10-12 wafers per hour for 8 inch wafers.

Bump structure: under-bump metallurgy. The UBM adheres well to the IC metallization and passivation, protecting the underlying bond pad. The phased Cr-Cu layer is an extremely reliable solder diffusion barrier for high lead solders (95Pb/Sn or 97Pb/Sn) but is not suitable for 63Sn/Pb solders (which can be formed by electroplating the solder over the evaporated UBM). The high Sn content of the eutectic Pb/Sn rapidly consumes the Cu in the Cr-Cu phased layer, resulting in poor or no adhesion.

Additionally, the Cu/Cr-Cu/Cu structure is a higher stress thin film than UBMs used in other processes. It is not recommended to use this process on probed wafers.

Bump structure: solder. The evaporative process provides for excellent metallurgical control. The high-lead solder material is ductile and allows for a predictable collapse of the bump upon assembly for ceramic packages. This process is predominantly limited to the use of high-lead solder alloys. Eutectic Pb/Sn solder cannot be evaporated due to the low vapor pressure of the Sn.

Alternative alloys. The evaporative process is limited to materials with high vapor pressures, since processing times to deposit materials such as tin (low vapor pressure material) would be prohibitively long in a manufacturing environment. Although the evaporative process exhibits excellent alloy control, typically, a maximum of two materials are deposited at one time. The option of using ternary or quaternary alloys is not economical.

The deposition of low alpha particle Pb/Sn solders is possible. However, the process would be extremely expensive due to the high amount of lead waste. Price premiums in the range of hundreds of dollars per wafer is not an uncommon estimate.[11]

Environmental considerations. When the solder is evaporated onto the wafer, the lead is also evaporated over the inside of the evaporation chamber, which can be quite large. This excess lead must be cleaned on a periodic basis.

Assembly with 63Sn/Pb solder. It is desirable to have eutectic 63Sn/Pb solder on the device to eliminate the extra costs of applying a fine pitch solder to the circuit board. However, eutectic 63Sn/Pb solder is not compatible with the evaporative solder deposition process for the reasons cited previously.

As shown in Fig. 4.8e, variations on this process are being offered to eliminate having to place a fine-pitch solder mask on the circuit board. Although a tin cap can be deposited on a high-lead bump to obtain a quasi-eutectic solder alloy at the top of the bump, this approach still presents several problems.

- The evaporative cost structure does not change significantly.
- Since most of the Pb bump will not be reflowed, the special "controlled collapse" feature of the bump is not taken advantage of. Therefore, the structure will be more sensitive to planarization issues. Similarly, the bump will not self-center, requiring much more accurate placement tolerances.
- A high contact pressure (9–15 g/bump) is needed during assembly of these devices.
- The Sn cap must wet to the board before the Sn goes into solution with the Pb. This represents a small process window for a successful operation.

Compatibility with probed wafers. Evaporative solder bumping is not recommended for use on probed wafers.

Option to preferentially lay down UBM without solder. It is not feasible to remove the metal mask between UBM and solder deposition. Therefore, solder will be deposited over all exposed UBM.

Production. Evaporated solder bumping is used in production at companies such as IBM and Motorola.

Electroplated Solder Bump Technology

Electroplating is a popular alternative to the evaporation process because of its lower equipment costs, facility and floor space requirements, and costs of processing.

Direct Chip Attach

Several plating processes are available. The traditional plating process for solder bump formation is adopted from the evaporated process and uses a Cr/Cr-Cu/Cu UBM with a high lead solder (only 3–5% Sn content). However, if a tin rich solder is required, such as with a eutectic 63Sn/Pb alloy, then the Sn will consume the Cu in the UBM, quickly degrading the integrity of the structure.

Therefore, one of the more accepted processes used for 63Sn/Pb solders involves the formation of a UBM by the deposition of an adhesion layer such as titanium/tungsten (TiW) followed by a thick solder wettable layer such as copper (Cu). If the solder has a high Sn content, then this thicker Cu layer must be used since the Sn can quickly consume the Cu, leaving a non-wettable adhesion layer (TiW) in contact with the eutectic solder alloy. This thick copper layer is sometimes called a *minibump* or *stud*. Companies that practice this process include Texas Instruments, Motorola, National Semiconductor, Aptos Semiconductor, and OKI Electric. Solder is deposited over the TiW/Cu-minibump UBM by electroplating. A schematic representation of this process is shown in Fig. 4.9.

The process flow is outlined as follows:

(a) Wafer cleaning. The wafer is cleaned to remove oxides and organic residue prior to metal deposition. Sputter cleaning as well as wet processes are used. As in the evaporated approach, the cleaning roughens the wafer passivation and bond pad surface to promote better adhesion of the UBM.

(b) UBM deposition. The UBM materials are typically TiW, Cu, and Au sequentially sputtered or evaporated over the entire wafer. The UBM adheres to the wafer passivation as well as to the bond pads. Theoretically, the UBM layer will provide an even current distribution to facilitate uniform plating. Figure 4.10a illustrates the blanket TiW layer as applied across the wafer.

For the creation of the minibump structure, the copper layer is plated over the bond pad to a height as determined by the patterned photoresist (4.10c). The minibump can measure between 10 and 25 μm in height within a total bump height of 85 to 100 μm.

(c) Electroplating of solder. A second mask process allows the formation of the plated solder bump. The photoresist is stripped after the bump is formed, leaving the UBM exposed on the wafer (Fig. 4.10d). The UBM is removed in one of two ways. In the first approach, a wet etch is used and the UBM is removed from the wafer with some undercutting around the bump. The solder is then reflowed into a sphere. In the second approach, the solder bumps are first reflowed, with the aim that any intermetallics formed within the bump structure will protect the bump by minimizing undercutting during the subsequent etching process. Figures 4.10e and 4.10f illustrate these steps.

Benefits and Drawbacks of Electroplated Solder Bumps

Cost considerations. Electroplating is typically less expensive than evaporated wafer bumping. Due to bump height uniformity and alloy control issues, future plating systems will migrate to the more costly single wafer fountain plating systems.

Bump structure: under-bump metallurgy. The typical plated UBM systems adhere well to the IC metallization and passivation, protecting the underlying bond pad. The UBM provides a low-resistance electrical path.

There are practical drawbacks to the plated solder bumping systems. The UBM systems used in plating typically have been derived from either gold plating systems or from high Pb plating requirements.

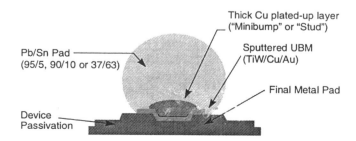

FIGURE 4.9 Cross section of the UBM, plated copper minibump and plated solder bump.

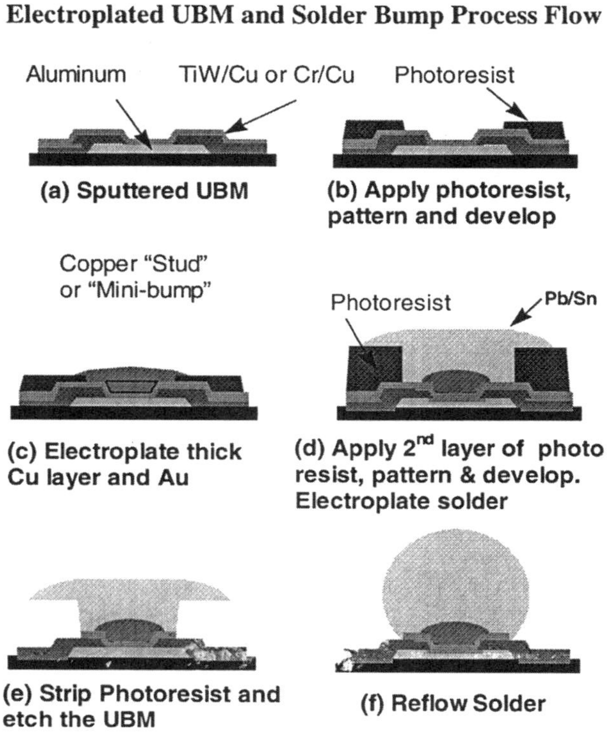

Electroplated UBM and Solder Bump Process Flow

FIGURE 4.10 Electroplated UBM with minibump and solder.

Variations of the Cr/Cr-Cu/Cu UBM systems (IBM, MCNC/Unitive) have shown poor results with 63Sn/Pb solders. This is due to the high Sn content of eutectic solders consuming the Cu from the Cr-Cu matrix, which results in a weaker interface with the remaining Cr-Cu, as the Cr is not wettable by the solder. The TiW with Cu minibump UBM is compatible with eutectic 63Sn/Pb solder. However, the Cu consumed by the Sn during multiple reflow operations creates brittle Cu/Sn intermetallics. If the Sn were to consume the Cu minibump, it would fall back onto the non-wettable TiW surface and de-wet from the UBM.

The TiW is a comparatively high-stress thin film that can reduce reliability. In addition, the copper minibump structure reduces reliability for the following reasons:

- Due to the high surface area and Cu volume of the minibump, intermetallic formation between the Cu and Sn are accelerated.
- The minibump UBM structure generates a large amount of stress to the wafer.
- The plating process can induce subsurface wafer stress.

Figure 8.11a depicts a SEM micrograph illustrating a failure mode that was generated through the combination of the wafer structure and the Cu minibump/electroplated bump structure. Inherent processing stresses, the wafer structure, and the differential cooling between the die and the substrate after reflow, can induce microcracks and crater the silicon under the solder bump.

The cratering of the silicon wafer occurred before underfill was applied since the underfill material can be seen within the crack as shown in Fig. 8.11b.

Bump structure: solder. There are several issues to be aware of when dealing with plated wafer solder bumping. Varying degrees of solder height and alloy uniformity can be obtained. This is due to non-uniform wafer current density distribution, the electric field between the anode plate and the wafer, and the plating solution flow. Since the height of the plated solder is limited by the height of the photoresist

Direct Chip Attach

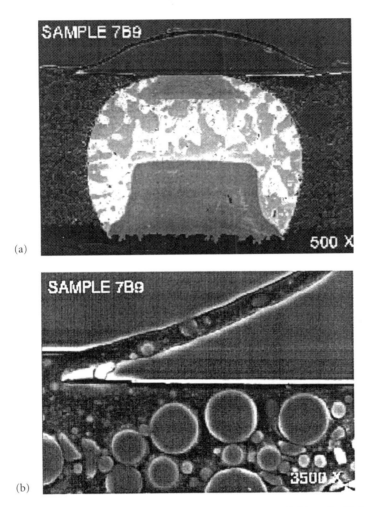

FIGURE 4.11 (a) Silicon cratering due to the plated copper minibump and (b) underfill found in the fractured section of (a).

and the pitch of the bumps, issues within the plating process can result in varying emergence times of the bump above the plating template. This condition can change the current density within that bump.

Extensive metallurgical control is required for consistency in solder deposition. Alloy control is a function of several bath parameters and current densities generated across the wafer.

These are significant issues in rack plating (lower cost) systems. Most of the current density issues across the wafer have been resolved with the single wafer plating systems, but these are costly (approximately $1.1M for a six-station fountain plating system that runs at 11 to 15 wafers/hr).

Variability in the chemical bath or in the current densities can create two problems.

1. Poor alloy control resulting in incomplete wetting of some solder bumps during the assembly process
2. Excessive height variation of the bumps

Other considerations arise in scaling the process to meet the requirements of larger wafer sizes. Additional issues involve solder voiding, which is more prevalent in plated bumping technologies. Higher solder voiding is due to the generation and entrapment of hydrogen gas as well as plating solutions within the deposited solder.

Alternative alloys. A maximum of two materials can typically be deposited at one time. This constrains the choices of bump material to high lead solders or, more recently, 63Sn/Pb solder. The option of using ternary or quaternary alloys is extremely limited.

Assembly with 63Sn/Pb solder. The use of the UBM/minibump structure enables the application of a 63Sn/Pb solder bump that is compatible with surface mount assembly. However, a limited number of thermal cycles is allowable with this type of structure.

Compatibility with probed wafers. Electroplated bumping is not recommended for use on probed wafers. As with the evaporative process, stress risers can be translated through the bump structure, degrading reliability.

Option to preferentially lay down UBM without solder. The electroplating process requires the wafer to be blanketed with the under-bump metals. This enables current distribution across the wafer, facilitating the plating process. The solder bumps are plated over the I/Os, effectively protecting the final metal from the subsequent etching step in which the interstitial metallization is removed from the wafer. There is no mechanism by which the UBM can remain untouched by the etchant without having a solder bump attached. Therefore, the plating process cannot both bump and preferentially protect features such as test structures with UBM only.

Production. Electroplating is one of the most common forms of bumping, and is used to bump die prior to inner lead bonding for TAB. Electroplated solder bumping is used in production at AT&T, Casio, Citizen Watch, Motorola, Nippon Denso and Sharp. Low-level production is being done at Cherry Semiconductor, Oki Electric, Fujitsu, NEC, and merchant vendors such as Aptos Semiconductor.[12] It should be noted that MCNC/Unitive offers a plated bumping process for high-lead solder bumps; however, they are not yet offering eutectic 63Sn/Pb bumps.[13]

Printed Solder Paste Bump Technology

The formation of the solder bump by printing solder paste is practiced in many forms by companies such as Delco Electronics (DE), Flip Chip Technologies (FCT) and their licensees, as well as Lucent Technologies and PacTech. The principal drawback in using printed solder deposition processes has been in achieving fine bump pitches. Various forms of solder printing can currently produce pitches of 0.010 in (250 mm). The process description provided in this section will be based upon the DE/FCT wafer bumping approach. FCT provides bump pitches of 0.008 in (200 mm), and a 0.006 in (150 mm) process will be introduced mid-1997. A schematic representation of a sputtered UBM and a bump formed with a solder paste is shown in Fig. 4.12.

The DE/FCT wafer bumping approach is termed *flex-on-cap (FOC)* and refers to the improved fatigue resistance of this structure over a minibump structure that had been used by Delco previously. The process flow is as follows:

(a) In-situ wafer clean. A back sputter clean of the wafer is performed to remove oxides and organic residue prior to metal deposition. The cleaning also serves to roughen the wafer passivation and surface of the bond pad to promote better adhesion of the UBM.

(b) UBM deposition. Three metals make up the UBM system. The first layer is sputtered Al, followed by sputtered layers of Ni and Cu. The Al forms a strong adhesion to the wafer passivation as well as to the Al bond pads. The Cu is used to keep the Ni from oxidizing and, unlike the plated minibump process, is not needed to allow the solder bump to adhere to the UBM. Figure 4.13a illustrates this step in the process.

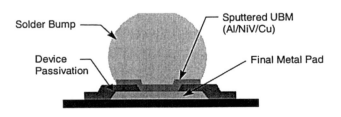

FIGURE 4.12 Cross section of a sputtered UBM and solder paste bump.

Direct Chip Attach

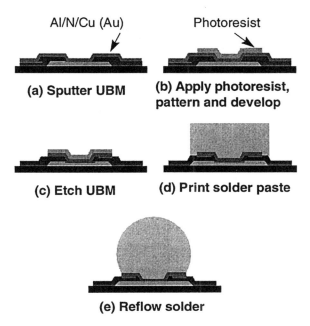

FIGURE 4.13 Sputtered UBM and solder paste bumping process.

(c) UBM patterning. A layer of photoresist is applied, patterned and developed (Fig. 4.13b). The Al/Ni/Cu layers are then etched away (Fig. 4.13c) except over the bond pad passivation openings and test structures. The resist is removed, leaving the tri-metal UBM over the bond pads.

(d) Solder paste deposition. Solder paste is printed onto the UBM using Delco Electronic's patent-pending process, and the bump is reflowed to form a sphere (Figs. 4.13d and 4.13e).

Benefits and Drawbacks of Printed Solder Bumps

Cost considerations. The solder paste bumping process is less expensive than the evaporative wafer bumping processes and competitive with plated bumping costs.

Bump structure: under-bump metallurgy. The sputtered Al layer provides excellent adhesion to the IC metallization and protects the underlying bond pad. The UBM also exhibits excellent adhesion to many passivation types including silicon nitride, silicon oxide, and several types of polyimide. In addition, the tri-metal UBM furnishes a low-resistance electrical path.

The Ni layer serves two functions: it is an excellent solder diffusion barrier (especially for 63Sn/Pb solders), and it provides a solder wettable surface after the Cu is consumed. The choice of materials and the physical structure of the UBM combine to provide a low-stress, highly reliable bump structure.

Bump structure: solder. The solder bump is extremely reliable, and its use has been well documented in automotive applications by Delco Electronics.[14] The 63Sn/Pb solder bump structure is robust enough to withstand more than 10 reflow cycles (customer requirements have not necessitated testing beyond this point).

The structure of the bump and the type of alloy used ensures a predictive amount of solder bump collapse. Depending on solder alloy, size of bump, and substrate attachment geometry, the deposited solder bump can experience a 10 to 30% collapse upon assembly. This feature provides for robust assembly processes with high yields.

The solder paste processing allows for excellent metallurgical control. A direct result of this material control is that assembly reflow temperatures are always predictable and consistent.

Alternative alloys. Solder paste deposition allows for a variety of solder alloys to be used. Non-binary solders are a marked benefit to this type of wafer bumping process. Ternary or quaternary alloys can address certain requirements that binary alloys cannot. FCT currently offers the following alloys:

(a) Lead-free alloys

CASTIN is a Cu/Sb/Ag/Sn alloy that reflows at 250–260° C (liquidus temperature begins at 211° C). It is a lead-free solder that can be used for ultra-low alpha particle requirements or to take advantage of a higher reflow temperature for a BGA assembly reflow hierarchy.

(b) Low alpha particle solder

An alpha particle is the nucleus of a helium atom (He_4^{++}). They are emitted through the decay of radioactive impurities or isotopes in the solder. A high-energy alpha particle (8 MeV) can generate up to 2.5×10^6 electron-hole pairs and cause soft errors in sensitive semiconductor devices such as memory chips. Since the main source of alpha particle generation in flip chip packaging is solder alloys that contain lead, it is advantageous to use low-alpha emission (<0.02 count/hr/cm^2) lead material or a lead-free alloy. Solder paste deposition provides for a cost-effective application of low alpha Pb/Sn solder.

(c) Other examples of non-binary alloys

Pb/In/Ag and Sn/Pb/Cd/In are also used in flip chip applications.

Environmental. Solder deposition processes are typically well controlled with little wasted material. The environmental impact of this type of solder bumping process is less than the evaporative or plated wafer bumping processes.

Assembly with 63Sn/Pb solder. The FCT process was designed specifically to accommodate a variety of solders including 63Sn/Pb. Since all of the eutectic 63Sn/Pb solder that is required for the joint resides on the chip, there are no additional cost adders that are incurred at the board level.

Compatibility with probed wafers. The Al/Ni/Cu UBM can be used on probed wafers without fear of affecting reliability.[15] This is due to both the UBM material set as well as the deposition process. There has never been a failure attributed to the probe marks left on the final metal pads. This is significant for companies that want to get yield information back to their fabs as fast as possible. Similarly, test flow within the fab is not affected by the fact that certain wafers are targeted for flip chip vs. wire bonded assembly.

Option to preferentially lay down UBM without solder. The FOC process provides the option of preferentially depositing either UBM material or a solder bump on the device. This is not an option with evaporated or electroplated solder bumping technologies.

Production. The Flex-on-Cap process has been used in high-volume production for many years. Delco Electronics bumps 300,000 devices per day. The process is robust and low in cost.

Solder Ball Bumping (SBB) Process

Solder ball bumping, or stud bump bonding (SBB), describes the process of using wire bonders to create a ball bond on the I/O pad with gold (Au) or solder wire. A slightly modified wire bonder is used to attach the wire to the bond pad, and the wire is broken off close to the stud. The wire is broken through mechanical, ultrasonic, or similar means. Reflow of the ball bond is typically done in a nitrogen oven to achieve better adhesion of the wire to the aluminum bond pad. The attached ball bump can be flattened *(coined)* to establish a uniform height for use with conductive adhesives or additional solder. The Au ball bump is normally not coined when used with isotropic conductive adhesives. Solder has been applied to the SBB with and without coining being performed.

A cost comparison between SBB and the other bumping technologies is difficult to assess. The SBB process typically has other unquantified costs that are greater than standard solder bumping processes. These include the conductive adhesive costs, high precision and slow assembly processes, and poorer reliability in many applications. It is estimated that stud bumping may be slightly less expensive than the electroplating process. Current costs of SBB are estimated at 60 to 70% of conventional wire bond costs.[16]

SBB has been used on microprocessors and memory devices. Much of this activity is done in Japan by companies such as Matsushita and Fujitsu.

Stud bumping: issues. There are significant issues to understand when evaluating this technology. Although SBB can be performed at a rate of eight bumps per second, it is not a batch process and is therefore not appropriate for high I/O count devices where the lower throughput would not justify its

Direct Chip Attach   4-19

cost savings. In addition, bumps cannot be applied over active circuitry, limiting SBB to peripheral I/O configurations.

When used in conjunction with conductive adhesives, stud bumping does not provide the same self-alignment properties found with solder bumps. For pitches of less than 150 μm, a placement accuracy of ≤5 mm is required. This requires the use of specialized flip chip assembly equipment rather than more conventional fine-pitch SMT assembly equipment. This approach often requires heat and pressure to be applied, which can considerably slow the assembly process.

A conventional solder SBB can have difficulty in sustaining a reliable joint during subsequent reflow operations. Figure 4.14a illustrates the stud bump bonded to an aluminum IC pad, and Fig. 4.14b shows the solder dewetting from the pad after reflow. However, much research is being conducted in this area. Special solder based wires that can wet directly to aluminum bond pads have been developed at Tanaka Denshi Kogyo[17] and Tohoku University. These alloys have had mixed success and are still limited in the reflow profiles and reliability results achieved to date. Corrosion of the aluminum bond pad is also a significant issue when using SBB techniques. In order to ensure that (a) the stud bump will not de-wet from the Al and (b) the Al bond pad will not corrode, a UBM can be applied over the Al bond pad. The addition of a UBM, however, will increase the cost of SBB to a point where it would offer neither a technological nor a cost advantage.

Electroless Nickel
The use of electroless nickel as an under-bump metallurgy has been vigorously pursued by many companies in the United States, Europe, and Asia since the late 1980s. This is because electroless nickel has the potential of being a lower-cost alternative to other bumping approaches. The number of process steps potentially required to create this UBM are few. Electroless nickel in its purest form does not require the use of photo masks and therefore eliminates processes such as patterning, developing, and stripping. Projected costs are estimated at approximately 60% of printing and "65 to 75% of conventional plated bump technologies."[18]

Electroless nickel can provide a UBM that can be used with conductive adhesives or various solder alloys. The process involves creating a UBM by building up nickel on the aluminum (Al) bond pad. Zincation is typically used to prepare the Al to receive the nickel. The Al bond pad is partially etched and activated by treatment with an alkaline zincate solution. There are alternative technologies which use special electroless nickel baths to eliminate the zincation step. Once the nickel is deposited, a layer of gold is applied over the nickel for oxide prevention.

Electroless nickel technology presents many challenges. For example, since electroless nickel only adheres to the bond pad and not to the passivation, corrosion can be an issue. The chemical baths must be well understood and controlled to avoid uneven or a lack of plating.

In addition, the wafer must be compatible with the process. Except for the Al bond pads, all exposed metal must be passivated. In addition, the wafers should not be back ground and should also be free of contaminants. This second issue is especially important since the introduction of foreign material can significantly effect the yield on an entire batch. For this reason, it is anticipated that electroless nickel will initially be used for low-value silicon enabling new flip chip markets.

 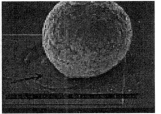

(a)   (b)

FIGURE 4.14 SEM micrographs of (a) Sn/Pb SBB and (b) stud bump after 300° C reflow in a nitrogen. Solder has dewetted from the Al bond pad.

Flip Chip Technologies is involved in an active research and development effort that is being conducted in conjunction with Delco Electronics (refer to the paper by Frank Stepniak for a detailed overview of the process).[19] FCT is sampling the technology as this is written and will have a production-ready electroless nickel process by press time. One of the benefits of this process is that various solder materials have already been characterized for use with the electroless nickel UBM. The eutectic 63Sn/Pb solder bump has been found to be compatible with this process. Initial reliability testing is showing excellent results of the electroless nickel and eutectic Sn/Pb structure.

4.5.2.1 Summary of Solder Deposition Processes

Table 4.3 summarizes the features of each solder deposition process (stud bumping is not listed) in terms of the under-bump metallurgy employed.

TABLE 4.3 UBM Deposition Systems

UBM	Evaporated (typical of C4)	Plating (two processes)	Solder Paste (typ. FCT)	Electroless nickel
Adhesion layer	CR	(1) TiW or (2) Cr/Cu	Al	Ni
Solder diffusion layer	Phased Cr-Cu	(1) Cu stud/minibump (2) Cr-Cu	Ni	Ni
Solder wettable layer	Cu	(1) Cu (2) Cu	Cu	Au
Oxide prevention	Au	Au	Cu	Au
Suitability for 63Sn/Pb	No	(1) Poor (2) No	Yes	Yes
Use with Probed Wafers	No	No	Yes	Mixed

*Two plating processes are represented, (1) one that uses the Cu minibump and (2) one that does not.

4.5.2.2 Assembly

Although flip chip technology has been in use for over 30 years, there are fewer than a dozen companies worldwide that use the process in production. Over 90% of these applications have been low lead count devices for watches, displays, vehicle modules and communications modules.[20] High-volume flip chip production has been captive, with companies having control of device design, wafer fabrication, and bumping, as well as test and assembly. Consequently, enabling technologies (such as underfill materials, dispensing equipment, efficient test methodologies, and high-throughput pick-and-place equipment) are not well characterized for new users with emerging applications.

Recent consortium activities have begun to focus on developing and characterizing flip chip assembly capabilities. This is to help promote an infrastructure for direct chip attach assembly. The main focus of these activities includes the following:

- Handling of the devices for assembly
- Placement equipment with required accuracy and throughput
- Application and cleaning of fluxes
- Application and characterization of underfills
- Reliable reworking methods
- Low-cost, fine-pitch board technology

Flip Chip Handling and Placement Equipment

The handling of flip chip devices depends on the type of assembly equipment used. Both waffle packs and tape and reel are used to convey flip chips prior to, and during, assembly. Tape and reel is mainly

Direct Chip Attach

used at captive flip chip manufacturing operations. For high-rate production, it is expected that tape and reel will be the principal method used for flip chip transport.

Tape carriers for flip chip devices need to incorporate certain characteristics that conventional tape does not address. Tape formed from polycarbonate materials offers superior impact and shrinkage properties at temperature and humidity extremes. Tight pocket tolerances minimize device movement, flatness avoids tombstoning or trampolining of the device, and conductivity provides ESD protection and low ionic contamination.[21]

The design of the flip chip tape carrier can incorporate pedestals that will support a bump-down peripheral array flip chip layout. Conversely, a ledge provided around the perimeter of the tape pocket can support a bump-down area array flip chip design. It is important for the designer to understand subsequent handling of the flip chip device to incorporate features into its layout that are conducive to efficient and cost effective transport. For example, in an area array flip chip device, the pads should be recessed approximately 0.015 to 0.020 inches to provide a region around the die that can sit on the ledge around the tape pocket. This is somewhat contradictory to a wire bond layout, where the pads are laid out closer to the edge of the device.

The assembly of flip chips onto substrates is relatively easy to accomplish when using assembly equipment designed for flip chip placement. One of the most notable features of flip chip assembly is the ability of the device to self-align to the circuit board pattern (surface tension driven). Solder-bumped flip chips will self-align if placed such that a portion of the solder bump makes contact with the wettable pad diameter. Typically, a 3σ placement accuracy of 1/4 the pad diameter will ensure successful alignment of the flip chip after joining.

To achieve the required placement accuracy, it is necessary for the placement machine to align off both the bumps on the flip chip and the pads on the substrate. For some substrate materials, local fiducials may prove to be adequate.

Several flip chip placement machines are currently available on the market. Placement rates for flip chip devices have been advertised as high as 2,000 chips per hour, but this is dependent on the specific fluxes and alignment methods employed. Some of the placement equipment manufacturers include Kulicke & Soffa, Research Devices, Zevatech, Universal, and Panasonic.

Fluxes and Underfills

Fluxes
The flux used during flip chip assembly must be evaluated on several levels. These include

- Its ability to aid in soldering the interconnect
- Its ability to be easily cleaned (or not cleaned if it is a no-clean type)
- Its interaction with any subsequent underfill operations

If the flux is adequately cleaned from the assembly, there is no impact on the underfill operation. However, if a no-clean flux system is used, any residual materials must not affect the adhesion of the underfill to the device and substrate surfaces. Because there are many flux systems that are compatible with flip chip technology, each application must be evaluated against the specific requirements of the assembly.

Underfills
Underfill materials protect the underside of the device from the environment and reduce stresses on the solder joint. The underfill process is a critical step in the assembly of large area flip chip devices or in the assembly of flip chip devices onto laminate substrates. "Underfill increases the reliability by increasing fatigue life and by reducing the junction temperature of the IC. The fatigue life is improved by lowering the strain due to thermal mismatch imposed on the solder joint."[22]

Similarly, the type of solder alloy chosen can also effect the thermal fatigue life of the bump. Table 4.4 shows the thermal fatigue life of several solder alloys as normalized to eutectic 63Sn/Pb. The test substrate was alumina, and the thermal cycle range was −50°C to +150°C.[23] It can be seen that solder alloys such as indium lead have twice the thermal fatigue life of a eutectic 63Sn/Pb. Similarly, the high lead bump

TABLE 4.4 Thermal Fatigue Life Comparison of Bump Alloys (Normalized to 63Pb/Sn with No Underfill)

Solder Alloy Composition	Reflow Temperature (°C)	Fatigue Life without Underfill	Fatigue Life with Underfill
63Sn/Pb	230	1.0	15
5Sn/Pb	360	1.2	Not tested
50In/Pb	260	2–3	>30
37In/Pb	290	2–3	>30
3.5 Ag/Sn	260	0.5	11
5Sb/Sn	280	0.3	11
Sn/Pb/Cd/In	230	1.0	13
Sn/Ag/Cu/Sb	260	1.0	13

employed in evaporative processes shows a 20% increase in fatigue life as compared to a 63Sn/Pb alloy. However, when underfill is used, the fatigue life for all solders increases by at least a factor of ten. It has been demonstrated that a 0.500 × 0.500-in device assembled on a laminate circuit board with the correct underfill will easily pass >1000 thermal cycles (–40 to 125° C).[24]

Figure 4.15 shows the failure rate of devices using 63Sn/Pb solder bumps with and without underfill. A large area 0.500 × 0.500-in test device was chosen to better illustrate the effects of the underfill (acceleration of failure mechanisms, etc.).

Figure 4.16 shows an SEM micrograph of the cross section of a solder joint that has experienced solder fatigue. This device was assembled without underfill. The failure mode occurs as expected, with the crack in the solder located close to the die, where the attachment area of the solder is the smallest.

Figure 4.17 shows the cross section of a solder joint that has been assembled using underfill. The solder fatigue is seen by the hairline fracture, also located close to the UBM. The difference between the severity of the solder cracks is due to the use of underfill. The underfill's material properties are not only critical for device reliability but are also an important consideration for manufacturability. An underfill must be dispensed, flowed, and cured within a reasonable period of time. The underfill is typically applied to two sides of an assembled die with a needle dispenser, allowing surface tension forces to draw the material under the device to the remaining sides. More than one pass may be necessary to adequately fill the gap between the device and the substrate.

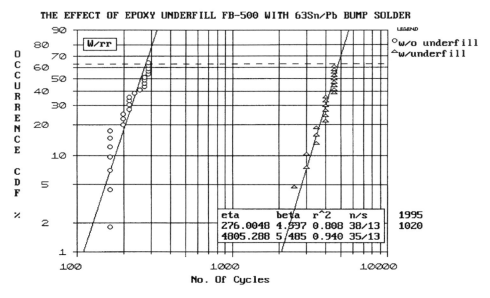

FIGURE 4.15 The effect of underfill on bump thermal fatigue life using an 0.500 × 0.500-in test die with 63Sn/Pb solder.

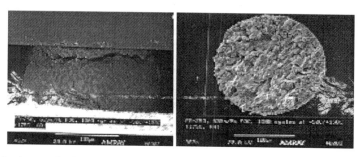

FIGURE 4.16 SEM micrograph of solder fatigue using a 63Sn/Pb bump assembled without underfill.

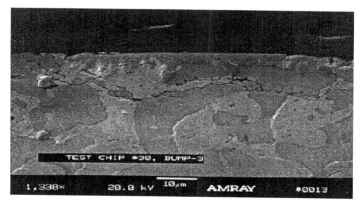

FIGURE 4.17 SEM micrograph of solder fatigue using a 63Sn/Pb bump assembled with underfill. Note how the hairline fracture reflects the stress relief that the underfill provides to the structure.

The flow properties are temperature dependent, with higher temperatures increasing flow rates. However, care must be taken that the temperature does not get too high, or the material will start to cure while still being flowed. The flow of the material must occur such that voids are not generated between the device and the substrate. The viscosity of the material must allow the surface tension forces to rapidly transfer the material through the substrate/device interface without creating air pockets. In addition, as bump height is reduced and bump densities are increased, new underfill materials will need to be created to address throughput and coverage considerations. Underfills that flow slowly will result in lengthy assembly times, with the underfill process becoming the manufacturing bottleneck.

Unfortunately, underfill material properties do not easily support both reliability and manufacturability considerations. This is because most underfills contain fillers that are used to achieve desired characteristics. Fillers that will provide a given coefficient of thermal expansion that help match the IC's CTE to the substrate's CTE can also hinder the flow rate. The filler system in the underfill must maintain uniform dispersion during application and operation since areas depleted of filler material, especially at the surface interface, will exhibit early failure.

The material must exhibit good adhesion to both the substrate and device surfaces to achieve the desired reliability. Therefore, the surface finish and passivation of the device, as well as the type of substrate used, must be evaluated to ensure that an underfill which will adhere to all material surfaces is chosen. This is significant, since many solder masks exhibit poor adhesion with some underfill materials. In addition, some wafer passivations, such as nitride, have presented a challenge to underfill development. Although initial adhesion appears good, underfills can experience delamination during thermal cycling.

In summary, the proper choice of fluxes and underfill materials is critical to ensure a reliable assembly while meeting production rate requirements. The manufacturer will need to choose the materials according to device type, device size, bump height and density, device passivation, substrate type and configuration, solder mask materials, and reliability requirements. There is no single combination of flux, solder resist, and underfill that will work well for all combinations of device types and board materials. However, much development in flip chip assembly materials is being conducted, and compatible material sets are being identified by a growing flip chip technology infrastructure.

Rework

Rework is an important issue for flip chip devices since there are presently no commercial "reworkable" underfills available (although development is progressing). Therefore, the assembly must be tested prior to applying underfill for those systems that are not disposable. Equipment that allows for the removal and replacement of die is available. However, this equipment is often designed for BGA devices, and modifications in air flow and device attachment are required. It is important to note that the device substrate site must be cleaned of residual solder after removal of the failed device to ensure that excess solder does not cause shorts or excessive height variation when the new device is attached. Some of the rework equipment manufacturers include SRT, AirVac, and Conceptronics.

Flip chip applications are now driving the flip chip assembly and test infrastructure. As the flip chip market and infrastructure grows, both equipment and material improvements will lead to new advancements in flip chip assembly technology.

4.5.2.3 Summary

It can be readily seen that the choice of flip chip bumping technology can impact component and assembly costs as well as overall device reliability. The solder joint, which consists of a UBM and a solder ball metallurgy, will effectively limit postprocessing choices in terms of temperature tolerances, direct SMT compatibility, geometries (height/diameter/pitch), assembly throughput, and overall manufacturing yields.

Costs associated with the bump deposition process must include

- capitalization
- licensing fees (if appropriate)
- merchant vendor pricing
- testing strategies
- the impact on assembly

Material choices such as lead-free solder, low alpha particle solder, or solders that will provide an assembly reflow hierarchy must also be investigated. The union of fluxes and underfills with a given passivation, solder, or circuit board material is not a trivial task, and initial experimentation will need to be performed.

It is important for design and packaging engineers to understand the material impacts of any wafer bumping system chosen. By selecting a bumping technology that addresses both reliability and manufacturability of the entire process, the overall system level cost is optimized.

With a growing merchant vendor infrastructure, the designer and packaging engineer will have much industry support. Flip chip technology is a highly reliable, cost-effective solution for a wide variety of applications. There is no question that it will be incorporated into a growing number of product types because of its ability to provide the competitive edge in design, manufacturing, reliability, and product pricing.

Historical data, as defined by high-volume production, system-level costs, or device/system reliability, can help define which technology choice is the most appropriate for a given application.

Acknowledgments

Special thanks to the following people at Flip Chip Technologies and Delco Electronics: Steve Wortendyke and Mechelle Jones, for assisting in the preparation of the manuscript; and Michael Varnau, Shing Yeh, and Alan Harvath, for providing Flip Chip Technologies with device, materials, reliability, and assembly information.

References

1. Adaptation from a 1994 IBM Microelectronic Presentation.
2. Elenius, P., "Flex on Cap—Solder Paste Bumping," 3rd International Symposium and Exhibition on Advanced Packaging Materials, Processes, Properties and Interfaces, Braselton, GA, March 1997.
3. Elenius, P., private communication with Motorola.
4. Tummala, R. R. and Rymaszewski, E., *Microelectronics Packaging Handbook,* Van Nostrand Reinhold, New York, 1989, p. 367.
5. Davis, E.M., Harding, W.E., Schwartz, R.S., Corning, J.J., "Solid Logic Technology: Versatile, High Performance Microelectronics," *IBM Journal of Research and Development,* Vol. 8, 1964.
6. Harvath, A., Delco Electronics, November, 1996, private communication; Totta, P. and Tummala, R. R., IMAPS 1st International Advanced Technology Workshop on Low Cost Flip Chip Technology, March, 1997.
7. Varnau, M. J. and Yeh, S., Flip Chip/BGA Workshop, Binghamton, NY, October 1996.
8. Delco Electronics internal evaluation program.
9. IPC J-STD-012, Implementation of Flip Chip and Chip Scale Technology, January 1996, p. 37.
10. Greer, S. E., *Proceedings of the 1996 IEEE 46th Electronic Components and Technology Conference,* Orlando, FL, 1996.
11. Elenius, P., In private discussion with IBM. Approximately three pounds of lead are deposited per 8-in wafer. A 4-ft spherical dome plus the sidewalls of an evaporator measure approximately 4000 in^2 total area. The deposition of 0.004 in of Pb results in 160 in^3 of material being deposited on all surfaces.
12. Prismark Partners, L.L.C., *Flip Chip Time Bulletin,* February 1996.
13. Magill, P., Private discussion regarding MCNC/Unitive offering of eutectic solder joints, IMAPS Advanced Technology Workshop on Low Cost Packaging, Ojai, CA, January 1997.
14. Although the reliability of the UBM/solder bump structure has been well documented internal to Delco Electronics, please also reference the following paper: Varnau, M. J. and Yeh, S., Flip Chip/BGA Workshop, Binghamton, NY, October 1996.
15. Varnau, M. J., "Impact of Wafer Probe Damage on Flip Chip Yields and Reliability," International Electronics and Manufacturing Technology Symposium, Austin TX, October 1996.
16. Prismark Partners, L.L.C., *Flip Chip Time Bulletin,* February 1996.
17. Ogashiwa, T., Akimoto, H., Shigyo, H., Murakami, Y., Inoue, A., Masumoto, T., *Japan Journal of Applied Physics* 1, Regulation Paper Short Notes (Japan), Vol. 31, No. 3, March 1992.
18. Prismark Partners, L.L.C., *Flip Chip Time Bulletin,* February 1996.
19. Stepniak, F., "Solder Flip Chips Employing Electroless Nickel: Weighing Increased Risks Against Reduced Costs," INTERpack 1997, Kohala, Hawaii, June 1997.
20. Prismark Partners, L.L.C., *Flip Chip Time Bulletin,* February 1996.
21. Kelley, S. and Becker, T. L., meeting with 3M at Nepcon West, Anaheim, CA, February 1997.
22. Varnau, M. J., Delco Electronics private communication.
23. Varnau, M. J. and Yeh, S., Flip Chip/BGA Workshop, Binghamton, NY, October 1996.
24. Yeh, S., Delco Electronics private communication.

4.6 Chip-Scale Packages

Chip-scale packages (CSPs) have been defined by IPC and EIA in "Implementation of Flip Chip and Chip Scale Technology" as IC die that have been enclosed in a package that is no larger than 1.2× the dimensions of the bare die itself. The packages have pitch dimensions <0.25 mm (<10 mil), making them valuable in space-limited applications such as memory products, PC cards, and MCMs. However, at this writing, there is very limited infrastructure to support CSP. It remains to be seen if improvements in test and burn-in support devices, standards, and sufficiently flat substrates will be successful. Many of these same problems affect COB and flip chip assembly. As with COB and flip chip, one of the major issues with CSP is the mismatch of the coefficient of thermal expansion (CTE) between the component and any substrate other than ceramic. Ikemizu et al. report that the solder joint reliability for CSPs on PCB is poorer than that for QFP due to CTE mismatch. Underfill will mitigate this issue, but it requires additional process steps.

Development of a package that can be assembled and successfully used without underfill will go a long way toward increasing the market penetration of these technologies. CSP is predicted to be that package. Other advantages of the CSP packages over COB and flip chip are

- Grid/termination pitch will be greater, allowing use of standard placement equipment.
- KGD will not be a issue.
- The die is environmentally protected.

Another issue that affects both CSP and flip chip is wafer bumping. Section 4.5.2 contains a comparative analysis of solder bumping methods.

Both EIA-JEDEC and EIA-J groups are working to establish standard package outlines and pitches for industry. Examples of their initial work for CSP and other small packages are documented in several JEDEC product outlines as follows:

- MO-207: Rectangular Chip-size, fine pitch ball grid array family (Jan. 1999)
- MO-209: Plastic thin shrink small outline no lead package (Oct. 1998)
- MO-210: Thin fine pitch ball grid array family, 0.80-mm pitch (Jan. 1999)
- MO-211: Die size ball grid array (Oct. 1998)

Like other product outlines, these product outlines carry this statement:

This registered outline has been prepared by the JEDEC JC-11 committee and reflects a product with anticipated usage in the electronics industry; changes are likely to occur.

This statement reflects the flux that exists regarding CSP at the time of this writing. Since it is certain the interest in CSP will continue, any reader considering the implementation of CSP will need to read the latest JEDEC product outlines.

These product outlines also reflect the different grid pitches in use and proposed as shown below:

	Grid termination pitches
MO-207	0.75 mm, 0.65mm, 0.50 mm
MO-209	0.65 mm, 0.50 mm
MO-210	0.80 mm
MO-211	0.50 mm, 0.40 mm
Note: (0.80 mm = 31.5 mil, 0.75 mm = 29.5 mil, 0.65 mm = 25.6 mil, 0.50 mm = 19.7 mil, 0.40 mm = 15.7mil)	

One intent of the package definitions is that standard placement equipment will be able to place the components. This is an important part of keeping the implementation of CSPs economically viable.

A board-level evaluation of eight types of CSPs was performed by Newman and Yuan. Characterized by package materials and internal die interconnect, their study evaluated solder joint defects and reliability issues after thermal cycling. They determined that optimization of solder paste screening parameters resulted in no observed solder joint opens, and that failure results for thermally cycled CSPs fell short of comparative results for BGAs.

Solder paste printing for CSP boards is similar to printing for BGAs. The stencils should be laser cut or electroformed for clean openings and accurate paste volumes. Burr reports the results of a study that shows that electroformed and laser-cut/nickel-plated stencils provide more consistent print volumes than chemically etched, bimetal, or laser-cut (without plating) stencils. He also reports that square aperture openings with 0.010-in radius corners provide better solder paste release than round openings, with metal squeegees.

In addition to the CSP packages defined by the JEDEC product outlines, various manufacturers are introducing CSPs that fit their needs best. As an example, Dallas Semiconductor Inc., in November, 1998, introduced the 1-Wire minimalist chip package, which essentially wraps small die in a durable film or packages them in a form-fit metal package. Using serial communication techniques, the packages have just two connections to the substrate upon which they are mounted. With just two bumped pads, their placement can be done with the same accuracy requirements and using the same equipment that will place 0603 and 0805 chip resistors and capacitors.

Cubic Memory Inc. will provide a patented "Vertical Integration Process" page, which will "vertically connect multiple semiconductor dice with the added capability of connecting dice horizontally" (http://www.cubicmemory.com/ Technology/VIP/). The process allows memory die to be packaged in a very compact format while claiming to reduce system delay and interconnect capacitance.

Bibliography

"Procurement standard for known good die (KGD) EIA/JESD49." EIA, Arlington, VA 1996.
"Semiconductor design standard for flip chip applications J-STD-026." EIA, Arlington, VA 1998.
"Performance standard for flip/chip scale bumps J-STD-028." EIA, Arlington, VA 1998.
Beddingfield C, Kost D. "Flip chip assembly of Motorola fast static RAM known good die." *Proceedings 47th Electronic Components and Technology Conference,* IEEE, 1997.
Burr DC. "Solder paste printing guidelines for BGA and CSP assemblies." *SMT Magazine,* Vol. 13 No. 1, Jan. 1999.
Carbin JW. "Better flip chip underfill throughput." *SMT Magazine,* Vol. 13, No. 1, Jan. 1999.
Carter T, Craig E. "Known good die comes of age." *Semiconductor Intl.,* Vol. 20, No. 12, 1997.
Conte A, Chillare S, Groover R. "Low-cost flip chip packaging technologies." *Electronic Packaging & Production,* Vol. 39, No. 1, Jan. 1999.
Gilg L. "Known Good Die." Journal of Electronic Testing: theory and applications, Vol. 10, No. 1–2, Feb.–April 1997.
Gilg L, Chung T. "Known good die versus chip size package: Options, availability, manufacturability and reliability." Surface Mount International, *Proceedings of the Technical Program,* 1996.
Gilg L, Chung T. "Known good die versus chip size package: options, availability, manufacturability and reliability." *Proceedings of the 1996 International Electronics Packaging Conference.* Int. Electron Packaging Soc., Edina, MN, 1996.
Glidden, W. "Selecting pick-and-place equipment for flip chip devices." *Electronics Packaging & Production,* Vol. 38, No. 2, Feb. 1998.
Gurnett K. "Will chip scale packaging replace known good die?" *Microelectronics Journal,* Vol. 28, No. 6–7, Aug.–Sept. 1997.
Min BY et al., "Burn-in and dynamic test carrier for KGD." *Proceedings of the Technical Program,* NEPCON West, Reed Exhibition Companies, Des Plaines, IL, 1994.
Novellino J. "What is known-good die and how do your get there?" *Electronic Design,* Vol. 45, No. 17, Aug. 18, 1997.
Patterson DS, Elenius P, Leal JA. "Wafer bumping technologies—a comparative analysis of solder deposition processes and assembly considerations." From http://www.flipchip.com, Flip Chip Technologies LLC, Phoenix, AZ, 1997.
Spletter P, Reber C. "To KGD or not to KGD? That is the question!." *Proceedings of the 1997 International Conference on Multichip Modules,* IEEE, 1997.
Vardaman J. "What does CSP cost?" *SMT magazine,* Vol. 11, No. 6, June 1997.
Wilson KT, Kleiner MH. "Testing strategies for KGD." *Proceedings of the Technical Program,* NEPCON West, Reed Exhibition Companies, Des Plaines, IL, 1994.
Wood, AG. "KGD definitions influence MCM applications." Surface Mount International Conference and Exposition, *Proceedings of the Technical Program,* 1994.

5
Circuit Boards

Glenn R. Blackwell
Purdue University

5.1 Introduction 5.1
5.2 Overview 5.2
5.3 Basic Circuit Board Design 5.16
5.4 Prototypes 5.17
5.5 DFM and DFT Issues 5.18
5.6 Board Materials 5.21
5.7 Circuit Design and Board Layout 5.23
5.8 Simulation.. 5.26
5.9 Standards.. 5.27

5.1 Introduction

This chapter assumes that the reader will be using one of the many ECAD programs to design the schematic and/or the circuit board. It does not address the issues surrounding manual board layout and fabrication.

Substrates and circuit boards are used in virtually every electronic assembly. There were many changes in circuit board standards and accepted practices in the last part of the 1990s. Included are:

- Release of IPC-2221, "Generic Standard on Printed Board Design," and IPC-2222, "Sectional Design Standard for Rigid Organic Boards." These replace the older IPC-D-275 standard on circuit boards.
- Release of IPC-2224, "Sectional Standard for Design of PWB's for PC Cards."
- Update of IPC-D-356A, "Bare Board Electrical Test Data Format."
- Publication of data from Hughes Aircraft and others that extends the original work done by Jennings in the 1970s on current-carrying capability of board conductors.

The changes in these IPC standards include the following:

- Upgraded thermal standards reflecting increased board densities
- Guidelines for V-groove scoring of panels
- Expansion of the sections on laminates, with new materials
- Updated fiducial requirements and dimensioning requirements
- A design and performance trade-off checklist
- Tables that allow determination of the number of SMT or THT parts that will fit in a board area

For the serious designer of circuit boards, all of these standards and documents should be close at hand.

The new information from Hughes and others regarding current capability is an extension of the work into finer conductor sizes. The finer conductors work hand in hand with smaller components and smaller via sizes in the continuing quest for ever-smaller circuit boards and products.

Where Else?

Virtually every chapter in this book has some topics that are affected by circuit board issues and has some topics that affect circuit board designs. The reader is encouraged to consider all topics that may affect designs. Chapter 6, "EMC and PCB Design," contains significant information for high-speed/transmission-line design as well as information on electromagnetic compatibility (EMC).

5.2 Overview*

5.2.1 Board Types, Materials, and Fabrication

Printed wiring boards (PWBs) can be divided in to four types of boards: (1) rigid boards, (2) flexible and rigid-flex boards, (3) metal-core boards, and (4) injection molded boards. The board that is most widely used is the rigid board. The boards can be further classified into single-sided, double-sided, or multilayer boards. The ever increasing packaging density and faster propagation speeds, which stem from the demand for high-performance systems, have forced the evolution of the boards from single-sided to double-sided to multilayer boards. On single-sided boards, all of the interconnections are on one side. Double-sided boards have connections on both sides of the board and allow wires to cross over each other without the need for jumpers. This was accomplished at first by Z-wires, then by eyelets, and at present by PTHs. The increased pin count of the ICs has increased the routing requirements, which led to multilayer boards. The necessity of a controlled impedance for the high-speed *traces*, the need for bypass capacitors, and the need for low inductance values for the power and ground distribution networks have made the requirement of power and ground planes a must in high-performance boards. These planes are possible only in multilayer boards. In multilayer boards, the PTHs can be buried (providing interconnection between inner layers), semiburied (providing interconnection from one of the two outer layers to one of the internal layers), or through vias (providing interconnection between the outer two layers).

The properties that must be considered when choosing PWB substrate materials are their mechanical, electrical, chemical, and thermal properties. Early PWBs consisted of copper-clad phenolic and paper laminate materials. The copper was patterned using resists and etched. Holes were punched in the laminate to plug the component leads, and the leads were soldered to the printed copper pattern. At present, the copper-clad laminates and the prepregs are made with a variety of different matrix resin systems and reinforcements [ASM 1989]. The most commonly used resin systems are fire resistant (FR-4) difunctional and polyfunctional epoxies. Their glass transition temperatures, T_g, range from 125 to 150° C. They have well understood processability, good performance, and low price. Other resins include high-temperature, one-component epoxy, polyimide, cyanate esters, and polytetrafluorocthylene (PTFE) (trade name Teflon®). Polyimide resins have high T_g, long-term thermal resistance, low coefficient of thermal expansion (CTE), long PTH life and high reliability, and are primarily used for high-performance multilayer boards with a large number of layers. Cyanate esters have low dielectric constants and high T_g and are used in applications where increased signal speed and improved laminate-dimensional stability are needed. Teflon has the lowest dielectric constant, low dissipation factor, and excellent temperature stability but is difficult to process and has high cost. It is used mainly in high-performance applications where higher densities and transmission velocities are required.

The most commonly used reinforcement is continuous filament electrical (E-) glass. Other high-performance reinforcements include high strength (S-) glass, high modulus (D-) glass, and quartz [Seraphim et al. 1989]. the chemical composition of these glasses determines the key properties of CTE and dielectric constant. As the level of silicon dioxide (SiO_2) in the glass increases, the CTE and the

---

*Adapted from Kollipara, R., and Tripathi, V., "Printed Wiring Boards," Chapter 71 in Whitaker, J., *The Electronics Handbook*, CRC/IEEE Press, Boca Raton, 1997.

dielectric constant decrease. The substrates of most rigid boards are made from FR-4 epoxy resin-impregnated F-glass cloth. Rolls of glass cloth are coated with liquid resin (A-stage). Then the resin is partially cured to a semistable state (B-stage or prepreg). The rolls are cut into large sheets and several sheets are stacked to form the desired final thickness. If the laminates are to he copper clad, then copper foils form the outside of the stack. The stack is then laminated and cured irreversibly to form the final resin state (C-stage) [ASM 1989]. The single-sided boards typically use phenolic or polyester resins with random mat glass or paper reinforcement. The double-sided boards are usually made of glass-reinforced epoxy. Most multilayer boards are also made of glass-reinforced epoxy. The internal circuits are made on single- or double-sided copper-clad laminates. The inner layers are stacked up with B-stage polymer sheets separating the layers. Rigid pins are used to establish layer-to-layer orientation. The B-stage prepreg melts during lamination and reflows. When it is cured, it glues the entire package into a rigid assembly [ASM 1989]. An alternative approach to pin-parallel composite building is a sequential buildup of the layers, which allows buried vias. Glass reinforced polyimide is the next most used multilayer substrate material due to its excellent handling strength and its higher temperature cycling capability. Other laminate materials include Teflon and various resin combinations of epoxy, polyimide, and cyanate esters with reinforcements. The dielectric thickness of the flat laminated sheets ranges from 0.1 to 3.18 mm (0.004 to 0.125 in), with 1.5 mm (0.059 in) being the most commonly used for single- and double-sided boards. The inner layers of a multilayer board are thinner with a typical range of 0.13 to 0.75 mm (0.005 to 0.030 in) [ASM 1989]. The number of layers could be 20 or more. The commonly used substrate board materials and their properties are listed in Table 5.1.

TABLE 5.1 Wiring Board Material Properties

Material	ε'_r	$\varepsilon''_r/\varepsilon'_r$	CTE ($\times 10^{-6}/°C$) x, y	z	T_g, °C
FR-4 epoxy-glass	4.0–5.0	0.02–0.03	16–20	60–90	125–135
Polyimide-glass	3.9–4.5	0.005–0.02	12–14	60	>260
Teflon®	2.1	0.0004–0.0005	70–120	—	—
Benzocyclobutene	2.6	0.0004	35–60		>350
High-temperature one-component epoxy-glass	4.45–4.45	0.02–0.022			170–180
Cyanate ester-glass	3.5–3.9	0.003–0.007			240–250
Ceramic	~10.0	0.0005	6–7		—
Copper	—	—	17		—
Copper/invar/copper	—	—	3–6		—

The most common conductive material used is copper. The substrate material can be purchased as copper-clad laminates or unclad laminates. The foil thickness of copper-clad laminates is normally expressed in ounces of copper per square foot. The available range is from 1/8 to 5 oz with 1-oz copper (0.036 mm or 0.0014 in) being the most commonly used cladding. The process of removing copper in the unwanted regions from the copper-clad laminates is called *subtractive* technology. If copper or other metal is added on to the unclad laminates, the process is called *additive* technology. Other metals used during electroplating may include Sn, Pb, Ni, Au, and Pd. Selectively screened or stenciled pastes that contain silver or carbon are also used as conducting materials [ASM 1989].

Solder masks are heat- and chemical-resistant organic coatings that are applied to the PWB surfaces to limit solderability to the PTHs and surface-mount pads; provide a thermal and electrical insulation layer isolating adjacent circuitry and components; and protect the circuitry from mechanical and handling damage, dust, moisture, and other contaminants. The coating materials used are thermally cured epoxies or ultraviolet curable acrylates. Pigments or dyes are added for color. Green is the standard color. Fillers and additives are added to modify rheology and improve adhesion [ASM 1989].

Flexible and rigid-flex boards are required in some applications. A flexible printed board has a random arrangement of printed conductors on a flexible insulating base with or without cover layers [ASM 1989].

Like the base material, the conductor, adhesive, and cover layer materials should be flexible. The boards can be single-sided, double-sided, or multilayered. However, multilayered boards tend to be too rigid and are prone to conductor damage. Rigid-flex boards are like multilayered boards with bonding and connections between layers confined to restricted areas of the wiring plane. Connections between the rigid laminated areas are provided by multiconductor layers sandwiched between thin base layers and are flexible. The metal claddings are made of copper foil, beryllium copper, aluminum, Inconel, or conductive polymer thick films, with copper foil being the most commonly used. Typical adhesives systems used include polyester, epoxy/modified epoxy, acrylic, phenolics, polyimide, and fluorocarbons. Laminates, which eliminate the use of adhesives by placing conductors directly on the insulator, are called *adhesiveless* laminates. Dielectric base materials include polyimide films, polyester films, aramids, reinforced composites, and fluorocarbons. The manufacturing steps are similar to those of the rigid boards. As insulating film or coating applied over the conductor side acts as a permanent protective cover. It protects the conductors from moisture, contamination, and damage and reduces stress on conductors during flexing. Pad access holes and registration holes are drilled or punched in an insulating film coated with an adhesive, and the film is aligned over the conductor pattern and laminated under heat and pressure. Often, the same base material is used as the insulating film. When coatings are used instead of films, they are screen printed onto the circuit, leaving pad areas exposed. The materials used are acrylated epoxy, acrylated polyurethane, and thiolenes, which are liquid polymers and are cured using ultraviolet (UV) radiation or infrared (IR) heating to form a permanent, thin, tough coating [ASM 1989].

When selecting a board material, the thermal expansion properties of the material must be a consideration. If the components and the board do not have a closely matched CTE, then the electrical and/or mechanical connections may be broken, and reliability of the board will suffer. The widely used epoxy-glass PWB material has a CTE that is larger than that of the encapsulating material (plastic or ceramic) of the components. When leaded devices in dual-in-line (DIP) package format are used, the mechanical forces generated by the mismatches in the CTEs are taken by the PTHs, which accommodate the leads and provide electrical and mechanical connection. When surface mount devices (SMDs) are packaged, the solder joint, which provides both the mechanical and electrical connection, can accommodate little stress without deformation. In such cases, the degree of component and board CTE mismatch and the thermal environment in which the board operates must be considered and if necessary, the board must be CTE tailored for the components to be packaged. The three typical approaches to CTE tailoring are constraining dielectric structures, constraining metal cores, and constraining low-CTE metal planes [ASM 1989, Seraphim et al. 1989].

Constraining inorganic dielectric material systems include cofired ceramic printed boards and thick-film ceramic wiring boards. Cofired ceramic multilayer board technology uses multiple layers of green ceramic tape into which via holes are made and on which circuit metallization is printed. The multiple layers are aligned and fired at high temperatures to yield a much higher interconnection density capability and much better controlled electrical properties. The ceramic used contains more than 90% of alumina and has a much higher thermal conductivity than that of epoxy-glass, making heat transfer efficient. The thick-film ceramic wiring boards are usually made of cofired alumina and are typically bonded to an aluminum thermal plane with a flexible thermally conducting adhesive when heat removal is required. Both organic and inorganic fiber reinforcements can be used with the conventional PWB resin for CTE tailoring. The organic fibers include several types of aramid fibers, notably Kevlar 29 and 49 and Technora HM-50. The inorganic fiber most commonly used for CTE tailoring is quartz [ASM 1989].

In the constraining metal core technology, the PWB material can be any one of the standard materials such as epoxy-glass, polyimide-glass, or Teflon-based materials. The constraining core materials include metals, composite metals, and low-CTE fiber-resin combinations, which have a low–CTE material with sufficient strength to constrain the module. The two most commonly used constraining core materials are copper-invar-copper (CIC) and copper-molybdenum-copper (CMC). The PWB and the core are bonded with a rigid adhesive. In the constraining low-CTE metal plane technology, the ground and power planes in a standard multilayer board are replaced by an approximately 0.15-mm thick CIC layers. Both epoxy and polyimide laminates have been used [ASM 1989].

Molded boards are made with resins containing fillers to improve thermal and mechanical properties. The resins are molded into a die or cavity to form the desired shape, including three-dimensional features. The board is metallized using conventional seeding and plating techniques. Alternatively, three-dimensional prefabricated films can also be transfer molded and further processed to form structures with finer dimensions. With proper selection of filler materials and epoxy compounds or by molding metal cores, the molded boards can be CTE tailored with enhanced thermal dissipation properties [Seraphim et al. 1989].

The standard PWB manufacturing primarily involves five technologies [ASM 1989].

1. *Machining.* This involves drilling, punching, and routing. The drill holes are required for PTHs. The smaller diameter holes cost more, and the larger aspect ratios (board thickness-to-hole diameter ratios) resulting from small holes make plating difficult and less reliable. They should be limited to thinner boards or buried vias. The locational accuracy is also important, especially because of the smaller features (pad size or annular ring), which are less tolerant to misregistration. The registration is also complicated by substrate dimensional changes due to temperature and humidity fluctuations, and material relaxation and shifting during manufacturing operations. The newer technologies for drilling are laser and water-jet cutting. CO_2 laser radiation is absorbed by both glass and epoxy and can be used for cutting. The typical range of minimum hole diameters, maximum aspect ratios, and locational accuracies are given in Table 5.2.

TABLE 5.2 Typical Limits of PWB Parameters

Parameter	Limit
Minimum trace width, mm	0.05–0.15
Minimum trace separation, mm	~0.25
Minimum PTH diameter, mm	0.2–0.85
Location accuracy of PTH, mm	0.015–0.05
Maximum aspect ratios	3.5–15.0
Maximum number of layers	≥20
Maximum board thickness, mm	≥7.0

2. *Imaging.* In this step, the artwork pattern is transferred to the individual layers. Screen printing technology was used early on for creating patterns for print-and-etch circuits. It is still being used for simpler single- and double-sided boards because of its low capital investment requirements and high volume capability with low material costs. The limiting factors are the minimum line width and spacings that can be achieved with good yields. Multilayer boards and fine line circuits are processed using photoimaging. The photoimageable films are applied by flood screen printing, liquid roller coating, dip coating, spin coating, or roller laminating of dry films. Electrophoresis is also being used recently. Laminate dimensional instability contributes to the misregistration and should be controlled. Emerging photoimaging technologies are direct laser imaging and imaging by liquid crystal light valves. Again, Table 5.2 shows typical minimum trace widths and separations.

3. *Laminating.* It is used to make the multilayer boards and the base laminates that make up the single- and double-sided boards. Prepregs are sheets of glass cloth impregnated with B-stage epoxy resin and are used to bond multilayer boards. The types of pressing techniques used are hydraulic cold or hot press lamination, with or without vacuum assist, and vacuum autoclave nomination. With these techniques, especially with autoclave, layer thicknesses and dielectric constants can be closely controlled. These features allow the fabrication of controlled-impedance multilayer boards of eight layers or more. The autoclave technique is also capable of laminating three-dimensional forms and is used to produce rigid-flex boards.

4. *Plating.* In this step, metal finishing is applied to the board to make the necessary electrical connections. Plating processes could be wet chemical processes (electroless or electrolytic plating) or dry plasma processes (sputtering and chemical vapor deposition). Electroless plating, which

does not need an external current source, is the core of additive technology. It is used to metallize the resin and glass portions of a multilayer board's PTHs with high aspect ratios (>15:1) and the three-dimensional circuit paths of molded boards. On the other hand, electrolytic plating, which requires an external current source, is the method of choice for bulk metallization. The advantages of electrolytic plating over electroless plating include greater rate of plating, simpler and less expensive process, and the ability to deposit a broader variety of metals. The newer plasma processes can offer pure, uniform, thin foils of various metals with less than 1-mil line widths and spacings and have less environmental impacts.

5. *Etching.* It involves removal of metals and dielectrics and may include both wet and dry processes. Copper can be etched with cupric chloride or other isotropic etchants, which limit the practical feature sizes to more than two times the copper thickness. The uniformity of etching is also critical for fine line feature circuits. New anisotropic etching solutions are being developed to extend the fine line capability.

The typical range of minimum trace widths, trace separations, printed through hole diameters, and maximum aspect ratios are shown in Table 5.2. The board cost for minimum widths, spacings, and PTH diameters, and small tolerances for the pad placements is high, and reliability may be lower.

5.2.2 Design of Printed Wiring Boards

The design of PWBs has become a challenging task, especially when designing high-performance and high-density boards. The designed board has to meet signal-integrity requirements [crosstalk, simultaneously switching output (SSO) noise, delay, and reflections], *electromagnetic compatibility (EMC)* requirements [meeting *electromagnetic interference (EMI)* specifications and meeting minimum *susceptibility* requirements], thermal requirements (being able to handle the mismatches in the CTEs of various components over the temperature range, power dissipation, and heat flow), mechanical requirements (strength, rigidity/flexibility), material requirements, manufacturing requirements (ease of manufacturability, which may affect cost and reliability), testing requirements (ease of testability, incorporation of *test coupons*), and environmental requirements (humidity, dust). When designing the circuit, the circuit design engineer may not have been concerned with all of these requirements when verifying the circuit design by simulation. But the PWB designer, in addition to wiring the board, must ensure that these requirements are met so that the assembled board functions to specification. Computer-aided design (CAD) tools are indispensable for the design of all but simple PWBs.

The PWB design process is part of an overall system design process. As an example, for a digital system design, the sequence of events may roughly follow the order shown in Fig. 5.1 [ASM 1989, Byers 1991]. The logic design and verification are performed by a circuit designer. The circuit designer and the signal-integrity and EMC experts should help the board designer in choosing proper parts, layout, and electrical rule development. This will reduce the number of times the printed circuit has to be reworked and shortens the product's time to market. The netlist generation is important for both circuit simulation and PWB design programs. The *netlist* contains the nets (a complete set of physical interconnectivity of the various parts in the circuit) and the circuit devices. Netlist files are usually divided into the component listings and the pin listings. For a list of the schematic capture software and PWB software packages, refer to Byers [1991].

Once the circuit is successfully simulated by the circuit designer, the PWB designer starts on the board design. The board designer should consult with the circuit designer to determine critical path circuits, line-length requirements, power and fusing requirements, bypassing schemes, and any required guards or shielding. First, the size and shape of the board are chosen from the package count and the number of connectors and switches in the design netlist. Some designs may require a rigidly defined shape, size, and defined connector placement; for example, PWB outlines for computer use. This may compromise the optimum layout of the circuit [Byers 1991].

Next, the netlist parts are chosen from the component library and placed on the board so that the interconnections form the simplest pattern. If a particular part is not available in the library, then it can

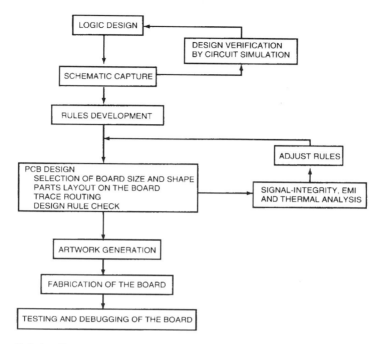

FIGURE 5.1 Overall design flow.

be created by the PCB program's library editor or purchased separately. Packages containing multiple gates are given their gate assignments with an eye toward a simpler layout. There are software packages that automate and speed up the component placement process. However, they may not achieve complete component placement. Problems with signal path length and clock skewing may be encountered, and so part of the placement may have to be completed manually. Placement can also be done interactively [Byers 1991].

Once the size and shape of the board are decided on, any areas on the board that cannot be used for parts are defined using fill areas that prevent parts placement within their perimeter. The proper parts placement should result in a cost-effective design and meet manufacturing standards. Critical lines may not allow vias, in which case they must be routed only on the surface. Feedback loops must be kept short, and equal line length requirements and shielding requirements, if any, should be met. Layout could be on a grid with components oriented in one direction to increase assembly speed, eliminate errors, and improve inspection. Devices should be placed parallel to the edges of the board [ASM 1989, Byers 1991].

The actual placement of a component may rely on the placement of a part before it. Usually, the connectors, user accessible switches and controls, and displays may have rigidly defined locations. Thus, these should be placed first. Next I/O interface chips, which are typically next to the connectors, are placed. Sometimes, a particular group of components may be placed as a block in one area (e.g., memory chips). Then, the automatic placement program may be used to place the remaining components. The algorithms can be based on various criteria. Routability, the relative ease with which all of the traces can be completed, may be one criterion. Overall total connection length may also be used to find an optimal layout. Algorithms based on the requirements for distances, density, and number of crossings can be formulated to help with the layout. After the initial placement, a placement improvement program can be run that fine tunes the board layout by component swapping, logic gate swapping, or (when allowed) even pin swapping. Parts can also be edited manually. Some programs support block editing, macros and changing of package outlines. The PWB designer may check with the circuit design engineer after a good initial layout is completed to make sure that all of the requirements will be met when the board is routed [Byers 1991].

The last step in the PWB layout process is the placement of the interconnection traces. Most PCB programs have an autorouter software that does most, if not all, of the trace routing. The software uses the wire list netlist and the placement of the parts on the board from the previous step as inputs and decides which trace should go where based on an algorithm. Most routers are point-to-point routers. The most widely used autorouter algorithm is the Lee algorithm. The Lee router works based on the cost functions that change its routing parameters. Cost functions may be associated with direction of traces (to force all of the tracks on a particular layer of the board vertically, the cost of the horizontal tracks can be set high), maximum trace length, maximum number of vias, trace density, or other criteria. In this respect, Lee's algorithm is quite versatile. Lee routers have a high trace completion rate (typically 90% or better) but are slow. There are other autorouters that are faster than the Lee router, but their completion rate may not be as high as the Lee router. These include Hightower router, pattern router, channel router, and gridless router [ASM 1989, Byers 1991]. In many cases, fast routers can be run first to get the bulk of the work done in a short time, and then the Lee router can be run to finish routing the remaining traces.

The autorouters may not foresee that placing one trace may block the path of another. Some routers, called *clean-up* routers, can get the offending traces out of the way by rip-up or shove-aside. The rip-up router works by removing the offending trace, completing the blocked trace, and proceeding to find a new path for the removed trace. The rip-up router may require user help in identifying the trace that is to be removed. Shove-aside routers work by moving traces when an uncompleted trace has a path to its destination but does not have enough room for the track. Manual routing and editing may be needed if the automatic trace routing is not 100% complete. Traces may be modified without deleting them (pushing and shoving, placing or deleting vias, or moving segments from one layer to another), and trace widths may be adjusted to accommodate a connection. For example, a short portion of a trace width may be narrowed so that it may pass between the IC pads without shorting them and is called *necking down* [Byers 1991].

Once the parts layout and trace pattern are found to be optimal, the board's integrity is verified by subjecting the trace pattern to *design rule check* for digital systems. A CAD program checks to see that the tracks, vias, and pads have been placed according to the design rule set. The program also makes sure that all of the nodes in each net are connected and that there are no shorted or broken traces [Byers 1991]. Any extra pins are also listed. The final placement is checked for signal integrity, EMI compliance, and thermal performance. If any problems arises, the rules are adjusted and the board layout and/or trace routing are modified to correct the problems.

Next, a netlist file of the PWB artwork is generated, which contains the trace patterns of each board layer. All of the layers are aligned using placement holes or datum drawn on each layer. The artwork files are sent to a Gerber or other acceptable format photoplotter, and photographic transparencies suitable for PWB manufacturing are produced. From the completed PWB design, solder masks, which apply an epoxy film on the PWB prior to flow soldering to prevent solder bridges from forming between adjacent traces, are also generated. Silkscreen mats could also be produced for board nomenclature. The PWB programs may also support drill file formats that numerically control robotic drilling machines used to locate and drill properly sized holes in the PWB [Byers 1991].

Finally, the board is manufactured and tested. Then the board is ready for assembly and soldering. Finished boards are tested for system functionality and specifications.

5.2.3 PWB Interconnection Models

The principal electrical properties of the PWB material are relative dielectric constant, loss tangent, and dielectric strength. When designing high-speed and high-performance boards, the wave propagation along the traces should be well controlled. This is done by incorporating power and signal planes in the boards and utilizing various transmission line structures with well defined properties. The electrical analog and digital signals are transferred by all of the PWB interconnects and the fidelity of these connections is dependent on the electrical properties of interconnects, as well as the types of circuits and signals. The interconnects can be characterized as electrically short and modeled as lumped elements in circuit simulators if the signal wavelength is large as compared to the interconnect length ($\lambda > 30 \ *l$). For digital signals, the corresponding criterion can be expressed in terms of signal rise time being large as compared to the propagation delay (e.g., $T_r > 10 \ T_d$). Propagation delay represents the time taken by the signal to travel from one end of the interconnect to the other end and is obviously equal to the interconnect length divided by the signal velocity, which in most cases corresponds to the velocity of light in the dielectric medium. Vias, bond wires, pads, short wires and traces, and bends in PWBs shown in Fig. 5.2 can be modeled as lumped elements, whereas, long traces must he modeled as distributed circuits or transmission lines.

The transmission lines are characterized by their characteristic impedances and propagation constants, which can also be expressed in terms of the associated distributed parameters R, L, G, and C per unit length of the lines [Magnuson, Alexander, and Tripathi 1992]. In general, the characteristic parameters are expressed as

$$\gamma \equiv \alpha + j\beta = \sqrt{(R + j\omega L)(G + j\omega C)}$$

$$Z_0 = \sqrt{\frac{(R + j\omega L)}{(G + j\omega C)}}$$

Propagation constant $\gamma \equiv \alpha + j\beta$ characterizes the amplitude and phase variation associated with an AC signal at a given frequency or the amplitude variation and signal delay associated with a digital signal. The characteristic impedance is the ratio of voltage to current associated with a wave and is equal to the

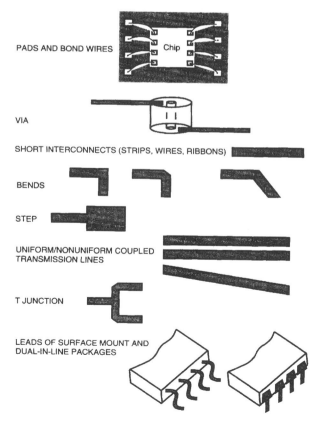

FIGURE 5.2 PWB interconnect examples.

impedance the lines must be terminated in for zero reflection. The signal amplitude, in general, decreases, and it lags behind in phase as it travels along the interconnect with a velocity, in general, equal to the group velocity. For example, the voltage associated with a wave at a given frequency can be expressed as

$$V = V_0 e^{-\alpha z} \cos(\omega t - \beta z)$$

where V_0 = amplitude at the input ($z = 0$)
      z = distance

For low loss and lossless lines, the signal velocity and characteristic impedance can be expressed as

$$\upsilon = \frac{1}{\sqrt{LC}} = \frac{1}{\sqrt{\mu_0 \varepsilon_0 \varepsilon_{reff}}}$$

$$Z_0 = \sqrt{\frac{L}{C}} = \frac{\sqrt{\mu_0 \varepsilon_0 \varepsilon_{reff}}}{C}$$

The effective dielectric constant ε_{reff} is an important parameter used to represent the overall effect of the presence of different dielectric media surrounding the interconnect traces. For most traces ε_{reff} is either equal to or approximately equal to the ε_r, the relative dielectric constant of the board material, except for the traces on the top layer where ε_{reff} is a little more than the average of the two media.

The line losses are expressed in terms of the series resistance and shunt conductance per unit length (R and G) due to conductor and dielectric loss, respectively. The signals can be distorted or degraded due to these conductor and dielectric losses, as illustrated in Fig. 5.3 for a typical interconnect. The resistance is, in general, frequency dependent, since the current distribution across the conductors is nonuniform and depends on frequency due to the skin effect. Because of this exclusion of current and flux from the inside of the conductors, resistance increases and inductance decreases with increasing frequency. In the high-frequency limit, the resistance and inductances can be estimated by assuming that the current is confined over the cross section one skin depth from the conductor surface. The skin depth is a measure of how far the fields and currents penetrate into a conductor and is given by

$$\delta = \sqrt{\frac{2}{\omega \mu \sigma}}$$

The conductor losses can be found by evaluating R per unit length and using the expression for γ or by using an incremental inductance rule, which leads to

$$\alpha_c = \frac{\beta \Delta Z_0}{2 Z_0}$$

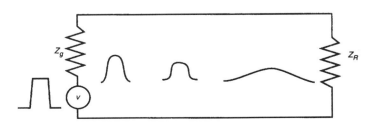

FIGURE 5.3 Signal degradation due to losses and dispersion as it travels along an interconnect.

where α_c is the attenuation constant due to conductor loss, Z_o is the characteristic impedance, and ΔZ_o is the change in characteristic impedance when all of the conductor walls are receded by an amount $\delta/2$. This expression can be readily implemented with the expression for Z_0. The substrate loss is accounted for by assigning the medium a conductivity σ which is equal to $\omega\varepsilon_0\varepsilon''_r$. For many conductors buried in near homogeneous medium,

$$G = C\frac{\sigma}{\varepsilon}$$

If the lines are not terminated in their characteristic impedances, there are reflections from the terminations. These can be expressed in terms of the ratio of reflected voltage or current to the incident voltage or current and are given as

$$\frac{V_{reflected}}{V_{incident}} = -\frac{I_{reflected}}{V_{incident}} = \frac{Z_R - Z_0}{Z_R + Z_0}$$

where Z_R is the termination impedance and Z_o is the characteristic impedance of the line. For a perfect match, the lines must be terminated in their characteristic impedances. If the signal is reflected, the signal received by the receiver is different from that sent by the driver. That is, the effect of mismatch includes signal distortion, as illustrated in Fig. 5.4, as well as ringing and increase in crosstalk due to multiple reflections resulting in an increase in coupling of the signal to the passive lines.

The electromagnetic coupling between the interconnects is the factor that sets the upper limit to the number of tracks per channel or, in general, the interconnect density. The time-varying voltages and currents result in capacitive and inductive coupling between the interconnects. For longer interconnects, this coupling is distributed and modeled in terms of distributed self- and mutual line constants of the multiconductor transmission line systems. In general, this coupling results in both the near- and far-end crosstalk as illustrated in Fig. 5.5 for two coupled microstrips. Crosstalk increases noise margins and degrades signal quality. Crosstalk increases with longer trace coupling distances, smaller separation between traces, shorter pulse rise and fall times, and larger magnitude currents or voltages being switched, and it decreases with the use of adjacent power and ground planes or with power and ground traces interlaced between signal traces on the same layer.

The commonly used PWB transmission line structures are microstrip, embedded microstrip, stripline, and dual stripline, whose cross sections are shown in Fig. 5.6. The empirical CAD oriented expressions for transmission line parameters, and the models for wires, ribbons, and vias are given in Table 5.3.

The traces on the outside layers (microstrips) offer faster clock and logic signal speeds than the stripline traces. Hooking up components is also easier in microstrip structure than in stripline structure. Stripline offers better noise immunity for RF emissions than microstrip. The minimum spacing between the signal traces may be dictated by maximum crosstalk allowed rather than by process constraints. Stripline allows

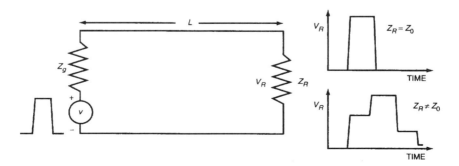

FIGURE 5.4 Voltage waveform is different when lines are not terminated in their characteristic impedance.

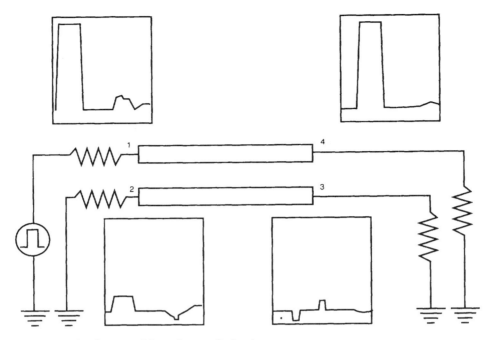

FIGURE 5.5 Example of near- and far-end crosstalk signal.

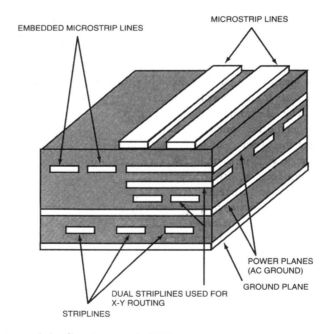

FIGURE 5.6 Example transmission line structures in PWBs.

closer spacing of traces than microstrip for the same layer thickness. Lower characteristic impedance structures have smaller spacing between signal and ground planes. This makes the boards thinner, allowing drilling of smaller diameter holes, which in turn allow higher circuit densities. Trace width and individual layer thickness tolerances of ±10% are common. Tight tolerances (±2%) can he specified, which would result in higher board cost. Typical impedance tolerances are ±10%. New statistical tech-

TABLE 5.3 Interconnect Models

Interconnect type	Model
 Round wire	$L \cong 0.002l\left[\ln\left(\dfrac{2l}{r} - 0.75\right)\right], \mu H; l > r$ l, r in cm.
 Straight rectangular bar or ribbon	$L \cong 0.002l\left[\ln\dfrac{2l}{b+c} + 0.5 + 0.2235\dfrac{b+c}{l}\right], \mu H$ b, c, l in cm.
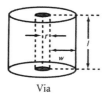 Via	$L = 0.002l\left[\ln\dfrac{2l}{r+W} - 1 + \xi\right], \mu H$ where $\xi = 0.25\left[\cos\left(\dfrac{r}{r+W}\dfrac{\pi}{2}\right) - 0.07\sin\left(\dfrac{r}{r+W}\pi\right)\right]$ l, r, W in cm.
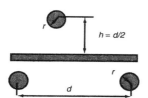 Round wire over a ground plane and parallel round wires	$Z_0 = \dfrac{120}{\sqrt{\varepsilon_r}}\cosh^{-1}\dfrac{d}{2r}, \text{ohm}$ $= \dfrac{120}{\sqrt{\varepsilon_r}}\ln\dfrac{d}{r}, \text{ohm}; d \gg r$
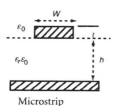 Microstrip	$\varepsilon_{r\text{eff}} = \dfrac{\varepsilon_r + 1}{2} + \dfrac{\varepsilon_r - 1}{2}\left[1 + \dfrac{10h}{W_e}\right]^{-a\cdot b}$ where $\dfrac{W_e}{h} = \dfrac{W}{h} + \dfrac{1.25}{\pi}\dfrac{t}{h}\left[1 + \ln\dfrac{2[h^4 + (2\pi W)^4]^{0.25}}{t}\right]$ $a = 1 + \dfrac{1}{49}\ln\left\{\dfrac{\left[\left(\dfrac{W_e}{h}\right)^4 + \left(\dfrac{W_e}{52h}\right)^2\right]}{\left[\left(\dfrac{W_e}{h}\right)^4 + 0.432\right]}\right\}$ $+ \dfrac{1}{18.7}\ln\left[1 + \left(\dfrac{W_e}{18.1h}\right)^3\right]$ $b = 0.564\left[\dfrac{\varepsilon_r - 0.9}{\varepsilon_r + 3}\right]^{0.053}$ $Z_0 = \dfrac{60}{\sqrt{\varepsilon_{r\text{eff}}}}\ln\left[\dfrac{F_1 h}{W_e} + \sqrt{1 + \left(\dfrac{2h}{W_e}\right)^2}\right]$ with $F_1 = 6 + (2\pi - 6)e^{-\left(30.666\frac{h}{W_e}\right)^{0.7528}}$

(continues)

TABLE 5.3 Interconnect Models (continued)

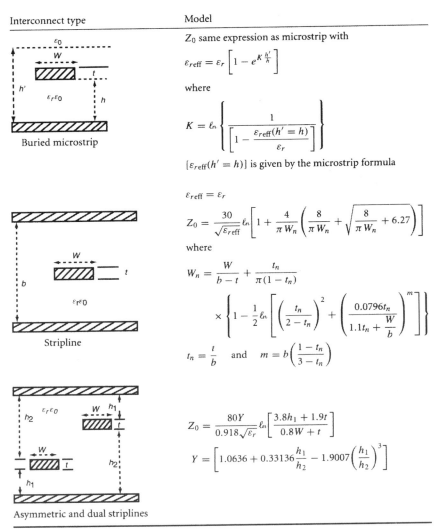

Interconnect type	Model
Buried microstrip	Z_0 same expression as microstrip with $$\varepsilon_{reff} = \varepsilon_r \left[1 - e^{K\frac{h'}{h}}\right]$$ where $$K = \ell_n \left\{ \frac{1}{\left[1 - \frac{\varepsilon_{reff}(h' = h)}{\varepsilon_r}\right]} \right\}$$ $[\varepsilon_{reff}(h' = h)]$ is given by the microstrip formula
Stripline	$$\varepsilon_{reff} = \varepsilon_r$$ $$Z_0 = \frac{30}{\sqrt{\varepsilon_{reff}}} \ell_n \left[1 + \frac{4}{\pi W_n}\left(\frac{8}{\pi W_n} + \sqrt{\frac{8}{\pi W_n} + 6.27}\right)\right]$$ where $$W_n = \frac{W}{b-t} + \frac{t_n}{\pi(1-t_n)}$$ $$\times \left\{1 - \frac{1}{2}\ell_n\left[\left(\frac{t_n}{2-t_n}\right)^2 + \left(\frac{0.0796 t_n}{1.1 t_n + \frac{W}{b}}\right)^m\right]\right\}$$ $$t_n = \frac{t}{b} \quad \text{and} \quad m = b\left(\frac{1-t_n}{3-t_n}\right)$$
Asymmetric and dual striplines	$$Z_0 = \frac{80Y}{0.918\sqrt{\varepsilon_r}} \ell_n \left[\frac{3.8 h_1 + 1.9 t}{0.8 W + t}\right]$$ $$Y = \left[1.0636 + 0.33136 \frac{h_1}{h_2} - 1.9007 \left(\frac{h_1}{h_2}\right)^3\right]$$

niques have been developed for designing the line structures of high-speed wiring boards [Mikazuki and Matsui 1994].

The low dielectric constant materials improve the density of circuit interconnections in cases where the density is limited by crosstalk considerations rather than by process constraints. Benzocyclobutene is one of the low dielectric constant polymers with excellent electrical, thermal, and adhesion properties. In addition, water absorption is lower by a factor of 15 compared to conventional polyimides. Teflon also has low dielectric constant and loss, which are stable over wide ranges of temperature, humidity, and frequency [ASM 1989, Evans 1994].

In surface mount technology, the components are soldered directly to the surface of a PWB, as opposed to through-hole mounting. This allows efficient use of board real estate, resulting in smaller boards and simpler assembly. Significant improvement in electrical performance is possible with the reduced package parasitics and the short interconnections. However, the reliability of the solder joint is a concern if there are CTE mismatches [Seraphim et al. 1989].

In addition to establishing an impedance-reference system for signal lines, the power and ground planes establish stable voltage levels for the circuits [Montrose 1995]. When large currents are switches, large voltage drops can be developed between the power supply and the components. The planes minimize the voltage drops by providing a very small resistance path and by supplying a larger capacitance and lower inductance contribution when two planes are closely spaced. Large decoupling capacitors are also added between the power and ground planes for increased voltage stability. High-performance and high-density boards require accurate computer simulations to determine the total electrical response of the components and complex PWB structures involving various transmission lines, vias, bends, and planes with vias.

Defining Terms

Design rules: A set of electrical or mechanical rules that must be followed to ensure the successful manufacturing and functioning of the board. These may include minimum track widths and track spacings, track width required to carry a given current, maximum length of clock lines, and maximum allowable distance of coupling between a pair of signal lines.

Electromagnetic compatibility (EMC): The ability of a product to coexist in its intended electromagnetic environment without causing or suffering functional degradation or damage.

Electromagnetic interference (EMI): A process by which disruptive electromagnetic energy is transmitted from one electronic device to another via radiated or conducted paths or both.

Netlist: A file of component connections generated from a schematic. The file lists net names and the pins, which are a part of each net in the design.

Schematic: A drawing or set of drawings that shows an electrical circuit design.

Suppression: Designing a product to reduce or eliminate RF energy at the source without relying on a secondary method such as a metal housing.

Susceptibility: A relative measure of a device or system's propensity to be disrupted or damaged by EMI exposure.

Test coupon: Small pieces of board carrying a special pattern, made alongside a required board, which can be used for destructive testing.

Trace: A node-to-node connection, which consists of one or more tracks. A track is a metal line on the PWB. It has a start point, an end point, a width, and a layer.

Via: A hole through one or more layers on a PWB that does not have a component lead through it. It is used to make a connection from a track on one layer to a track on another layer.

References

ASM. 1989. *Electronic Materials Handbook*, vol. 1, Packaging, Sec. 5, Printed Wiring Boards, pp. 505–629 ASM International, Materials Park, OH.

Beckert, B.A. 1993. Hot analysis tools for PCB design. *Comp. Aided Eng.* 12(1):44–49.

Byers, T.J. 1991. *Printed Circuit Board Design With Microcomputers*. Intertext, New York.

Evans, R. 1994. Effects of losses on signals in PWBs. *IEEE Trans. Comp. Pac., Man. Tech. Pt. B: Adv. Pac.* 17(2):2 17–222.

Magnuson, P.C, Alexander, G.C., and Tripathi, V.K. 1992. *Transmission Lines and Wave Propagation*, 3rd ed. CRC Press, Boca Raton, FL.

Maliniak, L. 1995. Signal analysis: A must for PCB design success. *Elec. Design* 43(19):69–82.

Mikazuki, T and Matsui, N. 1994. Statistical design techniques for high speed circuit boards with correlated structure distributions. *IEEE Trans. Comp. Pac., Man. Tech.* 17(1):159–165.

Montrose, M.I. 1995. *Printed Circuit Board Design Techniques for EMC Compliance*. IEEE Press, New York.

Seraphim, D.P., Barr, D.E., Chen, W.T., Schmitt, G.P., and Tummala, R.R. 1989. Printed-circuit board packaging. In *Microelectronics Packaging Handbook*, eds. Rao R. Tummala and Eugene J. Rymaszewski, pp. 853–921. Van Nostrand-Reinhold, New York.

Simovich, S., Mehrotra, S., Franzon, P., and Steer, M. 1994. Delay and reflection noise macromodeling for signal integrity management of PCBs and MCMs. *IEEE Trans. Comp. Pac., Man. Tech. Pt. B: Adv. Pac.* 17(1):15–20.

Further Information

Electronic Packaging and Production journal.

IEEE Transactions on Components, Packaging, and Manufacturing Technology—Part A.

IEEE Transaction on Components, Packaging, and Manufacturing Technology—Part B: Advanced Packaging.

Harper, C.A. 1991. *Electronic Packaging and Interconnection Handbook.*

PCB Design Conference, Miller Freeman Inc., 600 Harrison Street, San Francisco, CA 94107, (415) 905-4994. *Printed Circuit Design* journal.

Proceedings of the International Symposium on Electromagnetic Compatibility, sponsored by IEEE Transactions on EMC.

5.3 Basic Circuit Board Design: Overview and Guidelines

The beginning of the design process is the schematic capture process. Using the parts available in the libraries of the electronic computer aided design (ECAD) software, and possibly creating part definitions that are not in the libraries, the circuit is created. Large ECAD packages are hierarchical; that is, the user can, e.g., create a top-level diagram in which the major circuit blocks are indicated by labeled boxes. Each box can then be opened up to reveal any sub-boxes, eventually leading to the lowest/part-level boxes. A "rat's nest" is then created, which consists of directly connecting component terminations with other terminations (pin 6 on IC8 to R1, pin 22 on IC 17 to pin 6 on IC3, etc.), which tells the software where connections are needed.

At this point, the schematic capture program is capable of doing some error checking, e.g., whether the user has connected an output pin to ground, or two input pins together with no other input. Some software will also simulate (see Section 5.8, "Simulation") digital circuits if the user provides test vectors that describes machine states.

After entering the complete schematic into the ECAD program, the program will create a *netlist* of all circuit connections. The netlist is a listing of each set of interconnects on the schematic. If pin 6 of IC 19 fans out to 4 inputs, that net indicates all pins to which it is connected. The user can name each net in the list.

Following schematic capture and simulation, the software can rout the circuit board. Routing consists of consulting the rat's nest the user has defined, and from that a physical circuit board layout is created that will allow all signals to be routed appropriately. The user must define the board outline. Some systems will then automatically provide a component layout, but in most cases I/O connect, DFM, and DFT issues will dictate component placement. The software must then consider the component layout and rout the traces. If the circuitry includes special considerations such as RF or power circuits, the software may be able to accommodate that, or the user may have to rout those circuits manually. While many digital circuits can be routed automatically, it is most common for analog circuits to be routed manually. As the routing is being done, the software will perform design rule checking (DRC), verifying issues like trace and component spacing. If the user doesn't modify the design rules to reflect the actual manufacturing practices, the software may mark a number of DRC violations with *X*s, using its default rules. In all ECAD packages, the designer can modify the design rules through simple software commands.

In Fig. 5.7, the ECAD software has placed an *X* at each location where the default design rules have been violated. When the software did design rule checking, it determined that the component pads and traces were closer together than allowed by the rules. This is of great help to the board designer in making sure that clearances set by the manufacturing and design groups are not violated.

The user will frequently need to manually change some of the autorouting work. If rules were set up by any of the design groups but not entered into the software, it may violate some of those rules that the manufacturing and/or test group expected to have followed. For example, SMT tantalum capacitors with large bodies and relatively small pads may get their pads placed too close to each other. After routing, it is necessary to double-check the resulting layout for exact equivalence to the original schematic. There may have been manual changes that resulted in the loss of a needed connection.

Circuit Boards

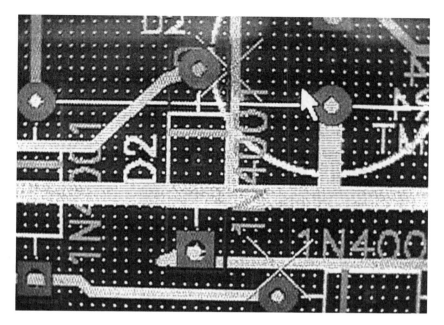

FIGURE 5.7 *X* marks the spot of DRC violations.

From this point, a number of files will be created for circuit board fabrication. The standard format for circuit board files has been in a *Gerber* format. The intent of the Gerber format was to create an interchange format for the industry, but it is actually a rather loose format that often results in non-readable files in systems other than the one in which it was created. To solve this problem, the industry has recently agreed upon a new interchange standard, GenCAM.

The GenCAM (generic computer aided manufacturing; GenCAM is a service mark of the IPC) standard is being developed under the guidance of the IPC and is intended to be integrated software that will have all necessary functional descriptions for printed circuit boards and printed circuit assemblies in a single file format. This will replace the need to have a collection of file formats, e.g.,

- Gerber files for layout information
- Excellon files for drilling
- IPC 356 net list files

GenCAM is in its early phases. As of this writing, it is in beta testing. Available is a conformance test module that allows developers to validate that their designs pass "certain tests of correctness." The most up-to-date information on GenCAM is available at http://www.gencam.org.

5.4 Prototyping

Any new design should be built as a prototype. The design may be for investigative purposes, or it may be too expensive to run it through the assembly line for single-digit quantities. If the assembly line is an option, the CAD design should be implemented. There are many circuit board vendors (search the World Wide Web for "circuit board fabrication") to whom one can e-mail the CAD files and receive a plated-through hole/via board in return in less than one week, for less than $1,000.

Most designs today do not lend themselves well to breadboarding. The additional parasitic inductance and capacitance that results from the use of a breadboard is not acceptable in analog circuits above 1 MHz, nor in high-speed digital circuits. If direct assembly is not an option, one of the most common prototyping techniques for analog circuits uses a copper-clad board as a ground plane. The ground pins of the components are soldered directly to the ground plane, with other components soldered together in the air above the board. This prototyping technique is known variously as the *dead-bug* technique (from the appearance of ICs mounted upside down with their leads in the air) or the *bird's nest* technique (Fig. 5.8). This technique has the favor of Robert Pease (see references) of National Semiconductor, a world-renowned guru of analog design. The ground pins of the ICs are bent over and soldered directly to the ground plane.

A variation on this technique includes using copper-clad perfboard with holes at 0.1-in centers. Another variation uses the *perfboard* but mounts the components on the non-clad side of the board. The holes are then used to mount the components, and point-to-point wiring is done on the copper-clad side of the board. At each hole to be used as a through hole, the copper must be removed from the area around the hole to prevent shorting the leads out.

When prototyping, whether using a fabricated board or a dead-bug technique, care should be taken in the use of IC sockets. Sockets include enough parasitic capacitance and inductance to degrade the performance of fast analog or digital circuits. If sockets must be used, the least effect is caused by sockets that use individually machined pins. Sometimes called *cage jacks*, these will have the lowest C and L.

Analog Devices' "Practical Analog Design Techniques" offers the following suggestions for analog prototyping:

- Always use a ground plane for precision or high frequency circuits.
- Minimize parasitic resistance, capacitance, and inductance.
- If sockets are required, use "pin sockets."
- Pay equal attention to signal routing, component placement, grounding, and decoupling in both the prototype and the final board design.

The guide also notes that prototype boards can be made using variations of a CNC mill. Double-sided copper-plated blank boards can be milled with traces using the same Gerber-type file used for the production fabrication of the boards. These boards will not have through-plated holes, but wires can be added and soldered at each hole location to create an electrical connection between the top and bottom layers. These boards will also have a great deal of copper remaining after the milling process, which can be used as a ground plane.

5.5 DFM and DFT Issues

The circuit board design must consider a myriad of issues, including but not limited to:

- Types of circuitry, such as analog, digital, RF, power, etc.
- Types of components, such as through hole, standard surface mount technology (SMT), BGAs, die-size components, power components, etc.
- Expected operating speeds and edge timing

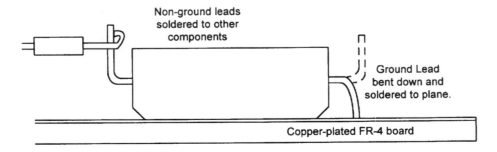

FIGURE 5.8 Example of a *dead-bug* prototype.

Circuit Boards

- Surface finish requirements
- Placement techniques to be used
- Inspection methods to be used
- Test methods to be used

The design of any board must consider these issues within the context of design for manufacturability (DFM) and design for testability (DFT) constraints. The DFM constraints that are defined by the component placement issues constitute one of the largest sets of considerations. The SMT placement machine is frequently the most expensive piece of equipment on the assembly line. The speed with which it can work is a function not only of the sizes of the components but also of the board layout and whether that layout can be optimized for production.

The first and most basic consideration is to keep the board one-sided. In many modern designs, this is not possible, but for those designs in which it is possible, the assembly and test will be considerably easier and faster. Part of the simplification, of course, comes from not having to flip the board over for a second set of placement operations, and the possible requirement for a two-sided test fixture. The second basic consideration is the ability to panellize the design. When a four-up panel is designed, solder paste stencil operations drop by four in number, since one pass of the squeegee over the stencil will apply paste to four boards, and placement is faster since fiducial corrections and nozzle changes will occur only one-fourth as often as with the same layout in a one-up configuration (Fig. 5.9). Board transport time is also lessened.

With panellizing comes the need for singulation/depanelling, an extra process step that must be considered in the overall DFM criteria. The board fabricator must also include handling and depanelling clearance areas. Table 5.4 is an example of the time differences between assembling a one-up versus a four-up board of identical panel design. Each reader must calculate the difference in assembly/time costs for a particular facility.

DFT issues are concisely discussed in the Surface Mount Technology Association's *TP-101 Guidelines*. This publication specifies the best practices for the location, shape, size, and position tolerance for test pads. These issues are discussed further in Chapter 9, "Design for Test." The primary concerns at the board layout are that accurate tooling holes and appropriate test pad size and location are all critical to increasing the probability that the spring-loaded probes in a bed-of-nails test fixture will reliably make contact and transfer signals to and from the board under test. For panellized boards, tooling holes and fiducials must be present both on the main panel and on individual boards. If local fiducials are required for fine or ultra-fine pitch devices, they must, of course, be present on each board. To summarize TP-101's requirements:

- Test pads should be provided for all electrical circuit nodes.
- Tolerance from tooling hole to any test pad should be 0.002 in or less.
- There should be two 0.125-in unplated tooling holes per board/panel.

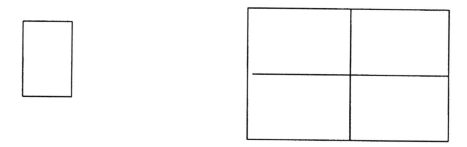

FIGURE 5.9 One-up vs. four-up circuit board layouts.

TABLE 5.4 Time Differences between Assembling a One-Up versus a Four-Up Board of Identical Panel Design

Machine times	
Number of placements per panel	100
Fiducial correction time	6 sec
Placement speed, components/hour	14,400 (4/sec)
Nozzle changes for the component mix	15 sec
Number of boards to be run	2,000
I/O board handling time	10 sec
Depanelling time/panel	45 sec
For one-up panels	
I/O board handling	10 sec
Fiducial correction	6 sec
Placement time	25 sec
Nozzle changes	15 sec
(100 placements ÷ 4/sec)	
Total time for one board	56 sec/board
Total time for four boards	224 sec
For four-up panels	
I/O board handling	10 sec
Fiducial correction	6 sec
Placement time	100 sec
Nozzle changes	15 sec
Depanelling time	45 sec
(400 placements ÷ 4/sec)	
Total time for four panels	176 sec
Total time for 2000 boards with one-up panel	
56 sec/board × 2000 = 112,000 sec = 31 hr 7 min	
Total time for 2000 boards with four-up panels (500 panels)	
176 sec/4 panels × 500 panels = 88,000 sec = 24 hr 27 min	

- The tolerance of the tooling hole diameter should be +0.003/–0.000 in.
- Test pad diameter should be at least 0.035 in for small probes and 0.040 in for large probes.
- Test pads should be at least 0.2 in away from components that are 0.2 in high or taller.
- Test pads should be at least 0.125 in away from the board edges.
- There should be a component-free zone around each test pad of at least 0.018 in.
- Test pads should be solder tinned.
- Test pads cannot be covered by solder resist or mask.
- Test pads on 0.100-in grid spacing are probed more reliably than pads on 0.75- or 0.050-in centers.
- Test pads should be distributed evenly over the surface of the board to prevent deformation of the board in the test fixture.

Other mechanical design rules include the need for fiducials and the location of any mounting holes beyond those that will be used for tooling holes. Fiducials are vision targets for the assembly machines, and two types may be needed—*global* fiducials and *local* fiducials.

Global fiducials are used by vision systems to accurately locate the board in the assembly machines. The vision system will identify the fiducials then calculate whether they are in the exact location described by the CAD data for the board. If they are not in that exact location, either because the board is not mechanically held by the machine in the exactly correct location or because the board artwork and fabrication have resulted in etching that does not exactly meet the CAD data, the machine system software will correct (offset) the machine operations by the amount of board offset. In this way, e.g., the placement machine will place a component in the best location on the board it is holding, even if the artwork and/or etch process created some "shrink" or "stretch" in the exact component pad locations.

Because large ICs with many leads, and fine-pitch ICs with small spaces between their leads, need to be placed very accurately, local fiducials will be placed on the board in the immediate vicinity of these parts. The placement machine will identify these fiducials (and any offset from the CAD data) during the placement operations and will correct for any offset, placing the part in the best location (Fig. 5.10).

As noted, fiducial shapes are intended to be identified by automated vision systems. Although circular fiducials are the most common, since different systems may be optimized for different shapes, it is best to consult the manufacturing engineers to determine the optimal shape for the fiducials.

5.6 Board Materials

For FR-4 boards, one major problem is presented by the through holes used for both THT leads and board layer connections. This can be easily seen by comparing the difference in the coefficients of thermal expansion (CTE) for the three axes of the FR-4 material, and the CTE of the copper-plated through hole:

X-axis CTE of FR-4	12–18 ppm/°C
Y-axis CTE of FR-4	12–18 ppm/°C
Z-axis CTE of FR-4	100–200 ppm/°C
CTE of copper	10–15 ppm/°C

This comparison clearly shows that, if the board is heated by either environmental temperature excursions or by heat generated by on-board components, the z-axis (thickness) of the FR-4 will expand at roughly six times the expansion rate of the copper barrel. Repeated thermal excursions and the resultant barrel cracking have been shown to be among the major failure modes of FR-4 printed circuit boards. Other rigid circuit board materials do not necessarily fare any better. Kevlar is often touted as a material that has a higher glass transition temperature than fiberglass and will therefore withstand heat better. While this is true, Kevlar has an even higher Z-axis CTE than FR-4! The problem is not solved. It should be noted that, in the manufacturing process, barrel cracking is less of a problem if the barrels are filled with solder.

5.6.1 Board Warpage

Circuit board warpage is a fact of life that must be minimized for successful implementation of newer part packages. Whereas THT devices were very tolerant in their ability to successful solder in the presence of board warp, surface mount devices, and especially newer large-area devices such as BGAs, are extremely non-tolerant of board warp. While JEDEC specifications call for warpage of 0.008 in (0.2 mm) or less, recent writings on BGAs indicate that substrate warpage of 0.004 in (0.1 mm) or less is required to

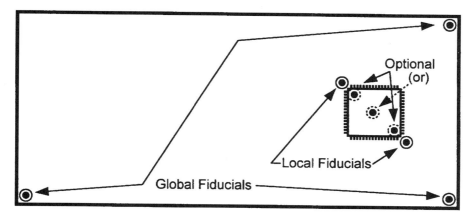

FIGURE 5.10 Examples of fiducial locations.

successfully solder plastic BGA packages. This is required because the BGA itself can warp, and with this type of area array package, the total allowed warp between the two cannot exceed 0.008 in.

5.6.2 Board Surface Finishes

After circuit board fabrication, the copper must be protected in some fashion to maintain its solderability. Bare copper left exposed to air will rapidly oxidize, and solder failures will become unacceptable after only a day's exposure. Tin-lead coating will protect the copper and preserve solderability. The traditional board surface finish is hot-air surface leveling (HASL). In the HASL finish process, a board finishing the solder plating process, commonly dipped in molten solder, is subjected to a blast from a hot air "knife" to "level" the pads. However, the result of HASL is anything but a leveled/even set of pad surfaces. As can be seen in Fig. 5.11, the HASL deposit does not result in a flat pad for placement. With today's component technologies, HASL is not the optimum pad surface finish.

One study showed the following results with HASL deposits, which give an indication of the problems with deposit consistency:

Pad type	Min. Solder Thickness (μin)	Max. Solder Thickness (μin)	Avg. Solder Thickness (μin)
THT	33	900	135
PLCC (50-mil pitch)	68	866	228
QFP (25-mil pitch)	151	600	284

Electroplating is another pad surface treatment and is typically followed by fusing the solder to provide a thin, even, coating that is suitable for fine-pitch placement. Electroplating does, however, leave a porous surface, and the following fusing process is necessary, since a porous surface is not desirable.

Alternative metallic (such as electroplated nickel/gold and electroless nickel/immersion gold) and organic surface (OSP) finishes can provide better planarity than HASL. This is particularly important with smaller pitch and high-lead-count components.

An organic solderability preservative (OSP) coating of bare copper pads is a non-plating option that coats the copper with an organic coating that retards oxide formation. In Peterson's 1996 study, he found the following advantages to OSP coating compared to HASL:

- A reduced cost of 3 to 10% in bare board cost was realized.
- The HASL process could not produce zero defects.
- Solder skips occurred with both coverings but are "much more" noticeable with OSP due to the color difference between the unplated pad and the tinned lead.
- Vision and X-ray systems should operate more efficiently with OSP coatings.
- OSP has a distinct advantage in not contributing to coplanarity problems in fine-pitch assembly.

In addition to provide a planar surface, the OSP coating process uses temperatures below 50° C, so there is no thermal shock to the board structure and therefore no opportunity to cause board warp. The major concern with OSPs is that they will degrade with time and with each heat cycle to which the board is subjected. Designers and manufacturing engineers must plan the process cycle to minimize the number of reflow and/or curing cycles. The OSP manufacturers can provide heat degradation information.

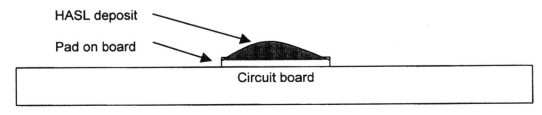

FIGURE 5.11 HASL deposit.

Circuit Boards

Build-up technologies (BUT) is an additive process to create high-performance multilayer printed wiring boards. It is more expensive and slower than the subtractive/etch process, but the final results are more accurate.

Microvias can be produced with plasma, laser, and photo-dielectric techniques, as well as with chemical and mechanical techniques. Microvias can be as small as 1 mil and are produced by laser etch. They can be placed in component pads due to their small size and lessened wicking ability.

5.7 Circuit Design and the Board Layout

5.7.1 Grounds

Grounds should be run separately for each section of the circuitry on a board and brought together with the power ground at only one point on the board. On multilayer boards, the ground is frequently run in its own plane. This plane should be interrupted no more than necessary, and with nothing larger than a via or through hole. An intact ground plane will act as a shield to separate the circuitry on the opposite sides of the board, and it will also not act as an antenna—unlike a trace, which will act as an antenna. On single-sided analog boards, the ground plane may be run on the component side. It will act as a shield from the components to the trace/solder side of the board.

A brief illustration of the issue of inter-plane capacitance and crosstalk is presented in Fig. 5.12, which shows the two common copper layout styles in circuit board design (cutaway views).

To define a capacitor C (two conductors separated by an insulator), where copper is the conductor, air is an insulator, and FR-4 is an insulator,

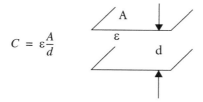

where C = value of the capacitor (capacitance)
         ε = dielectric constant

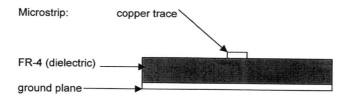

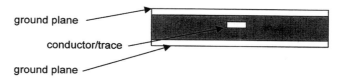

FIGURE 5.12 Copper layout techniques.

A = area of the "plates"
d = distance between the plates

For a circuit board, the area A of the plates can be approximated as the area of the conductor/trace and a corresponding area of the ground plane. The distance d between the plates would then be a measurement normal to the plane of the trace.

The impedance of a capacitor to the flow of current is calculated as

$$X_c = \frac{1}{2\pi f C}$$

where X_c = capacitive reactance (impedance)
   f = frequency of operation
   C = value of the capacitor or the capacitance of the area of the circuit board trace

It can be seen that as frequencies become higher, the impedances become lower, allowing current to flow between the plates of the capacitor.

At high frequencies, SMT, and especially fine pitch technology (FPT, lead pitch of 25 mils or less), demand copper traces that are ever closer together, diminishing d in the capacitor equation. Diminishing d leads to a larger value of C. A larger value of C leads to a smaller value of X_C, allowing more *crosstalk* between traces.

5.7.2 Board Partitioning

As mentioned, the grounds for functionally different circuits should be separated on the board. This leads directly to the concept that the circuits themselves should be partitioned. A board layout should attempt to keep the various types of circuits separate (Fig. 5.13). The power bus traces are run on the opposite side of the board so that they do not interrupt the ground plane, which is on the component side of the board. If it is necessary to run the power bus on the component side of the board, it is best to run it along the periphery of the board so that it creates minimal disturbance to the ground plane.

Analog Design

The analog design world relies less on CAD systems' autorouters than does digital design. As generally described in the "Simulation" section of this chapter, analog design also follows the stages of schematic capture, simulation, component layout, critical signal routing, non-critical signal routing, power routing, and a "copper pour" as needed. The copper pour is a CAD technique to fill an area of the board with copper, rather than having that area etched to bare FR-4. This is typically done to create a portion of a ground plane.

Part of this process involves both electrical and mechanical design rules being incorporated into the overall board design. The electrical design rules come from the electrical designer and will enable the

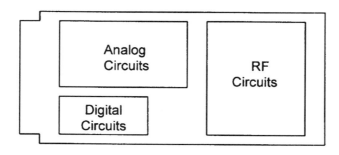

FIGURE 5.13 Partitioning of a circuit board.

Circuit Boards

best-performing trace and plane layout to be accomplished. As mentioned earlier, the mechanical design rules will incorporate issues such as:

- Tooling hole locations
- Fiducial locations
- No acute angles in the layout

The first rule of design in the analog portion of the board is to have a ground that is separate from any other grounds such as digital, RF, or power. This is particularly important since analog amplifiers will amplify *any* signal presented to them, whether it is a legitimate signal or noise, and whether it is present on a signal input or on the power line. Any digital switching noise picked up by the analog circuitry through the ground will be amplified in this section.

At audio frequencies (20 to 20 kHz), there are very few board layout issues, except in the power traces. Since every trace has some inductance and some capacitance, high-current traces may contribute to crosstalk. Crosstalk occurs primarily across capacitively coupled traces and is the unwanted appearance on one trace of signals and/or noise from an adjacent trace.

As frequencies exceed 100 kHz, coupling occurs more easily. To minimize unwanted coupling, traces should be as short and wide as possible. This will minimize intertrace capacitance and coupling. Adequate IC bypassing with 0.1 µF or 0.01 µF capacitors from each power pin to ground, is necessary to prevent noise from being coupled, regardless of the circuit style used. The bypass capacitors should be physically placed within 0.25 in of the power pin.

Another issue with analog circuit layout is the feedback paths that can occur if input lines and output lines are routed close to each other. While it is important to keep components as close together as possible, and thereby keep signal lengths as short as possible, feedback will occur if inputs and outputs are close.

When creating the analog component layout, it is important to lay out the components in the order they will be in the signal path. In this fashion trace lengths will be minimized. The ground plane rules covered in that section of this chapter should be followed in analog circuit design.

Generally, all analog signal traces should be routed first, followed by the routing of the power supply traces. Power supply traces should be routed along the edge of the analog section of the board.

Digital Design

The first rule of design in the digital portion of the board is to have a ground that is separate from any other grounds such as analog, RF, or power. This is to primarily protect other circuits from the switching noise created each time a digital gate shifts state. The faster the gate, and the lower its the transition time, the higher the frequency components of its spectrum.

Like the analog circuits, digital circuits should either have a ground plane on the component side of the board or a separate "digital" ground that connects only to the other grounds at one point on the board.

At some particular speed, the circuit board traces need to be considered as transmission lines. The normal rule of thumb is that this occurs when the two-way delay of the line is more than the rise time of the pulse. The "delay" of the line is really the propagation time of the signal. Since we assume that electricity travels down a copper wire or trace at the speed of light, we have a beginning set of information to work with. Ala Einstein, $c = 3 \times 10^8$ m/sec, or 186,280 miles/sec, in a vacuum. This translates to approximately 1.017 ns/ft or 85 ps/in (and 3.33 ns/m). In any other medium, the speed is related to the speed in vacuum by:

$$S_m = 1.017 \sqrt{\varepsilon_r}$$

where S_m = speed of transmission in a medium other than space
       ε_r = relative dielectric coefficient for the medium

For FR-4 ε_r is approximately equal to 4. This means the speed of transmission is approximately 1.017 ns/ft, or ≈2 ns/ft, which inverted leads to the common transmission rate figure of 6 in/ns.

From these two approximations, and remembering that the concern is the two-way (i.e., with reflection) transmission time, it can be seen that a trace must be considered as a transmission line if a signal with a 1 ns rise time is being sent down a trace that is 3 in long or longer (round trip = 6 in or longer). This would be considered the critical length for a 1 ns rise time pulse. On the other hand, a pulse with a 0.1 µs rise time can be sent over a 300-in trace before it is reflected like a transmission line.

If the trace is shorter than the critical length, the reflection occurs while the output gate is still driving the line positively or negatively, and the reflection has virtually no effect on the line. During this transition time, other gates' inputs are not expecting a stable signal and will ignore any reflection noise. If, however, the trace is longer than the critical length, the reflection will occur after the gate output has stabilized, leading to noise on the line. These concerns are discussed further in Chapter 6.

Blankenhorn notes symptoms that can indicate when a design has crossed the boundary into the world of high-speed design:

- A product works only when cards are plugged into certain slots.
- A product works with one manufacturer's component but not with another manufacturer's component with the same part number (this may indicate other problems besides or in addition to high-speed issues).
- One batch of boards works, and the next batch does not, indicating dependency on microscopic aspects of board fabrication.
- With one type of package (e.g., a QFP), the board works, but with a different package (e.g., a PLCC), the board does not.
- The design was completed without any knowledge of high-speed design rules and, in spite of the fact that the simulation works, the board will not.

RF Design

RF circuits cover a broad range of frequencies, from 0.5 kHz to >2 GHz. This handbook will not attempt to cover the specifics of RF design, due to the complexity of these designs. But see Chapter 6 for EMI/RFI issues in PCB design. Generally, RF designs are in shielded enclosures. The nature of RF means that it can be transmitted and received, even when the circuit is intended to perform all of its activities on the circuit board. This is the reason for the FCC Type 15 disclaimer on all personal computers.

As in the case of light, the frequencies of RF circuits lead to very short bandwidths, which means that RF can "leak" through the smallest openings. This is the reason for the shielded enclosures, and the reason that RF I/O uses through pass-through filters rather than direct connections.

All RF layouts have the transmission line concerns noted in Section 5.8. Additionally, controlling both trace and termination impedance is very important. This is done by controlling the trace width and length, the distance between the trace and the ground and power planes, and the characteristics of the components at the sending and receiving ends of the line.

5.8 Simulation

A traditional board-level digital design starts with the creation of a schematic at the level of components and interconnects. The schematic is passed to the board designer in the form of a netlist. The design is laid out, prototyped, and then verified. Any errors found in the prototype are corrected by hand wiring. If management decided not to undergo the expense of redesigning the board, every production board must be corrected in the same fashion.

Simulation is intended to precede the prototype stage and find problems before they are committed to hardware. In this scenario, the schematic design/netlist is passed to the simulator, along with definitions of power supplies and required stimulus. The simulator has libraries of parts that ideally will include all parts used in the schematic. If not, the design team must create the part specifications. Depending on whether the part is analog or digital, the part models describe the maximum and minimum part

Circuit Boards

performance specifications for criteria such as amplification, frequency response, slew rate, logical function, device timing constraints such as setup and hold times, and line and load regulation.

At the beginning of simulation, the simulator first "elaborates" the design. In this function, the simulator combines the design's topological information from the netlist with the library's functional information about each device. In this fashion, the simulator builds up a circuit layout along with timing and stimulus information. The results of the simulation are then presented in both graphical and textual form.

Simulators will allow the designer to determine whether "typical" part specs are used, or whether the simulation will use maximum or minimum specs. This can be useful in determining, e.g., whether a design will perform correctly in the presence of low battery voltage. One of the most difficult things for the simulator to determine is timing delays and other effects of trace parasitic inductance and capacitance. This will be better handled after the board layout is complete. Simulators can also perform in a min:max mode for digital timing, returning dynamic timing results for circuits built with off-the-shelf parts whose timing specs may be anywhere in the spec range.

Fault simulation is another area primarily aimed at digital circuits. The purpose of fault simulation is to determine all the possible faults that could occur, such as a solder short between a signal trace and ground. The fault simulator will apply each fault to the board, then analyze whether the fault can be found by ATE equipment at the board's outputs. It will assist in locating test pads at which faults can be found. It will also report faults that cannot be found. Fault simulators can be used with automatic test pattern generators (ATPGs, discussed further in Chapter 9) to create test stimuli.

At the high end of consumer products' simulation are the high-speed PCs. Clock speeds now exceed 450 MHz, and bus speeds are at 100 MHz as of this writing, with the new Rambus scheduled to operate at frequencies up to 400 MHz with edge rates in the range of 200 to 400 ps. To support these speeds, an industry-wide effort created the I/O Buffer Information Specification, or IBIS. IBIS simulation models can now be purchased for the newest PC chips. The IBIS standard also includes transmission line descriptions so that subsystem suppliers could describe their systems for signal integrity simulation. This part of IBIS is termed the *electrical board description,* or EBD. This standard provides a way of describing the transmission lines from components to/from interconnects.

5.9 Standards Related to Circuit Board Design and Fabrication

5.9.1 Institute for Interconnecting and Packaging Electronic Circuits (IPC)[*]

IPC-A-22	UL Recognition Test Pattern
IPC-T-50	Terms and Definitions for Interconnecting and Packaging Electronic Circuits
IPC-L-108	Specification for Thin Metal-Clad Base Materials for Multilayer Printed Boards
IPC-L-109	Specification for Resin Preimpregnated Fabric (Prepreg) for Multilayer Printed Boards
IPC-L-115	Specification for Rigid Metal-Clad Base Materials for Printed Boards
IPC-MF-150	Metal Foil for Printed Wiring Applications
IPC-CF-152	Composite Metallic Materials Specification for Printed Wiring Boards
IPC-FC-232	Adhesive Coated Dielectric Films for Use as Cover Sheets for Flexible Printed Wiring
IPC-D-300	Printed Board Dimensions and Tolerances
IPC-D-310	Guidelines for Phototool Generation and Measurement Techniques
IPC-D-317	Design Guidelines for Electronic Packaging Utilizing High-speed Techniques
IPC-D-322	Guidelines for Selecting Printed Wiring Board Sizes Using Standard Panel Sizes
IPC-MC-324	Performance Specification for Metal Core Boards

[*]Institute for Interconnecting and Packaging Electronic Circuits, 2215 Sanders Rd., Northbook, IL 60062-6135.

IPC-D-325	Documentation Requirements for Printed Boards	
IPC-D-330	Design Guide	
IPC-PD-335	Electronic Packaging Handbook	
IPC-D-350	Printed Board Description in Digital Form	
IPC-D-351	Printed Board Drawings in Digital Form	
IPC-D-352	Electronic Design Data Description for Printed Boards in Digital Form	
IPC-D-354	Library Format Description for Printed Boards in Digital Form	
IPC-D-356	Bare Board Electrical Test Information in Digital Form	
IPC-AM-361	Specification for Rigid Substrates for Additive Process Printed Boards	
IPC-AM-372	Electroless Copper Film for Additive Printed Boards	
IPC-D-422	Design Guide for Press-Fit Rigid Printed Board Backplanes	
IPC-A-600	Acceptability of Printed Boards	
IPC-TM-650	Test Methods Manual	
Method 2.1.1	Microsectioning	
Method 2.1.6	Thickness, Glass Fabric	
Method 2.6.3	Moisture and Insulation Resistance, Rigid, Rigid/Flex and Flex Printed Wiring Boards	
IPC-ET-652	Guidelines for Electrical Testing of Printed Wiring Boards	
IPC-CM-770	Printed Board Component Mounting	
IPC-SM-780	Component Packaging and Interconnecting with Emphasis on Surface Mounting	
IPC-SM-782	Surface Mount Land Patterns (Configurations and Design Rules)	
IPC-SM-785	Guidelines for Accelerated Reliability Testing of Surface Mount Solder Attachments	
IPC-S-804	Solderability Test Methods for Printed Wiring Boards	
IPC-S-815	General Requirements for Soldering Electronic Interconnections	
IPC-CC-830	Qualification and Performance of Electrical Insulating Compound for Printed Board Assemblies	
IPC-SM-840	Qualification and Performance of Permanent Polymer Coating (Solder Mask) for Printed Boards	
IPC-100001	Universal Drilling and Profile Master Drawing (Qualification Board Series #1)	
IPC-100002	Universal Drilling and Profile Master Drawing (Qualification Board Series #2)	
IPC-100042	Master Drawing for Double-Sided Printed Boards (Qualification Board Series #1)	
IPC-100043	Master Drawing for 10 Layer Multilayer Printed Boards (Qualification Board Series #1)	
IPC-100044	Master Drawing for 4 Layer Multilayer Printed Boards (Qualification Board Series #1)	
IPC-100046	Master Drawing for Double-Sided Printed Boards (Qualification Board Series #2)	
IPC-100047	Master Drawing for 10 Layer Multilayer Printed Boards (Qualification Board Series #2)	

5.9.2 Department of Defense

Military[*]

DOD-STD-100	Engineering Drawing Practices
MIL-STD-1686	Electrostatic Discharge Control Program for Protection of Electrical and Electronic Parts, Assemblies, and Equipment (Excluding Electrically-Initiated Explosive Devices)
MIL-STD-2000	Soldering Requirements for Soldered Electrical and Electronic Assemblies
MIL-D-8510	Drawings, Undimensioned, Reproducibles, Photographic and Contact, Preparation of

---

[*]Standardization Documents Order Desk, Building 4D, 700 Robbins Ave., Philadelphia, PA 19111-5094.

MIL-P-13949	Plastic Sheet, Laminated, Copper-Clad (For Printed Wiring)
MIL-G-45204	Gold Plating (Electrodeposited)
MIL-I-46058	Insulating Compound, Electrical (for Coating Printed Circuit Assemblies)
MIL-P-81728	Plating, Tin-Lead (Electrodeposited)

Federal[*]

QQ-A-250	Aluminum and Aluminum Alloy Plate Sheet
QQ-N-290	Nickel Plating (Electrodeposited)
L-F-340	Film, Diazo Type, Sensitize D, Moist and Dry Process Roll and Sheet
QQ-S-635	Steel

Other Documents

American Society for Testing and Materials (ASTM)[†]
  ASTM B-152   Copper Sheet, Strip, Plate, and Rolled Bar

Electronic Industries Alliance (EIA)[‡]
  JEDEC Publ. 95   Registered and Standard Outlines for Solid State Products

Underwriters Laboratories (UL)[§]
  UL 746E   Standard Polymeric Materials, Materials used in Printed Wiring Boards

American National Standards Institute (ANSI)[**]
  ANSI-Y14.5   Dimensioning and Tolerancing

References

Practical Analog Design Techniques. Analog Devices, Norwood, MA, 1995.
Blankenhorn JC, "High-Speed design." *Printed Circuit Design,* vol. 10 no. 6, June 1993.
Brooks D, "Brookspeak." A sometimes-monthly column in *Printed Circuit Design,* Brooks' writings cover many circuit-related aspects of circuit board design, such as noise, coupling capacitors, and crossover.
Edlund G, "IBIS model accuracy." *Printed Circuit Design,* vol. 15 no. 5, May 1998.
Maxfield C, "Introduction to digital simulation." *Printed Circuit Design,* vol. 12 no. 5, May, 1995.
Messina BA, "Timing your PCB Design." *Printed Circuit Design,* vol. 15 no. 5, May 1998.
Pease RP, *Troubleshooting Analog Circuits.* Butterworth-Heinemann, New York, 1991.
Peterson JP, "Bare copper OSP process optimization." *SMT,* vol. 10 no. 7, July 1996.
Wang PKU, "Emerging Trends in Simulation." *Printed Circuit Design,* vol. 11 no. 5, May 1994.

---

[*]Standardization Documents Order Desk, Building 4D, 700 Robbins Ave., Philadelphia, PA 19111-5094.
[†]American Society for Testing and Materials, 1916 Race St., Philadelphia, PA 19103.
[‡]Electronic Industries Alliance (formerly Electronic Industries Association), 2500 Wilson Blvd., Arlington, VA 22201-3834.
[§]Underwriters Laboratories, 333 Pfingsten Ave., Northbrook, IL 60062.
[**]American National Standards Institute, 11 West 42nd St., New York, NY 10036.

EMC and Printed Circuit Board Design

Mark I. Montrose
Montrose Compliance Services, Inc.

6.1 Printed Circuit Board Basics 6.2
6.2 Transmission Lines and Impedance Control 6.31
6.3 Signal Integrity, Routing, and Termination..................... 6.44
6.4 Bypassing and Decoupling 6.61

Information in this chapter is intended for those who design and layout printed circuit boards (PCBs). It presents an overview on the fundamentals of PCB design related to electromagnetic compatibility (EMC). Electromagnetic compatibility and compliance engineers will find the information presented helpful in solving design problems at both the PCB and system level. This chapter may be used as a reference document for any design project.

A minimal amount of mathematical analysis is presented herein. The reference section provides numerous publications containing EMC theory and technical aspects of PCB design that are beyond the scope of this presentation.

Controlling emissions has become a necessity for acceptable performance of an electronic device for both the civilian and military environment. It is more cost-effective to design a product with suppression on the PCB than to "build a better box." Containment measures are not always economically justified and may degrade as the EMC life cycle of the product is extended beyond the original design specification. For example, end users usually remove covers from enclosures for ease of access during repair or upgrade. Sheet metal covers, particularly internal subassembly covers that act as partition shields, in many cases are never replaced. The same is true for blank metal panels or faceplates on the front of a system that contains a chassis or backplane assembly. Consequently, containment measures are compromised. Proper layout of a PCB with suppression techniques incorporated also assists with EMC compliance at the levels of cables and interconnects, whereas box shielding (containment) does not.

Why worry about EMC compliance? After all, isn't speed is the most important design parameter? Legal requirements dictate the maximum permissible interference potential of digital products. These requirements are based on experience in the marketplace, related to emission and immunity complaints. Often, these same techniques will aid in improving signal quality and signal-to-noise ratio performance.

When designing a PCB, techniques that were used several years ago are now less effective for proper signal functionality and compliance. Components have become faster and more complex. Use of custom gate array logic, application specific integrated circuits (ASICs), ball grid arrays (BGAs), multichip modules (MCMs), flip chip technology, and digital devices operating in the subnanosecond range present new and challenging opportunities for EMC engineers. The same challenge exists for I/O interconnects, mixed logic families, different voltage levels, analog and digital components, and packaging requirements. The design and layout of a printed circuit board for EMI suppression at the source must always be optimized while maintaining system wide functionality. This is a job for both the electrical design engineer and PCB designer.

To design a PCB, use of simulation software is becoming mandatory during the development cycle. Simulation software will not be discussed herein, as the requirement for performance, features, and integration between platforms and vendors frequently change.

In an effort to keep costs down, design for manufacturing (DFM) and design for testing (DFT) concerns must be addressed. For very sophisticated PCBs, DFM and test points may have to give way to functional requirements. The PCB designer must be aware of all facets of PCB layout during the design stage, besides placing components and routing traces to prevent serious functionality concerns from developing. A PCB that fails EMC tests will force a relayout of the board.

The subject of EMC and PCBs is very complex. Material for this chapter was extracted from two books published by IEEE Press, written by this author: *Printed Circuit Board Design Techniques for EMC Compliance*, 1996, and *EMC and the Printed Circuit Board—Design, Theory and Layout Made Simple*, 1999.

6.1 Printed Circuit Board Basics

Developing products that will pass legally required EMC tests is not as difficult as one might expect. Engineers often strive to design elegant products. However, elegance sometimes must give way to product safety, manufacturing, cost, and, of course, regulatory compliance. Such abstract problems can be challenging, particularly if the engineer is unfamiliar with design or manufacturing in other specialized fields of engineering. This chapter examines only EMC related aspects of a PCB and areas of concern during the design cycle. Details on manufacturing, test, and simulation of logic design will not be discussed herein. In addition, concepts and design techniques for the containment of RF energy are also not discussed within this chapter.

Fundamental concepts examined include

1. Hidden RF characteristics of passive components
2. How and why RF energy is created within the PCB
3. Fundamental principals and concepts for suppression of RF energy
4. Stackup layer assignments (PCB construction)
5. Return path for RF current
6. Design concepts for various layout concerns related to RF energy suppression

It is desirable to suppress RF energy internal to the PCB rather than to rely on containment by a metal chassis or conductive plastic enclosure. The use of planes (voltage and/or ground), internal to the PCB assembly, is one important design technique of suppressing common-mode RF energy created within the PCB, as is proper use of capacitors for a specific application.

6.1.1 Hidden RF Characteristic of Passive Components

Traditionally, EMC has been considered "black magic." In reality, EMC can be explained by mathematical concepts. Some of the relevant equations and formulas are complex and beyond the scope of this chapter. Even if mathematical analysis is applied, the equations become too complex for practical applications. Fortunately, simple models can be formulated to describe how EMC compliance can be achieved.

Many variables exist that cause EMI. This is because EMI is often the result of exceptions to normal rules of passive component behavior. A resistor at high frequency acts like a series combination of inductance with the resistor in parallel with a capacitor. A capacitor at high frequency acts like an inductor and resistor in series combination with the capacitor plates. An inductor at high frequencies performs like an inductor with a capacitor in parallel across the two terminals along with some resistance in the leads. The expected behavior of passive components for both high and low frequencies is illustrated in Fig. 6.1.

The reason we see a parasitic capacitance across resistor and inductor leads is that both devices are modeled as a two port component, with both an input and output port. Because we have a two-port device, capacitance is always present between the leads. A capacitor is defined as two parallel plates with

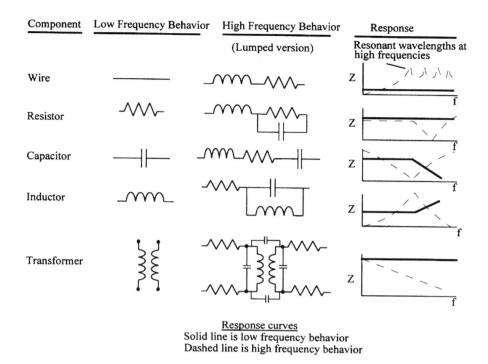

FIGURE 6.1 Component characteristics at RF frequencies.

a dielectric material separating the two plates. For passive components, the dielectric is air. The terminals contain electric charges, the same as if these leads were parallel plates. Thus, for any device, whether it is between the leads of a component, between a component and metal structure (chassis), a PCB and a metal enclosure, two parallel PCB traces or "any" electrical item, relative to another electrical item, parasitic capacitance will be present.

The capacitor at high frequencies does not function as a pure capacitor, because it has changed its functional (operational) characteristics when viewed in the frequency domain (ac characteristics). The capacitor will include lead-length inductance at frequencies above self-resonance. Section 6.4 presents details on capacitor usage. These details include why lead-length inductance is a major concern in today's products. One cannot select a capacitor using dc characteristics and then expect it to be a perfect component when RF energy, which is an ac component, is impressed across the terminals. Similarly, an inductor at high frequencies changes its magnitude of impedance due to parasitic capacitance at high frequencies, which occurs between the two leads and the individual windings.

To be a successful designer, one must recognize the limitations of passive component behavior. Use of proper design techniques to accommodate for these hidden features becomes mandatory, in addition to designing a product to meet a marketing functional specification.

The behavioral characteristics observed with passive components are referred to as the "hidden schematic."[1,2,10] Digital engineers generally assume that components have a single-frequency response in the time domain only, or dc. Consequently, passive component selection for use in the time domain, without regard to the characteristics exhibited in the frequency domain, or ac, will cause significant functional problems to occur, including EMC noncompliance.

To restate the complex problem present, consider the field of EMC as "everything that is not on a schematic or assembly drawing." This statement explains why the field of EMC is considered in the realm of black magic.

Once the hidden behavior of components is understood, it becomes a simple process to design a product that passes EMC and signal integrity requirements without difficulty. Hidden component behavior must take into consideration the switching speed of all active components, along with their unique characteristics, which also have hidden resistive, capacitive, and inductive elements. We now examine each passive device separately.

Wires and PCB Traces

One does not generally consider internal wiring, harnesses, and traces of a product as efficient radiators of RF energy. Every component has lead-length inductance, from the bond wires of the silicon die to the leads of resistors, capacitors, and inductors. Each wire or trace contains hidden parasitic capacitance and inductance. These parasitic components affect wire impedance and are frequency sensitive. Depending on the *LC* value (self-resonant frequency) and the length of the PCB trace, a self-resonance may occur between a component and trace, thus creating an efficient radiating antenna.

At low frequencies, wire is primarily resistive. At higher frequencies, the wire takes on the characteristics of an inductor. This impedance changes the relationship that the wire or trace has with grounding strategies, leading us to the use of ground planes and ground grids. The major difference between a wire and a trace is that wire is round, while a trace is rectangular. The impedance, Z, of wire contains both resistance, R, and inductive reactance, defined by $X_L = 2\pi f L$. Capacitive reactance, $X_c = 1/(2\pi f C)$, is not a part of the high-frequency impedance response of a wire. For dc and low-frequency applications, the wire (or trace) is essentially resistive. At higher frequencies, the wire (or trace) becomes the important part of this impedance equation. This is because we have in the equation the variable, f, or frequency. Above 100 kHz, inductive reactance ($j2\pi L$), becomes greater than resistance value. Therefore, the wire or trace is no longer a low-resistive connection but, rather, an inductor. As a rule of thumb, any wire or trace operating above the audio frequency range is inductive, not resistive, and may be considered an efficient antenna to propagate RF energy.

Most antennas are designed to be efficient radiators at one-fourth or one-half wavelength (λ) of a particular frequency of interest. Within the field of EMC, design recommendations are to design a product that does not allow a wire, or trace, to become an unintentional radiator below $\lambda/20$ of a particular frequency of interest. Inductive and capacitive elements can result in efficiencies through circuit resonance that mechanical dimensions do not describe.

For example, a 10-cm trace has R = 57 mΩ. Assuming 8 nH/cm, we achieve an inductive reactance of 5 mΩ at 100 kHz. For those traces carrying frequencies above 100 kHz, the trace becomes inductive. The resistance becomes negligible and no longer is part of the equation. This 10-cm trace is calculated to be an efficient radiator above 150 MHz ($\lambda/20$ of 100 kHz).

Resistors

Resistors are one of the most commonly used components on a PCB. Resistors also have limitations related to EMI. Depending on the type of material used for the resistor (carbon composition, carbon film, mica, wire-wound, etc.), a limitation exists related to frequency domain requirements. A wire-wound resistor is not suitable for high-frequency applications due to excessive inductance in the winding. Film resistors contain some inductance and are sometimes acceptable for high-frequency applications due to lower lead-length inductance.

A commonly overlooked aspect of resistors deals with package size and parasitic capacitance. Capacitance exists between both terminals of the resistor. This parasitic capacitance can play havoc with extremely high-frequency designs, especially those in the gigahertz range. For most applications, parasitic capacitance between resistor leads is not a major concern, compared to lead-length inductance present.

One major concern for resistors lies in the overvoltage stress condition to which the device may be subjected. If an ESD event is presented to the resistor, interesting results may occur. If the resistor is a surface-mount device, chances are this component will arc over, or self-destruct, as a result of the event. This destruction occurs because the physical distance between the two leads of a surface mount resistor is usually smaller than the physical distance between leads of a through-hole device. Although the wattage of the resistors may be the same, the ESD event arcs between two ends of the resistor. For resistors with radial or axial leads, the ESD event will see a higher resistive and inductive path than that of surface mount because of additional lead length inductance. Thus, ESD energy may be kept from entering the circuit, protected by both the resistor's hidden inductive and capacitive characteristics.

Capacitors

Section 6.4 presents a detailed discussion of capacitors. This section, however, will provide a brief overview on the hidden attributes of capacitors.

Capacitors are generally used for power bus decoupling, bypassing, and bulk applications. An actual capacitor is primarily capacitive up to its self-resonant frequency where $X_L = X_c$. This is described by the formula $X_c = 1/(2\pi f C)$ where X_c is capacitive reactance (unit of ohms), f is frequency in hertz, and C is capacitance in Farads. To illustrate this formula, an ideal 10 µf electrolytic capacitor has a capacitive reactance of 1.6 Ω at 10 kHz, which theoretically decreases to 160 µΩ at 100 MHz. At 100 MHz, a short circuit condition would occur, which is wonderful for EMI. However, physical parameters of electrolytic capacitors include high values of equivalent series inductance and equivalent series resistance that limits the effectiveness of this particular type of capacitor to operation below 1 MHz.

Another aspect of capacitor usage lies in lead-length inductance and body structure. This subject is discussed in Section 6.3 and will not be examined at this time. To summarize, parasitic inductance in the capacitor's wire bond leads causes X_L to exceed X_c above self-resonance. Hence, the capacitor ceases to function as a capacitor for its intended function or application.

Inductors

Inductors are used for EMI control within a PCB. For an inductor, inductive reactance increases linearly with increasing frequency. This is described by the formula $X_L = 2\pi f L$, where X_L is inductive reactance (ohms), f is frequency in hertz, and is L inductance (henries). As the frequency increases, the magnitude of impedance increases.

For example, an "ideal" 10 mH inductor has inductive reactance of 628 Ω at 10 kHz. This inductive reactance increases to 6.2 MΩ at 100 MHz. The inductor now appears to be an open circuit at 100 MHz to a digital signal operating at dc voltage transition levels. If we want to pass a signal at 100 MHz, great difficulty will occur, related to signal quality (time domain issue). Like a capacitor, the electrical parameters of an inductor limits this particular device to less than 1 MHz, as the parasitic capacitance between the windings and the two leads is excessively large.

The question now at hand is what to do at high frequencies when an inductor cannot be used. Ferrite beads become the saviors. Ferrite materials are alloys of iron/magnesium or iron/nickel. These materials have high permeability and provide for high impedance at high frequencies, with a minimum of capacitance that is always observed between windings in an inductor. Ferrites are generally used in high-frequency applications, because at low frequencies they are primarily resistive and thus impose few losses on the line. At high frequencies, they are reactive and frequency dependent. This is graphically shown in Fig. 6.2. In reality, ferrite beads are high-frequency attenuators of RF energy.

Ferrites are modeled as a parallel combination of a resistor and inductor. At low frequencies, the resistor is "shorted out" by the inductor, whereas at high frequencies, the inductive impedance is so high that it forces the current through the resistor.

Ferrites are "dissipative devices." They dissipate high-frequency RF energy as heat. This can only be explained by the resistive, not the inductive, effect of the device.

Transformers

Transformers are generally found in power supply applications, in addition to being used for isolation of data signals, I/O connections and power interfaces. Transformers are also widely used to provide common-mode (CM) isolation. Transformers use a differential-mode (DM) transfer mechanism across their input to magnetically link the primary windings to the secondary windings for energy transfer. Consequently, CM voltage across the primary is rejected. One flaw inherent in the manufacturing of transformers is signal source capacitance between the primary and secondary. As the frequency of the circuit increases, so does capacitive coupling; circuit isolation becomes compromised. If enough parasitic

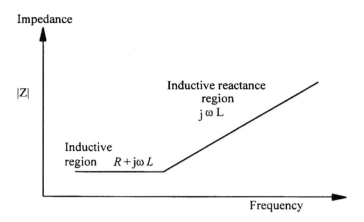

FIGURE 6.2 Characteristics of ferrite material.

capacitance exists, high-frequency RF energy, fast transients, ESD, lighting, and the like may pass through the transformer and cause an upset in the circuits on the other side of the isolation gap that received this transient event.

Depending on the type and application of the transformer, a shield may be provided between primary and secondary windings. This shield, connected to a ground reference source, is designed to prevent against capacitive coupling between the primary and secondary windings.

6.1.2 How and Why RF Energy is Created Within the PCB

We now investigate how RF energy is created within the PCB. To understand nonideal characteristics of passive components, as well as RF energy creation in the PCB, we need to understand Maxwell's equations. Maxwell's four equations describe the relationship of both electric and magnetic fields. These equations are derived from Ampere's law, Faraday's law, and two from Gauss's law.

Maxwell's equations describe the field strength and current density within a closed loop environment, requiring extensive knowledge of higher-order calculus. Since Maxwell's equations are extremely complex, a simplified discussion of physics is presented. For a rigorous analysis of Maxwell's equations, refer to the material listed in both the References and Bibliography. A detailed knowledge of Maxwell is not a prerequisite for PCB designers and layout engineers and is beyond the scope of this discussion. This important item of discussion is the fundamental concepts of how Maxwell works. Equation (6.1) presents his equations for reference only, illustrating that the field of EMC is based on complex math.

First Law : Electric Flux (from Gauss)
$$\nabla \cdot D = \rho \qquad \varphi e = \oint_s D \cdot ds = \int_v \rho \, dv = 0$$

Second Law : Magnetic Flux (from Gauss)
$$\nabla \cdot B = 0 \qquad \varphi m = \oint_s B \cdot ds = 0$$

Third Law : Electric Potential as E- Fields (from Faraday)
$$\nabla \times E = -\frac{\partial B}{\partial t} \qquad \oint E \cdot dl = -\int_s \frac{\partial B}{\partial t} \cdot ds$$

Fourth Law : Electric Current as H - Fields (from Ampere)
$$\nabla \times H = J + \frac{\partial D}{\partial t} \qquad \oint H \cdot dl = \int_s \left(J + \frac{\partial D}{\partial t} \right) \cdot ds = I_{total}$$

(6.1)

Maxwell's first equation is known as the divergence theorem, based on Gauss' law. This law says that the accumulation of an electric charge creates an electrostatic field, E. Electric charge is best observed between two boundaries, conductive and nonconductive. The boundary-condition behavior referenced in Gauss' law causes the conductive enclosure to act as an electrostatic shield (commonly called a Faraday cage, or Gaussian structure—which is a better description of the effect of Gauss' law). At the boundary, electric charges are kept on the inside of the boundary. Electric charges that exist on the outside of the boundary are excluded from internally generated fields.

Maxwell's second equation illustrates that there are no magnetic charges (no monopoles), only electric charges. These electric charges are either positively charged or negatively charged. Magnetic monopoles do not exist. Magnetic fields are produced through the action of electric currents and fields. Electric currents and fields emanate as a point source. Magnetic fields form closed loops around the circuit that generates these fields.

Maxwell's third equation, also called "Faraday's Law of Induction," describes a magnetic field traveling around a closed loop circuit, generating current. The third equation has a companion equation (fourth equation). The third equation describes the creation of electric fields from changing magnetic fields. Magnetic fields are commonly found in transformers or winding, such as electric motors, generators, and the like. The interaction of the third and fourth equation is the primary focus for electromagnetic compatibility. Together, they describe how coupled electric and magnetic fields propagate (radiate) at the speed of light. This equation also describes the concept of "skin effect," which predicts the effectiveness of magnetic shielding. In addition, inductance is described, which allows antennas to exist.

Maxwell's fourth equation is identified as "Ampere's law." This equation states that magnetic fields arise from two sources. The first source is current flow in the form of a transported charge. The second source describes how the changes in electric fields traveling in a closed loop circuit create magnetic fields. These electric and magnetic sources describe the function of inductors and electromagnetics. Of the two sources, the first source is the description of how electric currents create magnetic fields.

To summarize, Maxwell's equations describe the root causes of how EMI is created within a PCB by time-varying currents. Static-charge distributions produce static electric fields, not magnetic fields. Constant currents produce magnetic fields, not electric fields. Time-varying currents produce both electric and magnetic fields.

Static fields store energy. This is the basic function of a capacitor: accumulation of charge and retention. Constant current sources are a fundamental concept for the use of an inductor.

To "overly simplify" Maxwell, we associate his four equations to Ohm's law. The presentation that follows is a simplified discussion that allows one to visualize Maxwell in terms that are easy to understand. Although not mathematically perfect, this presentation concept is useful in presenting Maxwell to non-EMC engineers or those with minimal exposure to PCB suppression concepts and EMC theory.

$$\text{Ohm's law (time domain)} \qquad \text{Ohm's law (frequency domain)}$$
$$V = I * R \qquad\qquad V_{rf} = I_{rf} * Z \qquad\qquad (6.2)$$

where V = voltage
        I = current
        R = resistance
        Z = impedance $(R + jX)$

and subscript rf refers to radio frequency energy. To relate "Maxwell made simple" to Ohm's law, if RF current exists in a PCB trace that has a "fixed impedance value," an RF voltage will be created that is proportional to the RF current. Notice that, in the electromagnetics model, R is replaced by Z, a complex quantity that contains both resistance (real component) and reactance (a complex component).

For the impedance equation, various forms exist, depending on whether we are examining plane wave impedance or circuit impedance. For wire or a PCB trace, Eq. (6.3) is used.

$$Z = R + jX_L + \frac{1}{jXc} = R + j\omega L + \frac{1}{j\omega C} \qquad (6.3)$$

where $X_L = 2\pi fL$ (the component in the equation that relates only to a wire or PCB traces)
$Xc = 1/(2\pi fC)$ (not observed or present in a pure transmission line or free space)
$\omega = 2\pi f$

When a *component* has a known resistive and inductive element, such as a ferrite bead-on-lead, a resistor, a capacitor, or other device with parasitic components, Eq. (6.4) is applicable, as the magnitude of impedance versus frequency must be considered.

$$|Z| = \sqrt{R^2 + jX^2} = \sqrt{R^2 + j(X_L - X_c)^2} \qquad (6.4)$$

For frequencies greater than a few kilohertz, the value of inductive reactance typically exceeds R. Current takes the path of least impedance, Z. Below a few kilohertz, the path of least impedance is resistive; above a few kilohertz, the path of least reactance is dominant. Because most circuits operate at frequencies above a kilohertz, the belief that current takes the path of least resistance provides an incorrect concept of how RF current flow occurs within a transmission line structure.

Since current always takes the path of least impedance for wires carrying currents above 10 kHz, the impedance is equivalent to the path of least inductive reactance. If the load impedance connects to a wire, a cable, or a PCB trace, and the load impedance is much greater than the shunt capacitance of the transmission line path, inductance becomes the dominant element in the equation. If the wiring conductors have approximately the same cross-sectional shape, the path of least inductance is the one with the smallest loop area.

Each trace has a finite impedance value. Trace inductance is only one of the reasons why RF energy is developed within a PCB. Even the lead bond wires that connect a silicon die to its mounting pads may be sufficiently long to cause RF potentials to exist. Traces routed on a board can be highly inductive, especially traces that are electrically long. Electrically long traces are those physically long in routed length such that the round-trip propagation delayed signal on the trace does not return to the source driver before the next edge-triggered event occurs, when viewed in the time domain. In the frequency domain, an electrically long transmission line (trace) exceeds approximately $\lambda/10$ of the frequency that is present within the trace. If RF voltage occurs across an impedance, RF current is developed, per Ohm's law.

Maxwell's third equation states that a moving electrical charge in a trace generates an electric current that creates a magnetic field. Magnetic fields, created by this moving electrical charge, are also identified as magnetic lines of flux. Magnetic lines of flux can easily be visualized using the *right-hand rule*, graphically shown in Fig. 6.3. To understand this rule, make your right hand into a loose fist with your thumb pointing straight up. Current flow is in the direction of the thumb, upward, simulating current flowing in a wire or PCB trace. Your curved fingers encircling the wire point in the direction of the magnetic field or lines of magnetic flux. Time-varying magnetic fields create a transverse orthogonal electric field. RF emissions are a combination of both magnetic and electric fields. These fields will exit the PCB structure by either radiated or conducted means.

Notice that the magnetic field travels around a closed-loop boundary. In a PCB, RF currents are generated by a source driver and transferred to a load through a trace. RF currents must return to their source (Ampere's law) through a return system. Consequently, an RF current loop is developed. This loop does not have to be circular and is often a convoluted shape. Since this process creates a closed loop circuit within the return system, a magnetic field is developed. This magnetic field creates a radiated electric field. In the near field, the magnetic field component will dominate, whereas in the far field, the ratio of the electric to magnetic field (wave impedance) is approximately $120\pi\Omega$, or 377 Ω, independent of the source. Obviously, in the far field, magnetic fields can be measured using a loop antenna and a sufficiently sensitive receiver. The reception level will simply be $E/120\pi$ (A/m, if E is in V/m). The same applies to electric fields, which may be observed in the near field with appropriate test instrumentation.

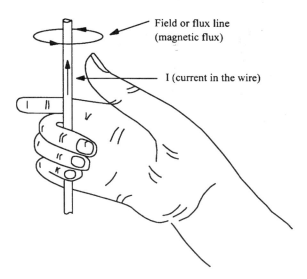

FIGURE 6.3 Right hand rule.

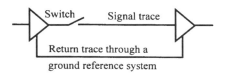

FIGURE 6.4 Closed-loop circuit.

Another simplified explanation of how RF exists within a PCB is depicted in Figs. 6.4 and 6.5. Here, we examine a simple circuit. The circuit on the left side in Fig. 6.5 represents the time domain. The circuit on the right represents the equivalent circuit in the frequency domain. According to Kirchhoff's and Ampere's laws, a closed-loop circuit must be present if the circuit is to work. Kirchhoff's voltage law states that the algebraic sum of the voltage around any closed path in a circuit must be zero. Ampere's law describes the magnetic induction at a point due to given currents in terms of the current elements and their positions relative to that point.

Consider a typical circuit with a switch in series with a source driver (Fig. 6.5). When the switch is closed, the circuit operates as desired; when the switch is opened, nothing happens. For the time domain, the desired signal component travels from source to load. This signal component must have a return

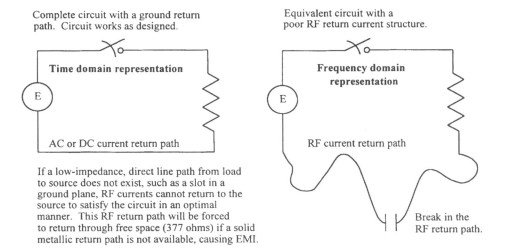

FIGURE 6.5 Representation of a closed-loop circuit.

path to complete the circuit, generally through a 0-V (ground) return structure (Kirchhoff's law). RF current must travel from source to load and return by the lowest impedance path possible, usually a ground trace or ground plane (also referred to as an *image plane*). RF current that exists is best described by Ampere's law. If a conductive path does not exist, free space becomes the return path.

Without a closed-loop circuit, a signal would never travel through a transmission line from source to load. When the switch is closed, the circuit is complete, and ac or dc current flows. In the frequency domain, we observe the current as RF energy. There are *not* two types of currents, time domain or frequency domain. There is only one current, which may be represented in *either* the time domain or frequency domain. The RF return path from load to source must also exist or the circuit would not work. Hence, a PCB structure must conform to Maxwell's equations, Kirchhoff voltage law, and Ampere's law.

6.1.3 Concept of Flux Cancellation (Flux Minimization)

To review *one* fundamental concept regarding how EMI is created within a PCB, we examine a basic mechanism of how magnetic lines of flux are created within a transmission line. Magnetic lines of flux are created by a current flowing through an impedance, either fixed or variable. Impedance in a network will always exist within a trace, component bond lead wires, vias, and the like. If magnetic lines of flux are present within a PCB, defined by Maxwell, various transmission paths for RF energy must also be present. These transmission paths may be either radiated through free space or conducted through cable interconnects.

To eliminate RF currents within a PCB, the concept of *flux cancellation* or *flux minimization* needs to be discussed. Although the term *cancellation* is used throughout this chapter, we may substitute the term *minimization*. Magnetic lines of flux travel within a transmission line. If we bring the RF return path parallel, and adjacent to its corresponding source trace, magnetic flux lines observed in the return path (counterclockwise field), relative to the source path (clockwise field), will be in the opposite direction. When we combine a clockwise field with a counterclockwise field, a cancellation effect is observed. If unwanted magnetic lines of flux between a source and return path are canceled or minimized, then radiated or conducted RF current cannot exist, except within the minuscule boundary of the trace. The concept of implementing flux cancellation is simple. However, one must be aware of many pitfalls and oversights that may occur when implementing flux cancellation or minimization techniques. With one small mistake, many additional problems will develop creating more work for the EMC engineer to diagnose and debug. The easiest way to implement flux cancellation is to use image planes.[6] Regardless of how well we design and lay out a PCB, magnetic and electric fields will always be present. If we minimize magnetic lines of flux, EMI issues cannot exist.

A brief summary of some techniques for flux cancellation is presented below, and discussed within this chapter.

- Having proper stackup assignment and impedance control for multilayer boards
- Routing a clock trace adjacent to a RF return path, ground plane (multilayer PCB), ground grid, or use of a ground or guard trace (single- and double-sided boards)
- Capturing magnetic lines of flux created internal to a component's plastic package into the 0-V reference system to reduce component radiation
- Carefully choosing logic families to minimize RF spectral energy distribution from component and trace radiation (use of slower edge rate devices)
- Reducing RF currents on traces by reducing the RF drive voltage from clock generation circuits, for example, transistor-transistor logic (TTL) versus complimentary metal oxide semiconductor (CMOS)
- Reducing ground noise voltage in the power and ground plane structure
- Providing sufficient decoupling for components that consume power when all device pins switch simultaneously under maximum capacitive load

- Properly terminating clock and signal traces to prevent ringing, overshoot, and undershoot
- Using data line filters and common-mode chokes on selected nets
- Making proper use of bypass (not decoupling) capacitors when external I/O cables shields are provided
- Providing a grounded heatsink for components that radiate large amounts of internal generated common-mode RF energy

As seen in this list, magnetic lines of flux are only part of the explanation of how EMI is created within a PCB. Other major areas of concern include

- Common-mode (CM) and differential-mode (DM) currents between circuits and interconnects
- Ground loops creating a magnetic field structure
- Component radiation
- Impedance mismatches

Remember that common-mode energy causes the majority of EMI emissions. These common-mode levels are developed as a result of not minimizing RF fields in the board or circuit design.

6.1.4 Common-Mode and Differential-Mode Currents

In any circuit, there are both common-mode CM and DM currents that together determine the amount of RF energy developed and propagated. Differential-mode signals carry data or signal of interest (information). Common-mode is a side effect of differential-mode and is most troublesome for EMC compliance. Common-mode and differential-mode current configurations are shown in Fig. 6.6.

The radiated emissions of differential-mode currents subtract and tend to cancel. On the other hand, emissions from common-mode currents add. For a 1-m length of cable, with wires separated by 0.050 in (e.g., typical ribbon cable spacing), a differential-mode current of 20 mA, or 8 µA common-mode current, will produce a radiated electric field at 3 m, 100 µV/m at 30 MHz. This level just meets the FCC Class B limit.[2,3] This is a ratio of 2500, or 68 dB, between the two modes. This small amount of common-mode current is capable of producing a significant amount of radiated emissions. A number of factors, such as physical distance to conducting planes and other structural symmetries can create common-mode currents. Much *less* common-mode current will produce the same amount of RF propagated energy than a *larger* amount of differential-mode currents. This is because common-mode currents do not cancel out within the RF return path.

When using simulation software to predict emissions from I/O interconnects that are driven from a PCB, differential-mode analysis is usually performed. It is *impossible* to predict radiated emissions based solely on differential-mode (transmission-line) currents. These calculated currents can severely underpredict the radiated emissions from PCB traces, since numerous factors, including parasitic parameters,

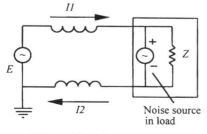 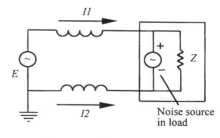

FIGURE 6.6 Common-mode and differential-mode current configurations.

are involved in the creation of common-mode currents from differential-mode voltage sources. These parameters usually cannot be anticipated and are dynamically present within a PCB in the formation of power surges in the planes during edge-switching transitions.

Differential-Mode Currents

Differential-mode (DM) current is the component of RF energy that is present on both the signal and return paths that are opposite to each other. If a 180° phase shift is established precisely, and the two paths are parallel, RF differential-mode fields will be canceled. Common-mode effects, however, may be created as a result of ground bounce and power plane fluctuation caused by components drawing current from a power distribution network.

Differential-mode signals

1. convey desired information
2. cause minimal interference, as the RF fields generated oppose each other and cancel out if properly set up

As seen in Fig. 6.6, differential-mode configuration, a circuit driver, E, sends out a current that is received by a load, identified as Z. Because there is outgoing current, return current must also be present. These two currents, source and return are traveling in opposite directions. This configuration represents standard differential-mode operation. We do not want to eliminate differential mode performance. Because a circuit board can only be made to emulate a perfect self-shielding environment (e.g., a coax), complete E-field capture and H-field cancellation may not be achieved. The remaining fields, which are not coupled to each other, are the source of common-mode EMI. In the battle to control EMI and crosstalk in differential mode, the key is to control excess energy fields through proper source control and careful handling of the energy-coupling mechanisms.

Common-Mode Currents

Common-mode (CM) current is the component of RF energy that is present on both the signal and return paths, usually in common phase. It is created by poor differential-mode cancellation, due to an imbalance between two transmitted signal paths. The measured RF field due to common-mode current will be the sum of the currents that exist in both the signal trace and return trace. This summation could be substantial and is the major cause of RF emissions, especially from I/O cables. If the differential signals are not exactly opposite and in phase, their currents will not cancel out.

Common-mode signals

1. are the major sources of RF radiation
2. contain no useful information

Common-mode current begins as the result of currents mixing in a shared metallic structure, such as the power and ground planes. Typically, this happens because currents are flowing through undesirable or unintentional return paths. Common-mode currents develop when return currents lose their pairing with their original signal path (e.g., splits or breaks in planes) or when several signal conductors share common areas of the return plane. Since planes have finite impedance, common-mode currents set up RF transient voltages within the planes. These RF transients set up currents in other conductive surfaces and signal lines that act as antennas to radiate EMI. The most common cause of coupling is the establishment of common-mode currents in conductors and shields of cables running to and from the PCB or enclosure. The key to preventing common-mode EMI is to understand and control the path of the power supply and return currents in the board. This is accomplished by controlling the position of the power and ground planes in the layer stackup assignment, and the currents within the planes. This is in addition to providing proper RF grounding to the case of the system or product.

In Fig. 6.6, current source $I1$ represents the flow of current from source E to load Z. Current flow $I2$ is current that is observed in the return system, usually identified as an image plane, ground plane, or

0-V reference. The measured radiated electric field of the common-mode currents is caused by the summed contribution of both $I1$ and $I2$ current produced fields.

With differential-mode currents, the electric field component is the difference between $I1$ and $I2$. If $I1 = I2$ exactly, there will be no radiation from differential-mode currents that emanate from the circuit (assuming the distance from the point of observation is much larger than the separation between the two current-carrying conductors), hence, no EMI. This occurs if the distance separation between $I1$ and $I2$ is electrically small. Design and layout techniques for cancellation of radiation emanating from differential-mode currents are easily implemented in a PCB with an image plane or RF return path, such as a ground trace. On the other hand, RF fields created by common-mode currents are harder to suppress. Common-mode currents are the main source of EMI. Fields due to differential-mode currents are rarely observed as a significant radiated electromagnetic field.

A PCB with a solid ground plane produces common-mode RF currents. This is because RF current encounters a finite inductance (impedance) in the ground plane material, usually copper. This inductance creates a voltage gradient between source and load, commonly identified as *ground-noise voltage*. This voltage, also termed *ground shift*, is a equivalent shift in the reference level within the planes. This reference shift is responsible for a significant amount of common-mode EMI. This voltage gradient causes a small portion of the signal trace current to flow through the distributed stray capacitance of the ground plane. This is illustrated in Fig. 6.7, where the following abbreviations are used:

L_s = partial self-inductance of the signal trace
M_{sg} = partial mutual-inductance between signal trace and ground plane
L_g = partial self-inductance of the ground plane
M_{gs} = partial mutual-inductance between ground plane and signal trace
C_{stray} = distributed stray capacitance of the ground plane
V_{gnd} = ground plane noise voltage

To calculate ground-noise voltage V_{gnd}, use Eq. (6.5) referenced to Figs. 6.6 and 6.7.

$$V_{gnd} = L_g \frac{dI_1}{dt} - M_{gs} \frac{dI_2}{dt} \tag{6.5}$$

To reduce the total ground-noise voltage, increase the mutual inductance between the trace and its nearest image plane. Doing so provides an additional path for signal return current to image back to its source.

Generally, common-mode currents are typically several orders of magnitude lower than differential-mode currents. It should be remembered that common-mode current is a by-product of differential-

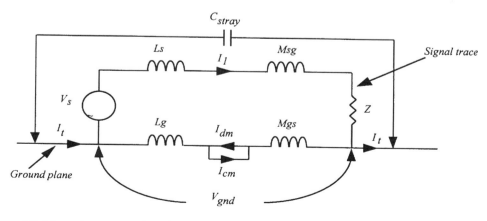

FIGURE 6.7 Schematic representation of a ground plane.

mode switching that does get cancelled out. However, common-mode currents (*I*1 and *Icm*) produce higher emissions than those created by differential-mode (*I*1 and *Idm*) currents. This is because common-mode RF current fields are additive, whereas differential-mode current fields tend to cancel. This was illustrated in Fig. 6.6.

To reduce *It* currents, ground-noise voltage must be reduced. This is best accomplished by reducing the distance spacing between the signal trace and ground plane. In most cases, this is not fully possible, because the spacing between a signal plane and image plane must be at a specific distance to maintain constant trace impedance of the PCB. Hence, there are prudent limits to distance separation between the two planes. The ground noise voltage must still be reduced. Ground-noise voltage can be reduced by providing additional paths for RF currents to flow.[4]

An RF current return path is best achieved with a ground plane for multilayer PCBs, or a ground trace for single- and double-sided boards. The RF current in the return path will couple with the RF current in the source path (magnetic flux lines traveling in opposite direction to each other). The flux that is coupled due to the opposite field will cancel each other out and approach zero (flux cancellation or minimization). This is seen in Fig. 6.8. If we have an optimal return path, differential-mode RF currents will be minimized.

If a current return path is not provided through a path of least impedance, residual common-mode RF currents will be developed. There will always be some common-mode currents in a PCB, for a finite distance spacing must exist between the signal trace and return path (flux cancellation almost approaches 100%). The portion of the differential-mode return current that is not canceled out becomes residual RF common-mode current. This situation will occur under many conditions, especially when a ground reference difference exists between circuits. This includes ground bounce, trace impedance mismatches, and lack of decoupling.

Consider a pair of parallel wires carrying a differential-mode signal. Within this wire, RF currents flow in opposite directions (coupling occurs). Consequently, RF fields created in the transmission line tend to cancel. In reality, this cancellation cannot be 100%, as a finite distance must exist between the two wires due to the physical parameters required during manufacturing of the board. This finite distance is insignificant related to the overall concept being discussed. This parallel wire set will act as a balanced transmission line that delivers a clean differential (signal-ended) signal to a load.

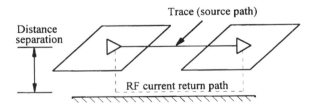

The closer we bring the RF return path to the RF source path enhanced coupling occurs. As a result, a cancellation effect, or minimization of flux lines is observed, producing less RF energy.

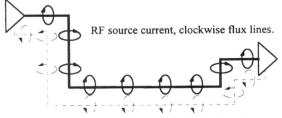

FIGURE 6.8 RF current return path and distance spacing.

EMC and Printed Circuit Board Design

Using this same wire pair, look at what happens when common-mode voltage is placed on this wire. No useful information is transmitted to the load, since the wires carry the same voltage. This wire pair now functions as a driven antenna with respect to ground. This driven antenna radiates unwanted or unneeded common-mode voltage with extreme efficiency. Common-mode currents are generally observed in I/O cables. This is why I/O cables radiate well. An illustration of how a PCB and an interconnect cable allow CM and DM current to exist is shown in Fig. 6.9.

6.1.5 RF Current Density Distribution

A 0-V reference, or return plane, allows RF current to return to its source from a load. This return plane completes the closed-loop circuit requirements for functionality. Current distribution in traces tends to spread out within the return structure as illustrated in Fig. 6.10. This distribution will exist in both the forward direction and the return path. Current distribution shares a common impedance between trace and plane (or trace-to-trace), which results in mutual coupling due to the current spread. The peak current density lies directly beneath the trace and falls off sharply from each side of the trace into the ground plane structure.

When the distance spacing is far apart between trace and plane, the loop area between the forward and return path increases. This return path raises the inductance of the circuit where inductance is proportional to loop area. This was shown in Fig. 6.8. Equation (6.6) describes the current distribution that is optimum for minimizing total loop inductance for both the forward and return current path. The current described in Eq. (6.6) also minimizes the total energy stored in the magnetic field surrounding the signal trace.[5]

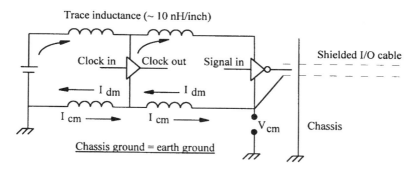

FIGURE 6.9 System equivalent circuit of differential- and common-mode returns.

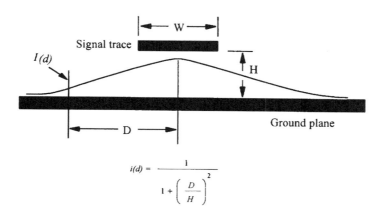

$$i(d) = \frac{1}{1 + \left(\dfrac{D}{H}\right)^2}$$

FIGURE 6.10 Current density distribution from trace to reference plane.

$$i(d) = \frac{Io}{\pi H} \times \frac{1}{1 + \left(\frac{D}{H}\right)^2} \tag{6.6}$$

where $i(d)$ = signal current density, (A/inch or A/cm)
Io = total current (A)
H = height of the trace above the ground plane (inches or cm)
D = perpendicular distance from the centerline of the trace (inches or cm)

The mutual coupling factor between source and return is highly dependent on frequency of operation and the skin depth effect of the ground plane impedance. As the skin depth increases, the resistive component of the ground plane impedance will also increase. This increase will be observed with proportionality at relatively high frequencies.[1–3]

A primary concern with current density distribution within a transmission line relates to crosstalk. Crosstalk is easily identified as EMI between PCB traces. The definition of EMI is *the creation of unwanted energy that propagates through a dielectric and causes disruption to other circuits and components*. Electromagnetic fields are propagated down a transmission line or PCB trace. Depending on configuration of the topology, RF flux lines will be developed based on the topologies shown within Fig. 6.11. With magnetic lines of flux, the current spread from the trace is described by Eq. (6.6), relative to the adjacent plane or adjacent lines (or signal routing layer). The distance that the current spread is best illustrated within Fig. 6.10. Current spread is typically equal to the width of an individual trace. For example, if a trace is 0.008 in (0.002 mm) wide, flux coupling to an adjacent trace will be developed if the adjacent trace is less than or equal to 0.008 in (0.002 mm) away. If the adjacent trace is routed greater than one trace width away, coupling of RF flux will be minimized. To implement this design technique, the section on the 3-W rule, described later in this chapter, details this layout technique.

6.1.6 Skin Effect and Lead Inductance

A consequence of Maxwell's third and fourth equations is skin effect related to a voltage charge imposed on a homogeneous medium where current flows, such as a lead bond wire from a component die or a PCB trace. If voltage is maintained at a constant dc level, current flow will be uniform throughout the transmission path. A finite period of time is required for uniformity to occur. The current first flows on the outside edge of the conductor and then diffuses inward.[11]

When the source voltage is not dc, but high frequency ac or RF frequencies, current flow tends to concentrate in the outer portion of the conductor, a phenomenon called *skin effect*. Skin depth is defined as the distance to the point inside the conductor at which the electromagnetic field, and hence current, is reduced to 37% of the surface value. We define skin depth (δ) mathematically by Eq. (6.7).

$$\delta = \sqrt{\frac{2}{\omega\mu\sigma}} = \sqrt{\frac{2}{2\pi f\mu\sigma}} = \frac{1}{\sqrt{\pi f\mu\sigma}} \tag{6.7}$$

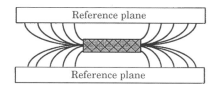

Microstrip topology (one reference plane only) or stripline topology (both reference planes)

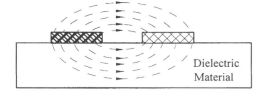

Coplanar strips (parallel lines)

FIGURE 6.11 Field distribution for microstrip and coplanar strips.

where ω = angular (radian) frequency ($2\pi f$)
μ = material permeability ($4\pi \times 10^{-7}$ H/m)
σ = material conductivity (5.82×10^7 mho/m for copper)
f = frequency (hertz)

Table 6.1 presents an abbreviated list of skin depth values at various frequencies for a 1-mil thick copper substrate (1 mil = 0.001 in = 2.54×10^{-5} m).

TABLE 6.1 Skin Depth for Copper Substrate

f	δ (copper)
60 Hz	0.0086 in (8.6 mil, 2.2 mm)
100 Hz	0.0066 in (6.6 mil, 1.7 mm)
1 kHz	0.0021 in (2.1 mil, 0.53 mm)
10 kHz	0.00066 in (0.66 mil, 0.17 mm)
100 kHz	0.00021 in (0.21 mil, 0.053 mm)
1 MHz	0.000066 in (0.066 mil, 0.017 mm)
10 MHz	0.000021 in (0.021 mil, 0.0053 mm)
100 MHz	0.0000066 in (0.0066 mil, 0.0017 mm)
1 GHz	0.0000021 in (0.0021 mil, 0.00053 mm)

As noted in the table, if any of the three parameters of Eq. (6.7) decreases, skin depth increases.

The wire's internal inductance equals the value of the internal dc resistance independent of the wire radius up to the frequency where the wire radius is on the order of a skin depth. Below this particular frequency, the wire's impedance *increases* as $\sqrt{f}$ or 10 dB/decade. Internal inductance is the portion of the magnetic field internal to the wire per-unit length, where the transverse magnetic field contributes to the per-unit-length inductance of the line. The portion of the magnetic lines of flux, external to the transmission line, contributes to a portion of the total per-unit-length inductance of the line, and is referred to as *external inductance*. Above this particular frequency, the wire's internal inductance *decreases* as $\sqrt{f}$ or –10 dB/decade.

For a solid round copper wire, the effective dc resistance is described by Eq. (6.8). Table 6.2 provides details on some of the parameters used in Eq. (6.7). Signals may be further attenuated by the resistance

TABLE 6.2 Physical Characteristics of Wire

Wire gage (AWG)	Solid wire diameter (mils)	Stranded wire diameter (mils)	Rdc—solid wire (Ω/1000 ft) @ 25°C
28	12.6	16.0 (19 × 40)	62.9
		15.0 (7 × 36)	
26	15.9	20.0 (19 × 38)	39.6
		21.0 (10 × 36)	
		19.0 (7 × 34)	
24	20.1	24.0 (19 × 36)	24.8
		23.0 (10 × 34)	
		24.0 (7 × 32)	
22	25.3	30.0 (26 × 36)	15.6
		31.0 (19 × 34)	
		30.0 (7 × 30)	
20	32.0	36.0 (26 × 34)	9.8
		37.0 (19 × 32)	
		35.0 (10 × 30)	
18	40.3	49.0 (19 × 30)	6.2
		47.0 (16 × 30)	
		48.0 (7 × 26)	
16	50.8	59.0 (26 × 30)	3.9
		60.0 (7 × 24)	

of the copper used in the conductor and by skin effect losses resulting from the finish of the copper surface. The resistance of the copper may reduce steady-state voltage levels below functional requirements for noise immunity. This condition is especially true for high-frequency differential devices, such as emitter coupled logic (ECL), where a voltage divider is formed by termination resistors and line resistance.

$$Rdc = \frac{L}{\sigma \pi r_\omega^2} \Omega \qquad (6.8)$$

where L = length of the wire
r_ω = radius (Table 6.2)
σ = conductivity

Units for L and r_ω must be consistent in English or metric units. As the frequency increases, the current over the wire cross section will tend to crowd closer to the outer periphery of the conductor. Eventually, the current will be concentrated on the wire's surface equal to the thickness of the skin depth as described by Eq. (6.9) when the skin depth is less than the wire radius.

$$\delta = \frac{1}{\sqrt{\pi f \mu_o \sigma}} \qquad (6.9)$$

where, at various frequencies
δ = skin depth
μ_0 = permeability of copper ($4\pi \times 10^{-7}$ H/m)
$\omega = 2\pi f$ (where f = frequency in hertz)
σ = the conductivity of copper (5.8×10^7 mho/m = 1.4736×10^6 mho/in)

Inductance of a conductor at high frequency is inversely proportional to the log of the conductor diameter, or the width of a flat conductor. For a round conductor, located above a return path, inductance is

$$L = 0.005 \times \ln\left(\frac{4h}{d}\right) \quad \mu\text{H/in or }\mu\text{H/cm} \qquad (6.10)$$

where d is the diameter, and h is the height above the RF current return path, in the same units (inches or centimeters). For flat conductors, such as a PCB trace, inductance is defined by

$$L = 0.005 \times \ln\left(\frac{2\pi h}{w}\right) \quad \mu\text{H/in or }\mu\text{H/cm} \qquad (6.11)$$

Due to the logarithmic relationship of the ratio h/d, or h/w, the reactive component of impedance for large-diameter wires dominates the resistive component above only a few hundred hertz. It is difficult to acquire a decrease in inductance by increasing the conductor diameter or size. Doubling the diameter, or width, by 100% will only decrease the inductance by 20%. The size of the wire would have to be increased by 500% for a 50% decrease in inductance. If a large decrease in inductance is required, alternative methods of design must be employed.[3] Thus, it is impractical to obtain a truly low-impedance connection between two points, such as grounding a circuit using only wire. Such a connection would permit coupling of voltages between circuits, due to current flow through an appreciable amount of common impedance.

6.1.7 Grounding Methods

Several types of grounding methods and terms have been devised, including the following types: digital, analog, safety, signal, noisy, quiet, earth, single-point, multipoint, and so on. Grounding methods must be specified and designed into a product—not left to chance. Designing a good grounding system is also cost-effective in the long run. In any PCB, a choice must be made between two basic types of grounding:

EMC and Printed Circuit Board Design

single-point versus multipoint. Interactions with other grounding methods can exist if planned in advance. The choice of grounding is product application dependent. It must be remembered that, if single-point grounding is to be used, be consistent in its application throughout the product design. The same is true for multipoint grounding; do not mix multipoint ground with a single-point ground system unless the design allows for isolation between planes and functional subsections.

Figures 6.12 through 6.14 illustrates three grounding methods: single-point, multipoint, and hybrid. The following text presents a detailed explanation of each concept, related to operational frequency and appropriate use.[1,11]

Single-Point Grounding

Single-point grounds are usually formed with signal radials and are commonly found in audio circuits, analog instrumentation, 60 Hz and dc power systems, and products packaged in plastic enclosures. Although single-point grounding is commonly used for low-frequency products, it is occasionally found in extremely high-frequency circuits and systems.

Use of single-point grounding on a CPU-motherboard or adapter (daughter card) allows loop currents to exist between the 0-V reference and chassis housing if metal is used as chassis ground. Loop currents create magnetic fields. Magnetic fields create electric fields. Electric fields generate RF currents. It is nearly

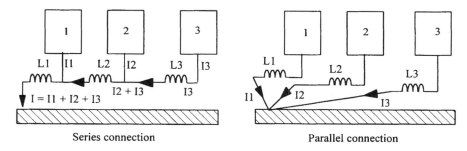

FIGURE 6.12 Single-point grounding methods. *Note:* this is inappropriate for high-frequency operation.

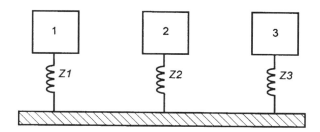

FIGURE 6.13 Multipoint grounding.

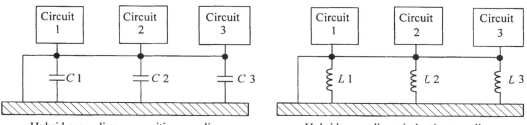

FIGURE 6.14 Hybrid grounding.

impossible to implement single-point grounding in personal computers and similar devices, because different subassemblies and peripherals are grounded directly to the metal chassis in different locations. These electromagnetic fields create a distributed transfer impedance between the chassis and the PCB that inherently develops loop structures. Multipoint grounding places these loops in regions where they are least likely to cause problems (RF loop currents can be controlled and directed rather than allowed to transfer energy inadvertently to other circuits and systems susceptible to electromagnetic field disturbance).

Multipoint Grounding

High-frequency designs generally require use of multiple chassis ground connections. Multipoint grounding minimizes ground impedance present in the power planes of the PCB by shunting RF currents from the ground planes to chassis ground. This low plane impedance is caused primarily by the lower inductance characteristic of solid copper planes. In very high-frequency circuits, lengths of ground leads from components must be kept as short as possible. Trace lengths add inductance to a circuit at approximately 12 to 20 nH per inch. This variable inductance value is based on two parameters: trace width and thickness. Inductance allows a resonance to occur when the distributed capacitance between the ground planes and chassis ground forms a tuned resonant circuit. The capacitance value, C, in Eq. (6.12) is sometimes known, within a specific tolerance range. Inductance, L, is determined by knowledge of the impedance of copper planes. Typical values of inductance for a copper plane, 10 × 10 in (25.4 × 25.4 cm), are provided in Table 6.3. The equations for solving this value of impedance are complex and beyond the scope of this chapter.

TABLE 6.3 Impedance of a 10 × 10 in (25.4 × 25.4 mm) Copper Metal Plane

Frequency (MHz)	Skin Depth (cm)	Impedance (Ω/sq)
1 MHz	6.6×10^{-3}	0.00026
10 MHz	2.1×10^{-3}	0.00082
100 MHz	6.6×10^{-4}	0.0026
1 GHz	2.1×10^{-4}	0.0082

$$f = \frac{1}{2\pi\sqrt{LC}} \qquad (6.12)$$

where f = resonant frequency (hertz)
 L = inductance of the circuit (henries)
 C = capacitance of the circuit (farads)

Equation (6.12) describes most aspects of frequency-domain concerns. This equation, although simple in format, requires knowledge of planar characteristics for accurate values of both L and C.

Examine Eq. (6.12) using Fig. 6.15. This illustration shows both capacitance and inductance that exist between a PCB and a screw-secured mounting panel. Capacitance and inductance are always present. Depending on the aspect ratio between mounting posts, relative to the self-resonant frequency of the power planes, loop currents will be generated and coupled (either radiated or conducted) to other PCBs located nearby, the chassis housing, internal cables or harnesses, peripheral devices, I/O circuits and connectors, or to free space.[1,11]

In addition to inductance in the planes, long traces also act as small antennas when routed microstrip, especially for clock signals and other periodic data pulses. By minimizing trace inductance and removing RF currents created in the transmission line (coupling RF currents present in the signal trace to the 0-V plane or chassis ground), significant improvement in signal quality and RF suppression will occur.

Digital circuits must be treated as high-frequency analog circuits. A good low-inductance ground is necessary on any PCB containing many logic circuits. The ground planes internal to the PCB (more than the power plane) generally provide a low-inductance return for the power supply and signal currents. This allows for creating a constant impedance transmission line for signal interconnects. When making

EMC and Printed Circuit Board Design 6-21

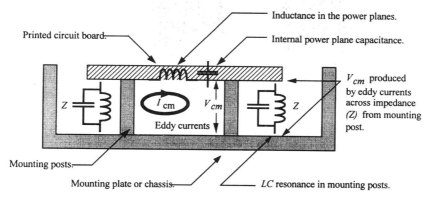

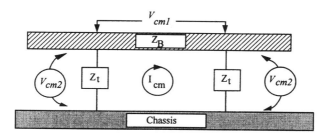

V_{cm2} is reduced by the mounting posts (ground stitch locations).
Resonance is thus controlled, along with enhanced RF suppression.

ELECTROMAGNETIC MODEL OF MULTI-POINT GROUNDING

FIGURE 6.15 Resonance in a multipoint ground to chassis.

ground plane to chassis plane connection, provide for removal of high-frequency RF energy present within the 0-V network with decoupling capacitors. RF currents are created by the resonant circuit of the planes and their relationship to signal traces. These currents are bypassed through the use of high-quality bypass capacitors, usually 0.1 µF in parallel with 0.001 µF at each ground connection, as will be reiterated in Section 6.4. The chassis grounds are frequently connected directly to the ground planes of the PCB to minimize RF voltages and currents that exist between board and chassis. If magnetic loops currents are small (1/20 wavelength of the highest RF generated frequency), RF suppression or flux cancellation or minimization is enhanced.

6.1.8 Ground And Signal Loops (Excluding Eddy Currents)

Ground loops are a major contributor to the propagation of RF energy. RF current will attempt to return to its source through any available path or medium: components, wire harnesses, ground planes, adjacent traces, and so forth. RF current is created between a source and load in the return path. This is due to a voltage potential difference between two devices, regardless of whether inductance exists between these points. Inductance causes magnetic coupling of RF current to occur between a source and victim circuit, increasing RF losses in the return path.[1,11]

One of the most important design considerations for EMI suppression on a PCB is ground or signal return loop control. An analysis must be made for each and every ground stitch connection (mechanical securement between the PCB ground and chassis ground) related to RF currents generated from RF noisy electrical circuits. Always locate high-speed logic components and frequency generating components as close as possible to a ground stitch connection. Placing these components here will minimize RF loops in the form of eddy currents to chassis ground and to divert this unwanted energy into the 0-V reference system.

An example of RF loop currents that could occur in a computer with adapter cards and single-point grounding is shown in Fig. 6.16. As observed, an excessive signal return loop area exists. Each loop will create a distinct electromagnetic field based on the physical size of the loop. The magnetic field developed within this loop antenna will create an electromagnetic field at a frequency that can easily be calculated. If RF energy is created from loop antennas present within a PCB layout, containment measures will probably be required. Containment may keep unwanted RF currents from coupling to other systems or circuits in addition to preventing the escape of this energy to the external environment as EMI. Internally generated RF loop currents are to always be avoided.

RF currents in power planes also have the tendency to couple, via crosstalk, to other signal traces, thus causing improper operation or functional signal degradation. If using multipoint grounding, consideration of loops becomes a major design concern.

In addition to inductance in the power and ground planes, traces act as small antennas. This is especially true for clock signals and other periodic data pulses routing microstrip. By minimizing trace inductance and removing RF currents created within in the trace, coupling RF currents from the signal trace to the ground planes or chassis ground provides significant improvement in signal quality. In addition, enhanced RF energy suppression will occur.

The smaller the magnetic loop area for RF return currents, the less the electric field gradient voltage potential difference. If the magnetic loop area is small, less than 1/20 wavelength of the highest RF generated frequency, RF energy is generally not developed.

6.1.9 Aspect Ratio—Distance Between Ground Connections

Aspect ratio refers to the ratio of a longer dimension to a shorter one. When providing ground stitch connections in a PCB using multipoint grounding to a metallic structure, we must concern ourselves with the distance spacing in all directions of the ground stitch location.[1,11]

RF currents that exist within the power and ground plane structure will tend to couple to other components, cables, peripherals, or other electronic items within the assembly. This undesirable coupling may cause improper operation, functional signal degradation, or EMI. When using multipoint grounding to a metal chassis, and providing a third wire ground connection to the ac mains, RF ground loops become a major design concern. This configuration is typical with personal computers. An example of a single point ground connection for a personal computer is shown in Fig. 6.16.

Because the edge rate of components is becoming faster, multipoint grounding is becoming mandatory, especially when I/O interconnects are provided in the design. Once an interconnect cable is attached to a connector, the device at the other end of the interconnect may provide an RF path to a third-wire ac ground mains connection, if provided to its respective power source. The power source for the load may be completely different from the power source from the driver (e.g., the negative terminal of a battery). A large ground loop between I/O interconnects can cause undesirable levels of radiated common-mode energy. How can we minimize RF loops that may occur within a PCB structure? The easiest way is to design the board with many ground stitch locations to chassis ground, if chassis ground is provided. The

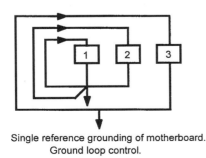
Single reference grounding of motherboard.
Ground loop control.

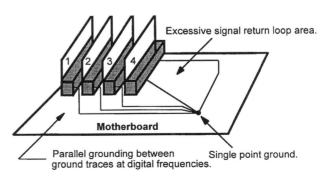

FIGURE 6.16 Ground loop control.

question that now exists is how far apart do we make the ground connections from each other, assuming the designer has the option of specifying this design requirement?

The distance spacing between ground stitch locations should not exceed $\lambda/20$ of the highest frequency of concern, including harmonics. If many high-bandwidth components are used, multiple ground stitch locations are typically required. If the unit is a slow edge rate device, connections to chassis ground may be minimized, or the distance between ground stitch locations increased.

For example, $\lambda/20$ of a 64 MHz oscillator is 23.4 cm (9.2 in). If the straight-line distance between any two ground stitch locations to a 0-V reference (in either the x- and/or y-axis) is greater than 9.2 in, then a potential efficient RF loop exists. This loop could be the source of RF energy propagation, which could cause noncompliance with international EMI emission limits. Unless other design measures are implemented, suppression of RF currents caused by poor loop control is not possible and containment measures (e.g., sheet metal) must be implemented. Sheet metal is an expensive band-aid that might not work for RF containment. An example of aspect ratio is illustrated in Fig. 6.17.

6.1.10 Image Planes

An image plane is a layer of copper or similar conductive material with the PCB structure. This layer may be identified as a voltage plane, ground plane, chassis plane, or isolated plane physically adjacent to a circuit or signal routing plane. Image planes provide a low-impedance path for RF currents to return to their source (flux return). This return completes the RF current path, thus reducing EMI emissions. The term *image plane* was popularized by German, Ott, and Paul[6] and is now used as industry standard terminology.

RF currents must return to their source one way or another. This path may be a mirror image of its original trace route or through another trace located in the near vicinity (crosstalk). This return path may be a power plane, ground plane, chassis plane, or free space. RF currents will capacitively or inductively couple themselves to any conductive medium. If this coupling is not 100%, common-mode RF current will be generated between the trace and its return path. An image plane internal to the PCB reduces ground noise voltage in addition to allowing RF currents to return to their source (mirror image) in a tightly coupled (nearly 100%) manner. Tight coupling provides for enhanced flux cancellation, which is another reason for use of a solid return plane without splits, breaks, or oversized through holes. An example of how an image plane appears to a signal trace is detailed in Fig. 6.18.[1,11]

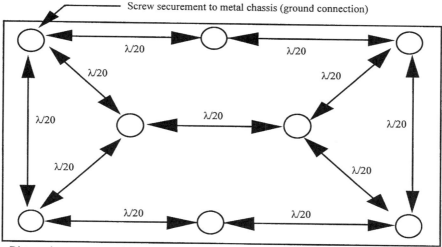

FIGURE 6.17 Example of aspect ratio.

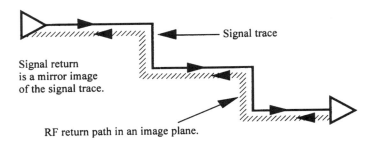

FIGURE 6.18 Image plane concept.

Regarding image plane theory, the material presented herein is based on a finite-sized plane, typical of most PCBs. Image planes are not reliable for reducing RF currents on I/O cables because approximating finite-sized conductive planes is not always valid. When I/O cables are provided, the dimensions of the configuration and source impedance are important parameters to consider.[7]

If three internal signal routing layers are physically adjacent in a multilayer stackup assignment, the middle routing layer (e.g., the one not immediately adjacent to an image plane) will couple its RF return currents to one or both of the other two signal planes. This coupling will cause undesired RF energy to be transferred through both mutual inductive and capacitive coupling to the other two signal planes. This coupling can cause significant crosstalk to occur. Flux cancellation performance is enhanced when the signal layers are adjacent to a 0-V reference or ground plane and not adjacent to a power plane, since the power distribution network generally contains more switching energy than the 0-V or return structure. This switching energy sometimes transfers or couples RF energy to other functional areas, whereas the 0-V or return structure is generally held at ground potential or is more stable than power. By being tied to ground potential, circuits and cable interconnects will not be modulated by switching currents.

For an image plane to be effective, *no signal or power traces can be located within this solid return plane.* Exceptions exist when a moat (or isolation) occurs. If a signal trace or even a power trace (e.g., +12 V trace) is routed in a solid +5 V plane, this plane is now fragmented into smaller parts. Provisions have now been made for a ground or signal return loop to exist for signal traces routed on an adjacent signal layer across this plane violation. This RF loop occurs by not allowing RF currents in a signal trace to seek a straight-line path back to its source. Split planes can no longer function as a solid return path to remove common-mode RF currents for optimal flux cancellation.

Figure 6.19 illustrates a violation of the image plane concept. An image plane that is not a solid structure can no longer function as an optimal return path to remove common-mode RF currents. The losses across plane segmentations may actually produce RF fields. Vias placed in an image plane do not degrade the imaging capabilities of the plane, except where ground slots are provided, discussed next. Vias do affect other functional parameters on a PCB layout. These functional parameters include:[1,11]

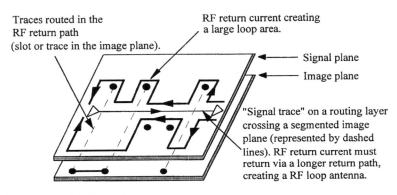

FIGURE 6.19 Image plane violation with traces.

EMC and Printed Circuit Board Design

- Reducing interplane capacitance; degrades decoupling effects
- Preventing RF return currents from traveling between routing planes
- Adding inductance and capacitance into a trace route
- Creating an impedance discontinuity in the transmission line

To ensure enhanced flux cancellation, or minimization within a PCB when using image planes, it becomes mandatory that all high-speed signal routing layers be adjacent to a solid plane, preferably at ground potential. The reason why ground planes are preferred over power planes is that various logic devices may be quite asymmetrical in their pull-up/pull-down current ratios. These switching components may not present an optimum condition for flux cancellation due to signal flux phase shift, greater inductance, poor impedance control and noise instability. The ground plane is also preferred, because this is where heavy switching currents are shunted to. TTL asymmetric drive is heaviest to ground, with fewer current spikes to the power plane. For ECL, the more noisy current spikes are to the positive voltage rail.

6.1.11 Slots in an Image Plane

Ground plane discontinuities are caused by through-hole components. Excessive through holes in a power or ground plane structure create the *Swiss cheese syndrome*.[8] The copper area between pins in the plane is reduced because the clearance area for many holes overlap (oversized through holes), leaving large areas of discontinuity. This is observed in Fig. 6.20. The return current flows on the image plane and does not mirror image to the signal trace on the adjacent layer due to this discontinuity. As seen in Fig. 6.20, return currents in the ground plane must travel around the slots or holes. This extra RF return path length creates a loop antenna, developing an electric field between the signal trace and return path. With additional inductance in the return path, there is reduced differential-mode coupling between the

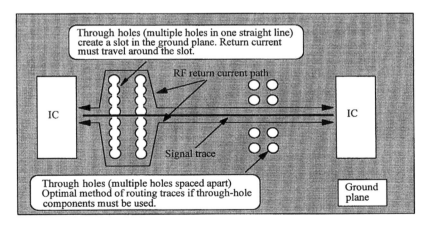

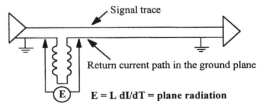

$E = L\, dI/dT = $ plane radiation

Equivalent circuit showing inductance in the return path. This inductance is approximately 1 nH/cm.

FIGURE 6.20 Ground loops when using through-hole components (slots in planes).

signal trace and RF current return plane (less flux cancellation or minimization). For through-hole components that have a space between pins (non-oversized holes), optimal reduction of signal and return current is achieved due to less lead length inductance in the signal return path owing to the solid image plane.[1,11]

If the signal trace is routed "around" through-hole discontinuities (left-hand side of Fig. 6.20), maintain a constant RF return path along the entire signal route. The same is true for the right-hand side of Fig. 6.20. There exists no image plane discontinuities here, hence, a shorter RF return path. Problems will arise when the signal trace travels through the middle of slotted holes in the routed PCB, when a solid plane does not exist due to the oversized through-hole area. When routing traces between through-hole components, use of the 3-W rule must be maintained between the trace and through-hole clearance area to prevent coupling RF energy between the trace and through-hole pins.

Generally, a slot in a PCB with oversized or overlapping holes will not cause RF problems for the majority of signal traces that are routed between through-hole device leads. For high-speed, high-threat signals, alternative methods of routing traces between through-hole component leads must be devised.

In addition to reducing ground-noise voltage, image planes prevent ground loops from occurring. This is because RF currents want to tightly couple to their source without having to find another path home. Loop control is maintained and minimized. Placement of an image plane adjacent to each and every signal plane removes common-mode RF currents created by signal traces. Image planes carry high levels of RF currents that must be sourced to ground potential. To help remove excess RF energy, all 0-V reference and chassis planes must be connected to chassis ground by a low-impedance ground stitch connection.[6,9]

There is one concern related to image planes. This deals with the concept of skin effect. Skin effect refers to current flow that resides in the first skin depth of the material. Current does not, and cannot, significantly flow in the center of traces and wires—and is predominantly observed on the external surface of the conductive media. Different materials have different skin depth values. The skin depth of copper is extremely shallow above 30 MHz. Typically, this is observed at 0.0000066 in (0.0017 mm) at 100 MHz. RF current present on a ground plane cannot penetrate 0.0014 in (0.036 mm) thick copper. As a result, both common-mode and differential-mode currents flow only on the top (skin) layer of the plane. There is no significant current flowing internal to the image plane. Placing an additional image plane beneath a primary reference plane would not provide additional EMI reduction. If the second reference plane is at voltage potential (the primary plane at ground potential) a decoupling capacitor is created. These two planes can now be used as both a decoupling capacitor and dual image planes.[1]

6.1.12 Partitioning

Proper placement of components for optimal functionality and EMC suppression is important for any PCB layout. Most designs incorporate functional subsections or areas by logical function. Grouping functional areas together minimizes signal trace length, routing, and creation of antennas. This makes trace routing easier while maintaining signal quality. Figure 6.21 illustrates functional grouping of subsections (or areas) on a typical CPU motherboard.[1,9,11]

Extensive use of chassis ground stitch connections is observed in Fig. 6.21. High-frequency designs require new methodologies for bonding ground planes to chassis ground. Multipoint grounding techniques effectively partitions common-mode eddy currents emanating from various segments in the design and keeps them from coupling into other segments. Products with clock rates above 50 MHz generally require frequent ground stitch connections to chassis ground to minimize effects of common-mode eddy currents and ground loops present between functional sections. At least four ground points surround each subsection. These ground points illustrate best-case implementation of aspect ratio. Note that a chassis bond connection, screw or equivalent, is located on both ends of the dc power connector (item P) used for powering external peripheral devices. RF noise generated in the PCB or peripheral power subsystem must be ac shunted to chassis ground by bypass capacitors. Bypass capacitors reduce coupling of power-supply-generated RF currents into both signal and data lines. Shunting of RF currents on the

EMC and Printed Circuit Board Design

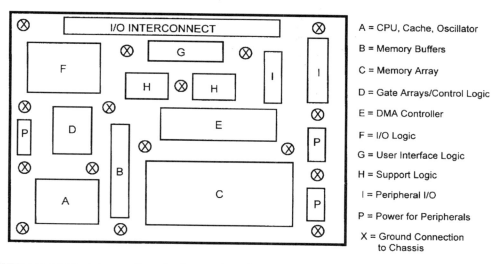

FIGURE 6.21 Multipoint grounding—implementation of partitioning.

power connector optimizes signal quality for data transfer between motherboard and external peripheral devices, in addition to reducing both radiated and conducted emissions.

Most PCBs consist of functional subsections or areas. For example, a typical personal computer contains the following on the motherboard: CPU, memory, ASICs, bus interface, system controllers, PCI bus, SCSI bus, peripheral interface (fixed and floppy disk drives), video, audio, and other components. Associated with each area are various bandwidths of RF energy. Logic families generate RF energy at different portions of the frequency spectrum. The higher the frequency component of the signal (faster edge rate), the greater the bandwidth of RF spectral energy. RF energy is generated from higher frequency components and time variant edges of digital signals. Clock signals are the greatest contributors to the generation of RF energy, are periodic (50% duty cycle), and are easy to measure with spectrum analyzers or receivers. Table 6.4 illustrates the spectral bandwidth of various logic families.

TABLE 6.4 Sample Chart of Logic Families Illustrating Spectral Bandwidth of RF Energy

Logic Family	Published Rise/Fall Time (approx.) t_r, t_f	Principal Harmonic Content $f = (1/\pi t_r)$	Typical Frequencies Observed as EMI (10th Harmonic) $f_{max} = 10*f$
74L	31–35	10 MHz	100 MHz
74C	25–60	13 MHz	130 MHz
CD4 (CMOS)	25 NS	13 MHz	130 MHz
74HC	13–15 ns	24 MHz	240 MHz
74	10–12 ns	32 MHz	320 MHz
(Flip-flop)	15–22 ns	21 MHz	210 MHz
74LS	9.5 ns	34 MHz	340 MHz
(Flip-flop)	13–15 ns	24 MHz	240 MHz
74H	4–6 ns	80 MHz	800 MHz
74S	3–4 ns	106 MHz	1.1 GHz
74HCT	5–15 ns	64 MHz	640 MHz
74ALS	2–10 ns	160 MHz	1.6 GHz
74ACT	2–5 ns	160 MHz	1.6 GHz
74F	1.5–1.6 ns	212 MHz	2.1 GHz
ECL 10K	1.5 ns	212 MHz	2.1 GHz
ECL 100K	0.75 ns	424 MHz	4.2 GHz

To prevent RF coupling between different bandwidth areas, functional partitioning is used. Partitioning is another word for the physical separation between functional sections. Partitioning is product specific and may be achieved using multiple PCBs, isolation, various topology layouts, or other creative means.

Proper partitioning allows for optimal functionality and ease of routing traces while minimizing trace lengths. It allows smaller RF loop areas to exist while optimizing signal quality. The design engineer must specify which components are associated with each functional subsection. Use the information provided by the component manufacturer and design engineer to optimize component placement prior to routing any traces.

6.1.13 Critical Frequencies ($\lambda/20$)

Throughout this chapter, reference is made to critical frequencies or high-threat clock and periodic signal traces that have a length greater than $\lambda/20$. The following is provided to show how one calculates the wavelength of a signal and the corresponding critical frequency. A summary of miscellaneous frequencies and their respective wavelength distance is shown in Table 6.5 based on the following equations.

$$f_{MHz} = \frac{300}{\lambda(m)} = \frac{984}{\lambda(ft)}$$

$$\lambda(m) = \frac{300}{f_{MHz}}$$

$$\lambda(ft) = \frac{984}{f_{MHz}} \tag{6.13}$$

TABLE 6.5 1/20 Wavelength at Various Frequencies

Frequency of Interest	$\lambda/20$ Wavelength Distance
10 MHz	1.5 m (5 ft)
27 MHz	0.56 m (1.8 ft)
35 MHz	0.43 m (1.4 ft)
50 MHz	0.3 m (12 in)
80 MHz	0.19 m (7.52 in)
100 MHz	0.15 m (6 in)
160 MHz	9.4 cm (3.7 in)
200 MHz	7.5 cm (3 in)
400 MHz	3.6 cm (1.5 in)
600 MHz	2.5 cm (1.0 in)
1000 MHz	1.5 cm (0.6 in)

6.1.14 Fundamental Principles and Concepts for Suppression of RF Energy

Fundamental Principles

The fundamental principals that describe EMI and common-mode noise created within a PCB are detailed below. These fundamental principles deal with energy transferred from a source to load. Common-mode currents are developed between circuits, not necessarily only within a power distribution system. Common-mode currents, by definition, are associated to both the power and return structure in addition to being created by components. To minimize common-mode currents, a metallic chassis is commonly provided. Since the movement of a charge occurs through an impedance (trace, cable, wire, and the line), a voltage will be developed across this impedance due to the voltage potential difference between two points. This voltage potential difference will cause radiated emissions to develop if electrically long transmission lines, I/O cables, enclosure apertures, and slots are present.

The following principles are discussed later in this chapter.

1. For high-speed logic, higher-frequency components will create higher fundamental RF frequencies based on fast edge-time transitions.
2. To minimize distribution of RF currents, proper layout of PCB traces, component placement, and provisions to allow RF currents to return to their source must be provided in an efficient manner. This keeps RF energy from being propagated throughout the structure.
3. To minimize development of common-mode RF currents, proper decoupling of switching devices, along with minimizing ground bounce and ground noise voltage within a plane structure, must occur.
4. To minimize propagation of RF currents, proper termination of transmission line structures must occur. At low frequencies, RF currents are not a major problem. At higher frequencies, RF currents will be developed that can radiate easily within the structure or enclosure.
5. Provide for an optimal 0-V reference. An appropriate grounding methodology needs to be specified and implemented early in the design cycle.

Fundamental Concepts

One fundamental concept for suppressing RF energy within a PCB deals with flux cancellation or minimization. As discussed earlier, current that travels in a transmission line or interconnect causes magnetic lines of flux to be developed. These lines of magnetic flux develop an electric field. Both field structures allow RF energy to propagate. If we cancel or minimize magnetic lines of flux, RF energy will not be present, other than within the boundary between the trace and image plane. Flux cancellation or minimization virtually guarantees compliance with regulatory requirements.

The following concepts that must be understood to minimize radiated emissions.

1. Minimize common-mode currents developed as a result of a voltage traveling across an impedance.
2. Minimize distribution of common-mode currents throughout the network.

Flux cancellation or minimization within a PCB is necessary because of the following sequence of events.

1. Current transients are caused by the production of high-frequency signals based on a combination of periodic signals (e.g., clocks) and nonperiodic signals (e.g., high-speed data busses) demanded from the power and ground plane structure.
2. RF voltages, in turn, are the product of current transients and the impedance of the return path provided (Ohm's law).
3. Common-mode RF currents are created from the RF voltage drop between two devices. This voltage drop builds up on inadequate RF return paths between source and load (insufficient differential-mode cancellation of RF currents).
4. Radiated emissions will propagate as a result of these common-mode RF currents.

To summarize what is to be presented later,

- Multilayer boards provide superior signal quality and EMC performance, since signal impedance control through stripline or microstrip is observed. The distribution impedance of the power and ground planes must be dramatically reduced. These planes contain RF spectral current surges caused by logic crossover, momentary shorts, and capacitive loading on signals with wide buses. Central to the issue of microstrip (or stripline) is understanding flux cancellation or flux minimization that minimizes (controls) inductance in any transmission line. Various logic devices may be quite asymmetrical in their pull-up/pull-down current ratios.
- Asymmetrical current draw in a PCB causes an imbalanced situation to exist. This imbalance relates to flux cancellation or minimization. Flux cancellation will occur through return currents present within the ground or power plane, or both, depending on stackup and component technology. Generally, ground (negative) returns for TTL is preferred. For ECL, positive return is use, since ECL generally runs on 5.2 V, with the more positive line at ground potential. At low

frequencies, CMOS current draw is very low, creating little difference between ground and voltage planes. One must look at the entire equivalent circuit before making a judgment.

Where three or more image planes are solid provided in a multilayer stackup assembly (e.g., one power and two ground planes), optimal flux cancellation or minimization is achieved when the RF return path is adjacent to a solid image plane, at a common potential throughout the entire trace route. This is one of the basic fundamental concepts of implementing flux cancellation within a PCB.[1,11]

To briefly restate this important concept related to flux cancellation or minimization, it is noted that not all components behave the same way on a PCB related to their pull-up/pull-down current ratios. For example, some devices have 15 mA pull-up/65 mA pull-down. Other devices have 65 mA pull-up/pull-down values (or 50%). When many components are provided within a PCB, asymmetrical power consumption will occur when all devices switch simultaneously. This asymmetrical condition creates an imbalance in the power and ground plane structure. The fundamental concept of board-level suppression lies in flux cancellation (minimization) of RF currents within the board related to traces, components, and circuits referenced to a 0-V reference. Power planes, due to this flux phase shift, may not perform as well for flux cancellation as ground planes due to the asymmetry noted above. Consequently, optimal performance may be achieved when traces are routed adjacent to 0-V reference planes rather than adjacent to power planes.

6.1.15 Summary

These are the key points on how EMC is created within the PCB:

1. Current transients are developed from the production of high-frequency periodic signals.
2. RF voltage drops between components are the products of currents traveling through a common return impedance path.
3. Common-mode currents are created by unbalanced differential-mode currents, which are created by an inadequate ground return/reference structure.
4. Radiated emissions are generally caused by common-mode currents.

Section 6.1 References

1. Montrose, M. I. 1999. *EMC and the Printed Circuit Board—Design, Theory and Layout Made Simple.* Piscataway, NJ: IEEE Press.
2. Paul, C. R. 1992. *Introduction to Electromagnetic Compatibility,* New York, NY: John Wiley & Sons, Inc.
3. Ott, H. 1988. *Noise Reduction Techniques in Electronic Systems.* 2nd ed. New York: John Wiley & Sons.
4. Dockey, R. W., and R. F. German. 1993. "New Techniques for Reducing Printed Circuit Board Common-Mode Radiation." *Proceedings of the IEEE International Symposium on Electromagnetic Compatibility,* New York: IEEE, pp. 334–339.
5. Johnson, H. W., and M. Graham. 1993. *High Speed Digital Design.* Englewood Cliffs, NJ: Prentice Hall.
6. German, R. F., H. Ott, and C. R. Paul. 1990. "Effect of an image plane on PCB radiation." *Proceedings of the IEEE International Symposium on Electromagnetic Compatibility,* New York: IEEE, pp. 284–291.
7. Hsu, T. 1991. "The Validity of Using Image Plane Theory to Predict Printed Circuit Board Radiation." *Proceedings of the IEEE International Symposium on Electromagnetic Compatibility,* New York: IEEE, pp. 58–60.
8. Mardiguian, M. 1992. *Controlling Radiated Emissions by Design.* New York: Van Nostrand Reinhold.
9. Montrose, M. I. 1991. "Overview of Design Techniques for Printed Circuit Board Layout Used in High Technology Products." *Proceedings of the IEEE International Symposium on Electromagnetic Compatibility,* New York: IEEE, pp. 61–66.

10. Gerke, D., and W. Kimmel, 1994. "The Designers Guide to Electromagnetic Compatibility." *EDN* January 10.
11. Montrose, M. I. 1996. *Printed Circuit Board Design Techniques for EMC Compliance.* Piscataway, NJ: IEEE Press.

6.2 Transmission Lines and Impedance Control

6.2.1 Overview on Transmission Lines

With today's high-technology products and faster logic devices, PCB transmission line effects become a limiting factor for proper circuit operation. A trace routed adjacent to a reference plane forms a simple transmission line. Consider the case of a multilayer PCB. When a trace is routed on an outer PCB layer, we have the microstrip topology, although it may be asymmetrical in construction. When a trace is routed on an internal PCB layer, the result is called *stripline topology*. Details on the microstrip and stripline configurations are provided in this section.

A transmission line allows a signal to propagate from one device to another at or near the speed of light within a medium, modified (slowed down) by the capacitance of the trace and by active devices in the circuit. If a transmission line is not properly terminated, circuit functionality and EMI concerns can exist. These concerns include voltage droop, ringing, and overshoot. All concerns will severely compromise switching operations and system signal integrity. Transmission line effects must be considered when the round-trip propagation delay exceeds the switching-current transition time. Faster logic devices and their corresponding increase in edge rates are becoming more common in the subnanosecond range. A very long trace in a PCB can become an antenna for radiating RF currents or cause functionality problems if proper circuit design techniques are not used early in the design cycle.

A transmission line contains some form of energy. Is this energy transmitted by electrons, line voltages and currents, or by something else? In a transmission line, electrons do not travel in the conventional sense. An *electromagnetic field* is the component that is present within and around a transmission line. The energy is carried along the transmission line by an electromagnetic field.

When dealing with transmission line effects, the impedance of the trace becomes an important factor when designing a product for optimal performance. A signal that travels down a PCB trace will be absorbed at the far end if, and only if, the trace is terminated in its characteristic impedance. If proper termination is not provided, most of the transmitted signal will be reflected back in the opposite direction. If an improper termination exists, multiple reflections will occur, resulting in a longer signal-settling time. This condition is known as ringing, discussed in Section 6.4.

When a high-speed electrical signal travels through a transmission line, a propagating electromagnetic wave will move down the line (e.g., a wire, coaxial cable or PCB trace). A PCB trace looks very different to the signal source at high signal speeds from the way it does at dc or at low signal speeds. The characteristic impedance of the transmission line is identified as Z_o. For a lossless line, the characteristic impedance is equal to the square root of L/C, where L is the inductance per unit length divided by C, the capacitance per unit length. Impedance is also the ratio of the line voltage to the line current, in analogy to Ohm's law. When we examine Eq. (6.14), we see subscripts for the line voltage and the line current. The ratio of line voltage to line current is constant with respect to the line distance x only for a matched termination. The (x) subscript indicates that variations in V and I will exist along the line, except for special cases. Other mathematical formulas are present for different units of measurements, also shown in Eq. (6.14).

$$Z_o = \sqrt{\frac{L_o}{C_o}} = \frac{V_x}{I_x} \ \Omega; \quad L_o = C_o \times Z_o^2 \ \text{pH/in}; \quad L = 5 \times \ln\frac{2 \times \pi \times H}{W} \ \text{nH/in} \quad (6.14)$$

where Z_o = characteristic impedance
 L_o = inductance of the transmission line per unit length

C_o = capacitance of the transmission line per unit length
H = height of the transmission line above a reference source
W = width of the transmission line (PCB trace)

We now examine characteristic impedance. As a data signal transitions from a low to a high state, or vice versa, and propagates down a PCB trace, the impedance it encounters (the voltage to current ratio) is equal to a specific characteristic impedance. Once the signal has propagated up and down, trace reflections, if any, have died or become a non-issue related to signal integrity when the quiescent state is achieved. The characteristic impedance of the trace now has no effect on the signal. The signal becomes dc and the line behaves as a typical wire.

Various techniques are available for creating transmission lines for the transference of RF energy between a source and load within a PCB. Another word for transmission line, when referenced to a PCB, is "trace." Two basic topologies are available for developing a transmission line structure. Each topology has two basic configurations: microstrip (single and embedded) and stripline (single and dual). Another topology available is coplanar, which is implemented in both microstrip and stripline configuration.

Note: None of the equations provided in the next section for microstrip and stripline is applicable to PCBs constructed of two or more dielectric materials, excluding air, or fabricated with more than one type of laminate. All equations are extracted from IPC-D-317A, *Design Guidelines for Electronic Packaging Utilizing High-Speed Techniques.*[*]

6.2.2 Creating Transmission Lines in a PCB

Different logic families have different characteristic source impedances. Emitter-coupled logic (ECL) has a source and load impedance of 50 Ω. Transistor-transistor logic (TTL) has a source impedance range of 70 to 100 Ω. If a transmission line is to be created within a PCB, the engineer must seek to match the source and load impedance of the logic family being used.[1,2,4,5]

Most high-speed traces must be impedance controlled. Calculations to determine proper trace width and separation to the nearest reference plane must occur. Board manufacturers and CAD programs can easily perform these calculations. If necessary, board fabricators can be consulted for assistance in designing the PCB, or a computer application program can be used to determine the most effective approach relative to trace width and distance spacing between planes for optimal performance. These approximate formulas may not be fully accurate because of manufacturing tolerances during the fabrication process. These formulas were simplified from exact models. Stock material may have a different thickness value and a different dielectric constant. The finished etched trace width may be different from a desired design requirement, or any number of manufacturing issues may exist. The board vendors know what the real variables are during construction and assembly. These vendors should be consulted to provide the real or actual dielectric constant value, as well as the finished etched trace width for both base and crest dimensions, as detailed in Fig. 6.22.

Microstrip Topology

Microstrip is one topology used to provide trace-controlled impedance on a PCB for digital circuits. Microstrip lines are exposed to both air and a dielectric referenced to a planar structure. The approximate formula for calculating the characteristic impedance of a surface microstrip trace is provided in Eq. (6.15) for the configuration of Fig. 6.23. The capacitance of the trace is described by Eq. (6.16).

$$Z_o = \left(\frac{87}{\sqrt{\varepsilon_r + 1.41}}\right) \ln\left(\frac{5.98H}{0.8W + T}\right) \qquad (6.15)$$

---

[*]Within the IPC standards, typographical and mathematical errors exist in the section related to impedance calculation. Before applying the equations detailed within IPC-D-317, study and identify all errors before literal use. Equations presented herein have been independently verified for accuracy.

EMC and Printed Circuit Board Design

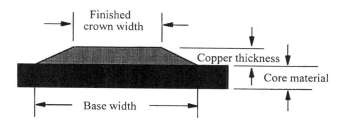

FIGURE 6.22 Finished trace width dimensions after etching.

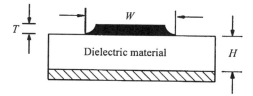

FIGURE 6.23 Surface microstrip topology.

$$C_o = \frac{0.67(\varepsilon_r + 1.41)}{\ln\left(\frac{5.98H}{0.8W + T}\right)} \quad \text{pF/in} \qquad (6.16)$$

where Z_o = characteristic impedance (Ω)
W = width of the trace
T = thickness of the trace
H = distance between signal trace and reference plane
C_o = intrinsic capacitance of the trace (pF/unit distance)
ε_r = dielectric constant of the planar material

Note: Use consistent dimensions for the above (inches or centimeters).

Equation (6.15) is typically accurate to ±5% when the ratio of W to H is 0.6 or less. When the ratio of W to H is between 0.6 and 2.0, accuracy drops to ±20%.

When measuring or calculating trace impedance, the width of the line should technically be measured at the middle of the trace thickness. Depending on the manufacturing process, the finished line width after etching may be different from that specified by Fig. 6.23. The width of the copper on the top of the trace may be etched away, thus making the trace width smaller than desired. Using the average between top and bottom of the trace thickness, we find that a more typical, accurate impedance number is possible. With respect to the measurement of a trace's width, with a ln (natural logarithm) expression, how much significance should we give to achieving a highly accurate trace impedance value for the majority of designs? Most manufacturing tolerances are well within 10% of desired impedance.

The propagation delay of a signal that is routed microstrip is described by Eq. (2.4). This equation has a variable of only ε_r, or dielectric constant. This equation states that the propagational speed of a signal within this transmission line is related only to the effective permittivity of the dielectric material. Kaupp derived this equation for the propagation delay function under the square root radical.[10]

$$t_{pd} = 1.017\sqrt{0.475\varepsilon_r + 0.67} \quad \text{(ns/ft)}$$

$$t_{pd} = 85\sqrt{0.475\varepsilon_r + 0.67} \quad \text{(ps/in)} \qquad (6.17)$$

Embedded Microstrip Topology

The embedded microstrip is a modified version of standard microstrip. The difference lies in providing a dielectric material on the top surface of the copper trace. This dielectric material may include another routing layer, such as core and prepreg. This material may also be solder mask, conformal coating, potting, or other material required for functional or mechanical purposes. As long as the material provided contains the same dielectric constant, with a thickness of 0.008 to 0.010 in (0.0020 to 0.0025 mm) air or the environment will have little effect on the impedance calculations. Another way to view embedded microstrip is to compare it to a single, asymmetric stripline with one plane infinitely far away.

Coated microstrip uses the same conductor geometry as uncoated except that the effective relative permittivity will be higher. *Coated microstrip* refers to placing a substrate on the outer microstrip layer. This substrate can be solder mask, conformal coating, or another material, including another microstrip layer. The dielectric on top of the trace may be asymmetrical to the host material. The difference between coated and uncoated microstrip is that the conductors on the top layer are fully enclosed by a dielectric substrate. The equations for embedded microstrip are the same as those for uncoated microstrip, with a modified permittivity, ε_r'. If the dielectric thickness above the conductor is more than a few thousandths of an inch, ε_r' will need to be determined either by experimentation or through use of an electromagnetic field solver. For "very thin" coatings, such as solder mask or conformal coating, the effect is negligible. Masks and coatings will drop the impedance of the trace by several ohms.

The approximate characteristic impedance for embedded microstrip impedance is provided by Eq. (6.18). For embedded microstrip, particularly those with asymmetrical dielectric heights, knowledge of the base and crown widths after etching will improve accuracy. These formulas are reasonable as long as the thickness of the upper dielectric material $[B - (T + H)]$ is greater than 0.004 in (0.001 mm). If the coating is thin, or if the relative dielectric coefficient of the coating is different (e.g., conformal coating), the impedance will typically be between those calculated for both microstrip and embedded microstrip.

The characteristic impedance of embedded microstrip is shown in Eq. (6.18) for the configuration shown in Fig. 6.24. The intrinsic capacitance of the trace is defined by Eq. (6.19).

$$Z_o = \left(\frac{87}{\sqrt{\varepsilon_r' + 1.41}}\right) \ln\left(\frac{5.98H}{0.8W + T}\right) \; \Omega \tag{6.18}$$

where $\varepsilon_r' = \varepsilon_r \left\{ 1 - e^{-\frac{1.55B}{H}} \right\}$

$$C_o = \left(\frac{1}{H + T}\right) \times \ln\left(1 - \frac{0.6897(\varepsilon_r + 1.41)}{\sqrt{\varepsilon_r}}\right) \; \text{pF/in} \tag{6.19}$$

where Z_o = characteristic impedance (Ω)
  C_o = intrinsic capacitance of the trace (pF/unit distance)

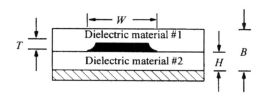

NOTE: Thickness of the dielectric material may asymmetric

FIGURE 6.24 Embedded microstrip topology.

W = width of the trace
T = thickness of the trace
H = distance between signal trace and reference plane
B = overall distance of both dielectrics
ε_r = dielectric constant of the planar material
$0.1 < W/H < 3.0$
$0.1 < \varepsilon_r < 15$

Note: Use consistent dimensions for the above (inches or centimeters).

The propagation delay of a signal routed embedded microstrip is given in Eq. (6.20). For a typical embedded microstrip, with FR-4 material and a dielectric constant that is 4.1, propagation delay is 0.35 ns/cm or 1.65 ns/ft (0.137 ns/in). This propagation delay is the same as single stripline, discussed next, except with a modified ε_r'.

$$t_{pd} = 1.017\sqrt{\varepsilon_r'} \quad \text{(ns/ft), or}$$

$$t_{pd} = 85\sqrt{\varepsilon_r'} \quad \text{(ps/in)} \tag{6.20}$$

where $\varepsilon_r' = \varepsilon_r \left[1 - e^{\left(\frac{-1.55B}{H}\right)}\right]$; $0.1 < W/H < 3.0$; $1 < \varepsilon_r < 15$

Single Stripline Topology

Stripline topology refers to a trace that is located between two planar conductive structures with a dielectric material that completely surrounds the trace (Fig. 6.25). Consequently, stripline traces are routed internal to the board and are not exposed to the external environment.

Stripline, compared to microstrip, has several advantages. Namely, stripline captures magnetic fields and minimizes crosstalk between routing layer. Stripline also provides an RF current reference return path for magnetic field flux cancellation. The two reference planes will capture any energy that is radiated from a routed trace. Reference planes prevent RF energy from radiating to the outside environment. Radiated emissions will still exist, but not from the traces. Emissions will be propagated from the physical components that are located on the outside layers of the board. In addition, the bond lead wires inside a component's package may be optimal in length for developing radiated RF emissions. Even with a perfect layout and design for suppression of EMI energy within a PCB, component radiation may be far greater than any other aspect of the design requiring containment of the fields by a metal chassis or enclosure.

When measuring or calculating trace impedance, the microstrip section should be consulted for a discussion of why we should measure trace impedance of the line at the middle of the trace thickness after etching.

The approximate characteristic impedance for single stripline impedance is provided in Eq. (6.21) for the illustration in Fig. 6.25. Intrinsic capacitance is presented by Eq. (6.22). Note that Eqs. (6.21 and

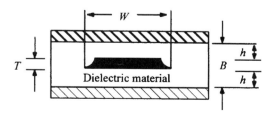

FIGURE 6.25 Single stripline topology.

6.22) are based on variables chosen for an optimal value of height, width, and trace thickness. During actual board construction, the impedance may vary by as much as ±5% from calculated values.

$$Z_o = \left(\frac{60}{\sqrt{\varepsilon_r}}\right) \ln\left(\frac{1.9B}{0.8W + T}\right) \; \Omega \tag{6.21}$$

$$C_o = \frac{1.41\varepsilon_r}{\ln\left(\frac{3.8h}{0.8W + T}\right)} \; \text{pF/in} \tag{6.22}$$

where Z_o = characteristic impedance (Ω)
 W = width of the trace
 T = thickness of the trace
 B = distance between both reference planes
 h = distance between signal plane and reference plane
 C_o = intrinsic capacitance of the trace (pF/unit distance)
 ε_r = dielectric constant of the planar material
 $W/(H - T) < 0.35$
 $T/H < 0.25$

Note: Use consistent dimensions for the above (inches or centimeters).

The propagation delay of signal stripline is described by Eq. (6.23), which has only ε_r as the variable.

$$tpd = 0.017\sqrt{\varepsilon_e} \; \text{(ns/ft), or}$$

$$tpd = 85\sqrt{\varepsilon_e} \; \text{(ps/in)} \tag{6.23}$$

Dual or Asymmetric Stripline Topology

A variation on single stripline topology is dual or asymmetric stripline, which increases coupling between a circuit plane and the nearest reference plane. When the circuit is placed approximately in the middle one-third of the interplane region, the error caused by assuming the circuit to be centered will be quite small and will fall within the tolerance range of the assembled board.

The approximate characteristic impedance for dual stripline is provided by Eq. (6.24) for the illustration of Fig. 6.26. This equation is a modified version of that used for single stripline. Note that the same approximation reason for single stripline is used to compute Z_o.

$$Z_o = \left(\frac{80}{\sqrt{\varepsilon_r}}\right) \ln\left[\frac{1.9(2H + T)}{0.8W + T}\right]\left[1 - \frac{H}{4(H + D + T)}\right]$$

$$Z_o = \left(\frac{80}{\sqrt{\varepsilon_r}}\right) \ln\left[\frac{1.9(2H + T)}{0.8W + T}\right]\left[1 - \frac{H}{4(H1)}\right] \tag{6.24}$$

$$C_o = \frac{2.82\varepsilon_r}{\ln\left[\frac{2(H - T)}{0.268W + 0.335T}\right]} \tag{6.25}$$

where Z_o = characteristic impedance (Ω)
 W = width of the trace
 T = thickness of the trace
 D = distance between signal planes

EMC and Printed Circuit Board Design

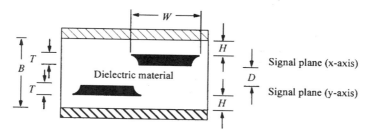

FIGURE 6.26 Dual or asymmetric stripline topology.

H = dielectric thickness between signal plane and reference plane
C_o = intrinsic capacitance of the trace (pF/unit distance)
ε_r = dielectric constant of the planar material

$W/(H - T) < 0.35$
$T/H < 0.25$

Note: Use consistent dimensions for the above (inches or centimeters).

Equation (6.24) can be applied to the asymmetrical (single) stripline configuration when the trace is not centered equally between the two reference planes. In this situation, H is the distance from the center of the line to the nearest reference plane. The letter D would become the distance from the center of the line being evaluated to the other reference plane.

The propagation delay for the dual stripline configuration is the same as that for the single stripline, since both configurations are embedded in a homogenous dielectric material.

$$t_{pd} = 1.017\sqrt{\varepsilon_r} \text{ (ns/ft), or}$$

$$t_{pd} = 85\sqrt{\varepsilon_r} \text{ (ps/in)} \tag{6.26}$$

Note: When using the dual stripline, both routing layers must be routed orthogonal to each other. This means that one routing layer is provided for the x-axis traces, while the other layer is used for y-axis traces. Routing these layers at a 90° angle prevents crosstalk from being developed between the two planes, especially when wide busses or high-frequency traces can cause data corruption to the alternate routing layer.

The actual operating impedance of a line can be significantly influenced (e.g., ≈30%) by multiple high-density crossovers of orthogonally routed traces, increasing the loading on the net and reducing the impedance of the transmission line. This impedance change occurs because orthogonally routed traces include a loaded impedance circuit to the image plane, along with capacitance to the signal trace under observation. This is best illustrated by Fig. 6.27.

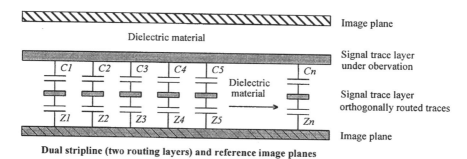

FIGURE 6.27 Impedance influence on dual stripline routing planes.

Differential Microstrip and Stripline Topology

Differential traces are two conductors physically adjacent to each other throughout the entire trace route. The impedance for differentially routed traces is not the same as a single-ended trace. For this configuration, line-to-ground (or reference plane) impedance is sometimes only considered, as if the traces were routed single-ended. This concern should also be with the line-to-line impedance between the two traces operating in differential mode.

For Fig. 6.28, differential traces are shown. If the configuration is microstrip, the upper reference plane is not provided. For stripline, both reference planes are provided with equal, center spacing between the parallel traces and the two planes.

When calculating differential impedance, Z_o (Z_{diff}), only trace width W should be adjusted to alter Z_{diff}. The user should not adjust the distance spacing between the traces, identified as D, which should be the minimal spacing specified by the PCB vendor that is manufacturable.[3] Technically, as long as the routed length of the differential pair is approximately the same, taking into consideration the velocity of propagation of the electromagnetic field within the transmission line, extreme accuracy on matched trace lengths need not occur. The speed of propagation is so fast that a minor difference in routed lengths will not be observed by components within the net. This fact is true for signals that operate below 1 GHz.

The reason for routing differential traces coplanar is for EMI control—flux cancellation. An electromagnetic field propagates down one trace and returns on the other. With closer spacing between differentially trace, enhanced suppression of RF energy occurs. This is the basis of why ground traces perform as well as they do when routed as a differential trace. Routing differential traces on different layers using the same trace width is an acceptable practice for signal integrity but not for EMC compliance. When routing differential traces on different layers, one must be aware at all times where the RF return current flows, especially when jumping layers.

It is also important to note that, when routing differential traces, the trace impedance requirement for various logic families may be different. For example, all traces on a particular routing layer may be designed for 55 Ω. The differential impedance for a trace pair may have to be specified at 82 Ω. This means that one must use a different trace width for the paired traces, per layer, which is not always easy to implement by the PCB designer.

$$Z_{diff} \approx 2 \times Z_o \left(1 - 0.48 e^{-0.96\frac{D}{h}}\right) \Omega \quad \text{microstrip}; \quad Z_{diff} \approx 2 \times Z_o \left(1 - 0.347 e^{-2.9\frac{D}{h}}\right) \Omega \quad \text{stripline}$$

(6.27)

where

$$Z_o = \frac{87}{\sqrt{\varepsilon_r + 1.41}} \ln\left[\frac{5.95 H}{0.8 W + T}\right] \Omega \quad \text{microstrip}$$

$$Z_o = \frac{60}{\sqrt{\varepsilon_r}} \ln\left[\frac{1.9 B}{0.8 W + T}\right] \Omega \quad \text{stripline}$$

(6.28)

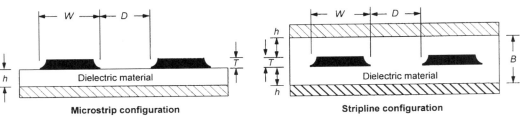

FIGURE 6.28 Differential trace routing topology.

where B = plane separation
W = width of the trace
T = thickness of the trace
D = trace edge-to-edge spacing
h = distance spacing to nearest reference plane

Note: Use consistent dimensions for the above (inches or centimeters).

6.2.3 Relative Permittivity (Dielectric Constant)

What is the electrical propagation mode present within a transmission line (PCB traces), free space or any conducting media or structure? A transmission line allows a signal to propagate from one device to another at or near the speed of light, modified (slowed down) by the capacitance of the traces and by active devices in the circuit. This signal contains some form of energy. Is this energy classified as electrons, voltage, current, or something else? In a transmission line, electrons do not travel in the conventional sense. An *electromagnetic field* is the component that is present within and around a transmission line. This energy is carried within the transmission line by an electromagnetic field. Propagation delay increases in proportion to the square root of the dielectric constant of the media in which the electromagnetic field propagates through. This slowing of the signal is based on the relative permittivity, or dielectric constant of the material.[1]

Electromagnetic waves propagate at a speed that depends on the electrical properties of the surrounding medium. Propagation delay is typically measured in units of picoseconds/inch. Propagation delay is the inverse of velocity of propagation (the speed at which data is transmitted through conductors in a PCB). The dielectric constant varies with several material parameters. Factors that influence the relative permittivity of a given material include the electrical frequency, temperature, extent of water absorption (also forming a dissipative loss), and the electrical characterization technique. In addition, if the PCB material is a composite of two or more laminates, the value of ε_r may vary significantly as the relative amount of resin and glass of the composite is varied.[4]

The relative dielectric constant, ε_r, is a measure of the amount of energy stored in a dielectric insulator per unit electric field. It is a measure of the capacitance between a pair of conductors (trace–air, trace–trace, wire–wire, trace–wire, etc.) within the dielectric insulator, compared to the capacitance of the same conductor pair in a vacuum. The dielectric constant of a vacuum is 1.0. All materials have a dielectric constant greater than 1. The larger the number, the more energy stored per unit insulator volume. The higher the capacitance, the slower the electromagnetic wave travels down the transmission line.

In air, or vacuum, the velocity of propagation is the speed of light. In a dielectric material, the velocity of propagation is slower (approximately 0.6 times the speed of light for common PCB laminates). Both the velocity of propagation and the effective dielectric constant are given by Eq. (6.29).

$$V_p = \frac{C}{\sqrt{\varepsilon_r}} \quad \text{(velocity of propagation)}$$

$$\varepsilon_r' = \left(\frac{C}{V_p}\right)^2 \quad \text{(dielectric constant)}$$

(6.29)

where $C = 3 \times 10^8$ m/sec, or about 30 cm/ns (12 in/ns)
ε_r' = effective dielectric constant
V_p = velocity of propagation

The effective relative permittivity, ε_r', is the relative permittivity that is experienced by an electrical signal transmitted along a conductive path. Effective relative permittivity can be determined by using a time domain reflectometer (TDR) or by measuring the propagation delay for a known length line and calculating the value.

The propagation delay and dielectric constant of common PCB base materials are presented in Table 6.6. Coaxial cables often use a dielectric insulator to reduce the effective dielectric insulator inside the cable to improve performance. This dielectric insulator lowers the propagation delay while simultaneously lowering the dielectric losses.

TABLE 6.6 Propagation Delay in Various Transmission Media

Medium	Propagation Delay (ps/in)	Relative Dielectric Constant
Air	85	1.0
FR–4 (PCB), microstrip	141–167	2.8–4.5
FR–4 (PCB), stripline	180	4.5
Alumina (PCB), stripline	240–270	8–10
Coax (65% velocity)	129	2.3
Coax (75% velocity)	113	1.8

FR-4, currently the most common material used in the fabrication of a PCB, has a dielectric constant that varies with the frequency of the signal within the material. Most engineers have generally assume that ε_r was in the range of 4.5 to 4.7. This value, referenced by designers for over 20 years, has been published in various technical reference manuals. This value was based on measurements taken with a 1-MHz reference signal. Measurements were not made on FR-4 material under actual operating conditions, especially with today's high-speed designs. What worked over 20 years ago is now insufficient for twenty-first century products. Knowledge of the correct value of ε_r for FR-4 must now be introduced. A more accurate value of ε_r is determined by measuring the *actual* propagation delay of a signal in a transmission line using a TDR. The values in Table 6.7 are based on a typical, high-speed edge rate signal.

TABLE 6.7 Dielectric Constants and Wave Velocities within Various PCB Materials

Material	ε_r (at 30 MHz)	Velocity (in/ns)	Velocity (ps/in)
Air	1.0	11.76	85.0
PTFE/glass (Teflon™)	2.2	7.95	125.8
RO 2800	2.9	6.95	143.9
CE/custom ply (Canide ester)	3.0	6.86	145.8
BT/custom ply (Beta-triazine)	3.3	6.50	153.8
CE/glass	3.7	6.12	163.4
Silicon dioxide	3.9	5.97	167.5
BT/glass	4.0	5.88	170.1
Polyimide/glass	4.1	5.82	171.8
FR–4 glass	4.5	5.87	170.4
Glass cloth	6.0	4.70	212.8
Alumina	9.0	3.90	256.4

Note: Values measured at TDR frequencies using velocity techniques.
Values were not measured at 1 MHz, which provides faster velocity values. Units for velocity differ due to scaling and are presented in this format for ease of presentation.
Source: IPC–2141, *Controlled Impedance circuit Boards and High Speed Logic Design*, Institute for Interconnecting and Packaging Electronics Circuits. © 1996. Reprinted with permission.

Figure 6.29 shows the "real" value of ε_r for FR-4 material based on research by the Institute for Interconnecting and Packaging Electronics Circuits Organization (IPC). This chart was published in document IPC-2141, *Controlled Impedance Circuit Boards and High Speed Logic Design*. The figure shows the frequency range from 100 kHz to 10 GHz for FR-4 laminate with a glass-to-resin ratio of approximately 40:60 by weight, respectively. The value of ε_r for this laminate ratio varies from about 4.7 to 4.0 over this large frequency range. This change in the magnitude of ε_r is due principally to the frequency

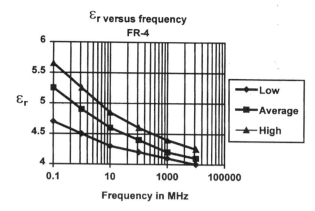

FIGURE 6.29 Actual dielectric constant values for FR-4 material.

response of the resin, and is reduced if the proportion of the glass-to-resin ratio in the composite is increased. In addition, the frequency response will also be changed if an alternative resin system is selected. Material suppliers typically quote values of dielectric properties determined at 1-MHz, not at actual system frequencies that now easily exceed 100 MHz.[5]

If a TDR is used for measuring the velocity of propagation, it must use a frequency corresponding to the actual operating conditions of the PCB for comparing dielectric parameters. The TDR is a wideband measurement technique using time domain analysis. The location of the TDR on the trace being measured may affect measurement values. IPC-2141 provides an excellent discussion of how to use a TDR for propagational delay measurements.

The dielectric constant of various materials used to manufacture a PCB is provided in Table 6.7. These values are based on measurements using a TDR and not on published limited-basis reference information. In addition, the dielectric constant for these materials changes with temperature. For FR-4 glass epoxy, the dielectric constant will vary as much as ±20% in the temperature range 0–70° C. If a stable substrate is required for a unique application, ceramic or Teflon may be a better choice than FR-4.

For microstrip topology, the effective dielectric constant is always lower than the number provided by the manufacturer of the material. The reason is that part of the energy flow is in air or solder mask, and part of the energy flows within the dielectric medium. Therefore, the signal will propagate faster down the trace than for the stripline configuration.

When a stripline conductor is surrounded by a single dielectric that extends to the reference planes, the value of ε_r' may be equated to that of ε_r for the dielectric measured under appropriate operating conditions. If more than one dielectric exists between the conductor and reference plane, the value of ε_r' is determined from a weighted sum of all values of ε_r for contributing dielectrics. Use of an electromagnetic field solver is required for a more accurate ε_r' value.[6-8] For purposes of evaluating the electrical characteristics of a PCB, a composite such as a reinforced laminate, with a specific ratio of compounds, is usually regarded as a homogeneous dielectric with an associated relative permittivity.

For microstrip with a compound dielectric medium consisting of board material and air, Kaupp[10] derived an empirical relationship that gives the effective relative permittivity as a function of board material. Use of an electromagnetic field solver is required for a more accurate answer.[6]

$$\varepsilon_r' = 0.475\varepsilon_r + 0.67 \quad \text{for } 2 < \varepsilon_r < 6 \tag{6.30}$$

In this expression, ε_r' relates to values determined at 25 MHz, the value used by Kaupp when he developed this equation.

Trace geometries also affect the electromagnetic field within a PCB structure. These geometries also determine if the electromagnetic field is radiated into free space or if it will stay internal to the assembly. If the electric field stays local to, or in the board, the effective dielectric constant becomes greater and signals propagate more slowly. The dielectric constant value will change internal to the board based on where the electric field shares its electrons. For microstrip, the electric field shares its electrons with free space, whereas stripline configurations capture free electrons. Microstrip permits faster propagation of electromagnetic waves. These electric fields are associated with capacitive coupling, owing to the field structure within the PCB. The greater the capacitive coupling between a trace and its reference plane, the slower the propagation of the electromagnetic field within the transmission line.

6.2.4 Capacitive Loading of Signal Traces

Capacitive input loading affects trace impedance and will increase with gate loading (additional devices added to the routed net). The unloaded propagation delay for a transmission line is defined by $t_{pd} = \sqrt{L_o C_o}$. If a lumped load, C_d, is placed in the transmission line (includes all loads with their capacitance added together), the propagation delay of the signal trace will increase by a factor of[1,2]

$$t'_{pd} = t_{pd}\sqrt{1 + \frac{C_d}{C_o}} \quad \text{ns/length} \quad (6.31)$$

where t_{pd} = unmodified propagation delay, nonloaded circuit
t'_{pd} = modified propagation delay when capacitance is added to the circuit
C_d = input gate capacitance from all loads added together
C_o = characteristic capacitance of the line/unit length

For C_o, units must be per unit length, not the total line capacitance.

For example, let's assume a load of five CMOS components are on a signal route, each with 10 pF input capacitance (total of C_d = 50 pF). With a capacitance value on a glass epoxy board, 25 mil traces, and a characteristic board impedance Z_o = 50 Ω (t_r = 1.65 ns/ft), there exists a value of C_o = 35 pF. The modified propagation delay is:

$$t'_{pd} = 1.65 \text{ ns/ft} \sqrt{1 + \frac{50}{35}} = 2.57 \text{ ns/ft} \quad (6.32)$$

This equation states that the signal arrives at its destination 2.57 ns/ft (0.54 ns/cm) later than expected. The characteristic impedance of this transmission line, altered by gate loading, Z'_o, is:

$$Z'_o = \frac{Z_o}{\sqrt{1 + \frac{C_d}{C_o}}} \quad (6.33)$$

where Z_o = original line impedance (Ω)
Z'_o = modified line impedance (Ω)
C_d = input gate capacitance—sum of all capacitive loads
C_o = characteristic capacitance of the transmission line

For the example above, trace impedance decreased from 50 to 32 Ω.

$$Z'_o = \frac{50}{\sqrt{1 + \frac{50}{35}}} = 32 \text{ Ω}$$

Typical values of C_d are 5 pF for ECL inputs, 10 pF for each CMOS device, and 10 to 15 pF for TTL. Typical C_o values of a PCB trace are 2 to 2.5 pF/in. These C_o values are subject to wide variations due to the physical geometry and the length of the trace. Sockets and vias also add to the distributed capacitance

EMC and Printed Circuit Board Design

(sockets ≈ 2 pF and vias ≈ 0.3 to 0.8 pF each). Given that $t_{pd} = \sqrt{L_o \times C_o}$ and $Z_o = \sqrt{L_o/C_o}$, C_o can be calculated as

$$C_o = 1000\left(\frac{t_{pd}}{Z_o}\right) \text{ pF/length} \tag{6.34}$$

This loaded propagation delay value is one method that may be used to decide if a trace should be treated as a transmission line ($2 \times t'_{pd} \times$ trace length $> t_r$ or t_f) where t_r is the rising edge of the signal and t_f the falling edge.

C_d, the distributed capacitance per length of trace, depends on the capacitive load of all devices including vias and sockets, if provided. To mask transmission line effects, slower edge times are recommended. A heavily loaded trace slows the rise and fall times of the signal due to an increased time constant ($\tau = ZC$) associated with increased distributed capacitance and filtering of high-frequency components from the switching device. Notice that impedance, Z, is used, and not R (pure resistance) in the time constant equation. This is because Z contains both real resistance and inductive reactance. Inductive reactance, ($j\omega L$), is much greater than R in the trace structure at RF frequencies, which must be taken into consideration. Heavily loaded traces seem advantageous, until the loaded trace condition is considered in detail.

A high C_d increases the loaded propagation delay and lowers the loaded characteristic impedance. The higher loaded propagation delay value increases the likelihood that transmission line effects will not be masked during rise and fall transition intervals. Lower loaded characteristic impedance often exaggerates impedance mismatches between the driving device and the PCB trace. Thus, the apparent benefits of a heavily loaded trace are not realized unless the driving gate is designed to drive large capacitive loads.[9]

Loading alters the characteristic impedance of the trace. As with the loaded propagation delay, a high ratio between distributed capacitance and intrinsic capacitance exaggerates the effects of loading on the characteristic impedance. Because $Z_o = \sqrt{L_o/(C_o + C_d)}$, the additional load, C_d, adds capacitance. The loading factor $\sqrt{1 + C_d/C_o}$ divides into Z_o, and the characteristic impedance is lowered when the trace is loaded. Reflections on a loaded trace, which cause ringing, overshoots, and switching delays, are more extreme when the loaded characteristic impedance differs substantially from the driving device's output impedance and the receiving device's input impedance. The units of measurements used for capacitance and inductance is in per *inch* or *centimeter* units. If the capacitance used in the L_o equation is picofarads per inch, the resulting inductance will be in picohenries per inch.

With knowledge of added capacitance lowering the trace impedance, it becomes apparent that, if a device is driving more than one line, the active impedance of each line must be determined separately. This determination must be based on the number of loads and the length of each line. Careful control of circuit impedance and reflections for trace routing and load distribution must be given serious consideration during the design and layout of the PCB.

If capacitive input loading is high, compensating a signal may not be practical. Compensation refers to modifying the transmitted signal to enhance the quality of the received signal pulse using a variety of design techniques. For example, use of a series resistor, or a different termination method to prevent reflections or ringing that may be present in the transmission line, is one method to compensate a distorted signal. Reflections in multiple lines from a single source must also be considered.

The low impedance often encountered in the PCB sometimes prevents proper Z_o (impedance) termination. If this condition exists, a series resistor in the trace is helpful (without corrupting signal integrity). Even a 10-Ω resistor provides benefit; however, a 33-Ω resistor is commonly used.

Section 6.2 References

1. Montrose, M. 1999. *EMC and the Printed Circuit Board Design—Design, Theory and Layout Made Simple*. Piscataway, NJ: IEEE Press.
2. Montrose, M. I. 1996. *Printed Circuit Board Design Techniques for EMC Compliance*. Piscataway, NJ: IEEE Press.

3. National Semiconductor, 1996. *LVDS Owner's Manual.*
4. IPC-D-317A. 1995, January. *Design Guidelines for Electronic packaging Utilizing High-Speed Techniques.* Institute for Interconnecting and Packaging Electronic Circuits (IPC).
5. IPC-2141. 1996, April. *Controlled Impedance Circuit Boards and High Speed Logic Design.* Institute for Interconnecting and Packaging Electronic Circuits.
6. Booton, R. C. 1992. *Computational Methods for Electromagnetics and Microwaves.* New York, NY: John Wiley & Sons, Inc.
7. Collin, R. R. 1992. *Foundations for Microwave Engineering.* Second Edition. Reading MA: Addison-Wesley Publishing Company.
8. Sadiku, M. 1992. *Numerical Techniques in Electromagnetics.* Boca Raton, FL: CRC Press, Inc.
9. Motorola, Inc. *Transmission Line Effects in PCB Applications* (#AN1051/D).
10. Kaupp, H. R. 1967, April. "Characteristics of Microstrip Transmission Lines," *IEEE Transactions.* Vol. EC-16, No. 2.

6.3 Signal Integrity, Routing and Termination

6.3.1 Impedance Matching—Reflections and Ringing

Reflections are an unwanted by-product in digital designs. Ringing within a transmission line contains both overshoot and reflection before stabilizing to a quiescent level, and it is a manifestation of the same effect. Overshoot is an excessive voltage level above the power rail or below the ground reference. Overshoot, if severe enough, can overstress devices and cause damage or failure. Excessive voltage levels below ground reference are still overshoot. Undershoot is a condition in which the voltage level does not reach the desired amplitude for both maximum and minimum transition levels. Components must have sufficient tolerance that allows for proper voltage level transition requirements. Termination and proper PCB and IC package design can control overshoot.

For an unterminated transmission line, ringing and reflected noise are the same. This can be observed with measurement equipment at the frequency associated as a quarter-wavelength of the transmission line, as is most apparent in an unterminated, point-to-point trace. The driven end of the line is commonly tied to ac ground with a low-impedance (5- to 20-Ω) load. This transmission line closely approximates a quarter-wavelength resonator (stub shorted on one end, open on the other). Ringing is the resonance of that stub.

As signal edges become faster, consideration must be given to propagation and reflection delays of the routed trace. If the propagation time and reflection within the trace are longer than the edge transition time from source to load, an *electrically long trace* will exist. This electrically long trace can cause signal integrity problems, depending on the type and nature of the signal. These problems include crosstalk, ringing, and reflections. EMI concerns are usually secondary to signal quality when referenced to electrically long lines. Although long traces can exhibit resonances, suppression and containment measures implemented within the product may mask EMI energy created. Therefore, components may cease to function properly if impedance mismatches exist in the system between source and load. Reflections are frequently both a signal integrity and an EMI issue, when the edge time of the signals constitutes a significant percentage of the propagation time between the device load intervals. This percentage depends on which logic family is used and speed of operation of the circuit. Solutions to reflection problems may require extending the edge time (slowing the edge rate) or decreasing the distance between load device intervals.

Reflections from signals on a trace are one source of RF noise within a network. Reflections are observed when impedance discontinuities exist in the transmission line. These discontinuities consist of

- Changes in trace width
- Improperly matched termination networks
- Lack of terminations

EMC and Printed Circuit Board Design

- T-stubs or bifurcated traces*
- Vias between routing layers
- Varying loads and logic families
- Large power plane discontinuities
- Connector transitions
- Changes in impedance of the trace

When a signal travels down a transmission line, a fraction of the source voltage will initially propagate. This source voltage is a function of frequency, edge rate, and amplitude. Ideally, all traces should be treated as transmission lines. Transmission lines are described by both their characteristic impedance, Z_o, and propagation delay, t_{pd}. These two parameters depend on the inductance and capacitance per unit length of the trace, the actual interconnect component, the physical dimensions of the interconnect, the RF return path, and the permittivity of the insulator between them. Propagation delay is also a function of the length of the trace and dielectric constant of the material. When the load impedance at the end of the interconnect equals that of the characteristic impedance of the trace, no signal is reflected.

A typical transmission line is shown in Fig. 6.30. Here we notice that

- Maximum energy transfer occurs when $Z_{out} = Z_o = Z_{load}$.
- Minimum reflections will occur when $Z_{out} = Z_o$ and $Z_o = Z_{load}$.

If the load is not matched to the transmission line, a voltage waveform will be reflected back toward the source. The value of this reflected voltage is

$$V_r = V_o \left(\frac{Z_L - Z_o}{Z_L + Z_o} \right) \tag{6.35}$$

where V_r = reflected voltage
 V_o = source voltage
 Z_L = load resistance
 Z_o = characteristic impedance of the transmission path

When Z_{out} is less than Z_o, a negative reflected wave is created. If Z_L is greater than Z_o, a positive wave is observed. This wave will repeat itself at the source driver if the impedance is different from line impedance, Z_o.

Equation (6.35) relates the reflected signal in terms of voltage. When a portion of the propagating signal reflects from the far end, this component of energy will travel back to the source. As it reflects back, the reflected signal may cross over the tail of the incoming signal. At this point, both signals will propagate simultaneously in opposite directions, neither interfering with the other.

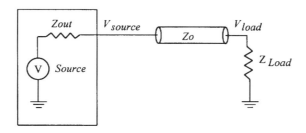

FIGURE 6.30 Typical transmission line system.

---

*A bifurcated trace is a single trace that is broken up into two traces routed to different locations.

We can derive an equation for the reflected wave. The reflection equation, Eq. (6.36), is for the fraction of the propagating signal that is reflected back toward the source.

$$\text{Percent reflection} = \left(\frac{Z_L - Z_o}{Z_L + Z_o}\right) \times 100 \qquad (6.36)$$

This equation applies to any impedance mismatch, regardless of voltage levels. Use Z_o for the signal source of the mismatch and Z_L for the load. To improve the noise margin budget requirements for logic devices, positive reflections are acceptable as long as they do not exceed V_{Hmax} of the receive component.

A forward-traveling wave is initiated at the source in the same manner as the incoming backward-traveling wave, which is the original pulse returned back to the source by the load. The corresponding points in the incoming wave are reduced by the percentage of the reflection on the line. The process of repeated reflections can continue as re-reflections at both the source and load. At any point in time, the total voltage (or current) becomes the *sum* of all voltage (or current) sources present. It is for this reason that we may observe a 7-V signal on the output of a source driver, while the power supply bench voltage is operating at 5 V. The term *ringback* is the effect of the rising edge of a logic transition that meets or exceeds the logic level required for functionality, and then recrosses the threshold level before settling down. Ringback is caused by a mismatch of logic drivers and receivers, poor termination techniques and impedance mismatches of the network.[9]

Sharp transitions in a trace may be observed through use of a time domain reflectometer (TDR). Multiple reflections caused by impedance mismatches are observed by a sharp jump in the signal voltage level. These abrupt transitions usually have rise and fall times that can be comparable to the edge transition of the original pulse. The time delay from the original pulse to the occurrence of the first reflection can be used to determine the location of the mismatch. A TDR determines discontinuities within a transmission line structure. An oscilloscope observes reflections. Both types of discontinuities are shown in Fig. 6.31. Although several impedance discontinuities are shown, only one reflection is illustrated.[1,4]

If discontinuities in a transmission line occur, reflections will be observed. These reflections will cause signal integrity problems to occur. How much energy will be reflected is based on the following.

1. For the reflected signal that is returned back to the source,

$$\tau_{reflected} = E_{reflected}/E_{incident} = (R_{load} - R_{source})/(R_{load} + R_{source})$$

2. For the reflected signal that is traveling beyond the discontinuity,

$$\tau_{transmitted} = E_{transmitted}/E_{incident} = (R_{load}^2)/(R_{load} + R_{source})$$

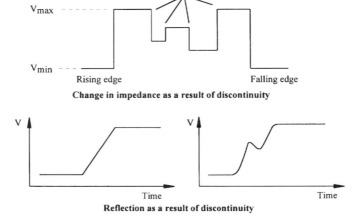

FIGURE 6.31 Discontinuities in a transmission line.

EMC and Printed Circuit Board Design

It is interesting to note that

- When $R_{load} < R_{source}$, the reflection back to the source is inverted in polarity.
- When $R_{load} = 0$, theoretically, the reflection back to the source is 100% and inverted.
- When $R_{load} \gg R_{source}$, theoretically, the reflection is 100% and not inverted.

We say *theoretically* because the mathematics of the problem is for optimal, based case configuration. Within any transmission line, losses occur, and variable impedance values are present. With these variable parameters, the resultant reflection will approach, but never exactly match, the theoretical value.

6.3.2 Calculating Trace Lengths (Electrically Long Traces)

When creating transmission lines, PCB designers need the ability to quickly determine if a trace routed on a PCB can be considered electrically long during component placement. A simple calculation is available that determines whether the *approximate* length of a routed trace is electrically long under typical conditions. When determining whether a trace is electrically long, think in the *time domain*. The equations provided below are best used when doing *preliminary* component placement on a PCB. For extremely fast edge rates, detailed calculations are required based on the actual dielectric constant value of the core and prepreg material. The dielectric constant determines the velocity of propagation of a transmitted wave.

Assume typical velocity of propagation of a signal within a transmission line is 60% the speed of light. We can calculate the maximum permissible unterminated line length per Eq. (6.37). This equation is valid when the two-way propagation delay (source-load-source) equals the signal rise time transition or edge rate.[1,2]

$$l_{max} = \frac{t_r}{2t'_{pd}} \tag{6.37}$$

where t_r = edge rate (ns)
t'_{pd} = propagation delay (ns)
l_{max} = maximum routed trace length (cm)

To simplify Eq. (6.37), determine the real value of the propagation delay using the actual dielectric constant value based at the frequency of interest. This equation takes into account propagation delay and edge rate. Equations (6.38) and (6.39) are presented for determining the maximum round-trip routed electrical line length before termination is required. The one-way length, from source to load, is one-half the value of l_{max} calculated. The factor used in the calculation is for a dielectric constant of 4.6, typical of FR-4.

$$l_{max} = 9 \times t_r \quad \text{for microstrip topology (cm)}$$

$$l_{max} = 3.5 \times t_r \quad \text{for microstrip topology (inches)} \tag{6.38}$$

$$l_{max} = 7 \times t_r \quad \text{for stripline topology (cm)}$$

$$l_{max} = 2.75 \times t_r \quad \text{for stripline topology (inches)} \tag{6.39}$$

For example, if a signal transition is 2 ns, the maximum round-trip, unterminated trace length when routed on microstrip is

$$l_{max} = 9 \times t_r = 18 \text{ cm (7 inches)}$$

When this same signal is routed stripline, the maximum unterminated trace length of this same 2-ns signal edge becomes

$$l_{max} = 7 \times t_r = 14 \text{ cm (5.5 inches)}$$

These equations are also useful when we are evaluating the propagational time intervals between load intervals on a line with multiple devices.

To calculate the constant for l_{max}, use the following example.

Example

$$k = x\left(\frac{a}{t_{pd}}\right)$$

where k = constant factor for transmission line length determination
   a = 30.5 for cm, 12 for inches
   x = 0.5 (converts transmission line to one way path)
   $t_{pd} = 1.017\sqrt{0.475\varepsilon_r + 0.67}$ for microstrip, $1.017\sqrt{\varepsilon_r}$ for stripline

For example, for $\varepsilon_r = 4.1$,

$$l_{max} = 8.87 \text{ for microstrip (cm) or } 3.49 \text{ (inches)}$$
$$l_{max} = 6.99 \text{ for stripline (cm) or } 2.75 \text{ (inches)}$$

If a trace or routed interval is longer than l_{max}, then termination should be implemented, as signal reflections (ringing) may occur in this electrically long trace. Even with good termination, a finite amount of RF currents can still be in the trace. For example, use of a series termination resistor will

- minimize RF currents within the trace
- absorb reflections and ringing
- match trace impedance
- minimize overshoot
- reduce RF energy generated by slowing the edge rate of the clock signal

During layout, when placing components that use clock or periodic waveform signals, locate them such that the signal traces are routed for the best straight-line path possible, with minimal trace length and number of vias in the route. Vias will add inductance and discontinuities to the trace, approximately 1 to 3 nH each. Inductance in a trace may also cause signal integrity concerns, impedance mismatches, and potential RF emissions. Inductance in a trace allows this wire to act as an antenna. The faster the edge rate of the signal transition, the more important this design rule becomes. If a periodic signal or clock trace must traverse from one routing plane to another, this transition should occur at a component lead (pin escape or breakout) and not anywhere else. If possible, reduce inductance presented to the trace by using fewer vias.

Equation (6.40) is used to determine if a trace or loading interval is electrically long and requires termination.

$$l_d < l_{max} \tag{6.40}$$

where l_{max} is the calculated maximum trace length, and l_d is the length of the trace route as measured in the actual board layout. Keep in mind that l_d is the *round-trip* length of the trace.

Ideally, trace impedance should be kept at ±10% of the desired impedance of the complete transmission line structure. In some cases, ±20 to 30% may be acceptable only after careful consideration has been given to signal integrity and performance. The width of the trace, its height above a reference plane, dielectric constant of the board material, plus other microstrip and stripline constants determine the

EMC and Printed Circuit Board Design 6-49

impedance of the trace. It is always best to maintain constant impedance control at all times in any dynamic signal condition.

6.3.3 Routing Layers

PCB designers have to determine which signal layers to use for routing clocks and periodic signals. Clock and periodic signals must be routed on either one layer, or on an adjacent layer separated by a single reference plane. An example of routing a trace between layers is shown in Fig. 6.32. Three issues must be remembered when selecting routing layers: which layers to use for trace routing, jumping between designated layers, and maintaining constant trace impedance.[1,2] Figure 6.32 is a representative example of how to route sensitive traces.

1. The designer needs to use a solid image or reference plane adjacent to the signal trace. Minimize trace length while maintaining a controlled impedance value of the transmission line. If a series termination resistor is used, connect the resistor to the pin of the driver without use of a via between the resistor and component. Place the resistor directly next to the output pin of the device. After the resistor, now place a via to the internal stripline layers.
2. Do not route clock or other sensitive traces on the outer layers of a multilayer board. The outer layers of a PCB are generally reserved for large signal buses and I/O circuitry. Functional signal quality of these signal traces could be corrupted by other traces containing high levels of RF energy using microstrip topology. When routing traces on outer levels, a change in the distributed capacitance of the trace, as the trace relates to a reference plane may occur, thus affecting performance and possible signal degradation.
3. If we maintain constant trace impedance and minimize or eliminate use of vias, the trace will not radiate any more than a coax. When we reference the electric field, E, to an image plane, magnetic

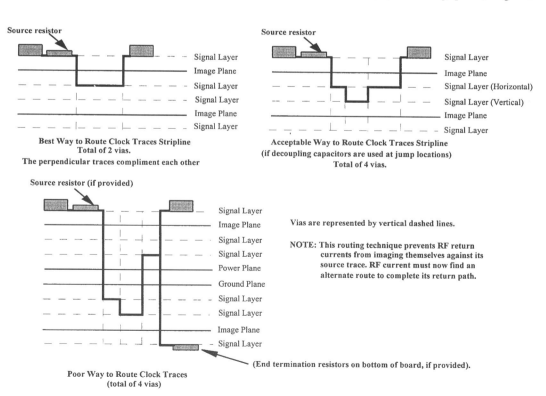

FIGURE 6.32 Routing layers for clock signals.

lines of flux present within the transmission line are cancelled by the image, thus minimizing emissions. A low-impedance RF return path adjacent to the rerouted trace performs the same function as the braid, or shield of a coax.

Three phenomena by which planes, and hence PCBs, create EMI are enumerated below. Proper understanding of these concepts will allow the designer to incorporate suppression techniques on any PCB in an optimal manner.

1. Discontinuities in the image plane due to use of vias and jumping clock traces between layers. The RF return current will divert from a direct line RF return path, creating a loop antenna. Once we lose image tracking, distance separation is what creates the loop antenna.
2. Peak surge currents injected into the power and ground network (image planes) due to components switching signal pins simultaneously. The power and ground planes will bounce at switching frequency. This bouncing will allow RF energy to propagate throughout the PCB, which is what we do not want.
3. Flux loss into the annular keep-out region of the via if 3-W routing is not provided for the trace route. Distance separation of a trace from a via must also conform to 3-W spacing. The 3-W rule is discussed in Section 6.3.5. This requirement prevents RF energy (magnetic lines of flux) present within a transmission line (trace) from coupling into the via. This via may contain a static signal, such as reset. This static signal may now repropagate RF energy throughout the PCB into areas susceptible to RF disruption.

6.3.4 Layer Jumping—Use of Vias

When routing clock or high-threat signals, it is common practice to via the trace to a routing plane (e.g., horizontal or x-axis) and then via this same trace to another plane (e.g., vertical or y-axis) from source to load. It is generally assumed that, if each and every trace is routed adjacent to an RF return path, there will be tight coupling of common-mode RF currents along the entire trace route. In reality, this assumption is partially incorrect.

As a signal trace jumps from one layer to another, RF return current must follow the trace route. When a trace is routed internal to a PCB between two planar structures, commonly identified as the power and ground planes, or two planes with the same potential, the return current is shared between these two planes. The only time return current can jump between the two planes is at a location where decoupling capacitors are located. If both planes are at the same potential (e.g., 0-V reference) the RF return current jump will occur at where a via connects both 0-V planes to a component, assigned to that via.

When a jump is made from a horizontal to vertical layer, the RF return current *cannot* fully make this jump. This is because a discontinuity was placed in the trace route by a via. The return current must now find an alternate low-inductance (impedance) path to complete its route. This alternate path may not exist in a position that is immediately adjacent to the location of the layer jump, or via. Therefore, RF currents on the signal trace can couple to other circuits and pose problems as both crosstalk and EMI. Use of vias in a trace route will always create a concern in any high-speed product.

To minimize development of EMI and crosstalk due to layer jumping, the following design techniques have been found effective:

1. Route all clocks and high-threat signal traces on only one routing layer as the initial approach concept. This means that both x- and y-axis routes are in the same plane. (*Note:* This technique is likely to be rejected by the PCB designer as being unacceptable, because it makes autorouting of the board nearly impossible.)
2. Verify that a solid RF return path is adjacent to the routing layer, with no discontinuities in the route created by use of vias or jumping the trace to another routing plane.

If a via must be used for routing a sensitive trace, high-threat or clock signal between the horizontal and vertical routing layer, incorporate ground vias at *each and every* location where the signal axis jumps are executed. The ground via is always at 0-V potential.

A ground via is a via that is placed directly adjacent to each signal route jump from a horizontal to a vertical routing layer. Ground vias can be used only when there are more than *one* 0-V reference planes internal to the PCB. This via is connected to all ground planes (0-V reference) in the board that serves as the RF return path for signal jump currents. This via essentially ties the 0-V reference planes together adjacent and parallel to this signal trace location. When using two ground vias per signal trace via, a continuous RF path will be present for the return current throughout its entire trace route.*

What happens when only one 0-V reference (ground) plane is provided and the alternate plane is at voltage potential as commonly found with four-layer PCB stackups? To maintain a constant return path for RF currents, the 0-V (ground) plane should be allowed to act as the primary return path. The signal trace must be routed against this 0-V plane. When the trace routes against the power plane, after jumping layers, use of a *ground trace* is required on the power plane. This ground trace must connect to the ground plane, by vias, at both ends of the ground trace routing. This trace must also be parallel to the signal trace as close as possible. Using this configuration, we can now maintain a constant RF return path throughout the entire trace route (see Fig. 6.33).

How can we minimize use of ground vias when layer jumping is mandatory? In a properly designed PCB, the first traces to be routed must be clock signals, which must be *manually routed.* Much freedom is permitted in routing the first few traces anywhere within by the PCB designer, e.g., clocks and high-threat signals. The designer is then able to route the rest of the board using the shortest trace distance routing possible (shortest Manhattan length). These early routed trace must make a layer jump adjacent to the *ground pin via* of any component. This layer jump will co-share this component's ground via. This ground via being referenced will perform the function of providing return path for the signal trace as well as 0-V reference to the component, allowing RF return current to make a layer jump, detailed in Fig. 6.34.

6.3.5 Trace Separation and the 3-W Rule

Crosstalk occurs between traces on a PCB. Crosstalk is the unwanted transference of RF energy from one transmission path to another. These paths include, among other things, PCB traces. This undesirable effect is associated with not only clock or periodic signals but also with other system critical nets. Data, address, control lines, and I/O may be affected by crosstalk and coupling. Clocks and periodic signals create the majority of crosstalk problems and can cause functionality concerns (signal integrity) with other functional sections of the assembly. Use of the 3-W rule will allow a designer to comply with PCB layout requirement without having to implement other design techniques.[1,2] These design techniques take up physical real estate and may make routing more difficult.

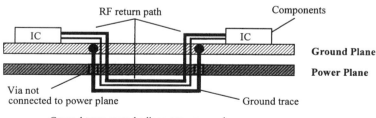

Ground trace routed adjacent to power plane connected to main ground plane by via to guarantee undisturbed RF return path.

Four layer PCB with trace routed on top and bottom layer

FIGURE 6.33 Routing a ground trace to ensure a complete RF return path.

---

*Use of ground vias was first identified and presented to industry by W. Michael King. Ground vias are also described in Refs. 1, 2, and 5.

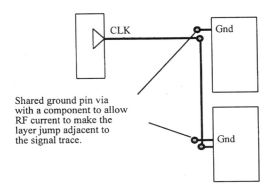

Optimal routing of traces with sensitive signals making a layer jump, assuring a constant RF retun path by sharing the ground pin of a component.

FIGURE 6.34 Routing a ground trace to ensure a complete RF return path.

The basis for use of the 3-W rule is to minimize coupling between transmission lines or PCB traces. The rule states that *the distance separation between traces must be three times the width of a single trace measured from centerline to centerline*. Otherwise stated, *the distance separation between two traces must be greater than two times the width of a single trace*. For example, a clock line is 6 mils wide. No other trace can be routed within a minimum of 2 × 6 mils of this trace, or 12 mils, edge-to-edge. As observed, much real estate is lost in areas where trace isolation occurs. An example of the 3-W rule is shown in Fig. 6.35.

Note that the 3-W rule represents the approximate 70% flux boundary at logic current levels. For the approximate 98% boundary, 10-W should be used. These values are derived form complex mathematical analysis, which is beyond the scope of this book.

Use of the 3-W rule is mandatory for *only* high threat signals, such as clock traces, differential pairs, video, audio, the reset line, or other system critical nets. Not all traces within a PCB have to conform to 3-W routing. It is important to determine which traces are to be classified as critical. Before using this design technique, it is important to determine exactly which traces must be routed 3-W.

As shown in the middle drawing of Fig. 6.35, a via is located between two traces. This via is usually associated with a third routed net and may contain a signal trace that is susceptible to electromagnetic disruption. For example, the reset line, a video or audio trace, an analog level control trace, or an I/O interface may pick up electromagnetic energy, either inductively or capacitively. To minimize crosstalk corruption to the via, the distance spacing between adjacent traces must include the angular diameter and clearance of the via. The same requirement exists for this distance spacing between a routed trace rich in RF spectral energy that may couple to a component's breakout pin (pin escape) to this routed trace.

Use of the 3-W rule should not be restricted to clock or periodic signal traces. Differential pairs (balanced, ECL, and similar sensitive nets) are also prime candidates for 3-W. The distance between paired traces must be 1-W for differential traces and 3-W from each of the differential pair to adjacent traces. For differential traces, power plane noise and single ended signals can capacitively, or inductively, couple into the paired traces. This can cause data corruption if those traces, not associated with the differential pair are physically closer than 3-W. An example of routing differential pair traces within a PCB structure is shown in Fig. 6.36.

6.3.6 Trace Termination

Trace termination plays an important role in ensuring optimal signal integrity as well as minimizing creation of RF energy. To prevent trace impedance matching problems and provide higher quality signal transfer between circuits, termination may be required. Transmission line effects in high-speed circuits

EMC and Printed Circuit Board Design

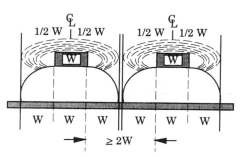

The distance spacing between both traces must have a mimimum overlap of 2W

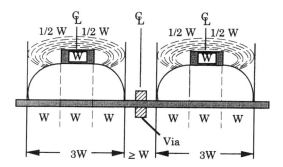

For the via, add annular keep-out diameter which includes both the via and annular (anti-pad) clearance

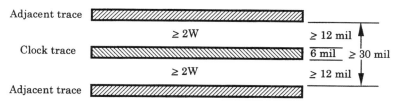

Top down view
3W spacing without a via between the traces

FIGURE 6.35 Designing with the 3-W rule.

and traces must always be considered. If the clock speed is fast, e.g., 100 MHz, and components are, for example, FCT series (2-ns edge rate typical), reflections from a long trace route could cause the receiver to double clock on a single edge transition. This is possible because it takes a finite time for the signal to propagate from source to load and return. If the return signal does not occur before the next edge transition event, signal integrity issues arise. Any signal that clocks a flip-flop is a possible candidate for causing transmission line effects regardless of the actual frequency of operation.

Engineers and designers sometimes daisy chain signal and clock traces for ease of routing. Unless the distance is small between loads (with respect to propagation length of the signal rise time or edge transition), reflections may result from daisy-chained traces. Daisy chaining affects signal quality and EMI spectral energy distribution, sometimes to the point of nonfunctionality or noncompliance. Therefore, radial connections for fast edge signals and clocks are preferred over daisy chains. Each component must have its respective trace terminated in its characteristic impedance. Various types of termination are shown in Fig. 6.37. Parallel termination at the end of a trace route is feasible only when the driver can tolerate the total current sink of all terminated loads.

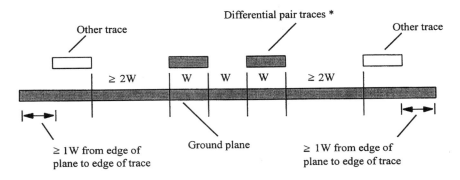

FIGURE 6.36 Parallel differential pair routing and the 3-W rule.

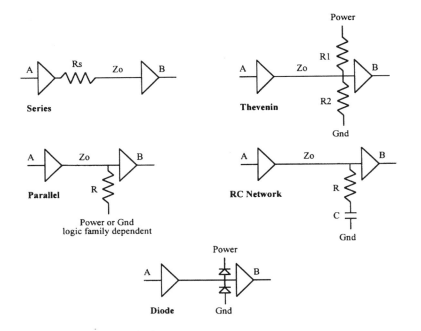

FIGURE 6.37 Common termination methods.

The need to terminate a transmission line is based on several design criteria. The most important criterion is when the existence of an electrically long trace within a PCB is present. When a trace is electrically long, or when the length exceeds one-sixth of the electrical length of the edge rate, the trace requires termination. Even if a trace is short, termination may still be required if the load is capacitive or highly inductive to prevent ringing within the transmission line structure.

Termination not only matches trace impedance and removes ringing but will sometimes slow down the edge rate transition of the clock signal. Excessive termination could degrade signal amplitude and integrity to the point of nonfunctionality. Reducing either dI/dt or dV/dt present within the transmission line will also reduce RF emissions generated by high amplitude voltage and current levels.

The easiest way to terminate is to use a resistive element. Two basic configurations exist: source and load. Several methodologies are available for these configurations. A summary of these termination methods is shown in Table 6.8. Each method is discussed in depth in this section.

EMC and Printed Circuit Board Design

TABLE 6.8 Termination Types and Their Properties

Termination Type	Added Parts	Delay Added	Power Required	Parts Values	Comments
Series termination resistor	1	Yes	Low	$R_s = Z_o - R_o$	Good DC noise margin
Parallel termination resistor	1	Small	High	$R = Z_o$	Power consumption is a problem
Thevenin network	2	Small	High	$R = 2 * Z_o$	High power for CMOS
RC network	2	Small	Medium	$R = Z_o$ $C = 20\text{–}600$ pF	Check bandwidth and added capacitance
Diode network	2	Small	Low	—	Limits overshoot; some ringing at diodes

Table 6.8 and Fig. 6.37 © Motorola, Inc., reprinted by permission.[9]

6.3.7 Series Termination

Source termination provides a mechanism whereby the output impedance of the driver and resistor matches the impedance of the trace. The reflection coefficient at the source will be zero. Thus, a clean signal is observed at the load. In other words, the resistor absorbs the reflections.

Series termination is optimal when a lumped load or a single component is located at the end of a routed trace. A series resistor is used when the driving device's output impedance, R_o, is less than Z_o, the loaded characteristic impedance of the trace. This resistor must be located directly at the output of the driver without use of a via between the component and resistor (Fig. 6.38). The series resistor, R_s, is calculated by Eq. (6.41).

$$R_s = Z_o - R_o \tag{6.41}$$

where R_s = series resistor
Z_o = characteristic impedance of the transmission line
R_o = output resistance of the source driver

For example, if $R_o = 22\ \Omega$ and trace impedance, $Z_o = 55\ \Omega$, $R_s = 55 - 22 = 33\ \Omega$. Use of a 33-$\Omega$ series resistor is common in today's high-technology products. The series resistor, R_s, can be calculated to be greater than or equal to the source impedance of the driving component and lower than or equal to the line impedance, Z_o. This value is typically between 15 and 75 Ω (usually 33 Ω).

Series termination minimizes the effects of ringing and reflection. Source resistance plays a major role in allowing a signal to travel down a transmission line with maximum signal quality. If a source resistor is not present, there will be very little damping. The system will ring for a long time (tens of nanoseconds). PCI drivers are optimal for this function because they have extremely low output impedance. A series resistor at the source that is about two-thirds of the transmission line impedance will remove ringing. A target value for a *slightly underdamped system* (to make edges sharper) is to have $R_s = 2/3 Z_o$. A wavefront of slightly more than half the power supply voltage proceeds down the transmission line and doubles at the open circuit far end, giving the voltage level desired at the load. The reflected wavefront is almost completely absorbed in the series resistor. Sophisticated drivers will attempt to match the transmission line impedance so that no external components are necessary.

When $R_s + R_o = Z_o$, the voltage waveform at the output of the series resistor is at one-half the voltage level sourced by the driver, assuming that a perfect voltage divider exists. For example, if the driver provides a 5-V output, the output of the series resistor, V_b, will be 2.5 V. The reason for this is described

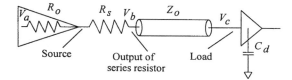

FIGURE 6.38 Series termination circuit.

by Eq. (6.42). If the receiver has high-input impedance, the full waveform will be observed immediately when received, while the source will receive the reflected waveform at $2 \times t_{pd}$ (round-trip travel).

$$\Delta V_b = \Delta V_a \left(\frac{Z_o}{R_o + R_s + Z_o} \right) \qquad (6.42)$$

6.3.8 End Termination

End termination is used when multiple loads exist within a single trace route. Multiple-source drivers may be connected to a bus structure or daisy chained. The last physical device on a routed net is where the load termination must be positioned.

To summarize,

1. The signal of interest travels down the transmission line at full voltage and current level without degradation.
2. The transmitted voltage level is observed at the load.
3. The termination will remove reflections by matching the line, thus damping out overshoot and ringback.

There is a right way and a wrong way when placing end terminators on a PCB. This difference is shown in Fig. 6.39. Regardless of the method chosen, termination must occur at the *very end of the trace*. For purposes of discussion, the RC method is shown in this figure.[1]

6.3.9 Parallel Termination

For parallel termination, a single resistor is provided at the end of the trace route (Fig. 6.40). This resistor, R, must have a value equal to the required impedance of the trace or transmission line. The other end

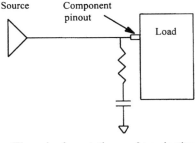
Wrong implementation - end termination
(Schematic representation)

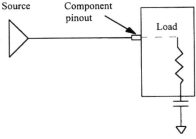
Correct location of end termination
(Schematic representation)

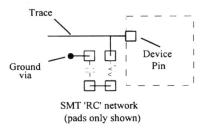
Wrong implementation - end termination
(Actual layout representation)

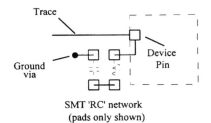
Correct implementation - end termination
(Actual layout representation)

FIGURE 6.39 Locating end terminators on a PCB.

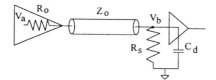

FIGURE 6.40 Parallel termination circuit.

of the resistor is tied to a reference source, generally ground. Parallel termination will add a small propagation delay to the signal due to the addition of resistance, which is part of the time constant equation, $\tau = Z_o C$, present in the network. This equation includes the total impedance of the transmission line. The total impedance, Z_o, is the result of the termination resistor, line impedance, and source output impedance. The variable C in the equation is both the input shunt capacitance of the load and distributed capacitance of the trace.

One disadvantage of parallel termination is that this method consumes dc power, since the resistor is generally in the range of 50 to 150 Ω. In applications of critical device loading or where power consumption is critical (for example, battery-powered products such as notebook computers), parallel termination is a poor choice. The driver must source current to the load. An increase in drive current will cause an increase in dc consumption from the power supply, an undesirable feature in battery-operated products.

Parallel termination is rarely used in TTL or CMOS designs. This is because a large drive current is required in the logic HI state. When the source driver switches to Vcc, or logic HI, the driver must supply a current of Vcc/R to the termination resistor. When in the logic LOW state, little or no drive current is present. Assuming a 55 Ω transmission line, the current required for a 5-V drive signal is 5 V/55 Ω = 91 mA. Very few drivers can source that much current. The drive requirements of TTL are different for logic LOW as compared to logic HI. CMOS sources the same amount of current in both the LOW and HI logic states.

Since parallel termination creates a dc current path when the driver is logic HI, excessive power dissipation and V_{OH} degradation (noise margin) occur. A driver's output is always switching. The dc current consumed by the termination resistor is always present. At higher frequencies, the ac switching current becomes the major component of the circuit function. When using parallel termination, one should consider how much V_{OH} degradation is acceptable by the receivers.

When parallel termination is provided, the net result observed on an oscilloscope should be nearly identical to that of series, Thevenin, or RC, since a properly terminated transmission line should respond the same regardless of which method is used.

When using simple parallel termination, a single pull-down resistor is provided at the load. This allows fast circuit performance when driving distributed loads. This resistor has a Z_o value equal to the characteristic impedance of the trace and source driver. The other end of the resistor is tied to a reference point, usually ground. For ECL logic, the reference is power. The voltage level on the trace is described by Eq. (6.43). On PCB stackups that include Omega layers, parallel termination is commonly found. An Omega layer is a single layer within a multilayer stackup assignment that has resistors built into the copper plane using photoresist material and laser etched for a the desired resistance value. This termination method is extremely expensive and found in only high-technology products where component density is high and large pin-out devices leave no physical room for hundreds, or even thousands, of discrete termination resistors.

$$\Delta V_a = \Delta V_b \left(\frac{Z_o}{R_o + Z_o} \right) \quad (6.43)$$

6.3.10 Thevenin Network

Thevenin termination has one advantage over parallel termination. Thevenin provides a connection that has one resistor to the power rail and the other resistor to ground (Fig. 6.41). Unlike parallel termination, Thevenin permits optimizing the voltage transition points between logic HI and logic LOW. When using

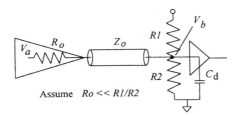

FIGURE 6.41 Thevenin termination circuit.

Thevenin termination, an important consideration in choosing the resistor values is to avoid improper setting of the voltage reference level of the loads for both the HI and LOW logic transition points. The ratio of $R1/R2$ determines the relative proportions of logic HI and LOW drive current.

Designers commonly, but arbitrarily, use a 220/330-Ω ratio (132 Ω parallel) for driving bus logic. Determining the resistor ratio value may be difficult to do if the logic switch point for various families are different. This is especially true when both TTL and CMOS are used. A 1:1 resistor ratio (e.g., 110/110 Ω will create a 55-Ω value, the desired characteristic Z_o of the trace) limiting the line voltage to 2.5 V, thus allowing an invalid transition level for certain logic devices. Hence, Thevenin termination is optimal for TTL logic, not CMOS.

The Thevenin equivalent resistance must be equal to the characteristic impedance of the trace. Thevenin resistors provide a voltage division. To determine the proper voltage reference desired, use Eq. (6.44).

$$V_{ref} = \frac{R2}{R1 + R2} \text{ V} \qquad (6.44)$$

where V_{ref} = desired voltage level to the input of the load
V = voltage source from the power rail
R1 = pull-up resistor
R2 = pull-down resistor

For the Thevenin termination circuit,

- $R1 = R2$: The drive requirements for both logic HI and LOW are identical.
- $R2 > R1$: The LOW current requirements are greater than the HI current requirement. This setting is appropriate for TTL and CMOS devices.
- $R1 > R2$: The HI current requirements are greater than the LOW current requirement.

With these constraints, I_{OHmax} or I_{OLmax} must never be exceeded per the device's functional requirements. This constraint must be present, as TTL and CMOS sinks (positive) current in the LOW state. In the high state, TTL and CMOS source (negative) current. Positive current refers to current that enters a device, while negative current is the current that leaves the component. ECL logic devices source (negative) current in both logic states.

With a properly chosen termination ratio for the resistors, an optimal dc voltage level now will exist for both logic HI and LOW states. The advantage of using parallel termination over Thevenin is parallel's use of one less component. If we compare the results of parallel to Thevenin, both termination methods provide identical results. Signal integrity of a transmitted electromagnetic wave always appear identical when properly terminated, regardless of termination method chosen.

6.3.11 RC Network

The RC (also known as ac) termination method works well in both TTL and CMOS systems. The resistor matches the characteristic impedance of the trace, identical to parallel. The capacitor holds the dc voltage level of the signal. The source driver does not have to provide current to drive an end terminator.

Consequently, ac current (RF energy) flows to ground during a switching state. In addition, the capacitor allows RF energy (which is an ac sine wave, not the dc logic level of the signal) to pass through to the load device. Although a propagation delay is presented to the signal due to both the resistor and capacitor's time constant, less power dissipation exists than parallel or Thevenin termination. From the viewpoint of the circuit, all termination methods produce identical results. The main difference lies in power dissipation, with the RC network consuming far less power than the other two.

The termination resistor must equal the characteristic impedance, Z_o, of the trace, while the capacitor is generally very small (20 to 600 pF). The time constant must be greater than twice the loaded propagation delay (round-trip travel time). This time constant is greater than twice the loaded propagation delay, because a signal must travel from source to load and return. It takes one time constant each way for a total of two time constants. If we make the time constant slightly greater than the total propagation delay within the routed transmission line, reflections will be minimized or eliminated. It is common to select a time constant that is three times the propagation delay of the round trip signal. RC termination finds excellent use in buses containing similar layouts.

To determine the proper value of the resistor and capacitor, Eq. (6.45) provides a simple calculation, which includes round-trip propagation delay $2 \times t'_{pd}$.

$$\tau = R_s C_s \tag{6.45}$$

where $\tau > 2 \times t'_{pd}$ for optimal performance

Figure 6.42 shows the results of RC termination. The lumped capacitance (C_d plus C_s) affects the edge rate of the signal, causing a slower signal to be observed by the load.

If the round-trip propagation delay is 4 ns, RC must be > 8 ns. Calculate C_s using the known round-trip propagation delay value using the value of ε_r that is appropriate for the dielectric material provided. *Note:* The self-resonant characteristic of the capacitor is critical during evaluation. It is also important to avoid inserting additional equivalent series inductance (ESL) into the circuit.

When selecting a capacitor value for a very long transmission line, if the ratio of line delay to signal rise time is greater than 1, there is no advantage of using RC over Thevenin or parallel termination. For random data transitions, if the value of C is large, an increase in peak I_{OH} and I_{OL} levels is required from the driver. This increase could be as much as twice that required than if series, Thevenin or parallel is used. Series termination requires the same peak I_{OH} and I_{OL} levels. If the driver cannot source enough current, then during the initial line transition (at least two rise times plus the capacitor charging time), the signal will not meet V_{OH} and V_{OL} levels. For this reason, RC termination is not optimal for use with *random data* signals on an electrically long transmission line.

If the ratio of line delay to signal rise time is moderate, approximately 1/3 to 1/2, RC termination provides some useful benefits. Under this condition, reduced ringing behavior occurs. Ringing is not completely eliminated; however, the effects are significant. An increase in peak driver current will also be observed from the driver.

6.3.12 Diode Network

Diode termination is commonly used with differential or paired networks. A schematic representation was shown previously in Fig. 6.37. Diodes are often used to limit overshoot on traces while providing low-power dissipation. The major disadvantage of diode networks lies in their frequency response to

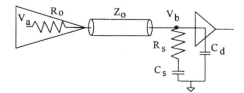

FIGURE 6.42 RC network circuit.

high-speed signals. Although overshoots are prevented at the receiver's input, reflections will still exist in the transmission line, as diodes do not affect trace impedance or absorb reflections. To gain the benefits of both techniques, diodes may be used in conjunction with the other methods discussed herein to minimize reflection problems. The main disadvantage lies in large current reflections that occur with this termination network. One should be aware, however, that when a diode clamps a large impulse current, this current can propagate into the power and ground plane, thus increasing EMI as power or ground bounce.

For certain applications, Schottky diodes are preferred. When using fast switching diodes, the diode switching time must be at least four times as fast as the signal rise time. When the line impedance is not well defined, as in backplane assemblies, use of diode termination is convenient and easy to use. The Schottky diode's low forward voltage value, V_f, is typically 0.3 to 0.45 V. This low voltage level clamps the input signal voltage, V_f, to ground. For the high voltage value, the clamp level is dependent on the maximum voltage rating of the device. When both diodes are provided, overshoot is significantly reduced for both positive and negative transitions. Some applications may not require both diodes to be used simultaneously.

The advantages of using diode termination include

- Impedance-matched lines are not required, including controlled transmission-line environments.
- The diodes may be able to replace termination resistors if the edge rate is slow.
- Clamping action reduces overshoots and enhances signal integrity.

Most of the discussion on diode termination deals with zero-time transitions when, in fact, ramp transitions must be considered. Diode termination is only perfect when used at the input of the load device. During the first half of the signal transition time (from 0 to 50%), the voltage level of the signal is doubled by the open circuit impedance at the receiving end of the transmission line. This open circuit impedance will produce a 0 to 100% voltage amplitude of the signal at the receiving end. During the last half of the signal transition (50 to 100%), the diode becomes a short circuit device. This result in a half-amplitude triangle-shaped pulse reflected back to the transmitter. The initial signal return back to the source has a phase amplitude that must be taken into consideration with the incoming signal. This phase difference may lead to either increasing or decreasing the amplitude of the desired signal. The source and return signal is either added or subtracted together, based on the phase of the signal related to each other. If additional logic devices are located throughout a routed net, such as a daisy-chain configuration, false triggering my occur to these intermediate devices. If the signal edge transition changes quickly, the first reflection to reach a receiving device may be out of phase, with respect to the desired voltage level.

When using diode terminations, it is common to overlook the package lead-length inductance when modeling or performing system analysis.

Section 6.3 References

1. Montrose, M. 1999. *EMC and the Printed Circuit Board Design—Design, Theory and Layout Made Simple.* Piscataway, NJ: IEEE Press.
2. Montrose, M. I. 1996. *Printed Circuit Board Design Techniques for EMC Compliance.* Piscataway, NJ: IEEE Press.
3. Paul, C. R. 1984. *Analysis of Multiconductor Transmission Lines.* New York: John Wiley & Sons, Inc.
4. Witte, Robert. 1991. *Spectrum and Network Measurements.* Englewood Cliffs, NJ: Prentice-Hall, Inc.
5. Johnson, H. W., and M. Graham. 1993. *High Speed Digital Design.* Englewood Cliffs, NJ: Prentice Hall.
6. Paul, C. R. 1992. *Introduction to Electromagnetic Compatibility.* New York: John Wiley & Sons, Inc.
7. Ott, H. 1988. *Noise Reduction Techniques in Electronic Systems.* 2nd ed. New York: John Wiley & Sons.

8. Dockey, R. W., and R. F. German. 1993. "New Techniques for Reducing Printed Circuit Board Common-Mode Radiation." *Proceedings of the IEEE International Symposium on Electromagnetic Compatibility.* IEEE, pp. 334-339.
9. Motorola, Inc. 1989. *Transmission Line Effects in PCB Applications* (#AN1051/D).

6.4 Bypassing and Decoupling

Bypassing and decoupling refers to energy transference from one circuit to another in addition to enhancing the quality of the power distribution system. Three circuit areas are of primary concern: power and ground planes, components, and internal power connections.

Decoupling is a means of overcoming physical and time constraints caused by digital circuitry switching logic levels. Digital logic usually involves two possible states, "0" or "1." Some conceptual devices may not be binary but ternary. The setting and detection of these two states is achieved with switches internal to the component that determines whether the device is to be at logic LOW or logic HIGH. There is a finite period of time for the device to make this determination. Within this window, a margin of protection is provided to guarantee against false triggering. Moving the logic switching state near the trigger level creates a degree of uncertainty. If we add high-frequency noise, the degree of uncertainty increases and false triggering may occur.

Decoupling is also required to provide sufficient dynamic voltage and current level for proper operation of components during clock or data transitions when all component signal pins switch simultaneously under maximum capacitive load. Decoupling is accomplished by ensuring a low-impedance power source is present in both the circuit traces and power planes. Because decoupling capacitors have increasingly low impedance at high frequencies up to the point of self-resonance, high-frequency noise is effectively diverted, while low-frequency RF energy remains relatively unaffected. Optimal implementation is achieved using a capacitor for a specific application: bulk, bypass, and decoupling. All capacitor values must be calculated for a specific function. In addition, properly select the dielectric material of the capacitor package. Do not leave it to random choice from past usage or experience.

Three common uses of capacitors follow. Of course, a capacitor may also be used in other applications such as timing, wave shaping, integration, and filtering.

1. *Decoupling.* Removes RF energy injected into the power distribution network from high frequency components consuming power at the speed the device is switching at. Decoupling capacitors also provide a localized source of dc power for devices and components and are particularly useful in reducing peak current surges propagated across the board.
2. *Bypassing.* Diverts unwanted common-mode RF noise from components or cables from one area to another. This is essential in creating an ac shunt to remove undesired energy from entering susceptible areas, in addition to providing other functions of filtering (bandwidth limiting).
3. *Bulk.* Used to maintain constant dc voltage and current levels to components when all signal pins switch simultaneously under maximum capacitive load. It also prevents power dropout due to dI/dt current surges generated by components.

An ideal capacitor has no losses in its conductive plates and dielectric. Current is always present between the two parallel plates. Because of this current, an element of inductance is associated with the parallel plate configuration. Because one plate is charging while its adjacent counterpart is discharging, a mutual coupling factor is added to the overall inductance of the capacitor.

6.4.1 Review of Resonance

All capacitors consist of an LCR circuit where L = inductance related to lead length, R = resistance in the leads, and C = capacitance. A schematic representation of a capacitor is shown in Fig. 6.43. At a calculable frequency, the series combination of L and C becomes resonant, providing very low impedance

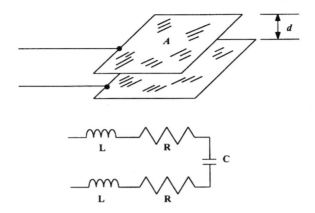

Leads internal to the capacitor actually consist of both inductance and resistance

L = approximately 10 nH (equivalent series inductance - ESL).
R = <1 ohm (equivalent series resistance - ESR).

FIGURE 6.43 Physical characteristics of a capacitor with leads.

and effective RF energy shunting at resonance. At frequencies above self-resonance, the impedance of the capacitor becomes increasingly inductive, and bypassing or decoupling becomes less effective. Hence, bypassing and decoupling are affected by the lead-length inductance of the capacitor (including surface mount, radial, or axial styles), the trace length between the capacitor and components, feed-through pads (or vias), and so forth. Before discussing bypassing and decoupling of circuits on a PCB, a review of resonance is provided.

Resonance occurs in a circuit when the reactive value difference between the inductive and capacitive vector is zero. This is equivalent to saying that the circuit is purely resistive in its response to ac voltage. Three types of resonance are common.

1. Series resonance
2. Parallel resonance
3. Parallel C–series RL resonance

Resonant circuits are frequency selective, since they pass more or less RF current at certain frequencies than at others. A series LCR circuit will pass the selected frequency (as measured across C) if R is high and the source resistance is low. If R is low and the source resistance is high, the circuit will reject the chosen frequency. A parallel resonant circuit placed in series with the load will reject a specific frequency.

Series Resonance

The overall impedance of a series RLC circuit is $Z = \sqrt{R^2 + (X_L - X_c)^2}$. If an RLC circuit is to behave resistively, the value can be calculated as shown in Fig. 6.44 where $\omega(2\pi f)$ is known as the *resonant angular frequency*.

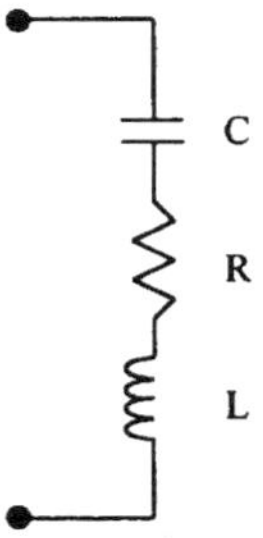

$$X_L = X_C$$

$$\omega L = \frac{1}{\omega C}$$

$$\omega = \frac{1}{\sqrt{LC}}$$

FIGURE 6.44 Series resonance.

EMC and Printed Circuit Board Design

With a series RLC circuit at resonance,

- Impedance is at minimum.
- Impedance equals resistance.
- The phase angle difference is zero.
- Current is at maximum.
- Power transfer (IV) is at maximum.

Parallel Resonance

A parallel RLC circuit behaves as shown in Fig. 6.45. The resonant frequency is the same as for a series RLC circuit. With a parallel RLC circuit at resonance,

- Impedance is at maximum.
- Impedance equals resistance.
- The phase angle difference is zero.
- Current is at minimum.
- Power transfer (IV) is at minimum.

Parallel C–Series RL Resonance (Antiresonant Circuit)

Practical resonant circuits generally consist of an inductor and variable capacitor in parallel, since the inductor will possess some resistance. The equivalent circuit is shown in Fig. 6.46. The resistance in the inductive branch may be a discrete element or the internal resistance of a nonideal inductor.

At resonance, the capacitor and inductor trade the same stored energy on alternate half cycles. When the capacitor discharges, the inductor charges, and vice versa. At the antiresonant frequency, the tank circuit presents a high impedance to the primary circuit current, even though the current within the tank is high. Power is dissipated only in the resistive portion of the network.

The antiresonant circuit is equivalent to a parallel RLC circuit whose resistance is Q^2R.

6.4.2 Physical Characteristics

Impedance

The equivalent circuit of a capacitor was shown previously in Fig. 6.43. The impedance of this capacitor is expressed by Eq. (6.46).

$$\omega = \frac{1}{\sqrt{LC}}$$

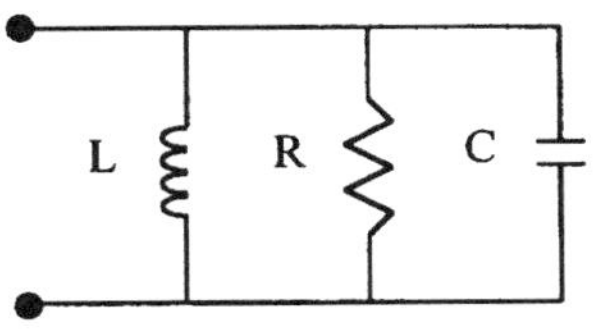

FIGURE 6.45 Parallel resonance.

$$w = \sqrt{\frac{1}{LC} - \left(\frac{R}{L}\right)^2} \approx \frac{1}{\sqrt{LC}} \quad [R << \omega_o L]$$

$$Q = \frac{1}{\omega_o CR} = \frac{X_C}{R} = \frac{X_L}{R}$$

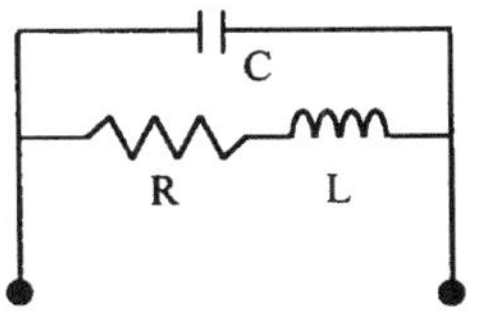

FIGURE 6.46 Parallel C–series resonance.

$$|Z| = \sqrt{R_s^2 + \left(2\pi fL - \frac{1}{2\pi fC}\right)^2} \qquad (6.46)$$

where Z = impedance (Ω)
R_s = equivalent series resistance, ESR (Ω)
L = equivalent series inductance, ESL (H)
C = capacitance (F)
f = frequency (Hz)

From this equation, |Z| exhibits its minimum value at a resonant frequency f_o such that

$$f_o = \frac{1}{2\pi\sqrt{LC}} \qquad (6.47)$$

In reality, the impedance equation [Eq. (6.46)] reflects hidden parasitics that are present when we take into account ESL and ESR.

Equivalent series resistance (ESR) is a term referring to resistive losses in a capacitor. This loss consists of the distributed plate resistance of the metal electrodes, the contact resistance between internal electrodes, and the external termination points. Note that skin effect at high frequencies increases this resistive value in the leads of the component. Thus, the high-frequency "ESR" is higher in equivalence than dc "ESR."

Equivalent series inductance (ESL) is the loss element that must be overcome as current flow is constricted within a device package. The tighter the restriction, the higher the current density and higher the ESL. The ratio of width to length must be taken into consideration to minimize this parasitic element.

Examining Eq. (6.46), we have a variation of the same equation with ESR and ESL, shown in Eq. (6.48).

$$|Z| = \sqrt{(ESR)^2 + (X_{ESL} - X_C)^2} \qquad (6.48)$$

where $X_{ESL} = 2\pi f(ESL);$ $X_C = \frac{1}{2\pi fC}$

For certain types of capacitors, with regard to dielectric material, the capacitance value varies with temperature and dc bias. Equivalent series resistance varies with temperature, dc bias, and frequency, while ESL remains fairly unchanged.

For an ideal planar capacitor where current uniformly enters from one side and exits from another side, inductance will be practically zero. For those cases, Z will approach R_s at high frequencies and will not exhibit an inherent resonance, which is exactly what a power and ground plane structure within a PCB does. This is best illustrated by Fig. 6.47.

The impedance of an "ideal" capacitor decreases with frequency at a rate of –20 dB/decade. Because a capacitor has inductance in its leads, this inductance prevents the capacitor from behaving as desired, described by Eq. (6.47).

It should be noted that long power traces in two-sided boards that are not routed for idealized flux cancellation are, in effect, extensions of the lead lengths of the capacitor, and this fact seriously alters the self-resonance of the power distribution system.

Above self-resonance, the impedance of the capacitor becomes inductive, increasing at +20 dB/decade as detailed in Fig. 6.48. Above the self-resonant frequency, the capacitor ceases to function as a capacitor. The magnitude of ESR is extremely small and as such does not significantly affect the self-resonant frequency of the capacitor.

The effectiveness of a capacitor in reducing power distribution noise at a particular frequency of interest is illustrated by Eq. (6.49).

$$\Delta V(f) = |Z(f)| \cdot \Delta I(f) \qquad (6.49)$$

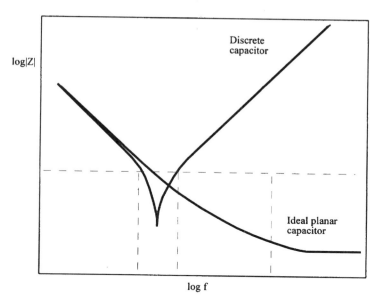

FIGURE 6.47 Theoretical impedance frequency response of ideal planar capacitors.

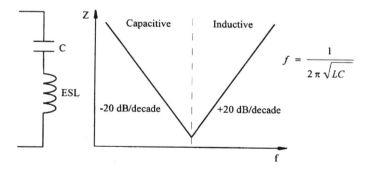

FIGURE 6.48 Effects of lead length inductance within a capacitor.

where ΔV = allowed power supply sag
ΔI = current supplied to the device
f = frequency of interest

To optimize the power distribution system by ensuring that noise does not exceed a desired tolerance limit, $|Z|$ must be less than $\Delta V/\Delta I$ for the required current supply. The maximum $|Z|$ should be estimated from the maximum ΔI required. If $\Delta I = 1$ A and $\Delta V = 3.3$ V, then the impedance of the capacitor must be less than 0.3 Ω.

For an ideal capacitor to work as desired, the device should have a high C to provide a low impedance at a desired frequency, and a low L so that the impedance will not increase at higher frequencies. In addition, the capacitor must have a low R_s to obtain the least possible impedance. For this reason, power and ground planes structures are optimal in providing low-impedance decoupling within a PCB over discrete components.

Energy Storage

Decoupling capacitors ideally should be able to supply all the current necessary during a state transition of a logic device. This is described by Eq. (6.50). Use of decoupling capacitors on two-layer boards also reduces power supply ripple.

$$C = \frac{\Delta I}{\Delta V/\Delta t}; \text{ that is, } \frac{20 \text{ mA}}{100 \text{ mV}/5 \text{ ns}} = 0.001 \text{ μF or } 1000 \text{ pF} \qquad (6.50)$$

where ΔI = current transient
ΔV = allowable power supply voltage change (ripple)
Δt = switching time

Note that for ΔV, EMI requirements are usually more demanding than chip supply needs.

The response of a decoupling capacitor is based on a sudden change in demand for current. It is useful to interpret the frequency domain impedance response in terms of the capacitor's ability to supply current. This charge transfer ability is also for the time domain function that the capacitor is generally selected for. The low-frequency impedance between the power and ground planes indicates how much voltage on the board will change when experiencing a relatively slow transient. This response is an indication of the time-average voltage swing experienced during a faster transient. With low impedance, more current is available to the components under a sudden change in voltage. High-frequency impedance is an indication of how much current the board can initially supply in response to a fast transient. Boards with the lowest impedance above 100 MHz can supply the greatest amount of current (for a given voltage change) during the first few nanoseconds of a sudden transient surge.

6.4.3 Resonance

When selecting bypass and decoupling capacitors, calculate the charge and discharge frequency of the capacitor based on logic family and clock speed of the circuit (self-resonant frequency). One must select a capacitance value based on the reactance that the capacitor presents to the circuit. A capacitor is capacitive up to its self-resonant frequency. Above self-resonance, the capacitor becomes inductive. This minimizes RF decoupling. Table 6.9 illustrates the self-resonant frequency of two types of ceramic capacitors, one with standard 0.25-in leads and the other surface mount. The self-resonant frequency of SMT capacitors is always higher, although this benefit can be obviated by connection inductance. This higher self-resonant frequency is due to lower lead-length inductance provided by the smaller case package size and lack of long radial or axial lead lengths.

TABLE 6.9 Approximate Self-Resonant Frequencies of Capacitors (Lead Length Dependent)

Capacitor Value	Through Hole* 0.25-in Leads	Surface Mount** (0805)
1.0 μf	2.6 MHz	5 MHz
0.1 μf	8.2 MHz	16 MHz
0.01 μf	26 MHz	50 MHz
1000 pF	82 MHz	159 MHz
500 pF	116 MHz	225 MHz
100 pF	260 MHz	503 MHz
10 pF	821 MHz	1.6 GHz

* For through hole, L = 3.75 nH (15 nH/in).
** For surface mount, L = 1 nH.

In performing SPICE testing or analysis on various package-size SMT capacitors, all with the same capacitive value, the self-resonant frequency changed by only a few megahertz between package sizes while keeping all other measurement constants unchanged. SMT package sizes of 1210, 0805, and 0603 are common in today's products using various types of dielectric material. Only the lead inductance is different between packaging, with capacitance value remaining constant. The dielectric material did not play a significant part in changing the self-resonant frequency of the capacitor. The change in self-resonant frequency observed between different package sizes, based on lead length inductance in SMT packaging, was negligible, and varied by ±2 to 5 MHz.

EMC and Printed Circuit Board Design

When actual testing was performed in a laboratory environment on a large sample of capacitors, an interesting phenomenon was observed. The capacitors were self-resonant at the frequency analyzed, as expected. Based on a large sample size, the self-resonant frequency varied considerably. The self-resonant frequency varied because of the tolerance rating of the capacitor. Because of the manufacturing process, capacitors are provided with a tolerance rating of generally ±10%. More expensive capacitors are in the ±2 to 5% range. Since the physical size of the capacitor is fixed, due to the manufacturing process used, the value of capacitance can change owing to the thickness and variation of the dielectric material and other parameters. With manufacturing tolerance for the capacitance part of the component, the actual self-resonant frequency will change based on the tolerance rating of the device. If a design requires an exact value of decoupling, the use of an expensive precision capacitor is required. The resonance equation easily illustrates this tolerance change.

Leaded capacitors are nothing more than surface-mount devices with leads attached. A typical leaded capacitor has on the average approximately 2.5 nH of inductance for every 0.10 in of lead length above the surface of the board. Surface mount capacitors average 1 nH lead-length inductance.

An inductor does not change resonant response like a capacitor. Instead, the magnitude of impedance changes as the frequency changes. Parasitic capacitance around an inductor can, however, cause parallel resonance and alter response. The higher the frequency of the circuit, the greater the impedance. RF current traveling through this impedance causes a RF voltage. Consequently, RF current is created in the device as related to Ohm's law, $V_{rf} = I_{rf} \times Z_{rf}$. As examined above, one of the most important design concerns when using capacitors for decoupling lies in lead-length inductance. SMT capacitors perform better at higher frequencies than radial or axial capacitors because of lower internal lead inductance. Table 6.10 shows the magnitude of impedance of a 15-nH inductor versus frequency. This inductance value is caused by the lead lengths of the capacitor and the method of placement of the capacitor on a typical PCB.

TABLE 6.10 Magnitude of Impedance of a 15-nH Inductor vs. Frequency

Frequency (MHz)	Z (Ω)
0.1	0.01
0.5	0.05
1.0	0.10
10.0	1.0
20.0	1.9
30.0	2.8
40.0	3.8
50.0	4.7
60.0	5.7
70.0	6.6
80.0	7.5
90.0	8.5
100.0	9.4

Figure 6.49 shows the self-resonant frequency of various capacitor values along with different logic families. It is observed that capacitors are capacitive until they approach self-resonance (null point) before going inductive. Above the point where capacitors go inductive, they proportionally cease to function for RF decoupling; however, they may still be the best source of charge for the device, even at frequencies where they are inductive. This is because the internal bond wire from the capacitor's plates to its mounting pad (or pin) must be taken into consideration. Inductance is what causes capacitors to become less useful at frequencies above self-resonance for decoupling purposes.

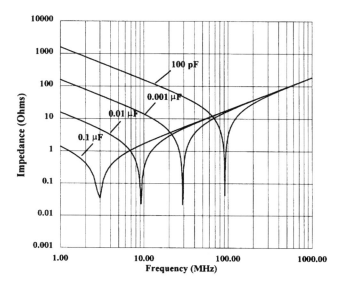

FIGURE 6.49 Self-resonant frequency of through-hole capacitors. Capacitors provided with 30-nH series inductance (trace plus lead length).

Certain logic families generate a large spectrum of RF energy. The energy developed is generally higher in frequency than the self-resonant frequency range that a decoupling capacitor presents to the circuit. For example, a 0.1 µF capacitor will usually not decouple RF currents for an *ACT* or *F* logic device, whereas a 0.001 µF capacitor is a more appropriate choice due to the faster edge rate (0.8 to 2.0 ns minimum) typical of these higher-speed components.

Compare the difference between through-hole and surface-mount capacitors (SMT). Since SMT devices have much less lead length inductance, the self-resonant frequency is higher than through-hole. Figure 6.50 illustrates a plot of the self-resonant frequency of various values of ceramic capacitors. All capacitors in the figure have the same lead length inductance for comparison purposes.

Effective decoupling is achieved when capacitors are properly placed on the PCB. Random placement, or excessive use, of capacitors is a waste of material and cost. Sometimes, fewer capacitors strategically placed perform best for decoupling. In certain applications, two capacitors in parallel are required to provide greater spectral bandwidth of RF suppression. These parallel capacitors must differ by two orders of magnitude or value (e.g., 0.1 and 0.001 µF) or 100× for optimal performance. Use of parallel capacitors is discussed later in this chapter.

6.4.4 Power and Ground Planes

A benefit of using multilayer PCBs is the placement of the power and ground planes adjacent to each other. The physical relationship of these two planes creates one large decoupling capacitor. This capacitor usually provides adequate decoupling for low-speed (slower edge rate) designs; however, additional layers add significant cost to the PCB. If components have signal edges (t_r or t_f) slower than 10 ns (e.g., standard TTL logic), use of high-performance, high self-resonant frequency decoupling capacitors is generally not required. Bulk capacitors are, still needed however to maintain proper voltage levels. For performance reasons, values such as 0.1 to 10 µF is appropriate for device power pins.

Another factor to consider when using power and ground planes as a primary decoupling capacitor is the self-resonant frequency of this built-in capacitor. If the self-resonant frequency of the power and ground planes is the same as the self-resonant frequency of the lumped total of the decoupling capacitors installed on the board, there will be a sharp resonance where these two frequencies meet. No longer will there be a wide spectral distribution of decoupling. If a clock harmonic is at the same frequency as this

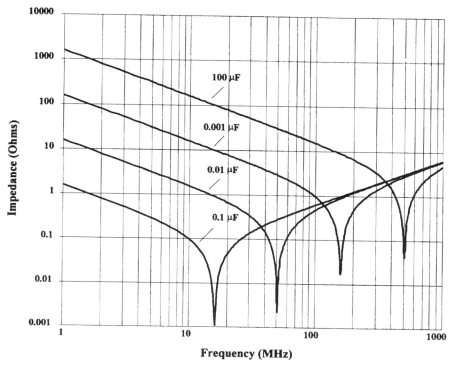

FIGURE 6.50 Self-resonant frequency of SMT capacitors (ESL = 1 nH).

sharp resonance, the board will act as if little decoupling exists. When this situation develops, the PCB may become an unintentional radiator with possible noncompliance with EMI requirements. Should this occur, additional decoupling capacitors (with a different self-resonant frequency) will be required to shift the resonance of the PCB's power and ground planes.

One simple method to change the self-resonant frequency of the power and ground planes is to change the physical distance spacing between these planes. Increasing or decreasing the height separation, or relocation within the layer stackup, will change the capacitance value of the assembly. Equations (6.52) and (6.53) provide this calculation. One disadvantage of using this technique is that the impedance of the signal routing layers may also change, which is a performance concern. Many multilayer PCBs generally have a self-resonant frequency between 200 and 400 MHz.

In the past, slower-speed logic devices fell well below the spectrum of the self-resonant frequency of the PCB's power and ground plane structure. Logic devices used in newer, high-technology designs approach or exceed this critical resonant frequency. When both the impedance of the power planes and the individual decoupling capacitors approach the same resonant frequency, severe performance deterioration occurs. This degraded high-frequency impedance will result in serious EMI problems. Basically, the assembled PCB becomes an unintentional transmitter. The PCB is not really the transmitter; rather, the highly repetitive circuits or clocks are the cause of RF energy. Decoupling will not solve this type of problem (due to the resonance of the decoupling effect), requiring system-level containment measures to be employed.

6.4.5 Capacitors in Parallel

It is common practice during a product design to make provisions for parallel decoupling of capacitors with the intent of providing greater spectral distribution of performance and minimizing ground bounce. Ground bounce is one cause of EMI created within a PCB. When parallel decoupling is provided, one must not forget that a third capacitor exists—the power and ground plane structure.

When dc power is consumed by component switching, a momentary surge occurs in the power distribution network. Decoupling provides a localized point source charge, since a finite inductance exists within the power supply network. By keeping the voltage level at a stable reference point, false logic switching is prevented. Decoupling capacitors also minimize radiated emissions by providing a very small loop area for creating high spectral content switching currents instead of having a larger loop area created between the component and a remote power source.

Research on the effectiveness of multiple decoupling capacitors shows that parallel decoupling may not be significantly effective and that, at high frequencies, only a 6-dB improvement may occur over the use of a single large-value capacitor.[4] Although 6 dB appears to be a small number for suppression of RF current, it may be all that is required to bring a noncompliant product into compliance with international EMI specifications. According to Paul,

> Above the self-resonant frequency of the larger value capacitor where its impedance increases with frequency (inductive), the impedance of the smaller capacitor is decreasing (capacitive). At some point, the impedance of the smaller value capacitor will be smaller than that of the larger value capacitor and will dominate thereby giving a smaller net impedance than that of the larger value capacitor alone.[4]

This 6-dB improvement is the result of lower lead-length and device-body inductance provided by the capacitors in parallel. There are now two sets of parallel leads from the internal plates of the capacitors. These two sets provide greater trace width than would be available if only one set of leads were provided. With a wider trace width, there is less lead-length inductance. This reduced lead-length inductance is a significant reason why parallel decoupling capacitors work as well as they do.

Figure 6.51 shows a plot of two bypass capacitors, 0.01 μF and 100 pF, both individually and in parallel. The 0.01 μF capacitor has a self-resonant frequency at 14.85 MHz. The 100 pF capacitor has its self-resonant frequency at 148.5 MHz. At 110 MHz, there is a large increase in impedance due to the parallel combination. The 0.01 μF capacitor is inductive, while the 100 pF capacitor is still capacitive. We have both L and C in resonance; hence, an antiresonant frequency effect is exactly what we do not want in a PCB if compliance to EMI requirements is mandatory.

As shown in the figure, between the self-resonant frequency of the larger value capacitor, 0.01 μF, and the self-resonant frequency of the smaller value capacitor, 100 pF, the impedance of the larger value capacitor is essentially inductive, whereas the impedance of the smaller value capacitor is capacitive. In

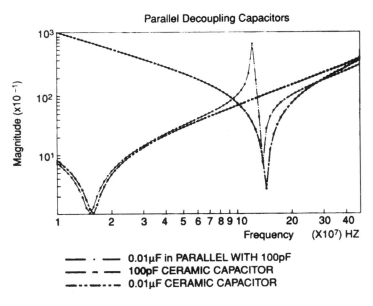

FIGURE 6.51 Resonance of parallel capacitors.

this frequency range, there exists a parallel resonant LC circuit, and we should therefore expect to find an infinite impedance of the parallel combination. Around this resonant point, the impedance of the parallel combination is actually larger than the impedance of either isolated capacitor.

In Fig. 6.51, observe that, at 500 MHz, the impedances of the individual capacitors are virtually identical. The parallel impedance is only 6 dB lower. This 6-dB improvement is only valid over a limited frequency range from about 120 to 160 MHz.

To further examine what occurs when two capacitors are used in parallel, examine a Bode plot of the impedance presented by two capacitors in parallel (Fig. 6.52). For this Bode plot, the frequency responses of the magnitude at the various break frequencies are[4]

$$f_1 = \frac{1}{2\pi\sqrt{LC_1}} < f_2 = \frac{1}{2\pi\sqrt{LC_2}} < f_3 = \frac{1}{2\pi\sqrt{LC_3}} = 2f_2 \qquad (6.51)$$

By shortening the lead lengths of the larger value capacitor (0.01 µF), we can obtain the same results by a factor of 2. For this reason, a single capacitor may be more optimal in a specific design than two, especially if minimal lead-length inductance exists.

To remove RF current generated by digital components switching all signal pins simultaneously (and it is desired to parallel decouple), it is common practice to place two capacitors in parallel (e.g., 0.1 µF and 0.001 µF) immediately adjacent to each power pin. If parallel decoupling is used within a PCB layout, one must be aware that the capacitance values should differ by two orders of magnitude, or 100×. The total capacitance of parallel capacitors is not important. Parallel reactance provided by the parallel capacitors (due to self-resonant frequency) is the important item.

To optimize the effects of parallel bypassing, and to allow use of only one capacitor, reduction in capacitor lead-length inductance is required. A finite amount of lead-length inductance will always exist when installing the capacitor on the PCB. Note that the lead length must also include the length of the via connecting the capacitor to the planes. The shorter the lead length from either single or parallel decoupling, the greater the performance. In addition, some manufacturers provide capacitors with significantly reduced "body" inductance internal to the capacitor.

6.4.6 Power and Ground Plane Capacitance

The effects of the internal power and ground planes inside the PCB are not considered in Fig. 6.51. However, multiple bypassing effects are illustrated in Fig. 6.53. Power and ground planes have very little lead-length inductance equivalence and no equivalent series resistance (ESR). Use of power planes as a decoupling capacitor reduces RF energy at frequencies generally in the higher frequency ranges.

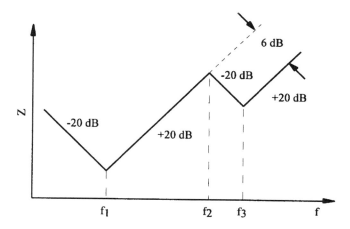

FIGURE 6.52 Bode plot of parallel capacitors.

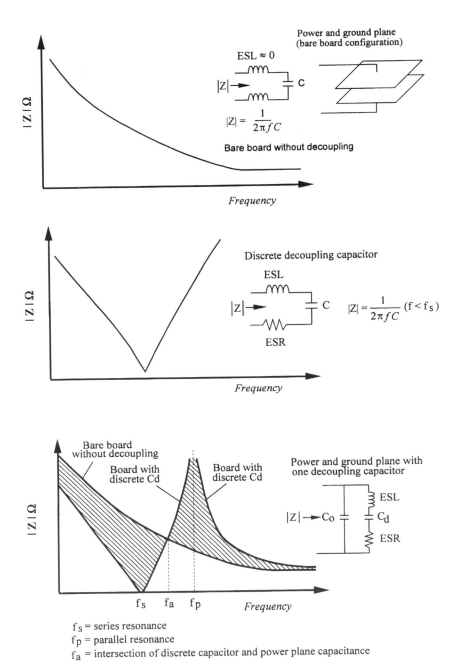

FIGURE 6.53 Decoupling effects of power ground planes with discrete capacitors.

On most multilayer boards, the maximum inductance of the planes between two components is significantly less than 1 nH. Conversely, lead-length inductance (e.g., the inductance associated with the traces connecting a component to its respective via plus the via themselves) is typically 2.5 to 10 nH or greater.[8]

Capacitance will always be present between a voltage and ground plane pair. Depending on the thickness of the core, the dielectric constant, and the placement of the planes within the board stackup, various values of internal capacitance can exist. Network analysis, mathematical calculations, or modeling

EMC and Printed Circuit Board Design

will reveal the actual capacitance of the power planes. This is in addition to determining the impedance of all circuit planes and the self-resonant frequency of the total assembly as potential RF radiators. This value of capacitance is easily calculated by Eqs. (6.52) and (6.53). These equations may be used to estimate the capacitance between planes, since planes are finite, have multiple holes, vias, and the like. Actual capacitance is generally less than the calculated value.

$$C = \frac{\varepsilon_o \varepsilon_r A}{d} = \frac{\varepsilon A}{d} \quad (6.52)$$

where ε = permittivity of the medium between capacitor plates (F/m)
A = area of the parallel plates (m²)
d = separation of the plates (m)
C = capacitance between the power and ground planes (pF)

Introducing relative permittivity ε_r of the dielectric material, and the value of ε_o, the permittivity of free space, we obtain the capacitance of the parallel-plate capacitor, namely, the power and ground plane combination.

$$C = 8.85 \frac{A \varepsilon_r}{d} \text{ (pF)} \quad (6.53)$$

where ε_r = the relative permittivity of the medium between the plates, typically ≈ 4.5 (varies for linear material, usually between 1 and 10)
ε_o = permittivity of free space, $1/36\pi \times 10^{-9}$ F/m = 8.85×10^{-12} F/m = 8.85 pF/m

Equations (6.52) and (6.53) show that the power and ground planes, when separated by 0.01 in of FR-4 material, will have a capacitance of 100 pF/in².

Because discrete decoupling capacitors are common in multilayer PCBs, we must question the value of these capacitors when low-frequency, slow edge rate components are provided, generally in the frequency range less below 25 MHz. Research into the effects of power and ground planes along with discrete capacitors reveals interesting results.[6]

In Fig. 6.53, the impedance of the *bare board* closely approximates the ideal decoupling impedance that would result if only pure capacitance, free of interconnect inductance and resistance, could be added. This ideal impedance is given by $Z_c = 1/j\omega C_o$. The discrete capacitor becomes zero at the series resonant frequency, f_s, and infinite at the parallel resonance frequency, f_p, where n = number of discrete capacitors provided, C_d is the discrete capacitor, and C_o is the capacitance of the power and ground plane structure, conditioned by the source impedance of the power supply.[4,6]

$$f_s = \frac{1}{2\pi\sqrt{LC}} \qquad f_p = f_s \sqrt{1 + \frac{nC_d}{C_o}} \quad (6.54)$$

For frequencies below series resonance, discrete decoupling capacitors behave as capacitors with an impedance of $Z = 1/(j\omega C)$. For frequencies near series resonance, the impedance of the loaded PCB is actually less than that of the ideal PCB. However, at frequencies above f_s, the decoupling capacitors begin to exhibit inductive behavior as a result of their associated interconnect inductance. Thus, the discrete decoupling capacitors function as inductors above series resonance. The frequency at which the magnitude of the board impedance is the 'same with or without the decoupling capacitors,'[6] where the unloaded PCB intersects that of the loaded, nonideal PCB is described by

$$f_a = f_s \sqrt{1 + \frac{nC_d}{2C_o}} \quad (6.55)$$

For frequencies above f_a, additional n decoupling capacitors provide no additional benefit, as long as the switching frequencies of the components are within the decoupling range of the power and ground plane structure. The bare board impedance remains far below that of the board that is loaded with discrete capacitors. At frequencies near the loaded board pole, parallel resonant frequency, the magnitude of the loaded board impedance is extremely high. Decoupling performance of the loaded board is far worse than that of the unloaded board without additional decoupling. The analysis clearly indicates that minimizing the series inductance of the decoupling capacitor connection is crucial to achieving ideal capacitor behavior over the widest possible frequency range, which in the time domain corresponds to the ability to supply charge rapidly. Lowering the interconnect inductance increases the series and parallel-resonance frequency, thereby extending the range of ideal capacitor behavior.[4,6]

Parallel resonances correspond to poles in the board impedance expression. Series resonances are null points. When multiple capacitors are provided, the poles and zeros will alternate so that there will be exactly one parallel resonance between each pair of series resonances. A parallel resonance will always exist between two series resonances.

Although good distributive capacitance exists when using a power and ground plane structure, adjacent close stacking of these two planes plays a critical part in the overall PCB assembly. If two sets of power and ground planes exist, for example, +5 V/ground and +3.3 V/ground, both with different dielectric spacing between the two planes, it is possible to have multiple decoupling capacitors built internal to the board. With proper selection of layer stackup, both high-frequency and low-frequency decoupling can be achieved without the use of any discrete devices. To expand on this concept, a technology known as *buried capacitance* is finding use in high-technology products that require high-frequency decoupling.

6.4.7 Buried Capacitance

Buried capacitance[*] is a patented manufacturing process in which the power and ground planes are separated by a 0.001-in (0.25-mm) dielectric. With this small dielectric spacing, decoupling is effective up to 200 to 300 MHz. Above this frequency range, use of discrete capacitors is required to decouple components that operate above the cutoff frequency of the buried capacitance. The important item to remember is that the closer the distance spacing between the power and ground planes, the better the decoupling performance. It is to be remembered that, although buried capacitors may eliminate the employment and cost of discrete components, use of this technology may far exceed the cost of all discrete components that were removed.

To better understand the concept of buried capacitance, consider the power and ground planes as pure capacitance at low frequencies with very little inductance. These planes can be considered to be an equal-potential surface with no voltage gradient except for a small dc voltage drop. This capacitance is calculated simply as area divided by thickness times permittivity. For a 10-in-square board, with 2 mil FR-4 dielectric between the power and ground planes, we have 45 nF (0.045 µF).

At some frequency, a full wave will be observed between the power and ground planes along the edge length of the PCB. Assuming velocity of propagation to be 6 in/ns (15.24 cm/ns), observe that the frequency will be 600 MHz for a 10 × 10-in board. At this frequency, the planes are not at equal potential, for the voltage measured between two points can differ greatly as we move the test probe around the board. A reasonable transition frequency is one-tenth of 600 MHz or 60 MHz. Below this frequency, the planes can be considered as pure capacitance.

Knowing the velocity of propagation and capacitance per square area, calculate the plane inductance. For a 2-mil-thick dielectric, capacitance is 0.45 nF/in^2, velocity of propagation = 6 in/ns, and inductance is 0.062 nH/sq. This inductance is a spreading inductance, similar in interpretation to spreading resistance,

---

[*]Buried capacitance is a registered trademark of HADCO Corp. (which purchased Zycon Corp., developers of this technology).

which is a very small number. This small number is the primary reason why power planes are mainly pure capacitance.

With known inductance and capacitance, calculate unit impedance as $Z_o = \sqrt{L/C}$, which is 0.372 Ω-in. A plane wave traveling down a long length of a 10-in-wide board will see 0.372/10 = 0.0372 Ω impedance, again, a small number.

Decoupling capacitance is increased, because the distance spacing between the planes (d) is in the denominator. Inductance is constant, because the velocity of propagation remains constant. The total impedance is thus decreased. The power and ground planes are the means of distributing power. Reducing the dielectric thickness is effective at increasing decoupling capacitance and transporting high-frequency power through a lower-impedance distribution system.

6.4.8 Calculating Power and Ground Plane Capacitance

Capacitance between a power and ground plane is described by

$$C_{pp} = k\frac{\varepsilon_r A}{d} \qquad (6.56)$$

where C_{pp} = capacitance of parallel plates (pF)
 ε_r = dielectric constant of the board material (vacuum = 1, FR-4 material = 4.1 to 4.7)
 A = common area between the parallel plates (square inches or cm)
 d = distance spacing between the plates (inches or cm)
 k = conversion constant (0.2249 for inches, 0.884 for cm)

One caveat in solving this equation is that the inductance caused by the antipads (holes for through vias) in the power and ground planes can minimize the theoretical effectiveness of this equation.

Because of the efficiency of the power planes as a decoupling capacitor, use of high, self-resonant frequency decoupling capacitors may not be required for standard TTL or slow-speed logic. This optimum efficiency exists, however, only when the power planes are closely spaced—less than 0.01 inch with 0.005 inch preferred for high-speed applications. If additional decoupling capacitors are not properly chosen, the power planes will go inductive below the lower cut-in range of the higher self-resonant frequency decoupling capacitor. With this gap in resonance, a pole is generated, causing undesirable effects on RF suppression. At this point, RF suppression techniques on the PCB become ineffective, and containment measures must be used at a much greater expense.

6.4.9 Lead-Length Inductance

All capacitors have lead and device body inductance. Vias also add to this inductance value. Lead inductance must be minimized at all times. When a signal trace plus lead-length inductance are combined, a higher impedance mismatch will be present between the component's power/ground pins and the system's power/ground plane. With trace impedance mismatch, a voltage gradient is developed between these two sources, creating RF currents. RF fields cause RF emissions; hence, decoupling capacitors must be designed for minimum inductive lead length, including via and pin escapes (or pad connections from the component pin to the point where the device pin connects to a via).

In a capacitor, the dielectric material determines the magnitude of the zero point for the self-resonant frequency of operation. All dielectric material is temperature sensitive. The capacitance value of the capacitor will change in relation to the ambient temperature provided to its case package. At certain temperatures, the capacitance may change substantially, resulting in improper performance, or no performance at all when used as a bypass or decoupling element. The more stable the temperature rating of the dielectric material, the better performance of the capacitor.

In addition to the sensitivity of the dielectric material to temperature, the equivalent series inductance (ESL) and the equivalent series resistance (ESR) must be low at the desired frequency of operation. ESL acts like a parasitic inductor, whereas ESR acts like a parasitic resistor, both in series with the capacitor. ESL is not a major factor in today's small SMT capacitors. Radial and axial lead devices will always have large ESL values. Together, ESL and ESR degrade a capacitor's effectiveness as a bypass element. When selecting a capacitor, one should choose a capacitor family that publishes actual ESL and ESR values in its data sheet. Random selection of a standard capacitor may result in improper performance if ESL and ESR are too high. Most vendors of capacitors do not publish ESL and ESR values, so it is best to be aware of this selection parameter when choosing capacitors used in high-speed, high-technology PCBs.

Because surface-mount capacitors have essentially little ESL and ESR, their use is preferred over radial or axial types. Typically, ESL is <1.0 nH, and ESR should be 0.5 Ω or less. For decoupling capacitors, capacitance tolerance is not as important as the temperature stability, dielectric constant, ESL, ESR, and self-resonant frequency.[1,2]

6.4.10 Placement

Power Planes

Multilayer PCBs generally contain one or more pair of voltage and ground planes. A power plane functions as a low-inductance capacitor that minimizes RF currents generated from components and traces. Multiple chassis ground stitch connections to all ground planes minimize voltage gradients between the board and chassis, and between/among board layers. These gradients also are a major source of common-mode RF fields. This is in addition to sourcing RF currents to chassis ground. In many cases, multiple ground stitch connections are not always possible, especially in card cage designs. In such situations, care must be taken to analyze and determine where RF loops will occur and how to optimize grounding of the power planes.

Power planes that are positioned next to ground planes provide for flux cancellation in addition to decoupling RF currents created from power fluctuations of components and noise injected into the power and ground planes. Components switching logic states cause a current surge during the transition. This current surge places a strain on the power distribution network. An image plane is a solid copper plane at voltage or ground potential located adjacent to a signal routing plane. RF currents generated by traces on the signal plane will mirror image themselves in this adjacent metal plane. This metal plane must not be isolated from the power distribution network.[1,2,9] To remove common-mode RF currents created within a PCB, all routing (signal) layers must be physically adjacent to an image plane. (Refer to Chapter 4 for a detailed discussion of image planes.)

Decoupling Capacitors

Before determining where to locate decoupling capacitors, the physical structure of a PCB must be understood. Figure 6.54 shows the electrical equivalent circuit of a PCB. In this figure, observe the loops that exist between power and ground caused by traces, IC wire bonds, lead frames of components, socket pins, component interconnect leads, and decoupling capacitor. The key to effective decoupling is to minimize R_2, L_2, R'_2, L'_2, R_3, L_3, R'_3, L'_3, R_4, L_4, R'_4, and L'_4. Placement of power and ground pins in the center of the component helps reduce R_4, L_4, R'_4, and L'_4. Basically, the impedance of the PCB structure must be minimized. The easiest way to minimize the resistive and inductive components of the PCB is to provide a solid plane. To minimize the inductance from component leads, use of SMT, ball grid arrays, and flip chips is preferred. With less lead bond lengths from the die to PCB pad, overall impedance is reduced.

Figure 6.54[1,2] makes it clear that EMI is a function of loop geometry and frequency. Hence, the smallest closed-loop area is desired. We acquire this small area by placing a local decoupling capacitor, C_{pcb}, for current storage adjacent to the power pins of the IC. It is mandatory that the decoupling loop impedance be much lower than the rest of the power distribution system. This low impedance will cause the high-frequency RF energy developed by both traces and components to remain almost entirely within this small loop area. Consequently, lower EMI emissions are observed.

If the impedance of the loop is smaller than the rest of the system, some fraction of the high-frequency RF component will transfer, or couple, to the larger loop formed by the power distribution system. With

EMC and Printed Circuit Board Design

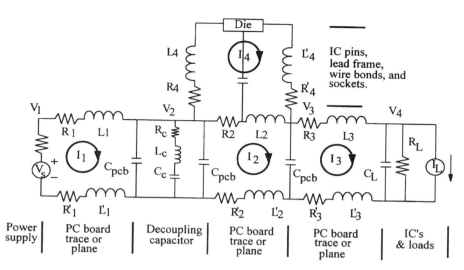

FIGURE 6.54 Equivalent circuit of a PCB.

this situation, RF currents are developed and, hence, higher EMI emissions. This situation is best illustrated in Fig. 6.55.

To summarize,

The important parameter when using decoupling capacitors is to minimize lead-length inductance and to locate the capacitors as close as possible to the component.

Decoupling capacitors must be provided for every device with edges faster than 2 ns and should be provided, placement wise, for *every component*. Making provisions for decoupling capacitors is a necessity, because future EMI testing may indicate a requirement for these devices. During testing, it may be possible to determine that there may be excess capacitors in the assembly. Having to add capacitors to an assembled board is difficult, if not impossible. Today, CMOS, ECL, and other fast logic families require additional discrete decoupling capacitors besides the power and ground planes.

If a decoupling capacitor must be provided to a through-hole device after assembly, retrofit can be performed. Several manufacturers provide a decoupling capacitor assembly using a flat, level construction that resides between the component and PCB. This flat pack shares the same power and ground pins of the components. Because these capacitors are flat in construction, lead length inductance is much less compared to capacitors with radial or axial leads. Since the capacitor and component share power and ground pins, R_2, L_2, R'_2, and L'_2 are also reduced. Some lead-length inductance will remain, which cannot be removed. The most widely used board level retrofit capacitors are known as Micro-Q™. Other manufacturers provide similar products. An example of this type of capacitor is detailed in Fig. 6.56. Other configurations exist in pin grid array (PGA) form factor. For PGA retrofit decoupling capacitors,

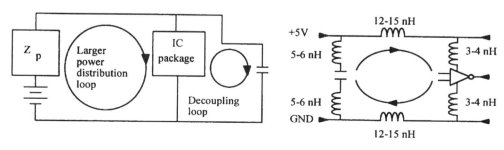

FIGURE 6.55 Power distribution model for loop control.

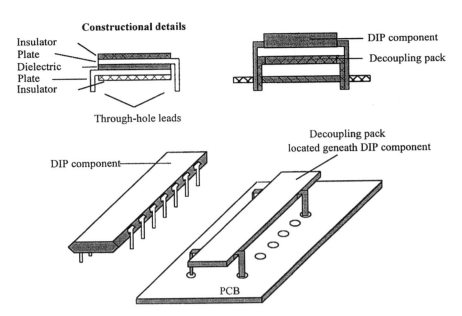

FIGURE 6.56 Retrofit decoupling capacitor, DIP mounting style.

unique assemblies are available based on the particular pinout configuration of the device requiring this part.

A retrofit capacitor has a self-resonant frequency generally in the range of 10 to 50 MHz, depending on the capacitance of the device. Since DIP style leads are provided, higher frequency use cannot occur, owing to excessive lead-length inductance. Although sometimes termed a "retrofit" device, the improved decoupling performance of these capacitors, compared to axial leaded capacitors on two-layer boards, makes them suitable for initial design implementation.

Poor planning during PCB layout and component selection may require use of Micro-Q* devices. As yet, no equivalent retrofit exists for SMT packaged components.

When selecting a capacitor, consider not only the self-resonant frequency, but the dielectric material as well. The most commonly used material is Z5U (barium titanite ceramic). This material has a high dielectric constant. This dielectric constant allows small capacitors to have large capacitance values with self-resonant frequencies from 1 to 20 MHz, depending on design and construction. Above self-resonance, performance of Z5U decreases, as the loss factor of the dielectric becomes dominant, which limits its usefulness to approximately 50 MHz.

Another dielectric material commonly used is NPO (strontium titanite). This material has better high-frequency performance owing to a low dielectric constant. NPO is also a more temperature-stable dielectric. The capacitance value (and self-resonant frequency) is less likely to change when the capacitor is subjected to changes in ambient temperature or operating conditions.

Placement of 1 nF (1000 pF) capacitors (capacitors with a very high self-resonant frequency) on 1-inch center grid throughout the PCB may provide additional protection from reflections and RF currents generated by both signal traces and the power planes, especially if a high-density PCB stackup is used. It is not the exact location that counts in the placement of these additional decoupling capacitors. A lumped model analysis of the PCB will show that the capacitors will still function as desired, regardless of where the device is actually placed for overall decoupling performance. Depending on the resonant structure of the board, values of the capacitors placed in the grid may be as small as 30 to 40 pF.[1,2,7]

---

*Micro-Q is a trademark of Circuit Components Inc. (formerly Rogers Corporation).

VLSI and high-speed components (e.g., F, ACT, BCT, CMOS, ECL logic families) may require use of decoupling capacitors in parallel. As slew rates of components become steeper, a greater spectral distribution of RF energy is created. Parallel capacitors generally provide optimal bypassing of power plane noise. Multiple paired sets of capacitors are placed between the power and ground pins of VLSI components located around all four sides. These high-frequency decoupling capacitors are typically rated 0.1 µF in parallel with 0.001 µF for 50 MHz systems. Higher clock frequencies use a parallel combination of 0.01 µF and 100 pF components.

While the focus in this chapter is on multilayer boards, single- and double-sided boards also require decoupling. Figure 6.57 illustrates correct and incorrect ways of locating decoupling capacitors for a single- or double-sided assembly. When placing decoupling capacitors, ground loop control must be considered at all times. When using multilayer boards with internal power and ground planes, placement of the decoupling capacitor may be anywhere in the vicinity of the component's power pins,[6] although this implementation may actually cause the PCB to become more RF active. This requirement is based on whether the component has mounting pads vias straight down to the power/ground plane or whether a routed trace connects to the discrete capacitor. Location of the capacitor is not critical during placement for the previous statement because of the lumped distributed capacitance of the power planes. The components themselves must via down to the power and ground plane, the same as the decoupling capacitor.[6]

Another function of a decoupling capacitor is to provide localized energy storage, thereby reducing power supply radiating loops. When current flows in a closed-loop circuit, the RF energy produced is proportional to IAF, where I is the current in the loop, A is the area of the loop, and F is the frequency of the current. The type of logic family selected predetermines current and frequency. It is necessary to minimize the area of the logic current loop to reduce radiation. Minimal loop area can be accomplished by taking care in placement of decoupling capacitors. A good example of a large loop area is shown in Fig. 6.57.

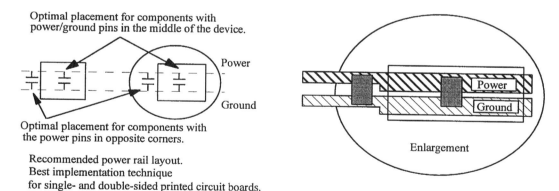

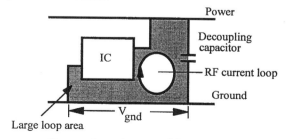

FIGURE 6.57 Placement of decoupling capacitors, single- and double-sided board.

In Fig. 6.57, V_{gnd} is *LdI/dt* induced noise in the ground trace flowing in the decoupling capacitor. This V_{gnd} drives the ground structure of the board, contributing to the overall common-mode voltage level. One should minimize the ground path that is common with a board's ground structure and decoupling capacitors.

6.4.11 Selection of a Decoupling Capacitor

Clock circuit components must be given emphasis to be RF decoupled. This is due to the switching energy generated by components injected into the power and ground distribution system. This energy will be transferred to other circuits or subsections as common-mode or differential-mode RF. Bulk capacitors such as tantalum and high-frequency ceramic monolithic are both required, each for a different application. Furthermore, monolithic capacitors must have a self-resonant frequency higher than the clock harmonics requiring suppression. Typically, one selects a capacitor with a self-resonant frequency in the range of 10 to 30 MHz for circuits with edge rates of 2 ns or less. Many PCBs are self-resonant in the 200 to 400 MHz range. Proper selection of decoupling capacitors, along with the self-resonant frequency of the PCB structure, acting as one large capacitor, will provide enhanced EMI suppression. Surface mount devices have a much higher self-resonant frequency by approximately two orders of magnitude (or 100×) as the result of less lead-length inductance. Aluminum electrolytic capacitors are ineffective for high-frequency decoupling and are best suited for power supply subsystems or power line filtering.

It is common to select a decoupling capacitor for a particular application, usually the first harmonic of a clock or processor. Sometimes, a capacitor is selected for the third or fifth harmonic, since this is where the majority of RF current is produced. There also need to be plenty of larger discrete capacitors—bulk and decoupling. The use of common decoupling capacitor values of 0.1 µF in parallel with 0.001 µF can be too inductive and too slow to supply charge current at frequencies above 200 to 300 MHz.

When performing component placement on a PCB, one should make provisions for adequate high-frequency RF decoupling capacitors. One should also verify that all bypass and decoupling capacitors chosen are selected based on intended application. This is especially true for clock generation circuits. The self-resonant frequency must take into account all significant clock harmonics requiring suppression, generally considered the fifth harmonic of the original clock frequency. Finally, capacitive reactance (self-resonant reactance in ohms) of decoupling capacitors is calculated per Eq. (6.57).

$$X_c = \frac{1}{2\pi f C} \tag{6.57}$$

where X_c = capacitance reactance (ohms)
    f = resonant frequency (hertz)
    C = capacitance value

6.4.12 Calculating Capacitor Values (Wave-Shaping)

Capacitors can also be used to wave-shape differential-mode RF currents on individual traces. These parts are generally used in I/O circuits and connectors and are rarely used in clock networks. The capacitor, C, alters the signal edge of the output clock line (slew rate) by rounding the time period the signal edge takes to transition from logic state 0 to logic state 1. This is illustrated in Fig. 6.58.

In examining Fig. 6.58, observe the change in the slew rate (clock edge) of the desired signal. Although the transition points remain unchanged, the time period, t_r, is different. This elongation or slowing down of the signal edge is a result of the capacitor charging and discharging. The change in transition time is described by the equations and illustration presented in Fig. 6.59. Note that a Thevenin equivalent circuit is shown without the load. The source voltage, V_b, and series impedance are internal to the driver or clock generation circuit. The capacitive effect on the trace, seen in the figure, is a result of this capacitor being located in the circuit. To determine the time rate of change of the capacitor detailed in Fig. 6.58, the equations in Fig. 6.59 are used.

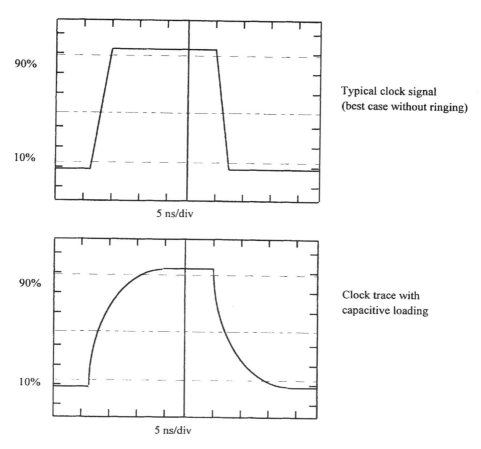

FIGURE 6.58 Capacitive effects on clock signals.

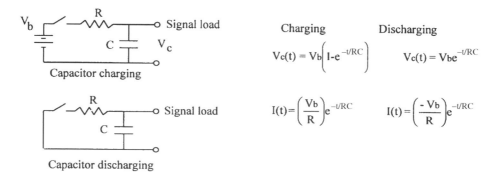

FIGURE 6.59 Capacitor equations.

When a Fourier analysis is performed on this signal edge (conversion from time to frequency domain), a significant reduction of RF energy is observed along with a decrease in spectral RF distribution. Hence, there is improved EMI compliance. Caution is required during the design stage to ensure that slower edge rates will not adversely affect operational performance.

The value for a decoupling capacitor can be calculate in two ways. Although the capacitance is calculated for optimal filtering at a particular resonant frequency, use and implementation depend on installation, lead length, trace length, and other parasitic parameters that may change the resonant frequency of the capacitor. The installed value of capacitive reactance is the item of interest. Calculating the value of capacitance will be in the ballpark and is generally accurate enough for actual implementation.

Before calculating a decoupling capacitor value, the Thevenin impedance of the network should be determined. This impedance value should be equal to these two resistors placed in parallel. Using a Thevenin equivalent circuit, assume that $Z_s = 150$ Ω, and $Z_L = 2.0$ kΩ.

$$Z_t = \frac{Z_s \times Z_L}{Z_s + Z_L} = \frac{150 \times 2000}{2150} = 140 \text{ Ω} \tag{6.58}$$

Method 1

Equation (6.59) is used to determine the capacitance value, knowing the edge rate of the clock signal.

$$t_r = k\, R_t C_{max} = 3.3 \times R_t \times C_{max} \tag{6.59}$$

where t_r = edge rate of the signal (the faster of either the rising or falling edge)
   R_t = total resistance within the network
   C_{max} = maximum capacitance value to be used
   k = one time constant

Note: C is in nanofarads if t_r is in nanoseconds; C in picofarads if t_r is in picoseconds.

The capacitor must be chosen so that $t_r = 3.3 \times R \times C$ equals an acceptable rise or fall time for proper functionality of the signal; otherwise, baseline shift may occur. Baseline shift refers to the steady-state voltage level that is identified as logic LOW or logic HIGH for a particular logic family. The number 3.3 is the value for the time constant of a capacitor charging, based on the equation $\tau = RC$. Approximately three (3) time constants equal one (1) rise time. Since we are interested in only one time constant for calculating capacitance value, this value of k is 1/3 t_r, which becomes 3.3$_r$ when incorporated within the equation.

For example, if the edge rate is 5 ns and the impedance of the circuit is 140 Ω, we can calculate the maximum value of C as

$$C_{max} = \frac{0.3 \times 5}{140} = 0.01 \text{ nF or } 10 \text{ pF} \tag{6.60}$$

A 60-MHz clock with a period of 8.33 ns on and 8.33 ns off, $R = 33$ Ω (typical for an unterminated TTL part), has an acceptable $t_r = t_f = 2$ ns (25% of the on or off value). Therefore,

$$\left(C = \frac{0.3 \times t_r}{R_t}\right) \quad C = \frac{0.3(2 \times 10^{-9})}{33} = 18 \text{ pF} \tag{6.61}$$

Method 2

- Determine highest frequency to be filtered, f_{max}.
- For differential pair traces, determine the maximum tolerable value of each capacitor. To minimize signal distortion, use Eq. (6.62).

$$C_{max} = \frac{100}{f_{max} \times R_t} \qquad \frac{1}{2\pi f_{max} \times \dfrac{C}{2}} \geq 3 \times R_t \tag{6.62}$$

where C is in nanofarads and f is in megahertz.

To filter a 20-MHz signal with $R_L = 140\ \Omega$, the capacitance value would be

$$C_{min} = \frac{100}{20 \times 140} = 0.036 \text{ nF or 36 pF} \qquad (6.63)$$

with negligible source impedance Z_c.

When using bypassing capacitors implement the following:

- If degradation of the edge rate is acceptable, generally three times the value of C, increase the capacitance value to the next highest standard value.
- Select a capacitor with proper voltage rating and dielectric constant for intended use.
- Select a capacitor with a tight tolerance level. A tolerance level of +80/−0% is acceptable for power supply filtering but is inappropriate as a decoupling capacitor for high-speed signals.
- Install the capacitor with minimal lead-length and trace inductance.
- Verify that the circuit still works with the capacitor installed. A capacitor too large in value can cause excessive signal degradation.

6.4.13 Selection of Bulk Capacitors

Bulk capacitors provide dc voltage and current to components when devices are switching all data, address, and control signals simultaneously under maximum capacitive load. Components have a tendency to cause current fluctuations on the power planes. These fluctuations can cause improper performance of components due to voltage sags. Bulk capacitors provide energy storage for circuits to maintain optimal voltage and current requirements. Bulk capacitors play no significant role in EMI control.

Bulk capacitors (tantalum dielectric) should be used in addition to high, self-resonant frequency decoupling to provide dc power for components and power plane RF modulation. Place one bulk capacitor for every two LSI and VLSI components, in addition to the following locations:

- Power entry connector from the power supply
- All power terminals on I/O connectors for daughter cards, peripheral devices, and secondary circuits
- Near power consuming circuits and components
- The farthest location from the input power connector
- High-density component placement remote from the dc input power connector
- Adjacent to clock generation circuits

When using bulk capacitors, calculate the voltage rating such that the nominal level equals 50% of the capacitor's actual rating to prevent self-destruction, should a voltage surge occur. For example, with power = 5 V, use a capacitor with a minimum of a 10-V rating.

Table 6.11 shows the typical quantity of capacitors required for some popular logic families. This chart is based on the maximum allowable power drop that is equal to 25% of the noise immunity level of the circuit being decoupled. Note that, for standard CMOS logic, this table is conservative, since the trace wiring to the components cannot provide the required peak current without an excessive voltage drop.

Memory arrays require additional bulk capacitors due to the extra current required for proper operation during a refresh cycle. The same is true for VLSI components with large pin counts. High-density pin grid array (PGA) modules also must have additional bulk capacitors provided, especially when all signal, address and control pins switch simultaneously under maximum capacitive load.

Use Eq. (6.64) to calculate the peak surge current consumed by many capacitors, it is observed that more is not necessarily better. An excessive number of capacitors can draw a large amount of current, thereby placing a strain on the power supply.

Selection of a capacitor for a particular application, based on past experience with slow-speed digital logic, generally will not provide proper bypassing and decoupling when used with high-technology, high-

TABLE 6.11 Number of Decoupling Capacitors for Selected Logic Families

Logic Family	Peak Transient Current Requirement (mA)		Number of Decoupling Capacitor for a Fan-Out of 5 Gates + 10 cm Trace Length
	Gate Overcurrent (mA)	1 Gate Drive (mA)	
CMOS	1	0.3	0.6
TTL	16	17	2.6
LS-TTL	8	2.5	2.0
HCMOS	15	5.5	1.2
STTL	30	5	1.8
FAST	15	5.5	1.8
ECL	1	1.2	0.9

speed designs. Consideration of resonance, placement on the PCB, lead-length inductance, existence of power and ground planes, and the like must all be included when selecting a capacitor or capacitor combination.

For bulk capacitors, the following procedures are provided to determine optimal selection.[1,2,10]

Method 1

1. Determine maximum current (ΔI) consumption anticipated on the board. Assume that all gates switch simultaneously. Include the effect of power surges by logic crossover (cross-conduction currents).
2. Calculate maximum amount of power supply noise permitted (ΔV) by devices for functionality purposes. Factor in a safety margin.
3. Determine maximum common-path impedance acceptable to the circuit.

$$Z_{cm} = \frac{\Delta V}{\Delta I} \qquad (6.64)$$

4. If solid planes are used, allocate the impedance, Z_{cm}, to the connection between the power and ground structure.
5. Calculate the impedance of the interconnect cable, L_{cable}, from the power supply to the PCB. Add this value to Z_{cm} to determine the frequency below which the power supply wiring is adequate ($Z_{total} = Z_{cm} + L_{cable}$).

$$f = \frac{Z_{total}}{2\pi L_{cable}} \qquad (6.65)$$

6. If the switching frequency is below the calculated f of Eq. (4.20), the power supply wiring is fine. Above f, bulk capacitors, C_{bulk}, are required. Calculate the value of the bulk capacitor for an impedance value Z_{total} at frequency f.

$$C_{bulk} = \frac{1}{2\pi f Z_{total}} \qquad (6.66)$$

Method 2

A PCB has 200 CMOS gates (G), each switching 5 pF (C) loads within a 2-ns time period. Power supply inductance is 80 nH.

$$\Delta I = GC\frac{\Delta V}{\Delta t} = 200(5 \text{ pF})\frac{5 \text{ V}}{2 \text{ ns}} = 2.5 \text{ A (worst-case peak surge)}$$

$$\Delta V = 0.200 \text{ V (from noise margin budget)}$$

$$Z_{total} = \frac{\Delta V}{\Delta I} = \frac{0.20}{2.5} = 0.08 \text{ }\Omega$$

$$L_{cable} = 80 \text{ nH}$$

$$f_{ps} = \frac{Z_{total}}{2\pi L_{cable}} = \frac{0.08 \text{ }\Omega}{2\pi 80 \text{ nH}} = 159 \text{ kHz}$$

$$C = \frac{1}{2\pi f_{ps} Z_{total}} = 12.5 \text{ }\mu\text{F}$$

(6.67)

Capacitors commonly found on PCBs for bulk purposes are generally in the range of 10 to 100 µF.

Capacitance required for preventing creation of RF currents due to the switching energy of components can be determined by knowing the resonant frequency of the logic circuits to be decoupled. The hardest part in calculating this resonant value is knowing the inductance of the capacitor's leads (ESL). If ESL is not known, an impedance meter or network analyzer may be used to measure the ESL value. The drawback of using an impedance meter is that low-frequency instruments may not catch higher-frequency responses. ESL can also be approximated by knowing only the capacitance value and self-resonant frequency parasitics.

6.4.14 Designing a Capacitor Internal to a Component's Package

Technology has progressed to the point where the majority of radiated emissions need not be caused by poor PCB layout, trace routing, impedance mismatches, or power supply corruption. Radiated emissions are the result of using digital components. What do we mean by the statement that digital components are the primary source of RF energy? The answer is simple. RF energy is produced by the Fourier spectra created by switching transistors internal to the silicon wafer that is physically glued down inside a protective enclosure. This enclosure is commonly identified as the *package*, which consists of either plastic or ceramic material.

Recent advances in integrated circuit (IC) components such as microprocessors, digital signal processors, and applications-specific integrated circuits (ASICs) have turned them into significant sources of electromagnetic noise. In recent years, clock rates have increased from 25 and 33 MHz to 200 through 500 MHz and beyond. With these clock rates, we have a corresponding internal dynamic power dissipation increase due to switching currents that may exceed 10 W.

The silicon die demands current from a power distribution network, which must drive a transmission line at certain levels of voltage and current. In addition, technology has progressed to the point where millions of transistors are provided within a single die or wafer. Manufacturing technology has also approached 0.18 micron line width, which allows for faster edge rate devices and *die shrink*. Die shrink is where the number of silicon dies on a wafer increases, thus improving the yield and total number of devices from a single process batch. The cost of the product decreases when an increase in the number of functional units occurs. Because of smaller line widths, propagation delay between the individual gates within the component package becomes shorter, along with a corresponding faster edge rate. The faster the edge rate, the greater the ability of the device to create radiated emissions. With faster internal edges, switching effects will cause greater losses across the lead bond inductance internal to the package.

With faster edge rates, dc current is demanded from the power distribution network at a quicker rate. This faster switching speed bounces the power distribution network, creating an imbalance in the differential-mode current between power and ground. With an imbalance in the differential mode, common-mode currents are produced. Common-mode currents are observed during EMI tests radiating from cable assemblies, interconnects, or PCB components.

To address component-level problems, EMC engineers and component manufacturers must advance state-of-the-art principles in implementing suppression techniques for ICs, especially decoupling. Design and cost margins also play an important part in determining how an EMC solution will be implemented.

As presented earlier, decoupling capacitors provide an instantaneous point source of charge for the time period that a device switches. A decoupling capacitor must not only provide a voltage charge at the speed that the device is switching at, but it must also recharge quickly. The self-resonant value of the capacitor depends on various parameters, which include not only the capacitance value but also ESL and ESR.

A component vendor can use various techniques to implement a decoupling capacitor internal to the component package. One approach is to implement a built-in decoupling capacitor before affixing the silicon die into the package, illustrated in Fig. 6.60.

Two layers of metal film, separated by a thin layer of a dielectric material, will form a high-quality parallel plate capacitor. Since the applied voltage is extremely low, the dielectric layer can be very thin. This thin dielectric results in adequate capacitance for a very small area. The effective lead length inductance approaches zero. The resonant frequency of the parallel plate configuration will be very high. The cost to manufacturers to implement this technique will be minimal compared to the overall cost of the IC. In addition to improved performance, overall PCB cost may be reduced, because use of discrete decoupling capacitors may not be required.

Another technique that will provide decoupling within a component package is brute force. High-density, high-technology components often feature SMT capacitors located directly inside the device package. The use of discrete components is common in multichip modules. Depending on the inrush peak current surge of the silicon die, an appropriate capacitor is selected based on the current charge the device requires in addition to providing decoupling of differential-mode currents at the self-resonant frequency of the component. Even with this internal capacitor, additional externally located discrete capacitors may be required. An example of how a discrete capacitor is provided in these modules is shown in Fig. 6.61.

For a synchronous design, CMOS power dissipation is a capacitive discharge effect. A device that consumes 2400 mW, 3.6 V, running at 100 MHz, has an effective load capacitance of approximately 2000 pF, generally observed right after a clock event. If we allow a 10% drop in voltage, this means we require 20 nF of bulk capacitance for optimal performance. A gate capacitance of 6 to 7 $\mu F/m^2$ provides an area of approximately 3 mm^2. This dimension is huge for a high-density component. In addition, the capacitive value is not very large.

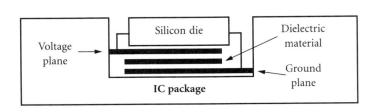

FIGURE 6.60 Decoupling capacitor internal to a silicon package.

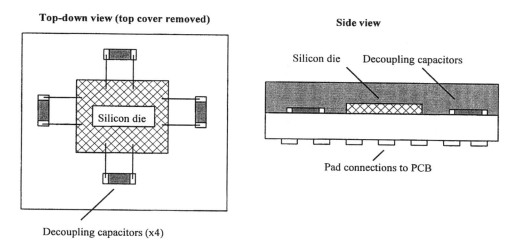

FIGURE 6.61 Locating decoupling capacitors internal to multichip module packaging.

CMOS gates provide distributed capacitance, both by coupling to the supply rails driving them and by the series capacitance of their own input transistors. This internal capacitance does not come close to the required value for functional operation. Silicon dies do not permit extra silicon to be available for use as bulk capacitance (floor space), since deep submicron designs consume routing space and are required to support the oxide layers of the assembly.

6.4.15 Vias and Their Effects in Solid Power Planes

Use of vias in solid power planes will decrease the total capacitance available based on the number of vias and the amount of real estate that has been etched out from the planes. A capacitor works by virtue of energy storage contained within a metallic structure. With less metal (copper plane), current density distribution is decreased. As a result, less area exists to support the number of electrons that create the current density distribution. Figure 6.62 illustrates the value of capacitance between parallel power planes in two configurations: solid planes, and planes with 30% of the area removed by vias and clearance pads.

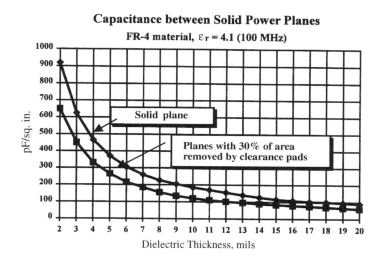

FIGURE 6.62 Effects of vias in power and ground planes.

Section 6.4 References

1. Montrose, M. 1999. *EMC and the Printed Circuit Board Design—Design, Theory and Layout Made Simple.* Piscataway, NJ: IEEE Press.
2. Montrose, M. I. 1996. *Printed Circuit Board Design Techniques for EMC Compliance.* Piscataway, NJ: IEEE Press.
3. Ott, H. 1988. *Noise Reduction Techniques in Electronic Systems.* 2nd ed. New York: John Wiley & Sons.
4. Paul, C. R. 1992. "Effectiveness of Multiple Decoupling Capacitors," *IEEE Transactions on Electromagnetic Compatibility,* May, vol. EMC-34, pp. 130–133.
5. Mardiguian, M. 1992. *Controlling Radiated Emissions by Design.* New York: Van Nostrand Reinhold.
6. Drewniak, J. L., T. H. Hubing, T. P. Van Doren, D. M. Hockanson. "Power Bus Decoupling on Multilayer Printed Circuit Boards." *IEEE Transactions on EMC* 37(2), 155–166.
7. Montrose, M., 1991. "Overview on Design Techniques for PCB Layout Used in High Technology Products." *Proceedings of the IEEE International Symposium on Electromagnetic Compatibility,* 61–66.
8. Van Doren, T. P., J. Drewniak, T. H. Hubing. "Printed Circuit Board Response to the Addition of Decoupling Capacitors," Tech. Rep. #TR92-4-007, University of Missouri, Rolla EMC Lab., (September 30, 1992).
9. Montrose, M. 1996. "Analysis on the Effectiveness of Image Planes within a Printed Circuit Board." *Proceedings of the IEEE International Symposium on Electromagnetic Compatibility,* 326–331.
10. Johnson, H. W., and M. Graham. 1993. *High Speed Digital Design.* Englewood Cliffs, NJ: Prentice Hall.

Additional Bibliography/Suggested Readings

Brit. D.S., et al. 1997. "Effects of Gapped Ground Planes and Guard Traces on Radiated EMI." *Proceeding of the IEEE International Symposium on Electromagnetic Compatibility,* New York: IEEE, pp. 159–164.

Brown, R. et. al. 1973. *Lines, Waves and Antennas.* New York: Ronald Press Company.

Coombs, C. 1996. *Printed Circuits Handbook.* New York, NY: McGraw-Hill.

DiBene, J. T., and Knighten, J. L. 1997. "Effects of Device Variations on the EMI potential of High Speed Digital Integrated Circuits." *Proceeding of the IEEE International Symposium on Electromagnetic Compatibility.* New York: IEEE, 208–212.

Hubing, T., et. al. 1995. "Power Bus Decoupling on Multilayer Printed Circuit Boards." *IEEE Transactions on EMC* 37,2: 155–166.

IPC-2141. *Controlled Impedance Circuit Boards and High Speed Logic Design.* April, 1996. Institute for Interconnecting and Packaging Electronic Circuits.

IPC-D-317A. *Design Guidelines for Electronic Packaging Utilizing High-Speed Techniques.* January, 1995. Institute for Interconnecting and Packaging Electronic Circuits.

IPC-TM-650. *Characteristics Impedance and Time Delay of Lines on Printed Boards by TDR.* April, 1996. Institute for Interconnecting and Packaging Electronic Circuits.

Kaupp, H.R. 1967. "Characteristics of Microstrip Transmission Lines," *IEEE Transactions,* Vol. EC-16, No. 2, April.

Kraus, John. 1984. *Electromagnetics.* New York: McGraw-Hill Inc.

Mardiguian, M. 1992. *Controlling Radiated Emissions by Design.* New York: Van Nostrand Reinhold.

Montrose, M. I. 1996. "Analysis of the Effectiveness of Clock Trace Termination Methods and Trace Lengths on a Printed Circuit Board." *Proceedings of the IEEE International Symposium on Electromagnetic Compatibility.* New York: IEEE, 453–458.

Montrose, M. I. 1998. "Time and Frequency Domain Analysis of Right Angle Corners on Printed Circuit Board Traces." *Proceedings of the IEEE International Symposium on Electromagnetic Compatibility.* New York: IEEE, 551–556.

Motorola, Inc. *MECL System Design Handbook (#HB205),* Chapters 3 and 7.

Motorola, Inc. *ECL Clock Distribution Techniques (#AN1405).*

Motorola, Inc. *Low Skew Clock Drivers and Their System Design Considerations (#AN1091).*

Paul, C. R., K. White, and J. Fessler. 1992. "Effect of Image Plane Dimensions on Radiated Emissions," *Proceeding of the IEEE International Symposium on Electromagnetic Compatibility,* New York: IEEE, 106–111.

Paul, C. R. "Effectiveness of Multiple Decoupling Capacitors," *IEEE Transactions on Electromagnetic Compatibility,* May 1992, vol. EMC-34, pp. 130–133.

Williams, Tim. 1996. *EMC for Product Designers.* 2nd ed. Oxford, England: Butterworth-Heinemann.

7
Hybrid Assemblies

Janet K. Lumpp
University of Kentucky

7.1 Introduction 7.1
7.2 Ceramic Substrates 7.1
7.3 Thick Film 7.2
7.4 Thin Film 7.6
7.5 Chip Resistors and Multilayer Ceramic Capacitors 7.8
7.6 Component and Assembly Packages 7.9
7.7 Buried Passive Circuit Elements 7.10
7.8 Bare Die Assembly 7.11
7.9 Multichip Module Technology 7.12

7.1 Introduction

As the term *hybrid* implies, many technologies are combined to build a hybrid circuit assembly. The substrate or package may be a simple ceramic platform, complex multilayer cofired ceramic, semiconductor wafer, or laminate structure. An interconnect is fabricated by depositing and patterning multiple layers of conductors and insulators with or without buried passive components. Resistors, capacitors, and inductors are patterned, and they overlap the conductor traces of the interconnect. Discrete components and integrated circuits are added as surface mount components or bare die using solder, wire bonding, TAB, flip chip, and die attach technologies. In a traditional sense, however, a hybrid circuit is a thick or thin film circuit on a ceramic substrate with bare die and packaged surface mount components sealed in a metal or ceramic package. The same technologies are utilized in multiple chip modules, hermetic packaging, and chip scale packaging.

7.2 Ceramic Substrates

In traditional hybrid assemblies, a ceramic substrate serves as a platform to mechanically and thermally support the interconnect, devices, and components that form the circuit on one or both sides of the substrate. Electrically, the substrate must be a low dielectric constant insulator to isolate the various conductor traces and pads. As described below, the substrate may also contain buried cofired interconnect structures and passive components. Typical ceramic compositions include aluminum oxide (alumina), aluminum nitride, and beryllium oxide (beryllia). The purity and fabrication techniques determine the electrical resistivity, thermal conductivity, thermal expansion coefficient, dielectric constant, loss tangent, surface roughness, density, and mechanical strength of the different grades of material. Table 7.1 compares the properties of 96% and 99% alumina. Aluminum nitride and beryllia are more expensive substitutes for alumina, but have substantially higher thermal conductivities and aluminum nitride matches the thermal expansion coefficient of silicon more closely than alumina.

Ceramic substrates are fabricated from powders by pressing or casting techniques. Hot pressing produces high-density substrates in smaller batches than tape casting. *Tape casting* or doctor blade

TABLE 7.1 Typical Properties of Alumina

Property	96% Alumina	99% Alumina
Electrical Properties		
Resistivity	10^{14} Ω–cm	10^{16} Ω–cm
Dielectric constant	9.0 at 1 MHz, 25° C	9.2 at 1 MHz, 25° C
	8.9 at 10 GHz, 25° C	9.0 at 10 GHz, 25° C
Dissipation factor	0.0003 at 1 MHz	0.0001 at 1 MHz, 25° C
	0.0006 at 10 GHz, 25° C	0.0002 at 10 GHz, 25° C
Dielectric strength	16 V/mm at 60 Hz, 25° C	16 V/mm at 60 Hz, 25° C
Mechanical Properties		
Tensile strength	1760 kg/cm^2	1760 kg/cm^2
Density	3.92 g/cm^3	3.92 g/cm^3
Thermal conductivity	0.89 W/° C–in at 25° C	0.93 W/°C–in at 25° C
Coefficient of thermal expansion	6.4 ppm/° C	6.4 ppm/°C
Surface Properties		
Surface roughness, average	0.50 mm center line average (CLA)	0.23 mm CLA
Surface flatness, average (Camber)	0.002 cm/cm	0.002 cm/cm

Source: from Sergent, J.E., "Hybrid Microelectronics Technology," in Whitaker, J., *The Electronics Handbook,* CRC/IEEE Press, 1997, by permission.

processing produces a thin sheet of green tape from a slurry of ceramic powder, organic binders, and solvents. A doctor blade is set at a fixed height over a table, the slurry is dispensed behind the blade, and a plastic carrier film passes under the blade to cast a layer of slurry onto the carrier film. The solvents evaporate from the slurry and the flexible green sheet is rolled onto a spindle for further processing. The roll of green tape can be punched into substrates of various sizes and shapes and laminated to create three-dimensional packages. The term *green tape* refers to a powder and organic composite prior to firing. During firing the organic components evaporate or are burned out, and the powder sinters into a monolithic component. Heating and cooling cycles are carefully controlled to prevent warpage and camber and to produce uniform shrinkage from the green to fired state.

7.3 Thick Film

7.3.1 Materials

A thick film ink or paste consists of three types of ingredients:[1] functional material, binder, and vehicle. The functional material is the metal or metal oxide powder that will make the fired ink act as a conductor, resistor, or dielectric. The binder material is a glass powder with thermal, mechanical, and electrical properties tailored to provide adhesion to the substrate, sintering of the functional material particles, high density, and thermal expansion match between film layers and the substrate. The vehicle components are organic solvents, plasticizers, and binders that control the rheology of the ink during the printing process. The vehicle materials suspend and carry the functional material and glass binder powders into the desired lines and patterns without allowing them to flow together on the surface of the substrate. Although crucial to the thick film process, the vehicle compounds are only temporary components of the ink and are subsequently burned out of the films during the firing cycle. Only the functional material and binder are present in the final form of a thick film circuit.

Functional Materials

A wide variety of functional materials are available for each category of inks.[2] Dielectric inks, for example, include capacitor materials with high dielectric constants, low dielectric constant insulator layers, and low melting point sealing glasses. Resistor inks are the most homogenous group based primarily on ruthenium dioxide as the functional material. Conductors serve the widest range of functions within the circuit; however, it is desirable to print the fewest number of layers, so compromises must be made in choosing an alloy to satisfy all the demands on the conductor layers. Interconnection is the overall purpose of the conductor films and in this role the film must also function as a resistor, capacitor, or inductor termination, wire bond pad, component lead pad, die bond pad, and via fill. Each of these applications has distinct requirements for adhesion and continuity with solder, gold wire, aluminum wire, conductive adhesive, and other thick film layers.

Conductor inks are evaluated for properties such as resistivity, solderability, solder leach resistance, line resolution, compatibility with substrate and other films, stability, migration resistance, wire bonding, die bonding, and cost. Noble metal inks include gold, silver, platinum, palladium, and alloys of silver-palladium, gold-palladium, and gold-platinum. Base metal inks include copper, molybdenum, tungsten, nickel, and aluminum. Gold conductors have long been associated with high-reliability circuits for military and telecommunications applications. Pure gold inks are excellent for high conductivity, corrosion resistance, wire bonding, and die bonding but are difficult to solder. When alloyed with palladium or platinum, gold films become solderable. Pure silver inks are less expensive than gold and are solderable, but they suffer from migration and poor solder leach resistance. Silver-palladium inks are the most common conductor inks for hybrid circuits, offering the good properties of silver with increased solder leach resistance, decreased silver migration, and wire bonding. Thick film copper has gained acceptance as processing improvements have been achieved to allow nitrogen firing with complete vehicle removal. Table 7.2 compares thick film conductor compositions and their compatibility with different bonding processes. The refractory metal inks are used primarily as cofired buried layers and via fill in multilayer cofired ceramic substrates and packages as described below.

TABLE 7.2 Process Compatibility of Thick Film Conductors

Material	Process				
	Au Wire Bonding	Al Wire Bonding	Eutectic Bonding	Sn/Pb Solder	Epoxy Bonding
Au	Y	N	Y	N	Y
Pd/Au	N	Y	N	Y	Y
Pt/Au	N	Y	N	Y	Y
Ag	Y	N	N	Y	Y
Pd/Ag	N	Y	N	Y	Y
Pt/Ag	N	Y	N	Y	Y
Pt/Pd/Ag	N	Y	N	Y	Y
Cu	N	Y	N	Y	N

Resistor inks contain conductive oxides mixed in various proportions with the binder glass to produce inks of different resistivities and thermal coefficients of resistance (TCR). Common oxides include RuO_2, IrO_2, $Bi_2Ru_2O_7$, and $Pb_2Ru_2O_5$ for circuits fired in air. Nitrogen firing with copper conductors requires alternative resistor inks containing SnO_2, In_2O_3, $SrRuO_3$, LaB_6, $TiSi_2$, or TaN. Thick film resistor inks are available in sheet resistances ranging from 10 to 10^7 Ω/sq allowing resistors from a few ohms to hundreds of megohms to be fabricated. A recent field of resistor development has addressed the need for surge suppression in telecommunications and consumer electronics.[3]

The resistivity of a thick film ink is specified as a *sheet resistance* in units of ohms per square (Ω/sq). The thickness of the printed film is assumed to be constant at 25 microns and the available design dimensions are length and width. The length of a line (L) is divided by the width (W) to determine the number of squares (N), which is then multiplied by the sheet resistivity (R_s) to determine the resistor value (R). In a serpentine design, the corner squares count as one-half square each for calculating the total resistance of the design.

$$R = \frac{R_s L}{W} = R_s N$$

For example, using a 1000-Ω/sq ink a 750-Ω resistor can be designed using 7.5 squares. Choosing a width of 0.020 in, the resistor length should be 0.150 in. The actual length will also include overlap with the conductor termination pads that are part of the interconnect design. Rectangular, serpentine, and top-hat designs are most common for designing various resistor values with convenient end points and to incorporate trimming.

To allow for processing variations, thick film resistors are typically designed to 80% of the desired value and then trimmed after firing to increase the resistance to the design value. Trimming is accomplished by probing the resistor at the conductor terminations and simultaneously removing material with a laser until the desired resistor value is achieved. The notch cut by the laser decreases the width of the film and thereby increases the resistor value. It is not always possible to isolate each resistor in an interconnect pattern, and it may be necessary to trim several resistors to tune the circuit to the final operating characteristics. *Laser trimming* is highly automated, and several resistors can be trimmed per minute. Thin film resistors are similarly trimmed to adjust the final value; however, the more precise photolithography patterning produces a smaller error due to fabrication. Tuning can also be incorporated into a design by patterning a ladder of resistors of different incremental values in series with the primary resistor area. The ladder is shorted together by conductor film and the trimming processing involves breaking the shorting bars with a laser to route the current through the resistor segments, thereby increasing the resistance value. Resistor chains consume real estate while providing flexibility.

Dielectric inks combine glass and ceramic powders with the binder glass to adjust the dielectric constant, loss tangent, melting point, density, and viscosity of the film. A low dielectric constant is preferred for all applications except capacitor films. High density is desired to prevent shorting between conductor layers at crossovers and in multilayer interconnects. Pinholes should be avoided as well as flowing of the dielectric during firing of subsequent layers. Ceramic additives promote crystallization of the glass to increase density, melting point, and viscosity. Dielectric layers in multilayer interconnects include vias that must remain open to allow the via fill conductor ink to complete the three-dimensional network. A new dielectric ink composition that allows photodefinable via patterning is available as an alternative to using a mask pattern to define the vias in the screen printing process. A blanket coating of the photodefinable dielectric is printed, exposed through a mask, and developed to remove the via plugs. Another new via patterning method is diffusion patterning using a printed patterning paste that converts areas of the dielectric into a water soluble layer that is rinsed off to open the vias.[4]

Vehicle

The organic vehicle component of a thick film ink includes a range of high molecular weight polymers, solvents, and other viscosity modifiers. As described in the screen printing section, the rheological properties of the ink are critical to successful printing and line resolution. Solvents are removed during the low-temperature drying process along with low boiling point and low vapor pressure organic compounds. Certain combinations of resistor and conductor ink compositions are designed to be cofired by allowing the second layer to be printed after the first layer has only been dried, thereby reducing the number of process steps and energy consumption.

Binder

Binder glasses serve as a permanent bonding material responsible for holding the functional material particles in contact with each other and interacting with the substrate to maintain adhesion through multiple firings and during the lifetime of the circuit. The compositions of these glasses are critical (and usually proprietary) for fine tuning properties such as glass transition temperature, thermal expansion coefficient, dielectric constant, dielectric loss tangent, resistivity, thermal conductivity, viscosity as a function of temperature, and mechanical strength. Lead borosilicate and bismuth borosilicate glasses are common basic glass compositions. Additives such as ceramic powders influence the high-temperature properties and dielectric properties.

During the high-temperature portion of the firing cycle, the binder glass particles sinter or melt in the case of liquid phase sintering. The binder maintains contact between the functional material particles to form the conductive pathways through the film. Adhesion is achieved by the glass binder in the film wetting the substrate surface.

7.3.2 Screen Printing

Screen printing or *stencil printing* is an old ink-transfer process drawing on the artistic fabric designs produced by silk screening various colors in sequence with appropriate alignment to create a full-color image. Using the modern-day example of a T-shirt design, the different ink colors of a thick film circuit are also different electrical circuit elements. To describe the technical aspects of the screen printing process, it is useful to describe the equipment, the process sequence, and the large number of variables contributing to process control. Screen printing equipment includes the screens or stencils, squeegee blades, alignment fixtures, and substrate holder.

Screens and stencils contain the pattern for a specific layer and allow ink to be transferred through the openings in the screen or stencil onto the substrate. Screens are assembled from finely woven stainless steel wire mesh stretched and mounted onto a metal frame. The wire diameter and wire spacing determine the size of the individual openings in the mesh. The number of openings per linear inch is the *mesh count* or *mesh number*. The mesh can be woven in a simple over one under one pattern or in more complex weaves and is then mounted on the frame to align the mesh with the sides of the frame or at various angles. Screen fabrics are typically mounted at 45 or 22.5° to avoid aligning the wires with the predominant lines of the circuit pattern. After mounting the mesh to the frame, the screen surface is coated with a light sensitive emulsion to block the openings in the mesh. The emulsion can be either a positive or negative sensitivity meaning the exposed regions will either become soluble or be polymerized, respectively. A mask is prepared with an appropriate positive or negative image of the desired layer pattern, and the mask is placed over the emulsion aligning the pattern with the frame. The emulsion is then exposed to ultraviolet light through the mask to polymerize or depolymerize regions of the emulsion coating. The exposed emulsion is rinsed to remove the unwanted areas and allowed to dry. The emulsion still covers the majority of the screen and, in the areas where the emulsion has been removed, the ink will flow through the screen onto the substrate as shown in Fig. 7.1. Depending on the manufacturer's specifications, the rinsing process may require only warm water or a specific developer/stripper solution. The emulsion patterning process is analogous to the deposition and patterning of photoresist in thin film processing.

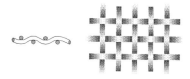

a. Plain weave mesh, side view and top view.

b. Emulsion coated screen

c. Patterned emulsion coated screen.

FIGURE 7.1 Screen preparation for thick film printing.

Stencil preparation involves a thin metal sheet bonded to the same type of metal frame used for screen assembly. The metal sheet is similarly patterned to create openings for the printing process. The openings are etched either by wet chemical etching of photoresist defined patterns or by laser cutting.[5] The exact cross-sectional shapes of the openings created by wet etching and laser cutting are not the same and, depending on the resolution required, laser processing may be the better choice. Stencils are more durable than emulsion-coated screens and therefore are more common for large-volume production and in applications where the dimensions are critical and should not vary from the beginning to the end of the batch. Stencils are more expensive to assemble and pattern than screens, and design changes can be incorporated more readily by patterning a new screen than by fabricating a new stencil. Screens can be recycled by stripping off the emulsion and recoating the mesh surface. If the mesh has stretched or been damaged, it can be cut off and replaced. Stencils can only be recycled in the sense that the metal sheet can be reclaimed and the frame reused. Screens allow annular features to be printed, because the screen supports the center circular region of emulsion, whereas a stencil cannot suspend an island piece to block the flow of ink.

Alignment is an extremely critical element of screen/stencil printing, because it influences the thickness of the ink, uniformity of the thickness over the area of the substrate, and orientation of various layers of ink. The screen or stencil frame is mounted in the printer and is aligned with the substrate to ensure that the pattern is oriented with the edges of the substrate, overlaps existing layers on the substrate, and is parallel to the surface of the substrate. Proper alignment is adjusted manually or automatically and is verified by visual inspection. Fiducial marks included in the pattern and on the substrate assist in visual alignment. A simple test for coplanarity is to place a piece of paper between the screen and substrate and pull on the paper to determine if it is firmly held between the two surfaces. If the paper tends to rotate about a point, then the screen and substrate are not parallel to each other. Test cycles can also be run by printing onto a clear plastic sheet placed over the substrate. The printed areas can be inspected for proper ink volume and the alignment with the previous layers can be verified.

The substrate is held in place by vacuum chucks and alignment pins that protrude into corresponding holes in the substrate or make contact with the edges of the substrate. Substrates are placed in the printing equipment manually or by automatic feed systems. In high-volume, high-precision assembly facilities, the automatic visual alignment system checks each substrate and makes minor adjustments before each printing cycle.

Squeegee blades are available in different durometer or stiffness grades and are mounted in a holder such that one edge points down toward the screen/stencil. A line of ink is dispensed in front of the squeegee blade between the blade and the beginning of the pattern. During the printing cycle, the blade presses down on the screen/stencil and moves across parallel to the plane of the substrate. The force of the blade causes the screen or stencil to flex bringing a narrow section of the screen/stencil in contact with the substrate temporarily. As the squeegee moves along, the entire pattern is transferred to the substrate similar to a slit exposing a piece of film. As the squeegee pushes the ink across the top side of the screen/stencil, the ink should roll smoothly like a travelling wave to fill in the openings in the screen/stencil without leaving gaps or voids. The applied shear stress reduces the viscosity of the ink to allow it to flow into the openings and through to the substrate. The screen/stencil must make contact with the substrate to achieve transfer from the filled openings to the substrate below. The vehicle compounds in the ink formulation are responsible for the change in viscosity as a function of shear stress.

The thickness of ink transferred to the substrate depends on several variables, including the thickness of the emulsion, speed of the squeegee blade, and snap off distance. Process control of a screen/stencil printing operation has to accommodate a large number of process variables,[6] not all of which are well characterized or can be adjusted in real time. The stencil, screen, emulsion, and squeegee surfaces suffer from mechanical wear after repeated use. The wear rate depends on the squeegee pressure, velocity, and composition. Stainless steel squeegee blades are replacing rubber blades to reduce wear. The rheological properties of the ink change over time as the ink sits exposed to the environment and is kneaded back and forth across the screen/stencil.

7.4 Thin Film

7.4.1 Deposition

Whereas thick film screen printing accomplishes both deposition and patterning in one step, thin film technology is primarily blanket coating followed by photolithography patterning. A variety of deposition techniques are available spanning the full range of cost and complexity, with different techniques better suited to metal, resistor, and insulator materials. The simplest deposition technique is evaporation. In a vacuum chamber, a heater melts a metal source and the metal vapor coats the heated substrate. Materials with high melting points and low vapor pressures are more readily deposited by sputtering. Direct-current sputtering uses ionizable gases such as argon to bombard the source with ions, removed fragments of material, and deposit the sputtered material onto the heated substrate. The source material acts as the cathode in dc sputtering. If the desired film material is nonconductive, then it will make a poor cathode. RF sputtering is the next option using an alternating field to prevent charge buildup on the source. All of these techniques can produce pure single component films or can be combined with reactive gases to produce compound films. Multilayer and alloyed films can also be deposited from multiple pure sources exposed in sequence or mixed sources. An excellent summary table listing evaporation and sputtering deposition techniques for various materials is included in Ref. 7.

Thin film deposition is carried out in vacuum to prevent oxidation of metal films and to increase the mean free path of the vapor phase species. Heating the substrate adds energy to the growing film to improve adhesion, promote microstructure development, and relieve stress. Reactive sputtering produces oxide and nitride films to serve as insulators, capacitor dielectrics, and resistors. Sputtering and evaporation equipment is designed to hold multiple substrates for batch processing and carousels of different sources covered by automated shutters to control film composition. A quartz crystal microbalance monitors the film growth rate during each deposition run. Other thin film deposition techniques include chemical vapor deposition (CVD), laser assisted chemical vapor deposition (LCVD), pulsed laser deposition (PLD), sol-gel, metallorganic decomposition, electroless plating, and electroplating (see Fig. 7.2).

7.4.2 Materials

Metal films for thin film interconnections include copper, gold, and aluminum with adhesion layers of titanium or chromium. Thin film conductors serve all the same functions as their thick film counterparts as passive component terminations, die bond pads, solder bond pads, and wire bond pads. Pure chromium and ruthenium serve as resistor films. Alloys of nickel and chromium (nichrome) are used as an adhesion promotion layer or as a resistor film by adjusting the composition to vary the sheet resistance. Similarly, the composition of tantalum, tantalum nitride, and tantalum-oxynitride films is varied to produce resistor films. When compared to thick film resistors, however, thin film materials offer a smaller range of sheet resistance values of 10 to 2000 Ω/sq. Thin film capacitors consist of silicon monoxide, tantalum pentoxide, or manganese oxide films. High-density, pinhole-free layers are important for producing high-reliability capacitors and insulating layers.

FIGURE 7.2 A thin film circuit. *Source:* from Sergent, J.E., "Hybrid Microelectronics Technology," in Whitaker, J., *The Electronics Handbook,* CRC/IEEE Press, 1997, by permission.

7.4.3 Photolithography

Patterning of thin film interconnects and devices is identical to semiconductor fabrication. The deposited film is coated with photoresist, and the photoresist is soft baked, exposed through a mask, developed, and hard baked. The underlying film is etched through the openings in the photoresist, and then the photoresist is stripped. Lift-off patterning changes the order of the photolithography and film deposition steps. The film is deposited over the patterned photoresist, and the excess film material on top of the photoresist lifts off when the photoresist is stripped off of the substrate. Photoresist materials are positive or negative light-sensitive polymers, and the mask design must complement the mode of the photoresist so that the correct areas are removed in the etching process. Depending on the film composition and the feature resolution, the etching step may be a wet chemical etch involving acid baths or a dry chemical etch such as plasma etching or reactive ion etching. The pros and cons of these techniques include environmental, equipment cost, line resolution, and throughput issues.

7.5 Chip Resistors and Multilayer Ceramic Capacitors

Thick film technology is also utilized to manufacture surface mount discrete components such as resistors and capacitors. Chip resistors are fabricated in large arrays on a ceramic substrate by printing conductor pads and resistor lengths. Each isolated resistor is trimmed and sealed. The large substrate is scribed and snapped apart into individual resistor components. Solder can also be added to the pads or wraparound terminations can be fabricated so that the components can be placed in any orientation during the surface mount assembly phase. The advantages of chip resistors are

- They can be trimmed to desired value (Fig. 7.3).
- They are small in size because each value is printed with the highest practical sheet resistance and fewest squares.
- They require fewer printing cycles.
- Thick film resistors can be incorporated into printed circuit board or thin film circuits.

Multilayer ceramic (MLC) capacitors are fabricated using cofire technology by printing conductor patterns on ceramic green tape dielectrics, stacking layers, laminating, and cofiring. The conductor

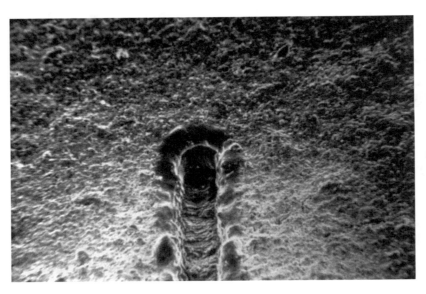

FIGURE 7.3 A laser-trimmed thick film resistor. *Source:* from Sergent, J.E., "Hybrid Microelectronics Technology," in Whitaker, J., *The Electronics Handbook*, CRC/IEEE Press, 1997, by permission.

patterns produce a vertical interdigitated sequence of terminals connected to two opposite edges of the stack. The large multilayer substrate is cut into individual capacitors, and the edge terminations are metallized to produce miniature parallel plate capacitor stacks. Wraparound terminations are common to eliminate the need for a specific orientation in the surface mount assembly phase. Semiconductor thin film capacitors are also available as surface mount components.

7.6 Component and Assembly Packages

Individual integrated circuits and multichip modules are enclosed in packages to provide mechanical, chemical, electrical, and thermal protection. Ceramic and metal packages can be hermetically sealed to isolate the circuit from corrosive conditions such as humid or salt water environments. Plastic packages are nonhermetic but provide an inexpensive method to conformally encase die, wire bonds, and leadframes. A typical plastic package is injection molded around a single IC that has been attached and wire bonded to a metal leadframe. After molding, the leadframe is cut apart from the strip or reel, and the leads are trimmed and bent to the desired configuration. *Through-hole components* have long, straight leads that are inserted through the printed circuit board, trimmed, and clinched prior to wave soldering. *Surface mount components* have shorter *I* leads or bent gull-wing or *J* leads. Gull-wing leads have the advantage of ease of solder joint inspection, while *J* leads consume less surface area on the substrate. The lead spacing corresponds to a set of pads patterned into the conductor film, and the component is soldered to the pads by printing solder paste, placing the components in the proper orientation, and reflowing the solder.

Multilayer ceramics packages with embedded three-dimensional interconnects are fabricated using a combination of thick film and ceramic green sheet technologies. In the flexible "green" condition, tapes can be unrolled, punched, printed, and rolled again in batch processing style. Vias are punched and via fill inks are printed, as well as conductor traces. Finally, the individual substrates or package layers are stamped from the roll. Multilayer units are stacked and laminated under pressure and heat with careful alignment of the vias and lines as shown in Fig. 7.4. Green sheets and stacks are fired to remove the binder materials, sinter the ceramic particles, and sinter the embedded thick film layers. The high firing temperatures required to sinter the ceramic limits the selection of metals available for via fill and interconnection. Molybdenum and tungsten are the most commonly cofired metals, and the ink compositions have been adjusted to accommodate the shrinkage that occurs during firing and to match the TCE of the ceramic. Three-dimensional package cavities are assembled from punched sheets and picture frame type rings, and the cofired embedded metallizations create a hermetically sealed interconnect.

Low-temperature cofired ceramic (LTCC) materials have been developed in recent years to reduce the processing costs associated with high-temperature firing and allow higher conductivity metals, such as silver and copper, to be used as cofire metallizations. The ceramic layers are actually glass ceramic mixtures with thermal expansion coefficients designed to match silicon for multichip module (MCM) applications. The thermal conductivities of LTCC materials, however, are lower than high density alumina or aluminum nitride.

Metal packages are fabricated by forming, milling, and sealing techniques with glass sealed pins inserted through the base or sides of the package for hermetic feed-throughs. A hybrid assembly is bonded to the base of the package with an adhesive, and the feed-through pins are connected to the circuit by wire bonds. Metal lids are sealed over the top of a ceramic or metal package by brazing, laser welding, or glass sealing to complete the package.

Most plastic, ceramic, and metal packages have peripheral leads with the exception of pin grid arrays. More recently, ball grid array (BGA) packages and chip scale packages (CSP) have addressed the needs for higher I/O counts and higher-density hybrid circuits. The primary advantage of arrays over peripheral leads is the larger number of I/O per unit area. As I/O increase, a peripheral lead package must increase in size to provide more fan out, whereas a two-dimensional array of pins or balls can accommodate the increasing number of I/O and maintain adequate spacing between pads. Escape routing must be included in the interconnect pattern under an array device, and a similar three-dimensional interconnect is

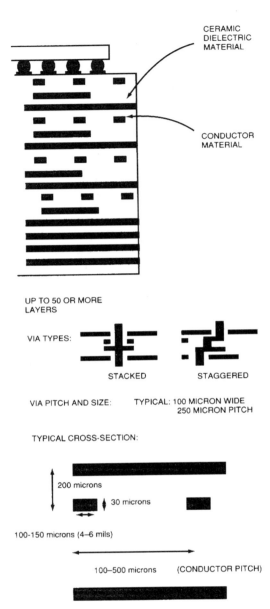

FIGURE 7.4 Cross-sectional views of multilayer cofired ceramic substrates. *Source:* from Sergent, J.E., "Hybrid Microelectronics Technology," in Whitaker, J., *The Electronics Handbook*, CRC/IEEE Press, 1997, by permission.

fabricated in the layers of the ball grid array package or ceramic cofired pin grid array package to route the peripheral pads of the IC to the balls or pins. Ball grid array attachment is similar to flip chip technology; however, the balls and pad spacings are generally larger for BGA devices. A minimalist approach to packaging is the chip scale package that provides protection, ease of handling, and testability but is only slightly larger than the die itself.

7.7 Buried Passive Circuit Elements

It is desirable to fabricate not only a three-dimensional interconnect within a multilayer substrate but also to bury or integrate passive circuit elements such as resistors, capacitors, and inductors.[8] The challenges of buried passives are

- Compatibility of the materials to be able to cofire several compositions in a single assembly

- Accuracy of component values where trimming is not possible
- Repair and rework of completed assemblies
- A range of component values available in high-temperature cofire, low-temperature cofire, thin film multilayer, thick film multilayer, and printed circuit (laminate) systems

Each substrate system requires individual development of materials and processes for each type of buried passive component. Some advantages of buried, embedded, or integrated passives are

- Decreased surface area dedicated to passive component placement
- Decreased signal delays due to shorter interconnects
- Increased reliability due to fewer solder joints[9,10]

7.8 Bare Die Assembly

Packaged components occupy a large percentage of real estate due to the fanout of the leadframe from the die bond pads to the solder bond pads. Direct attachment of bare die to a hybrid assembly saves space and is accomplished by wire bonding, TAB, and flip chip processes. Wire bonding and some TAB geometries require that the die be mechanically attached to the substrate. This is also true in laminate multichip modules (MCM-L), which require wire bonding of bare die. Die attach methods include adhesive, solder, and gold/silicon eutectic bonds. A related issue that hinders the use of bare die is known good die (KGD). Packaged ICs are tested by standard methods such as burn-in and functional testing to eliminate bad die. Individual bare die, however, cannot be burned-in without elaborate fixtures. KGD is a key player in the development of multichip modules.

7.8.1 Wire Bonding

Wire Bonding is the oldest process and perhaps the best understood. The most common techniques are gold ball bonding and aluminum wedge bonding. In gold ball bonding, the end of the wire is heated by an electric discharge such that the tip melts and forms a ball. The ball is pressed onto a pad on the die or substrate and forms a metallurgical bond. Either thermocompression (heat and pressure) or thermosonic (heat and ultrasonic vibrations) provide energy to form the bond. With the wire coming off the top of the ball, the bonder is free to move in any direction toward the second pad to complete the bonding cycle. At the second pad, the capillary tool again uses thermocompression or thermosonic energy to form a metallurgical bond, and then the wire is broken off. The next bonding cycle begins with melting the tip of the wire. The tooling for ultrasonic wedge bonding is slightly different, with the wire fed in at an angle and passing under the foot of the tool. Ultrasonic energy, pressure, and sometimes heat are used to form the metallurgical bond between the wire and pad. After making the first bond, the tool must move inline with the wedge bond toward the second pad where the bonding motion is repeated and the wire is broken off. The advantages of ultrasonic wedge bonding are the low temperature, reduced cycle time, and reduced equipment requirements as compared to thermocompression ball bonding, which requires the die to be heated. Ball bonding has the distinct advantage that the second bond can be made in any direction relative to the first bond site.[11]

7.8.2 Tape Automated Bonding

Tape automated bonding (TAB) involves many processing steps but offers the opportunity to test bare die prior to final assembly. A TAB tape is a leadframe with individual leads fanning out from the spacing of the pads on the die to the spacing of the pads on the substrate. The leadframe is etched from a metal sheet, in which case the leads are shorted together by metal connections, and testing of the die is prohibited. Other lead frame types are attached to a plastic carrier or plated on to the carrier and supported by nonconducting bars allowing for burn-in and die testing in specially designed sockets. Bumps of metallization are deposited and plated onto the tips of the TAB leads or onto the die pads.

The compositions are designed to prevent diffusion and to make dissimilar metals compatible and bondable. *Inner lead bonding* occurs first to attach the leadframe to the bare die. Several bonding methods are available using single-point and gang bonding approaches. Single-point techniques include thermocompression, eutectic, reflow, and laser tools, whereas gang bonding uses a thermode to thermocompression bond all the leads at once. After inner lead bonding, the die can be encapsulated to protect the integrated circuitry from mechanical and chemical attack. At this point, the die can be tested if the leads are not shorted together. Special test pads are often patterned onto the TAB leadframe for the purpose of testing and then cut off during the subsequent excise and lead-forming steps. In preparation for outer *lead bonding,* the die is attached to the substrate by an adhesive or eutectic die attach. Outer lead bonding techniques are similar to inner lead bonding with one addition, mass bonding. Using solder reflow, many TAB devices can be bonded at the same time along with other surface mount components in a mass bonding solder reflow oven. TAB is an equipment intensive process requiring different tooling for each die size for gang bonding, testing, excise, and lead-forming operations. Single-point processes are more flexible for attaching different die sizes with one piece of equipment, but they accomplish fewer bonds per minute.[12]

7.8.3 Flip Chip Bonding

The ideal method for attaching bare die without giving up real estate to fan-out is by *flip chip* bonding. Solder bumps are deposited or plated onto the die and or substrate pads. The die is placed face down over the pads, and the solder is reflowed. During reflow, the surface tension of the molten solder causes self-alignment of the die pads over the substrate interconnect pads. Underfill materials are added to distribute mechanical stress through the underfill and solder to relieve stress on the solder joints. Fillers in the underfill material, such as ceramic powders, improve the thermal conductivity of the underfill and provide a heat path to the substrate. Encapsulants are also used to protect the die from mechanical and chemical damage. Prior to underfill and encapsulation, the die can be removed by melting the solder or shearing the solder joints, and the bumps and pads can be redressed for rework operations. The ability to rework a flip chip bond allows the die to be tested and burned in on a carrier substrate and then attached to the actual circuit assembly in the second bonding process.[13,14]

7.9 Multichip Module Technology*

As noted earlier, multichip module (MCM) technology may use a variety of assembly techniques. An example of the most common technology, MCM-L, using a fiberglass substrate and pin-grid array packaging, is shown in Fig. 7.5.

7.9.1 Multichip Module Technology Definitions

In its broadest sense, a multichip module is an assembly in which more than one integrated circuit (IC) is bare mounted on a common substrate. This definition is often narrowed in such a way as to imply that the common substrate provides a higher wiring density than a conventional printed circuit board (PCB). The main physical components are shown in Fig. 7.6 and can be described as follows:

1. The substrate technology provides the interconnect between the chips (ICs or die) and any discrete circuit elements, such as resistors, capacitors, and inductors.
2. The chip connection technology provides the means whereby signals and power are transferred between the substrate and the chips.
3. The MCM package technology is the housing that surrounds the MCM and allows signals, power, and heat to be conducted to the outside world.

---

*Adapted from Franzon, P.D.; "Multichip Module Technology," in *The Electronics Handbook,* ed. J.C. Whitaker, CRC/IEEE Press, 1996.

Hybrid Assemblies

FIGURE 7.5 Eight chips wire bonded into an MCM. *Source:* MicroModule Systems.

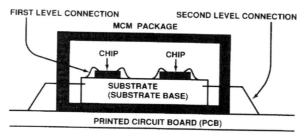

FIGURE 7.6 Physical technology components to an MCM.

What cannot be shown in Fig. 7.6 are several other components that are important to the success of an MCM.

1. The test technology used to ensure correct function of the bare die, the MCM substrate, and the assembled MCM
2. The repair technology used to replace failed die, if any are detected after assembly
3. The design technology

There are many different versions of these technology components. The substrate technology is broadly divided into three categories.

1. *Laminate MCMs* (MCM-L), as shown in (Fig. 7.7). Essentially fine-line PCBs, MCM-Ls are usually constructed by first patterning copper conductors on fiberglass/resin-impregnated sheets as shown in Fig. 7.8. These sheets are laminated together under heat and pressure. Connections between conductors on different sheets are made through via holes drilled in the sheets and plated. Recent developments in MCM-L technology have emphasized the use of flexible laminates. Flexible laminates have the potential for permitting finer lines and vias than fiberglass-based laminates.
2. *Ceramic MCMs* (MCM-C) are mostly based on pin grid array (PGA) technology. The basic manufacturing steps are outlined in Fig. 7.9. MCM-Cs are made by first casting a uniform sheet of prefired ceramic material, called *green tape*, then printing a metal ink onto the green tape, then

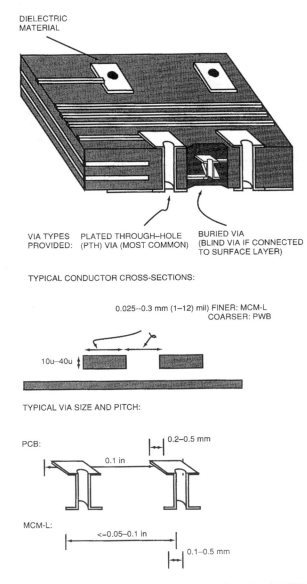

FIGURE 7.7 Typical cross sections and feature sizes in printed circuit board and MCM-L technologies.

punching and metal filling holes for the vias, and finally cofiring the stacked sheets together under pressure in an oven. In addition to metal, other inks can be used, including ones to print discrete resistors and capacitors in the MCM-C.

3. Thin-film (deposited) MCMs (MCM-Ds) are based on chip metallization processes. MCM-Ds have very fine feature sizes, giving high wiring densities (Fig. 7.10). MCM-Ds are made one layer at a time, using successive photolithographic definitions of metal conductor patterns and vias and the deposition of insulating layers, usually polyimide (Fig. 7.11). Often, MCM-Ds are built on a silicon substrate, allowing capacitors, resistors, and transistors to be built, cheaply, as part of the substrate.

Table 53.1 gives one comparison of the alternative substrate technologies. MCM-Ls provide the lowest wiring and via density but are still very useful for making a small assembly when the total wire count in the design is not too high. MCM-Ds provide the highest wiring density and are useful in designs

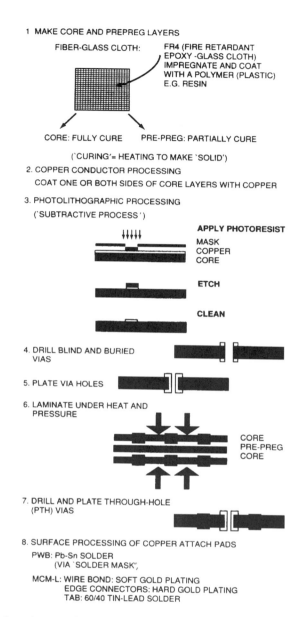

FIGURE 7.8 Basic manufacturing steps for MCM-Ls, using fiberglass/resin prepreg sheets as a basis.

TABLE 7.3 Rough Comparison of the Three Different Substrate Technologies; 1 mil = 1/1000 in ≈ 25 μm

	Min. Wire Pitch, μm	Via Size, μm	Approx. Cost/Part, $/in^2	Nonrecurring Costs, $
PCB	300	500	0.3	100
MCM-L	150	200	4	100–10,000
MCM-C	200	100	12	25,000
MCM-D	30	25	20	15,000

containing high-pin-count chips; however, it is generally the most expensive technology on a per-unit-area basis. MCM-Cs fall somewhere in between, providing intermediate wiring densities at intermediate costs.

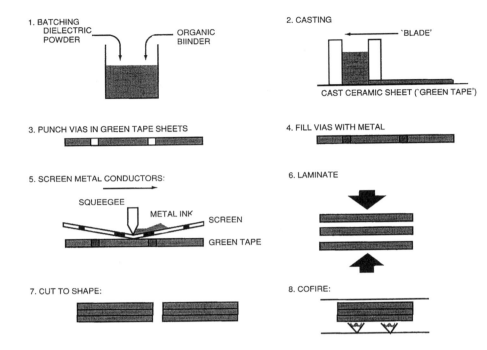

FIGURE 7.9 Basic manufacturing steps for MCM-Cs.

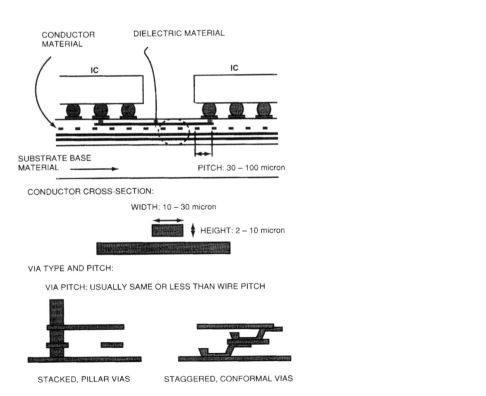

FIGURE 7.10 Typical cross sections and feature sizes in MCM-D technology.

Hybrid Assemblies

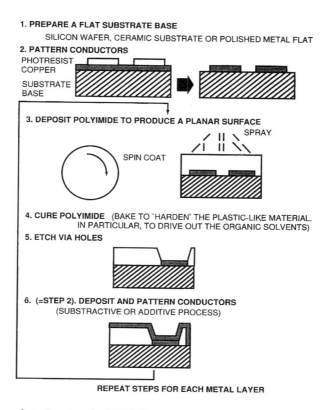

FIGURE 7.11 Basic manufacturing steps for MCM-Cs.

Current research promises to reduce the cost of MCM-Ls and MCM-D substrates. MCM-Ls based on flexible board technologies should be both cheaper and provide denser wiring than current fiberglass MCM-L. technologies. Although the nonrecurring costs of MCM-Ds will always be high, due to the requirement to make fine feature photolithographic masks, the cost per part is projected to decrease as more efficient manufacturing techniques are brought to fruition.

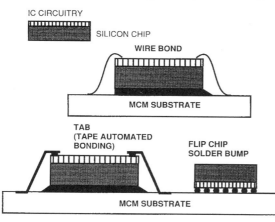

FIGURE 7.12 Chip attach options.

The chip-to-substrate connection alternatives are presented in Fig. 7.12. Currently, over 95% of the die mounted in MCMs or single-chip packages are wire bonded. Most bare die come suitable for wire bonding, and wire bonding techniques are well known. Tape automated bonding (TAB) is an alternative to wire bonding but has enjoyed only mixed success. With TAB assembly, the chip is first attached to the TAB frame inner leads. These leads are then shaped (called *forming*), after which the outer leads can be attached to the MCM. TAB has a significant advantage over wire bonding in that the TAB-mounted chips can be more easily tested than bare die. The high tooling costs, however, for making the TAB frames, make it less desirable in all but high-volume production chips. The wire pitch in wire bonding or TAB is generally 75 µm or more.

With flip-chip solder-bump attachment, solder bumps are deposited over the area of the silicon wafer. The wafer is then diced into individual die and flipped over the MCM substrate. The module is then placed in a reflow oven, where the solder makes a strong connection between the chip and the substrate. Flip-chip attachment is gaining in popularity due to the following unique advantages:

- Solder bumps can be placed over the entire area of the chip, allowing chips to have thousands of connections. For example, a 1-cm^2 chip could support 1600 solder bumps (at a conservative 250 µm pitch) but only 533 wire bonds.

- Transistors can be placed under solder bumps but cannot be placed under wire bond or TAB pads. The reason arises from the relative attachment process. Making a good wire bond or TAB lead attachment requires compression of the bond into the chip pad, damaging any transistors beneath a pad. On the other hand, a good soldered connection is made through the application of heat only. As a result, the chip can be smaller by an area equivalent to the total pad ring area. For example, consider a 100-mm^2 chip) with a 250-µm pad ring. The area consumed by the pad ring would total 10 mm^2, or 10% of the chip area. This area could be used for other functions if the chip were flipped and solder bumped. Alternatively, the chip could be made smaller and cheaper.

- The electrical parasitics of a solder bump are far better than for a wire bond or TAB lead. The latter generally introduce about 1 nH of inductance and 1 pF of capacitance into a circuit. In contrast, a solder bump introduces about 10 pH and 10 nF. The lower parasitic inductance and capacitance make solder bumps attractive for high-frequency radio applications, for example, in the 5.6-GHz communications band and in high-clock-rate digital applications; MCMs clocking at over 350 MHz are on the drawing board as of this writing.

- The costs of flip-chip are comparable to wire bonding. Solder-bumping a wafer today costs about $400, and assembly only requires alignment and an oven. In contrast, wire bonding costs about 1 cent per bond. A wafer containing one hundred 400-pin chips would cost the same with either approach.

Flip chip solder-bump, however, does have some disadvantages. The most significant follows from the requirement that the solder bump pads are larger than wire bond pads (60 to 80 µm vs. approximately 50 µm). The solder-bump pads need to be larger in order too make the bumps taller. Taller bumps can more easily absorb the stresses that arise when the module is heated and cooled during assembly. As the thermal coefficient of expansion (TCE) of silicon is usually different from that of the MCM substrate, these two elements expand and contract at different rates, creating stresses in the connecting bumps. For the same reason, it is better to place the bumps in the center of the chip, rather than around the edge. By reducing the distance over which differential expansion and contraction occur, the stresses are reduced.

The larger bump pad sizes require that the chips be designed specifically for solder-bumping, or that the wafers be postprocessed to distribute solder bumps pads over their surface, and we must wire (redistribute) these pads to the conventional wire-bond pads.

Another potential disadvantage arises from the lead content of the solder bump. Containing radioactive isotopes, most lead is a source of alpha particles that can potentially change the state of nearby transistors not shielded by aluminum. The effects of alpha particles are mainly of concern to dynamic random access memories (DRAMs) and dynamic logic.

7.9.2 Design, Repair, and Test

An important adjunct to the physical technology is the technology required to test and repair the die and modules. An important question that has to be answered for every MCM is how much to test the die before assembly vs. how much to rely on postassembly test to locate failed die. This is purely a cost question. The more the bare die are tested, the more likely that the assembled MCM will work. If the assembled MCM does not work, then it must either be repaired (by replacing a die) or discarded. The question reduces to the one that asks what level of bare die test provides a sufficiently high confidence that the assembled MCM will work, or is the assembled module cheap enough to throw away if a die is faulty.

In general, there are four levels of bare die test and burn-in, referred to as four levels of known good die (KGD). In Table 7.4, these test levels are summarized along with their impact. The lowest KCD level is to just use the same tests normally done at wafer level, referred to as the *wafer sort tests*. Here, the chips are normally subject to a low-speed functional test combined with some parametric measurements (e.g., measurement of transistor curves). With this KGD level, test costs for bare die are limited to wafer test costs only. There is some risk, however, that the chip will not work when tested as part of the MCM. This risk is measured as the test escape rate. With conventional packaging, the chips are tested again, perhaps at full speed, once they are packaged, making the test escape rate zero.

TABLE 7.4 Levels of Known Good Die and Their Impact

Known Good Die Level	Test Cost Impact	Test Escape Impact
Wafer level functional and parametric	Low	<1–2% fallout for mature ICs, possibly >5% fallout for new ICs
At-pin speed sorted	Medium	Min. fallout for new digital ICs
Burned-in	High	Burn-in important for memories
Dynamically burned-in with full testing	Highest	Min. memory fallout

If the MCM contains chips that must meet tight timing specifications (e.g., a workstation) or the MCM must meet high reliability standards (e.g., for aerospace), however, then higher KGD test levels are required. For example, a workstation company will want the chips to be speed sorted. Speed sorting requires producing a test fixture that can carry high-speed signals to and from the tester. The test fixture type usually used at wafer sort, generally referred to as a *probe card*, is usually not capable of carrying high-speed signals. Instead, it is necessary to build a more expensive, high-speed test fixture or temporarily mount the bare die in some temporary package for testing. Naturally, these additional expenses increase the cost of providing at-pin speed sorted die.

Some applications require even higher levels of known good die. Aerospace applications usually demand burn-in of the die, particularly for memories, so as to reduce the chance of infant mortality. There are two levels of burned-in KGD. In the lowest level, the die are stressed at high temperatures for a period of time and then tested. In the higher level, the rinse are continuously tested while in the oven.

How do you decide what test level is appropriate for your MCM? The answer is driven purely by costs; the cost of test vs. the cost of repair. For example, consider a four-chip MCM using mature ICs. Mature ICs tend to have very high yields, and the process engineers have learned how to prevent most failure modes. As a result, the chances are very small that a mature IC would pass the wafer-level functional test and fail in the MCM. With a test escape rate of 2%, there is only an 8% ($1 - 0.98^4$) chance that each MCM would need to have a chip replaced after assembly. If the repair costs $30 per replaced chip, then the average excess repair cost per MCM is $2.40. It is unlikely that a higher level of KGD would add only $0.60 to the cost of a chip. Thus, the lowest test level is justified.

On the other hand, consider an MCM containing four immature ICs. The functional and parametric wafer-level tests are poor at capturing speed faults. In addition, the process engineers have not had a chance to learn how to maximize the chip yield and how to best detect potential problems. If the test escape rate was 5%, there would be a 40% chance that a four-chip MCM would require a chip to be replaced. The average repair cost would be $12 per MCM in this scenario, and the added test cost of obtaining speed sorted die would be justified.

For high-speed systems, speed sorting also greatly influences the speed rating of the module. Today, microprocessors, for example, are graded according to their clock rate. Faster parts command a higher price. In an MCM, the entire module will be limited by the slowest chip on the module, and when a system is partitioned into several smaller chips, it is highly likely that a module will contain a slow chip if they are not tested.

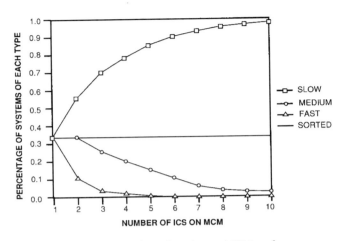

FIGURE 7.13 The effect of speed sorting on MCM performance.

For example, consider a set of chips that are manufactured in equal numbers of slow, medium, and fast parts. If the chips are speed sorted before assembly into the MCM, there should be 33% slow modules, 33% medium modules, and 33% fast modules. If not speed sorted, these rations change drastically as shown in Fig. 7.13. For a four-chip module assembled from unsorted die, there will be 80% slow systems, 19% medium systems, and only 1% fast systems. This dramatic reduction in fast modules also justifies the need for speed-sorted die.

Design technology is also an important component to the MCM equation. Most MCM computer-aided design (CAD) tools have their basis in the PCB design tools. The main differences have been the inclusion of features to permit the rise of the wide range of physical geometries possible in an MCM (particularly via geometries) as well as the ability to use bare die. A new format, the *die interchange format*, has been specifically developed to handle physical information concerning bare die (e.g., part locations).

There is more to MCM design technology, however, than just making small changes to the physical design tools (those tools that actually produce the MCM wiring patterns). Design correctness is more important in an MCM than in a PCB. For example, a jumper wire is difficult to place on an MCM to correct for an error. Thus, recent tool developments have concentrated on improving the designers ability to ensure that the multichip system is designed correctly before it is built. These developments include new simulation libraries that allow the designer to simulate the entire chip set before building the MCM, as well as tools that automate the electrical and thermal design of the MCM.

7.9.3 When to Use Multichip Modules

A number of scenarios, enumerated below, typically lead to consideration of an MCM alternative.

1. You must achieve a smaller form factor than is possible with single-chip packaging. Often, integrating the digital components onto an MCM-L provides the most cost-effective way to achieve the required size. MCM-D provides the greatest form factor reduction when the very smallest size is needed. For example, an MCM-D might be used to integrate a number of ICs so that they can fit into one existing socket. MicroModule Systems integrated four ICs in an MCM-D to make an Intel 286 to 486 PC upgrade component that was the same size as the original 286 package. If the design many discrete capacitors and resistors, then using an MCM-D technology with integrated capacitors and resistors might be the cheapest alternative. Motorola often uses MCM-Cs for this purpose, whereas AT&T uses MCM-Ds with integrated capacitors and resistors. In both cases, the capacitors and resistors are printed onto the MCM substrate using the same techniques previously used to make so-called hybrid components. Size reduction is currently the most common driver of commercial use of MCMs.
2. The alternative is a single die that would either be too large to manufacture or would be so large as to have insufficient yield. The design might be custom or semicustom. In this case, partitioning the die into a number of smaller die and using an MCM to achieve the interconnect performance of the large die is an attractive alternative. For example, Silicon Graphics once implemented a four-chip graphics coprocessor, using four application-specific integrated circuits (ASICs) on an MCM-C (essentially a multicavity pin grid array [PGA]). The alternative was a single, full custom

chip, which would have taken a lot longer to design. But they needed a single chip form factor to fit the coprocessor into the board space allocated.

Recently, teams of researchers have been investigating the possibility of building a *megachip* in a multichip module. The prospect is that a multichip module containing several chips can be built to provide the performance and size of a single, very large, highly integrated chip, but at a lower cost. Cost reduction arises due to the far better yields (percentage of die that work) of a set of small die over a larger die. This use of MCM technology is of particular interest to the designers of high-performance computing equipment. This concept is explored further in Dehkordi et al. [1995] and Franzon et al. [1995].

3. You have a technology mix that makes a single IC expensive or impossible, and electrical performance is important. For example, you need to interface a complementary-metal-oxide-semiconductor (CMOS) digital IC with a GaAs microwave monolithic integrated circuit (MMIC) or high bandwidth analog IC. Or you need a large amount of static random access memory (SRAM) and a small amount of logic. In these cases, an MCM might be very useful. An MCM-L might be the best choice if the number of interchip wires is small. A high-layer-count MCM-C or MCM-D might be the best choice if the required wiring density is large. For example, AT&T wanted to integrate several existing die into one new die. All of the existing die had been designed for different processes. They determined that it was cheaper to integrate the existing die into a small MCM rather than redesign all of the die into a common IC process.

4. You are pad limited or speed limited between two ICs in the single chip version. For example, many computer designs benefit by having, a very wide bus (256 bits or more) between the two levels of cache. In this case, an MCM allows a large number of very short connections between the ICs. For example, in the Intel Pentium Pro (P6) processor, there is a 236-bit bus between the central processing, unit and the second level cache chip. Intel decided that an MCM-C was required to route this bus between the two chips and achieve the required bus delay. The combination of MCM-D with flip-chip solder-bump technology provides the highest performance means of connecting several die. The greatest benefit is obtained if the die are specially designed for the technology.

5. You are not sure that you can achieve the required electrical performance with single-chip packaging. An example might be a bus operating above 100 MHz. If your simulations show that it will be difficult to guarantee the bus speed, then an MCM might be justified. Another example might be a mixed-signal design in which noise control might be difficult. In general, MCMs offer superior noise levels and interconnect speeds to single-chip package designs.

6. The conventional design has a lot of solder joints, and reliability is important to you. An MCM design has far fewer solder joints, resulting in less field failures and product returns. (Flip-chip solder bumps have shown themselves to be far more reliable than the solder joints used at board level.)

Although there are many other cases where the use of MCM technology makes sense, these are the main ones that have been encountered so far. If a MCM is justified, the next question might be to decide what ICs need to be placed on the MCM. A number of factors follow that need to be considered in this decision:

1. It is highly desirable that the final MCM package be compatible with single-chip packaging assembly techniques to facilitate manufacturability.
2. Although wires between chips on the MCM are inexpensive, off-MCM pins are expensive. The MCM should contain as much of the wiring as is feasible. On the other hand, an overly complex MCM with a high component count will have a poor final yield.
3. In a mixed-signal design, separating the noisy digital components from the sensitive analog components is often desirable. This can be done by placing the digital components on an MCM. If an MCM-D with an integrated decoupling capacitor is used, then the on-MCM noise might be low enough to allow both analog and digital components to be easily mixed. Be aware in this case,

however, that the on-MCM ground will have noise characteristics that differ from those of the PCB ground.

In short, most MCM system design issues are decided on by careful modeling of the system-wide cost and performance. Despite the higher cost of the MCM package itself, cost savings achieved elsewhere in the system often can be used to justify the use of an MCM.

7.9.4 Issues in the Design of Multichip Modules

The most important issue in the design of multichip modules is obtaining bare die. The first questions to be addressed in any MCM project relate to whether the required bare die are available in the quantities required, with the appropriate test level, with second sourcing, at the right price, etc. Obtaining answers to these questions is time consuming, as many chip manufacturers still see their bare die sales as satisfying a niche market. If the manufactured die are not properly available, then it is important to address alternative chip sources early.

The next most important issue is the test and verification plan. There are a number of contrasts with using a printed circuit board. First, as the nonrecurring engineering cost is higher for a multichip module, the desirability for first pass success is higher. Complete prefabrication design verification is more critical when MCMs are being used, so more effort must be spent on logic and electrical simulation prior to fabrication.

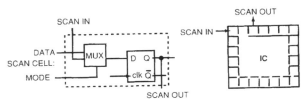

FIGURE 7.14 The use of boundary scan eases fault isolation in an MCM.

It is also important to determine, during design, how faults are going to be diagnosed in the assembled MCM. In a prototype, you wish to be able locate design errors before redoing the design. In a production module, you need to locate faulty die or wire bonds/solder bumps if faulty module is to be repaired (typically by replacing a die). It is more difficult to physically probe lines on an MCM, however, than on a PCB. A fault isolation test plan must be developed and implemented. The test plan must be able to isolate a fault to a single chip or chip-to-chip interconnection. It is best to base such a plan on the use of chips with *boundary scan* (Fig. 7.14). With boundary scan chips, test vectors cam be scanned in serially into registers around each chip. The MCM can then be run for one clock cycle, and the results scanned out. The results are used to determine which chip or interconnection has failed. If boundary scan is not available, and repair is viewed as necessary, then an alternative means for sensitizing between-chip faults is needed.

The decision as to whether a test is considered necessary is based purely on cost and test-escape considerations. Sandborn and Moreno [1994] and Ng, in Doane and Franzon [1992, Chap. 4], provide more information. In general, if an MCM consists only of inexpensive, mature die, repair is unlikely to be worthwhile. The cost of repair (generally $20 to $30 per repaired MCM + the chip cost) is likely to be more than the cost of just throwing away the failed module. For MCMS with only one expensive, low-yielding die, the same is true, particularly if it is confirmed that the cause of most failures is that die. On the other hand, fault diagnosis and repair is usually desirable for modules containing multiple high-value die. You do not wish to throw all of these die away because only one failed.

Thermal design is often important in an MCM. An MCM will have a higher heat density than the equivalent PCB, sometimes requiring a more complex thermal solution. If the MCM is dissipating more than 1 W, it is necessary to check whether any heat sinks and/or thermal spreaders are necessary.

Sometimes, this higher concentration of heat in an MCM can work to the designer's advantage. If the MCM uses one larger heat sink, as compared with the multiple heat sinks required on the single-chip packaged version, then there is the potential for cost savings.

Generally, electrical design is easier for an MCM than for a PCBs; the nets connecting the chips are shorter, and the parasitic inductances and capacitances are smaller. With MCM-D technology, it is possible to closely space the power and ground planes so as to produce am excellent decoupling capacitor. Electrical design, however, cannot be ignored in an MCMs; 300-MHz MCMs will have the same design complexity as 75-MHz PCBs.

MCMs are often used for mixed-signal (mixed analog/RF and digital) designs. The electrical design issues are similar for mixed signal MCMS as for mixed signal PCBs, and there is a definite lack of tools to help the mixed-signal designer. Current design practices tend to be qualitative. There is an important fact to remember that is unique to mixed signal MCM design. The on-MCM power and ground supplies are separated, by the package parasitics, from the on-PCB power and ground supplies. Many designers have assumed that the on-MCM and on-PCB references voltages are the same only to find noise problems appearing in their prototypes. For more information on MCM design techniques, the reader referred to Doane and Franzon [1992], Tummula and Rymaszewski [1989], and Messner et al. [1992].

7.9.5 The Future of MCM Packaging

MCM technology has some problems to be solved, as many newer technologies do. Since it involves no new techniques on the part of either the wafer fabs, hybrid circuit manufacturers, or the MCM assemblers, many of these problems are being solved. Although known good die (KGD) is a potential issue (see also Chapter 4), users of hybrid assemblies are familiar with the issues surrounding KGD. Users include the automotive, defense, consumer, and telecom industries. Since assembly of the final MCM package uses standard assembly technologies rather than the increased accuracy and test difficulties presented by CSP and ultrafine pitch, it is not necessary to change the final assembly process. The ability to take "standard" function silicon die and combine them to provide added functionality still is in use, and will continue to be, by users such as AT&T/Lucent, Delco/Delphi Electronics, Motorola, and NEC. Yoshimura et al. (1998)[15] describe the development of a new MCM for cellular phones by NEC Research and Development. Sasidhar et al. (1998)[16] describe the work NASA has done in developing a new flight computer utilizing a 3-D MCM stack.

However, MCM utilization is not increasing to the previously expected levels. The main drawbacks appear to be:

- "Standard" circuit assemblers hesitate to get into wire bonding.
- Mounting bare die on FR-4, as MCM-L devices do, can result in problems.
- The advent of flip chips and chip-scale packaging (CSP) of devices means that many assemblers can take advantage of very small packages without the need for wire bonding.

7.9.6 Definitions

MCM: multi-chip module
MCM-C: an MCM assembled on a ceramic substrate
MCM-D: an MCM assembled using thin-film lithography and deposition techniques
MCM-L: an MCM assembled on standard circuit board substrate material, such as FR-4.
Wire bonding: a standard IC assembly technique where gold or aluminum wires are bonded to the IC die then to the leadframe or substrate

References, Sections 7.1–7.8

1. R.W. Vest, "Materials Science of Thick Film Technology," *Amer. Cer. Soc. Bull.*, 65, 4, 631, 1986.
2. D.L. Hankey, A.S. Shaikh, S. Vasudevan, and C.S. Khadilkar, "Thick Film Materials and Processes," in *Hybrid Microelectronics Handbook*, 2/e, J.E. Sergent and C.A. Harper, Editors. McGraw-Hill, New York, 1995.

3. S. Vasudevan, "Effect of Design Parameters on Overstress Characterization of Thick Film Resistors for Lightning Surge Protection," *Proc. Intl. Symp. Microelec.*, 634, Philadelphia, PA, 1997.
4. S.J. Horowitz, J.P. Page, D.I. Amey, S.K. Ladd, J.E. Mandry and D. Holmes, "Design Trade-Offs of MCM-C Versus Ball Grid Array on Printed Wiring Boards," *Int'l. J. Microcircuits & Elec. Pkg.*, vol. 19, no. 4, 352, 1996.
5. W.E. Coleman, "Stencil Design and Applications for SMD, Through Hole, BGA and Flip Chips," *Advancing Microelectronics*, vol. 23, no. 1, 24, 1996.
6. J. Pan, G.L. Tonkay and A. Quintero, "Screen Printing Process Design of Experiments for Fine Line Printing of Thick Film Ceramic Substrates," *Proc. Intl. Symp. Microelec.*, 264, San Diego, CA, 1998.
7. A. Elshabini-Riad and F.D. Barlow, "Thin Film Materials and Processing" in *Hybrid Microelectronics Handbook,* 2/e, J.E. Sergent and C.A. Harper, Editors. McGraw-Hill, New York, 1995.
8. H. Kanda, R.C. Mason, C. Okabe, J.D. Smith and R. Velasquez, "Buried Resistors and Capacitors for Multilayer Hybrids," *Advancing Microelectronics*, vol. 23, no. 4, 19, 1996.
9. R.C. Frye, "Passive Components in Electronic Applications: Requirements and Prospects for Integration," *Int'l. J. Microcircuits & Elec. Pkg.*, vol. 19, no. 4, 483, 1996.
10. R.C. Frye, "Integrated Passive Components: Technologies, Materials, and Design," *Int'l. J. Microcircuits & Elec. Pkg.*, vol. 20, vol. 4, 578, 1997.
11. E.N. Larson, "Wire Bonding" in *Multichip Module Technologies and Alternatives: The Basics,* D.A. Doane and P.D. Franzon, Editors. Van Nostrand Reinhold, New York, 1993.
12. D.R. Haagenson, S.J. Bezuk and R.D. Pendse, "Tape Automated Bonding" in *Multichip Module Technologies and Alternatives The Basics,* D.A. Doane and P.D. Franzon, Editors. Van Nostrand Reinhold, New York, 1993.
13. C.C. Wong, "Flip Chip Connection Technology" in *Multichip Module Technologies and Alternatives The Basics,* D.A. Doane and P.D. Franzon, Editors. Van Nostrand Reinhold, New York, 1993.
14. K.J. Puttlitz, Sr., "Flip Chip Solder Bump (FCSB) Technology: An Example" in *Multichip Module Technologies and Alternatives The Basics,* D.A. Doane and P.D. Franzon, Editors. Van Nostrand Reinhold, New York, 1993.
15. T. Yoshimura, H. Shirakawa, Y. Tanaka; "Package Technology for Cellular Phones." *NEC Research and Development,* vol. 39 no. 3, July 1998.
16. K. Sasidhar, L. Alkalai, A. Chatterjee; "Testing NASA's 3D-Stack MCM Space Flight Computer." *IEEE Design and Test of Computers,* vol. 15, no. 3, July-Sept. 1998.

References, Section 7.9

Dehkordi, P., Ramamurthi, K., Bouldin, D., Davidson, H., and Sandborn, P. 1995. Impact of packaging technology on system partitioning: A case study. In *1995 IEEE MCM Conference,* pp. 144–151.

Doane, D.A., and Franzon, P.D., eds. 1992. *Multichip Module Technologies and Alternatives: The Basics.* Van Nostrand Reinhold, New York.

Franzon, P.D., Stanaski, A., Tekmen, Y., and Banerjia, S. 1996. System design optimization for MCM. *Trans. CPMT.*

Messner, G., Turlik, I., Balde, J.W., and Garrou, P.E., eds. 1992. *Thin Film Multichip Modules.* ISHM.

Sandborn, P.A., and Morena., H. 1994. *Conceptual Design of Multichip Modules and Systems.* Kluwer: Norwell, MA.

Tummula, R.R., and Rymaszewski, E.J., eds. 1989. *Microelectronics Packaging Handbook.* Van Nostrand Reinhold, Princeton, NJ.

Suggested Readings

A. Csendes, V. Szekely, M. Rencz; "Efficient Thermal Simulation Tool for ICs, Microsystem Elements, and MCMs." *Microelectronics Journal,* vol. 29, no. 4–5, Apr–May 1998.

A. Shibuya, I. Hazeyama, T. Shimoto, N. Takahashi; "New MCM Composed of D/L Base Substrate, High-density-wiring CSP and 3-D Modules." *NEC Research and Development,* vol. 39 no. 3, July 1998.

A. Dohya, M. Bonkohara, K. Tanaka; "Leadless Chip Assembly Technology for Low-Cost MCMs." *NEC Research and Development,* vol. 39 no. 3, July 1998.

P. Mannion; "Passives Tackle Next-Generation Integration Requirements." *Electronics Design,* vol. 47 no. 5, March 9, 1999.

E. Swartzlander; "VLSI, MCM, and WSI: A Design Comparison." *IEEE Design and Test of Computers,* vol. 15 no. 3, Jul-Sep 1998.

G. Katopis, W. Becker; "S-390 Cost Performance Considerations for MCM Packaging." *IEEE Transactions on Components Packaging & manufacturing Technology Part B—Advanced Packaging,* vol. 21 no. 3, Aug. 1998.

K. Fujii et al.; "Application of the MCM Technology into an MMIC T/R Module." *IEICE Transactions on Electronics,* no. 6 June 1998

E. Davidson; "Large Chip vs. MCM for a High-Performance System." *IEEE Micro.,* vol. 18 no. 4, July/Aug 1998.

J. Powell, B. Durwood; "Beyond the GHz Barrier with Thick Film Ceramic Substrates." *Electronic Engineering,* vol. 70 no. 858, Jun 1998.

R. Buck; "MCM-C Packaging Provides Cost-Effective Microprocessor." *Semiconductor International,* vol. 21 no. 8, July 1998.

R. Chahal, et al.; "Novel Integrated Decoupling Capacitor for MCM-L Technology." *IEEE Transactions on Components Packaging & Manufacturing Technology Part B—Advanced Packaging,* vol. 21 no. 2, May 1998.

8
Interconnects

Glenn R. Blackwell
Purdue University

8.1 General Considerations 8.1
8.2 Wires for Interconnection 8.3
8.3 Single-Point Interconnects 8.4
8.4 Connectors 8.5
8.5 Board Interconnects 8.10
8.6 Component Sockets 8.11
8.7 Fiber-Optic Interconnects and Connections 8.16
8.8 Coaxial Cable and Interconnects 8.49
8.9 Microwave Guides and Couplers 8.57

8.1 General Considerations

This chapter provides information on separable interconnect devices used for interconnecting components, circuit boards, and modules. The needs for these devices must be considered during the design phase.

Where Else?

Other information related to interconnecting circuit sections can be found in Chapter 2, "Surface Mount Technology"; Chapter 4, "Direct Chip Attach"; and Chapter 5, "Circuit Boards."

Interconnect devices must meet certain performance requirements in most or all of the following categories:

- Signal integrity
- Power loss
- Electrical characteristics: contact resistance, inductance and capacitance; voltage and power ratings; shielding; filtering
- Mechanical characteristics: contact count; contact spacing; contact forces; polarization/keying; wiping characteristics; shock and vibration; size; soldering technique; CTE with respect to substrate
- Environmental issues: soldering technique; cleaning materials; thermal; corrosion protection

Signal integrity is defined by loss of quality of the input signal. Ideally, the signal at the output of an interconnect should be equal in all characteristics with the signal at the input of the interconnect. In reality, signal degradation occurs. The user must define what level of degradation is acceptable or, alternatively, must define minimum acceptable signal levels at the output. Tests that indicate signal integrity include

- Voltage standing wave ratio (VSWR)
- Frequency response
- Rise-time of signal edges
- Current flow

Both THT and SMT technology connectors are available for most uses. It is possible to use THT connectors and sockets in SMT designs and processes by using the pin-in-paste process described in Chapter 2. The limiting factor on using THT sockets is that, for some connectors, the plastic housing is not designed to withstand reflow heating temperatures.

8.1.1 High-Speed Connector Problems

In today's designs, with clock rates over 100 MHz and rise times commonly 1 nanosecond (ns) or less, designers cannot ignore the role interconnections play in a logic design. Interconnect effects can play a significant part in the timing and noise characteristics of a circuit. The faster clock rates and rise times increase both capacitive and inductive coupling effects, which makes crosstalk problems greater. They also mean shorter time for reflections to decay before the data is clocked and read, which decreases the maximum line length that can be used for unterminated systems. This all means that one of the major interconnect challenges is to ensure signal integrity as the high-speed pulses move along the total interconnect path, from device to PCB, through the PCB to the backplane, and on out to any network connections which may be present.

An interconnect that must pass the short rise time of a high-speed signal pulse can be detrimental to maintaining signal integrity due to an unwanted reflection. The shorter the rise time, the greater the risk of degradation of the signal. The ideal situation is that the connector will have the appropriate termination characteristics to create no degradation of the signal.

8.1.2 Crosstalk

In addition to general signal degradation, another potential impact of a connector on signal integrity is increased crosstalk, which is caused by inductive and capacitive coupling between contacts and/or lines feeding contacts. This can cause noise to be induced in one contact/line from a switching signal in another contact/line. Crosstalk can be expressed in a percentage of the voltage swing of the signal, or in decibels (dB), where 0 dB represents 100% crosstalk, –20 dB represents 10% crosstalk, and –40 dB represents 1% crosstalk.

Two types of crosstalk can be generated: *backward* crosstalk and *forward* crosstalk. Backward, also called *near-end* crosstalk, is measured at the driving end of the connector and represents the sum of the capacitive and inductive coupling. Forward, also called *far-end* crosstalk, is measured at the receiving end of the connector and represents capacitive minus inductive coupling.

Crosstalk can also be *single-line* or *multiline*. Single-line crosstalk is the result of one driven line inducing a voltage on one quiet line. Multiline crosstalk is two or more driven lines inducing voltage(s) on one quiet line. In most systems, many lines/contacts will be in use at once, so multiline crosstalk is a more realistic figure.

To completely understand the crosstalk specifications of a connector, a user needs the following information:

- Is the crosstalk specification measured on the actual connector, or is it the result of a simulation?
- What is the value of the signal rise time used for the crosstalk measurement, and the digital technology used to generate it (TTL and CMOS rise times are typically 10 to 90% rise times, while ECL is typically a 20 to 80% rise time)?
- What is the signal pattern on the connector, i.e., the location of the ground lines relative to the measured lines?
- How many lines are being driven by the switched signal?
- What are the source and load impedances?
- Is the crosstalk specification for forward or backward, or the total?
- Is the crosstalk specification for single-line or multiline?
- If multiline, how many lines are being driven?

Interconnects 8-3

One common way to minimize crosstalk is to intersperse signal contacts with ground contacts. However, this has the disadvantage of minimizing signal contact density.

Levels of Interconnects

Granitz has defined six levels of interconnects:

- Level 1: chip pad to package leads, e.g., wire bonds
- Level 2: components to circuit board, e.g., PLCC socket
- Level 3: circuit board (board-to-board) connector, e.g., edge-card for mother board to daughter board connection
- Level 4: subassembly interconnects, e.g., ribbon cables
- Level 5: subassembly to I/O, e.g., BNC connector
- Level 6: system interconnects, e.g., Ethernet connectors

Each level of interconnection may be accomplished in several ways. For example, a level 4 interconnect between circuit boards may use a card edge connector, a ribbon cable, or even discrete wires. This chapter will include information on connector materials and types that can be used for levels 2 through 6.

8.2 Wires for Interconnection

Selection of wires for interconnection involves several considerations.

- Color coding
- Temperature rating
- Solid or stranded
- Individual, multiple-cable, flat cable
- Current-carrying capacity (ampacity)

Color codes are at the users discretion. Some cable types, such as flat cable, will have limited selections available, based on the manufacturer. While custom cables can be ordered, they can be difficult to cost justify. If, however, part of the design criteria includes developing the product to be suitable for field installation of the particular cable in question, distinct color coding can be very valuable in minimizing field errors.

Temperature rating of wires depends on both the assembly environment and the test environment, as well as the expected use environment. A need for high-temperature wires exists if the wires will go through the reflow process, if burn-in at elevated temperature is required, or if the use environment will exceed 60° C. Standard wire temperature rating is to 60° C, and insulation is available that is rated up to 500° C.

The decision for solid or stranded wire is made with consideration of flexibility and vibration in the user environment and assembly techniques. If flexibility is required, such as connecting the top and bottom sections of a hinged device like a laptop computer, then stranded wire is called for. If the environment will involve considerable vibration, such as automotive service, then stranded wire is called for. If the assembly technique is wirewrap, then the solid wire specifically made for wirewrap is called for. This wire is available in a variety of sizes, and the wirewrap pins used must accommodate the wirewrap wire to be used.

The choice among individual, multiple-cable, and flat cable may depend on both application (such as the laptop connection) and ease-of-assembly considerations. If it is expected that the wire/cable will be routinely subjected to flexing, such as in the laptop application, consult the cable manufacturers. Typically, the finer the base wire used, and the larger the number of strands used, the longer the cable will stand up to continued flexing. Wire selection issues may also include any expected field assembly or installation.

Current-carrying capacity (ampacity) defines minimum wire sizes. The chart below applies to both solid and stranded wire. Remember that larger diameter wires have smaller AWG sizes. The ampacities shown in the table are for wire lengths of 1 m or less and are much higher than the ampacities for longer wire runs.

TABLE 8.1 Wire Ampacity and Size

AWG wire size	Resistance, Ω per 1000 ft	Ampacity for low-temperature insulation	Ampacity for high-temperature insulation	Bare wire diameter mm (in)
30	100	2	4	0.254 (0.0100)
28	60	3	6	0.320 (0.0126)
26	40	4	7	0.404 (0.0159)
24	25	6	10	0.511 (0.0201)
22	14	8	13	0.643 (0.0253)
20	10	10	17	0.810 (0.0319)
18	6	15	24	1.024 (0.0403)

Note: Ω/1000 ft is approximate. Exact value depends on whether the conductor is solid or stranded, and if stranded, the type of stranding. Consult the manufacturers' data sheets for exact values.

Other considerations when using wires in a circuit include the inductance and resistance of the wire. While resistance concerns are normally answered by proper wire size selection, inductance is an issue in high-frequency or high-speed circuits that must take into account wire lengths and placement within the overall circuit. The user must also remember that any length wire (or lead) is potentially an antenna in high-frequency and/or high power applications.

8.3 Single-Point Interconnects

This section covers interconnects used for individual wires or individual THT component leads. It includes interconnect points for stranded and solid wire, bus bars, test points, terminals, pins, and jumpers. Many of these interconnects are available in both THT and SMT configurations. A common name for this type of interconnect is a *terminal*.

8.3.1 Terminals for THT Component Leads

There are several types of terminals that allow the direct insertion of THT component leads. They are used in breadboard and prototype assemblies, where they may need to be changed, to tune component values to a specific board or for a specific concern, e.g., frequency of operation. If the lead is to be soldered into the terminal, the terminal will have a cup-style opening in the top. If the lead is inserted in the terminal in a manner similar to a component socket, the terminal will have one of the many styles of spring-loaded contacts shown in Section 8.6.1.

8.3.2 Terminals for Wire Connections

Terminals for wire connections may have a cup-style top for insertion and soldering, may have a solid or bifurcated post for wrapping and soldering the wire, or may be designed to have the wire crimped to the post. They may be used to interconnect boards, to allow wire routing when a circuit board cannot be fully routed with its own traces, or to connect the board to external I/O or power.

The terminals are usually designed for THT mounting and, for additional strength, may be mechanically affixed to the board in addition to being soldered. In this case, the terminal and corresponding board hole may be designed for a press fit or may be crimped, rolled, or flared after insertion.

Solder-cup terminals are designed to allow stranded or solid wire or a THT part lead to be soldered directly to them. The THT parts would normally be two-lead parts with axial leads, such as a resistor or inductor. The terminal is inserted in a through hole in the board and soldered. The top of the terminal

Interconnects

has a hollowed-out cup into which the stripped end of a wire is inserted then soldered. For proper soldering, the stripped end of the wire must be inserted all the way to the bottom of the cup.

Turret terminals allows multiple wires to be wrapped around them, then be soldered to them. Normal good soldering practice is to have the wires wrap from one-half the way around the turret to far enough around to touch the incoming wire, but not allow the wire to overlap itself or another wire on the same turret. The terminal mounts in a through hole in the board, and the turret base may mount on the board as a press fit or with flange retention.

Bifurcated terminals are similar to turret terminals in that they are designed for wire attachment, but they have a slot or slit opening through which one or more wires are placed to be soldered.

Pin terminals come in a variety of styles, each designed to allow wire attachment. They are normally press fit into the board. In addition to being used on boards, they are also common on relays and on devices that are hermetically sealed. In this case, the terminal serves as a feed-through for the glass seal.

Wirewrap terminals are designed specifically for solid-wire wirewrap applications. Wirewrap pins are designed for one to four wires to be wrapped around them. They are square in cross section, which allows a properly wrapped wire to form a "gas-tight" connection at each corner of the wrap. The bottom of the terminal inserts into the board and may press fit into appropriate size holes, may be soldered to lands on the board, or may crimp to the board. The style chosen will depend on what assembly technique will best fit with the board design, the available assembly procedures, and the amount of retention strength needed. Press-fit pins normally have a compliant retention area that will accommodate slight variations in PTH diameter.

Wirewrap terminals are normally gold plated to minimize oxidation under each corner wrap. They can be used for a prototype or for a permanent assembly.

8.4 Connectors

Connectors for the purposes of this chapter include any multi-termination device which is not designed to be a component socket. Connector selection must consider electrical, mechanical, and environmental issues.

All connectors have two electrical sides. One side is designed to create a separable mating with a mating connector, such as the panel-mount receptacle and the wire-mount pin of a stereo connection. The other side is a "permanent" electrical connection that is designed not normally to be disconnected.

Connectors are considered to be used in mating pairs. In cases where a connector is used to mate with a series of contacts fixed on a circuit board, one-half of the pair is considered to be the contacts on the board, regardless of whether they are in a connector body, are pins soldered into the board, or consist of an etched pattern at the edge of the board (edge-card connector).

8.4.1 Electrical Considerations

Electrical considerations in connector selection include

- Contact current ratings
- Contact resistance
- Contact impedance; capacitance and inductance
- Millivolt drop of contacts
- Power rating
- Shielding and filtering requirements

Contact current ratings must consider the maximum ac and dc currents expected. Duty cycle should be considered if the application is such that there is absolutely no chance of a continuous flow of current. The current rating of the contacts in a connector, and of the complete connector, are based on an allowable temperature rise. Generally, and following UL guidelines, a 30° C temperature rise is used. The rise in the contacts is due to I^2R heating of the contacts caused by the currents flowing through each contact. Remember that the total of the connector temperature rise plus the maximum environmental temperature cannot exceed the maximum allowable temperature rating of the connector.

If several contacts are expected to carry currents large enough to cause heating, those contacts should be distributed as evenly as possible among the contacts. To prevent failures, never expect to operate all contacts at their maximum allowable current. If a contact fails mechanically due to overtemperature, there is a good chance the failure will cascade through the other contacts as the molding compound fails.

8.4.2 Mechanical Considerations

Mechanical considerations in connector selection include

- One- or two-part connector
- Number of contacts
- Wipe length and style of contacts
- Normal force required to seat connector
- Insertion/extraction force
- Number of expected connect/disconnect cycles
- Body material with regard to TCE and soldering technique
- Body style and profile
- Connector-to-cable attachment technique
- Polarization and keying requirements
- Shock and vibration
- Area and height of connector
- Assembly technology to be used for connecting cable
- Manual or automatic connector placement

One-part connectors are connectors that have their mating halves integral with a board or other structure. Edge-card connectors use etched contacts on the board as their mating half. These are sometimes called *integral* connectors. The connector provides an disconnect interface between the board and a cable, a set of discrete wires, or a mother board. The connector may also be designed to provide some board support (Fig. 8.1).

Two-part connectors consist of a male-female pair. As discussed in Section 8.4.4, the male connector may have pins and the female connector leaf-spring contacts. The connectors may create a separable link between the following:

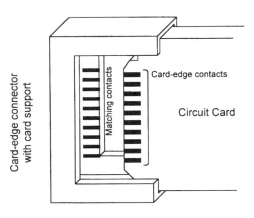

FIGURE 8.1 Edge card connector example, with card support.

- Two cables
- A circuit board and a cable
- Two boards (mother-daughter boards)

In the case of two cables, the cables may terminate at the permanent side of the connectors by solder, crimp, insulation displacement, or press-fit pin. For circuit board mounting, the connectors may be soldered or press fit to the board, or they may have wirewrap terminations. They may also have a non-circuit mechanical means of attachment, such as screws or rivets. The pin/contact set may exit the board normal to or parallel to the board surface. The male pin set is also commonly called a *header*. A variety of body (a.k.a. housing) styles are available (see the paragraph on body styles below).

The number of contacts is determined by the number of lines of I/O, power, and ground to be connected. A high number of contacts means a larger connector and may mean a need to consider connectors with smaller pin/contact pitch (center-to-center distance between pins or contacts, including edge-card contacts). Connectors carrying high-frequency, fast-rise-time signals may need interspersed ground conductors between each signal path.

Wipe length means the total linear movement of the connector through which a single mating pair of contacts will be in touch with each other. With long, unsupported connectors, flexing of the connector may mean that some center or end contacts may not mate. Appropriate wipe length and stiffening of the body will prevent this.

The *normal force* required is the force an individual contact beam exerts on a mating connector half, in a direction normal to the direction of the wiping motion. For connectors with more than one contact beam, the normal force is not additive. That is, if the normal force of a contact beam is 50 g, and there are 10 contacts in the connector, the normal force is still 50 g. Normal force must maintain contact pressure sufficient to preserve the electrical performance of the connector in the presence of oxide growth and expected environmental factors.

Insertion/extraction force is also called the *mating force* and is the total force necessary to connect and disconnect the connector halves. It increases as the number of contacts in the connector increase and is related to the normal force of the contacts. Above a certain force, special tools may be needed to connect/disconnect the connector.

Frequent removal/replacement operations at the same connector, or at connectors with large pin counts, may require the use of zero-insertion-force (ZIF) connectors, with their accompanying larger area requirement on the board. The ZIF connector has the advantages of reducing wear on the individual contacts and allowing minimization of the insertion/extraction force of the connector. The ZIF connector allows mechanical release of the contact forces, allowing easy insertion. The connector has a mechanical arm that then locks the contact in place. To disconnect, the arm is unlocked and the contacts disconnected. See Section 8.6.1 for an example of a ZIF socket.

The number of expected connect/disconnect cycles affects the quality of the contact type and material selected. If frequent disconnects are expected, the connector supplier must be aware of this during the selection process. This will require a more robust contact style, with consideration of the thickness of the plating on the contact, and may, as noted in the previous paragraph, suggest that a ZIF socket should be considered.

Body materials are determined primarily by the manufacturer, and frequently the only option open to the buyer is the temperature rating. As discussed in Section 8.4.1, the maximum temperature rating is the sum of the environmental temperature and any temperature rise caused by I^2R heating of the contacts, due to the currents flowing through the contacts. Typically, either thermoplastic or thermoset plastics are used. Issues also include use in a corrosive environment. The main purposes of the body material are to provide mechanical strength for the contacts, provide dielectric insulation between the contacts, and resist any environmental factors that could degrade the mechanical and dielectric strengths.

Body or housing styles are numerous. The simplest would be the molded body of the stereo cassette connector. In electronics use, the connector body may simply house the contacts, or it may provide card support, latches, and polarizing and keying. Systems in which connectors hold boards horizontally may need additional board stabilization if vibration is expected in the operating environment.

Body profile depends on two clearances. The clearance above the board and the cleaning clearance below the connector body, which is the clearance that is necessary to allow removal of flux residues and other contaminants following the soldering step of the assembly process. Like IC's themselves, low-profile connectors are harder to clean under and should lead the user to consider no-clean flux. Some connectors are designed with standoffs to allow for thorough cleaning after soldering. The other issue with the body is the size of the opening for the insertion of the mating leads, or its "lead-in."

The connector-to-board attachment technique may be THT or SMT technology and may include mechanical attachment for additional strength. THT connectors may be press fit or soldered to the board. SMT connectors must be soldered or welded to the board. Regardless of the soldering technology, connectors will have multiple insertion and removal forces placed on them during their lifetime. They may also be subject to vibration in their intended-use environment. Their solder fillets will experience far more stress and strain than fillets on components directly soldered to the board. Therefore, board layout considerations include the following:

1. SMT land areas should be as large as possible.
2. PTH diameter should match the diagonal measurement of the pins, ensuring that maximum mechanical contact is made within the hole.
3. Users with high-retentive-force connectors should consider mechanical attachment to the board, whether SMT or THT.

The mechanical attachment to the board may be screws, rivets, or other appropriate devices. If it is expected that the connector may need to be disassembled in the field after manufacture, an attachment technique should be chosen that will facilitate field repair.

THT connectors to be soldered should have some board-retention mechanism established during/after insertion and prior to wave soldering. The soldering pins may be crimped or have an interference fit into the hole. If paste-in-hole soldering is to be used, the connector body must be made of a material that is suitable for reflow soldering temperatures.

Press-fit THT connectors are also available, with compliant sections on each pin to ensure a gas-tight fit in each hole, as long as the pin-hole dimensions are designed and fabricated to create an interference fit. Additional mechanical attachments may also be needed. Press-fit connectors are installed by the use of an arbor press to create the force necessary to press all the pins into all the holes simultaneously.

SMT connectors have compliant leads that are designed to mount on a matching land pattern on the board. To increase mechanical strength, the connector may be epoxied to the board or may have a mechanical attachment. SMT body materials must also allow for reflow soldering, which also requires meeting the solderability, coplanarity, etc., requirements noted in the chapter on SMT assembly.

In terms of polarization/keying requirements, the user's assembly requirements may dictate that an indexing/polarizing key be used on the connector to ensure correct assembly orientation. This key can allow the same connector to be used for different applications with minimal risk of cross-connections. The key can be a slot with a matching plug in the connector halves, or it may be an asymmetrical body that will only allow mating in one orientation.

Shock and vibration affect both the strength of the solder connections and the ability of the connector halves to stay mated during the product lifetime. After consultation with the connector manufacturer, the most appropriate connector may be one that has a locking mechanism to hold the two halves together.

The area and height of connector affect the amount of board space necessary and the z-axis minimum height between the connector and surrounding boards, enclosure panels, etc.

Assembly technology to be used may be either hand or automated placement. Hand placement, while slow, is very flexible in being able to deal with variations in devices. Automated assembly has limitations, and selection of a connector may be constrained by the ability of the existing equipment to handle the new connector(s) or may require new placement tooling.

Interconnects 8-9

8.4.3 Environmental Considerations

Environmental considerations in connector selection include

- Soldering technique to be used
- Cleaning technology (if any) to be used
- Expected user environment

Appropriateness for the soldering technology means consideration of whether the connector is expected to be reflow soldered, wave soldered, or hand soldered. Wave soldering will require contact technology that will minimize solder wicking from the connector leads to the contacts. Wave or hand soldering will require a THT connector, where reflow soldering will require SMT connector, unless paste-in-hole (see Chapter 2) soldering techniques are used. The user must remember that, when using paste-in-hole techniques, while the connector will be THT, the body of the connector must be suitable for reflow temperatures. Like all devices, verifying and maintaining the solderability of the connector leads is important to the overall quality of the final board.

During reflow soldering, the user must consider that the amount of metal present in the combined leads and contacts of the connector will sink more heat than a standard component of the same lead count. A longer dwell time in preheat or soak areas of the profile may be required.

Cleaning the board after soldering must be considered. If no-clean flux is used, and the board will not be cleaned (not automatically true), then cleaning is not an issue, but flux residue may be. If the board is to be cleaned, then these issues should be considered:

- Cleaning fluids must be compatible with the metals and the connector housing materials.
- The connector may need to be sealed.
- An temporary contact insert may be needed to prevent cleaning fluids from contaminating the contact surfaces.
- A minimum standoff may be required above the board surface to ensure complete cleaning underneath the connector.
- No cleaning residue may be left on the contacts.

Handling techniques for connectors are similar to those for components. Handling must not allow finger oils to be deposited on the contacts. Solderability of the connector leads must be maintained. Storage must not allow damage to either the connector leads or the contacts.

The expected user environment may include corrosive dusts or vapors in the air. These issues must be known at the time of design so that appropriate protective design decisions can be made. If available connector body and contact materials are sufficient, no further action may be necessary. High dust areas may require a protective body around the connector or sealing of the enclosures. Corrosive vapors, such as those encountered by marine electronics used in salt water boats, may require conformal coating, sealing/gasketing of the enclosure, or other protective techniques. If the connector is expected to be used only during the initial assembly process or during nonroutine maintenance, it may be appropriate to apply a conformal coat over the connector if the connector on the board is a set of edge-card contacts.

8.4.4 Contact Types

Ideally, the force of the contact against the component termination will result in a contact that is gas tight. Vendors use a variety of different contact types and should be consulted during contact selection. The variety of considerations discussed earlier in this chapter should all be considered. An example of a contact pairing is shown in Fig. 8.2.

The solid pin contact shown is the male connector half, or header. The spring side of the pin will connect to a circuit board by a soldering, a press-fit, or a wire-wrap connection. The separable pin contacts are tin, tin-lead, or gold plated.

FIGURE 8.2 Contact example.

The female receptacles most commonly use leaf spring contacts. Contact spring materials are either tin, tin-lead, or gold plated. Tin and tin-lead are less expensive than gold but will oxidize faster. Their oxides are also more resistant to the wiping action of the contacts. The continued expansion and contraction of contacts and pins due to thermal changes will allow the formation of oxides. Additionally, there is a documented phenomenon in which oxides will grow within the contact area due to the presence of small electrical currents. Gold is very resistant to both types of oxidation.

For these reasons, it is important to understand the current and voltage levels of the circuits to be connected. Low currents can lead to the growth of oxides between contact leaves of the mating connectors. As mentioned above, this oxide growth will happen much more rapidly with tin and tin-lead platings.

Contact-type determination depends on the available types for the connector body and/or connector style chosen. Frequently the connector manufacturer uses a specific type contact for each connector variety made. Like connectors themselves, contacts vary in their cost, so the less expensive contacts are used in less expensive connectors, while more expensive (and typically more reliable) contacts are used in more expensive connectors.

Therefore, the contact type may not be optional after one has selected the connector usage, the plating material, and the body material. In general, leaf-type contacts are less expensive than multiple-wipe contacts. The variation in contact types will lead to different normal and retention forces, and users who anticipate high-vibration environments for connectorized assemblies should consult with the connector manufacturer about retention forces. Assembly forces range from 5 to 15 lb, depending primarily on the number of contacts and the retention force. Contacts may also be open or closed, depending on soldering issues.

Like any component, thermal demands and restrictions on and due to connectors occur in three areas:

- Reflow soldering
- Use during burn-in tests
- Expected environment for the final product

These concerns reflect themselves primarily in the body material selection, and in the contact material selection, recognizing that tin and tin-lead will oxide much faster than gold, especially at elevated temperatures.

8.5 Board Interconnects

Level 3 board interconnects may be done by wires or cables or by connectors attached to the board(s). One of the most common board interconnect techniques is the mother and daughter board configuration, used commonly in computers. A space-saving board interconnect technique, useful when disconnect are not anticipated on a regular basis, is the use of square pins on stacked parallel boards. They may either go from one board-mounted connector to another, or they may be swaged/soldered into one board and pass through mating holes in the other board. The closest spacing is obtained using the through-pin technique (Fig. 8.3).

The level 3 board-to-board connection techniques include pin-to-pin pitch of 0.100 in (2.54 mm) and 0.050 in (1.27 mm). Other pitches are available, including 0.156 in. As pin pitch gets smaller, with a corresponding decrease in spring-contact size, edge-card connectors become difficult to use. Control of the thickness and warpage of circuit cards is not good compared to the deflection range of the spring contacts. This may mean that edge-card connections may not be usable in some applications, including designs using thick multilayer boards.

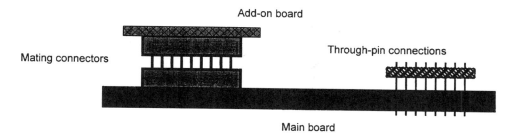

FIGURE 8.3 Examples of pin connectors.

8.6 Component Sockets

Sockets are level 2 interconnect devices that are available for components ranging from two-lead capacitors to 100+ pin THT and SMT devices. The socket is soldered to the board and the component held in place with one of a variety of spring elements. Advantages to using a socket, as opposed to soldering the component directly to the board, include

- It provides ease of component replacement due to failure, upgrade, test, EPROM, etc.
- It allows performance testing of several components as long as they are in the same package.
- It provides protection of the component from processing extremes.
- It is economically advantageous when boards must be shipped across economic zones.

Disadvantages of using sockets include:

- An additional assembly procedure is required.
- There is additional cost for the socket.
- Contacts become electrically noisy with time.
- Contacts constitute one more item to fail.
- Vibration environments may preclude the use of a socket.
- Additional lead resistance, capacitance, and inductance exist, compared to direct component attachment.

8.6.1 Socket Types

The most common types of sockets are shown in Table 8.2, with their corresponding EIA and MIL standards. Contact type designation is per IPC CM-770, "Guidelines for Printed Board Component Mounting." See also Fig. 8.4.

While most sockets are an assembly consisting of individual contacts enclosed in a metal or plastic housing and placed as a single discrete component, some contact types are available mounted on a carrier. The carrier holds the contacts in a standard pattern that is suitable for later insertion of an electrical or electronic component. The carrier-type contacts are placed and soldered on the board, and the carrier is then removed, resulting in individual contacts present on the board in some standard configuration. For some low-lead-count components with compliant leads, individual contacts can be placed and soldered onto plated-through holes without the use of a carrier.

Three classes of sockets are available: standard insertion force, low insertion force, and zero insertion force (ZIF). The low insertion force socket is the standard socket, equipped with a fixed spring element at each contact location. Pressure is required to seat the component into the socket or remove the component from the socket. The insertion/removal pressure is directly related to the single contact force times the number of contacts. This limits the maximum number of contacts for low insertion force

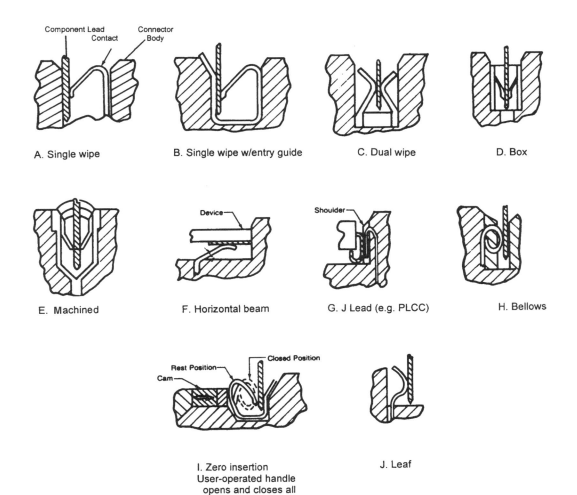

FIGURE 8.4 Contact types for sockets. Adapted from IPC CM-770, "Guidelines for Printed Circuit Board Component Mounting.

TABLE 8.2 Socket Types

Socket Type	Abbreviation	Contact Type	Standard
Discrete transistor	TO-5	E	MIL-S-83502
Power transistor	TO-3	F	MIL-S-12833
Dual in-line (may be 0.3, 0.4, 0.6, or 0.9 inch pin grid width)	DIP	A, B, C, D, E, I	EIA RS-415
Single in-line	SIP	A, B, C, D, E, J	EIA RS-415
Pin grid array	PGA	E	none
Leaded chip carrier	LCC	F, G	EIA RS-428
Plastic leaded chip carrier	PLCC	G	EIA RS-428
Leadless ceramic chip carrier	LCCC	G, special	EIA RS-428
Relay sockets		D, E	EIA RS-488
Crystal sockets	HC	D, E	EIA RS-192
Lamp and LED sockets		Lamp = J LED = E	EIA RS-488
Special sockets		C, D, E	EIA RS 488

sockets. Literature reports indicate that removal of some large ICs has resulted in damage to the IC package, in some cases rendering the IC unusable.

Zero insertion force sockets have a cam-and-lever assembly that allows the contact force to be removed when the component is inserted or removed from the socket. After insertion, the cam assembly is operated to bring the contacts against the component leads with appropriate force. Prior to removal, the cam assembly is operated to remove the force (Fig. 8.5). An area must be left clear around a ZIF socket to allow operation of the cam and arm assembly.

8.6.2 Contact Types

As in connectors, ideally, the force of the contacts against the component terminations will result in a contact that is gas tight. The contact types are as shown in Fig. 8.5.

Contact spring materials are either tin, tin-lead, or gold plated. Tin and tin-lead are less expensive than gold but will oxidize faster. The continued expansion and contraction of contacts and leads due to thermal changes will allow the formation of oxides. Additionally, there is a documented phenomenon in which oxides will grow within the contact area due to the presence of small electrical currents. Gold is very resistant to both types of oxidation.

8.6.3 Socket Selection

Among the items to consider when selecting a socket are:

- Appropriate for the component leads
- Through hole or surface mount
- Handling techniques
- Appropriate for the soldering technology to be used
- Cleaning technology to be used
- Contact/retention force, individual and overall
- Contact material, gold or tin-lead
- Contact type, A through J per IPC CM-770
- Thermal demands on the socket
- Body materials
- Body profile, where high- and low-profile bodies are available
- Mechanical polarization requirements
- Solder land or hole design

Appropriate for the component leads means consideration of the number of leads and the style of component leads. The component supplier may or may not have recommendations on sockets.

FIGURE 8.5 Zero-insertion force (ZIF) sockets (left, open; right, closed).

The decision to use THT or SMT sockets depends largely on the soldering technology to be used on the board. THT components may be mounted in either THT or SMT sockets, and SMT components likewise may use either, limited solely by availability. This in one way to convert technologies. For example, a DIP part may be mounted in a socket which is designed for reflow soldering. If SMT sockets are chosen, the board designer should pay close attention to IPC-C-406, "Design and Application Guidelines for Surface Mount Connectors."

Handling techniques for sockets are similar to those for components. Handling must not allow finger oils on the contacts. Solderability of the socket leads must be maintained. Storage must not allow damage to either the socket leads or the contacts.

Appropriate for the soldering technology means consideration of whether the socket is expected to be reflow soldered, wave soldered, or hand soldered. Wave soldering will require contact technology that will minimize solder wicking from the socket leads to the contacts. Reflow soldering will require that the body materials of the socket be suitable for reflow temperatures. Wave or hand soldering will require THT sockets, where reflow soldering will require SMT sockets, unless paste-in-hole (see Chapter 2) soldering techniques are used. The user must remember that, when using paste-in-hole techniques, while the socket will be THT, the body of the socket must be suitable for reflow temperatures. Like all devices, verifying and maintaining the solderability of the socket leads is important to the overall quality of the final board.

During reflow soldering, the user must consider that the amount of metal present in the combined leads and contacts will sink more heat than a standard component of the same lead count. A longer dwell time in pre-heat or soak areas of the profile may be required.

Cleaning the board after soldering must be considered. If no-clean flux is used, and the board will not be cleaned (not automatically true), then cleaning is not an issue, but flux residue may be. If the board is to be cleaned, then these issues should be considered:

- Cleaning fluids must be compatible with the metals and the socket housing materials.
- The socket may need to be sealed.
- An temporary contact insert may be needed to prevent cleaning fluids from contaminating the contact surfaces.
- A minimum standoff may be required above the board surface to ensure complete cleaning underneath the socket.
- No cleaning residue may be left on the contacts.

Contact force is individual contact force, whereas retention force is total force on the component. This force must be considered for removal operations, especially with components that have terminations on all four sides. This is important to minimize the risk of component damage in situations where it is expected that a removed component can be tested or inspected, or where the component may be used again. Frequent removal/replacement operations at the same socket, or large pin counts, may require the use of zero-insertion force sockets, with their accompanying larger area needed on the board.

Contact material selection affects both socket cost and contact longevity. Tin and tin-lead plated contacts are less expensive than gold but will oxidize much faster. Their oxides are also more resistant to the wiping action of the contacts against the component leads.

Contact type determination depends on the available types for the component to be socketed. Frequently, the socket manufacturer uses a specific type contact for each socket variety made. Like sockets themselves, contacts vary in their cost, so that the less expensive contacts are used in less expensive sockets, while more expensive contacts are used in more expensive sockets. Therefore, the contact type may not be optional after one has selected the socket usage, the plating material, and the body material. In general, leaf-type contacts (A, B, F, G, J) are less expensive than multiple-wipe contacts (C, D, E). The variation in contact types will lead to different retention forces, and user who anticipates high-vibration environments for socketed assemblies should consult with the socket manufacturer about retention forces.

Interconnects

Assembly forces range from 5 to 15 lb, depending primarily on the number of contacts and the retention force.

Like any component, thermal demands on sockets occur in three areas:

- Reflow soldering
- Use during burn-in tests
- Expected environment for the final product

These concerns reflect themselves primarily in the body material selection, and in the contact material selection, recognizing that tin and tin-lead will oxide much faster than gold, especially at elevated temperatures.

Body materials are determined primarily by the manufacturer, and the only option open to the buyer is the temperature rating. Typically, either thermoplastic and thermoset plastics are used.

Body profile depends on two clearances. The clearance above the board and the cleaning clearance below the socket body. Like ICs themselves, low-profile sockets are harder to clean under and should lead the user to consider no-clean flux. Some sockets are designed with standoffs to allow for thorough cleaning after soldering. The other issue with the body is the size of the opening for the insertion of the component leads, or its *lead-in*.

The user's assembly requirements may dictate that an indexing/polarizing leg or boss be used on the socket to ensure correct assembly orientation. Many sockets, however, are symmetrical, with only visual markings (dot, corner angle, tab, etc.) to indicate orientation. It is important that the board be marked with orientation information and that the controlling documentation thoroughly identify the visual markings and pin number identification of the socket and the component.

Regardless of the soldering technology to be used, sockets will have both insertion and removal forces placed on them during their lifetime. Their solder fillets will experience far more stress and strain than fillets on components directly soldered to the board. Therefore, board layout should ensure that:

- SMT land areas are as large as possible.
- PTH diameter should match the diagonal measurement of the pins, ensuring that maximum contact is made with the hole.
- Users with high retentive-force sockets should consider mechanical retention to the board, whether SMT or THT.

8.6.4 Examples of Specific Socket Types

Chip carrier sockets are designed for ceramic chip carriers (CCCs) and may be THT or SMT. The pitch of the socket leads will be the same as the pitch of the CCC leads/terminations. The hold-down is necessary to maintain constant contact with the component leads. THT sockets are available for CCCs. For large termination-count CCCs, the force necessary to hold the part down may distort the connector body, placing undue force on the solder joints. To prevent this, sockets are available that use four corner screws to transfer the forces from the hold down to a backup plate.

PLCC sockets do not require hold-downs, and in SMT designs they match the land pattern of the PLCC itself. Users must remember that this is one type of component for which damage to the component can occur during the component removal process. Special tools are used to remove PLCCs from their sockets. These tools insert into two corners of the socket, with hooks to fit under two corners of the PLCC, allowing the user to pull it up. It is not uncommon for the PLCC body to break during the removal process if care it not taken. See the examples of PLCC sockets under "SMT-mount sockets," below.

PGA sockets match the pin configurations of the components. Due to the number of pins, zero-insertion force (ZIF) sockets are commonly used for PGAs. Like other component types, ZIF PGA sockets are commonly used for test and burn-in applications.

Area array (e.g., *BGA*) *sockets* mate with the pad on the bottom of the array or with the individual balls on the BGA. Like the CCC socket, a hold-down is required to keep the part in place. Each pad location has a leaf-type contact and a corresponding SMT lead on the board. BGAs also have adapters that allow the BGA to be soldered to them. The adapter then has pin-grid array (PGA) terminations on the bottom, allowing the BGA to be plugged into a standard PGA socket.

SMT-mount sockets generally have their leads under the body of the socket, making visual alignment and inspection difficult. For this reason, there should be visual alignment markers on the board.

In the PLCC sockets shown in Fig. 8.6, note that the leads under the left socket are not visible from the top of the socket. As the bottom-side view on the right shows, they are hidden under the body of the socket. This makes post-solder visual or laser inspection very difficult, although x-ray will work.

DIP sockets (Fig. 8.7a) are used for the insertion of DIP ICs. Their bottom terminals may attach to the board using THT, SMT, or they may terminate in wire-wrap terminals (Fig. 8.7b), with the terminals designed to press-fit into proper sized holes in the board. See the further description of wire-wrap terminals earlier in the chapter.

8.7 Fiber-Optic Interconnects and Connections

Adapted from Englebert, J., Hassett S., Summers, T.; "Optical fiber and cable for telecommunications," in Whitaker, J.C., *The Electronics Handbook*, ECR/IEEE Press, Boca Raton, 1997.

8.7.1 Basic Principle of Operation

Every optical fiber used for telecommunications consists of three sections: core, cladding, and protective coating. The core is the area where the light is most concentrated. It consists of solid silica glass and a dopant (typically, germania) to change the index of refraction of the glass. The cladding concentrically surrounds the core. It is solid silica glass with no (or at least different) dopants from the core. The core and the cladding are manufactured as a single piece of glass and cannot be separated from one another. The refractive index characteristics of the core and the cladding are designed to contain the light in the core as the light travels down the fiber. The core has a higher index of refraction than the cladding.

FIGURE 8.6 PLCC sockets.

(a)

(b)

FIGURE 8.7 Examples of (a) high and low DIP sockets for through-hole soldering and (b) wire-wrap sockets.

The standard core diameters in use today are 8.3 μm (single-mode), 50 μm (multimode), and 62.5 μm (multimode). (Note: single-mode and multimode fiber will be discussed in more detail later in this chapter.) The cladding surrounding each of these cores is 125 μm. Core sizes of 85 and 100 μm have been used in early applications but are not typically used today.

The third section of an optical fiber is the outer protective coating. The typical diameter of a coated fiber is 245 μm. The protective coating is typically an ultraviolet (UV) light cured acrylate coating applied during the fiber manufacturing process. The coating provides physical and environmental protection for the fiber. Figure 8.8 provides a cross section of an optical fiber.

To understand how light travels down a fiber, it is necessary to understand the principles of how light behaves in various media. In a vacuum, light travels at approximately 3.0×10^8 m/s. In other media (air, water, glass, etc.), the light travels at some slower speed. The ratio of the speed of light in a vacuum to the speed of light in another medium is called the *index refraction* for that medium. It is typically represented by the letter n and is always greater than or equal to one. The larger the index of refraction, the slower the light travels in that medium.

When light passes from one medium to another medium that has a different index of refraction, the light is either reflected back into the first medium at the interface or it is refracted as it passes into the second medium. See Fig. 8.9. Behavior of the light depends on the angle of incidence and the indices of refraction of the two materials. As the light passes from one medium to another, the refraction of the light obeys Snell's law:

$$n_1 \sin \theta_1 = n_2 \sin \theta_2$$

Based on this relationship, as the angle of incidence θ_1 increases, θ_2 approaches 90°. When $\theta_2 = 90°$, θ_1 is called the *critical angle*. For angles of incidence greater than the critical angle, the light is reflected back into the first medium. The angle of reflection is equal to the angle of incidence.

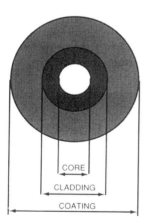

FIGURE 8.8 Cross section of a typical optical fiber.

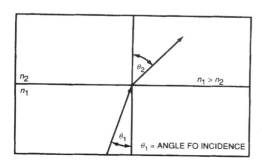

FIGURE 8.9 Refraction light.

To ensure that the light reflects and stays in the core, the light must enter the core through an imaginary acceptance cone that makes an angle θ_{NA} relative to the axis of the fiber. The size of the acceptance cone is a function of the refractive index difference between the core and the cladding. The acceptance cone is not specified for a typical optical fiber; instead, a numerical aperture (NA) is specified and is related to the acceptance cone by the following:

$$NA = \sin \theta_{NA} = \sqrt{n_1^2 - n_2^2}$$

The numerical aperture of an optical fiber is a measure of the maximum angle of the light entering the end of the fiber that will propagate in the core of the fiber, as shown in Fig. 8.10. The propagation of light down the core is called *total internal reflection*.

After light enters an optical fiber, it travels down the fiber in stable propagation states called *modes*. There can be from one to hundreds of modes depending on the type of fiber (see single-mode and multimode fiber). Each mode carries a portion of the light from the incoming signal. The approximate number of modes N_m is expressed by the following relationship:

$$N_m \approx \frac{\left(D \times NA \times \frac{\pi}{\lambda}\right)^2}{2}$$

where D = core diameter, µm
 NA = numerical aperture
 λ = wavelength of operation, µm

As one can see, the larger the core diameter and numerical aperture, and the shorter the wavelength, the greater the number of modes in an optical fiber. This gives a simplified reason for the different data rate capacities for different sizes of multimode fibers. The greater the number of modes, the greater the modal dispersion among those modes and, subsequently, the lower the data rate capacity.

The wavelength of operation can also affect the number of modes and, subsequently, the data rate capacity. Multimode fibers generally operate at one of two primary wavelengths: 850 nm and 1300 nm. Since the wavelength of operation affects the data rate capacity, higher data rate systems generally operate at the longer wavelength. Single-mode fibers generally operate at one of the primary wavelengths of 1310 and 1550 nm. For single-mode fibers, the previous equation on the number of modes does not hold true since at both wavelengths there is only one mode. Data rate capacities in single-mode tiber will be discussed later. As a note, all four of these wavelengths (850, 1300, 1310, and 1550 nm) are invisible to the naked eye and are in the near-infrared region of the electromagnetic spectrum.

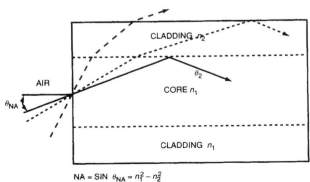

FIGURE 8.10 Numerical aperture (NA) properties.

Attenuation

Attenuation is the loss of optical power as the light propagates in the optical fiber. Mathematically, attenuation is expressed as

$$\text{Attenuation (dB)} = 10 \log\left(\frac{P_{in}}{P_{out}}\right)$$

By convention, attenuation is reported as a positive value.

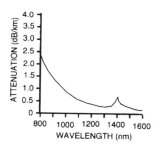

FIGURE 8.11 Spectral attenuation of a typical single-mode fiber.

As stated, the operating wavelengths of choice for single-mode fiber are 1310 and 1550 nm. The typical attenuation of a single-mode fiber as a function of wavelength is shown in Fig. 8.11. A typical spectral attenuation curve for multimode fiber has a similar shape but greater attenuation.

To be useful to the system designer, optical fiber attenuation is usually normalized per unit length. The attenuation over a specific length of fiber, determined in the preceding equation, is divided by the length of the fiber under measurement. The resultant value is called the *attenuation coefficient*, and is typically reported as decibels per kilometer. Optical fiber attenuation is caused by sources internal to the fiber and also by external sources. Fiber attenuation that is derived from internal sources is referred to as *intrinsic attenuation*. Fiber attenuation derived from external sources is referred to as *extrinsic attenuation*.

Intrinsic Attenuation

Two primary mechanisms contribute to intrinsic attenuation: Rayleigh scattering and absorption.

Rayleigh Scattering

Rayleigh scattering is the predominant loss mechanism in optical fiber. Light propagates in the core of the optical fiber and interacts with the atoms in the glass. The light waves elastically collide with the atoms, and light is scattered as a result. Rayleigh scattering is the result of these elastic collisions between the light wave and the atoms in the fiber. If the scattered light maintains an angle that supports continued forward propagation within the core, then no attenuation occurs. If the light is scattered at an angle that does not support continued propagation in the desired direction, then the light is diverted out of the core and attenuation occurs.

A portion of the scattered light is directed back toward the light source. As with the forward signal, the scattering angle must be sufficient to support wave propagation in the core back to the source. This Rayleigh scattered light in the backward direction can be used by an optical time domain reflectometer (OTDR) to analyze the attenuation of fibers and the optical loss associated with localized events in the fiber, such as splices. Rayleigh scattering accounts for about 96% of the intrinsic loss in the fiber.

Absorption

The optical fiber's core consists of very pure silicon dioxide with dopants to change the index of refraction. Pure glass is relatively transparent to light at the near-infrared wavelengths. However, some small quantity of impurities can enter the fiber during manufacturing. Hydroxide bonds can also exist within the fiber in minute quantities. The hydroxide bonds and other impurities will absorb light at various wavelengths in the near-infrared wavelength range. The light is absorbed by the impurities and is converted to vibrational energy or some other form of energy. The optical signal is attenuated in the process. Absorption due to the hydroxide bonds is noticeable near 1383 nm as revealed by the large increase in attenuation at that wavelength. See Fig. 8.11. Unlike Rayleigh scattering, absorption can be limited by controlling the amount of impurities in the glass.

Extrinsic Attenuation

Two external mechanisms contribute to the loss of optical power: *macrobending* and *microbending*. Both of the mechanisms cause a reduction of optical power. The two effects can be differentiated by the wavelengths of light that are attenuated.

Macrobending

If a bend is imposed on an optical fiber, strain is placed on the optical fiber along the region that is bent. The bending strain will affect the refractive index in that specific area and the angle of incidence of the light at the core/cladding interface. As a result, light propagating in the core can refract out since the critical angle may be exceeded. Such a bend, effected on a large scale, is referred to as a macrobend.

For a single-mode fiber, macrobend performance is linked to the mode field diameter of a fiber. (See single-mode fiber.) A fiber with a larger mode field diameter will be more susceptible to macrobending attenuation than a fiber with a smaller mode field diameter. Likewise, a specific fiber's susceptibility to macrobending depends on the wavelength of light in the fiber. A longer wavelength, for example, 1550 nm, will result in a larger mode field diameter. Therefore, longer wavelengths will be more susceptible to macrobending losses. A macrobend induced loss may severely attenuate a 1550-nm signal without affecting a 1310-nm signal. For multimode fiber, macrobending losses are independent of wavelength.

Microbending

Like macrobending, microbending will cause a reduction of optical power due to geometric changes in the core and very localized changes in the index of refraction due to strain in the glass. However, microbending is very localized, and the bend may not be clearly visible on inspection. The localized strain might be caused by localized contact of the fiber with a sharp, rigid object or an inconsistent stress applied around the fiber.

Microbending attenuation affects all optical wavelengths. One wavelength will not be much more susceptible to increased loss as is the case with macrobending in single-mode fibers. Mechanical damage to the cable, due to excessive crushing or impact forces, could result in microbending induced attenuation.

Data Rate Limitations

Data rates are limited by *dispersion*. Dispersion causes a light pulse to spread in time as it travels through a fiber. This spreading increases the potential that sequential pulses may overlap. Pulses launched close together (high data rates) that spread too much (high dispersion) result in errors. For example, if the input pulse is 10 ns long, the output pulse might be 11 ns long after 2 km.

There are three types of dispersion. Modal *(intermodal)* dispersion and chromatic *(intramodal)* dispersion are the two most common and have the most impact on typical telecommunications systems. Chromatic dispersion occurs in both single-mode and multimode fiber, whereas modal dispersion occurs only in multimode fiber. The third type of dispersion is *polarization mode dispersion*.

Modal Dispersion

The various modes of light in a multimode fiber follow different paths through the core of the fiber. These individual paths have different optical lengths. The modes following the shortest paths will arrive at the end of the fiber before the modes following the longer paths. A pulse of light, which may fill hundreds of modes in a multimode fiber, will therefore broaden in time as it travels through the fiber. This type of dispersion is called modal dispersion.

Chromatic Dispersion

Chromatic dispersion is determined by the fiber's material composition, structure, and design, and by the light source's operating wavelength and spectral width. Chromatic dispersion is measured in units of picoseconds of light pulse spread per nanometer of laser spectral width per kilometer of fiber length (ps/nm·km). Chromatic dispersion occurs because different wavelengths of light travel at different speeds. There are two components of chromatic dispersion: material and waveguide dispersions. The amount of material dispersion varies with the composition of the glass and is a function of the wavelength of the light source. No transmitter produces a pure light source of only one wavelength. Instead, light sources produce a range of wavelengths distributed around a center wavelength. This range is called the *spectral width*. See Fig. 8.12. Since each portion of this wavelength range travels at a different speed in glass, material dispersion results. This dispersion increases proportionally with distance.

In the case of single-mode fiber, waveguide dispersion results from a portion of the light traveling in the cladding as well as the core of the fiber depending on the index of refraction profile of the core. The fraction of the light in the cladding increases as the wavelength increases, so that the effective refractive index is a function of wavelength. Because each pulse consists of more than one wavelength, the pulse spreads out. This dispersion also increases proportionally with distance. The effect of chromatic dispersion can be minimized by using a better light source (smaller spectral width) with a center wavelength closer to the zero dispersion wavelength. See Fig. 8.13.

Modal dispersion is the dominant factor in multimode fiber but does not occur in single-mode fiber, where only one mode propagates. Chromatic dispersion is present in both fiber types. Parameters are specified for single-mode fiber in most industry standards to aid in calculating the data rate capacity for the fiber over a given length. Industry standards for multimode fiber, however, specify bandwidth instead of dispersion.

Bandwidth is a measure of the modal dispersion of multimode fiber. It is typically measured by the fiber manufacturer on the bare fiber and reported in normalized units of megahertz·kilometer. Because bandwidth is not linear, special care must be taken when estimating the data rate capacity for an installed system, since a number of fibers may be concatenated together to compose the system. Bandwidths of fibers that are concatenated together forming a long span cannot simply be added together then divided by the length to get the bandwidth of the system. The formulas for calculating system bandwidth are beyond the scope of this chapter but can be found in the references. Typically, the end equipment manufacturer should state the fiber bandwidth necessary far proper operation of a system over a given

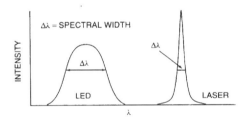

FIGURE 8.12 Spectral width of LED and laser sources.

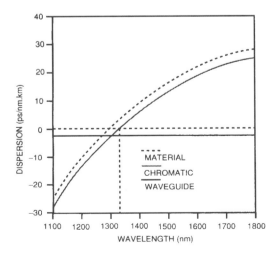

FIGURE 8.13 Dispersion as a function of wavelength.

length. Standards such as fiber distributed data interface (FDDI), asynchronous transfer mode (ATM), token ring, ethernet, fibre channel, etc., have set maximum distances to ensure proper operation over optical fiber.

Polarization Mode Dispersion (PMD)

Polarization mode dispersion is an optical effect that can spread or disperse an optical signal. Single-mode fiber is more sensitive to PMD than multimode fiber. PMD typically originates with narrow spectral width laser diode sources, which polarize the emitted light. The light is then coupled into the single-mode fiber where it is resolved into two orthogonally polarized components making up the fundamental mode. PMD occurs when the two polarized components begin to propagate at different speeds. This difference in propagation velocity represents a difference in the index of refraction between the two components and is referred to as the fiber's *birefringence*. The result is a distortion of the originally transmitted signal. The difference in propagation velocity between the two components is also wavelength dependent since the refractive index is a function of wavelength. Since both of the polarized components carry a portion of the transmitted power, any offset in the arrival time of the two components acts to disperse the original signal. See Fig. 8.14. PMD may limit the operational transmission capacity of digital systems and cause composite second-order distortions in analog systems. Additionally, it is important to note that PMD increases as the square root of the length, whereas chromatic dispersion increases linearly with length. Test methods and specifications for PMD are still under development.

Fiber Types

There are different fibers for a range of applications. They can be broken down into two categories: single-mode and multimode. Both fiber types act as the transmission media of light, but they operate in different ways, have different characteristics, and serve different applications. Single-mode fibers are used in applications where low attenuation and high data rates are required, such as on long spans where repeater spacing needs to be maximized. System cost, operating wavelength, and transmission speed also play a role in choosing between transmission media. Multimode fiber is better suited for shorter distance applications. Where costly transceivers are heavily concentrated, such as in a local area network (LAN), the primary cost of the system does not lie with the cable. In this case, multimode fiber is more economical because multimode fiber can be used with inexpensive connectors and light-emitting diode (LED) transmitters making the total system cost lower. This makes multimode fiber the ideal choice for a LAN where low attenuation characteristics are not a concern.

Multimode fiber can be broken down into two categories: *step index* and *graded index*. The step index fiber is manufactured with a definitive material demarcation between the cladding and the core. The core has an index of refraction that is higher than the index of refraction of the cladding. In other words, there is an abrupt change in the index of refraction at the interface between the core and the cladding. See Fig. 8.15. The typical core size for a step index multimode fiber starts at 100 μm. Such fibers are rarely used in telecommunication systems.

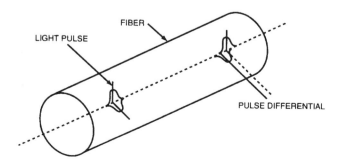

FIGURE 8.14 Polarization mode dispersion.

Interconnects

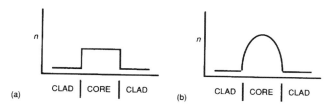

FIGURE 8.15 Multimode fiber characteristics: (a) step index multimode fiber, (b) graded index multimode fiber.

The graded index profile is more complicated. The index of refraction is constant across the cladding. Beginning at the core/cladding interface, the index of refraction gradually increases reaching a maximum at the center of the core and approximately forming a parabolic shape, as shown in Fig. 8.15. The typical core sizes of graded index multimode fibers used in telecommunication systems are 50 and 62.5 μm.

Graded index multimode fiber was designed to mitigate modal dispersion. In a step index fiber, all of the modes travel at the same speed. Because of the difference in physical path lengths, significant modal dispersion occurs. In a graded index fiber, the speed of the modes varies depending on the location of the mode in the core. Near the outside of the core (lower index of refraction), the mode travels faster than at the middle of the core (higher index of refraction). The effect is to better equalize the propagation times of the modes and reduce the modal dispersion.

As explained before, single-mode fiber allows only one mode of transmission, as long as the transmission wavelength is above a certain wavelength called the *cutoff wavelength*. The core is designed such that only one mode will travel down the fiber when operating above the cutoff wavelength. If the fibers were operating at a wavelength less than the cutoff wavelength value, then there would be multimode transmission. The fiber is designed so that the cutoff wavelength is the minimum wavelength which guarantees the propagation of one mode only.

This one mode travels down the fiber core and has a diameter associated with its power density. The power density is Gaussian shaped and extends into the cladding such that part of the light travels in the cladding near the core. The diameter of the power distribution in the fiber is called the *mode field diameter* (MFD). See Fig. 8.16. The MFD is dependent on the wavelength and has a strong influence on the bend sensitivity of a fiber.

Table 8.3 summarizes some of the parameters for multimode and single-mode fiber (typical values are shown).

The lowest dispersion for a conventional single-mode fiber is found near 1310 nm. At other wavelengths, the dispersion is significantly greater. However, the spectral attenuation curve shows that the lowest attenuation is near 1550 nm. To have both the lowest dispersion and the lowest attenuation at the same wavelength, dispersion-shifted fiber was developed with the zero dispersion point shifted to the 1550-nm region. The index of refraction profile of this type of fiber is sometimes referred to as the W profile because of its segmented shape. See Fig. 8.17. Dispersion-shifted single-mode fiber is typically used in long-distance applications where very low attenuation and dispersion are necessary.

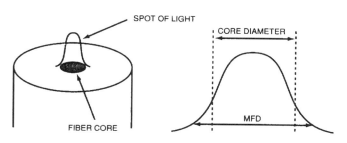

FIGURE 8.16 Mode field diameter (MFD).

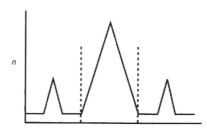

FIGURE 8.17 Dispersion shifted single-mode fiber index of refraction profile.

TABLE 8.3 Parameters for Multimode and Single-Mode Fiber

Parameter	Multimode (62.5/125 μm)	Dispersion-Unshifted Single-Mode
Core size	62.5 ± 3.0 μm	8.3 μm (typical)
Mode field diameter	—	8.8–9.3 μm (typical)
Numerical aperture	0.275 ± 0.015	0.13 (typical)
Cladding diameter	125.0 ± 2.0 μm	125.0 ± 1.0 μm
Attenuation	≤3.5 dB/km at 850 nm	≤0.5 dB/km at 1310 nm
	≤1.5 dB/km at 1300 nm	≤0.4 dB/km at 1550 nm
Bandwidth	≥160 MHz·km at 850 nm	—
	≥500 MHz·km at 1300 nm	
Dispersion	—	≤3.5 ps/nm·km for 1285–1330 nm

8.7.2 Special Use Fibers

Polarization Maintaining Fiber

Polarization maintaining fiber (PMF) is designed far use in fiber optic sensor applications, interferometric sensors, polarization sensitive components, and high data rate communication systems. In PMF, the light maintains a constant state of polarization as it travels down the fiber. One method of maintaining the polarization state is to manufacture a fiber core that is elliptical in shape. Another method is to purposely manufacture a fiber with inherent stresses on the fiber core. This creates a high birefringence leading to the polarization maintaining properties. Operational wavelengths of PMF range from 630 to 1550) nm.

Radiation-Hardened Fiber

When single-mode fibers are exposed to radiation, the attenuation increase due to radiation depends on the dopants and other materials in the fiber. The ionization from the radiation creates trapped corners, which create absorption bands at the most common operating wavelengths, 1310 and 1550 nm. The creation of the absorption bands at these wavelengths means the attenuation at these wavelengths increases. The radiation sensitivity of a fiber depends on fiber composition and the carrier life, which depends on dopant concentration. Some applications call for fibers to operate in the presence of radiation. A radiation-hardened fiber is one that initially has some radiation insensitivity but also recovers quickly from any radiation-induced effect. Fibers with a high concentration of germania as a dopant have been found to operate and recover fairly well in a radiation environment. It is believed that the high concentration of the dopant inhibits carrier formation within the glass. Some dopants, such as phosphorous, are more susceptible to radiation effects.

Fiber Nonlinearities

The information carrying capacity of an optical system is normally limited by two factors: attenuation and dispersion/bandwidth. Systems that use higher launch power to overcome attenuation limitations, and/or are optimized to minimize dispersion, are finding nonlinear dispersion effects to be the limiting factor.

In recent years, the output powers of transmitters used in optical fiber systems has dramatically increased. The use of erbium doped fiber amplifiers (EDFA) and Nd:YAG lasers has made output powers in excess of 18 dBm (60 mW) a reality. These higher optical powers lead to unique interactions between the optical fiber and the electromagnetic wave traversing the fiber.

The nonlinear effects between the light and the optical fiber are classified into two groups: *scattering effects* and *refractive index effects*. This section briefly addresses two scattering effects: stimulated Raman scattering and stimulated Brillouin scattering. This section also discusses two refractive index effects *(self-phase modulation/cross-phase modulation)* and four wave mixing.

Factors Affecting Nonlinear Behavior

Several factors influence nonlinear fiber behavior in optical fibers. The effective core area and the effective length of the system both influence the nonlinear behavior of the fiber. The core area determines the optical intensity in the core. The effective length, which depends on both the fiber length and the attenuation coefficient of the fiber, determines the length of interaction. A smaller effective core area and longer interaction length (longer fiber length and lower attenuation coefficient) make nonlinear fiber behavior more severe.

Stimulated Raman Scattering (SRS)

Stimulated Raman scattering is an interaction between the vibrations of the atoms in the glass and the optical signal. Light propagating in the fiber core can excite atoms to higher energy vibrational states. The energy removed from the light signal is scattered by the excited atoms as a light wave. The light can be scattered in the forward or reverse propagating direction. It is different from Rayleigh scattering in that the scattered wavelength may be different (shorter or longer) from the wavelength of the original signal. In general, the longer wavelength that is forward scattered is of concern in an optical fiber system. The scattered light represents a new optical wave in the core, and may interfere with other optical signals in the fiber that share the same wavelength.

Signal degradation for single-channel systems occurs at very high powers, typically in the 1 W range. The range of wavelengths of scattered light extends over a range of 120 nm. In optical fibers that carry multiple optical channels [wavelength division multiplexed (WDM) systems], optical channels operating at shorter wavelengths may provide gain to channels operating at longer wavelengths. For example, a 1310-nm signal might be scattered to provide gain to a signal over the wavelength range of 1310–1430 nm or longer. As a result, shorter wavelength optical channels experiencing SRS can interfere with channels operating at longer wavelengths. The optical power required for degradation of longer wavelength channels is reduced as the number of channels carried by the fiber increases. In general, SRS is not a significant concern in current optical fiber systems.

Stimulated Brillouin Scattering (SBS)

Stimulated Brillouin scattering is an interaction between acoustic waves and the optical signal in the fiber. Light can interact with electrons in the glass, resulting in the scattering of light and the generation of acoustic phonons. The interaction creates a backward propagating optical wave. Optical power from the forward propagating wave is diverted to a backward propagating wave. Above a specific threshold power, the optical power in the forward propagating wave is limited, and all additional power is scattered to the backward propagating wave. This effect can reduce the clarity of the received optical signal.

SBS may be a concern in some specific optical fiber systems. It can be observed at powers as low as several milliwatts, depending on the transmitter characteristics. SBS is aggravated by narrow spectral width transmitters, such as those used in externally modulated optical systems. Directly modulated systems are at less risk of SBS due to laser chirp and its concomitant broadened spectral width.

Several fiber related parameters can inhibit SBS or increase the threshold power at which SBS can occur. Variations in the speed of the acoustic wave over the length of a fiber can increase the power threshold to SBS. Parameters that affect the speed of sound in glass include fiber strain and changes in core dopant concentration.

Refractive Index Effects

The refractive index of glass is a function of both the wavelength of the light and the optical power density of the light propagating in the material. The effect of optical power on refractive index is typically very small, but current optical sources can produce output powers that are sufficient to cause a significant change in the refractive index. The following nonlinear effects are based on the change in refractive index with a variation of optical power.

Self-Phase Modulation (SPM)

At typical low powers, or with minimal power variations (analog systems with a small modulation index), the change in refractive index with optical power is very small. On systems that employ a large power variation, such as amplitude shift keyed (ASK) digital systems, the optical power can vary widely within a single pulse. As the optical power increases at the edge of the pulse, the refractive index increases. Likewise, as the optical power decreases at the trailing edge of the pulse, the refractive index decreases. The effect on the pulse is to produce a *chirp*, a variation of the wavelength dependent on power. Light at the leading edge of the pulse experiences a lengthening of the wavelength (red shift). Light at the trailing end of the pulse experiences a shortening of wavelength (blue shift). Except in special cases such as *solitons*, the net effect is the broadening of the signal pulse.

Self-phase modulation generates an effect similar to chromatic dispersion. The additional dispersion can limit the operating speed of a system. For example, a standard single-mode fiber, designed for operation at 1310 nm, might be limited in operating speed at 1550 nm due to self-phase modulation. The effect is more severe if the fiber is operated away from the zero dispersion wavelength of the fiber.

The soliton makes the phenomenon of self-phase modulation desirable. A soliton is a wave that can propagate an indefinite distance without suffering any dispersion. The wave is only attenuated. An optical soliton can be created by coupling the effect of self-phase modulation and positive dispersion. At the proper power, the combination of self-phase modulation and anomalous dispersion essentially creates a dispersionless optical pulse. The optical soliton will continue to be a popular area of study as optical systems strive far higher operating speeds over longer distances.

Cross-Phase Modulation

Two or more optical channels may be transmitted on a single fiber. The optical power of one channel can affect the local refractive index experienced by other optical signals at that location. Since each optical signal in the fiber has a different wavelength and therefore has a different refractive index, each signal will travel through the fiber at a different speed. Therefore, a pulse will be broadened as it begins to overlap a second, high power pulse, which is also broadened at the same location. As the pulse passes through the second pulse, they are recompressed to restore it close to its initial shape.

Four Wave Mixing

Four wave mixing is the creation of new optical wavelengths at the expense of the original two or more wavelengths in a single optical fiber. At least two discrete optical wavelengths, representing two optical channels in a wavelength division multiplexed system or two axial modes of a laser transmitter, will modulate the refractive index of the optical fiber. As a result, two modulated indices of refraction in the glass result, and the phase matching between the waves produces beating. The beating of the modulated index of refraction creates additional wavelengths at the expense of the power in the initial wavelengths. If more than two wavelengths are carried by the fiber, still more mixing products can be created.

The operational issues caused by four wave mixing are manifested by two mechanisms. First, there is a reduction in power of the operational channels since the resultant mixing channels acquire power at the expense of the operational channels. Second, the mixing product channels can exist at wavelengths that currently carry optical signal channels. Interference between the two channels can further reduce the power in the optical signal channel.

The current method available to defeat or reduce the risk of our wave mixing is to use a property of the optical fiber itself. Since four wave mixing is a phase matching effect, system operation away from the zero dispersion wavelength will ensure that a small amount of dispersion can be introduced to

Interconnects

minimize the phase matching. Additionally, the spacing of the optical channels can be increased to minimize the interaction length of the channels. For example, two optical channels could be operated at 1540 and 1560 nm on dispersion-shifted fiber with limited risk of four wave mixing.

8.7.3 Fiber Manufacturing

There are many ways to manufacture an optical fiber; this chapter discusses the most common methods. Three methods that have proven economical on a large scale are *outside vapor disposition* (OVD), *vapor axial deposition* (VAD), and *inside vapor disposition* (IVD). All three processes involve the deposition of soot produced when volatile feed materials are delivered in gaseous form to a target reaction area in the presence of oxygen. The deposited soot is then *sintered,* or consolidated, into a preform of high-purity glass.

The (OVD) process uses a glass mandrel, or seed rod. The rod is placed in a lathe, and the feed materials are delivered to the region near the rod surface by a gas burner. The flame source and the feed materials traverse longitudinally along the rod as the lathe turns. See Fig. 8.18. Layers of a porous glass soot are built up, and the chemical composition of each layer can be controlled. Once enough layers been deposited, the seed rod is removed, and the entire cylindrical soot tube is heated, collapsing the hole formed by the vacated seed rod, forming a solid, transparent preform.

The VAD process is similar to the OVD process except that the soot is deposited on the endface of a silica seed rod mounted vertically. The porous cylindrical soot rod created in this way is drawn upward. See Fig. 8.19. This type of process allows for several burners to be used in the deposition of the glass soot. Again, as in the OVD process, the seed rod is removed and the consolidation process forms a preform.

The IVD process is a departure from the previous two processes because instead of a seed rod, a seed tube is used. The tube is placed horizontally in a lathe and the gas burner moves along outside of the tube surface. The feed materials flow into the tube and the presinter soot is formed on the inside of the seed tube building inward toward the center of the tube. See Fig. 8.20. Typically, the seed tube is the outer part of the cladding glass. Concentrations of the feed materials flowing into the tube can be changed to alter the properties of the final product. The consolidation process forms a solid, high-purity preform ready for the drawing process.

In the drawing process, the preform, or *blank,* as it is also called, is mounted vertically in a draw tower and heated to about 2000° C at the bottom end. As the preform softens, the bottom end, or gob, begins to pull away leaving behind a thin strand of fiber, as illustrated in Fig. 8.21. The speed control, tension, and temperature of the draw are critical to obtaining consistent fiber dimensions. Vibration, particulate, and mechanical damage to the glass surface must be avoided to guarantee a high-strength glass. A coating is applied to the fiber during this process. Typically, the coating is a UV cured acrylate, which protects the glass surface to preserve its inherent strength. It also protects the fiber from microbending and simplifies handling of the fiber.

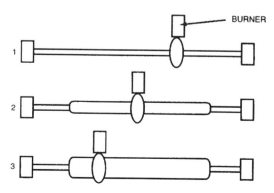

FIGURE 8.18 Outside vapor deposition laydown.

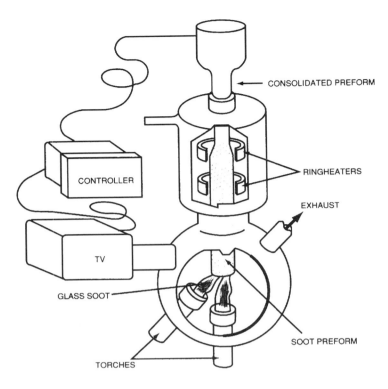

FIGURE 8.19 Vapor axial deposition process.

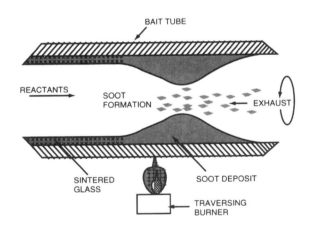

FIGURE 8.20 Inside vapor deposition laydown.

Variations in the feed materials during the laydown process determine what type of fiber will be manufactured. Changes in the dopant concentration in the core section and the size of the core determine whether the fiber will be intended for multimode or single-mode applications. Additional material can be added to the outside of the cladding to enhance certain properties of the fiber. For example, carbon can be deposited on the outside of the cladding to hermetically seal the surface of the glass.

The carbon layer in a hermetic fiber impedes the migration of hydrogen generated during undersea applications from penetrating into the fiber core, thus potentially affecting the attenuation characteristics. Titanium oxide can also be deposited to increase the fibers fatigue resistance parameter, which is related to the fiber's longevity.

Interconnects 8-29

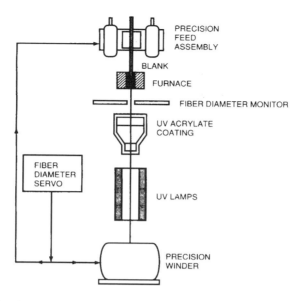

FIGURE 8.21 The fiber drawing process.

8.7.4 Optical Cable Design

The purpose of a cable design is to package the transmission medium, in this case the optical fiber, so that it can withstand environmental and mechanical stresses before, during, and after deployment. Optical cables must meet stringent performance levels to ensure reliable service. These performance levels are determined by the telecommunications industry such that the cable can survive in the intended environment.

As stated before, the advantages of optical fiber cable are many: light weight, small diameter, excellent transmission characteristics, and an enormous information carrying capacity over relatively long distances. Another advantage is the fact that a cable can be dielectric, which almost eliminates lightning and ground faults as an outside plant concern.

Cable Design Considerations

The specifics of any particular cable design depend on the application, but there are a few generalities. The purpose of any cable structure is to protect the optical fiber in a given application. Different applications expose the cable to different mechanical and environmental conditions, potentially causing stress on the cable. These stresses come from installation conditions, temperature fluctuations, handling, and so forth. Another purpose, of the cable structure is to package the fibers in groups to ease system administration and termination.

Even though each cable may be subjected to slightly different conditions in the field, optical cable is designed to meet certain requirements as established by the telecommunications industry. To facilitate testing to these requirements, standard *fiber optic test procedures* (FOTPs) were developed by the Electronic Industries Alliance and the Telecommunications Industry Association (EIA/TIA). These test methods simulate the stresses the cable might experience during any given cable deployment.

A common denominator for all cable designs is identification/color coding of fibers and fiber units. The industry standard for color coding is EIA/TIA-598-A, "Color Coding of Fiber Optic Cables." This document outlines the standard color coding and identification of the fiber units. Fibers are bundled in units to allow easier administration) and termination in the field. Considering that there could be more than 300 fibers in a cable, the installers have be able to distinguish the fibers from one another. The 12 colors specified are listed in Table 8.4.

TABLE 8.4 Standard Optical Cable Color Coding

Unit or Fiber	Color	Unit or Fiber	Color
1	Blue	7	Red
2	Orange	8	Black
3	Green	9	Yellow
4	Brown	10	Violet
5	Slate	11	Rose
6	White	12	Aqua

The purpose of the coloring is to provide unique fiber identification. The coloring does not enhance the mechanical properties of the fiber, and it must be compatible with fusion splicing techniques. One splicing technique involves using a light injection and detection device. This device operates by injecting light into a fiber through its coloring and coating. The light is then detected where the fiber is bent around a small mandrel. Basically, this device uses the principle of macrobending to inject and detect light in the fiber to optimize splice loss. If a coating of ink is too thick or opaque, however, the device still be ineffective.

Fibers usually are colored by one of two processes. The polyvinyl chloride-(PVC)-based ink process thermally dries the ink. A volatile solvent acts as a carrier far the ink pigment and is driven off by heat. The UV-based ink process applies a thin layer of ink on the fiber and cures the ink through exposure to an ultraviolet lamp.

The 12 colors of the individual fibers alone do not provide a totally unique identifier, so *units*, defined as fiber bundles, buffer tubes, or ribbons, are also color coded or printed to aid in field identification. The same colors apply to the units. For units 13–24, the colors repeat, but a colored dash is used to distinguish the fiber bundles. For any count over 24, a double dash is used. Ribbons offer a unique template for identification, since the surface of a ribbon allows alphanumeric printing. For ribbon units, a position number and/or the associated color can be printed on the ribbon.

Stranded Loose Tube Cable

A typical stranded loose tube cable is illustrated in Fig. 8.22. There are several major components used to enlist met the cable. Each component has a defined function.

Central Member

The central member provides the rigidity to resist contraction of the cable at low temperatures (sometimes called *antibuckling*). It also provides a structure around which to strand buffer tubes, fillers, and/or a limited number of twisted copper pairs. The central member may add to the overall tensile strength of the cable, and in some designs, can provide all of the tensile strength for the cable. The central member maybe either steel or glass reinforced plastic (GRP), depending on the customer's requirements. GRP central members are used when the customer wants an all-dielectric cable (no metallic components). If necessary, the central member may be overcoated with plastic so that the overcoated outer diameter will

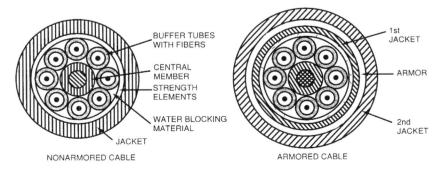

FIGURE 8.22 Stranded loose tube cable.

allow tubes to be stranded into a round core. Coating of a steel central member also inhibits corrosion of the steel.

Buffer Tube

The buffer tube mechanically decouples the fiber from the cable and the stresses applied to the cable. The buffer tube contains the fibers and a filling compound. The fibers freely float inside the buffer tubes to ensure that the fibers experience no strain when the cable undergoes environmental and mechanical stress. The filling compound inside the buffer tube can be a mineral oil-based gel. The purpose of the gel is to allow the fibers to float in the tube, decoupling the fibers from the tube itself and any mechanical stress the cable may experience. Field concerns with respect to the gel are that it must not drip from the buffer tubes at elevated temperatures. It must also be compatible with the other cable components, as well as termination hardware/closure components. The gel must also be dermatologically safe, nonnutritive to fungus, and easily removed. An additional purpose of the gel is to prevent the ingress of water or moisture into the tube, which upon freezing could cause microbending or mechanical damage.

The tubes provide mechanical protection for the fibers and are color coded for identification. Tube material should be selected to be flexible, yet have sufficient hoop strength to provide mechanical resistance to crush forces during installation and operation. Some buffer tubes have two layers with polycarbonate forming the first layer and polyester forming the outer layer. Other tubes maybe a single layer of polypropylene or polyester. The buffer tube and filling compound work in such a way as to allow the fibers to move and readjust to compensate for cable expansion and contraction. Since the buffer tubes are plastic, they expand and contract with temperature changes; therefore, a stiffening element in the cable, such as a central member, may be needed to mitigate contraction at cold temperatures.

Stranding

The buffer tubes are stranded around a central member to provide the necessary operating window far the fibers. The cable's operating window allows for contraction at low temperatures and elongation due to tensile loads without affecting the fibers. As the cable contracts at low temperatures, the excess fiber length (fiber length relative to tube length) increases. As the cable elongates at high tensile loads, the excess fiber length decreases. The excess fiber length inside the buffer tubes and the proper lay length of the stranded buffer tubes ensure the fibers remain in a stress-free state throughout the operating range of the cable. The stranded buffer tubes maybe held together by a binder. The minimum number of buffer tubes or fillers needed far a round cable is typically five.

Water Blocking Compound

A water blocking agent is applied to the cable core (stranded buffer tubes) to prevent water migration along the cable to prevent potential crushing forces due to freezing water. The water blocking compound can be either a water swellable powder, tape, or yarn, or it could be a water blocking gel. The water blocking gel is flooded into the stranded core of buffer tubes. The compound is usually a petroleum-based microcrystalline gel used as a physical barrier to prevent water penetration into the cable core. Water swellable *super absorbent polymers* (SAPs) can be used in the gel or in the form of a powder held within tapes or yarns. They serve the same purpose as the gel in the prevention of moisture ingress into the cable, but their operating principle is different. Whereas the gel acts as a physical barrier, the SAPs are water activated and swell inside the cable core. The swelling then acts as a blocking agent to prevent further water migration. Whether the blocking compound is a gel or a SAP in some form, the purpose is to prevent water migration down the cable core if the cable's sheath integrity is compromised.

Tensile Strength Elements

For those designs that do not rely solely on the central member for its tensile strength, high tensile strength yarns are typically helically wrapped around the cable core before a sheath is placed on the cable. Limits are established by the manufacturer as to the allowable stretch in the cable components before attenuation or reliability becomes an issue. The typical strength elements used in cable manufacture today are fiberglass yarns, aramid yarns, or a mixture of both. The yarns may be part of a composite material held together by an epoxy or a plastic, but the application of such components is still to wrap them around the cable core. Other designs include embedding the strength elements in the cable jacket.

These elements could be various sizes of steel wire or GRP rods. Embedding the strength elements in the sheath is typically employed in central tube cable manufacture and will be discussed later.

Jacket Material

The jacket is the first line of defense for the fibers against any mechanical and chemical influences on the cable. Several jacket material options could be selected. The best material depends on the application. The most commonly used cable jacket material for outdoor cables is polyethylene (PF).

Armoring

Steel armoring provides additional mechanical strength and resistance to rodents. Other armoring options will be discussed later in the armoring section.

Specialty Loose Tube Cables

There are some loose tube cables in operation, which are industry specialties. One such type of specialty cable is called an *aircore* cable. This is an unflooded stranded loose tube cable for indoor use and usually has a unique jacketing material such as polyvinyl chloride (PVC). This design is discussed in the premises cable section.

A second specialty application is a gas pressurized cable. These cables are similar to an aircore cable but are usually jacketed with PF. This allows for unrestricted outdoor use but the cable termination is very specialized. These cables are spliced in such a way that the cables can be pressurized to prevent water ingress. Additionally, the pressurization allows the customer to detect damage to the cable system by a resultant drop in pressure. These systems are somewhat costly to maintain and have limited use by today's customers.

Single-Tube Cable

The single, or central tube cable design is characterized by a single buffer tube in the center of the cable. Tensile and antibuckling strength elements may be helically wrapped around the buffer tube or placed longitudinally down the cable inside the outer jacket. Other design options include using the armor, if present, for antibuckling. The buffer tube is much the same as for a stranded loose tube cable except that the tube size is much larger. The single-core tube cable can be manufactured with single fibers, bundled fibers, or fiber ribbons. The largest bundled fiber cable could have as many as 96 fibers. The fibers are colored following the standard color coding schemes already discussed, but the units are groups of 12 fibers bound by color coded threads. A buffer tube is then extruded around all of the fiber units. An alternative fiber unit is the ribbon. Fibers are arranged in parallel in the standardized color code order, then bonded together by a UV cured acrylate or some other plastic material to form a multifiber ribbon. The number of fibers in a ribbon typically ranges from 2 to 24. The advantages of the ribbon are that it aids in the packing density of the cable, but more importantly, it aids in ease of installation in the field. The ribbons allow mass fusion and mass mechanical splicing. Printed identification on the ribbons makes installation easier since the installer does not have to depend on color coding, which can be lost or misinterpreted. The position or color is printed right on the ribbon coating allowing for easy identification. Typical single-tube cables are shown in Fig. 8.23.

Another ribbon cable design is the stranded ribbon cable. Buffer tube units are manufactured with ribbons inside, and these units are in turn stranded around a central element to form a cable core. The result is a stranded loose tube ribbon cable. The only difference between this cable and the standard loose tube cable is that the buffered units are larger and, instead of loose fibers in the tubes, there are ribbon stacks within the tubes. This design allows for a greater fiber packing density.

Slotted Rod

Another cable design that employs ribbons is the slotted rod design, sometimes called a *star design*. The considerations of this particular design are total system cost and fiber density. The slotted rod typically consists of a polyethylene or polypropylene form extruded over a central member. The PE form contains slots for the placement of ribbons. The rod is extruded so that the slots from helix lengthwise down the cable. The operating window is determined by the lay length of the slots and the slot depth, minus the

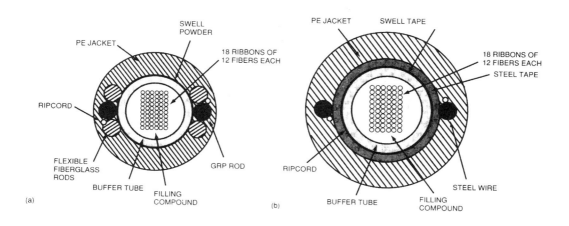

FIGURE 8.23 Types of cable: (a) all-electric single-tube cable, and (b) armored single-tube cable.

ribbon stack height. Care must be taken in choosing the lay of the slotted rod to allow for adequate cable bend performance without compromising the ribbon stack integrity, putting strain on the fibers, or causing attenuation problems. In a typical slotted rod construction, the central member acts as the primary tensile loading element in the cable. Ribbons used in slotted rod designs range from a 2-fiber ribbon to a 12-fiber ribbon. Buffer tubes can also be placed in the slots. The cable's fiber counts can range from 12 fibers to 1000 fibers. Black PE stripes in the slotted rod itself are typically used for slot identification. Figure 8.24 is a diagram of a typical slotted rod cable.

Sheath Options

Once the cable core (stranded tubes, single-tube, or slotted rod) is assembled, the cable then goes through a sheathing process. For cable designs that require a flooded cable, a flooding compound is applied to the cable core before jacketing. The purpose of the sheath is to provide mechanical, environmental, and chemical protection for the cable core. There are two basic sheath constructions: nonarmored and armored. Nonarmored sheaths have only a single jacket, whereas a typical armored cable has a combination of jacket and armor. Armoring provides additional mechanical strength and rodent resistance when required. Most armored cables are for direct buried applications.

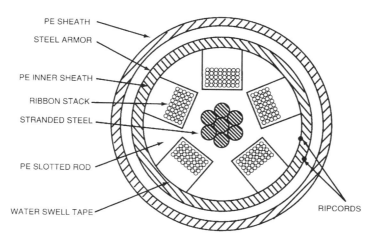

FIGURE 8.24 360-fiber slotted rod cable, 12-fiber ribbon—single armor.

Standard jacket materials for nonarmored cable are medium-density and high-density polyethylene. Other materials available are linear low-density polyethylene (LLDPE), polyvinyl chloride (PVC), various fluoropolymers, polyamide (nylon), flame-retardant polyethylene (FRPE), and polyurethane. In applications involving exposure to adverse environmental conditions, such as high temperatures, fire, or corrosive chemicals, the environment needs to be taken into account before the correct sheath option can be determined.

Although there are several colors of polyethylene, colors other than black will not provide the same level of protection against UV radiation. Additionally, color retention of all nonblack colors may decrease with UV exposure. Black is strongly recommended for use in cables deployed outside.

FRPE was developed to achieve flame retardance in a material without halogens (e.g., fluorine, chlorine, or bromine). The halogens retard the burning of the material but, if they are burned, they may emit corrosive gases, which could potentially cause equipment damage and produce life threatening substances. Instead, various combinations of olefinic polymers (such as polyethylene) with hydrated minerals such as alumina trihydrate and magnesium hydroxide are used. The hydrated minerals give off water at elevated temperatures and act to retard burning.

Other types of PE materials are used for their electrical properties. It was found that some specialized materials have significantly improved resistance to tracking in the presence of strong electrical fields. They are called *track resistance polyethylene* (TRPE). Tracking is an electrical phenomenon caused by water and impurities in the air that coat the surface of a cable. This condition could lead to localized surface conduction if the cable is placed in the presence of strong electric fields. Tracking causes grooves in the cable jacket, which eventually deepen until they wear through the jacket, destroying the cable's integrity. The track resistance of some TRPE materials is due to their lower electrical resistivity, which reduces localized electrical conduction and the formation of grooves on the surface of the PE. Tracking is normally not an issue except for cables installed near electrical systems above 115 kV; each high voltage application should be evaluated on a case-by-case basis.

The primary use of PVC is in an *air core*, or unfilled cable design. This air core design and PVC sheath provides flame retardancy. PVC sheaths should not be used with petroleum-based flooding compounds, because the flooding compounds draw out the PVC plasticizers. This may lead to embrittlement of the PVC over time. There may be other limitations in temperature performance ranges of PVC jacketed cables.

The primary use of fluoropolymers is for high-temperature environments or for chemical resistance. Polyamides (nylon) are primarily used as a gasoline, kerosene, oil, and jet fuel barrier. Although it is not as good as a fluoropolymer jacket, it is not as costly. Nylon has also been used as a sheathing option for resistance to termites, but its effectiveness is questioned by some.

Polyurethane used to be the standard jacket for some cable manufacturers. It was also used as a flame-retardant sheath option and had excellent low-temperature flexibility. Today, it is used only in specialty applications.

Armoring Options

Armoring of a cable has proven to provide additional compressive strength, and aids in cable cut through resistance. Most manufacturers use a steel tape to provide armoring protection for their cables. The tape itself may be an electrolytic chrome coated steel and both sides are coated with an ethylene acrylic acid copolymer, which inhibits corrosion of the steel tape. The tape width is selected for a particular cable core size to meet a minimum tape overlap determined by the manufacturer's process. The tape is processed through a corrugator before it is formed around the cable core. The corrugations allow for better flexibility of the completed cable. Depending on the manufacturing process, the armor overlap may be sealed by bonding the copolymer coating at the overlap with pressure, heat, and/or glue.

Stainless-steel tape is another armoring option. Stainless steel is more corrosion resistant than the steel tape discussed previously, and there is a perception that it provides more strength. However, it is more expensive, and standard steel armor is considered to be as effective for strength and corrosion resistance.

Interconnects

Copper cladded steel tape is claimed to offer better lightning protection, but industry experts do not agree on its effectiveness. Copper tape is used as a moisture barrier in some cable designs, or as a high-temperature/steam protection option.

Another moisture barrier armoring option is the laminated aluminum polyethylene (LAP) sheath option. LAP is a noncorrugated aluminum tape. Typically, the tape overlap is sealed to form a moisture barrier. The sealing occurs with pressure and heat, which causes the tape laminate to bond together at the overlap. Because of the cost, processability, and ease of sealing, the LAP option is sometimes preferred for cables that are used in pressurized cable systems.

A dielectric version of armoring is common in the European aerial cable market for ballistic protection. The *aramid armored* cable contains an annular ring of aramid yarns wound around the cable enclosed in an outer jacket. The yarns may provide some protection against shotgun blasts. This application can also be applied to buried cables where the yarns, combined with a large cable diameter, may provide some rodent resistance. The effectiveness of aramid armor against shotgun blasts or rodents is questionable.

Dual Layer (or High-Density) Loose Tube Cable

Typically, stranded loose tube cables consist of a single layer of buffer tubes stranded around a central member. As an extension of this design a second layer of buffer tubes can be stranded around an inner layer of buffer tubes. This inner layer usually has 5 to 9 stranded tubes; the outer layer typically has 11 to 15 tubes, allowing fiber counts up to 288 fibers. See Fig. 8.25.

Self-Supporting Aerial Cables

One type of self-supporting aerial cable is the *figure-8* cable. Standard aerial cables require lashing to an existing aerial support messenger in the field. The figure-8 cable eliminates the need for lashing the cable to an independently installed messenger. The cable and the messenger can be installed in a single step. The key element in the figure-8 cable is the messenger. The asphalt flooded stranded steel messenger chosen for figure-8 processing should meet the ASTM A640 standard. To simplify the sag and tension issues, the messenger can be designed to be the limiting factor, not the optical fiber cable core. Many figure-8 cables are designed to have zero fiber strain up to 60% of the minimum rated breaking strength of the messenger when installed. [The maximum load allowed on a messenger by the National Electric Safety Code (NESC) is 60%.]

The jacketing material chosen for figure-8 cables is also a consideration far the cable designer. The materials need to be able to stand up to the rigors of the outdoor environment. UV radiation, wind and ice loading, creep, and *galloping* are just some of the considerations that a designer should keep in mind. PE is typically the material of choice. Figure 8.26 is a cross-sectional drawing of the figure-8 cable design.

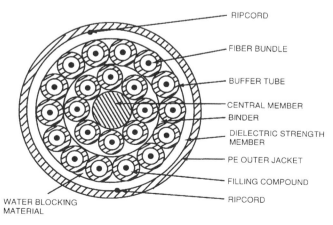

FIGURE 8.25 288-fiber stranded loose tube cable.

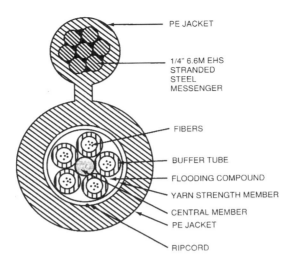

FIGURE 8.26 Typical construction of aerial cable.

Given the right circumstances, the cross section of a figure-8 cable could act as an airfoil, leading to galloping. Galloping is a high-amplitude, low-frequency vibration resulting from the wind acting on the installed cable. This mostly affects the termination and anchoring pole hardware.

Concentric Self-Supporting Cable

Another aerial cable is the concentric self-supporting cable. Because of its all-dielectric construction, these cables are generally used near high-voltage systems or in areas prone to lots of lightning strikes. The concentric design allows for an annular ring of strength elements within a round cable. Typically, the strength materials are aramid yarns. This cable still needs to meet the same requirements as a figure-8 cable (60% of the rated breaking strength) when installed. This design lends itself to being all-dielectric with the elimination of any steel elements in the cable. Concentric self-supporting cables are susceptible to galloping and acolia, vibrations. Acolian vibration is a low-amplitude, high-frequency vibration induced by steady winds.

Cables for Extremely Cold Environments

The proper choice of materials and design parameters allows for very cold temperature performance. Linear low density polyethylene is a common choice far a cable jacket material down to –60° C, although medium-density polyethylene will also perform given the correct contraction window of the cable.

Tapered Cable

Telecommunications customers desire an inexpensive option to loose tube cable to facilitate fiber drop points. Some outside plant systems have service areas such that, at specific locations along the route (which could be one long cable), fibers are dropped off. After the drop point, those particular fibers are unused. This represents a large cost to the customer. The solution is a loose tube cable with predetermined dropoff points. At each dropoff point, the buffer tubes are replaced by filler rods or an empty buffer tube. The tapered fiber cable design is the result, and this minimizes the fiber splicing. Fibers within a buffer tube are accessed at specific taper points along the cable route. The remaining fibers and buffer tubes are left undisturbed and pass through to the next splice point.

Applications/Considerations of Outdoor Cables

The outside plant cable is the backbone of the optical fiber system. The outside plant cable can connect main processing and switching centers with remote centers hundreds of kilometers away. Alternatively, the outside plant cable may distribute services from a remote location to a home. In each case, the cable is likely to be exposed to the rigors of the outdoors. The specific type of outside plant cable employed in an installation may depend on the fiber count to be used as well as the cable diameter, cost, mechanical performance characteristics, and other issues.

In general, outside plant cables are categorized as either all-dielectric (duct) cables or armored cables. Armored cables are used when the cable may be subject to attack by rodents, such as gophers and squirrels, or when the cable will be direct buried in rocky soil. An armored cable may also be used in outside plant applications where additional compressive resistance is required. Although the armored cable has greater compressive strength and offers resistance to gnawing rodents, the metallic armoring will require bonding and grounding.

All-dielectric (duct) cables are used in applications that are not subject to rodent attack and will not be direct buried in rocky soil. For example, a duct cable may be used when the cable will be placed in a conduit. All-dielectric cables may also be preferred in high lightning and high-voltage areas due to the absence of conductive materials in the cable.

Outdoor Installations

Aerial applications offer unique considerations. Domestically, the NESC sets guidelines limiting the strain on the supporting messenger or strength elements to 60% of their rated breaking strength. Weight of the cable is a concern for sag and tension calculations. The weight of the cable and environmental conditions determine the maximum pole spacing for the aerial system. As stated in the self-supporting cable section, some cables can act as airfoils simply from the cable's profile. If the cable acts as an airfoil, several scenarios could develop. Given a strong enough wind, the cable could experience additional tensile loading from the wind pushing on the cable. The wind could also subject the cable to galloping and aeolian vibration. Although these two phenomena may not affect the cable's optical performance to a great degree, it is a consideration for fatigue of the pole termination hardware. The cable's profile could also affect the ice loading a cable might experience. Ice formation or freezing rain accumulation will also add to the tensile forces the cable experiences. Aerial cables are also susceptible to lightning strikes because of their installation height. Dielectric cables minimize the lightning risk, but metallic termination hardware could take a hit and still affect the aerial system at that point. Lastly, aerial cables are also an inviting target for bored hunters who would like to try their marksmanship. Some self-supporting aerial designs provide a sacrificial aramid yarn layer around the cable core. This is not an armoring, but it may provide some protection from bird shot.

Direct buried applications might seem a safer way to deploy a cable, but even this application has its challenges. For cables that are directly plowed into the ground, crush is a concern when the trenches are backfilled. Ground heaves during the winter months test the tensile strength of the cables, and rodents looking for food and water find a buried cable to be an inviting opportunity. Soil conditions with regard to chemical analysis as well as microbe content may have adverse effects on the cable's sheath material. Ground strikes from lightning have two affects on buried cable. An indirect lightning strike may be close enough to the cable to cause pinholes in the jacket of an armored cable. This in turn may cause sheath-to-ground faults, which are usually of no practical consequence. The second effect would be a direct hit to the cable. Grounding, strength of hit, type of cable, soil conditions, and so forth dictate whether the cable would survive.

Inner duct installations might have the fewest worries of all. As long as the cable's minimum bend radius and tensile strength limit are not exceeded, there are few worries when installing the cable or after it is installed. Concerns about jacket material compatibility with the pulling lubricant should be answered before installation. The fill ratio is the ratio of the cable(s) cross-sectional area to the cross-sectional area of the inner duct. Typical installations limit the fill ratio to 50%, but exceeding 50% may not be a concern as long as the tensile strength limit and minimum bend radius limitations are observed. Prior planning before any duct installation to determine the pulling plan, distances, mid-assist points, and so on, should be performed to ensure the pull is well executed.

Ease of mid-span entry into the cable is a concern far future dropoff points in the system routing. Some designs, like SZ stranded loose tube cables, lend themselves to easy mid-span access. Other designs may tend to be more difficult.

Premises Cables

Outside plant optical fiber cables transport digital or analog information tens to hundreds of kilometers through harsh environments. The optical signals are useful only if they can be connected to sensitive optoelectronic hardware that can process and route the information to the end user. Optoelectronic hardware is typically installed in an indoor, climate-controlled environment that is isolated from the severe conditions of the outside plant. The premises indoor environment presents different performance requirements for the optical fiber cables. As a result, optical fiber cables used in the premises environment must be designed to meet different criteria.

Although premises optical fiber cables are installed indoors, they must be mechanically rugged and demonstrate excellent optical performance over a broad temperature range. Inside plant cables must be easy to terminate and connect to end equipment. The cables must be flexible to afford easy routing in cable trays, ladder racks, and cable organizational hardware. The cables should be available in high fiber counts to allow simple connections between the high fiber count of the outside plant trunk cable and the end equipment. The cables should have small diameters to maximize the use of indoor space. Finally, the cables must also be flame resistant and suitable for indoor use, as indicated by an Underwriter's Laboratory (UL) or equivalent listing.

Before discussing the design and materials used in the cable, an overview of the applications of premises cables is required.

National Electrical Code

The National Electrical Code (NEC) is a document issued every three years in the U.S. by the National Fire Protection Association. The document specifies the requirements for premises wire and cable installations to minimize the risk and spread of fire in commercial buildings and individual dwellings. Section 770 of the NEC defines the requirements for optical fiber cables. Although the NEC is advisory in nature, its content is generally adopted by local building authorities.

The NEC categorizes indoor spaces into two general areas: plenums and risers. A plenum is an indoor space that handles air as a primary function. Examples of plenum areas are fan rooms and air ducts. A riser is an area that passes from one floor of a building to another floor. Elevator shafts and conduit that pass from one floor to another floor are examples of risers. Indoor areas that are not classified as plenums or risers are also governed by the NEC but are not specifically named.

These different installation locations (plenums and risers) require different degrees of minimum flame resistance. Since plenum areas provide a renewable source of oxygen and distribute environmental air, the flame resistance requirements for cables installed in plenum areas are the most stringent. Likewise, riser cables must demonstrate the ability to retard the vertical spread of fire from floor to floor. Other indoor cables (not plenum or riser) must meet the least demanding fire resistance standards.

Optical fiber cables that can be installed in plenums and contain no conductive elements are listed as Type OFNP. OF refers to *optical fiber*, N refers to *nonconductive*, and P refers to the *plenum* listing. These cables are generally referred to as plenum cables. Likewise, optical fiber cables that can be installed in riser applications and contain no conductive elements are listed as Type OFNR. The first three letters maintain the same meaning as the preceding example with the plenum cable, and the R refers to the riser listing. These cables are generally referred to as riser cables. Optical fiber cables that can be installed indoors in nonplenum and nonriser applications and contain no conductive elements are listed as Type OFN. These cables maybe referred to as general purpose cables and meet the least demanding of the fire resistance standards.

Premises cables must be subjected to and pass standardized flame tests to acquire these listings. The tests are progressively more demanding as the application becomes more demanding. Therefore, plenum cables must pass the most stringent requirements, whereas general purpose cables must meet less demanding criteria.

To obtain a Type OFN listing, the cable must pass a standardized flame test, such as the UL 1581 flame test. The UL 1581 test specifies the maximum burn length the cable can experience to pass the test. Type OFN cables can be used in general purpose indoor applications.

Similar to the Type OFN listing is the Type OFN-LS listing. To obtain this listing, the cable must pass the UL. 1685 test. This test is similar to the UL 1581 test, but the UL 1685 test includes a smoke-generation test. The cable must also produce limited smoke as defined in the standardized test to achieve the listing. A cable with a Type OFN-LS listing is still restricted to general purpose indoor applications only.

To obtain a Type OFNR listing, the cable must pass a standardized flame test, such as the UL 1666 flame test. The UL 1666 test contains specifications for allowable flame propagation and heat generation. The test is more stringent than the UL 1581 test. Type OFNR cables can be used in riser and general purpose applications.

To obtain a Type OFNP listing, the cable must pass a standardized flame test, such as the NFPA 262-1990 (UL 910) flame test. The NFPA 262 test contains a specification for allowable flame propagation and a specification for the average and peak smoke that can be produced during the test. The plenum test is the most severe of all flame tests. Type OFNP cables can be used in plenum, riser, and general purpose applications.

Section 770 of the NEC contains a variety of other requirements on the placement and use of optical fiber cables in premises environments. The section also contains several exceptions that permit the use of unlisted cables under specific conditions. The NEC and the local building code should always be consulted during the design of the premises cable plant.

Cable Types

Two general cable design types are used in premises applications: tight buffered cables and loose tube cables. The loose tube premises cable is similar to those discussed in the section on outside plant cable. The tight buffered cable is a unique design developed primarily to meet the requirements of premises applications.

Tight Buffered Cables

Tight buffered cables have several beneficial characteristics that make them well suited for premises applications. Tight buffered cables are generally small, lightweight, and flexible in comparison to outside plant cables. As a result, they are relatively easy to install and easy to terminate with optical fiber connectors. Tight buffered cables are capable of passing the most stringent flame and smoke generation tests. Light buffered cables are normally listed as Type OFNP or Type OFNR.

Loose Tube Premises Cables

Loose tube cables offer high packing densities ideal for intrabuilding backbone applications and transition splicing to high fiber count outside plant cables. The cables demonstrate superior environmental and mechanical performance characteristics. Some loose tube premise cables are designed far use in both outside plant and premises applications. Variations of the loose tube cable that are halogen free or that possess water-blocking characteristics are also available.

Tight Buffered Premises Cables: General Construction and Overview

Tight buffered cables were the first generation of premises cables. The name *tight buffered* is derived from the layer of thermoplastic or elastomeric material that is tightly applied over the fiber coating. This method contrasts sharply with the loose tube design cable, in which 250-µm fibers are loosely contained in an oversized buffer tube.

A 250- or 500-µm coated optical fiber is overjacketed with a thermoplastic or elastomeric material, such as PVC, polyamide (nylon), or polyester, to an outer nominal diameter of approximately 900-µm. All tight buffered cables will contain these upjacketed fillers. The 900 µm coating on the fiber makes it easier to terminate with an optical fiber connector. The additional material also makes the fiber easier to handle due to its larger size.

The 900-µm tight buffered fiber is the fundamental building block of the tight buffered cable. Several different tight buffered cable designs can be manufactured. The design of the cable depends on the specific application and desires of the user.

In one cable design, a number of 900-μm optical fibers are stranded around a central member, a serving of strength member (such as aramid yarn) is applied to the core, and a jacket is extruded around the core to form the cable. See Fig. 8.27. As an alternative, the fibers in the cable can be allocated into smaller groups, called subunits. Each subunit contains the same number, of 900-μm in tight buffered fibers, a strength member, and a subunit jacket. Each of these individually jacketed groups of fibers (subunits) are stranded around another central member and the composite cable is enclosed in a common outer jacket. See Fig. 8.28.

Lower fiber count cables are necessary to connect the optical end equipment to the optical fiber distribution system. They can also be used to bring fiber to a workstation. Tight buffered interconnect cables with 900-μm fibers are again well suited far this application. One or two 900-μm tight buffered fibers are independently jacketed with a serving of strength member to connect a tight buffered interconnect cable commonly called jumpers. See Figs. 8.29 and 8.30. To applications requiring higher fiber counts, a number of these individually jacketed fibers can be stranded around a central member to construct a fan-out cable (Fig. 8.31).

Loose Tube Premises Cables—General Construction and Overview

The components and construction of the loose tube design cable are similar to those used in outside plant cables. Differences in materials are discussed in the Material Selection section of this chapter. Loose tube cable design and construction are discussed in the Outside Plant Cable section.

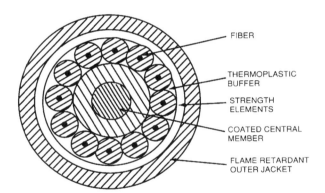

FIGURE 8.27 12-fiber tight buffered cable.

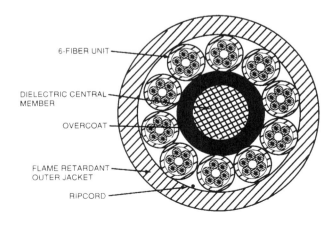

FIGURE 8.28 A 60-fiber unitized tight buffered cable.

Interconnects

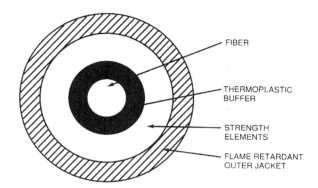

FIGURE 8.29 Single-fiber interconnect cable.

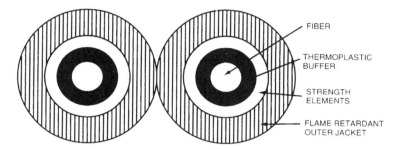

FIGURE 8.30 A two-fiber zipcord table.

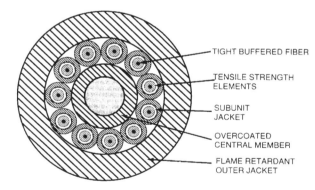

FIGURE 8.31 Tight buffered fan-out cable.

Material Selection

Materials are chosen for premises cables based on flame resistance, mechanical performance, chemical content chemical resistance, and cost, among other factors. The selection of material is based on the application for which the cable is designed.

Tight Buffered Cables

900-μm Tight Buffer Material

The tight buffer material can consist of PVC polyester, polyamide (nylon), or other polymers. The buffer material must provide stable performance over the operating temperature range of the cable. The buffering material must be compatible with other components in the cable. The material should also enhance the flame performance of the completed cable. For example, a plenum cable may use a tight buffer material with a greater flame resistance than the tight buffer material of a general purpose or riser-rated cable.

The tight buffer material and fiber coating must be easy to remove for termination. In some cases, only the tight buffer material needs to be removed with the fiber coating remaining intact. This would be useful in splicing operations with light injection and detection (LID) systems. Some options are available to reduce the composite coating strip force or allow only the thermoplastic buffer to be removed. One option includes the use of a slip layer, applied to the surface of the coating prior to tight buffering. This reduces the force required to remove the buffering material. A second option involves the placement of aramid yarn between the fiber coating and the tight buffer material. The former method reduces the strip force required to remove the composite coating prior to termination. The latter allows the thermoplastic buffer to be removed while maintaining the fiber coating intact.

Strength Member

Tensile strength is provided by the incorporation of high modulus strength elements. These strength elements could be aramid yarns; fiberglass yarns; or rigid, fiberglass or aramid reinforced plastic rods. Aramid or fiberglass yarns serve primarily as strength components. Reinforced plastic rods can provide antibucking protection for the cable, if required to enhance environmental performance. They can also provide tensile strength.

Jacket Material

The cable jacket can be constructed from PVC, fluoropolymers, polyurethane, FRPE, or other polymers. The specific jacket material used for a cable will depend on the application for which the cable is designed. Standard, indoor cables for riser or general purpose applications may use PVC for its rugged performance and cost effectiveness. Indoor cables designed far plenum applications may use fluoropolymers or filled PVCs due to the stringent flame resistance and low smoke requirements. Filled PVCs contain chemical flame and smoke inhibitors. They provide a more flexible cable that is generally more cost effective than fluoropolymer jacketed cables. Polyurethane is used on cables that will be subjected to extensive and unusually harsh handling but do not require superior flame resistance. FRPE provides a flame-resistant cable that is devoid of halogens. Halogen-free cables and cable jackets do not produce corrosive, halogen acid gases when burned.

Loose Tube Premises Cables

The materials used in loose tube premises cable construction are similar to that of the outside plant loose tube cable but with a few significant differences. First, premises cables do not employ a gel filling in the core interstitial areas. These gel fillings are generally hydrocarbon based, which degrade the flame resistance of the cables. The cables may employ noncombustible water-blocking agents, such as water absorbent tapes and yarns, if water blocking is desired. The cables will often employ a flame resistant tape to enhance the flame resistance characteristics of the cable. Finally, the cable jacket will be constructed from a flame resistant material, typically a PVC. Flame resistant polyethylene can also be used to provide a cable jacket that is halogen free, flame resistant, and also burns with limited smoke generation.

Applications of Cables

Premises cables are generally deployed in one of three areas: backbone, horizontal, and interconnect. Higher fiber count tight buffered cables can be used as intrabuilding backbones that connect a main cross connect or data center to intermediate cross connects and telecommunications closets. Likewise, lower fiber count cables can link an intermediate crisis connect to a horizontal cross connect to multiple workstations.

Interconnect cables patch optical signals from the optical end equipment to the hardware that contains the main distribution lines. These cables can also provide optical service to workstations or transfer an optical signal from one patch panel to a second patch panel.

Special Cables

Submarine Cables

Optical fiber cables connect every continent on Earth (except Antarctica) in the form of submarine cables. These cables are specially designed far the rigors experienced in a submarine environment: rough coral bottoms, strong currents and tides, sharks and other predators, boating and fishing traffic, and so forth. The typical approach to reduce the risks associated with these hazards is to strand steel rods around the cable. This steel rod armoring increases cable strength, provides mechanical protection, and adds to the overall density of the cable to ensure that it sinks to the bottom. Additional protection can be provided by burying the cable with an undersea plow. Some of the design considerations for submarine cables areas follows: (1) a hermetic barrier to prevent hydrogen from reaching the optical fibers, which can lead to unwanted attenuation, and (2) resistance to crushing at depths up to 5 km or more. Both of these are provided by a pressure tube and by steel wires nested in such a way to form a steel vault. The ultimate tensile strength of a typical submarine cable is greater than 60 kN.

Optical Ground Wire (OPGW)

The concept of integrating optical fiber cables with electrical power transmission systems has been very attractive to electric utility companies. One or more ground wires (or shield wires) are typically placed along the top of aerial high-voltage transmission to provide an area of protection from lightning strikes, follower currents, systems or phase to ground faults. Utilities have replaced some of these ground wires with a composite optical ground wire (OPGW), which integrates an optical fiber communication cable with a ground wire. Typically, OPGW cable design incorporates all-aluminum pipe roll formed, continuously welded, and drawn over the optical cable unit providing permanently airtight moisture proof protection. Aluminum clad steel wires and/or aluminum alloy wires surround the pipe giving the cable its primary strength. The optical cable unit consists of optical fibers stranded around a central strength member. The fibers are surrounded by an acrylate filler material and a polymer overcoat. Aramid filler yarn is stranded around the unit followed by a flame retardant tape. All of this is contained inside the aluminum pipe. See Fig. 8.32 for a typical cross section.

Ruggedized Cable

Ruggedized optical cable is a field-deployable, tight buffered cable used in applications where the cable is temporarily deployed on the ground to support military or other short-term communications requirements in a field environment. The cable is typically a low fiber count, multimode, all-dielectric, connectorized cable. The jacket material is typically polyurethane to give the cable low temperature flexibility, extra protection against abrasion, and resistance to petroleum products. The cable may also be flame rated. The connectors are specially designed to withstand exposure to the environment and other rigors associated with field deployment.

Industrial-Use Cables

Cables for the industrial environment call be the same as those mentioned in the section on premises cables. Industrial applications involve the transmission of RS-232/RS-485 data over optical fiber cables. Applications also include closed circuit television and transmission of voice, data, or video signals in

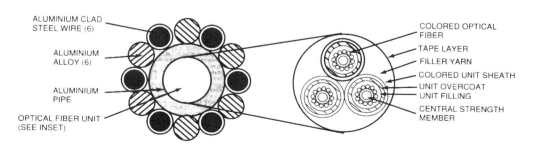

FIGURE 8.32 Optical ground wire. *Source:* Aloca Fujikura, Ltd.

areas sensitive to electromagnetic and radio frequency interference (EMI and RFI). The dielectric nature of optical fiber makes them virtually immune to EMI and RFI, providing great advantages for the system designer in an industrial environment.

Composite Cables

Composite cables consist of both optical fiber and copper twisted pairs. These cables are useful when an end user is planning far an upgrade to optical fiber in the future. The user may want to perform only one installation. The copper twisted pairs call be used now, with the application switched over to the fiber later by changing the end equipment. Alternatively, the media can continue to be used with each supporting a different application, such as voice over the copper and data or video over the fiber. Both copper and fiber components are individually jacketed to make termination easier. The cable's small diameter makes it ideal for space-constrained environments. The cables are typically flame rated for indoor use. See Fig. 8.33 for a typical cross section.

8.7.5 Termination of Optical Fiber

Optical fiber cables are not installed in continuous, point-to-point links for every application. Optical fibers must be terminated at desired locations, spliced into long continuous spans, and connected to optoelectronic hardware. The fibers are joined by one of two methods: splicing or connectorizing.

Splicing

Splicing is the permanent joining of two separate fibers. The core of the optical fiber is aligned and joined to the core of another optical fiber to create a single, continuous length of fiber. The splicing of the fiber can be accomplished by two methods—mechanical splicing or fusion splicing.

Mechanical Splicing

In mechanical splicing, the ends of two separate fibers are physically held in place with the end faces of the fibers butted together. An index matching gel occupies the space between the two fibers to minimize the amount of light that is reflected due to the air gap between the two end faces. The internal components of the splice housing must be meticulously sized to ensure that the fiber cores are properly aligned. Core alignment is required to ensure that splice loss is minimized.

Fusion Splicing

Fusion splicing is the physical joining of two fibers by heating the fiber to its softening point. The heat source is usually a pair of electrodes. The two optical fibers are then mated in the softened phase. After the electrical heat source is removed, the glass quickly cools and again becomes solid, forming a continuous length of fiber.

Fusion splicing has significant benefits over mechanical splicing. Since no air gap is present in a completed fusion splice, no index matching gel is necessary, and no light is reflected or lost due to

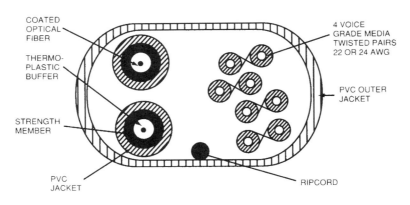

FIGURE 8.33 Composite tight buffered cable.

refractive index differences. The fusion splice generally produces a lower loss splice than mechanical splicing. Most importantly, the fusion splicer can incorporate a variety of active alignment techniques. These techniques automatically align the fiber cores to control the fusing process. Active alignment minimizes the splice loss and reduces the risk of operator error.

Two methods of active fiber alignments can be employed by fusion splicers. One method, profile alignment, aligns the two fibers based on the location of the fiber core or cladding. Lights and cameras are used to center the cores or claddings of the two fibers relative to each other prior to initiating the fusion process. A second system, light injection/detection, or "LID" system, injects light into the fiber. The light transmitted to the second fiber is detected by a receiver. The splicer automatically aligns the fibers to maximize the light coupled from one fiber to the next. When light transmission is maximized, the fiber is automatically spliced. The fuse time can also be automatically optimized by the fusion splicer to ensure the minimum splice loss is attained.

Mass Fusion Splicing

Most splicers can splice only one fiber at a time. This can slow productivity when large fiber count cables are being spliced. For fibers placed into multi-fiber ribbon matrices, the entire ribbon can be spliced at once. Mass fusion splicers can simultaneously splice from 2 to 12 fibers in one step. Unlike single-fiber splicing, active alignment systems are currently not available. As a result, splice losses are slightly higher for mass fusion splicers than for single fiber splicers. Likewise, additional tooling must be purchased to cleave and strip the ribbonized fibers. Mass fusion splicing does however, increase efficiency by splicing more fibers in a shorter period of time.

Connectors

Connectors are devices that are installed on fiber ends to allow interconnection and remating of optical fibers. When two connectors are mated, they mechanically align two fiber endfaces. The connector also allows the fiber to be easily placed in input and output ports of transceivers. Connectors are designed to be disconnected and reconnected easily by hand.

Optical fiber connectors are available in a variety of designs. Some connectors mate into threaded sleeves, while others employ a spring-loaded method to cause fiber end face contact. Most connectors incorporate a key, which ensures that the fiber end faces are consistently mated in the same position. While the majority of connectors are designed for one fiber, some designs can house up to 12 optical fibers. The most common connector designs are displayed in Fig. 8.34.

The fibers in two mated connectors may have an air gap creating a source of Fresnel reflections. The reflected light can be diverted back to the optical source or be doubly reflected at a second reflective point. The reflected light can affect the operation of the light source or cause signal degradation. As a result, the quantity of light reflected at the interface, or *reflectance*, is controlled by the connector polishing process. In turn, connectors are characterized by their reflectance.

Testing

Standard optical mechanical, and environmental test procedures are defined in industry standard *fiber optic test* procedures (FOTPs). These FOTPs are published by TIA/EIA as the TIA/EIA-455 series of documents. They are readily referenced by standards organizations and customers. Each FOTP spells out exactly how the test is to be conducted so that every organization conducts the tests the same way. The FOTP, however, does not prescribe the pass/fail criteria. These criteria would be prescribed by the fiber or cable specification or industry standard (see references). A list of all FOTPs can be obtained from TIA/EIA.

Optical Fiber Tests

The optical fiber tests are normally performed by the fiber manufacturer because cabling has no effect on most of these parameters. The fiber manufacturer controls each of these items (or a surrogate) to ensure they remain within the specification range. While there are additional tests performed to ensure the operability of optical fibers, the tests in Table 8.5 are the most practical for the end user.

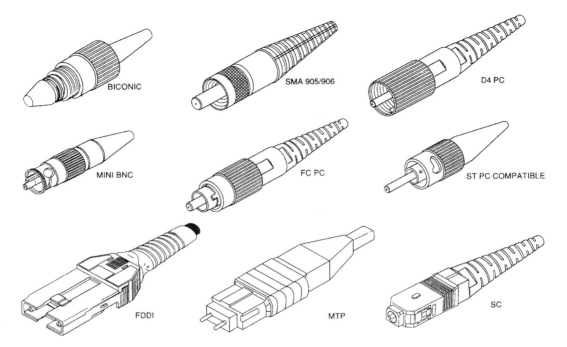

FIGURE 8.34 Common connector types.

Optical Cable Tests

Each optical cable design normally undergoes a series of mechanical and environmental tests to ensure the cable will perform as designed once deployed in the field. The tests are designed to simulate the installation, operating, and environmental conditions the cable will experience over its lifetime, Table 8.6 lists a brief description of several tests typically performed on each cable design. In most tests the attenuation is monitored either continuously at the extreme portion of the test, or after the test is complete and the cable has recovered. Typical results for a properly designed cable would indicate minimal attenuation change during or after each test. Depending on the test the physical integrity of the cable may also be examined. While the tests in Table 8.6 are generally performed on a cable that represents a design family of cables, attenuation coefficient testing is generally performed on every cable that is manufactured. Other final cable tests include cable dimensions, cable construction, and cable length.

Attenuation

The cabled fiber attenuation coefficient is one of the most important parameters far the customer since many optical systems are attenuation limited, The maximum attenuation coefficient should be supplied by the cable manufacturer for every fiber in every cable.

There are three ways of determining the attenuation coefficient. The first way is by using an optical time domain reflectometer (OTDR). The OTDR sends a series of pulses of light down the optical fiber and measures the power of the backscattered light returning to the OTDR. By using the timing of those backscattered pulses, the OTDR's internal algorithms then calculate the attenuation and length, and display a "trace" of the results on a screen. The OTDR can "see" nearly the entire length of the fiber giving the user the ability to detect any point discontinuities or measure the attenuation of discreet events along the length of the span. Use of an OFDR to measure the attenuation coefficient is described in FOTP-61; measuring length with an OTDR is described in FOTP-60.

The second method for determining the attenuation coefficient is the cutback method. This method involves measuring the power transmitted through the length of fiber under test then referencing that measurement against a short piece of fiber cut from the fiber under test. The cutback method is described in FOTP-78.

TABLE 8.5 Common Tests for Optical Fiber

Test	FOTP	Purpose
Core diameter	FOTP–43 and –58	Determine the size of the core in a multimode fiber to ensure compatibility with other fibers of the same core size as well as with end equipment.
Mode field diameter	FOTP–167	Measure of the spot size of light propagating in a single-mode fiber to ensure compatibility with other fibers of similar core size as well as with end equipment. Differences in mode field diameters of two fibers being spliced together can affect splice loss.
Cladding diameter	FOTP–176	Determine the size of the cladding. The consistency of cladding diameter can affect connector and mechanical splice performance.
Core–clad concentricity	FOTP–176	Distance between the center of the core and the center of the cladding. High values can affect splice and connector losses.
Core noncircularity	FOTP–176	Measures the roundness of multimode cores. High values can have a slight effect on splice and connector losses.
Cladding noncircularity	FOTP–176	Measures the roundness of the cladding. High values can have a slight effect on splice and connector losses.
Fiber cutoff wavelength	FOTP–80	Measures the minimum wavelength at which a single-mode fiber will support the propagation of only one mode. If the system wavelength is below the cutoff wavelength, multimode operation may occur introducing modal dispersion and higher attenuation. The difference in fiber and cable cutoff wavelength is due to the deployment of fiber during the test. Cabling can shift the cutoff wavelength to a lower value.
Cable cutoff wavelength	FOTP–170	
Curl	FOTP–111 (underdevelopment)	Measures the curvature of a short length of fiber in an unsupported condition. Excessive curl can affect splice loss in passive alignment fusion splicers such as mass fusion splicers.
Coating diameter	FOTP–173	Measures the outside diameter of a coated fiber. Out of spec values can affect cable manufacturing and potentially cable performance.
Numerical aperture	FOTP–47 and –177	Measures the numerical aperture of a fiber. Ensures compatibility with other fibers as well as end equipment.
Proof test	FOTP–31	Ensures the minimum strength of a fiber. Every fiber is normally subjected to the proof test.
Attenuation coefficient	FOTP–61 and –78	Measured by the fiber and cable manufacturers and reported to the customer in units of dB/km.
Bandwidth	FOTP–30 and –5	Measured by the fiber manufacturer and reported to the customer by the cable manufacturer in units of MHz·km.

TABLE 8.6 Common Tests for Optical Cable Design

Test	FOTP	Purpose
Temperature cycling	FOTP–3	Simulates environmental conditions once the cable is deployed.
Impact	FOTP–25	Simulates an object being dropped on the cable for a sudden and brief impact.
Tensile	FOTP–33	Measures the performance of the cable at its rated tensile load simulating installation by pulling.
Compressive load	FOTP–41	Measures cable performance while under a compressive or crushing force.
Cable twist	FOTP–85	Measures the ability of the cable to perform when under a twist condition.
Cycle flex or bend	FOTP–104	Measures the ability of the cable to perform even when subjected to a bend, and withstand repeated bending during installation.
Water penetration	FOTP–82	Measures the ability of an outdoor cable to prevent the ingress of water along the length of the cable.
Filling and flooding compound flow	FOTP–81	Measures the resistance to flow of compound flow filling and flooding compounds at elevated temperatures.

The third method is similar to the cutback test but a separate fiber reference is substituted for the piece of fiber cut from the fiber under test. This method is described in FOTP-53.

8.7.6 Summary

Optical fiber and optical cable has dramatically changed telecommunications services over the last 15 years. As the information age matures, the installation of optical cables will reach further into businesses, schools, and homes bringing a wider variety of services to the end user. As a designer or installer of optical systems, whether for a local area network or a nationwide telecommunications system, the ability to understand, choose, and specify the proper optical cables to fit a given application will become increasingly more important.

8.7.7 Defining Terms[*]

Absorption. In an optical fiber, that portion of attenuation resulting from conversion of optical power into heat. *Note:* components consist of tails of the ultraviolet and infrared absorption bands; impurities (e.g., the OH ion and transition metal ions); and defects (e.g., results of thermal history and exposure to nuclear radiation).

Attenuation. In an optical fiber, the diminution of average optical power. *Note:* in an optical fiber, attenuation results from absorption, scattering, and other radiation. Attenuation is generally expressed in decibels. However, attenuation is often used as a synonym for attenuation coefficient, expressed in decibels per kilometer.

Cladding. The optical material surrounding the core of an optical fiber.

Core. (1) The central region of all optical fiber through which light is transmitted. (2) The central region of an optical filler that has an index of refraction higher than the surrounding cladding material.

Dispersion. A term used to describe the chromatic, or wavelength, dependence of a parameter as opposed to the temporal dependence. The term is used, for example, to describe the process by which an electromagnetic signal is distorted because the various wavelength components of that signal have different propagation characteristics. The term is also used to describe the relationship between refractive index and wavelength.

Dispersion-shifted fiber. A single-mode fiber that has nominal zero-dispersion wavelength of 1550 nm with applicability in the 1500–1600 nm range, and has a dispersion coefficient that is a monotonically increasing function of wavelength. Also known as an EIA Class IVb fiber.

Dispersion-unshifted fiber. A single-mode filler that has a nominal zero-dispersion wavelength in the 1300 nm transmission window and has a dispersion coefficient that is a monotonically increasing function of wavelength. Also known as dispersion-unmodified, nonshifted, or EIA Class IVa fiber.

Insertion loss. The total optical power loss caused by insertion of an optical component such as a connector, splice, or coupler.

Macrobending. In an optical fiber, optical attenuation due to macroscopic deviations of the axis from a straight line; distinguished from microbending.

Microbending. In an optical fiber, sharp curvatures involving local axial displacements of a few micrometers and spatial wavelengths of a few millimeters or less. Such bends may result from fiber coating, cabling, packaging, installation, and so forth. *Note:* microbending can cause significant radiative losses and mode coupling.

Mode. In an optical fiber, one of those electromagnetic field distributions that satisfies Maxwell's equations and the boundary conditions. The field pattern of a mode depends on the wavelength, refractive index, and waveguide geometry.

Mode field diameter (MFD). Mode field diameter is a measure of the width of the guided optical power's intensity distribution in a single-mode fiber.

---

[*]Definitions adapted from NSI/EIA, 1989, *Fiber Optic Terminology,* EIA-440-A, Electronic Industries Alliance.

Optical time domain reflectometer (OTDR). A measurement device used to characterize a fiber wherein all optical pulse is transmitted through the fiber and the resulting light scattered and reflected back to the input is measured as a function of time. Useful in estimating attenuation coefficient as a function of distance, and identifying defects and other localized losses.

Rayleigh scattering. Light scattering by refractive index fluctuations (inhomogeneities in material density or composition) that are small with respect to wavelength. The scattered power is inversely proportional to the fourth power of the wavelength.

Reflection. The abrupt change in direction of a light beam at an interface between two dissimilar media so that the light beam returns into the medium from which it originated. Reflection from a smooth surface is termed *specular*, whereas reflection from a rough surface is termed *diffuse*.

Refractive index (of a medium) (n). Denoted by *n*, the ratio of the velocity of light in vacuum to the velocity in the medium. Its synonym is the index of refraction.

Spectral width. A measure of the wavelength extent of a spectrum. *Note:* one method of specifying the spectral line width is the full width at half-maximum (FWHM); specifically, the difference between the wavelengths at which the magnitude drops to one-half of its maximum value.

Total internal reflection. The total reflection that occurs when light strikes an interface at angles of incidence (with respect to the normal) greater than the critical angle.

8.8 Coaxial Cable and Interconnects

Adapted from Cozad, K.W.; "Coax/Transmission Line," Chapter 19 in Whitaker, J.C., *The Electronics Handbook,* CRC/IEEE Press, Boca Raton, 1997.

8.8.1 Introduction

Coaxial *transmission lines* provide the simplest and most versatile method for the transmission of RF and microwave energy (see Fig. 8.35). The most popular types consist of a cylindrical metallic inner conductor surrounded by a dielectric material and then enclosed by a cylindrical metallic outer conductor. The dielectric material is used to maintain the inner conductor concentrically within the outer conductor. This dielectric material is typically polyethylene (PE), polyproplene (PP), or tetrafluoroethylene (TUE). Most flexible coaxial cables are then coated with a protective jacket made of polyethylene or polyvinyl chloride (PVC).

8.8.2 General Applications

Coaxial transmission lines are used primarily in two areas: (1) short interconnections between RU electronic equipment and (2) longer connections between the signal source and its antenna. These applications usually occur in completely different environments that can significantly impact the performance of the transmission line. These will be discussed separately later in this chapter.

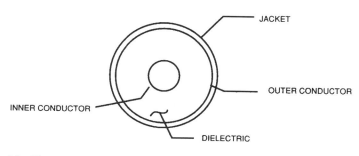

FIGURE 8.35 Coaxial cable cross section.

8.8.3 Theory

For purposes of this discussion, the coaxial transmission lines will be considered as *uniform transmission lines*. This implies that the line is a system of two conductors that has identical cross sections for all positions along its length. From electromagnetic field theory, coaxial transmission lines will have the following general characteristics:

1. An infinite number of field configurations is possible. Each of these is called a *mode*.
2. In a sinusoidal steady-state condition, the modes will be in a general form of waves traveling in the $\pm z$ direction (along the axis of the line) (Fig. 8.36).
3. The propagation constant γ is different for each mode.
4. There is a frequency, called the cutoff frequency, where the propagation constant is 0. Below this frequency, there is no propagation of that mode.
5. There is one mode for which the electromagnetic fields are transverse to the transmission line axis. This mode is called the *transverse electromagnetic* mode (TEM). The TEM has a cutoff frequency of 0 Hz. This is the mode that is primarily used for coaxial transmission line propagation.

Some standard characteristics of coaxial lines:

1. Characteristic impedance $Z_0 = (60/\sqrt{\varepsilon_r})\ln(D/d)$, where
   D = inner diameter of the outer conductor
   d = outer diameter of the inner conductor
   ε_r = dielectric constant of medium (relative to air = 1)
2. Capacitance per unit length $C = \dfrac{2\pi\varepsilon}{\ln(D/d)}$.
3. Inductance per unit length $L = \mu\ln(D/d)$, where
   ε = permittivity of the medium
   μ = permeability of the medium
4. Phase velocity $V_p = 1\sqrt{LC} = 3\times 10^8/\sqrt{\varepsilon_r}$ m/s.
5. Relative phase velocity $= 1/\sqrt{\varepsilon_r}$.
6. Attenuation constant α is usually given in a decibel loss per unit length, and it is a function of the conductivity of the metals used for the two conductors, the dielectric losses of the insulating medium, and the frequency of operation. This is usually determined by experimental measurements and is provided by the coaxial line supplier.

Since coaxial lines are typically used to transfer energy from a source to a load (receiver or antenna), it is important that this transfer is as efficient as possible. The attenuation constant is one factor in determining the efficiency as this represents a direct loss of energy within the line. Another significant factor is the level of impedance mismatch between the line and the load. The greater the difference between the characteristic impedance of the line and the input impedance of the load, the higher the reflected level of energy at the connection. This reflected energy reduces the amount of signal transmitted

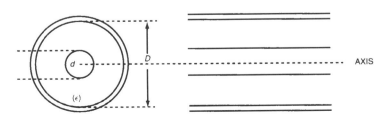

FIGURE 8.36

Interconnects

to the load and results in a lower efficiency of the system. The reflected signal also combines with the forward signal and produces impedance changes along the line as the combined voltages vary.

In a simplified form,

$$Z(s) = \frac{Z_0 \chi (1 + V^-/V^+)}{(1 - V^-/V^+)} = \frac{Z_0 (1 + V^-/V^+)}{(1 - V^-/V^+)}$$

where $Z(s)$ = resultant impedance at a point on the line
  V^- = reflected voltage component
  V^+ = toward voltage component

At the load, if $V^- = 0$, then $Z_L = Z_0$, and the line is said to be matched with its load. If $V^- \neq 0$, then it is said be mismatched. Since V^-/V^+ is determined by Z_L/Z_0, then we can define the voltage reflection coefficient Γ_0 at the load as

$$\Gamma_0 = \frac{(Z_L - Z_0)}{(Z_L + Z_0)}$$

The voltage standing wave ratio (VSWR) is then defined as

$$\text{VSWR} = \frac{V_{max}}{V_{min}} = \frac{(1 + \Gamma_0)}{(1 - \Gamma_0)}$$

or

$$\Gamma_0 = \frac{(\text{VSWR} - 1)}{(\text{VSWR} + 1)}$$

Table 8.7 compares the various characteristics to the efficiency of the transmission line.

8.8.4 Power Handling

The power handling of coaxial lines are based on voltage breakdown and the maximum operating temperature allowed on the inner conductor. Both of these limitations are primarily limited by the type of dielectric used as the insulating medium between the inner and outer conductors.

Voltage breakdown, or peak power handling, is frequency independent and typically determined from actual testing of the cable. A dc test voltage is applied to the cable and it is then checked for arcing and/or current leakage. The peak power rating is determined from the following equation:

$$P_{pk} = \frac{\left(\frac{EP \times 0.707 \times 0.7}{SF}\right)^2}{Z_0}$$

where P_{pk} = cable power rating
  EP = dc test voltage
  0.707 = RMS factor
  0.7 = dc to RF factor
  SF = safety factor on voltage
  Z_0 = characteristic impedance

TABLE 8.7

VSWR	Γ_0	%Power Reflected	% Power Transmitted	Transmission Loss, dB	Return Loss, dB
1.00	0.00	0.0	100.0	0.000	—
1.05	0.02	0.0	100.0	0.001	32.3
1.10	0.05	0.3	99.7	0.013	26.4
1.15	0.07	0.5	99.5	0.022	23.1
1.20	0.09	0.8	99.2	0.035	20.8
1.25	0.11	1.2	98.8	0.052	19.1
1.30	0.13	1.7	98.3	0.074	17.7
1.35	0.15	2.3	97.7	0.101	16.5
1.40	0.17	2.9	97.1	0.128	15.6
1.45	0.18	3.2	96.8	0.141	14.7
1.50	0.20	4.0	96.0	0.177	14.0
1.55	0.22	4.8	95.2	0.214	13.3
1.60	0.23	5.3	94.7	0.237	12.7
1.65	0.25	6.3	9.7	0.282	12.2
1.70	0.26	6.8	93.2	0.306	11.7
1.75	0.27	7.3	92.7	0.329	11.3
1.80	0.29	8.4	91.6	0.381	10.9
1.85	0.30	9.0	91.0	0.410	10.5
1.90	0.31	9.6	90.4	0.438	10.1
1.95	0.32	10.2	89.8	0.467	9.8
2.00	0.33	10.9	98.1	0.501	9.5
2.50	0.43	18.5	81.5	0.881	7.4
3.00	0.50	25.0	75.0	1.249	6.0
3.50	0.56	30.9	69.1	1.603	5.1
4.00	0.60	36.0	64.0	1.938	4.4
4.50	0.64	40.5	59.5	2.255	3.9
5.00	0.67	44.4	55.6	2.553	3.5
5.50	0.69	47.9	52.1	2.834	3.2
6.00	0.71	5.0	49.0	3.100	2.9
6.50	0.73	53.78	46.2	3.351	2.7
7.00	0.75	56.2	43.7	3.590	2.5
7.50	0.76	58.5	41.5	3.817	2.3
8.00	0.78	60.5	39.5	4.033	2.2
8.50	0.79	62.3	37.7	4.240	2.1
9.00	0.80	64.0	36.0	4.437	1.9
9.50	0.81	65.5	34.5	4.626	1.8
10.00	0.82	66.9	33.1	4.807	1.7

As previously discussed, when the transmission line is not matched to its load, the reflected signal combines with the forward signal to produce voltage maximums. Because of this, the peak power ratings must be derated for the VSWR present on the line. Typical deratings are given in Table 8.8.

TABLE 8.8

Modulation	Peak Power Derating
AM	$P_{max} = \dfrac{P_{pk}}{(1+M)^2 \text{VSWR}}$
FM (CW)	$P_{max} = \dfrac{P_{pk}}{\text{VSWR}}$
TV (NTSC)	$P_{max} = \dfrac{P_{pk}}{2.09 \text{VSWR}}$

where P_{max} = derated peak power
P_{pk} = peak power rating of cable
M = amplitude modulation index (100% = 1.0)
2.09 = modulation derating factor for aural/visual = 0.2

The average power ratings of transmission lines are determined by the safe long-term operating temperature allowable for the dielectric material chosen as an insulating material. The inner conductor is the generator of heat, and the maximum permissible inner conductor temperature varies with the type of dielectric. It is recommended that the system engineer closely review the manufacturer's specifications for this prior to completing a system design and installation.

8.8.5 Short Interconnection Applications

Coaxial transmission lines used for short interconnections between equipment, typically inside buildings, have traditionally been called RU cables. These cables are typically 0.5 in or smaller in diameter and use solid insulating material. The outer conductor is usually a copper or aluminum braid and the inner conductor is copper or copper clad steel. The primary benefits of these cables are that they are relatively inexpensive and are very easy to handle in terms of weight and flexibility. More commonly used cables are shown in Fig. 8.37 and listed in Table 8.9.

TABLE 8.9

Cable Designation	Nominal Z_0	V_p	Diameter, in	Nominal Atten. at 50 MHz, dB/100 ft	Nominal Atten. at 200 MHz, dB/100 ft	Nominal Atten. at 700 MHz, dB/100 ft	Max. Op. Voltage, RMS
RG-8A/U	52	0.66	0.405	1.6	3.2	6.5	4000
RG-8/X	50	0.78	0.242	2.5	5.4	11.1	600
RG-213/U	50	0.66	0.405	1.6	3.2	6.5	5000
RG-58/U	53.5	0.66	0.195	3.1	6.8	14.0	1900
RG-58A/U	50	0.66	0.195	3.3	7.3	17.0	1900
RG-58C/U	50	0.66	0.195	3.3	7.3	17.0	1900
RG-11A/U	75	0.66	0.405	1.3	2.9	5.8	5000
RG-59B/U	75	0.66	0.242	2.4	4.9	9.3	2300
RG-62B/U	93	0.84	0.242	2.0	4.2	8.6	750
RG-71/U	93	0.84	0.245	1.9	3.8	7.3	750
RG-141A/U	50	0.695	0.190	2.7	5.6	11.0	1400
RG-178B/U	50	0.695	0.70	10.5	19.0	37.0	1000
RG6A/U	75	0.66	0.332	1.9	4.1	8.1	2700

FIGURE 8.37

Although these cables provide adequate performance in most installations, the increasing use of closer spaced channels to provide more efficient use of frequency spectrums is resulting in the need to take a closer look at cable performance. For cables used in equipment rack jumpers, shielding effectiveness to prevent signal leakage into other electronic components can be critical. Cables using braided outer conductors can be especially susceptible to leakage from the separating of the braid wire after numerous bends or tight twists. For these critical installations., cables using a solid outer conductor will provide improved performance. Examples of these cables are listed in Table 8.10.

TABLE 8.10

Cable Designation	Nominal Z_0	V_p	Diameter, in	Nominal Atten. at 50 MHz, dB/100 ft	Nominal Atten. at 200 MHz, dB/100 ft	Nominal Atten. at 700 MHz, dB/100 ft	Max. Op. Voltage, RMS
FSJ2-50	50	0.84	0.290	1.27	2.58	4.97	6,400
FSJ2-50	50	0.83	0.415	0.848	1.73	3.37	13,200
FSJ4-50B	50	0.81	0.520	0.73	1.50	2.97	15,600
LDF2-50	50	0.88	0.44	0.736	1.50	2.93	15,600
LDF4-50A	50	0.88	0.63	0.479	0.983	1.92	40,000
ETS1-50T	50	0.82	0.29	1.27	2.56	4.89	6,400
ETS2-50T	50	0.83	0.415	0.856	1.77	3.48	13,200
FSJ1-75	75	0.78	0.29	1.3	2.68	5.30	3,300
FSJ4-75A	75	0.81	0.52	0.673	1.39	2.75	10,000
LDF4-75A	75	0.88	0.63	0.435	0.896	1.76	26,000

Solid outer conductors also prevent the deterioration of performance due to moisture ingress. Corrosion between the strands of the braid results in higher attenuation and loss of electrical contact. In outdoor use, proper review of the system requirements for long-term reliability will help prevent premature failure or extensive maintenance of the transmission system.

8.8.6 Connectors

The previous sections have focused on the RF cables themselves. However, use of the cable can be dependent on the availability of connectors to interface between the cable and the electric components and antennas. A list of commonly used connectors for RF cables is shown in Fig. 8.38. These connectors use either a bayonet-style coupling for ease of assembly or a screw-type coupling for better protection from moisture. The BNC connector, a bayonet style, is used extensively with test equipment when the cables must be removed and reattached numerous times during testing. However, the reliability begins to deteriorate after continued reattachment and periodic recalibration of the test setup is highly recommended. For highly sensitive measurements, one of the screw type connectors may provide improved results.

8.8.7 Intermodulation Products

With the cable and connector now assembled, another performance criterion can be reviewed. In the newer, high-channel communication systems, the effects of connector attachment and assembly must be included. A primary cause of signal distortion due to these connections is intermodulation products. They result from signals containing two or more frequency components transmitted through a nonlinear passive device. The nonlinearity produces harmonics and signals due to the intermodulation of the various frequencies. These nonlinearities result from the use of magnetic materials or imperfect contact joints. Therefore, connector designs to be used where intermodulation products will result in significant

Interconnects

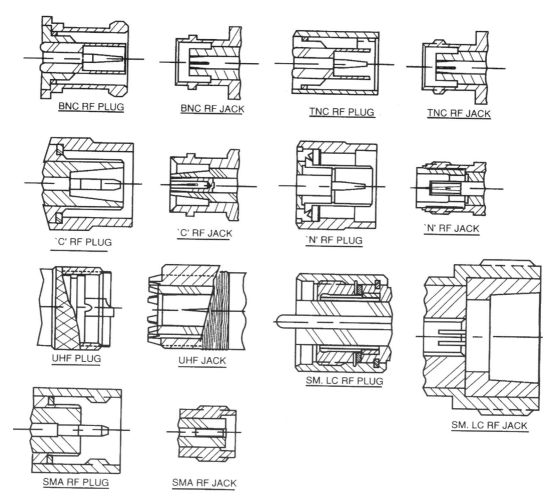

FIGURE 8.38

deterioration of the transmission system performance should be reviewed for material used in fabrication and contact joints that provides high contact pressures that can be maintained over the life of the system.

8.8.8 Long Interconnection Applications

For interconnections between a source and its antenna (Table 8.11), the length of coaxial line may be several hundred feet and experience extreme environmental conditions. Higher-power transmitters and the need for more efficient transmission lines require cable sizes much larger than the RU cables previously discussed. For high-power broadcast stations, the use of rigid coaxial transmission line is very common.

Semiflexible coaxial cables use solid outer conductors that are fabricated from copper strips that have been seam welded into continuous cylinders. For flexibility, the outers are corrugated to permit bending and to account for thermal expansion when exposed to various temperatures. Solid foam dielectric insulating materials or a spiral wound dielectric spacer is used to maintain the *concentricity* between the inner and outer conductors. Table 8.11 gives examples of common sizes and types. For even higher power handling and lower attenuation rigid coaxial transmission line can be used. These can range from 3-1/8 to 8-3/16 inches in diameter and are used primarily for FM and TV broadcasting where lengths up to 2000 ft may be necessary.

TABLE 8.11 Cable Designations and Characteristics

Cable Designation	Nominal Z_0	V_p	Diameter, in	Nominal Atten. at 50 MHz, dB/100 ft	Nominal Atten. at 200 MHz, dB/100 ft	Nominal Atten. at 700 MHz, dB/100 ft	Max. Op. Voltage, RMS
LDF5-50A	50	0.89	1.09	0.257	5.35	1.07	91
LDF7-50A	50	0.88	1.98	1.156	0.328	0.664	315
HJ5-50	50	0.916	1.11	0.260	0.540	1.04	90
HJ7-50A	50	0.921	1.98	0.145	0.290	0.576	305
HJ8-50B	50	0.933	3.01	0.098	0.208	0.450	640
HJ11-50	50	0.92	4.00	0.078	0.166	0.351	1100
HJ9-50	50	0.931	5.20	0.054	0.115	0.231	1890

Since it is not feasible to bend the rigid line, it is supplied in sections and must be assembled on site. The outer conductors are bolted together using a contact flange and the inners are spliced in a manner similar to the RU cable connector designs. A male connector, called a *bullet,* is inserted into one inner conductor and then the second inner conductor. The outer conductors are then pulled together and the flanges bolted.

For these larger cables and rigid lines, one of the most important requirements for long-term reliability is proper installation. It must be remembered that the environmental forces acting on lines installed outdoors are continuously changing. This results in extreme stress on the cable due to thermal changes, corrosion, and atmospheric contaminants that are not typically present with indoor installations. Following the manufacturer's recommended procedures for installation and maintenance can prevent catastrophic failures within the transmission system (see Table 8.12 and Figs. 8.39 and 8.40).

TABLE 8.12 Rigid Coaxial Transmission Lines

Rigid Line Type, in (Ω)	Atten.			Average Power, kW		
	57 MHz	201 MHz	600 MHz	57 MHz	201 MHz	600 MHz
3-1/8 (50)	0.072	0.138	0.257	70	39	22
4-1/16 (50)	0.050	0.094	0.165	140	62	36
6-1/8 (50)	0.039	0.074	0.127	300	150	71
6-1/8 (75)	0.034	0.066	0.120	250	110	60
8-3/16 (75)	0.029	0.053	0.089	350	185	90

8.8.9 Defining Terms

Attenuation. The decrease in magnitude of a signal as it travels through a transmitting medium.
Attenuation constant. The relative rate of attenuation per unit length of a transmitting medium.

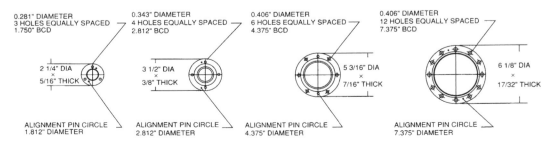

FIGURE 8.39 Electronic Industries Alliance (EIA) flanges.

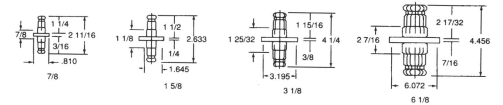

FIGURE 8.40 Male connectors for rigid lines.

Characteristic impedance. In theory, the ratio of the applied voltage to the resultant current in a transmission line of infinite length at the point the voltage is applied.
Concentric. Two or more items having the same center.
Dielectric. A nonconducting (insulating) medium.
EIA. The Electronic Industries Alliance (formerly Electronic Industries Association).
Jacket. The protective outer covering for a cable to prevent abrasion or corrosion.
Transmission line. An arrangement of two or more conductors used to transfer a signal from one point to another.
Velocity of propagation. The speed of transmission of electrical energy within a transmission medium typically compared to the speed in free space.

8.9 Microwave Guides and Couplers

Adapted from Steer, M.B.; Passive Microwave Devices," Chapter 21 in Whitaker, J.C., *The Electronics Handbook*, CRC/IEEE Press, Boca Raton, 1997.

8.9.1 Introduction

Wavelengths in air at microwave and millimeter-wave frequencies range from 1 m at 300 MHz to 1 mm at 300 GHz and are comparable to the physical dimensions of fabricated electrical components. For this reason circuit components commonly used at lower frequencies, such as resistors, capacitors, and inductors, are not readily available. The relationship between the wavelength and physical dimensions enables new classes of distributed components to be constructed that have no analogy at lower frequencies. Components are realized by disturbing the field structure on a transmission line, resulting in energy storage and thus reactive effects. Electric (E) field disturbances have a capacitive effect, and the magnetic (H) field disturbances appear inductive. Microwave components are fabricated in waveguide, coaxial lines, and strip lines. The majority of circuits are constructed using strip lines as the cost is relatively low and they are highly reproducible due to the photolithographic techniques used. Fabrication of waveguide components requires precision machining but they can tolerate higher power levels and are more easily realized at millimeter-wave frequencies (30–300 GHz) than either coaxial or microstrip components.

8.9.2 Characterization of Passive Elements

Passive microwave elements are defined in terms of their reflection and transmission properties for an incident wave of electric field or voltage. In Fig. 8.41a, a traveling voltage wave with phasor V_1^+ is incident at port 1 of a two-port passive element. A voltage V_1^- is reflected and V_2^- is transmitted. In the absence of an incident voltage wave at port 2 (the voltage wave V_2^-; is totally absorbed by Z_0), at port 1 the element has a voltage reflection coefficient

$$\Gamma_1 = \frac{V_1^-}{V_1^+} \tag{6.1}$$

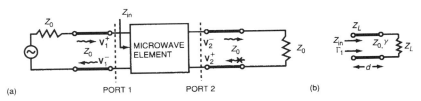

FIGURE 8.41 Incident, reflected, and transmitted traveling voltage waves at (a) a passive microwave element and (b) a transmission line.

and transmission coefficient

$$T = \frac{V_2^-}{V_1^+} \tag{8.2}$$

More convenient measures of reflection and transmission performance are the return loss and insertion loss as they are relative measures of power in transmitted and reflected signals. In decibels,

$$\text{return loss} = -20 \log \Gamma_1 \quad \text{insertion loss} = -20 \log T \tag{8.3}$$

The input impedance at port 1, Z_{in}, is related to Γ_1 by

$$Z_{in} = Z_0 \frac{1 + \Gamma_1}{1 - \Gamma_1} \tag{8.4}$$

The reflection characteristics are also described by the *voltage standing wave ratio (VSWR)*, a quantity that is more easily measured. The VSWR is the ratio of the maximum voltage amplitude on a transmission line ($|V_1^+| + |V_1^-|$) to the minimum voltage amplitude ($|V_1^+| - |V_1^-|$). Thus,

$$VSWR = \frac{1 + |\Gamma_1|}{1 - |\Gamma_1|} \tag{8.5}$$

These quantities will change if the loading conditions are changed. For this reason scattering parameters are used which are defined as the reflection and transmission coefficients with a load referred to as a *reference impedance*. Simple formulas relate the S parameters to other network parameters [Vendelin et al. 1990, pp. 16–17]. Thus,

$$S_{11} = \Gamma_1 \quad S_{21} = T \tag{8.6}$$

S_{11} and S_{12} are similarly defined when a voltage wave is incident at port 2. For a multiport,

$$S_{pq} = \frac{V_p^-}{V_q^-} \tag{8.7}$$

with all of the ports terminated in the reference impedance. S parameters are the most convenient network parameters to use with distributed circuits as a change in line length results in a phase change. Also, they are the only network parameters that can be measured directly at microwave and millimeter-wave frequencies. Most passive devices, with the notable exception of ferrite devices, are reciprocal, and so $S_{pq} = S_{qp}$. A lossless passive device also satisfies the unitary condition: $\Sigma_q |S_{pq}|^2 = 1$, which is a statement of power conservation indicating that all power is either reflected or transmitted. A passive

Interconnects   8-59

element is fully defined by its *S* parameters together with its reference impedance, here Z_0. In general, the reference impedance at each port can be different.

Circuits are designed to minimize the reflected energy and maximize transmission at least over the frequency range of operation. Thus, the return loss is high, and the VSWR $\approx$ 1 for well designed circuits. Individual elements may have high reflection, and the interaction of elements is used in design.

A terminated transmission line such as that in Fig. 8.41b has an input impedance

$$Z_{in} = Z_0 \frac{Z_L + jZ_o \tanh \gamma d}{Z_0 + jZ_L \tanh \gamma d} \tag{8.8}$$

Thus, a short section ($\gamma d \ll 1$) of short-circuited ($Z_L = 0$) transmission line looks like an inductor, and a capacitor if it is open circuited ($Z_L = \infty$). When the line is a half-wavelength long, an open circuit is presented at the input to the line if the other end is short circuited.

8.9.3 Transmission Line Sections

The simplest microwave circuit element is a uniform section of transmission line that can be used to introduce a time delay or a frequency-dependent phase shift. Other line segments for interconnections include bends, corners, twists, and transitions between lines of different dimensions. (See Fig. 8.42.) The dimensions and shapes are designed to minimize reflections and so maximize return loss and minimize insertion loss.

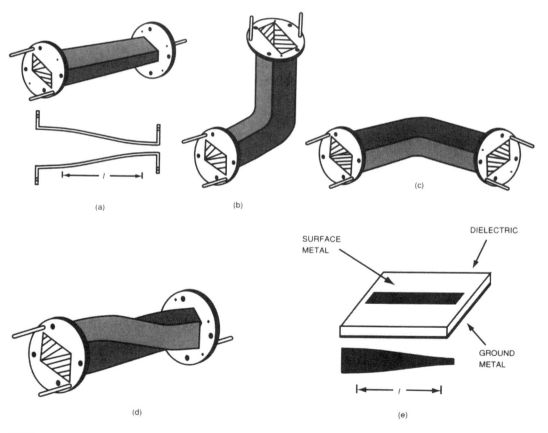

FIGURE 8.42 Sections of transmission lines used for interconnecting components: (a) waveguide tapered section, (b) waveguide E-plane bend, (c) waveguide H-plane bend, (d) waveguide twist, and (e) microstrip taper.

8.9.4 Discontinuities

The waveguide discontinuities shown in Figs. 8.43a through 8.43f illustrate most clearly the use of E- and H-field disturbances to realize capacitive and inductive components. An E-plane discontinuity (Fig. 8.43a) can be modeled approximately by a frequency-dependent capacitor. H-plane discontinuities (Figs. 8.43b and 8.43c resemble inductors, as does the circular iris of Fig. 8.43d. The resonant waveguide iris of Fig. 8.43e disturbs both the E- and H-fields and can be modeled by a parallel LC resonant circuit near the frequency of resonance. Posts in waveguide are used both as reactive elements (Fig. 8.43f) and to mount active devices (Fig. 8.43g). The equivalent circuits of microstrip discontinuities (Figs. 8.43k through 8.43o) are again modeled by capacitive elements if the E-field is interrupted, and by inductive elements if the H-field (or current) is interrupted. The stub shown in Fig. 8.43j presents a short circuit to the through transmission line when the length of the stub is $\lambda_g/4$. When the stubs are electrically short («$\lambda_g/4$), they introduce shunt capacitances in the through transmission line.

8.9.5 Impedance Transformers

Impedance transformers are used to interface two sections of line with different *characteristic impedances*. The smoothest transition and the one with the broadest bandwidth is a tapered line as shown in Figs. 8.42a and 8.42e. This element tends to be very long, and so step terminations called quarter-wave

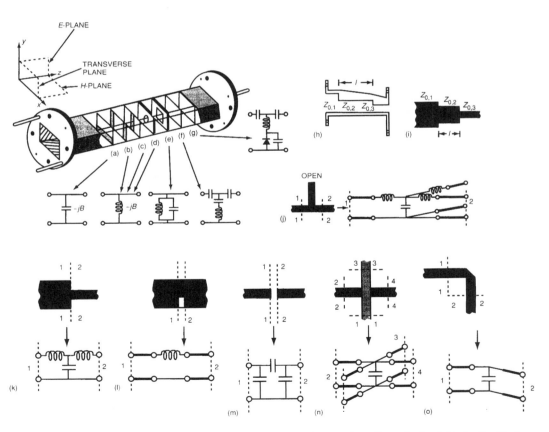

FIGURE 8.43 Discontinuities. Waveguide discontinuities: (a) capacitive E-plane discontinuity, (b) inductive H-plane discontinuity, (c) symmetrical inductive H-plane discontinuity, (d) inductive circular iris discontinuity, (e) resonant window discontinuity, (f) capacitive post discontinuity, (g) diode post mount, and (h) quarter-wave impedance transformer. Microstrip discontinuities: (i) quarter-wave impedance transformer, (j) open microstrip stub, (k) step, (l) notch, (m) gap, (n) crossover, and (o) bend.

impedance transformers (see Figs. 8.43h and 8.43i) are sometimes used, although their bandwidth is relatively small centered on the frequency at which $l = \lambda_g/4$. Ideally, $Z_{0,2} = \sqrt{Z_{0,1} Z_{0,3}}$.

8.9.6 Terminations

In a termination, power is absorbed by a length of lossy material at the end of a shorted piece of transmission line (Figs. 8.44a and 8.43c). This type of termination is called a matched load as power is absorbed and reflections are very small, irrespective of the characteristic impedance of the transmission line. This is generally preferred, as the characteristic impedance of transmission lines varies with frequency, and particularly so for waveguides. When the characteristic impedance of a line does not vary much with frequency, as is the case with a coaxial line, a simpler smaller termination can be realized by placing a resistor to ground (Fig. 8.44b).

8.9.7 Attenuators

Attenuators reduce the level of a signal traveling along a transmission line. The basic construction is to make the line lossy but with a characteristic impedance approximating that of the connecting lines so as to reduce reflections. The line is made lossy by introducing a resistive vane in the case of a waveguide (Fig. 8.44d), replacing part of the outer conductor of a coaxial line by resistive material (Fig. 8.44e), or covering the line by resistive material in the case of a microstrip line (Fig. 8.44f). If the amount of lossy material introduced into the transmission line is controlled, a variable attenuator is achieved—for example, Fig. 8.44d.

8.9.8 Microwave Resonators

In a lumped element resonant circuit, stored energy is transferred between an inductor, which stores magnetic energy, and a capacitor, which stores electric energy, and back again every period. Microwave resonators function the same way, exchanging energy stored in electric and magnetic forms but with the energy stored spatially. Resonators are described in terms of their quality factor,

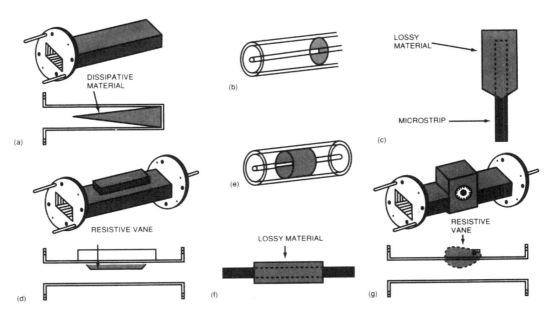

FIGURE 8.44 Terminations and attenuators: (a) waveguide matched load, (b) coaxial line resistive termination, (c) microstrip matched load, (d) waveguide fixed attenuator, (e) coaxial fixed attenuator, (f) microstrip attenuator, and (g) waveguide variable attenuator.

$$Q = 2\pi f_0 \left(\frac{\text{maximum energy stored in the resonator at } f_0}{\text{power lost in the cavity}} \right) \quad (8.9)$$

where f_0 is the resonant frequency. The Q is reduced and thus the resonator bandwidth is increased by the power lost due to coupling to the external circuit so that the loaded Q

$$Q_L = 2\pi f_0 \left(\frac{\text{maximum energy stored in the resonator at } f_0}{\text{power lost in the cavity}} \right)$$

$$= \frac{1}{1/Q + 1/Q_{\text{ext}}} \quad (8.10)$$

where Q_{ext} is called the external Q. Q_L accounts for the power extracted from the resonant circuit and is typically large. For the simple response shown in Fig. 8.45a, the half-power (3-dB) bandwidth is f_0/Q_L.

Near resonance, the response of a microwave resonator is very similar to the resonance response of a parallel or series R, L, C resonant circuit (Figs. 8.45f and 8.45g). These equivalent circuits can be used over a narrow frequency range.

Several types of resonators are shown in Fig. 8.45. Figure 8.45b is a rectangular cavity resonator coupled to an external coaxial line by a small coupling loop. Figure 8.45c is a microstrip patch reflection resonator. This resonator has large coupling to the external circuit. The coupling can be reduced and photolithographically controlled by introducing a gap as shown in Fig. 8.45d for a microstrip gap-coupled transmission line reflection resonator. The Q of a resonator can be dramatically increased by using a high dielectric constant material as shown in Fig. 8.45e for a dielectric transmission resonator in microstrip. One simple application of a cavity resonator is the waveguide wavemeter (Fig. 8.45h). Here, the resonant frequency of a rectangular cavity is varied by changing the physical dimensions of the cavity with a null of the detector indicating that the frequency corresponds to the resonant cavity frequency.

8.9.9 Tuning Elements

In rectangular waveguide, the basic adjustable tuning element is the sliding short shown in Fig. 8.46a. Varying the position of the short will change resonance frequencies of cavities. It can be combined with hybrid tees to achieve a variety of tuning functions. The post in Fig. 8.43f can be replaced by a screw to

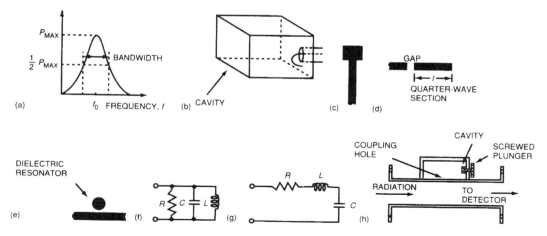

FIGURE 8.45 Microwave resonators: (a) resonator response, (b) rectangular cavity resonator, (c) microstrip patch resonator, (d) microstrip gap-coupled reflection resonator, (e) transmission dielectric transmission resonator in microstrip, (f) parallel equivalent circuits, (g) series equivalent circuits, and (h) waveguide wavemeter.

Interconnects

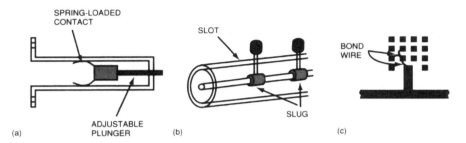

FIGURE 8.46 Tuning elements: (a) waveguide sliding short circuit, (b) coaxial line slug tuner, and (c) microstrip stub with tuning pads.

obtain a screw tuner, which is commonly used in waveguide filters. Sliding short circuits can be used in coaxial lines and in conjunction with branching elements to obtain stub tuners. Coaxial slug tuners are also used to provide adjustable matching at the input and output of active circuits. The slug is movable and changes the characteristic impedance of the transmission line. It is more difficult to achieve variable tuning in passive microstrip circuits. One solution is to provide a number of pads as shown in Fig. 8.46c, which, in this case, can be bonded to the stub to obtain an adjustable stub length. Variable amounts of phase shift can be inserted by using a variable length of line called a line stretcher, or by a line with a variable propagation constant. One type of waveguide variable phase shifter is similar to the variable attenuator of Fig. 8.44d with the resistive material replaced by a low-loss dielectric.

8.9.10 Hybrid Circuits and Directional Couplers

Hybrid circuits are multiport components that preferentially route a signal incident at one port to the other ports. This property is called *directivity*. One type of hybrid is called a *directional coupler*, the schematic of which is shown in Fig. 8.47a. Here, the signal incident at port 1 is coupled to ports 2 and 3, whereas very little is coupled to port 4. Similarly, a signal incident at port 2 is coupled to ports 1 and 4 but very little power appears at port 3. The feature that distinguishes a directional coupler from other types of hybrids is that the power at the output ports (here ports 2 and 3) is different. The performance of a directional coupler is specified by three parameters:

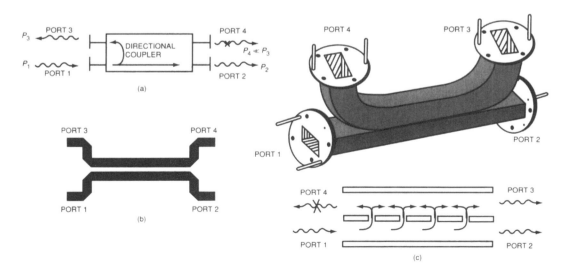

FIGURE 8.47 Directional couplers: (a) schematic, (b) backward-coupling microstrip directional coupler, and (c) forward-coupling waveguide directional coupler.

$$\text{Coupling factor} = P_1/P_3$$

$$\text{Directivity} = P_3/P_4 \qquad (8.11)$$

$$\text{Isolation} = P_1/P_4$$

Microstrip and waveguide realizations of directional couplers are shown in Figs. 8.47b and 8.47c, where the microstrip coupler couples in the backward direction and the waveguide coupler couples in the forward direction. The powers at the output ports of the hybrids shown in Figs. 8.48 and 8.49 are equal, and so the hybrids serve to split a signal into two equal half power signals, as well as having directional sensitivity.

8.9.11 Filters

Filters are combinations of microwave passive elements designed to have a specified frequency response. Typically, a topology of a filter is chosen based on established lumped element filter design theory. Then, computer-aided design techniques are used to optimize the response of the circuit to the desired response.

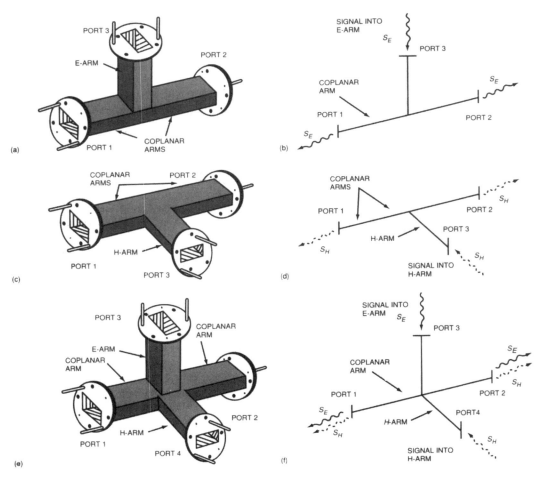

FIGURE 8.48 Waveguide hybrids: (a) E-plane tee and (b) its signal flow, (c) H-plane tee and (d) its signal flow, and (e) magic tee and (f) its signal flow. The negative sign indicates $180^{\to 0}$ phase reversal.

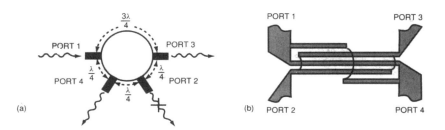

FIGURE 8.49 Microstrip hybrids: (a) rat race hybrid, and (b) Lange coupler.

8.9.12 Ferrite Components

Ferrite components are nonreciprocal in that the insertion loss for a wave traveling from port A to port B is not the same as that from port B to port A.

Circulators and Isolators

The most important type of ferrite component is a circulator (Figs. 8.50a and 8.50b). The essential element of a circulator is a piece of ferrite, which when magnetized becomes nonreciprocal, preferring progression of electromagnetic fields in one circular direction. An ideal circulator has the scattering matrix

$$[S] = \begin{bmatrix} 0 & 0 & S_{13} \\ S_{21} & 0 & 0 \\ 0 & S_{32} & 0 \end{bmatrix} \tag{8.12}$$

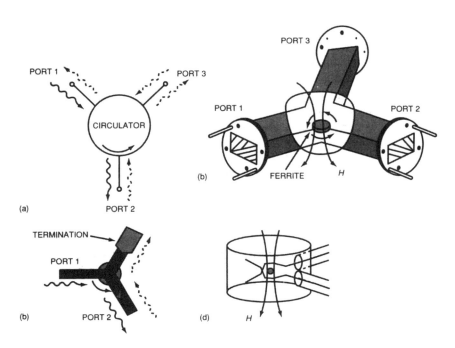

FIGURE 8.50 Ferrite components: (a) schematic of a circulator, (b) a waveguide circulator, (c) a microstrip isolator, and (d) a YIG tuned bandpass filter.

In addition to the insertion and return losses, the performance of a circulator is described by its isolation, which is its insertion loss in the undesired direction. An isolator is just a three-port circulator with one of the ports terminated in a matched load as shown in the microstrip realization of Fig. 8.50c. It is used in a transmission line to pass power in one direction, but not in the reverse direction. It is commonly used to protect the output of equipment from high reflected signals. The heart of isolators and circulators is the nonreciprocal element. Electronic versions have been developed for microwave monolithic integrated circuits (MMICs). A four-port version is called a *duplexer* and is used in radar systems and to separate the received and transmitted signals in a transceiver.

YIG Tuned Resonator

A magnetized YIG (yttrium iron garnet) sphere, shown in Fig. 8.50d, provides coupling between two lines over a very narrow bandwidth. The center frequency of this bandpass filter can be adjusted by varying the magnetizing field.

8.9.13 Passive Semiconductor Devices

A semiconductor diode can be modeled by a voltage-dependent resistor and capacitor in shunt. Thus, an applied dc voltage can be used to change the value of a passive circuit element. Diode optimized to produce a voltage variable capacitor are called varactors. In detector circuits, a diode voltage variable resistance is used to achieve rectification and, through design, produce a dc voltage proportional to the power of an incident microwave signal. A controllable variable resistance is used in a p-i-n diode to realize an electronic switch.

8.9.14 Defining Terms

Characteristic impedance.   Ratio of the voltage and current on a transmission line when there are no reflections.
Insertion loss.   Power lost when a signal passes through a device.
Reference impedance.   Impedance to which scattering parameters are referenced.
Return loss.   Power lost upon reflection from a device.
Voltage standing wave ratio (VSWR).   Ratio of the maximum voltage amplitude on a line to the minimum voltage amplitude.

Reference

Vendelin, G.D., Pavio, A.M., and Rohde, UL. 1990. *Microwave Circuit Design Using Linear and Nonlinear Techniques.* Wiley, New York.

Suggested Readings

The following books provide good overviews of passive microwave components:

Microwave Engineering Passive Circuits, by PA. Rizzi, Prentice-Hall, Englewood Cliffs, NY, 1988.
Microwave Devices and Circuits, by S.Y. Liao, 3rd ed. Prentice-Hall, Englewood Cliffs, NY, 1990.
Microwave Theory, Components and Devices, by I.A. Seeger, Prentice-Hall, Englewood Cliffs, NJ, 1986.
Microwave Technology, by E. Pehl, Artech House, Dedham, MA, 1985.
Microwave Engineering and Systems Applications, by E.A. Wolff and R. Kaul, Wiley, New York, 1988.
Microwave Engineering, by T.K. Ishii, 2nd ed. Harcourt Brace Jovanovich, Orlando, FL, 1989.

Microwave Circuit Design Using Linear and Nonlinear Techniques by G.D. Vendelin, A.M. Pavio, and U.L. Rohde, Wiley, New York, 1990, provides a comprehensive treatment of computer-aided design techniques for both passive and active microwave circuits.

The monthly journals IEEE *Transactions on Microwave Theory Techniques,* IEEE *Microwave and Guided Wave Letters,* and *IEEE Transactions on Antennas and Propagation* publish articles on modeling and design of microwave passive circuit components. Articles in the first two journals are more circuit and component oriented, whereas the third focuses on field theoretic analysis. These are published by the Institute of Electrical and Electronics Engineers, Inc. For subscription or ordering contact: IEEE Service Center, 445 Hoes Lane, P.O. Pox 1331, Piscataway, NI 08855-1331.

Articles can also be found in the biweekly magazine *Electronics Letters* and the bimonthly magazine *IEE Proceedings Part H-Microwave, Optics and Antennas*. Both are published by the Institute of Electrical Engineers and subscription inquiries should be sent to IEE Publication Sales, P.O. Box 96, Stenage, Herts. SG1 2SD, United Kingdom. Telephone number (0438) 313311.

The *International Journal of Microwave and Millimeter-Wave Computer-Aided Engineering* is a quarterly journal devoted to the computer-aided design aspects of microwave circuits and has articles on component modeling and computer-aided design techniques. It has a large number of review-type articles. For subscription information contact John Wiley & Sons, Inc., Periodicals Division, P.O. Box 7247-8491, Philadelphia, Pennsylvania 19170-8491.

References

1. *Reference Data for Engineers,* 8th ed. Prentice-Hall, Englewood Cliffs, NJ, 1993.
2. *The ARRL 1999 Handbook.* American Radio Relay League, 1999.
3. *Design and Application Guidelines for Surface Mount Connectors,* IPC-C-406. IPC, Lincolnwood, IL 1990.
4. Bristow JP, Liu Y, Marta T, Bounnak S, Johnson K. "Cost-effective optoelectronic packaging for multichip modules and backplane level optical interconnects." *Proceedings of the International Photo-Optical Instrumentation Engineers,* Bellingham, WA, 1996.
5. Cohen T, Patel G, Stokoe P. "At the crossroads of high-speed PCB interconnect technology." *Electronic Packaging and Production,* vol. 37, no. 16, December 1997.
6. Dudinski FE. Socketing CSPs." *Connector Specifier,* April 1998.
7. Granitz RF. "Levels of packaging." Instrumentation and Control Systems, 8–92.
8. Infantolino W. "Evaluation and optimization of a surface mount connector design with respect to solder creep." Advances in *Electronic Packaging,* Society of Mechanical Engineers, vol. 4-1, 1993.
9. Lubowe AG, Million TP. "MACII optical connectors for printed circuit board input/output." *Proceedings of the International Photo-Optical Instrumentation Engineers,* Bellingham, WA, 1996
10. Markstein HW. "Ensuring signal integrity in connectors, cables and backplanes." *Electronic Packaging and Production,* vol. 36, no. 10, October 1996.
11. Sugiura N. "Effect of power and ground pin assignment and inner layer structure on switching noise." *IEICE Transactions on Electronics,* vol. 9, no. 5, May 1995.
12. Sugiura N, Yasuda K, Oka H. "High-speed signal transmission characteristics for improved coaxial printed board connector in a rack system." *Intl. J. of Microcircuits & Electronic Packaging,* vol. 18, no. 2, 1995.
13. Vendelin GD, Pavio AM, and Rohde UL; *Microwave Circuit Design Using Linear and Nonlinear Techniques.* Wiley, New York, 1990.
14. Whitaker JC, DeSantis E, Paulson CR; *Interconnecting Electronic Systems.* CRC Press, Boca Raton, 1992.

9
Design for Test

Glenn R. Blackwell
Purdue University

9.1 Introduction 9.1
9.3 Testing Philosophies.. 9.1
9.3 Scan Test for Digital Devices... 9.4
9.4 General Electrical Design 9.15
9.5 Design for Test Fixtures... 9.18

9.1 Introduction

Design for test (DFT), also known as *design for testability*, is the first part of the test techniques, the second part being *testing*, as described in Chapter 12. This chapter covers primarily hardware test, with software test issues presented only as they relate to testing of companion hardware.

Where Else?

Other chapters with relevant information include Chapter 12, "Testing of Assemblies"; Chapter 2, "Surface Mount Technology Overview"; Chapter 4, "Chip on Board Assemblies"; and Chapter 6, "EMC and PCB Design."

9.2 Testing Philosophies

In an ideal world, the design and manufacturing processes would be so robust that testing would not be necessary. Until that world comes along, test is necessary. Although test adds no inherent value to a product, practically, it adds to the overall product value since, if done properly, it will find faults early enough in the manufacturing stage to allow their cost-effective correction, and the elimination of faulty units will result in a more successful marketing effort with higher sales. Design for test attempts to minimize the cost and maximize the success and value of the test process. Implementations of high-density circuits, high clock speeds, large data buses, surface mount technology (SMT), and high-density interconnects (HDI) all contribute to modern electronics but all make test more difficult. Additionally, for PC-based data acquisition, signal conditioning and signal acquisition are major issues. These are discussed further in Section 9.4.1.

Test is necessary in circuits to detect all possible faults that could prevent proper circuit operation. Faults that occur within components are checked at incoming inspections and at in-circuit test and are known as *functional* or *component* faults. Faults that occur during the manufacturing process are checked at shorts/opens, manufacturing defect analyzer (MDA), and in-circuit tests and are known as *manufacturing* defects. Faults that occur as a result of the interaction of specification limits of "good" components are known as *design* defects, as are faults that result from poor designs. Faults that occur as a result of software problems on a properly assembled board with good components are known as *software* faults. The design and test teams must understand which type of faults are most likely to occur in their product and design to minimize these expected faults. Again, the test system should only be in place to find random/chaotic faults, not repetitive faults that could have been designed out, whether through improvement in product design or improvement in process design.

Design for test is intended to be part of the concurrent engineering process wherein circuit design, manufacturing issues, and test issues are all considered by the design team during the design phase of the project. This is sometimes know as adding *testability* to a product. While adding testability increases design time, it is intended to reduce overall time-to-market while reducing the costs of manufacturing test and maintenance testing. A 1987 Mitre Corporation report showed that the system design phase of product development represented only 15% of a product's total life-cycle costs while having a 70% impact on a product's operation and support costs over the product's total life.

9.2.1 Bare Board Tests

It is important to note that, while most of this chapter deals with the testing of loaded boards, bare board test is the first step in the series of tests used in circuit board assembly. Placing and reflowing components on defective bare boards is wasteful of both time and money. Working with the board supplier to determine the supplier's level of faults will help determine if bare board testing should be implemented in the user's facility. If bare board testing in necessary, it is important to determine that the tests used will provide 100% fault coverage. This coverage must include double-sided testing, or voids in vias and/or through holes will not be detected.

Bare board testing must include tests for continuity of traces, vias, and through holes to rule out any opens. It must also include isolation tests to rule out shorts between adjacent traces and vias. The testing must follow requirements as spelled out in IPC-ET-652:

Continuity test threshold (maximum allowed resistance):
    Class 1, general electronic devices: 50 Ω
    Class 2, dedicated service devices: 20 Ω
    Class 3, high-reliability devices: 10 Ω
    Generally, the lower the test resistance is, the better. Many users will set their allowable much lower than the IPC standard, and 5 Ω is common.

Isolation test threshold (minimum allowed resistance):
    Class 1, general electronic devices: 500 kΩ
    Class 2, dedicated service devices: >2 MΩ
    Class 3, high-reliability devices: >2 MΩ
    Generally, the higher the test isolation is, the better. Many users will set their allowable much higher than the IPC standard, and 100 MΩ is common.

Isolation test voltage:
    High enough to provide the necessary minimum resistance test current, but low enough to prevent arc-over between adjacent conductive features. Since 1/16 in (0.0625 in) air gap is sufficient for 120 Vac applied, calculations can determine the maximum voltage.

Bare board testing must consider all nets shown on the original board CAD file, and this must result in 100% electrical test of the board. There must also be a method of marking either good or bad boards by the test system. For the isolation tests, it is not necessary to test each net to all other nets. Once adjacent nets are identified, testing will be faster if only physically adjacent nets are tested against one another.

9.2.2 Loaded Board Tests

Simplistically, test considerations require:

- One test node per circuit net
- Test fixture probe spacing of 0.080 in (2 mm) minimum
- Probe-to-device clearance of 0.030 in (0.9 mm) minimum

- All test node accessible from one side of the board
- A test node on any active unused pins
- Provision of extra gates to control and back-drive clock circuits
- Insertion of extra gates or jumpers in feedback loops and where needed to control critical circuit paths
- Unused inputs tied to pull-up or pull-down resistors so that individual devices may be isolated and back-driven by the ATE system
- Provision of a simple means of initializing all registers, flip-flops, counters, and state machines in the circuit
- Testability built into microprocessor-based boards

Realistically, most modern designs prohibit meeting all these requirements either due to device and/or circuit complexity, the demand for miniaturization of the circuit, or both. The design team must remember that quality should be designed in, not tested in. No single method of test should be mandated by quality. Test decisions should be made based on decisions made in conjunction with the optimization of circuit characteristics.

The most common types of electronic loaded board test, along with some of the disadvantages and advantages of each, are shown in Table 9.1

TABLE 9.1 Test Method Comparison

	Shorts/Opens	MDA	In-Circuit	Functional
Typical use	Go/no-go	Manufacturing defect detection	Manufacturing and component defect detection	Performance and spec-compliance testing
Go/no-go decision time	Very fast	Fast	Slow	Fast
Fault sensitivity	Shorts/opens	Manufacturing defect	Manufacturing and component defects	Mfg., component, design & software defects
Fault isolation level	Node	Component	Component	To spec, not any specific component or software line.
Multi-fault isolation	Good	Good	Good	Poor
Repair guidance	Good	Good	Good	Poor
Programming cost and time	Low	Low	Medium	High
Equipment costs	Low	Low	Medium	High

At the end of the circuit board/product assembly process, the board/product must meet its specifications, which were determined by initial product quality definitions. Generally, the following statements must then be true:

- All components work to spec.
- All components are soldered into their correct location on the assembly.
- All solder joints were properly formed.
- No combination of component specification limits and design criteria results in performance that fails to meet the overall performance specifications of the assembly/product.
- All product functions meet spec.

Ideally, all tests that find faults will allow for rapid identification and isolation of those faults. It is important to understand that testing has two major aspects: control and observation. To test any product or system, the test equipment must put the product/system into a known state with defined inputs, then observe the outputs to see if the product/system performs as its specifications say it should. Lack of control or observability will lead to tests that have no value.

Tests can be divided into high, medium, and low-level tests. These distinctions in no way indicate the value of the tests but, rather, their location in and partnership with the test process. High-level tests require specialized components that have test hardware and/or software built in. Medium-level test may not require specialized hardware but require specialized software for optimum testing. Low-level tests require no special on-board hardware or software in the product and are performed by external test equipment/systems that have been preprogrammed with all the necessary criteria to validate correct operation of the assembly/product.

Tests can also be divided into certain physical strategies:

- Incoming inspection to verify individual component specs
- Production in-circuit tests (ICT)
- Functional tests on the assembly/product

9.2.3 Vector Tests

Vector tests are primarily designed to test the functionality of a digital device before it becomes part of an assembly. A test vector is a set of input conditions that result in defined output(s). A test vector may include timing functions; e.g., to test a counter, a test vector could be designed to exercise the inputs by clearing and enabling the device then cycling the count input ten times, with an expected result of the output lines having a value equal to ten. A set of test vectors may also need to be developed for a device if has a set of possible functions that cannot all be exercised by one vector.

Test vectors for programmable devices such as PLDs, CPLDs, and FPGAs can be generated automatically, using an automatic test program generator (ATPG). ATPG refers to the addition of partial or full scan into a design, and it is available to generate test programs for common ATE systems from companies like Teradyne and Hewlett-Packard. Implementation of ATPG and scan will replace all flip flops, latches, and cross-coupled gates with scannable flip flops. Additionally, the circuit will be modified to prevent the asserting of sets and resets. The fault coverage provided by ATPG systems is claimed to be over 90%. Some systems will also generate *DFT reports* to allow determination of whether the initial design lends itself well to DFT concepts and good fault coverage. This allows the designer to focus on any logic elements that are preventing good fault coverage and to perform design changes to improve coverage.

However, ATPG will not work for all devices. Pinout limitations can prevent adequate coverage, and economic or space limitations may prevent the additional gates necessary to support full scan testing. In these cases, test vectors can be written manually to obtain maximum fault coverage. Scan testing also will not confirm if the overall design is correct—only that the chip was manufactured/programmed correctly. Scan also does not check the timing of a device, and a static timing analysis must be performed. Synopsis notes that logic whose fault effects pass into a memory element, or logic that requires the outputs of the memory to set up a fault, are said to be *in the shadow* of the memory. This memory *shadow* causes a reduction in fault coverage. Vendors in the ATPG arena include Synopsis, Tekmos, and Flynn Systems.

9.3 Scan Test for Digital Devices

There are a number of scan tests that can be designed into a digital product. Generally, these tests are designed to develop a test vector, or set of test vectors, to exercise the part through its functions. Full-scan tests are custom tests developed for each part individually. Boundary scan, a.k.a. JTAG or 1149.1 test, is a more closely defined series of tests for digital devices. It primarily exercises the devices' I/O lines.

The Joint Test Automation Group (JTAG) was formed by a group of European and American companies for the express purpose of developing a boundary-scan standard. JTAG Rev 2.0 was the document developed by this group in 1988.

Design for Test

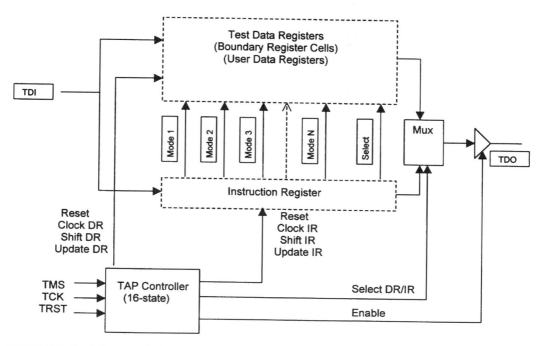

FIGURE 9.2 Block diagram of a boundary-scan device.

to note that the logic required by the 1149.1 architecture may not be used as part of the application logic. This is an advantage, since the test logic can be accessed to scan data and perform some on-line tests while the IC is in use.

The TAP responds to the test clock (TCK) and test mode select (TMS) inputs to shift data through the instruction register, or to shift data to a selected data register from the test data input (TDI) to the test data output (TDO) lines. The output control circuit allows multiplexing the serial output of the instruction or a selected data register such as ID to the TDO line. The instruction register holds test commands. The boundary-scan register allows testing of both the wiring interconnects as well as the application logic of the IC. The single-bit bypass register allows a scan path through the IC when no testing is in progress. The optional IDCODE register is a 32-bit shift register that provides device ID and manufacturer information, such as version number.

9.3.2 Test Access Port

The TAP is a finite state machine with 16 defined states. The serial input via the TMS line allows it to transition through its predefined states during scan and test operations. The serial technique was chosen to minimize the number of IC pins that would be required to implement 1149.1. A duplication of the TAP functions using a "no-state" machine (combinational logic) would require nine lines rather than four. And since 1149.1 test port signals are also intended to be bussed among ICs, boards, backplanes, and onto multichip modules (MCMs), a minimum number of lines is preferred for the bus.

TAP State Diagram

Shown in Fig 9.3 is the 16-state diagram of the 1149.1 architecture, with its six steady states indicated with an *ss*. The steady states are

- Test logic reset TLRST, resets test logic
- Run test/idle RT/IDLE, runs self-tests or idles test logic
- Shift data register SHIFT-DR, shifts data from TDI to TDO

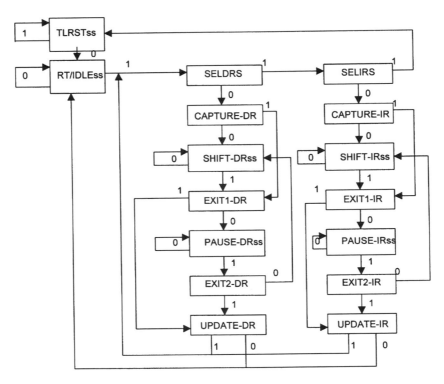

FIGURE 9.3 State diagram of 11491 test access port.

- Shift instruction SHIFT-IR, shifts instruction from TDI to TDO
- Pause data register PAUSE-DR, pauses data shifting
- Pause Instruction Register PAUSE-IR, pauses instruction shifting

The TDO line is enabled to allow data output during either SHIFT state. At this time, data or instructions are clocked into the TAP architecture from TDI on the rising edge of TCK, and clocked out through TDO on the falling edge of TCK.

The ten temporary states allow transitions between steady states and allow certain required test actions:

- Select data register scan SELDRS,
- Capture data register CAPTURE-DR
- Exit 1 data register EXIT1-DR
- Exit 2 data register EXIT2-DR
- Update data register UPDATE-DR
- Select Instruction register scan SELIRS
- Capture instruction register CAPTURE-IR
- Exit 1 instruction register EXIT1-IR
- Exit 2 instruction register EXIT2-IR
- Update instruction register UPDATE-IR

Test Access Port Operation Modes

As described earlier, the TAP uses five well defined operational modes: reset, idle, data scan, instruction scan, and run test. These are 1149.1 required operational modes that ensure that ICs from different manufacturers will operate together with the same test commands.

At power up, or during normal operation of the host IC, the TAP is forced into the test-logic-reset (TLR) state by driving the TMS line high and applying five or more TCKs. In this state, the TAP issues a reset signal that resets all test logic and allows normal operation of the application logic. When test access is required, a protocol is applied via the TCS and TCK lines, which instructs the TAP to exit the TSR state and shift through and to the desired states. From the run test/idle (RTI) state, the TAP will perform either an instruction register scan or a data register scan.

Note that the sequential states in the IR scan and the DR scan are identical. The first operation in either sequence is a capture operation. For the data registers, the capture-DR state is used to parallel load the data into the selected serial data path. If the boundary-scan register (BSR) is selected, the external data inputs to the application logic are captured. In the IR sequence, the capture-IR state is used to capture status information into the instruction register.

From the capture state, the shift sequence continues to either the shift or exit1 state. In most operations, the TAP will sequence to the shift state after the capture state is complete so that test data or status information can be shifted out for inspection and new data shifted in. Following the shift state, the TAP will return to either the RTI state via the exit1 and update states or enter the pause state via exit1. The pause state is used to temporarily suspend the shifting of data through either the selected data or instruction register while another required operation, such as refilling a tester memory buffer, is performed. From the pause state, sequence shifting can resume by re-entering the shift state via the exit2 state, or be terminated by entering the RTI state via the exit2 and update states.

Reset Mode*

The reset mode forces the TAP into the TLR state by either a control input the TMS pin or by activation of the optional TRST input. In the reset mode, the TAP outputs a reset signal to the test architecture to keep it in a reset/inactive state. When the IC is in its operational mode, the TAP should be in TLR to prevent the test logic from interacting with the application logic. There are, however, situations when certain test modes can be used during application logic operation.

Idle Mode

In this condition, the test architecture is in a suspended state where no test or scan operations are in progress. The architecture is not reset.

Date Scan Mode

The data scan mode must be accessed by sequencing the TAP through a series of states entered via the SELDRS state shown in the state diagram figure. During this mode, the TAP instructs a data register to perform a predefined series of test steps that consist of a capture step, a shift step, and an update step. The capture step causes the selected data register to parallel load with test data. The shift step causes the selected data register to serial-shift data from TDI to TDO. The update step causes the selected data register to output the test data it received during the shift step. If required, a pausing step can be used to suspend the transfer of data during the shifting step. Data register selection is determined by the instruction in the instruction register.

Instruction Scan Mode

The TAP enters this mode by being sequenced through a series of states entered via the SELIRS state shown on the state diagram. In the instruction scan mode, the instruction register receives commands from the TAP to perform a predefined set of test steps consisting of a capture step, a shift step, and an update step. Similar to the actions in the data scan mode, the capture step causes the instruction register to parallel load with the status information. The shift step causes the instruction register to shift data from TDI to TDO. The update step causes the instruction register to parallel output the instruction data it received during the shift step. If required, a pause step can be used to suspend the transfer of data during the shift step.

---

*The reset mode is an active-low mode that requires a pull-up on the TMS line to ensure the TAP remains in the TLR state when the TMS is not externally driven.

Run Test Mode

The TAP enters this mode by loading a self-test instruction into the instruction register, then transitioning the TAP into the RT/IDLE state. When the TAP enters the RT/IDLE state, the self-test starts and continues during this state. The self-test terminates when the TAP is brought out of this state. The TAP stays in this state for the number of TCK cycles required by the self-test operation being performed.

Software Commands

The 1149.1 standard provides a set of required and standard software command functions plus many possible functions that are optional or that can be defined by the manufacturer following 1149.1 guidelines. The required commands are

- EXTEST
- SAMPLE/PRELOAD
- BYPASS

The standard optional commands are

- INTEST
- RUNBIST
- IDCODE
- USERCODE

The following are the instructions defined by the 1149.1 standard. Optional instructions may be defined and included in the architecture.

- *Extest:* defined as all logic zeros, this instruction disables the application logic of the IC, placing the boundary-scan register between TDI and TDO. In this mode, the solder interconnects and combinational logic between ICs on the board may be tested. For example, if four 1149.1-standard ICs are connected in parallel, giving all of them the Extest instruction allows an external test system to place a word at the input of the first IC. If all solder connections are correctly made, and all circuit traces are fully functional, that same word will pass through all four ICs and be present on the outputs for the last IC.
- *Sample/Preload:* this instruction provides two modes. In sample, the application logic operates normally, and the TAP can sample system data entering and leaving the IC. In preload, test data can be loaded into the boundary register before executing other test instructions.
- *Bypass:* defined as all logic ones, this instruction allows a direct scan path from TDI to TDO and the application logic operates normally. To ensure proper operation of the IC, the TDI input is required to have external pull-up, so that if there is no external drive, logic ones will be input during an instruction scan, thereby loading in the bypass instruction.
- *Runbist:* this optional instruction places the TAP into the RT/IDLE state, and a user-defined self-test is performed on the application logic, placing the test results in a user-defined data register, which is between TDI and TDO. On completion of the self-test, the user-defined data register can be accessed to obtain test results.
- *Intest:* this optional instruction allows internal testing of the application logic. The boundary-scan register is selected between TDI and TDO. The ICs inputs can now be controlled, and its outputs can be observed. Boundary-scan cells with controllability must be used as part of the hardware. These type-2 and type-3 cells will be described later.
- *Idcode:* this optional instruction allows the application logic to operate normally and selects the IDCODE register between TDI and TDO. Any information in the IDCODE register, such as device type, manufacturer's ID, and version, are loaded and shifted out of this register during data scan operations of the TAP.

- *Usercode:* this optional instruction allows the application logic to operate normally and selects the IDCODE register between TDI and TDO. User-defined information, such as PLC software information, is loaded and shifted out during the data scan operations of the TAP.
- *Highz:* this optional instruction disables the application logic and the bypass register is selected between TDI and TDO. The outputs are forced into a high-impedance state and data shifts through the bypass register during data scan operations. This instruction is designed to allow in-circuit testing (ICT) of neighboring ICs.
- *Clamp:* similar to highz, this optional instruction disables the application logic and causes defined test data to be shifted out of the logic outputs of the IC. This instruction is designed to allow scan testing of neighboring ICs.

TAP Hardware

- *Instruction Register (IR):* this is a multi-bit position register used to store test instruction. As shown in Fig 9.4, it consists of a shift register, an output latch, and a decode logic section(s). The stored instruction regulates the operation of the test architecture and places one of the data registers between TDI and TDO for access during scans. During an instruction scan operation, the shift register captures parallel status data then shifts the data serially between TDI and TDO. The status inputs are user-defined inputs, which must always include a logic one and a logic zero in the least two significant bits. This allows stuck at-fault locations on the serial data path between ICs to be detected and fixed.

   The IR output latch section consists of one latch for each bit in the shift register. No change is made in these latches during an instruction scan. At the completion of the scan, these latches are updated with data from the shift register.

   The IR decode latch section receives the latched instruction from the output latch, decodes the instruction, then outputs control signals to the test architecture. This configures the test architecture and selects a data register to operate between TDI and TDO.

   When the TAP is given a TLRST command, it outputs a reset signal to the IR. This causes the shift register and output latch to initialize into one of two conditions. If the test architecture

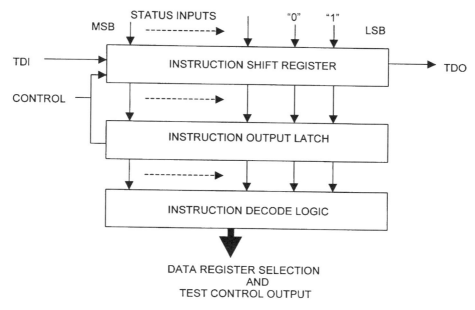

FIGURE 9.4 IEEE 11491 instruction register

includes an identification register, the shift register and output latch initialize to the IDCODE instruction that selects the identification register between TDI and TDO. If no identification register is present, they initialize to the bypass instruction that selects the bypass register between TDI and TDO. Both of these instructions enable immediate normal operation of the application logic.

Data Registers

The data registers are defined to be all registers connected between TDI and TDO, with the exception of the instruction register. IEEE 1149.1 defines no upper limit to the number of data registers that can be included in the hardware. The minimum is two data registers, defined as the bypass and boundary-scan registers. The data registers are defined in the following sections.

Identification Register
This optional register is a 32-bit data register that is accessed by the IDCODE instruction. When accessed, it will provide information that has been loaded into it, such as the IC's manufacturer, version number, and part number. It can also use an optional usercode instruction to load and shift out user-defined information. If a data scan operation is initiated from the TAP's TLRST state, and the IC includes this register, it will be selected, and the first bit shifted out of TDO will be a logic one. If the IC does not include this register, the bypass register will be selected, and the first bit shifted out of TDO will be a logic zero. In this manner, it is possible to determine if the IC includes an identification register by examining the first bit shifted out.

Bypass Register
The required bypass register is a single-bit shift register (single scan cell) that is selected between TDI and TDO when the instruction register is loaded with a bypass, clamp, or highz instruction. When selected, the bypass register loads to a logic zero when the TAP passes through the CAPTURE-DR state, then shifts data from TDI to TDO during the SHIFT-DR state. When the TAP is in bypass mode, this register allows the scan path length through an IC to be reduced to a single bit.

Boundary-Scan Register
This required shift register forms a test collar of boundary-scan cells (BRCs) around the application logic, with one cell for each input and output pin. It is selected between TDI and TDO when the instruction register loads an extest, intest, or sample/preload instruction. During normal IC operation, the presence of this register is transparent to the normal operation of the application logic, with the addition of one additional layer of logic delay. In the boundary test mode, the normal operation is halted and a test I/O operation is enabled via the TAP, the boundary-scan register, and the other registers. The series of BRCs that make up this register are each associated with one of the input or output pins of the application logic and any associated control pins. During data scan operations, data is shifted through each cell of this register from TDI to TDO. The test function performed by this register is defined by the test instruction loaded into the instruction register. BRCs can be of three different types.

9.3.3 Boundary-Scan Register Cells

As mentioned, there are three types of BRCs that can be used by IC designers. These are an observe-only cell, an observe-and-control cell, and an observe-and-control cell with output latch. An example of each is given in the paragraphs below.

The observe-only (OO) cell has two data inputs: DI data in from the application logic pins, and SI scan in (Fig. 9.5). It has two control inputs, SHIFTDR shift data register and CLKDR clock data register. It has one data output, SO scan out. During boundary-scan operations, the SHIFT-DR and CLK-DR inputs are controlled by the TAP to load data into the cell from DI then shift data through the cell from SI to SO.

When the TAP passes through the CAPTURE-DR state, the SHIFT-DR input causes the shift register to transfer data from SI to SO. The OO cells should by used on IC designs which requiring monitoring

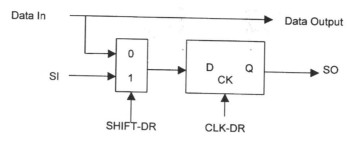

FIGURE 9.5 Observe-only cell.

of their state only during testing. It is also important to note that all the boundary cells add some delay to the lines they are inserted in, but OO cells have the shortest delay of all three types of cells. They would therefore be the preferred cell for lines that cannot tolerate longer additional delays.

The observe-and-control (OC) cell (Fig. 9.6) is identical to the observe-only cell, with the addition of a DO data output signal and a MODE control signal. The MODE signal comes from the IR and controls DO to output either the DI input of the Q output of the shift register. During instructions that allow normal operation of the application logic, MODE causes the system input data to flow directly from DI to DO.

During instructions that enable boundary testing, MODE causes the test data from the shift register to be output on DO. OC cells are obviously used on IC inputs that require both monitoring and control during boundary scan testing. The two drawbacks of the OC cell are that it adds more delay to the line it is monitoring, and it adds some ripple to the signal on that line.

The observe-and-control cell with latch (OCL) is identical to the observe-and-control cell, with the addition of an output latch on the shift register output and a latch UPDATE control signal (Fig. 9.7). The latch prevents DO from rippling as data is shifted through the shift register during scan operations. When the TAP passes through the UPDATE-DR state at the end of a scan operation, it outputs the UPDATE signal which causes the latch to output data from the shift register.

This type cell is used on all IC output and tristate control lines and any lines that require ripple-free control during testing. While this cell does not add ripple to the monitored line like the OC cell does, it does have the longest signal delay of any of the boundary register cells.

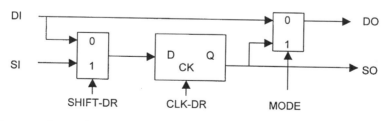

FIGURE 9.6 Observe-and-control cell.

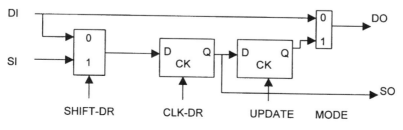

FIGURE 9.7 Observe-and-control cell with latch cell.

9.3.4 Boundary-Cell Applications

Shown in Fig. 9.8 is an example of a boundary scan register and its simplified I/O around the application logic of an IC. This example IC has four input pins, two output pins, and one bidirectional I/O control line. It shows all three types of boundary scan cells in use. The boundary scan cells are connected so that the application I/O is unaffected or so there is a scan path from TDI to TDO.

IEEE 1149.1 requires that the application-logic input pins on the IC be capable of being monitored by the boundary-scan register. In the above example, an OO cell is on INPUT4, allowing for monitoring of that line. As noted previously, observe-only cells have the shortest delay time of the three types of boundary register cells. An OC cell is used on INPUT2 and INPUT3, allowing for monitoring and control of those lines. An OCL cell is shown on INPUT4, allowing that line to be latched, with the accompanying reduction in ripple that latching provides.

The output pin OUTPUT1 is a two-state output which is required by 1149.1 to be observable and controllable, with latched output to prevent ripple. These conditions are met by using an OCL cell on this output. OUTPUT2 is shown as a tristate output, with lo, high, and high-z output states. This type of output is required by 1149.1 to be observable and controllable, which requires an additional OCL cell at this output. One OCL cell is used to control hi and low, while a second OCL cell is used to enable or disable OUPUT2.

The bidirectional I/O pin is also required to be fully observable and controllable. This pin acts like a tristate output, with the addition of input capability. This requires three OCL cells for compliance with these requirements of 1149.1.

9.3.5 Support Software

One of the major areas in which there has been a lack of standardization is in the area of boundary-scan software. IEEE 1149.1b-1994 described a boundary scan descriptive language (BSDL), modeled after VHDL. Even with this availability, many users do not have complete conformance to 1149.1. In many

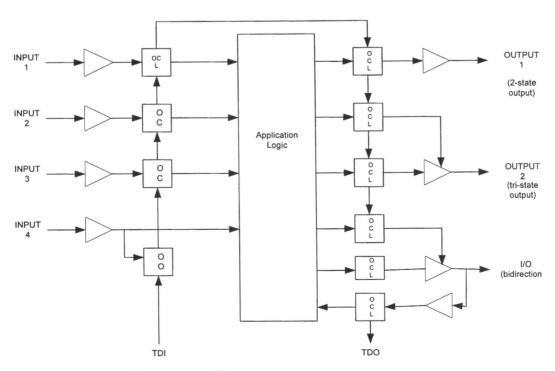

FIGURE 9.8 Example of a boundary-scan IC.

Design for Test   9-15

developments, it has been downstream tools, such as automatic test systems, that have been the first to find nonconforming issues in a design. By this time, corrections and changes are very expensive and time consuming. As more and more electronic design automation (EDA) tools present themselves as being 1149.1 compatible, this problem appears to be growing worse.

One group that is attempting to assist users with conformance is Hewlett-Packard's Manufacturing Test Division. To assist both IC and EDA users who anticipate using HP testers, HP has made available their back-end (test) tools on the World Wide Web (http://hpbscancentral.invision1.com/). This system has the following two major elements:

- A BSDL compilation service that will determine if a BSDL description is syntactically correct and will make numerous checks for semantic violations.
- An automatic test pattern generation that will derive a set of test vectors in a "truth table" format directly from a BSDL description. HP says this generation will exercise facets of the implementation including:
    - All transitions in the 1149.1 TAP state diagram
    - Mandatory Instructions and their registers
    - IDCODE/USERCODE readout
    - Loading of all non-private instructions and verification of target registers
    - Manipulation of the boundary register and all associated system pins

9.4 General Electrical Design

Discussed more fully in Chapter 12, the available test techniques need to be understood to make informed choices during the application of DFT principles in the design phase. Each test mode has its own positive and negative aspects, and decisions must be based on consideration of the issues of cost, development time, and fault coverage. It is unlikely that all test techniques are available in house, so purchase issues must also be considered. MacLean and Turino discuss in detail various aspects of each test technique (see Table 9.1).

TABLE 9.2 Test Techniques

Technique	Cost	Test Speed	Fault Coverage	Software Development
ICA/MDA	low	very fast	opens, shorts, tolerances, unpowered tests on passives	very fast
ICT	low	very fast	opens, shorts, missing components, parametric passive part function test, powered tests to exercise active components, E²PROM programming	fast
Functional	high	fast	test to the final functional specifications of the product, will not identify bad components	slow, expensive
Vectorless	low	fast	opens, shorts, missing	
Vision	high	slow	shorts, missing components, off-location components, insufficient solder joints, some opens, component alignment	by vendor/manufacturer of system
2-D X-ray	medium	slow/medium	solder adequacy, some opens, shorts, component alignment, limited use on multilayer boards	
3-D X-ray	high	slow	solder adequacy, opens, shorts	

ICA = in-circuit analyzer, MDA = manufacturing defect analyzer, ICT = in-circuit tester

Further descriptions of the test techniques can be found in Chapter 12. Using the characteristics of each technique as part of the DFT process, below are examples of issues that must be considered for each technique.

ICA/MDA
- Test pads at least 0.030 inch diameter
- Test pad grid at least 0.080 inch (smaller is possible, at higher cost)
- Test pads no closer than 0.050 inch to any part body
- Test pad for each net, ideally all on the bottom side
- Incorporate bed-of-nails fixture design needs, e.g., edge clearance prohibitions, into board layout

ICT
- Grids and spacings as for ICA/MDA
- Bed-of-nails fixture design as for ICA/MDA
- Provide tristate control
- Provide clock isolation
- Provide integration with IEEE1149.1 boundary-scan tests, if present (see also Section 4.3)

Functional
- Provide specification definitions
- Provide appropriate field connector(s)
- Preload product operating software
- Emulate or provide all field inputs and outputs

Vectorless
- Verify that the vectorless probe(s) have appropriate access to the parts to be tested
- Verify that the fields created by the vectorless probes will not affect any programmable parts

Vision
- Decide if 100% inspection is feasible, based on line speed and inspection time
- If 100% inspection is not feasible, develop a plan to allow either complete inspection of every X boards (e.g., if inspection speed is 1/5 line speed, every 5 boards) or inspect a certain proportion of each board, inspecting alternate areas of each board so that 100% coverage occurs every X boards (e.g., inspect 20% of each board, providing 100% inspection of all board areas every 5 boards)
- Provide component spacing to allow appropriate angles for laser or other light sources to reach all solder joints

X-Ray
- As with vision testing, determine if 100% inspection is feasible
- Generate board layout to prevent shadowing of one side by the other on two-sided boards
- For laminography techniques, be certain board warp is either eliminated or detected, so that the cross-section information is appropriate

General DFT considerations must be applied, regardless of the type of board or test method to be used. The ideal considerations include the following:

- A test pad for all circuit nodes. In this consideration, any used or unused electrical connection is a node, including unused pins of ICs and connectors, 0-Ω resistors used as jumpers, fuses and switches, as well as power and ground points. As circuit speeds rise, these consideration become more critical. For example, above 10 MHz there should be a power and ground test pad within one inch of each IC. This requirement also assumes there will be adequate decoupling on the board at each IC.

- Distribute test points evenly across the circuit board. As mentioned earlier, each probe exerts 7 to 16 oz of pressure on the board. When large circuit boards have 500 to 1000 test points, the amount of pressure that must be exerted to maintain contact can range into the hundreds of pounds. The board must be evenly supported, and the vacuum or mechanical compression mechanism must exert its pressure evenly as well. If the board or the test fixture flexes, some probes may not make contact or may not be able to pierce flux residues.
- Do not use traces or edge card connector fingers as test points. Test pads should be soldered, and any traces that are to be probed should be as wide as possible (40 mil minimum) and solder-plated. Gold-plated connector fingers must not be probed, as there is a risk of damage to the plating.
- In any board redesign, maintain test pad locations. Changes in test pad locations not only may require changes in the test fixture, they make it difficult to test several variations of a given design. This requirement is especially important in high-mix, low-volume applications, where both the cost of a new fixture and the time/cost to change fixtures for each variety of board are prohibitive.

Additionally, these active-component guidelines should be followed in all designs:

- Clock sources must be controllable by the tester.
- Power-on reset circuits must be controllable by the tester.
- Any IC control line that the tester must be able to control must not be hard wired to ground or V_{cc}, or to a common resistor. Pull-up and pull-down resistors must be unique to each IC or must be isolated via a multiplexer or other isolation circuit.
- In addition to the test methods, general electrical design for test requires consideration the types of circuits on the board. Analog, digital, microprocessor (and other software-controlled circuits), RF, and power circuits all require their own test considerations.

9.4.1 General Data Acquisition Considerations

The use of personal computer-based test systems normally starts with data acquisition. There are five steps to selecting a data acquisition system:

- Identify the inputs to be acquired.
- Identify the signal characteristics of the inputs.
- Determine the signal conditioning necessary.
- Select an appropriate signal conditioning device, whether stand-alone or PC-based, along with appropriate support software.
- Select the appropriate cabling to minimize noise pickup and signal loss.

Frequently, the test system design begins during the design phase, and further decisions are made after the prototype stage. One common pitfall to the decisions made for the test system is that they are based on measurements made during the prototype stage. During this stage, measurements are often made with directly-connected scopes and DMMs. Translating this directly to a data acquisition (DAQ) systems creates problems. The measurements are no longer taken directly over short probe wires/cables. Now the measurement system is remote, and the front end of the DAQ board may not electrically mimic the performance of the input of the scope and DMM. Two major issues cause the problems: there are much longer, impedance-variable connection cables, and there are no instrumentation amplifiers at the front end of the DAQ board to dig the signals out of environmental noise.

Signal conditioning may require amplification, isolation, filtering, linearization, and an excitation voltage. Amplification of low-level signals from transducers should be performed as close as possible to the measurement source, and only high-level signals transmitted to the PC. In some systems the DAQ is in the PC and an instrumentation amplifier must be used. The instrumentation amplifier can have a combination of high gain, over 100, and high common-mode rejection ratio (CMRR) over 100 dB, equal to 100,000:1. A standard differential amplifier input is not sufficient if this level of performance is necessary.

High CMRR is needed because of the amount of electrical pollution present in all environments today. The signal detected from any source is a combination of a differential-mode voltage (DMV) and a common-mode voltage (CMV). The DMV is the signal of interest. The CMV is noise that is present on both inputs. CMV may be balanced, in which case the noise voltage present on the two inputs is equal, or it may be unbalanced, in which case the noise voltage is not equal on the two inputs. Instrumentation amplifier specifications will address these two issues. Additionally, areas with high CMV also may approach the specification for maximum input voltage without damage. An oscilloscope will help to determine both the magnitude and the frequency spectrum of CMV. Digital scopes with fast Fourier transform (FFT) capabilities will make this assessment easier. Exceeding the maximum voltage spec may not only put the DAQ board/module at risk; if the board is PC-mounted, it may also put the PC at risk unless the DAQ board is isolated.

Connection techniques will aid in minimizing CMV. Twisted-pair cables are a necessity for carrying low-level signals while canceling some acquired noise, and shielding the cables will also keep the acquisition of noise voltages through the cabling to a minimum. They will not, however, do anything about CMV acquired at the source.

Isolation is necessary for any PC-mounted board that is connected to the external environment. Any mistake in the field (has anyone ever made a connection mistake?) may apply a fatal voltage to the DAQ card and, without isolation, that same voltage may be presented to the PC bus. Goodbye to the PC and the test system. Isolation is also necessary if the DAQ board and the signal input have grounds that are at different potentials. This difference will lead to a ground loop that will cause incorrect measurements and, if large enough, can damage the system. Well isolated input modules can withstand 500-V surges without damage.

To summarize, the voltages of concern for a DAQ system are:

- NMV: The actual measurement signal.
- CMV: Noise voltage, whether acquired at the source or during transmission of the signal along a cable.
- NMV + CMV: The total must not exceed full-scale input for the DAQ to meet its performance specifications.
- NMV + CMV: The total must not exceed the maximum input voltage to avoid damage to the DAQ.

The other pesky element of "field" measurements, as opposed to lab measurements, is *crosstalk*. Essentially, crosstalk is the electrical signal on one line of a cable being superimposed on a parallel line. Most prevalent in flat cables, crosstalk can lead to measurement errors. As the rise time of signals decreases, the impedance between lines decreases, leading to more crosstalk. The use of a multiplexer (mux) at the front end of a DAQ system is a good way to save money, but it can also lead to problems, because multiplexers rarely have an instrumentation amplifier at their front end. When the multiplexer is connected directly to the signal source, the lack of high CMRR and high capacitance common on most mux inputs can lead to crosstalk generation and susceptibility.

9.5 Design for Test Fixtures

9.5.1 Design for Bed-of-Nails Fixtures

An overall consideration for any test using a bed-of-nails (BON) fixture is the total stress on the board. The number of test points on boards is increasing, double-sided probing is common, and no-clean flux demands more aggressive spring forces. Additionally, there is a growing move to thinner, more flexible substrates, such as PCMCIA cards (Is it true the acronym really means "people can't memorize computer industry acronyms?"). A related concern is that board flex may mask manufacturing defects (solder opens), resulting in fatal escapes. A spring-loaded probe develops 7 to 12 ounces of pressure on its test point. A large complex board with 800 to 1500 test points may therefore be subject to 400 to 800 pounds of stress from the probes. Proper support for the board is obviously necessary to prevent damage to solder terminations and traces caused by bending and/or twisting of the board.

Two studies have been performed that indicate the seriousness of this problem. Delco Electronics (now Delphi-Delco Electronics) performed a board-flex test by purchasing five fixtures from three different fixture manufacturers. The test was performed with 6 × 8 × 0.063-in thick circuit boards, with 400 4-oz probes. The board layout with flex-measurement points is as shown in Fig. 9.9.

The corner points marked "R" were the reference and hold-down points. The densest area of probes was in the locations marked 7 and 8. As seen in Table 9.3 for three of the fixtures, the maximum flex was 0.071 inches.

TABLE 9.3 Board Flex Measurements with Vacuum Hold-Down (Numbers in Inches)

Board #	Vendor #1	Vendor #2	Vendor #3
1	0.024	0.000	0.036
2	0.030	0.037	0.030
3	0.046	0.000	0.021
4	0.052	0.033	0.000
5	0.014	0.000	0.000
6	0.038	0.036	0.022
7	0.053	0.044	0.026
8	0.071	0.030	0.029
9	0.063	0.039	0.040
10	0.020	0.036	0.000
11	0.034	0.032	0.000
12	0.027	0.000	0.012
13	0.039	0.025	0.014
14	0.050	0.000	0.015
Min | Max	0.014 | 0.071	0.000 | 0.044	0.000 | 0.040
Avg.	0.058	0.044	0.040

The second test that indicates concern over board flex and was performed by Murata Electronics (UK). Their test was designed to study the strength of types COG, X7R, and Y5V ceramic chip capacitors in the presence of board flex, whether that flex was caused by environmental stress in use or in-circuit fixtures. After mounting and reflow soldering the capacitors, the boards were flexed, and changes in capacitance were measured. Capacitance will change in the presence of flex, and a crack can result in a decrease in capacitance to as little as 50 to 30% of the initial unflexed value. Results of the test showed the following:

- All three dielectric types will show fatal component cracking if flex is in excess of 0.117 in.
- X7R and Y5V types showed fatal cracking at 0.078 in of flex.
- Size matters; EIA 1812 X7R capacitors showed 15% failures at 0.078 in of flex.

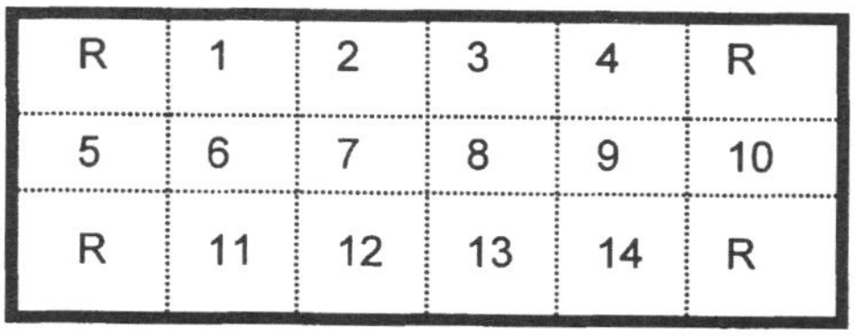

R	1	2	3	4	R
5	6	7	8	9	10
R	11	12	13	14	R

FIGURE 9.9 Board used for flex test

- EIA 1812 Y5V capacitors showed 20% failures at 0.078" of flex.
- EIA 2220 X7R showed 40% failures at 0.078" of flex.
- EIA 2220 Y5V showed 60% failures at 0.078" of flex.

Combining Delco's 0.071-in worst-case flex, and Murata's failure rates at 0.078-in flex, it is apparent that board flex needs to be minimized and that it does occur and must be investigated for BON fixtures. Add to this the fatal escape problem when boards flex during probing, and the need to minimize flex is obvious.

This also points out that test points should only be used as necessary. While this seems to fly in the face of some of the requirements, some thoughtful use of test pads will allow the designer avoid over-probing. For example, consider the test pad locations shown in Fig. 9.10.

Test pad C, although located near a component, may not be needed if the component and the board operate at relatively low clock speeds. Test pads A, C, and D, located at the ends of their respective traces, facilitate both bare-board tests for shorts and opens as well as in-circuit analysis and tests.

These overall circuit requirements for a BON fixture have been discussed:

- One test node per circuit net
- Test fixture probe spacing of 0.080 in (2 mm) minimum
- Probe-to-device clearance of 0.030 in (0.9 mm) minimum
- All test node access from one side of the board
- A test node on any active unused pins

Additionally, the fixture itself also adds the following constraints on the board design:

- Probes should be evenly spaced across the board to equalize stress across the board, as discussed above.
- Tall parts will create potential problems with probing, and/or with probes not correctly landing on test pads.

Shown in Fig. 9.11 is an example of a board with tolerances identified that must be taken into account during test pad layout. Note that the test point/pad/via location must accommodate the combined tolerances (a.k.a. tolerance buildup) that may exist in the PCB manufacturing process as well as in the BON fixture itself. PCB tolerances may come from artwork shrink or stretch as well, from etch issues, and from tooling hole tolerances. The sum of these various tolerances is the reason most fixture and probe manufacturers recommend a minimum of 0.030 in (0.9 mm) test pad/via sizes. As test pad size decreases, the chance for fixture-induced test failures increases as a result of probes missing pads.

The tolerances noted above are primarily driven by the need to accurately locate the test probes onto the test pads when the BON test fixture is activated. The industry term for probe accuracy is *pointing accuracy.*

The use of microvias may make these locations unsuitable for probing unless the surrounding pad is made a suitable size. A 4-mil microvia is not an appropriate test pad unless the doughnut around the via is enlarged appropriately. Test point locations must also allow for sufficient clearance from any

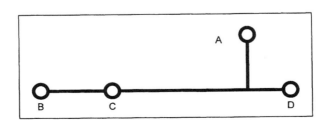

FIGURE 9.10 Test pad locations.

Design for Test

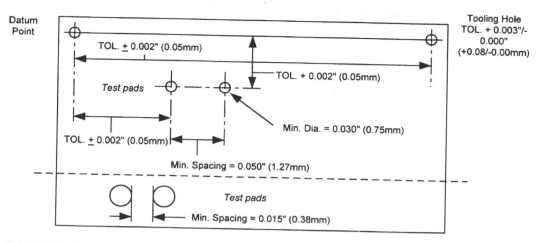

FIGURE 9.11 Examples of test probe contact locations.

component part body for the probe contact point and must prevent the probe landing on any part of a soldered joint. Any unevenness of the contact point can force the probe sideways, with the resulting risk of damage. A damaged probe will not retract or extend properly.

Test pad spacing of 0.050 to 0.100 in and test pad size of 0.030 in minimum have been discussed. It should be noted that manufacturers are capable of building fixtures which can test boards with 0.010-in lead/termination pitch.

Test fixtures commonly have the spring-loaded probes connected to the test system using wire-wrapped connections to the bottom of the probes (Fig. 9.12). These wire-wrap fixtures have a number of advantages.

- They are easy to wire.
- They are easy to change wiring connections.
- They use standard single-ended spring-loaded probes.
- They are easy to assemble and disassemble.

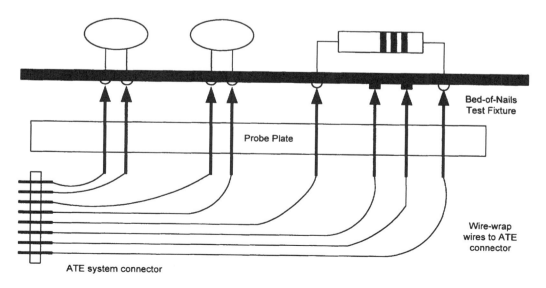

FIGURE 9.12 Examples of test probe contact locations.

However, the standard wire-wrap fixture also has a number of disadvantages:

- Impedance of connections is not easily controllable.
- Wire routing can significantly affect test performance.
- If multiple test fixtures are used in testing high-speed boards, test results can vary due to different impedances in different fixtures.

One answer to these disadvantages is to change to a wireless fixture (see Fig. 9.13). The advantages of a wireless fixture come from the use of a printed circuit board to make the connections from the probes to the ATE system connector. The circuit board can have controlled impedances and identical characteristics among several fixtures. However, there are, of course, disadvantages as well.

- One incurs design and fabrication expense of the PCB.
- It is difficult to change
- The use of double-ended probes means increased probe expense and less reliable probes
- The initial cost is high.

9.5.2 Design for Flying Probe Testers

Flying probe testers (a.k.a. *moving probe* and *x-y probe* testers) use two or more movable test probes that are computer controlled and can be moved to virtually any location over the DUT. Two-sided flying probe systems are available. The accuracy of the x-y positioning system makes flying probe testers suitable for very fine-pitch SMT applications. Flying probe testers also do not suffer from probe-density limitations, since only one probe is applied to either side the DUT at any given time.

The basic operation of a flying probe system uses point-to-point measurements between a pair of probes. As shown in Fig. 9.14, one probe will contact a point on the DUT, and the other probe will contact another point on the network to be measured. If a continuity measurement is being made, the probes are on opposite ends of the trace to be measured.

Like all other measurement systems, the flying probe system has its advantages and disadvantages.

Advantages

- It is not limited by BON probe density.
- It is able to probe very fine pitch pads.
- It is able to use capacitive testing techniques.

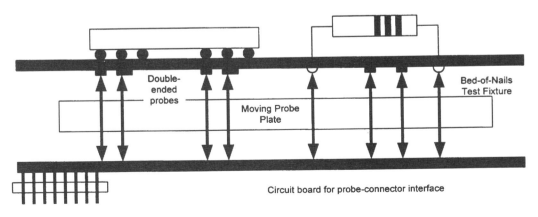

FIGURE 9.13 Wireless BON fixture.

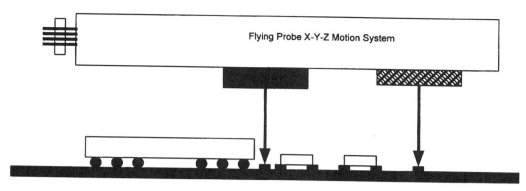

FIGURE 9.14 Flying probe measurement system.

Disadvantages

- It is slower than BON fixture system due to movement time of probes.
- It is more expensive than BON fixture.

In summary, DFT requires the consideration of a number of issues for successful testing with minimal expense and maximum probability of success. The necessity of a test engineer on the design team is certain for successful implementation of test. The many choices to be made require the test engineer's presence.

Definitions and Acronyms

Backdriving. An digital-logic in-circuit test technique that applies defined logic levels to the output pins of digital devices. This application will also be applied to the next devices inputs.

Bed-of-Nails (BON) Test Fixture. A test fixture consisting of spring-loaded test probes that are aligned in a test holder to make contact with test pads on the circuit assembly. The circuit board/assembly is typically pulled down onto the probes using vacuum applied by the fixture. See also clamshell fixture.

Clamshell Fixture. A type of BON that mechanically clamps the UUT between two sets of tests rather than using a vacuum pull-down. Commonly used when both sides of the UUT must be probed.

Design for Test (DFT). A part of concurrent engineering, DFT involves the test engineering function at the design phase of a project, to develop test criteria and be certain that the necessary hardware and software are available to allow all test goals to be met during the test phase of the project.

Failure. The end of the ability of a part or assembly to perform its specified function.

Fault. Generally, a device, assembly, component, or software that fails to perform in a specified manner. May be in the components used in an assembly, as a result of the assembly process, as a result of a software problem, or as a result of a poor design in which interaction problems occur between components that individually meet spec.

Fault Coverage. The percentage of overall faults/failures that a given test will correctly detect and identify.

First-Pass Yield. The number of assemblies that pass all tests without any rework.

Functional Test. Using simulated or real inputs and outputs to test the overall condition of a circuit assembly.

Guard, Guarding, Guarded Circuit. A test technique in which a component is isolated from parallel components or interactive components. In analog circuits, guarding places equal voltage potentials at each end of parallel components to the DUT to prevent current flowing in the parallel component. In digital circuits, guarding disables the output of a device connected to the DUT, commonly through activating a tristate output.

In-Circuit Analysis. See manufacturing defects analyzer

In-Circuit Test. Conducts both unpowered and powered tests to test both correctness of assembly, and functionality of individual components. Will identify bad components.

Infant Failures, Infant Mortality. Synonymous terms used to refer to failures that occur early in a the life product/assembly.

Manufacturing Defect Analyzer. A system that tests for shorts, opens, passive component values, and semiconductor junctions on an unpowered board.

ATE. Automatic test equipment, a system that may conduct any of a number of tests on a circuit assembly. A generic term that may include in-circuit tests, functional tests, or other test formats, under software control.

ATPG. Automatic test program generation, computer generation of test programs based on the circuit topology.

BON. Bed-of-nails test fixture.

CND. Cannot duplicate, see NFF.

COTS. Commercial off-the-shelf, a term used by the U.S. military services.

DFT. Design for test.

DUT. Device under test.

ESS. Environmental stress screening.

NFF. No fault found, a term used when a field failure cannot be replicated.

NTF. No trouble found, see NFF.

ODAS. Open data acquisition standard.

RTOK. Retest OK, see NFF.

UUT. Unit under test.

World Wide Web Related Sites

Evaluation Engineering Magazine, http://www.evaluationengineering.com
Everett Charles Technologies, http://www.ectinfo.com
Open Data Acquisition Association, http://www.opendaq.org
Test & Measurement World Magazine, http://www.tmworld.com

References

IEEE 1149.1 *(JTAG) Testability Primer, New for 1997.* Texas Instruments, http://www.ti.com/sc/docs/jtag/jtag2.htm, 1998.

BSDL/IEEE 1149.1 Verification Service. Hewlett-Packard Manufacturing Test Division, http://hpbscanentral.invision1.com/, 1997.

Chip Type Monolithic Ceramic Capacitor Bending Strength Technical Data. Murata Electronics (UK) Ltd., 1994.

Coombs, C.; *The Printed Circuit Handbook,* 4th edition. McGraw-Hill, New York, 1996.

MacLean K.; "Step-by-Step SMT: Step 2, Design for Test," http://www.smtmag.com/stepbystep/step2.html, 1998.

Turino, J.; *Design to Test.* Van Nostrand Reinhold, New York, 1990.

Walter, J.; "In-Circuit Test Fixturing." Internal report, Delco Electronics, Kokomo, 1991.

10
Adhesive and Its Application*

Ray Prasad
Ray Prasad Consultancy, Inc.

10.1 Introduction 10.1
10.2 Ideal Adhesive for Surface Mounting 10.2
10.3 General Classification of Adhesives 10.3
10.4 Adhesives for Surface Mounting 10.4
10.5 Conductive Adhesives for Surface Mounting 10.5
10.6 Adhesive Application Methods 10.8
10.7 Curing of Adhesive 10.13
10.8 Evaluation of Adhesives with Differential Scanning Calorimetry 10.19
10.9 Summary 10.25

10.1 Introduction

The use of adhesives is widespread in the electronics industry. This chapter will aid the reader in making correct decisions with respect to the use, selection, application, and curing of adhesives.

An adhesive in surface mounting is used to hold passive components on the bottom side of the board during wave soldering. This is necessary to avoid the displacement of these components under the action of the wave. When soldering is complete, the adhesive no longer has a useful function.

The types of components most commonly glued to the bottom side of Type III and Type II SMT boards are rectangular chip capacitors and resistors, the cylindrical transistors known as metal electrode leadless faces (MELFs), small outline transistors (SOTs), and small outline integrated circuits (SOICs). These components are wave soldered together with the through-hole technology (THT) devices.

Rarely, adhesive is also used to hold multileaded active devices such as plastic leaded chip carriers (PLCCs) and plastic quad flat packs (PQFPs) on the bottom side for wave soldering. Wave soldering of active devices with leads on all four sides is generally not recommended for reasons of reliability and excessive bridging. The problem of bridging can be minimized by passing the board over the solder wave at 45° to board flow, but reliability (because of, e.g., popcorning and flux seepage into the package) remains a concern. If active devices, including SOICs, are to be wave soldered, they must be properly qualified.

In another uncommon application, an adhesive holds both active and passive components placed on solder paste on the bottom or secondary side of the board during reflow soldering to allow the simultaneous reflow soldering of surface mount components on both sides.

---
*This chapter is reprinted with permission from Prasad, R.P., *Surface Mount Technology: Principles and Practice*, Chapman & Hall, New York, 1997.

This chapter focuses on adhesive types, and on dispensing methods and curing mechanisms for adhesives. Adhesives are also discussed in Chapter 2, Section 2.6

10.2 Ideal Adhesive for Surface Mounting

Reference 1 provides some general guidelines for selecting nonconductive and conductive adhesives. In this chapter, our focus will be on nonconductive adhesives, which are the most widely used. Electrically conductive adhesives are used for solder replacement, and thermally conductive adhesives are used for heat sink attachment. We will review them briefly in Section 10.5.

Many factors must be considered in the selection of an adhesive for surface mounting. In particular, it is important to keep in mind three main areas: precure properties, cure properties, and postcure properties.

10.2.1 Precure Properties

One-part adhesives are preferred over two-part adhesives for surface mounting because it is a nuisance to have to mix two-part adhesives in the right proportions for the right amount of time. One-part adhesives, eliminating one process variable in manufacturing, are easier to apply, and one does not have to worry about the short working life (pot life) of the mixture. The single-part adhesives have a shorter shell life, however. The terms *shelf life* and *pot life* can be confusing. Shelf life refers to the usable life of the adhesive as it sits in the container, whereas pot life, as indicated above, refers to the usable life of the adhesive after the two main components (catalyst and resin) have been mixed and catalysis has begun.

Two-part adhesives start to harden almost as soon as the two components are mixed and hence have a short pot life, even though each component has a long shelf life. Elaborate metering and dispensing systems are generally required for automated metering, mixing in the right proportions, and then dispensing the two-part adhesives.

Colored adhesives are very desirable, because they are easy to spot if applied in excess such that they contact the pads. Adhesive on pads prevents the soldering of terminations, hence it is not allowed. For most adhesives, it is a simple matter to generate color by the addition of pigments. In certain formulations, however, pigments are not allowed, because they would act as catalysts for side reactions with the polymers—perhaps drastically altering the cure properties. Typical colors for surface mount adhesives are red or yellow, but any color that allows easy detection can be used.

The uncured adhesive must have sufficient green strength to hold components in place during handling and placement before curing. This property is similar to the tackiness requirement of solder paste, which must secure components in their places before reflow. To allow enough time between dispensing of the adhesive and component placement, it should not cure at room temperature. The adhesive should have sufficient volume to fill the gap without spreading onto the solderable pads. It must be nontoxic, odorless, environmentally safe, and nonflammable.

Some consideration must also be given to storage conditions and shelf life. Most adhesives will have a longer shelf life if refrigerated. Finally, the adhesive must be compatible with the dispensing or stenciling method to be used in manufacturing. This means that it must have the proper viscosity. Adhesives that require refrigeration must be allowed to equilibrate to ambient temperature before use to ensure accurate dispensing. The issue of changes in viscosity with temperature is discussed in Section 10.6.3.

10.2.2 Cure Properties

The cure properties relate to the time and temperature of cure needed to accomplish the desired bond strength. The shorter the time and the lower the temperature to achieve the desired result, the better the adhesive. The specific times and temperatures for some adhesives are discussed in Section 10.7.

The surface mount adhesive must have a short cure time at low temperature, and it must provide adequate bond strength after curing to hold the part in the wave. If there is too much bond strength, reworking may be difficult; too little bond strength may cause loss of components in the wave. However, high bond strength at room temperature does not mean poor reworkability, as discussed in Section 10.2.3.

The adhesive should cure at a temperature low enough to prevent warpage in the substrate and damage to components. In other words, it is preferable that the adhesive be curable below the glass transition temperature of the substrate (126° C for FR-4). However, a very short cure time above the glass transition temperature is generally acceptable. The cured adhesive should neither increase strength too much nor degrade strength during wave soldering.

To ensure sufficient throughput, a short cure time is desired, and the adhesive cure property should be more dependent on the cure temperature than the cure time. Low shrinkage during cure, to minimize the stress on attached components, is another cure property. Finally, there should be no outgassing in adhesives, because this phenomenon will entrap flux and cause serious cleaning problems. Voids also may result from rapid curing of adhesive. (See Section 10.7.1.)

10.2.3 Postcure Properties

Although the adhesive loses its function after wave soldering, it still must not degrade the reliability of the assembly during subsequent manufacturing processes such as cleaning and repair/rework. Among the important postcure properties for adhesives is reworkability. To ensure reworkability, the adhesive should have a relatively low glass transition temperature. The cured adhesives soften (i.e., reach their T_g) as they are heated during rework. For fully cured adhesives, T_g in a range of 75° to 95° C is considered to accommodate reworkability.

Temperatures under the components often exceed 100° C during rework, because the terminations must reach much higher temperatures (>183° C) for eutectic tin-lead solder to melt. As long as the T_g of the cured adhesive is below 100° C, and the amount of adhesive is not excessive, reworkability should not be a problem. As discussed in Section 10.8, differential scanning calorimetry (DSC) can be used to determine the T_g of a cured adhesive.

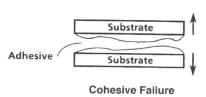

FIGURE 10.1 Schematic representation of a *cohesive* failure in adhesive.

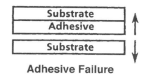

FIGURE 10.2 Schematic representation of an *adhesive* failure at the adhesive-substrate bond line.

Another useful indicator of reworkability is the location of the shear line after rework. If the shear line exists in the adhesive bulk, as shown in Fig. 10.1, it means that the weakest link is in the adhesive at the reworking temperature. The failure would occur as shown in Fig. 10.1, because one would not want to lift the solder mask or pad during rework. In the failure mechanism shown in Fig. 10.2, on the other hand, there is hardly any bond between the substrate and adhesive. This can happen due to contamination or undercuring of the adhesive.

Other important postcure properties for adhesives include nonconductivity, moisture resistance, and noncorrosivity. The adhesive should also have adequate insulation resistance and should remain inert to cleaning solvents. Insulation resistance is generally not a problem, because the building blocks of most adhesives are insulative in nature, but insulation resistance under humidity should be checked before final selection of an adhesive is made.

The test conditions for adhesives should include postcure checking of surface insulation resistance (SIR). It should be noted that SIR test results can flag not only poor insulative characteristics but also an adhesive's voiding characteristics. (See Section 10.7.1.)

10.3 General Classification of Adhesives

Adhesives can be based on electrical properties (insulative or conductive), chemical properties (acrylic or epoxy), curing properties (thermal or UV/thermal cure), or physical characteristics after cure (thermoplastic or thermosetting). Of course, these electrical, chemical, curing, and physical properties are interrelated.

Based on conductivity, adhesives can be classified as insulative or conductive. They must have high insulation resistance, since electrical interconnection is provided by the solder. However, the use of conductive adhesives as a replacement for solder for interconnection purposes is also being suggested. Silver fillers are usually added to adhesives to impart electrical conduction. We discuss conductive adhesives briefly in Section 10.5, but this chapter focuses on nonconductive (insulative) adhesives, because they are most widely used in the wave soldering of surface mount components.

Surface mount adhesives can be classified as *elastomeric, thermoplastic,* or *thermosetting*. Elastomeric adhesives, as the name implies, are materials having great elasticity. These adhesives may be formulated in solvents from synthetic or naturally occurring polymers. They are noted for high peel strength and flexibility, but they are not used generally in surface mounting.

Thermoplastic adhesives do not harden by cross-linking of polymers. Instead, they harden by evaporation of solvents or by cooling from high temperature to room temperature. They can soften and harden any number of times as the temperature is raised or lowered. If the softening takes place to such an extent that they cannot withstand the rough action of the wave and may be displaced by the wave, however, thermoplastic adhesives cannot be used for surface mounting.

Thermosetting adhesives cure by cross-linking polymer molecules a process that strengthens the bulk adhesive and transforms it from a rubbery state (Fig. 10.3, left) to a rigid state (Fig. 10.3, right). Once such a material has cured, subsequent heating does not add new bonds. Thermosetting adhesives are available as either one- or two-part systems.

10.4 Adhesives for Surface Mounting

Both conductive and nonconductive adhesives are used in surface mounting. Conductive adhesives are discussed in Section 10.5. The most commonly used nonconductive adhesives are epoxies and acrylics. Sometimes, urethanes and cyanoacrylates are chosen.

10.4.1 Epoxy Adhesives

Epoxies are available in one- and two-part systems. Two-part adhesives cure at room temperature but require careful mixing in the proper proportions. This makes the two-part systems less desirable for production situations. Single-component adhesives cure at elevated temperatures, with the time for cure depending on the temperature. It is difficult to formulate a one-part epoxy adhesive having a long shelf life that does not need high curing temperature and long cure time. Epoxies in general are cured thermally and are suitable for all different methods of application. The catalysts for heat curing adhesives are epoxides. An epoxide ring contains an oxygen atom bonded to two carbon atoms that are bonded to each other. The thermal energy breaks this bond to start the curing process.

The shelf life of adhesives in opened packages is short, but this may not be an important issue for most syringing applications since small (5 g) packages of adhesives are available. Any unused adhesive can be discarded without much cost impact if its shelf life has expired. For most one-part epoxies, the shelf life at 25° C is about two months. The shelf life can be prolonged, generally up to three months,

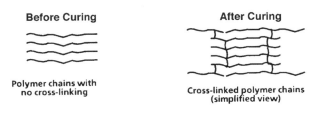

FIGURE 10.3 A schematic representation of the curing mechanism in a thermosetting adhesive.

by storage in a refrigerator at low temperature (0° C). Epoxy adhesives, like almost all adhesives used in surface mounting, should be handled with care, because they may cause skin irritation. Good ventilation is also essential.

10.4.2 Acrylic Adhesives

Like epoxies, acrylics are thermosetting adhesives that come as either single- or two-part systems but with a unique chemistry for quick curing. Acrylic adhesives harden by a polymerization reaction like that of the epoxies, but the mechanism of cure is different. The curing of adhesive is accomplished by using long-wavelength ultraviolet (UV) light or heat. The UV light causes the decomposition of peroxides in the adhesive and generates a radical or odd-electron species. These radicals cause a chain reaction to cure the adhesive by forming a high-molecular-weight polymer (cured adhesive).

The acrylic adhesive must extend past the components to allow initiation of polymerization by the UV light. Since all the adhesive cannot be exposed to UV light, there may be uncured adhesive under the component. Not surprisingly, the presence of such uncured adhesive will pose reliability problems during subsequent processing or in the field in the event of any chemical activity in the uncured portion. In addition, uncured adhesives cause outgassing during solder and will form voids. Such voids may entrap flux.

Total cure of acrylic adhesives is generally accomplished by both UV light and heat to ensure cure and to also reduce the cure time. These adhesives have been widely used for surface mounting because of faster throughput and because they allow in-line curing between placement and wave soldering of components. Like the epoxies, the acrylics are amenable to dispensing by various methods.

The acrylic adhesives differ from the epoxy types in one major way. Most, but not all, acrylic adhesives are anaerobic (i.e., they can cure in the absence of air). To prevent natural curing, therefore, they should not be wrapped in airtight containers. To avoid cure in the bag, the adhesive must be able to breathe. The acrylic adhesives are nontoxic but are irritating to the skin. Venting and skin protection are required, although automated dispensing often can eliminate skin contact problems. Acrylic adhesives can be stored for up to six months when refrigerated at 5° C and for up to two months at 30° C.

10.4.3 Other Adhesives for Surface Mounting

Two other kinds of nonconductive adhesives are available: urethanes and cyanoacrylates. The urethane adhesives are typically moisture sensitive and require dry storage to prevent premature polymerization. In addition, they are not resistant to water and solvents after polymerization. These materials seldom are used in surface mounting.

The cyanoacrylates are fast bonding, single-component adhesives generally known by their commercial names (Instant Glue, Super Glue, Crazy Glue, etc.). They cure by moisture absorption without application of heat. The cyanoacrylates are considered to bond too quickly for SMT, and they require good surface fit. An adhesive that cures too quickly is not suitable for surface mounting, because some time lapse between adhesive placement and component placement is necessary. Also, cyanoacrylates are thermoplastic and may not withstand the heat of wave soldering.

10.5 Conductive Adhesives for Surface Mounting

10.5.1 Electrically Conductive Adhesives

Electrically conductive adhesives have been proposed for surface mounting as a replacement for solder to correct the problem of solder joint cracking.[1] It is generally accepted that leadless ceramic chip carriers (LCCCs) soldered to a glass epoxy substrate are prone to solder joint cracking problems due to mismatch in the coefficients of thermal expansion (CTEs) between the carriers and the substrate. This problem exists in military applications, which require hermetically sealed ceramic packages.

Most electrically conductive adhesives are epoxy-based thermosetting resins that are hardened by applying heat. They cannot attain a flowable state again, but they will soften at their glass transition temperature. Nonconductive epoxy resin serves as a matrix, and conductivity is provided by filler metals. The metal particles must be present in large percentage so that they are touching each other, or they must be in close proximity to allow electron tunneling to the next conductive particle through the nonconductive epoxy matrix. Typically, it takes 60 to 80% filler metals, generally precious metals such as gold or silver, to make an adhesive electrically conductive. This is why these adhesives are very expensive. To reduce cost, nickel-filled adhesives are used. Copper is also used as a filler metal, but oxidation causes this metal to lose its conductivity.

The process for applying conductive adhesive is very simple. The adhesive is screen printed onto circuit boards to a thickness of about 2 mils. Nonconductive adhesive is also used to provide the mechanical strength for larger components. For smaller chip components, nonconductive adhesive is not needed for mechanical strength.

After the adhesive has been applied and the components placed, both conductive and nonconductive adhesives are cured. Depending on the adhesive, the heat for curing is provided by heating in a convection oven, by exposing to infrared or ultraviolet radiation, or by vapor phase condensation. Curing times can vary from a few minutes to an hour, depending on the adhesive and the curing equipment. High-strength materials cure at 150° to 180° C, and lower-strength, highly elastomeric materials cure at 80° to 160° C.[2]

Since flux is not used the use of conductive adhesive is truly a no-clean process. In addition, the conductive adhesives work well with all types of board finishes such as tin-lead, OSP (organic solderability protection), and gold or palladium.

Electrically conductive adhesives have been ignored by the surface mounting industry for many reasons. The military market is still struggling with the solder joint cracking problem, and the exotic substrates selected thus far instead of adhesives have not been completely successful. However, there is not much reliability data for adhesives either, and since plastic packages with leads are used in commercial applications, the cracking problem does not exist in that market sector. The higher cost of conductive adhesives is also a factor.

The process step savings provided by conductive adhesive may be exaggerated. The only meaningful process step saving between solder reflow and conductive epoxy is elimination of cleaning when adhesive is used.

There are other reasons for not using conductive adhesives for surface mounting. As we indicate throughout this book, very few assemblies are entirely surface mount (Type I); most are a mixture of surface mount and through-hole mount (Type II). Electrically conductive adhesives, however, do not work for interconnection of through holes and therefore cannot be used for mixed assemblies. In any event, since the electrical conductivity of these adhesives is lower than that of solder, they cannot replace solder if high electrical conductivity is critical. When conductive adhesives can be used, component placement must be precise, for twisted components cannot be corrected without the risk that they will smudge and cause shorts.

Repairs may also be difficult with electrically conductive adhesives. A conductive adhesive is hard to remove from a conductive pad, yet complete removal is critical, because the new component must sit flush. Also, it is not as convenient to just touch up leads with conductive adhesive as can be done with solder. Probably the biggest reason for the very limited use of conductive adhesives, however, is unfamiliarity for mass application. The conductive adhesives do have their place in applications such as hybrids and semiconductors. It is unlikely, however, that they will ever replace solder, a familiar and relatively inexpensive material.

Anisotropic Electrically Conductive Adhesives

Anisotropic adhesives are also electrically conductive and are intended for applications where soldering is difficult or cannot be used. This process for their use was developed by IBM to provide electrical connections between tungsten and copper. Now, there are many domestic companies (such as Alpha

Metal, Amp, and Sheldahl) and many Japanese companies that supply these adhesives. Examples of anisotropic adhesive applications include ultra fine pitch tape-automated bonding (TAB) and flip chips.

This type of adhesive is also referred to as *z-axis adhesive*, because it conducts electricity only in the Z axis (i.e., in the vertical direction between the pad on the board and the lead) and remains insulative in the X-Y horizontal plane. Since it is insulative in the X-Y direction, bridging with adjacent conductors is not a concern.

The anisotropic adhesives use a low concentration of conductive filler metals in a nonconductive polymer matrix. The filler metals do not touch each other in the matrix so as to prevent conductivity in the X-Y direction. The filler metals are either completely metallic or nickel-plated spherical elastomer plastics. The nickel-plated plastic spheres are considered more resilient.

As shown in Fig. 10.4, the electricity is conducted between the lead and pad through each particle. This means that anisotropic adhesives cannot carry high current, and hence their application is limited to low-power devices.

The electrical contact is made between the lead and the pad by curing the adhesive under pressure. The conductive surfaces must be in contact during the entire cure cycle. The adhesives come in both thermoplastic and thermosetting epoxy matrix. The thermosetting adhesives require much higher pressure (several kilograms) during the cure cycle than the thermoplastic adhesives, which require about one-tenth as much pressure.[3]

10.5.2 Thermally Conductive Adhesives

In addition to nonconductive and electrically conductive adhesives, thermally conductive adhesives are used. High-performance devices, particularly microprocessors, typically generate a large amount of heat. Since the power in electronic devices continues to increase over time, removal of this heat is necessary for the microprocessor to deliver peak performance.

Most microcomputer system designs employ forced-air cooling to remove heat from high-power packages. Such designs require a heat sink to be attached to these packages. The heat sinks are attached by various interface materials such as thermally conductive tapes and adhesives (Fig. 10.5, top) and thermal grease (Fig. 10.5, bottom).

The thermally conductive adhesives include epoxies, silicones, acrylics, and urethanes. These are generally filled with metallic or ceramic filler to enhance their thermal conductivity. These fillers increase the thermal conductivity of these adhesives from about 0.2 W/m-K to more than 1.5 W/m-K.

The upper diagram in Fig. 10.5 illustrates the structure of these materials. Like the nonconductive and electrically conductive adhesives discussed earlier, the thermally conductive adhesives also need to be cured to develop the necessary bond strength between the heat sink and the package.

Curing can be done thermally by baking in an oven or chemically by the use of an accelerator. A single-part adhesive typically needs thermally activated curing. Dual-part adhesives, where one part is the resin and the other part is the hardener, have to be mixed according to a predesigned ratio. Curing is initiated

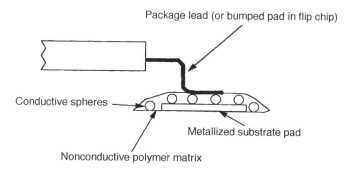

FIGURE 10.4 Lead attachment with anisotropic conductive adhesive.

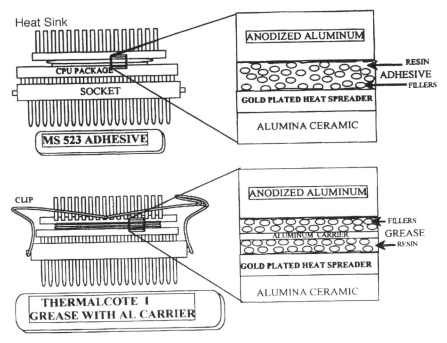

FIGURE 10.5 Heat sink attachment to high-power packages by thermally conductive adhesive and tapes (top) and thermal grease (bottom). *Source:* courtesy of Dr. Raiyomand Aspandiar, Intel Corp.

as soon as mixing occurs but, to develop greater bond strength, elevated temperature curing is generally required.

When heat sinks are attached to component packages with adhesives, as opposed to thermal grease and thermal tapes, they do not require a secondary mechanical attachment. The adhesive serves as both the heat transfer interface as well as the mechanical bonding interface. A critical property of the adhesive to consider during evaluation is the modulus of rigidity. The adhesive should absorb the stresses generated by the expansion mismatches of the two bonded surfaces without debonding or cracking.

10.6 Adhesive Application Methods

The commonly used methods of applying adhesives in surface mounting are pin transfer and syringing or pressure transfer. Proper selection of a method depends on a great number of considerations such as type of adhesive, volume or dot size, and speed of application. No matter which method is used, the following guidelines should be followed when dispensing adhesives:

1. Adhesives that are kept refrigerated should be removed from the refrigerator and allowed to come to room temperature before their containers are opened.
2. The adhesive should not extend onto the circuit pads. Adhesive that is placed on a part should not extend onto the component termination.
3. Sufficient adhesive should be applied to ensure that, when the component is placed, most of the space between the substrate and the component is filled with adhesive. For large components, more than one dot may be required.
4. It is very important that the proper amount of adhesive be placed. As mentioned earlier, too little will cause loss of components in the solder wave, and too much will either cause repair problems or flow onto the pad under component pressure, preventing proper soldering.
5. Figure 10.6 can be used as a general guideline for dot size requirements.

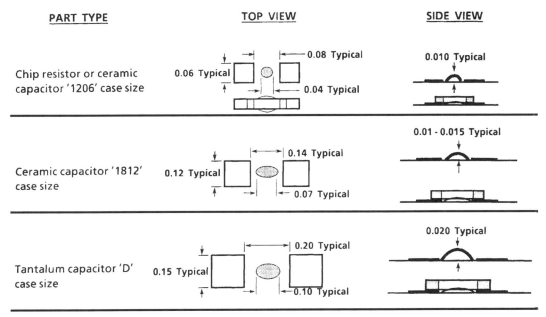

FIGURE 10.6 Guidelines for adhesive dot sizes for surface mount components. *Note:* diameter of dot size is typically about half the gap between lands.

6. Unused adhesives should be discarded.
7. If two-part adhesives are used, it will be necessary to properly proportion the "A" (resin) and "B" (catalyst) materials and mix them thoroughly before dispensing. This can be done either manually or automatically. Two-part adhesives are not commonly used for surface mounting, because they introduce additional process and equipment variables.

10.6.1 Stencil Printing

Like solder paste application, either screens or stencils can be used to print adhesive at the desired locations. Stenciling is more common than screening for adhesive printing, just as it is for solder paste printing. Stencils can deposit different heights of adhesive but screens cannot.

Stencil printing uses a squeegee to push adhesive through the holes in the stencil onto the substrate where adhesive is required. The stencils are made using an artwork film of the outer layer showing the locations at which adhesive needs to be deposited. Chapter 7 discusses the stencil printing process and equipment for paste application; this process applies to the screening of adhesives as well. Stencil printing is a very fast process. It allows the deposition of adhesives on all locations in one stroke. Thickness and size of adhesive dots are determined by the thickness of the stencil or of the wire mesh and the emulsion on the screen.

Stenciling of adhesive is cumbersome, hence it is not a very common production process. Cleaning the stencils after printing is a difficult task. Also, care is required to prevent smudging of adhesives onto adjacent pads, to preserve solderability.

10.6.2 Pin Transfer

Pin transfer, like stenciling, is a very fast dispensing method because it applies adhesive en masse. Viscosity control is very critical in pin transfer to prevent tailing, just as it is for stenciling. The pin transfer system can be controlled by hardware or by software.

In hardware-controlled systems, a grid of pins, which is installed on a plate on locations corresponding to adhesive locations on the substrate, is lowered into a shallow adhesive tray to pick up adhesive. Then, the grid is lowered onto the substrate. When the grid is raised again, a fixed amount of adhesive sticks to the substrate because the adhesive has greater affinity for the nonmetallic substrate surface than for the metallic pins. Gravity ensures that an almost uniform amount of adhesive is carried by the pins each time. Hardware-controlled systems are much faster than their software-controlled counterparts, but they are not as flexible. Some control in the size of dots can be exercised by changing the pin sizes, but this is very difficult.

Software-controlled systems offer greater flexibility at a slower speed, but there are some variations. For example, in some Japanese equipment, a jaw picks up the part, the adhesive is applied to the part (not the substrate) with a knife rather than with a pin, and then the part is placed on the substrate. This software-controlled system applies adhesive on one part at a time in fast succession. Thus, when there is a great variety of boards to be assembled, software-controlled systems are preferable.

For prototyping, pin transfer can be achieved manually, using a stylus as shown in Fig. 10.7. Manual pin transfer provides great flexibility. For example, as shown in Fig. 10.8, a dot of adhesive may be placed on the body of a large component if this is necessary to ensure an adequate adhesive bond.

10.6.3 Syringing

Syringing, the most commonly used method for dispensing adhesive, is not as fast as other methods, but it allows the most flexibility. Adhesive is placed inside a syringe and dispensed through a hollow needle by means of pressure applied pneumatically, hydraulically, or by an electric drive. In all cases, the system needs to control the flow rate of the adhesive to ensure uniformity.

Adhesive systems that can place two to three dots per second are generally an integral part of the pick-and-place equipment. Either dedicated adhesive dispensing heads with their own X-Y table can be used,

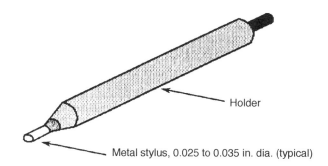

FIGURE 10.7 A stylus for manual pin transfer of adhesive.

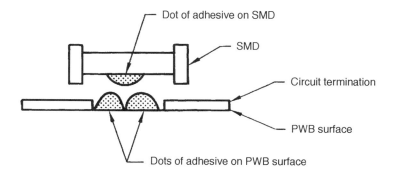

FIGURE 10.8 Guideline for adhesive dots on large components.

Adhesive and Its Application

or the component placement X-Y table can be shared for adhesive placement. The latter option is cheaper but slower, because the placement head can be either dispensing adhesive or placing components, but not both.

The adhesive dispensing head as an integral part of the pick-and-place system is shown in Fig. 10.9, which shows the syringe out of its housing. This figure shows a typical size of syringe with 5 to 10 g of adhesive capacity. The dispensing head can be programmed to dispense adhesive on desired locations. The coordinates of adhesive dots are generally downloaded from the CAD systems, not only to save time in programming the placement equipment but also to provide better accuracy. Some pick-and-place equipment automatically generate an adhesive dispensing program from the component placement

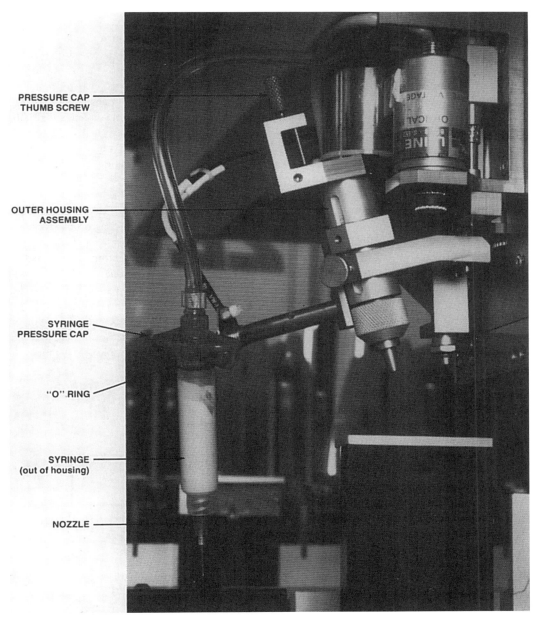

FIGURE 10.9 An automatic adhesive dispenser as an integral part of the pick-and-place equipment.

program. This is a very effective method for varying dot size or even for placing two dots, as is sometimes required.

If the adhesive requires ultraviolet cure, the dot of adhesive should be placed or formed so that a small amount extends out from under the edges of the component, but away from the terminations and the leads, as shown in Fig. 10.10. The exposed adhesive is necessary to initiate the ultraviolet cure.

The important dispensing parameters, pressure and timing, control the dot size and to some extent tailing, which is a function of adhesive viscosity. By varying the pressure, the dot size can be changed. Stringing or tailing (dragging of the adhesive's "tail" to the next location over components and substrate surface) can cause serious problems of solder skips on the pads. Stringing can be reduced by making some adjustments to the dispensing system. For example, smaller distance between the board and the nozzle, larger-diameter nozzle tips, and lower air pressure help reduce the incidence of stringing. If pressure is used for dispensing, which is commonly the case, any change in viscosity and restriction to flow rate will cause the pressure to drop off, resulting in a decrease in flow rate and a change in dot size.

The viscosity of adhesive also plays a role in stringing. For example, higher-viscosity adhesives are more prone to stringing than lower-viscosity adhesives. However, a very low viscosity may cause dispensing of excessive amounts of adhesive. Since viscosity changes with temperature, a change in ambient temperature can have a significant impact on the amount of adhesive that is dispensed. From a study conducted at IBM by Meeks[4] (see Fig. 10.11), it was shown that only a 5° C change in ambient temperature could influence the amount of adhesive dispensed by almost 50% (from 0.13 to 0.190 g). All other dispensing variables, such as nozzle size, pressure, and time, remained the same. Temperature-controlled housing or the positive displacement method (to dispense a known amount of adhesive) should be used to prevent variation in dot size due to change in ambient temperature.

Skipping of adhesive is another common problem in adhesive dispensing. The likely causes of skipping are clogged nozzles, worn dispenser tips, and circuit boards that are not flat.[5] The nozzles generally clog if adhesive is left unused for a long time (from a few hours to a few days, depending on the adhesive). To avoid clogging of nozzles, either the syringe should be discarded after every use, or a wire can be put inside the nozzle tip. A very high viscosity can also cause skipping. When using automatic dispensers in pick-and-place systems, care should be exercised to keep gripper tips clean of adhesive. If contact with adhesive occurs during component placement, the gripper tips should be cleaned with isopropyl alcohol. Also, when using the pick-and-place machine for dispensing adhesive, a minimum amount of pressure should be used to bring the component leads or terminations down onto the pads. Once the components have been placed on the adhesive, lateral movement should be avoided. These precautions are necessary to ensure that no adhesive gets onto the pads. Dispensing can also be accomplished manually for

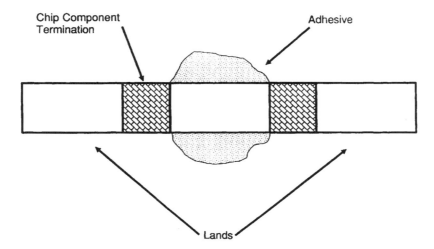

FIGURE 10.10 The adhesive extension required on the sides of components using UV-curing adhesive.

Adhesive and Its Application

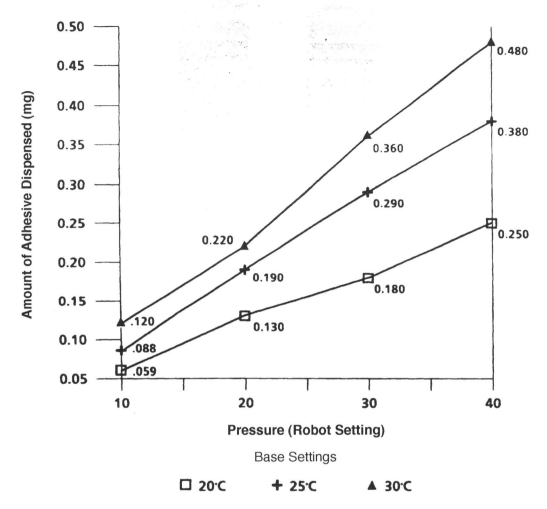

FIGURE 10.11 The impact of temperature on the amount of adhesive dispensed.[4]

prototyping applications using a semiautomatic dispenser (Fig. 10.12). Controlling dot size uniformity is difficult with semiautomated dispensers.

10.7 Curing of Adhesive

Once adhesive has been applied, components are placed. Now the adhesive must be cured to hold the part through the soldering process. There are two commonly used methods of cure: thermal cure and a combination of UV light and thermal cure. We discuss these curing processes in turn.

10.7.1 Thermal Cure

Most epoxy adhesives are designed for thermal cure, which is the most prevalent method of cure. Thermal cure can be accomplished simply in a convection oven or an infrared (IR) oven without added investment in a UV system. The IR or convection ovens can also be used for reflow soldering. This is one of the reasons for the popularity of thermal cure, especially in IR ovens. The single-part epoxy adhesives require a relatively longer cure time and higher temperatures. When using higher temperatures, care should be taken that boards do not warp and are properly held.

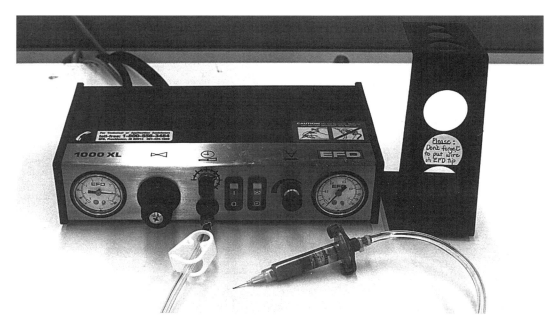

FIGURE 10.12 A semiautomatic dispenser for adhesive application

Thermal Cure Profile and Bond Strength

As shown in Fig. 10.13, the adhesive cure profile depends on the equipment. Batch convection ovens require a longer time, but the temperature is lower. In-line convection (and IR) ovens provide the same results in less time, since the curing is done at higher temperatures in multiple heating zones.

Different adhesives give different cure strengths for the same cure profile (time and temperature of cure), as shown in Fig. 10.14, where strength is depicted as the force required to shear off a chip capacitor, cured in a batch convection oven, from the board at room temperature. For each adhesive, the cure time is fixed at 15 min. The cure temperature in a batch convection oven was varied, and strength was measured with a Chatillon pull test gauge. Graphs similar to Fig. 10.14 should be developed for evaluating the cure profile and corresponding bond strength of an adhesive for production applications. The curing can be done in either an in-line convection or an IR oven.

In adhesive cure, temperature is more important than time. This is shown in Fig. 10.15. At any given cure temperature, the shear strength shows a minor increase as the time of cure is increased. However, when the cure temperature is increased, the shear strength increases significantly at the same cure time. For all practical purposes, the recommended minimum and maximum shear strengths for cured adhesive

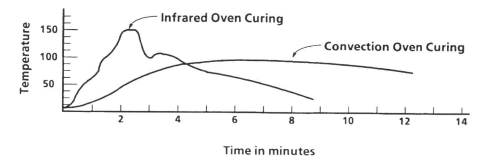

FIGURE 10.13 Adhesive cure profiles in in-line IR and batch convection ovens.

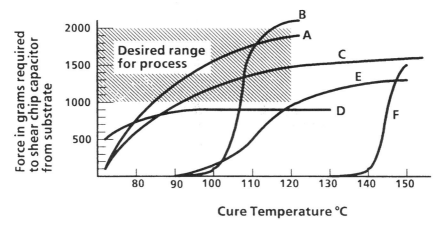

FIGURE 10.14 Cure strength for adhesives A through F at different temperatures for 15 min in a batch convection oven.

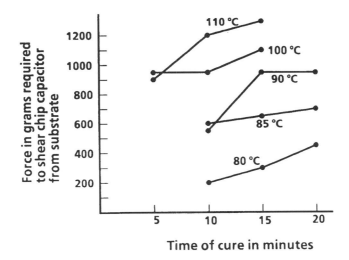

FIGURE 10.15 Cure strength of adhesive C, an epoxy adhesive, at different times and temperatures.

are 1000 and 2000 g, respectively. However, higher bond strengths (up to 4000 g) have been found not to cause rework problems, since adhesive softens at rework temperatures.

This point about cure temperature being more important than time is also true for in-line convection or infrared curing, as shown in Table 10.1. As the peak temperature in the IR oven is raised by raising panel temperatures, the average shear strength increases drastically. Table 10.1 also shows that additional curing takes place during soldering. Thus, an adhesive that is partially cured during its cure cycle will be fully cured during wave soldering.

Most of the adhesive gets its final cure during the preheat phase of wave soldering. Hence, it is not absolutely essential to accomplish the full cure during the curing cycle. Adequate cure is necessary to hold the component during wave soldering, however, and if one waits for the needed cure until the actual soldering, it may be too late. Chips could fall off in the solder wave.

It should be kept in mind that the surface of the substrate plays an important role in determining the bond strength of a cured adhesive. This is to be expected because bonding is a surface phenomenon. For example, a glass epoxy substrate surface has more bonding sites in the polymer structure than a surface covered by a solder mask.

TABLE 9.1 Impact of Cure Temperature on the Bond Strength of Epoxy Adhesive. The table also shows that additional curing takes place during wave soldering. *Source:* courtesy of Chris Ruff, Intel Corporation.

Belt Speed (ft/min)	Peak Temperature (°C)	Mean Shear Strength (g)		Approximate Rework (Removal) Time (s)
		After IR Cure	After Wave Soldering	
4.0	150	3000	3900	4–6
5.0	137	2000	3900	4–5
6.0	127	1000	3700	3–5

As illustrated in Fig. 10.16, different solder masks give different bond strengths with the same adhesive using identical curing profiles. Hence, an adhesive should be evaluated on the substrate surface that is to be used. If a new kind of solder mask is used, or if the solder mask is changed, an adhesive that was acceptable before may have only marginal shear strength under the new conditions and may cause the loss of many chip components in the wave.

Adhesive Cure Profile and Flux Entrapment

One additional requirement for adhesive thermal cure is also very important. The cure profile should be such that voids are not formed in the adhesive—this is an unacceptable condition. The voids will entrap flux, which is almost impossible to remove during cleaning. Such a condition is a serious cause for concern, especially if the flux is aggressive in nature. The circuit may corrode and cause failure if the flux

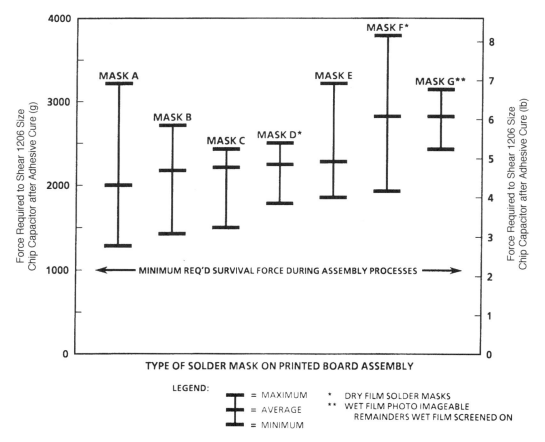

FIGURE 10.16 Bond strength of an epoxy adhesive with different types of solder masks. Note that letters A–G do not represent the same materials as in Figs. 10.14 and 10.15.

is not completely removed. Most often, the cause of voids is fast belt speed to meet production throughput requirements. A rapid ramp rate during cure has also been found to cause voiding in the adhesive.

Voiding may not be caused by rapid ramp rate alone, however. Some adhesives are more susceptible to voiding characteristics than others. For example, entrapped air in the adhesive may cause voiding during cure. Voids in adhesive may also be caused by moisture absorption in the bare circuit boards during storage. Similarly, susceptibility to moisture absorption increases if a board is not covered with solder mask. During adhesive cure, the evolution of water vapor may cause voids. Whatever the cause, voids in adhesive are generally formed during the cure cycle.

Baking boards before adhesive application and using an adhesive that has been centrifuged to remove any air after packing of the adhesive in the syringe certainly helps to prevent the formation of voids, as does the use of solder mask. Nevertheless, the most important way to prevent flux entrapment due to formation of voids in the adhesive during cure is to characterize the adhesive and the cure profile. This must be done before an adhesive with a given cure profile is used on products. How should an adhesive be characterized? We discussed earlier the precure, cure, and postcure properties of an ideal adhesive. Characterization of the adhesive cure profile to avoid void formation should also be added to the list.

There are two important elements in the cure profile for an adhesive, namely initial ramp rate (the rate at which temperature is raised) and peak temperature. The ramp rate determines its susceptibility to voiding, whereas the peak temperature determines the percentage cure and the bond strength after cure. Both are important, but controlling the ramp rate during cure is more critical.

Figure 10.17 shows the recommended cure profile for an epoxy adhesive in a four-zone IR oven (Fig. 10.17a), in a 10-zone IR oven (Fig. 10.17b), and in a ten-zone convection oven (Fig. 10.17c). A zone is defined as having both top and bottom heaters, thus a ten-zone oven has ten top heaters and ten bottom heaters. The cure profile will vary from oven to oven, but the profiles shown in Fig. 10.17 can be used as a general guideline.

As shown in the figure, the average ramp rate for the in-line IR or in-line convection oven for epoxy adhesive cure is about 0.5° C/s. This figure is obtained by dividing the rise in temperature by the time

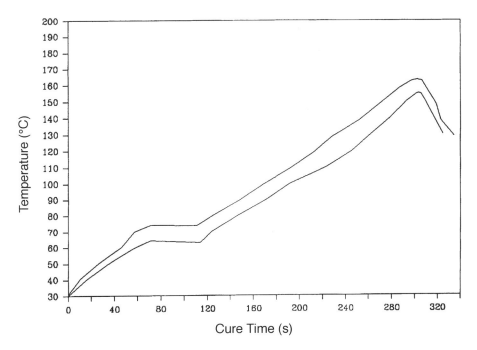

FIGURE 10.17 (a) Cure profile for an epoxy adhesive in a four-zone IR oven. Source: courtesy of Chris Ruff, Intel Corp.

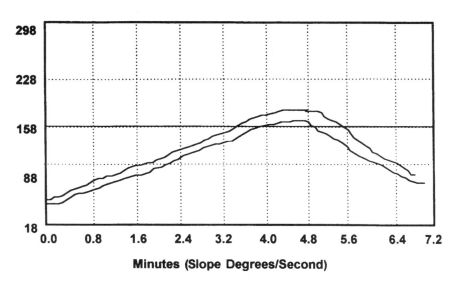

FIGURE 10.17 (b) Cure profile for an epoxy adhesive in a ten-zone IR oven. *Source:* courtesy of Dudi Amir, Intel Corp.

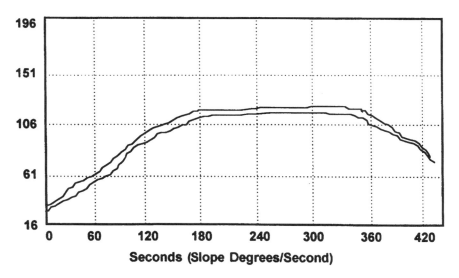

FIGURE 10.17 (c) Cure profile for an epoxy adhesive in a ten-zone convection oven. *Source:* courtesy of Dudi Amir, Intel Corp.

during the heating cycle: temperature of 160° C – 30° C = 130° C ÷ 300 s in the four-zone IR oven (Fig. 10.17a); temperature of 160° C – 30° C = 130° C ÷ 270 s in the ten-zone IR oven (Fig. 10.17b); temperature of 130° C – 30° C = 100° C ÷ 200 s in the ten-zone convection oven (Fig. 10.17c). Looking closely at Fig. 10.17c, the initial ramp rate in the convection oven in the beginning is higher (temperature of 120° C – 30° C = 90° C ÷ 120 = 0.75° C/s), but the rate is only 0.5° C if we take into account the peak temperature.

In addition to the ramp rate, it is important to note that there is no significant difference in the total cure time between the four-zone (Fig. 10.17a) and ten-zone (Figs. 10.17b and 10.17c) ovens. It is about six minutes for each oven. Therefore, it is important to keep in mind that switching to ovens with more zones should be done for reasons other than increasing throughput. A higher number of zones does make it easier to develop the thermal profile, however.

For comparison purposes, the ramp rate in a typical batch convection oven is only 0.1° C/s for the peak cure temperature of 100° C for 15 min. Such a low ramp rate in a batch oven may be ideal for preventing void formation, but it is not acceptable for production.

For the ovens discussed above, the ramp rate of 0.5° C/s translates into a belt speed of about 30 in/min. All the profiles shown in Fig. 10.17 were developed at 30 in/min. Increasing the belt speed will increase the ramp rate, but it will also increase the potential for void formation. For example, we found that a belt speed of 42 in/min pushed the average ramp rate to 0.8° C/s, but it caused voids in the adhesive. Increasing the ramp rate above 0.5° C/s may be possible in some ovens, but the adhesive must be conclusively characterized to show that higher ramp rates do not cause voids.

Cleanliness tests, including the surface insulation resistance (SIR) test, should be an integral part of the adhesive characterization process. Cleanliness tests other than SIR tests are necessary, because SIR tests are generally valid only for water-soluble fluxes and may not flag rosin flux entrapment problems. Another way to determine the voiding characteristics of an adhesive for a given cure profile is to look for voids visually during the initial cure profile development and adhesive qualification phases.

Figure 10.18 offers accept/reject criteria for voids after cure; again, these are only guidelines. The SIR or other applicable cleanliness tests should be the determining factors for the accept/reject criteria applied to voids, ramp rate, and belt speed. In addition, data on acceptable bond strength to prevent loss of chips in the wave without compromising reworkability must be collected for the final profile. The bond strength requirement was discussed in the preceding section. It should be pointed out that the cure profile and the voiding characteristics of thermal and UV adhesive will differ, since the UV adhesive is intended for a faster cure.

10.7.2 UV/Thermal Cure

The UV/thermal cure system uses, as the name implies, both UV light and heat. The very fast cure that is provided may be ideal for the high throughput required in an in-line manufacturing situation. The adhesives used for this system (i.e., acrylics) require both UV light and heat for full cure and have two cure "peaks," as discussed later in connection with differential scanning calorimetry (DSC).

The UV/thermal adhesive must extend past the components to allow initiation of polymerization by the UV light, which is essentially used to "tack" components in place and to partially cure the adhesive. Final cure is accomplished by heat energy from IR or convection or a combination of both.

It is not absolutely necessary to cure the UV/thermal adhesives by both UV light and heat. If a higher temperature is used, the UV cure step can be skipped. A higher cure temperature may also be necessary when the adhesive cannot extend past the component body (as required for UV cure; see Fig. 10.10) because of lead hindrance of components such as SOTs or SOICs.

For UV/thermal systems, it is important to have the right wattage, intensity, and ventilation. A lamp power of 200 W/in at a 4-in distance using a 2 kW lamp generally requires about 15 s of UV cure. Depending on the maximum temperature in an IR or convection oven, a full cure can be accomplished in less than two minutes.

10.8 Evaluation of Adhesives with Differential Scanning Calorimetry

Adhesives are received in batches from the manufacturer, and batch-to-batch variations in composition are to be expected, even though the chemical ingredients have not been changed. Some of these differences, however minute they may be, may affect the cure properties of an adhesive. For example, one particular batch may not cure to a strength that is adequate to withstand forces during wave soldering after it has been exposed to the standard cure profile. This can have a damaging impact on product yield.

Adhesives can be characterized by the supplier or by the user, as mutually agreed, to monitor the quality of adhesive from batch to batch. This is done by determining whether the adhesive can be fully cured when subjected to the curing profile. The equipment can be programmed to simulate any curing

Gross voiding (40 to 70%), unacceptable

Moderate voiding (25 to 40%), unacceptable

Gross porosity (5 to 20%), unacceptable

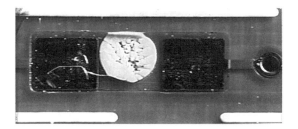

Moderate porosity (2 to 5%), minimum unacceptable

Minor porosity (0 to 2%), acceptable

FIGURE 10.18 Accept/reject criteria for voids in adhesive, to prevent flux entrapment during wave soldering. *Source:* courtesy of Chris Ruff, Intel Corp.

profile. It also can measure the glass transition temperature of the adhesive after cure to determine whether the reworkability properties have changed.

In the sections that follow, we discuss the results of adhesive characterization of epoxy and acrylic adhesives based on evaluations done at Intel.[6] The thermal events of interest for surface mount adhesives are the glass transition temperature and the curing peak.

10.8.1 Basic Properties of DSC Analysis

Differential scanning calorimetry is a thermal analysis technique used for material characterization, such as ascertaining the curing properties of adhesives. Typical DSC equipment is shown in Fig. 10.19. The output from a DSC analysis is either an isothermal DSC curve (heat flow versus time at a fixed temperature) or an isochronal DSC curve (heat flow versus temperature at a fixed heating rate).

Basically, in the DSC method of analysis, the heat generated by the sample material (e.g., an adhesive) is compared to the heat generated in a reference material as the temperature of both materials is raised at a predetermined rate.

The DSC controller aims to maintain the same temperature for both the sample and the reference material. If heat is generated by the sample material at a particular temperature, the DSC controller will reduce the heat input to the sample in comparison to the heat input to the reference material, and vice versa. This difference of heat input is plotted as the heat flow to or from the sample (Y axis) as a function of temperature (X axis).

The reference material should undergo no change in either its physical or chemical properties in the temperature range of study. If the sample material does not evolve heat to or absorb heat from the ambient, the plot will be a straight line. If there is some heat-evolving (exothermic) or heat-absorbing (endothermic) event, the DSC plot of heat flow versus temperature will exhibit a discontinuity.

As seen in Fig. 10.20, the glass transition temperature is represented as a change in the value of the baseline heat flow. This occurs because the heat capacity (the quantity of heat needed to raise the temperature of the adhesive by 1° C) of the adhesive below the glass transition temperature is different from the heat capacity above T_g.

As also seen in Fig. 10.20, curing of the adhesive is represented by an exothermic peak. That is, the adhesive gives off heat when it undergoes curing. This cure peak can be analyzed to give the starting temperature of the cure and the extent of cure for a particular temperature profile. A fusion peak is also shown in Fig. 10.20. However, adhesives do not undergo fusion or melting at the temperatures of use because these temperatures are too low.

The DSC curve shown in Fig. 10.20 is an isochronal curve; that is, heat flow is measured as the temperatures of the sample and the reference material are increased at a constant rate. Isothermal DSC curves can also be generated at any particular temperature. Isothermal curves depict heat flow versus time at a constant predetermined temperature.

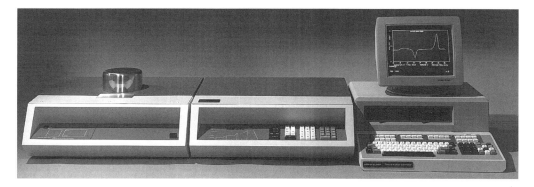

FIGURE 10.19 A differential scanning calorimeter. *Source:* courtesy of Perkin-Elmer.

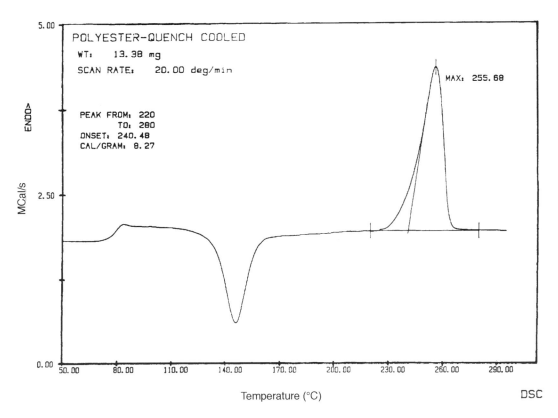

FIGURE 10.20 A typical DSC curve illustrating the various thermal events occurring in the sample under investigation.

Both isochronal and isothermal DSC curves are very useful in characterizing surface mount adhesive curing rates and glass transition temperatures. Results on the characterization of an epoxy adhesive (thermal cure) and an analytic adhesive (UV/thermal cure) are presented next.

10.8.2 DSC Characterization of an Epoxy Adhesive

Figure 10.21 shows an isochronal curve for an uncured epoxy adhesive (adhesive A, Fig. 10.14) subjected to a heating profile in the DSC furnace from 25° to 270° C at a heating rate of 75° C/minute. The results from Fig. 10.21 indicate the following:

1. There is an exothermic peak corresponding to the curing of the adhesive. The onset temperature of this peak is 122° C, and it reaches a minimum at 154° C.
2. The heat liberated during the curing of the adhesive is 96.25 calories per gram of adhesive.
3. The adhesive starts curing before it reaches the maximum temperature of 155° C (the peak temperature for most epoxy adhesives in an infrared oven).

When the same adhesive in the uncured state is cured in an IR oven and analyzed with DSC, the results are different, as shown in Fig. 10.22, which compares the isochronal DSC curves of the adhesive in the uncured and cured states. Before cure, the adhesive is in a fluid state. Its T_g is below room temperature because, by definition, this is the point at which the adhesive transforms from a rigid state to a glassy or fluid state. After cure, the T_g of the adhesive will increase because of cross-linking in the carbon chains as discussed earlier (see Fig. 10.3).

Adhesive and Its Application

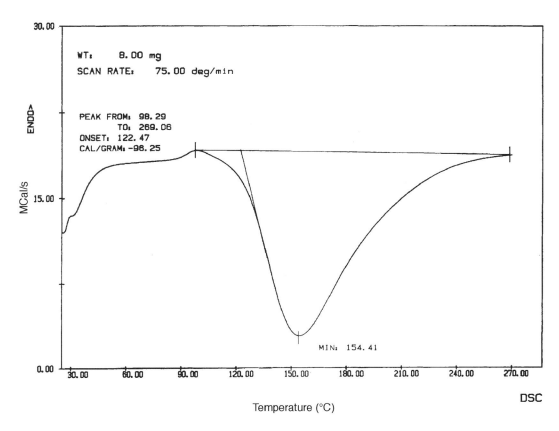

FIGURE 10.21 The isochronal DSC curve for an uncured epoxy adhesive. The curing exothermic peak is clearly visible.

An increase in heat flow occurs during the glass transition of an adhesive. Since the magnitude of this change is quite small, the Y axis in the DSC curves shown in Fig. 10.22 must be expanded, as shown in Fig. 10.23, to reveal the T_g effect. Figure 10.23 is a portion of the DSC curve from Fig. 10.22 (broken curve) replotted with the Y axis scaled up. From Fig. 10.23, it is apparent that the onset of the T_g occurs at 73° C, and the midpoint of the T_g occurs at 80° C. It is appropriate to characterize T_g from the DSC curves as the temperature value at the midpoint of the transition range. (The onset value depends on the baseline slope construction, whereas the midpoint value is based on the inflection point on the DSC curve and is therefore independent of any geometric construction.)

Since the T_g in the cured state is quite low, it will be very easy to rework the adhesive after IR cure. After IR cure, however, the adhesive and the SMT assembly must be subjected to another heat treatment, namely, wave soldering. This further heating will increase the T_g of the adhesive, as evident is Fig. 10.24, which shows a higher T_g (85° C).

The DSC curve in Fig. 10.24 is for the same epoxy adhesive characterized in Fig. 10.23 but after wave soldering. The 5° C increase in the T_g after wave soldering implies that additional physical curing of the adhesive occurs during the wave soldering step. A small increase in the T_g after wave soldering is not a real problem, because the adhesive still can be reworked. The requirement for SMT adhesives is that, after all soldering steps, the T_g must be below the melting point of the solder (i.e., 183° C).

Figure 10.24 also shows that the adhesive starts decomposing at about 260° C, as evidenced by the small wiggles at that temperature. The DSC curve shown in Fig. 10.24 was run with the encapsulated sample placed in a nitrogen atmosphere. In other atmospheres, the decomposition temperature might be different.

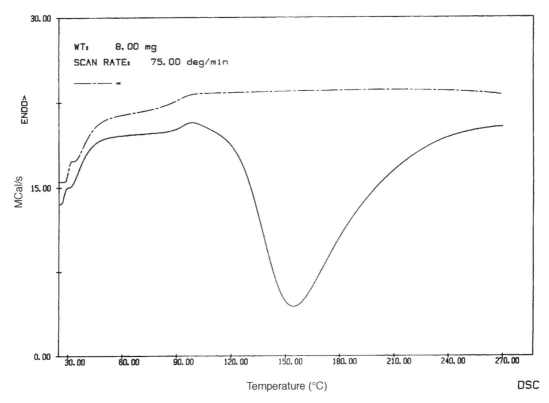

FIGURE 10.22 A comparison of the isochronal DSC curves for the epoxy adhesive in the cured (broken line) and uncured (solid line) states. The exothermic curing peak is absent for the cured adhesive.

10.8.3 DSC Characterization of an Acrylic Adhesive

As mentioned earlier, acrylic adhesives require UV light and heat, but they can be cured by heat alone if the temperature is high enough. However, as discussed in Section 8.4.2, the catalysts for UV curing are the peroxides (photoinitiators). Figure 10.25 shows the isochronal DSC curve of an acrylic adhesive (adhesive G) that had been cured in DSC equipment with a simulated IR oven curing profile. Also shown in Fig. 10.25 for comparison is the isochronal DSC curve for the same adhesive in an uncured state. From Fig. 10.25 we can draw two conclusions.

1. The uncured adhesive shows two cure peaks, one at 150° C and another at 190° C (solid line in Fig. 10.25). The lower temperature peak is caused by the curing reaction induced by the photoinitiator catalyst in the UV adhesive. In other words, the photoinitiators added to the UV adhesive start the curing reaction at ambient temperature. They reduce the curing cycle time in the UV adhesives by "tacking" components in place. For complete cure, thermal energy is required.
2. DSC analysis offers two ways to distinguish a UV adhesive from a thermal adhesive. An uncured UV adhesive will show two curing peaks (solid line in Fig. 10.25). If, however, it is first thermally cured and then analyzed by DSC, it will show only one peak (broken line in Fig. 10.25). This characteristic is similar to the curing characteristics of an uncured thermal adhesive, as discussed earlier. In other words, a low-temperature thermally cured WV adhesive will look like an uncured thermal adhesive.

Can an analytic UV adhesive be fully cured without UV? Yes, but, as shown in Fig. 10.26, the curing temperature must be significantly higher than that used for other adhesive cures. Such a high temperature may damage the temperature-sensitive through-hole components used in mixed assemblies.

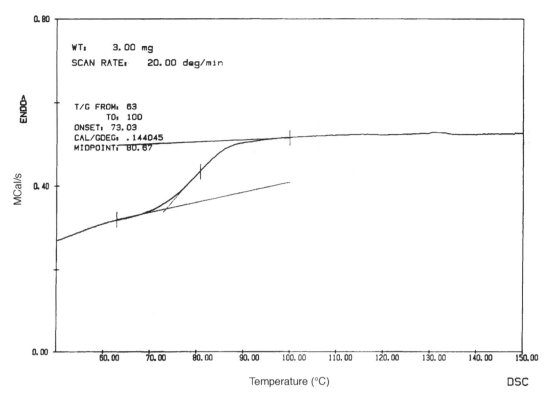

FIGURE 10.23 Part of the isochronal DSC curve showing the temperature range at which the T_g occurs for an epoxy adhesive first cured in an infrared oven. The T_g onset is at 73° C, and the midpoint is at 80° C.

Figure 10.26 compares the isochronal DSC curves for an adhesive in the uncured state and in the fully cured state. The adhesive was fully cured by heating it to 240° C, which is much higher than the 155° C maximum curing temperature in the IR oven. Figure 10.26 shows that the adhesive sample cured at 240° C does not exhibit any peak (broken line). This indicates that to achieve full cure of this adhesive, the material must be heated above the peak temperature of about 190° C. Again, as in Fig. 10.25, the solid line in Fig. 10.26 is for the same adhesive in an uncured state.

When UV acrylic adhesives are subjected to heat only, they do not fully cure. The partial cure can meet the bond strength requirement for wave soldering, however. Does this mean that UV adhesives can be thermally cured, and one does not need to worry about the undercure as long as the bond strength requirements are met? Not necessarily. A partial cured adhesive is more susceptible to absorption of deleterious chemicals during soldering and cleaning than a fully cured adhesive and, unless the partially cured adhesive meets all other requirements, such as insulation resistance, it should not be used.

10.9 Summary

Adhesive plays a critical role in the soldering of mixed surface mount assemblies. Among the considerations that should be taken into account in the selection of an adhesive are desired precure, cure, and postcure properties. Dispensing the right amount of adhesive is important as well. Too little may cause loss of devices in the wave, and too much may be too hard to rework or may spread on the pad, resulting in solder defects. Adhesive may be dispensed in various ways, but syringing is the most widely used method.

Epoxies and acrylics are the most widely used types of adhesive. Adhesives for both thermal and UV/thermal cures are available, but the former are more prevalent. The cure profile selected for an

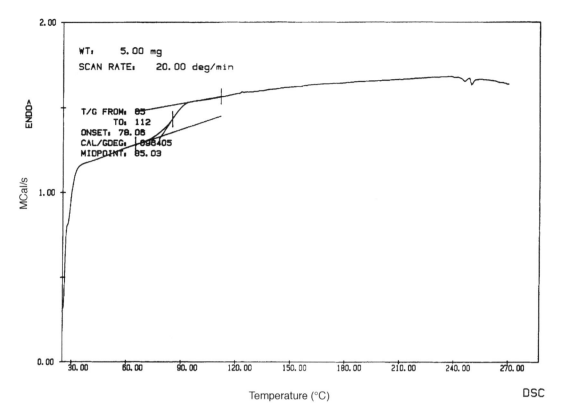

FIGURE 10.24 Part of the isochronal DSC curve showing the temperature range at which the T_g occurs for cured epoxy adhesive in an IR oven. The T_g onset is at 78° C, and the midpoint is at 85° C. The adhesive decomposes at 260° C.

adhesive should be at the lowest temperature and shortest time that will produce the required bond strength. Bond strength depends on the cure profile and the substrate surface, but the impact of temperature on bond strength is predominant.

Bond strength should not be the sole selection criterion, however. Consideration of voiding characteristics may be even more significant to ensure that flux entrapment and cleaning problems after wave soldering are not encountered. One way to prevent voiding is to control the ramp rate during the cure cycle of the adhesive and to fully characterize the adhesive cure profile before the adhesive is used on products. An integral part of that characterization process should be to visually look for gross voids and to confirm their impact by conducting cleanliness tests such as the test for surface insulation resistance. In addition, the cure profile should not affect the temperature-sensitive through-hole components used in mixed assemblies.

Minor variations in an adhesive can change its curing characteristics. Thus, it is important that a consistent quality be maintained from lot to lot. One of the simplest ways to monitor the curing characteristics of an adhesive is to analyze samples using differential scanning calorimetry (DSC). This can be performed either by the user or the supplier, per mutual agreement. DSC also serves other purposes. For example, it can be used to distinguish a thermal cure adhesive from a UV/thermal cure adhesive and a fully cured adhesive from a partially cured or uncured adhesive. DSC is used for characterizing other materials needed in the electronics industry as well.

References

1. Specifications on Adhesives, available from IPC, Northbrook, IL:

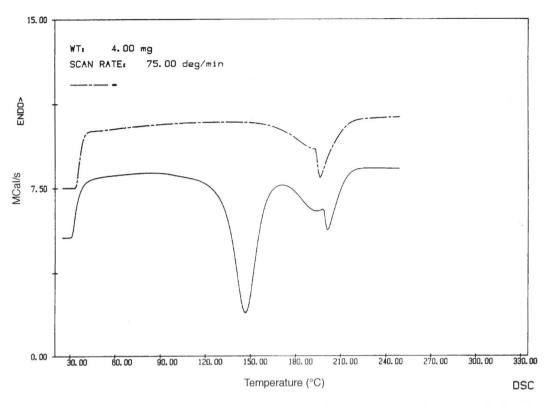

FIGURE 10.25 A comparison of the DSC isochronal curves for adhesive G, an acrylic (UV cure) adhesive. The broken curve represents already thermally cured adhesive at a peak temperature of 155° C. The presence of the peak in the broken curve means that the adhesive is only partially cured. The solid curve represents the uncured state of adhesive G.

- IPC-SM-817. *General Requirements for Dielectric Surface Mount Adhesives.*
- IPC-3406. *Guidelines for Conductive Adhesives.*
- IPC-3407. *General Requirements for Isotropically Conductive Adhesives.*
- IPC-3408. *General Requirements for Anisotropically Conductive Adhesives.*

2. Pound, Ronald. "Conductive epoxy is tested for SMT solder replacement." *Electronic Packaging and Production,* February 1985, pp. 86-90.
3. Zarrow, Phil, and Kopp, Debra. "Conductive Adhesives." *Circuit Assembly,* April 1996, pp. 22–25.
4. Meeks, S. "Application of surface mount adhesive to hot air leveled solder (HASL) circuit board evaluation of the bottom side adhesive dispense." Paper IPC-TR-664, presented at the IPC fall meeting, Chicago, 1987.
5. Kropp, Philip, and Eales, S. Kyle. "Troubleshooting guide for surface mount adhesive." *Surface Mount Technology,* August 1988, pp. 50–51.
6. Aspandiar, R., Intel Corporation, internal report, February, 1987.

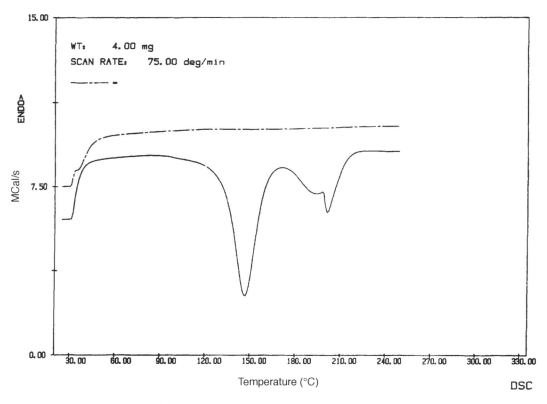

FIGURE 10.26 A comparison of the DSC isochronal curves for a UV-cured acrylic (adhesive G). The broken line curve represents the already thermally cured adhesive at a peak temperature of 240° C. The absence of peak in the broken curve means that the adhesive is fully cured. The solid curve represents the uncured state of adhesive G.

11
Thermal Management

Glenn R. Blackwell
Purdue University

11.1 Introduction 11.1
11.2 Overview 11.1
11.3 Fundamentals of Heat 11.3
11.4 Engineering Data 11.6
11.5 Heat Transfer 11.8
11.6 Heat Removal/Cooling 11.16

11.1 Introduction

Thermal management is critical to the long-term survival and operation of modern electronics. This chapter will introduce the reader to the principles of thermal management and then provide basic information on heat removal.

Where Else?

See also Chapter 15, "Reliability," which deals with the effect heat has on overall device and product reliability.

11.2 Overview

The life of all materials, including semiconductors, varies logarithmically with the reciprocal of the absolute temperature and is expressed by the Arrhenius equation:

$$L = A(\varepsilon^{b/T} - 1)$$

where, L = expected life
A = constant for the material
ε = emissivity
b = a constant related to the Boltzmann constant
T = absolute temperature, Kelvins

When solved, the Arrhenius equation predicts that the life of a device is halved for every 20° C rise in temperature. The equation can also be solved to predict the life at an elevated temperature compared to life at "normal" temperature.

$$\frac{L_{hot}}{L_{cold}} = 2^{-\Delta T/20}$$

where, $\Delta T = T_{hot} - T_{cold}$

All failure mechanisms and failure indications show increased activity at elevated temperatures. These include

- Increased leakage currents
- Increased oxide breakdown
- Increased electromigration
- Accelerated corrosion mechanisms
- Increased stresses due to differences in thermal expansion due to differing coefficient of thermal expansion (CTE) among materials

Further examination of the equation will also tell us that device life will increase dramatically as we cool devices. However, issues such as brittleness and practicality enter and, with the exception of supercomputers and the like, most attempts to reduce temperature beyond what we normally think of as room temperature will involve a greater expenditure of money than the value received.

The power rating of an electronic device, along with the expected operating conditions and ambient temperature(s), determines the device's need for heat dissipation. This chapter deals with the fundamentals of heat, the fundamentals of heat transfer, and heat removal techniques.

The packaging system of any electronics component or system has four major functions:

1. Mechanical support
2. Electrical interconnection for power and signal distribution
3. Protection of the circuitry from the expected environmental operating conditions
4. Thermal management to maintain internal product/system and device temperature to control the thermal effects on circuits and system performance

The latter function may be to prevent overheating of the circuitry, or it may be to keep the internal temperature up to an acceptable operating level if operation is expected in subzero environments, such as unconditioned aircraft compartments. The latter issue will not be covered in this chapter.

To aid the electrically-trained engineer in better understanding heat transfer concepts, these analogies may be helpful:

Electrical	Thermal
Voltage	ΔT
Current	Heat flux
E/I (Ω)	Degree/watt
Capacitance	Thermal mass

While designers of "standard" digital circuits may not have heat management concerns, most analog, mixed-signal, RF, and microprocessor designers are faced with them. In fact, as can be seen in the evolution of, e.g., Pentium® (Intel Corp.) microprocessors, heat management can become a major portion of the overall design effort. It may need to be considered at all packaging levels.

- Level 1, the individual device/component package, e.g., IC chip carrier
- Level 2, the carrier/package on a module substrate or circuit board
- Level 3, the overall board and/or board-to-board interconnect structures, e.g., daughter boards plugged into a mother board
- Level 4, the mother board or single main board in its individual chassis, housing, or enclosure
- Level 5, if present, the cabinet or enclosure that houses the entire product or system

At the component level, the drive to minimize both silicon-level feature size and package size and to maximize the number of transistors on an IC serves to increase the heat generation and reduce the size of the package with which to accomplish the dissipation. This trend shows no sign of slackening. Lower supply voltages have not helped, since it is I²R heating that generates the bulk of the heat. Surface mount power transistors and voltage regulators add to the amount of heat that needs to be transferred from the component to the board or ambient air. Power densities for high-performance chips are currently on the order of 10 to 70 W/cm², and for modules on the order of 0.5 to 15 W/cm². High-speed circuits also add to these issues. High speeds and fast edges require high-power circuitry, as do typical microwave circuits.

All of these issues require that designers understand heat management issues. That is the intent of this chapter. More information on relating these issues to component and product reliability is presented in Chapter 15.

While they are interesting topics, this chapter will not cover the very specialized techniques such as water and inert fluid cooling of semiconductor devices. However, the references include articles on these and other cooling topics.

11.3 Fundamentals of Heat[*]

In the commonly used model for materials, heat is a form of energy associated with the position and motion of the material's molecules, atoms and ions. The position is analogous with the state of the material and is potential energy, whereas the motion of the molecules, atoms, and ions is kinetic energy. Heat added to a material makes it hotter, and vise versa. Heat also can melt a solid into a liquid and convert liquids into gases, both changes of state. Heat energy is measured in calories (cal), British thermal units (Btu), or joules (J). A calorie is the amount of energy required to raise the temperature of one gram (1 g) of water one degree Celsius (1° C) (14.5 to 15.5° C). A Btu is a unit of energy necessary to raise the temperature of one pound (1 lb) of water by one degree Fahrenheit (1° F). A joule is an equivalent amount of energy equal to work done when a force of one newton (1 N) acts through a distance of one meter (1 m). Thus, heat energy can be turned into mechanical energy to do work. The relationship among the three measures is: 1 Btu = 251.996 cal = 1054.8 J.

11.3.1 Temperature

Temperature is a measure of the average kinetic energy of a substance. It can also be considered a relative measure of the difference of the heat content between bodies. Temperature is measured on either the Fahrenheit scale or the Celsius scale. The Fahrenheit scale registers the freezing point of water as 32° F and the boiling point as 212° F. The Celsius scale or centigrade scale (old) registers the freezing point of water as 0° C and the boiling point as 100 °C.

The Rankine scale is an absolute temperature scale based on the Fahrenheit scale. The Kevin scale is an absolute temperature scale based on the Celsius scale. The absolute scales are those in which zero degree corresponds with zero pressure on the hydrogen thermometer. For the definition of temperature just given, zero °R and zero K register zero kinetic energy.

The four scales are related by the following:

$$°C = 5/9(°F - 32)$$
$$°F = 9/5(°C) + 32$$
$$K = °C + 273.16$$
$$°R = °F + 459.69$$

11.3.2 Heat Capacity

Heat capacity is defined as the amount of heat energy required to raise the temperature of one mole or atom of a material by 1° C without changing the state of the material. Thus, it is the ratio of the change

---

[*]Adapted from Besch, David, "Thermal Properties," in *The Electronics Handbook,* J. Whitaker, ed., CRC/IEEE Press, 1996.

in heat energy of a unit mass of a substance to its change in temperature. The heat capacity, often called *thermal capacity*, is a characteristic of a material and is measured in cal/g per °C or Btu/lb per °F,

$$c_p = \frac{\partial H}{\partial T}$$

11.3.3 Specific Heat

Specific heat is the ratio of the heat capacity of a material to the heat capacity of a reference material, usually water. Since the heat capacity of water is 1 Btu/lb and 1 cal/g, the specific heat is numerically equal to the heat capacity.

11.3.4 Thermal Conductivity

Heat transfers through a material by conduction resulting when the energy of atomic and molecular vibrations is passed to atoms and molecules with lower energy. In addition, energy flows due to free electrons.

$$Q = kA\frac{\partial T}{\partial l}$$

where, Q = heat flow per unit time
k = thermal conductivity
A = area of thermal path
l = length of thermal path
T = temperature

The coefficient of thermal conductivity, k, is temperature sensitive and decreases as the temperature is raised above room temperature.

11.3.5 Thermal Expansion

As heat is added to a substance, the kinetic energy of the lattice atoms and molecules increases. This, in turn, causes an expansion of the material that is proportional to the temperature change, over normal temperature ranges. If a material is restrained from expanding or contracting during heating and cooling, internal stress is established in the material.

$$\frac{\partial l}{\partial T} = \beta_L l \quad \text{and} \quad \frac{\partial V}{\partial T} = \beta_V V$$

where, l = length
V = volume
T = temperature
β_L = coefficient of linear expansion
β_V = coefficient of volume expansion

11.3.6 Solids

Solids are materials in a state in which the energy of attraction between atoms or molecules is greater than the kinetic energy of the vibrating atoms or molecules. This atomic attraction causes most materials to form into a crystal structure. Noncrystalline solids are called *amorphous*, and they include glasses, a majority of plastics, and some metals in a semistable state resulting from being cooled rapidly from the liquid state. Amorphous materials lack a long range order.

Thermal Management

Crystalline materials will solidify into one of the following geometric patterns:

- Cubic
- Tetragonal
- Orthorhombic
- Monoclinic
- Triclinic
- Hexagonal
- Rhombohedral

Often, the properties of a material will be a function of the density and direction of the lattice plane of the crystal.

Some materials will undergo a change of state while still solid. As it is heated, pure iron changes from body-centered cubic to face-centered cubic at 912° C with a corresponding increase in atomic radius from 0.12 nm to 0.129 nm due to thermal expansion. Materials that can have two or more distinct types of crystals with the same composition are called *polymorphic*.

11.3.7 Liquids

Liquids are materials in a state in which the energies of the atomic or molecular vibrations are approximately equal to the energy of their attraction. Liquids flow under their own mass. The change from solid to liquid is called *melting*. Materials need a characteristic amount of heat to be melted, called the *heat of fusion*. During melting, the atomic crystal experiences a disorder that increases the volume of most materials. A few materials, like water, with *stereospecific* covalent bonds and low packing factors attain a denser structure when they are thermally excited.

11.3.8 Gases

Gases are materials in a state in which the kinetic energies of the atomic and molecular oscillations are much greater than the energy of attraction. For a given pressure, gas expands in proportion to the absolute temperature. For a given volume, the absolute pressure of a gas varies in proportion to the absolute pressure. For a given temperature, the volume of a given weight of gas varies inversely as the absolute pressure. These three facts can be summed up into the Gas Law:

$$PV = RT$$

where, P = absolute pressure
V = specific volume
T = absolute temperature
R = universal gas constant

Materials need a characteristic amount of heat to transform from liquid to solid, called the *heat of vaporization*.

11.3.9 Melting Point

Solder is an important material used in electronic systems. The tin-lead solder system is the most used solder compositions. The system's equilibrium diagram shows a typical eutectic at 63% Sn. Alloys around the eutectic are useful for general soldering. High-Pb-content solders have up to 10% Sn and are useful as high-temperature solders. High-Sn solders are used in special cases such as in high corrosive environments. Some useful alloys are listed in Table 11.1.

TABLE 11.1 Useful Solder Alloys and Their Melting Temperatures

Percent Sn	Percent Pb	Percent Ag	°C
60	40	–	190
60	38	2	192
10	90	–	302
90	10	–	213
95	5	5	230

11.4 Engineering Data

Graphs of resistivity and dielectric constant vs. temperature are difficult to translate to values of electronic components. The electronic design engineer is more concerned with how much a resistor changes with temperature and if the change will drive the circuit parameters out of specification. The following defines the commonly used terms for components related to temperature variation.

11.4.1 Temperature Coefficient of Capacitance

Capacitor values vary with temperature due to the change in the dielectric constant with temperature change. The temperature coefficient of capacitance (TCC) is expressed as this change in capacitance with a change in temperature.

$$TCC = \frac{1}{C}\frac{\partial C}{\partial T}$$

where, TCC = temperature coefficient of capacitance
C = capacitor value
T = temperature

The TCC is usually expressed in parts per million per degree Celsius (ppm/°C). Values of TCC may be positive, negative, or zero. If the TCC is positive, the capacitor will be marked with a P preceding the numerical value of the TCC. If negative, N will precede the value. Capacitors are marked with NPO if there is no change in value with a change in temperature. For example, a capacitor marked N1500 has a –1500/1,000,000 change in value per each degree Celsius change in temperature.

11.4.2 Temperature Coefficient of Resistance

Resistors change in value due to the variation in resistivity with temperature change. The temperature coefficient of resistance (TCR) represents this change. The TCR is usually expressed in parts per million per degree Celsius (ppm/°C).

$$TCR = \frac{1}{R}\frac{\partial R}{\partial T}$$

where, TCR = temperature coefficient of resistance
R = resistance value
T = temperature

Values of TCR may be positive, negative, or zero. TCR values for often used resistors are shown in Table 11.2. The last three TCR values refer to resistors imbedded in silicon monolithic integrated circuits.

11.4.3 Temperature Compensation

Temperature compensation refers to the active attempt by the design engineer to improve the performance and stability of an electronic circuit or system by minimizing the effects of temperature change. In

TABLE 11.2 TCR Values for Often-Used Resistors

Resistor Type	TCR, ppm/°C
Carbon composition	+500 to +2000
Wire wound	+200 to +500
Thick film	+20 to +200
Thin film	+20 to +100
Base diffused	+1500 to +2000
Emitter diffused	+600
Ion implanted	±100

addition to utilizing optimum TCC and TCR values of capacitors and resistors, the following components and techniques can also be explored:

- Thermistors
- Circuit design stability analysis
- Thermal analysis

Thermistors

Thermistors are semiconductor resistors that have resistor values that vary over a wide range. They are available with both positive and negative temperature coefficients and are used for temperature measurements and control systems as well as for temperature compensation. In the latter, they are used to offset unwanted increases or decreases in resistance due to temperature change.

Circuit Analysis

Analog circuits with semiconductor devices have potential problems with bias stability due to changes in temperature. The current through junction devices is an exponential function as follows:

$$i_D = I_S\left(e^{\frac{qVD}{nkT}} - 1\right) = 1$$

where, i_D = junction current
I_S = saturation current
v_D = junction voltage
q = electron charge
n = emission coefficient
k = Boltzmann's constant
T = temperature, in kelvins

Junction diodes and bipolar junction transistor currents have this exponential form. Some biasing circuits have better temperature stability than others. The designer can evaluate a circuit by finding its fractional temperature coefficient,

$$TC_F = \frac{1}{v(T)}\frac{\partial v(T)}{\partial T}$$

where, $v(T)$ = circuit variable
TC_F = temperature coefficient
T = temperature

Commercially available circuit simulation programs are useful for evaluating a given circuit for the result of temperature change. SPICE, for example, will run simulations at any temperature, with elaborate models included for all circuit components.

Thermal Analysis

Electronic systems that are small or that dissipate high power are subject to increases in internal temperature. Thermal analysis is a technique in which the designer evaluates the heat transfer from active devices that dissipate power to the ambient.

Defining Terms

Eutectic: alloy composition with minimum melting temperature at the intersection of two solubility curves.

Stereospecific: directional covalent bonding between two atoms.

11.5 Heat Transfer

11.5.1 Fundamentals of Heat Transfer

In the construction analysis of VLSI-based chips and packaging structures, all modes of heat transfer must be taken into consideration, with natural and forced air/liquid convection playing the main role in the cooling process of such systems. The temperature distribution problem may he calculated by applying the (energy) conservation law and equations describing heat conduction, convection, and radiation (and, if required, phase change). Initial conditions comprise the initial temperature or its distribution, whereas boundary conditions include adiabatic (no exchange with the surrounding medium, i.e., surface isolated, no heat flow across it), isothermal (constant temperature) or/and miscellaneous (i.e., exchange with external bodies, adjacent layers or surrounding medium). Material physical parameters, and thermal conductivity, *specific heat*, thermal coefficient of expansion, and *heat transfer coefficients*, can be functions of temperature.

Basic Heat Flow Relations, Data for Heat Transfer Modes

Thermal transport in a solid (or in a stagnant fluid: gas or liquid) occurs by *conduction* and is described in terms of the *Fourier equation*, here expressed in a differential form as

$$q = -k\nabla T(x, y, z)$$

where, q = heat flux (power density) at any point, x, y, z, in W/m²
k_i = thermal conductivity of the material of conducting medium (W/m-degree), here assumed to be independent of x, y, z (although it may be a function of temperature)
T = temperature, °C, K

In the one-dimensional case, and for the transfer area A (m) of heat flow path length L (m) and thermal conductivity k not varying over the heat path, the temperature difference ΔT(°C, K) resulting from the conduction of heat Q (W) normal to the transfer area, can be expressed in terms of a *conduction thermal resistance* θ (degree/W). This is done by analogy to electrical current flow in a conductor, where heat flow Q (W) is analogous to electric current I (A) and temperature T (°C, K) to voltage V (V), thus making thermal resistance θ analogous to electrical resistance R(Ω) and thermal conductivity k (W/m-degree) analogous to electrical conductivity, σ (1/Ω–m).

$$\theta = \frac{\Delta T}{Q} = \frac{L}{kA}$$

Expanding for multilayer (n layer) composite and rectilinear structure,

$$\theta = \sum_{i=1}^{n} \frac{\Delta l_i}{k_i A_i}$$

where, Δl_i = thickness of the ith layer, m
k_i = thermal conductivity of the material of the ith layer, W/m degree
A_i = cross-sectional area for heat flux of the ith layer, m²

In semiconductor packages, however, the heat flow is not constrained to be one dimensional, because it also spreads laterally. A commonly used estimate is to assume a 45° heat spreading area model, treating the flow as one dimensional but using an effective area A_{eff} that is the arithmetic mean of the areas at the top and bottom of each of the individual layers Δl_i of the flow path. Assuming the heat generating source to be square, and noting that with each successive layer A_{eff} is increased with respect to the cross-sectional area A_i for heat flow at the top of each layer, the thermal (spreading) resistance θ_{sp} is expressed as follows:

$$\theta_{sp} = \frac{\Delta l_i}{kA_{eff}} = \frac{\Delta l_i}{kA_i\left(\frac{1+2\Delta l_i}{\sqrt{A_i}}\right)}$$

On the other hand, if the heat generating region can be considered much smaller than the solid to which heat is spreading, then the semi-infinite heat sink case approach can be employed. If the *heat flux* is applied through a region of radius R, then either $\theta_{sp} = 1/\pi kR$ for uniform heat flux and the maximum temperature occurring at the center of the region, or $\theta_{sp} = 1/4kR$ for uniform temperature over the region of the heat source [Carstaw and Jaeger, 1967].

The preceding relations describe only static heat flow. In some applications, however (e.g., switching), it is necessary to take into account transient effects. When heat flows into a material volume V (m³) causing a temperature rise, thermal energy is stored there, and if the heat flow is finite, the time required to effect the temperature change is also finite, which is analogous to an electrical circuit having a capacitance that must be charged to enable a voltage to occur. Thus, the required power/heat flow Q to cause the temperature ΔT in time ΔT is given as follows:

$$Q = pc_p v\frac{\Delta T}{\Delta t} = C_\theta \frac{\Delta T}{\Delta t}$$

where, C_θ = thermal capacitance, W–s/degree
p = density of the medium, g/m³
c_p = specific heat of the medium, W-s/g-degree

Again, we can make use of electrical analogy, noting that thermal capacitance C_θ is analogous to electrical capacitance C (F).

A rigorous treatment of multidimensional heat flow leads to a time-dependent heat flow equation in a conducting medium, which, in Cartesian coordinates and for Q_V (W/m³) being the internal heat source/generation, is expressed in the form of

$$k\nabla^2 T(x, y, z, t) = -Q_V(x, y, z, t) + pc_p\frac{\partial T(x, y, z, t)}{\partial t}$$

An excellent treatment of analytical solutions of heat transfer problems has been given by Carslaw and Jaeger [1967]. Although analytical methods provide results for relatively simple geometries and idealized boundary/initial conditions, some of them are useful [Newell, 1975, Kennedy, 1960]. However, thermal analysis of complex geometries requires multidimensional numerical computer modeling limited only by the capabilities of computers and realistic CPU times. In these solutions, the designer is normally interested in the behavior of device/circuit/package over a wide range of operating conditions, including temperature dependence of material parameters, finite dimensions and geometric complexity of individual layers, nonuniformity of thermal flux generated within the active regions, and related factors.

Figure 11.1 displays temperature dependence of thermal material parameters of selected packaging materials, whereas Table 11.3 summarizes values of parameters of insulator, conductor, and semiconductor materials, as well as gases and liquids, needed for thermal calculations, all given at room temperature. Note the inclusion of the thermal coefficient of expansion $\beta(°C^{-1}, K^{-1})$, which shows the expansion and contraction ΔL of an unrestrained material of the original length L_0, while heated and cooled according to the following equation:

$$\Delta L = \beta L_0 (\Delta T)$$

Convection heat flow, which involves heat transfer between a moving fluid and a surface, and *radiation heat flow*, where energy is transferred by electromagnetic waves from a surface at a finite temperature, with or without the presence of an intervening medium, can be accounted for in terms of heat transfer coefficients $h(W/m^2\text{-degree})$. The values of the heat transfer coefficients depend on the local transport phenomena occurring on or near the package/structure surface. Only for simple geometric configurations can these values be analytically obtained. Little generalized heat transfer data is available for VLSI type conditions, making it imperative to create the ability to translate real-life designs to idealized conditions (e.g., through correlation studies). Extensive use of empirical relations in determining heat transfer correlations is made through the use of dimensional analysis in which useful design correlations relate the transfer coefficients to geometrical/flow conditions [Furkay, 1984].

For *convection,* both free-air and forced-air (gas) or liquid convection have to be considered, and both flow regimes must be treated: *laminar flow* and *turbulent flow*. The detailed nature of convection flow is heavily dependent on the geometry of the thermal duct, or whatever confines the fluid flow, and it is nonlinear. What is sought here are crude estimates, however, just barely acceptable for determining whether a problem exists in a given packaging situation, using as a model the relation of *Newton's law of cooling* for convection heat flow Q_c (W).

$$Q_c = h_c A_S (T_S - T_A)$$

where, h_c = average convective heat transfer coefficient, W/m²-degree
A_S = cross-sectional area for heat flow through the surface, m²
T_S = temperature of the surface, °C, K
T_A = ambient/fluid temperature, °C, K

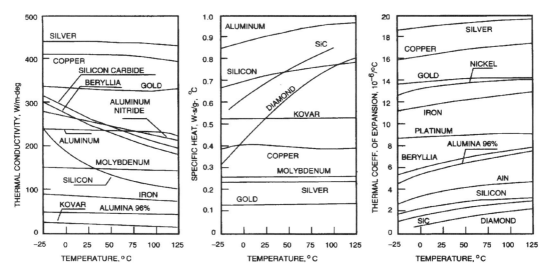

FIGURE 11.1 Temperature dependence of thermal conductivity k, specific heat c_p, and coefficient of thermal expansion (CTE) β of selected packaging materials.

TABLE 11.3 Selected Physical and Thermal Parameters of Some of the Materials Used in VLSI Packaging Applications (at Room Temperature, $T = 27°C$, 300 K)[a]

Material Type	Density, p, g/cm^3	Thermal Conductivity, k, W/m–°C	Specific Heat, cp, W–s/g–°C	Thermal Coeff. of Expansion, β 10^{-6}/°C
Insulator Materials				
Aluminum nitride	3.25	100–270	0.8	4
Alumina 96%	3.7	30	0.85	6
Beryllia	2.86	260–300	1.02–0.12	6.5
Diamond (IIa)	3.5	2000	0.52	1
Glass–ceramics	2.5	5	0.75	4–8
Quartz (fused)	2.2	1.46	0.67–0.74	0.54
Silicon carbide	3.2	90–260	0.69–0.71	2.2
Conductor Materials				
Aluminum	2.7	230	0.91	23
Beryllium	1.85	180	1.825	12
Copper	8.93	397	0.39	16.5
Gold	19.4	317	0.13	14.2
Iron	7.86	74	0.45	11.8
Kovar	7.7	17.3	0.52	5.2
Molybdenum	10.2	146	0.25	5.2
Nickel	8.9	88	0.45	13.3
Platinum	21.45	71.4	0.134	9
Silver	10.5	428	0.234	18.9
Semiconductor Materials (lightly doped)				
GaAs	5.32	50	0.322	5.9
Silicon	2.33	150	0.714	2.6
Gases				
Air	0.00122	0.0255	1.004	3.4 × 10^3
Nitrogen	0.00125	0.025	1.04	10^2
Oxygen	0.00143	0.026	0.912	10^2
Liquids				
FC–72	1.68	0.058	1.045	1600
Freon	1.53	0.073	0.97	2700
Water	0.996	0.613	4.18	270

[a]Approximate values, depending on exact composition of the material. *Source:* Compiled based in part on Touloukian, Y.S. and Ho, C.Y. 1979. *Master Index to Materials and Properties.* Plenum Publishing, New York.

For *forced convection* cooling applications, the designer can relate the temperature rise in the coolant temperature $\Delta T_{coolant}$ (°C, K), within an enclosure/heat exchanger containing subsystem(s) that obstruct the fluid flow, to a volumetric flow rate G(m^3/s) or fluid velocity v(m/s) as

$$\Delta T_{coolant} = T_{coolant-out} - T_{coolant-in} = \frac{Q_{flow}}{pGc_p} = \frac{Q_{flow}}{pvAc_p} = \frac{Q_{flow}}{\dot{m}c_p}$$

where, $T_{coolant}$ = the outlet/inlet coolant temperatures, respectively, °C, K
Q_{flow} = total heat flow/dissipation of all components within the enclosure upstream of the component of interest, W
$\dot{m}$ = mass flow rate of the fluid, g/s

Assuming a fixed temperature difference, the *convective heat transfer* maybe increased either by obtaining a greater heat transfer coefficient h_c or by increasing the surface area. The heat transfer coefficient maybe increased by increasing the fluid velocity, changing the coolant fluid, or utilizing *nucleate boiling*, a form of immersion cooling.

For nucleate boiling, which is a liquid-to-vapor phase change at a heated surface, increased heat transfer rates are the results of the formation and subsequent collapsing of bubbles in the coolant adjacent to the heated surface. The bulk of the coolant is maintained below the boiling temperature of the coolant, while the heated surface remains slightly above the boiling temperature. The *boiling heat transfer* rate Q_b can be approximated by a relation of the following form:

$$Q_b = C_{sf}A_S(T_S - T_{sat})^n = h_b A_S(T_S - T_{sat})$$

where, C_{sf} = constant, a function of the surface/fluid combination, W/m²-K^n [Rohsenow and Harnett, 1973]
T_{sat} = temperature of the boiling point (saturation) of the liquid, °C, K
n = coefficient, usual value of 3
h_b = boiling heat transfer coefficient, $C_{sf}(T_S - T_{sat})^{n-1}$ degree

Increased heat transfer of surface area in contact with the coolant is accomplished by the use of *extended surfaces*, plates or pin fins, giving the heat transfer rate Q_f by a fin or fin structure as

$$Q_f = h_c A_\eta (T_b - T_f)$$

where, C_{sf} = full wetted area of the extended surfaces, m²
η = fin efficiency
T_b = temperature of the fin base, °C, K
T_f = temperature of the fluid coolant, °C, K

Fin efficiency η ranges from 0 to 1; for a straight fin η = tanh mL/mL, where $m = \sqrt{2h_c/k\delta}$, L = the fin length (m), and δ = the fin thickness (m) [Kern and Kraus, 1972].

Formulas for heat convection coefficients h_c can be found from available empirical correlation and/or theoretical relations and are expressed in terms of dimensional analysis with the dimensionless parameters: Nusselt number *Nu*, Rayleigh number *Ra*, Grashof number *Gr*, Prandtl number *Pr* and Reynolds number *Re*, which are defined as follows:

$$Nu = \frac{h_c L_{ch}}{k}, \quad Ra = Gr\, Pr, \quad Gr = \frac{g\beta p^2}{\mu^2} L_{ch}^3 \Delta T, \quad Pr = \frac{\mu c_p}{k}, \quad Re = \upsilon L_{ch}\frac{p}{\mu}$$

where, L_{ch} = characteristic length parameter, m
g = gravitational constant, 9.81 m/s²
μ = fluid dynamic viscosity, g/m-s
$\Delta T = T_s - T_A$, degree

Examples of such expressions for selected cases used in VLSI packaging conditions are presented next. Convection heat transfer coefficients h_c averaged over the plate characteristic length are written in terms of correlations of an average value of *Nu* vs. *Ra*, *Re*, and *Pr*.

1. *For natural (air) convection over external flat horizontal and vertical platelike surfaces,*

$$Nu = C(Ra)^n$$

$$h_c = (k/L_{ch})Nu = (k/L_{ch})C(Ra) = C'(L_{ch})^{3n-1}\Delta T^n$$

where, C, n = constants depending on the surface orientation (and geometry in general), and the value of the Rayleigh number (see Table 11.4)
$C' = kC[(g\beta p^2/\mu 2)Pr]^n$

TABLE 11.4 Constants for Average Nusselt Numbers for Natural Convection[a] and Simplified Equations for Average Heat Transfer Coefficients h_c (W/m²–degree) for Natural Convection to Air Over External Flat Surfaces (at Atmospheric Pressure)[b]

Configuration	L_{ch}	Flow Regime	C	n	h_c	C'(27° C)	C'(75° C)
Vertical plate	H						
		$10^4 < Ra < 10^9$ laminar	0.59	0.25	$C'(\Delta T L_{ch})^{0.25}$	1.51	1.45
		$10^9 < Ra < 10^{12}$ turbulent	0.13	0.33	$C'(\Delta T)^{0.33}$	1.44	1.31
Horizontal plate (heated side up)	WL/[2(W+ L)]						
		$10^4 < Ra < 10^7$ laminar	0.54	0.25	$C'(\Delta T/L_{ch})^{0.25}$	1.38	1.32
		$10^7 < Ra < 10^{10}$ turbulent	0.15	0.33	$C'(\Delta T)^{0.33}$	1.66	1.52
(heated side down)							
		$10^5 < Ra < 10^{10}$		0.25	$C'(\Delta T/L_{ch})^{0.25}$	0.69	0.66

Note: H, L, and W = height, length, and width of the plate, respectively
[a]Compilation based on two sources: McAdams, W.H. 1954. *Heat Transmission*. McGraw–Hill, New York, and Kraus, A.D. and Bar–Cohen, A. 1983. *Thermal Analysis and Control of Electronic Equipment*. Hemisphere, New York.
[b]Physical properties of air and their temperature dependence, with units converted to metric system, are depicted in Fig. 11.2. Source: Keith, F. 1973. *Heat Transfer*. Harper & Row, New York.

Most standard applications in electronic equipment, including packaging structures, appear to fall within the laminar flow region. Ellison [1987] has found that the preceding expressions for natural air convection cooling are satisfactory for cabinet surfaces; however, they significantly underpredict the heat transfer from small surfaces. By curve fitting the empirical data under laminar conditions, the following formula for natural convection to air for small devices encountered in the electronics industry was found:

$$h_c = 0.83 f (\Delta T/L_{ch})^n \text{ (W/m}^2\text{-degree)}$$

where $f = 1.22$ and $n = 0.35$ for vertical plate; $f = 1.00$ and $n = 0.33$ for horizontal plate facing upward, that is, upper surface $T_S > T_A$, or lower surface $T_S < T_A$; and $f = 0.50$ and $n = 0.33$ for horizontal plate facing downward, that is, lower surface $T_S > T_A$, or upper surface $T_S < T_A$.

2. *For forced (air) convection (cooling) over external flat plates:*

$$Nu = C(Re)^m(Pr)^n$$

$$h_c = (k/L_{ch})C(Re)^m(Pr)^n = C''(L_{ch})^{m-1}v^m$$

where, C, m, n = constants depending on the geometry and the Reynolds number (see Table 11.5)
$C'' = kC(p/\mu)^m (Pr)^n$
L_{ch} = length of plate in the direction of fluid flow (characteristic length)

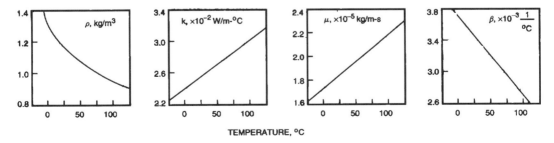

FIGURE 11.2 Temperature dependence of physical properties of air: density p, thermal conductivity k, dynamic viscosity μ, and CTE β. Note that specific heat c_p is almost constant for given temperatures, and as such is not displayed here.

TABLE 11.5 Constants for Average Nusselt Numbers for Forced Convection,[a] and Simplified Equations for Average Heat Transfer Coefficients h_c (W/m²–degree) for Forced Convection Air Cooling Over External Flat Surfaces and at Most Practical Fluid Speeds (at Atmospheric Pressure)[b]

Flow Regime		C	m	n	h_c for Forced Convection Air Cooling		
					h_c	C" (27° C)	C" (75° C)
$Re < 2 \times 10^5$	laminar	0.664	0.5	0.333	$C''(v/L_{ch})^{0.5}$	3.87	3.86
$Re < 3 \times 10^5$	turbulent	0.036	0.8	0.333	$C''v^{0.8}/(L_{ch})^{0.2}$	5.77	5.34

[a]Compilation based on four sources: Keith, F. 1973. *Heat Transfer.* Harper & Row, New York. Rohsenow, W.M. and Choi, H. 1961. *Heat, Mass, and Momentum Transfer.* Prentice–Hall, Englewood Cliffs, NJ. Kraus, A.D. and Bar–Cohen, A. 1983. *Thermal Analysis and Control of Electronic Equipment.* Hemisphere, New York. Moffat, R.J. and Ortega, A. 1988. Direct air cooling in electronic systems. In *Advances in Thermal Modeling of Electronic Components and Systems*, ed. A. Bar–Cohen and A.D. Kraus, Vol. 1, pp. 129–282. Hemisphere, New York.

[b]Physical properties of air and their temperature dependence, with units converted to metric system, are depicted in Fig. 11.2 (*Source:* Keith, F. 1973. *Heat Transfer.* Harper & Row, New York.)

Experimental verification of the these relations, as applied to the complexity of geometry encountered in electronic equipment and packaging systems, can be found in literature.

Ellison [1987] has determined that, for *laminar forced air flow,* better agreement between experimental and calculated results is obtained if a correlation factor f, which depends predominantly on air velocity, is used, in which case the heat convection coefficient h_c becomes

$$h_c = f(k/L_{ch})Nu$$

Although the correlation factor f was determined experimentally for laminar flow over 2.54 × 2.54 cm ceramic substrates producing the following values: $f = 1.46$ for $v = 1$ m/s, $f = 1.56 - v = 2$ m/s, f 1.6 $- v = 2.5$ m/s, $f = 1.7 - v = 5$ m/s, $f = 1.78 - v = 6$ m/s, $f = 1.9 - v = 8$ m/s, $f = 2.0 - v = 10$ m/s, and we expect the correlation factor to he somewhat different for other materials and other plate sizes, the quoted values are useful for purposes of estimation.

Buller and Kilburn [1981] performed experiments determining the heat transfer coefficients for laminar flow forced air cooling for integrated circuit packages mounted on printed wiring boards (thus for conditions differing from that of a flat plate), and correlated h_c with the air speed through use of the Colburn J factor, a dimensionless number, in the form of

$$h_c = J\rho c_p v (Pr)^{-2/3} = 0.387(k/L_{ch})Re^{0.54}Pr^{0.333}$$

where, $J = 0.387*(Re)^{-0.46}$

L_{ch} = redefined characteristic length, allows to account for the three-dimensional nature of the package, $[(A_F/C_F)(A_T/L)]^{0.5}$

W, H = width and height of the frontal area, respectively

A_F = W, H, frontal area normal to air flow

C_F = 2(W + H), frontal circumference

A_T = 2H(W + L) + (W + L), total wetted surface area exposed to flow

L = length in the direction of flow

Finally, Hannemann, Fox, and Mahalingham [1991] experimentally determined average heat transfer coefficient h_c for finned heat sink of the fin length L (m), for forced convection in air under laminar conditions at moderate temperatures and presented it in the form of

$$h_c = 4.37(v/L)^{0.5} (\text{W/m}^2\text{-degree})$$

Radiation heat flow between two surfaces or between a surface and its surroundings is governed by the Stefan Boltzmann equation providing the (nonlinear) *radiation heat transfer* in the form of

$$Q_r = \sigma A F_T (T_1^4 - T_2^4) = h_r A (T_1 - T_2)$$

where, Q_r = radiation heat flow, W
σ = Stefan-Boltzmann constant, 5.67×10^{-8} W/m²-K⁴
A = effective area of the emitting surface, m²
F_T = exchange radiation factor describing the effect of geometry and surface properties
T_1 = absolute temperature of the external/emitting surface, K
T_2 = absolute temperature of the ambient/target, K
h_r = average radiative heat transfer coefficient, $\sigma F_T (T_1^4 - T_2^4)/(T_1 - T_2)$, W/m²-degree

For two-surface radiation exchange between plates, F_T is given by

$$F_T = \frac{1}{\frac{1-\varepsilon_1}{\varepsilon_1} \frac{1-\varepsilon_2}{\varepsilon_2} \frac{A_1}{A_2} \frac{}{F_{1-2}} + 1}$$

where, $\varepsilon_1, \varepsilon_2$ = emissivity of materials 1 and 2
A_1, A_2 = radiation area of materials 1 and 2
F_{1-2} = geometric view factor [Ellison, 1987; Siegal and Howell, 1981]

The emissivities of common packaging materials are given in Table 11.6. The following approximations are useful in dealing with radiation:

- For a surface that is smaller compared to a surface by which it is totally enclosed (e.g., by a room or cabinet), $\varepsilon_2 = 1$ (no surface reflection) and $F_{1-2} = 1$, then $F_T = \varepsilon_1$.
- For most packaging applications in a human environment, $T_1 \approx T_2$, then $4\sigma F_T T_1^3$.
- For space applications, where the target is space with T_2 approaching 0 K, $Q_r = \sigma F_T T_1^4$.

TABLE 11.6 Emissivities of Some of Materials Used in Electronic Packaging

Aluminum, polished	0.039–0.057
Copper, polished	0.023–0.052
Stainless steel, polished	0.074
Steel, oxidized	0.80
Iron, polished	0.14–0.38
Iron, oxidized	0.31–0.61
Porcelain, glazed	0.92
Paint, flat black lacquer	0.96–0.98
Quartz, rough, fused	0.93–0.075

For purposes of physical reasoning convection and radiation heat flow can be viewed as being represented by (nonlinear) thermal resistances, *convective* θ_c and *radiational* θ_r, respectively,

$$\theta_{c,r} = \frac{1}{hA_S}$$

where h is either the convective or radiative heat transfer coefficients, or h_c or h_r, respectively, the total heat transfer, both convective and radiative, $h = h_c + h_r$. Note that, for nucleate boiling, ηh_c should be used.

11.5.2 Fundamentals of Device Package Thermal Characteristics

The common characteristic of all semiconductor packages is that they are composites, made up of several materials. These materials all have different coefficients of thermal expansion (CTE) and different thermal resistances.

The silicon die is, of course, the source of the heat that must be removed. The package surface serves to both trap the heat and to spread it from the die by conduction. The die is normally square or rectangular in shape.

11.6 Heat Removal/Cooling in the Design of Packaging Systems

As noted earlier, most electronic products and systems consist of several levels of packaging, and may include most or all of the defined levels 1 through 5. The goal of the heat removal technique(s) used is to move heat out of the junctions of power-dissipating devices while minimizing cost, noise, and overall complexity and size of the heat-removing devices.

Of the thermal resistances, θ, which have been discussed, the most commonly used basic model is shown in Fig. 11.3. For components that will be enclosed in some type of packaging, additional resistance will exist in the heat transfer path to the enclosure, through the enclosure, and from the enclosure to the outside air.

In this series model, the responsibility for θ_{jc} is with the package designer/supplier, while the remainder of the resistances are under the control of the board and system designers.

In this model, it must be remembered that the *ambient* temperature is rarely the room temperature in which the product or system is housed. Particularly in natural-convection cooled devices, the heat given off by the package and/or cooling fins can create a much higher temperature in the immediate vicinity. This means the *ambient* temperature into which the thermal resistances are working is not room temperature but a higher temperature. The calculations must take this real temperature into consideration if thermal problems are to be avoided. An example that shows this is the operation of an LM317T three-terminal voltage regulator without a heat sink. Housed in a TO-220 package, the rated junction-to-air thermal resistance $\theta_{ja} = 0.5$ W/cm². Using this information with an ambient temperature of 25° C = 77° F, an input-to-output voltage drop across the device of 5 V, and a current out of 0.5 A, shows that, theoretically, the junction temperature will be 125° C above ambient under these conditions:

$$T_j = T_a + (P_d \times \theta_{ja}) = 25°\text{C}[(5\text{ V} \times 0.5\text{ A}) \times 50°\text{ C/Watt}] = 25°\text{C} + 125°\text{C} = 150°\text{C} = T_j$$

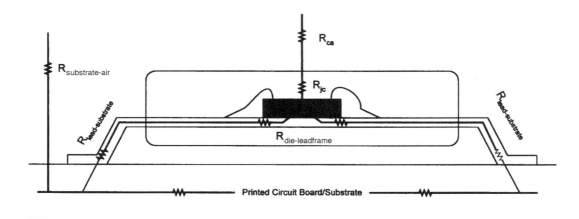

FIGURE 11.3 Thermal model of basic assembled IC and board.

Thermal Management

Since the maximum allowable $T_j = 150°$ C, if room temperature is truly $77°$ F = $25°$ C, the thermal protection circuit should just start shutting the circuit down at 0.5 A out. However, actual tests of a convection-cooled TO-220 package with no heat sink demonstrate that thermal-protection shutdown begins around 200 mA. The difference is that the ambient temperature immediately surrounding the heat sink is actually in excess of $75°$ C, which prevents the expected output current from being obtained.

$$T_j = T_a + (P_d \times \theta_{ja}) = 75° \text{C}[(5 \text{ V} \times 0.5 \text{ A}) \times 50° \text{C/Watt}] = 75° \text{C} + 125° \text{C} = 200° \text{C} = T_j = Oops...$$

However, at 200 mA,

$$T_j = T_a + (P_d \times \theta_{ja}) = 75° \text{C}[(5 \text{ V} \times 0.2 \text{ A}) \times 50° \text{C/Watt}] = 75° \text{C} + 50° \text{C} = 125 °\text{C} = T_j = OK$$

Therefore, the LM317 will not shut down at 200 mA, although the thermal protection circuits will be on the verge of beginning to control the output. These examples indicate the importance of defining *ambient* temperature. Note that the calculation examples are dependent on the ambient conditions. Trapped in an enclosed box, a 317 may shut down at under 200 mA.

11.6.1 Heat Removal Techniques

There are several heat removal techniques available to the circuit, board, and system designers.

- Enhanced thermal conductivity of the package itself
- Natural convective removal directly from the package
- Fan-forced convective removal directly from the package
- Conductive removal to a standard substrate/board
- Conductive removal to a thermally enhanced substrate/board

These techniques are frequently used in combination, and are discussed in the following sections. In many cases the designers will find the best combination of techniques that will allow for the maximum heat removal from the package. Again, heat is the major cause of component failures once they have been properly soldered onto a board, as described further in Chapter 15 on Reliability.

Enhanced Packages

Packages designed for enhanced thermal conduction include older through-hole designs such as TO-220 and TO-3 packages, the SMT D-PAK for power transistors, and SMT ICs in QFP and BGA packages incorporating integral heat spreaders or heat slugs. Heat spreaders are included in the top of the package to lower θ_{ja}, while heat slugs are attached directly to the die then molded into the top or bottom of the package and may either be used to increase convective heat transfer to the board or may be attached directly to the board or an external heat sink by solder, thermally conductive adhesive, or mechanical means.

While standard black-epoxy IC packages can be somewhat enhanced by the use of an external heat sink, the encapsulating epoxy is not a good heat conductor, and only minimal improvements can be made this way. The thermally enhanced packages are designed to increase conductive heat transfer from the die as well as conductive and/or convective heat transfer from the case to the air, heat sink, etc., and may be used as a stand-alone solution or have external enhancements. Examples of each are as follows:

Stand-alone	External enhancements
TO-220 and TO-3 packages used alone	Heat sink added to TO-220 and TO-3 packages
D-PAK soldered to a standard PCB	D-PAK soldered to a thermally enhanced PCB
BGA with heat spreader	BGA with heat sink attached to heat spreader
SO with heat slug	SO with heat slug soldered to enhanced board

Natural Convective Removal

This section considers removal of heat by convection from the package, from an enhanced package, and from a package with a heat sink. The thermal impedances of concern have been discussed.

Heat removal from the package is governed by the thermal resistances. In the case of heat removal from the package itself, it is only necessary to know the power dissipation of the package. This is calculated by determining the voltage across the device and the current through the device. The voltage across the device may be simply the power supply voltage, or it may be a $V_{in}-V_{out}$ voltage droop across the device, as in the case, e.g., of an LM7805 three-terminal regulator in a TO-220 package, as shown in Fig. 11.4.

If the maximum unregulated $V_{in} = 9$ V, and the regulated $V_{out} = 5$ V, then the voltage drop across the regulator is 4 V. This is the value used to calculate P_D, not the output voltage of 5 V. If the current through the regulator is 0.5 A, then:

$$P_D = V_{drop} \times 1 = 4 \text{ V} \times 0.5 \text{ A} = 2 \text{ W}$$

This 2 W would be the value of power to be dissipated and would be used to determine internal junction temperature T_j. If the thermal resistance from the internal junction to air θ_{ja} is 25° C/W, then,

$$T_j = T_A + (P_D \times \theta_{ja}) = 25°\text{ C} + (4 \text{ W} \times 25°\text{ C/W}) = 25°\text{ C} + 100°\text{ C} = 125°\text{ C}$$

This equation makes one common mistake and answers one common question. The mistake it makes is to assume that *ambient* temperature is the same as common *room* temperature. As will be seen later, even the use of a fan does not bring the immediate ambient temperature down to environmental ambient.

The most common natural convective heat removal addition to an IC is an external heat sink. While the correct application of heat sink information is not difficult, heat sinks users must not only consider θ_{jc}, θ_{cs}, and θ_{sa}, they must also consider the location of the die in the package if maximum cooling ability is required. Another item of consideration is the immediate ambient temperature, as discussed above with a bare device. However, the heat sink does have an advantage due to its typically large surface area, which means that larger air currents will flow through the fins of a large heat sink.

Typical textbook and data sheet calculations show the thermal resistances to be additive such that, if the user has a typical makeup of a power device, thermal grease or silicon thermal spacer, and heat sink, all the user has to do is add together the θ_{cs} of the silicon spacer/grease, and the θ_{sa} of the heat sink to arrive at an overall series resistance which can be used to predict temperature rise or conversely allowable power dissipation,

$$T_j = T_A + [(\theta_{jc} + \theta_{cs} + \theta_{sa}) \times P_D]$$

This assumes straight-line heat motion from case to sink. In reality, this does not happen. Kennedy did initial work on the concept of spreading resistance in thermal conduction. His work was done with circular sources and substrates. Subsequently, Lee, Song, and Au have adapted the spreading resistance concept to rectangular structures that more closely mimic the typical structures found in packaged components.

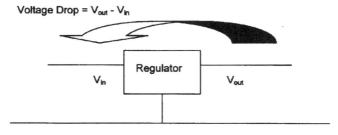

FIGURE 11.4 Voltage drop across a three-terminal regulator.

Thermal Management

The spreading resistance of the IC package-to-heat-sink thermal connection affects the efficiency of heat removal. The heat sink manufacturers' data sheets offer generalized information without any knowledge of the specific application in which the heat sink is to be used. The spreading resistance exists in any application where heat flows from one region to another, the two regions having different cross-sectional areas or having uneven temperature distribution over the surface area of concern. When a heat sink is attached to an IC, the area directly over the die is typically hotter than the remaining package area. To remove heat from this hotter area, some of the heat must flow laterally across either the package body or the heat sink base. This means there will be a hot spot directly over the die: For a given die power dissipation, the smaller the die, the hotter the temperature. However, since the same amount of heat must be dissipated by the package (and therefore the average temperature across the face of the package must be the same in either case), as the center of the package and heat sink get hotter with a smaller die, the temperatures at the edges of the package and the heat sink will decrease.

The following data are used to calculate the spreading resistance:

- The flat area of the package, A_p
- The flat area of the heat sink base, A_b
- The thickness of the heat sink base, t
- Heat sink thermal resistance, θ_{sa}
- Thermal conductivity of the heat sink base material, k

Additionally the total heat dissipation, Q, may be needed. The average temperature of the package surface may be calculated as $T_{avg} = \theta_{sa} \times Q$, and if we assume that $Q = P_{DISS}$, the power dissipation of the component, then $T_{avg} = (\theta_{sa} \times P_{DISS}) + T_A$ (see Fig. 11.5).

If the heat sink were the same size as the die heat source, spreading resistance would not be an issue. However, smaller heat sinks have a higher thermal resistance, and most heat sink applications closely match the area of the heat sink with the area of the package rather than with the area of the die.

One additional quantity is needed for determining maximum heat sink temperature. Since the heat flow is constricted in the immediate area of the heat source, a quantity *constriction resistance*, θ_c, is used to calculate the localized temperature rise at a given location above the average surface temperature.

$$\theta_c = \frac{\sqrt{A_p} - \sqrt{A_s}}{k\sqrt{\pi A_p A_s}} \times \frac{\theta_{cs} + \tanh(\lambda t)}{1 + \lambda k A_p \theta_{cs} \tanh(\lambda t)}$$

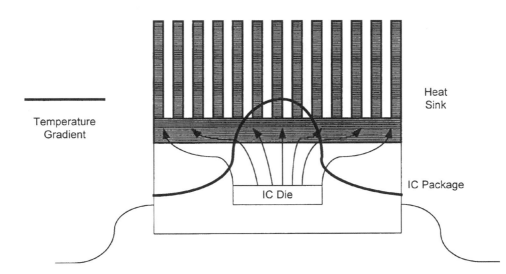

FIGURE 11.5 Thermal spreading resistance example.

To find λ,

$$\lambda = \frac{\pi^{3/2}}{\sqrt{A_p}} = \frac{1}{\sqrt{A_s}}$$

Studies by Lee, Song, and Au show that neither the shape of the heat source nor the shape of the heat sink plate (e.g., square or rectangular) affects these calculations by more than 5%. To use these equations in an example, consider a TO-3 package that will be bolted to an aluminum heat sink with base plate dimensions of 4 × 4 × 0.15 cm thick. The manufacturer states that the thermal resistance of this heat sink is 2° C/W, and that the thermal conductivity k of the aluminum is 200 W/mK. It is assumed that this is the average conductivity across the base of the device, that we will mount the TO-3 package centrally on the base, and that the area of the TO-3 base is 4.8 cm². Since our equations are based on the meter as the unit of measurement, our known quantities are as follows:

$$A_s = 4.8 \text{ cm}^2 = 4.8 \text{ E}^{-4} \text{ m}^2 \qquad A_p = 4 \times 4 \text{ cm} = 16 \text{ cm}^2 = 16 \text{ E}^{-4} \text{ m}^2$$

$$t = 0.15 \text{ cm} = 0.0015 \text{ m} = 15 \text{ E}^{-3} \text{ m} \qquad k = 200 \text{ W/mK}$$

$$\theta_{sa} = 2° \text{ C/W}$$

Now,

$$\lambda = \frac{\pi^{3/2}}{\sqrt{16 \text{ E}^{-4} \text{ m}^2}} + \frac{1}{\sqrt{4.8 \text{ E}^{-4} \text{ m}^2}} = 184.85 \text{ m}^{-1}$$

$$\tanh(\lambda t) = \tanh(184.85^{-1} \times 1.5 \text{ E}^{-3} \text{ m}) = 0.27$$

$$\theta_c = \frac{\sqrt{16 \text{ E}^{-4} \text{ m}^2} - \sqrt{4.816 \text{ E}^{-4} \text{ m}^2}}{200 \text{ W/mK}\sqrt{\pi \times 16 \text{ E}^{-4} \text{ m}^2 \times 4.816 \text{ E}^{-4} \text{ m}^2}} \times$$

$$\frac{184.85 \text{ m}^{-1} \times 200 \text{ W/mK} \times 16 \text{ E}^{-4} \text{ m}^2 \times 2° \text{ C/W} + 0.27}{1 + 184.85 \text{ m}^{-1} \times 200 \text{ W/mK} \times 16 \text{ E}^{-4} \text{ m}^2 \times 2° \text{ C/W} \times 0.27} = 0.214$$

Therefore, the actual thermal resistance is $\theta_{sa} + \theta_c$ = 2° C/W + 0.214° C/W = 2.214° C/W. This would be the correct number to use in T_j calculations with the combination of this specific heat sink and the TO-3 case.

Lee, Song, and Au also show that the effects of the specific die location in the package and the effects of boundary layer consideration on heat flow over the heat sink will result in little effects on these basic calculations.

Heat Sink Compounds

In the use of heat sinks, it is also important to remember that, when seen microscopically, neither the surface of the power component case nor the surface of the heat sink base is truly flat (Fig. 11.6). Since the gaps shown above between the case and the base are filled with air (a terrible conductor of heat), some means must be used to displace the air with a material that is a better conductor of heat. Therefore, the use must plan for the use of thermal paste, thermal grease, a silicon thermal spacer, thermal adhesive, or some other means. Note that, although many users tend to consider them to be interchangeable,

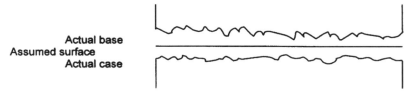

FIGURE 11.6 Typical component and heat sink base surfaces.

thermal paste has a thermal resistance between one-half and one-third that of thermal grease. Mounting pressure also has a direct effect on thermal resistance, and the interface manufacturer must be consulted to determine this effect as well as the proper mounting torque.

Examples of thermal paste include

- Dow Corning 340 compound
- General Electric G640 compound
- Thermalloy Thermalcoate
- Wakefield 120 compound

In certain applications, such as the use of a single, large heat sink or heat plate for more than one component, electrical insulation between the base and the case may also be necessary. In this case, Kapton, mica, Mylar, or silicon spacers will be necessary. Due to the hard surfaces of the Kapton, mica, and Mylar spacers, a thermal paste or thermal grease will be necessary to fill the air gaps, as discussed above. Beryllium oxide (BeO) is also used as an insulator and has a lower thermal resistance than aluminum and a high dielectric strength. It is thicker than the other insulators, resulting in decreased interface capacitance, which may be a factor in reducing EMI in power switching components. One downside to BeO is that it is very brittle, requiring care in handling. As with other spacers such as Mylar and Kapton, the manufacturer's recommendations for torque values on mounting screws/bolts must be followed. Table 11.7 lists examples of thermal resistances for various combinations of component cases and spacing materials, indicating the importance of the use of thermal paste.

TABLE 11.7 Example of the Effects of Various Thermal Spacers with and without Thermal Paste

Case Type	Case–to–Sink		BeO Insulator		Kapton Insul.		Mica Insulation	
	Dry	Paste	Dry	Paste	Dry	Paste	Dry	Paste
TO–3	0.3	0.1	0.6	0.15	0.8	0.5	0.5	0.4
TO–66	1.0	0.7	0.9	0.3	2.0	1.5	2.0	0.3
TO–220	1.2	0.9			4.5	2.5	3.5	1.5

Forced Convective Removal

As in natural convection cooling, the primary goal of thermal management using forced convection is to maintain the temperature of the electronic components in the system at a low enough temperature to minimize negative thermal effects. One of the major problems with most forced convection cooling systems is the difficulty in reaching all areas to be cooled with sufficient air volume and velocity.

The amount of air a fan or fans can force through an enclosure depends on the static pressure drops across the intake and discharge openings, across any filters in the system, across any internal ducting, heat sinks, etc. All of these items must be taken into account by the designer.

Convection heat transfer depends on a number of theoretical relations, including Newton's Law of Cooling, dimensionless numbers including Reynolds and Prandtls, transfer efficiencies, fin efficiencies, and so on and so forth. The author suggests that, rather than attempting to learn all these issues, designers of electronic systems for which convection transfer calculations are a concern avail themselves of one of the several Computational Fluid Dynamic (CFD)-based software packages that are specific to electronic convection heat transfer, such as Flotherm, Flotran, and others. If there is a thermal package associated with an ECAD package used for board design, check to see if it calculates flow and if it does a true 3-D analysis, which the CFD packages do. Most CFD tools can be used for IC package-level, board-level, and system enclosure-level computations.

One main measurement all designers of forced convection systems need to be aware of is the pressure drop with the enclosure. This drop is a function of both the frictional and the dynamic losses in the enclosure system, where frictional losses stem primarily from the equivalent duct length and average velocity, and dynamic losses stem primarily from changes in flow velocity due to contraction and expansion in the flow path, filters, etc. Fan flow rates are typically rated at open flow, with no back pressure, and with some defined low back pressure. Until the design knows the back pressure (static pressure) developed with the enclosure system, data sheet values of flow are of little value. Measurement of this quantity will be discussed later.

Initial determination of fan specifications must begin with calculation/approximation of the amount of heat that will be generated within the enclosure. The density of air must also be known. The basic heat transfer equation that applies for this case is

$$Q = C_p \times W \times \Delta T$$

where, Q = heat transfer amount
C_p = specific heat of air
ΔT = temperature rise in enclosure
W = mass flow = CFS × air density

Applying conversion factors to the basic equation allows simplification to an equation, which will provide an estimate of the amount of airflow necessary to dissipate a given quantity of heat at sea level.

$$\text{CFM} = \frac{1.74 \times Q}{\Delta T_c}$$

where Q is in watts and T is in °C.

Once airflow has been determined, the designer must determine the enclosure/system airflow resistance. This, along with the CFM, will allow fan selection to begin. Airflow resistance in a mechanical system is a function of static pressure and impedance provided by all the objects in between air intake and air discharge. In most systems, it is much easier to measure static pressure drop than it is to calculate it. Static pressure drop can be measured with a fan on either the intake or the discharge opening. Using the simplification that pressure will equalize within the enclosure, one would mount a fan on, e.g., the intake opening of the enclosure, then measure the pressure difference between the inside and the outside of the enclosure (Fig. 11.7).

The impedance follows a square law curve, where the static pressure drop changes as the square of changes in CFM. Therefore, this pressure measurement is only valid at the tested CFM.

Now that required flow and static pressure drop are known, it is possible to consult fan manufacturers' data sheets and being the selection of the fan. The most common types of fans used with electronic equipment are axial fans and squirrel cage blowers. Generally, axial fans are low-pressure units with flow rates in the range of 20 to 600 CFM. Squirrel cage blowers are higher-pressure units capable of moving 50 to 850 CFM. Squirrel cage fans will typically be noisier than axial fans, limiting their use to applications where axial fans cannot supply the required CFM and pressure. Additionally, axial fans are available as dc units capable of being powered off of an internal dc supply, but squirrel cage units typically require an ac supply. Most dc axial fans use solid state commutation, which reduces EMI. A brush-type motor should be tested for EMI prior to incorporation into a design.

For several reasons, the designer may want to consider the use of more than one fan. Reliability, in case of one fan failure, and economical use of the same fan used in other applications, are two reasons

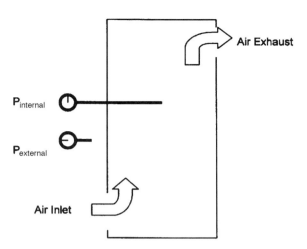

FIGURE 11.7 Static pressure drop measurement.

to consider a multiple-fan application. Multiple fans may be set in to blow in series, where the two (or more) fans are in a push-pull arrangement, or in parallel, with the two fans blowing side by side.

Series operation will result in increased pressure but not the doubling of pressure as one might expect. When fans are in series, they no longer have ambient pressure on one side and enclosure pressure on the other side, as in the initial test conditions. Series operation can be with two fans stacked at the input or output, or with one fan at the input and one fan at the output. Parallel operation will result in increased flow, but not necessarily doubled flow. Two-fan operation will result in doubled output (flow or pressure) only when the fans are both drawing from and discharging into free air. The results for multiple fan operation must be verified with the system enclosure as the test vehicle.

Designers must also remember that the equations presented earlier are valid only at sea level. For operations at elevated altitude, air density decreases, mass flow decreases for a given CFM, and therefore the CFM required at higher altitudes will be greater than the CFM at sea level to move the same mass of air through the enclosure and result in the same cooling effects. To consider the effects of altitude, the initial CFM equation must be modified to include barometric pressure as follows:

$$\text{CFM} = \frac{1.74 \times 29.9 Q}{p \Delta T}$$

where, Q = units of watts
T = units of °C
p = barometric pressure in inches of Hg at altitude

If the sea level calculation has already been done, the result can be modified for increased altitude by dividing the initial result by $(P_{altitude}/P_{sea\ level})$.

Example

For a given enclosure system, the internal power dissipation is 1200 W, the expected inlet air temperature is 35° C, the desired outlet temperature is no greater than 65° C, and the operational altitude may reach 6000 ft. Solving the equation yields:

$$\text{CFM} = \frac{1.74 \times 29.9 \times 1200}{24.4 \text{ in} \times (65 - 35°\text{ C})} = 85.6 \text{ CFM}$$

Or the sea level calculation gives $\text{CFM} = \dfrac{1.74 \times 1200}{30} = 69.9 \text{ CFM}$

Altitude correction gives $\dfrac{69.6 \text{ CFM}}{24.3 \text{ in}/29.9 \text{ in}} = 85.6 \text{ CFM}$

The initial static pressure determination should take place at the equivalent of the highest altitude at which the product is expected to operate. This will result in lower resistance and lower static pressure for a given CFM.

Conductive Removal to Standard Board/Substrate

Thermally Conductive Adhesives

The use of thermally conductive adhesives allows for a lowered conductive heat path to the board. This technique is adaptable to both standard and thermally enhanced substrates. It must also be remembered that the use of any adhesive in the assembly process will make rework and repair more difficult and may prevent them from being cost-effective.

Shown in Table 11.8 are process considerations for the application of thermal adhesives. It includes the variables the design and process engineers must investigate as they evaluate the use of a thermal adhesive. This chart is a starting point and does not include standard process issues such as which flux to use in the solder paste, what reflow profile to use, and which carrier to use for ICs. It assumes that the reader is already familiar with standard process steps and includes only those steps that have an impact on the successful use of thermal adhesives.

TABLE 11.8 Thermal Adhesive Process Considerations

Initial Considerations
Trace/pad thickness
Substrate warpage
Lead configuration(s)
Lead pitch
Lead surface area
Package standoff
Package/adhesive coverage necessary to meet thermal requirements
Adhesive Deposition
Before/after solder paste
Stencil or syringe
If syringe, time–pressure or positive–displacement?
Volume to be deposited
Required shape of deposit
Required volume of deposit
Required coverage area
Paste Deposition
Before or after adhesive deposition
Stencil or syringe
Volume to be deposited
Placement
Z–axis pressure
Pressure dwell time
Location
Adhesive Cure
Before or after reflow
Thermal or time cure
IR or convection
Oven settings needed
Expansion of adhesive (TCE) during cure
Post Cure and Post Placement
Ability to rework/repair

Regarding package and lead types, Heisler and Kyriacopoulos used design of experiment (DOE) techniques to examine different package types that were assembled to the board with thermal adhesives to enhance their thermal conductivity. They reported, during initial tests, a 90% acceptance rate with PLCC packages and a 30% acceptance rate with QFP packages. Subsequent work and testing to optimize the process resulted in better results. Among the test performed were ones to determine the volume of component area coverage based on the amount of adhesive dispensed. Their results indicate the following:

- For reliable formation of solder joints, no more than 50% of the area of the component bottom should be covered with thermal adhesive.
- Solder paste volume is the other determining factor in the reliable formation of solder joints. Thicker solder past deposition results in more reliable joint formation, since the adhesive prevents the component from *dropping* into the molten paste as reflow occurs.

This points out that it is very important to investigate thoroughly the affects thermal adhesives will have on each component being considered for this type of assembly. Like the addition of any process step, the use of thermal adhesives will cause a degradation in first-pass yield. And since the use of thermal adhesives has been studied less than many other process steps, and there will be a smaller in-house knowledge base, it is likely to require more investigative time and more process troubleshooting before it becomes a robust step in the process. This investigative process should include the following:

- Use of the appropriate θ to determine the theoretical effect of the addition of a thermal adhesive and the resultant decrease in T_j and T_c, using the manufacturer's data sheets. This calculation must include consideration of the thermal characteristics of the substrate.
- Theoretical calculation of the local temperature rise in the substrate as a result on the use of the thermal adhesive.
- Determination of actual (measured) T_c and from θ_{jc} the calculated T_j. For this determination, a known quantity of the thermal adhesive must be used. The assembly process need not mimic the expected actual process, although the volume and curing temperature and time must mimic the expected process.
- Determination of an appropriate assembly process. At this point, one of the main issues to investigate is that the thermal coefficient of expansions (TCEs) of the substrate, the component, and the thermal adhesive will all be different. It must be determined if the volume of thermal adhesive to be used will lift the component during the curing process.

In the above evaluations, it is important to determine the actual thermal effects on the component. If the T_j is only reduced by 1 to 2° C, the use of thermal adhesives is inconsequential, and another method of temperature reduction must be found. The use of a thermal pad, a heat sink, a thermal spreader, a heat slug, and/or convection cooling can all be considered.

Assuming thermal conduction to the substrate still needs to be enhanced, a silicon thermal pad may be used. It has the advantage of not needing any curing and of not causing any additional work during rework or repair. However, the thermal conductivity of a thermal paid may be lower than that of a thermal adhesive, and the thickness of the paid may not be available in the exact value necessary to fit under the component in question.

Conductive Removal to Enhanced Board/Substrate

Enhanced substrates are available with metal layers in the substrate. While these substantially increase thermal conductivity from components on the substrate, they also substantially increase cost. Substrates are available with internal layers of copper, invar, BeO, and graphite. All of these substrate cores reduce the thermal resistance of the substrate and also constrain the core of the board by reducing the CTE of the substrate. Used with heat slugs, they can dramatically reduce the thermal resistance of the component/board system (see Fig. 11.8).

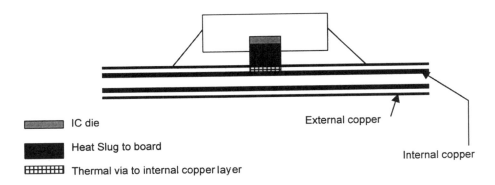

FIGURE 11.8 Thermally enhanced substrate.

As can be seen, there are a number of ways to reduce heat buildup in components. The designer's task is to select the techniques that make the most sense and are the most cost-effective for the project and product in question.

References

Assouad, Y, Caplot, M, Gautier, T; "Compact air-cooled heat sinks for power packages." Annual IEEE Semiconductor Thermal Measurement and Management Symposium, 1997.

Blanco, R, Mansingh, V; Thermal characteristics of buried resistors." Proceedings of the fifth IEEE SEMI-THERM Symposium, 1989.

Ellison, GN; "Thermal Analysis of microelectric packages and printed circuit boards using an analytical solution to the heat conduction equation." *Advances in Engineering Software*, vol. 22, no. 2, pp. 99–111, 1995.

Kennedy, DP; "Spreading Resistance in cylindrical semiconductor devices." *Journal of Applied Physics*, vol. 31, 1960

Lee, S, Song, S, Au,V; "Closed form equation for thermal constriction/spreading resistances with variable resistance boundary condition." *Proceedings of the 1994 IEPS Conference*, 1994.

Mansingh, et al.; "Three Dimensional computational thermal analysis of a pin grid array package." *Proceedings, Design Technology Conference, 1991.*

Markstein, HW; "Cooling electronic equipment enclosures." *Electronic Packaging and Production*, vol. 36 no. 5, May 1996.

Mei, YH, Liu, S; "Investigation into popcorning mechanisms for IC plastic packages: Defect initiation." *Application of Fracture Mechanics to Electronic Packaging ASME, EEP*, vol. 13, 1995.

Walker, AD, Marongiu, MJ; "Practical thermal management issues in the design of power supply components." *Power Conversion & Intelligent Motion*, vol. 23, no. 3, 1997.

12
Testing

Garry R. Grzelak
Teradyne Telecommunications

Glenn R. Blackwell
Purdue University

12.1 Introduction 12.1
12.2 Testing Philosophies 12.1
12.3 Test Strategies 12.4
12.4 Sources of Faults 12.4
12.5 Automatic Test Methods 12.9
12.6 Test Fixtures 12.15
12.7 Environmental Stress Screening 12.16
12.8 Test Software 12.19
12.9 Testing Software Programs 12.20

12.1 Introduction

In today's electronic world, testing is not an option for product designers and assemblers. This chapter is the second one that addresses the subject of test. The first is Chapter 9, "Design for Test."

Where Else?

Other information relating to issues discussed in this chapter can be found in Chapter 2, "SMT Overview," Chapter 6, "EMC and PCB Design," Chapter 9, "Design for Test," and Chapter 15, "Hardware Reliability and Failure Analysis."

12.2 Testing Philosophies

Testing can be considered both a value added and a non-value added part of the manufacturing process. When testing is used to monitor the manufacturing process and feed defect data back to help tune the process, it is considered to be value added. Testing is also considered to be value added when it provides a quality (defect-free, functionally performing unit) product to the customer. From a business case perspective, the overall cost to test must be less than the overall cost to manage defective product at the customer site. Non-value added testing includes redundant testing and testing that does not contribute to the overall quality to the customer.

Testing should be performed so as to allow any necessary rework to be accomplished judiciously and at the same quality level as the initial manufacturing process. This rework expectation is met if rework is done to production standards and if reworked products pass through the standard production testing after rework is finished. However, like all other facets of electronics manufacturing, the costs for all types of testing are increasing. Therefore, serious consideration of all factors must accompany the decision to test.

Testing is often performed in stages, with each progressive stage used to isolate and/or identify a particular defect in the fault spectrum. It is important to understand the range of faults in the fault spectrum and which faults will be considered to constitute an unacceptable quality level.

12.2.1 Types of testing

Four general categories of testing may be performed:

- In-circuit testing
- Functional testing
- Product performance testing
- Stress testing

In-circuit test consists of testing individual components while they are loaded on a circuit card. An ideal in-circuit test would be able to perform a test for each component or, stated another way, the failed test would identify a single component. In-circuit testing would utilize a variety of guarding techniques to achieve component isolation. In-circuit testing is largely used to make sure the assembly process has put the right parts into the right locations on the circuit board. By design, in-circuit testers add value to the assembly process by monitoring the manufacturing process.

Functional testing is used to determine whether a particular function of the board is working properly by toggling the circuitry through its legal states or conditions. Functional testing is designed to test the functions of each component or a group of components but not necessarily the entire function of the board. A simple example of a functional test is to make sure that an inverter inverts its signal properly to the correct logic levels.

If the inverter was run at the speed at which it was designed to function in the circuit, and it was subject to the same input signalling and output loading in the circuit as it was designed for, we can say that the inverter has passed a performance test. A performance test is used to exercise the circuitry in the capacity (speed and loading) in which the circuit was designed to function. Performance testing is sometimes called *hot box* testing, *subsystem* testing, or *system* testing.

Stress testing is used to understand where a product's failure point would be, as determined by introducing the product to various externally induced stresses. These stresses should be introduced in a way that would seem to accelerate the life of the components. Examples of stresses used to accelerate the life of electronic components include high- and low-temperature soaking, temperature cycling, vibration, and humidity. Stress testing is used on a sample of boards early in the product life cycle. The results of the data collected from stress testing can be used to improve the products robustness by focusing improvement tactics in the area of weakness, or to set stress screening levels for product quality.

Stress screening levels are set by derating (by perhaps 60% or 80%) the stress value that caused the failure. Stress screening levels can be applied to all products in the production cycle as a form of hurdle stress test that all products must pass to be considered as meeting the quality standard. An example of a stress screening profile might be to subject all the circuit boards to ten cycles of temperature cycling from $-20°$ to $+85°$ C, at a $9°$ C ramp rate per minute, allowing the product to soak for 1 hr at each of the temperature extremes while the units are powered up. Stress screening is discussed further in Section 12.7.

In-circuit, functional, and performance testing, and stress screening, are performed while the overall assembly process is taking place. These are not destructive to the product and serve the primary purpose of identifying defects that occur during the production processes. Defects found during these tests allow corrective action to be taken within the context of the applied quality control program. Rework, if cost-effective and meeting the criteria noted above, may correct defects found in this manner and add to the overall yield.

Whereas an in-circuit tester will typically identify a failure to an individual component, a functional or performance test will isolate the fault to a group of components; e.g., it will not result in a diagnostic message like "R3 open." A more highly skilled technician will need to perform a further analysis to determine the cause of the failure. Functional/performance testing systems must mimic all inputs and outputs to which the product will be connected. For example, an automotive engine control module will require the following inputs and outputs, among others:

- Engine RPM
- Engine manifold absolute pressure (MAP)
- Engine temperature
- Knock sensor
- Fuel injectors (or loads that mimic injectors)
- Ignition system

Once all the inputs and outputs are connected, whether they are the actual field I/O (performance) or simulations (functional) of the I/O, the test system will put the product through a series of tests designed to determine whether the product will perform to specification in its intended environment, to a high level of confidence.

Another form of performance testing employs a technique known as *dynamic reference testing*. In this approach, both the unit under test (UUT) and a known-good board are provided with the input stimuli simultaneously and have their outputs monitored simultaneously. The outputs from the two boards are exclusive-OR-ed and monitored to see if they continuously agree. The known-good board essentially acts as an infinite ROM for storing test patterns. Signature analysis techniques are frequently used to verify that the known-good board still functions properly.

One of the most crucial elements of test development is determining the defect mechanisms that the tests are designed to detect. The test techniques chosen, whether electrical, environmental, production or stress, must have a purpose tied to the product design, parts, manufacturing techniques, and expected performance.

Another crucial element is determining where defects can be found. Fluke presents a "Rule of Ten" in determining the relative costs of find failure at different points in a product's life cycle, as shown in Fig. 12.1.

Like all other issues, test is best handled in a concurrent engineering environment. One problem the test engineering member of the design team often faces, and shares with the manufacturing engineering member, is that design decisions that make their jobs more reliable are often contradictory to the customer and design demands. For example, both test and manufacturing would prefer wider traces and spaces on the circuit board, as well as large, well defined test pads at each circuit node. The typical customer demand for miniaturization is diametrically opposed to this.

In addition to finding and identifying verifiable faults, it is important that any automatic test system be investigated for unverifiable faults. Bateson gives an example of a test engineering manager who was

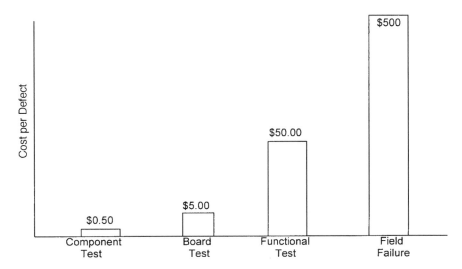

FIGURE 12.1 "Rule of Ten" in failure costs.

assigned to select a new in-circuit tester for a production line. After narrowing the field to three vendors, the engineer proceeded to evaluate each system by running a test case (benchmark) on all three systems. After the 75-board study, it was determined that vendor A's system caught 363 faults, Vendor B's caught 344 faults, and Vendor C's caught 351 faults. However, after verifying that the 75 boards actually contained 352 faults, it was determined that Vendor A caught 346 of the actual 352, Vendor B caught 341 of 352, and Vendor C caught 349 of 352. The extra faults detected by Vendor A's system were in fact false positives, i.e., unverifiable faults. This indicates that it is important to consider the validity of test systems for the particular fault classes that are expected to be present on the board to be tested.

12.3 Test Strategies

It is important to select an appropriate test strategy for a particular assembly before developing the tests. A balance must be established between the cost of the test development, the test coverage, and the time the test takes to run. Individual circumstances usually determine the drivers of your test strategies. For example, if the product life cycle is short and production quantities are low, the test engineer could be given a tight cost limit for development of the test. Obviously, the greater the number of UUTs to be tested, the greater the number required to amortize the test development costs. If your product is to be used in an application where someone's life or safety could be jeopardized by the failure of the electronics, then product functionality, test coverage, and reliability testing will be strong drivers. In today's typical high-volume manufacturing houses, the driver may be to fit the testing into a particular time specification of the overall manufacturing process. To keep up with production, a test may have to run in, for example, 23 seconds. Sometimes, the production cycle time requires a certain test time to avoid creating a bottleneck, as in a batch situation when boards are continuously coming out of the solder machines.

It is the purpose of the test strategy to maximize the combination cost, coverage, and test time for each assembly. One method to analyze this problem is to create a matrix. The matrix has the various test stages or processes on one axis and the likely fault spectrum on the other axis. The purpose of the test strategy matrix is to select the appropriate test process for the fault spectrum and test requirement of the UUT. You will note that the majority of the fault spectrum is detected in the in-circuit test stage. For example, you may note that system performance and stress screen for only 4% of the failures. This comes back to the question of what quality level you want to provide to the customer. It is often preferable to perform the system performance test and ESS rather than manage defects at the customer site.

One good question to ask is whether you could combine test stages (i.e., move the fault coverage for bad tweak to the in-circuit test or system performance test). In this way, you have reduced the test stages to three from four. Usually, this would be viewed as a cost savings in the overall test process.

One may ask why, in the test strategy, it is necessary to have different test stages at all. It comes back to the optimization question. At what test stage, using what skill level, will the test be optimized? A less skilled tester instead of a highly skilled technician can operate an in-circuit tester. If the in-circuit tester finds the majority of your faults, your overall rework cost would be lower than if you used a skilled technician to find the same faults.

Another factor to consider when exploring test strategies is the estimated cost of supporting the test program. This analysis would consider the cost to update the test program, how many times the program may need to be updated, and how likely the UUT is to be updated with an engineering change notice/order (ECN/ECO). Programs that can be easily modified and updated have an advantage when many changes and updates are typical of the manufacturing process.

12.4 Sources of Faults

The trends in electronics toward higher circuit operating speeds, miniaturization of circuits, and finer-pitch, higher lead-count parts have all contributed to increasing problems, both on the manufacturing floor and in implementing test procedures. Space saved through the use of surface mount parts is frequently not automatically available for test pads. Instead, it is either taken up by additional circuitry or used to reduce the size of the overall board/product even further. The faults on circuit board assemblies are exacerbated by these and other issues.

The faults present on circuit board assemblies can be grouped into four general categories:

- Component faults, whether outright performance failures or mechanical faults
- Manufacturing/process faults, typically in assembly, placement, or soldering
- Performance faults, which may be the result of design deficiencies, component faults, software errors, dynamic failures, or a combination of any or all of these, and which together cause the performance to fall outside allowable specifications
- Software faults, which may manifest themselves in a number of performance ways and which require their own test/verification considerations

Studies have shown a variety of fault distributions, but generally the distribution of non-software fault sources, is approximately as shown in Fig. 12.2.

One of the criteria for testing is to first determine where the faults are most likely to occur, what types of faults are expected or experienced, and then decide the best and most cost-effective testing to implement. For example, if the solder deposition (syringe or stencil) contributes to most faults, then the first test to consider would be solder paste presence and/or volume.

One of the most frustrating fault types both to categorize and to define are the faults variously known as *no-trouble found* (NTF), *no-fault found* (NFF), *cannot duplicate* (CND), *retest OK* (RTOK), and other like terms. While much of the discussion in this section deals with assembly faults, some consideration must be given to faults that do not turn up until the device is in the field. The inability to duplicate field failures can be very frustrating and is one reason why test definitions, as well as problem definitions, must be as rigorous as possible. It is important that field technicians gather information that is as complete as possible to allow them to better diagnose and define failures prior to returning defective devices to the factory. When repetitive field failures are occurring, and verified in the factory, it is apparent that in-house testing is not replicating field stress/environmental conditions adequately. Williams et al., 1998, report that the two most common causes of NTF reports in a Boeing 777 processor module were improperly seated connectors and changes in expected temperatures. The temperature shifts should indicate a change in in-house testing, while the connector problems may indicate the need for consideration of redesign.

12.4.1 Component Faults

As noted, component faults are either due to electrical/electronic performance deficiencies/failures of the component, or to mechanical failures. Performance deficiencies/failures depend on the type of component. Passive components such as resistors may not meet their accuracy or temperature coefficient (tempco) specifications, and ICs may be deficient in the same ways. Additionally, ICs may not meet other performance specifications, such as frequency response (for analog components) and

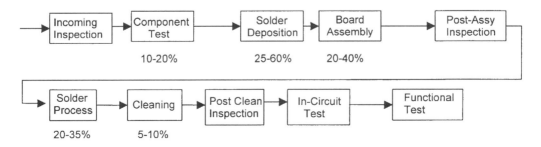

FIGURE 12.2 Fault sources in circuit board assembly.

risetime specifications (for digital devices). Performance tests are best conducted as part of incoming inspection.

Mechanical failure mechanisms may be from package failures, such as cracked ceramic resistor or capacitor bodies, moisture absorption failures in ICs, or termination faults such as the electrochemical effect of using two dissimilar metals, improper tinning of terminations/leads or coplanarity faults and bent leads on ICs. Package failures may be found during incoming inspection or, if parts and parts kits are stored in house, inspections may need to be performed just prior to assembly. A particularly insidious problem is that of moisture absorption by IC packages. If stored improperly in house, moisture absorption, which was not a problem with the part at incoming inspection, may become a problem at reflow. Chapter 4 discusses this problem further, particularly with regard to QFPs and BGAs.

A decision must be made either to purchase components at a specified quality level or test for the quality level yourself. And will this decision apply to passive components or just semiconductor components? Regardless of who does the testing, are all components tested against an acceptable quality level (AQL)? Will the supplier buy into being part of the supply line, or is the supplier just seen as a separate entity with no responsibility once the shipping packages are opened at the assembly facility? Decisions of this type are made based not only on pure economic issues but also on the expected use of the device and circuit use requirements. Incoming inspection may not be performed at all on parts for a clock-radio, but 100% inspection may be performed on parts that will be used in a heart pacemaker.

Table 12.1 shows the board-level defect rate that can be expected with certain component defect rates for a board with 175 components. For example, for a 1% component defect rate, 17.8% of the boards can be expected to have 0 faults, 30.4% of the boards can be expected to have 1 fault, 26.2% of the boards can be expected to have 2 faults, etc. It indicates the value of part mechanical and performance verification. Obviously, increased component count will cause the board defect rate to increase for any given component defect rate.

TABLE 12.1 Board Fault Distributions at 0.5, 1.0, and 2.0% Component Defect Rates

Faults/PCB	Component Defect Rate		
	0.5%	1.0%	2.0%
0	41.8%	17.8%	3.3%
1	36.3%	30.4%	11.3%
2	15.9%	26.2%	18.6%
3	4.7%	15.4%	21.1%
4	1.1%	6.9%	18.4%
5+	0.2%	3.4%	27.2%

Incoming inspection of components does not guarantee a product's reliability. The "bathtub" failure distribution curve, shown in Fig. 12.3, is still applicable for semiconductors and indicates why many manufacturers perform various types of stress testing, which will be discussed later. While the rates of

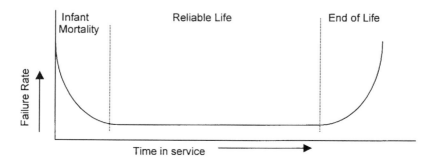

FIGURE 12.3 Distribution of electronic values with time—the bathtub curve.

infant mortality vary greatly among semiconductors, and the rates for the parts used in a particular design should be determined from the component manufacturer, infant mortality still must be considered in products that are expected to perform with high reliability.

12.4.2 Manufacturing Faults

This category of faults includes bare printed circuit board manufacturing defects, solder paste deposition errors, placement errors, reflow soldering errors, and cleaning system errors. Bare board errors can include

- Problems with planarity (levelness of the bare board)
- Problems with evenness of solder tinning
- Problems with through hole and/or via tinning
- Problems with solderability
- Inappropriate solder mask location
- Artwork shrink or stretch

Solder paste deposition errors can originate from

- Problems with solder paste volume deposited, either too much or too little
- Problems with the location of the solder paste deposit
- Problems with the solder paste chemistry, such as viscosity and flux, also solder paste age

Placement errors can include

- Wrong component placed
- Component placed in wrong location
- Component leads not placed on solder pads with the appropriate percentage of co-location
- Component leads not placed deep enough into solder paste (inappropriate down force)
- Component rotation incorrect
- Component lead problems not detected by camera system

Reflow solder problems can include

- Incorrect thermal profile
- Uneven heating on board as a result of large thermal masses on board
- Uneven heating resulting in tombstoning of components
- Uneven conveyor flow resulting in component movement

Cleaning system problems can include

- Incomplete washing of flux residue
- Incomplete washing of solder balls
- Incomplete drying
- Contaminated wash solution
- Non-hermetically sealed component packages

12.4.3 Performance Faults

As stated earlier, performance faults may be the result of design deficiencies, component faults, software errors, dynamic failures, or a combination of any or all of these which together cause the performance to fall outside the allowable specifications. These faults may show symptoms such as:

- Too narrow or too broad frequency response
- Signal distortion

- Race conditions
- Excessive radiated noise, causing EMI/RFI
- Inability to drive other circuits correctly
- Signal dropout
- Incorrect outputs for a given set of inputs

12.4.4 Software Faults

Software faults can cover such a wide spectrum of possible errors that it is impossible to cover all of them here. The faults can be something as widely recognized as the infamous "blue screen of death" on certain operating systems, to difficult-to-discover errors that are not immediately obvious but affect the user, such as an incorrect mathematical calculation result.

12.4.5 Fault Distributions

Since there is no way to predict the fault distribution on a new product running on a new line, an investigation of the defects is necessary as a first step in reducing the faults, once the line is up and running. This first step is preliminary to application of quality techniques as described in Chapter 1. As an example, consider the defects shown in Table 12.2. This is an example listing from an assembly line producing automotive engine control modules. The total production was 2300 boards, with a first pass yield of 1449 good boards, or 63% yield. As can be seen, average faults per board were 0.52, with an average of 1.4 faults per faulty board.

TABLE 12.2 Example Fault Distribution

Fault type	Faults per circuit assembly					Percent of faults
	1	2	3	4	5	
Solder shorts	250	88	52	31	9	35.9
Solder opens	103	29	32	16	3	15.3
Wrong component	17	8	5	4	3	3.2
Missing component	31	13	7	4	3	5.2
Bent leads	47	21	13	8	6	7.9
Solder balls remaining	80	45	23	15	1	14
Tombstones	2	1	0	0	1	~0
Component off location	81	44	34	19	8	15.5
Performance faults	19	9	5	3	2	3.3
Total faults	630	260	171	100	36	1197 total faults
Total PCBs with faults	630	130	57	25	9	851 total faulty boards

The predominant faults are solder shorts, solder opens, solder balls, and off-location components. With an understanding of the process steps and process machinery, and using the *design of experiment* techniques described in Chapter 1, the defect investigation would begin with a goal of improving first-pass yield.

Theoretical yield can be plotted as a function of overall defect rate, where $Y = e^{-DR}$ = yield, and DR = defect rate, which was 0.5/board in the above example. Graphically, it is as shown in Fig. 12.4.

If only overall defect rate is known, the Poisson distribution can be used to calculate the expected distribution of faults as follows:

$$P(d, x) = e^{-d}\frac{d_z}{x!}$$

which would lead to the predicted fault distribution shown in Fig. 12.5.

Testing

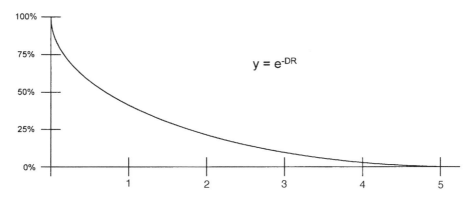

FIGURE 12.4 Defects per board.

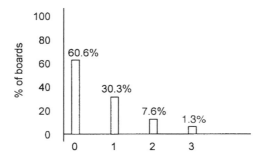

FIGURE 12.5 Number of defects per board.

If a product has multiple boards, e.g., a laptop computer, the overall product yield can be calculated from the defect rate.

$$\text{product yield \%} = (\text{board defect rate \%})^N$$

where N = number of boards

Therefore, if the board defect rate is 0.5 defects/board, the expected yield based on the number of boards in the product is as shown below.

Number of boards	Product yield
1	60.6%
2	36.7%
3	22.2%
4	13.5%
5	8.2%

From this, it is obvious that high first-pass yield gains greater and greater importance as the number of boards in the product increases.

12.5 Automated Test Methods

The test techniques to be described represent a spectrum of techniques that would rarely, if ever, all be used in the same facility or on the same product. It is one of the duties of the test team to determine the proper test techniques to be used to allow appropriate testing of the product under consideration.

Parts testing and bare circuit board testing, while important, are not covered in this chapter. These tests are best developed in conjunction with parts and board suppliers.

It is important to note, for the engineers who are developing test systems for the U.S. military, that customized test systems are not always required. The military has been accepting *commercial off-the shelf* (COTS) test systems for a number of years. Molnar, 1998, notes that the Navy has been actively engaged for more than 25 years in the development of COTS systems. He reports that new Navy initiatives are concentrating on improvements in training, extending calibration intervals, automation, and interoperability. These are valid goals for commercial users of test instrumentation as well. Thames, 1998, provides guidance in using COTS equipment in military applications.

12.5.1 Manual Testing

Employed in the prototype shop and in low-volume production, manual testing involves one or more employees using various test equipment by hooking that equipment to a board and diagnosing the faults on it. Typically, manual testing begins with functional testing then proceeds to component-level testing if functional problem are found. For manual testing, acquisition cost is quite low, but test execution time can be high, and the required skill level of the operator is very high for reliable tests.

12.5.2 PC-Based Testing

Sometimes called *rack-and-stack,* PC-based testing typically uses a rack of VXI-bus or IEEE-488-bus (called GPIB by Hewlett-Packard) test equipment to connect to the circuit board, or unit, under test (UUT). Appropriate software, such as TestPoint, VEE, and LabVIEW, is used to control the testing. Access to the UUT may be by techniques such as edge-card connectors, dedicated test connector, or bed-of-nails fixture. PC-based testing offers low initial cost and high flexibility. The setup time and testing time required rely heavily on the skill of the programmer. Programming time can be long if the programmer is not skilled in the use of the software.

As noted above, the military has developed its own acronym, which is sometimes used to describe the equipment used for this type of testing: *commercial off-the-shelf testers* (COTS). COTS may apply to PC-based testers or to combinations of other commercial test equipment.

When selecting hardware for PC-based testing, not only does any external test equipment such as a DMM or an oscilloscope need to be compatible with whatever test software and hardware is chosen, the data acquisition (DAQ) cards need to be compatible. The *Open Data Acquisition Standard* (ODAS), drafted by the Open Data Acquisition Association (http://www.opendaq.org), has established a software interface definition for PC-based DAQ cards. The standard defines the interface between application software and the device drivers for DAQ hardware such as PC plug-in (PCPI) cards. The association members include ComputerBoards, Data Translation, Hewlett-Packard, LABTECH, OMEGA Engineering, and Strawberry Tree. The goal of the association is to develop and maintain a universal, open standard allowing interoperability between DAQ hardware and software from multiple vendors.

The standard provides a specification for the five primary functions found on most DAQ cards:

1. Analog input
2. Analog output
3. Digital input
4. Digital output
5. Counter/timers

The standard is based on Microsoft 32-bit OM technologies, not any one vendor's software. The primary purpose for this is that most software uses the Windows interface, with its required *dynamic link libraries* (DLLs). Since one of the primary purposes of the standard is to allow use of several manufacturers' software and hardware products, the issue of multiple calls to a DLL had to be addressed. It is common that, when a new software package is installed, an older application may no longer run. This happens when the new application installs a version of a DLL that differs from what the first application uses. COM solves this problem with DLLs. The standard is designed to be a "thin" software layer that could be written on top of existing drivers.

The ODA standard is designed with a high degree of interoperability, so that any software can be connected to any hardware. This is accomplished by user-defined aliases for hardware devices. End users can swap ODAS-compatible hardware devices without changing the support software.

Users of ODAS-compatible products have a choice of using writing their own software using, e.g., Visual Basic, C++, or Delphi, or purchasing instrument-simulation/control software, e.g., HP VEE or LABTECH Notebook, and using that software to create an executable program.

12.5.3 Shorts Tester

A loaded-board shorts tester is used when shorts are one of the predominant (or the only) faults on the board. This typically applies to a commercially produced device with very rapid and high throughput, using a bed-of-nails fixture for UUT access. The shorts tester will identify both single and multiple shorts and produce location data for rework. The software will typically allow a board to be set up in a matter of hours, and it is easily changed for board modifications. Shorts testers require little in terms of operator skills. As would be expected, shorts testers have little or no parametric capability and no functional test capability.

12.5.4 Cable and Backplane Testing

Test systems designed for cabinet and interconnect wiring continuity testing can be used to check for the following problems with wire and cable:

- Verify point-to-point wiring in cables, harnesses, and wire-wrap panels
- Perform dc voltage measurements
- Perform resistance measurements to detect termination resistors
- Perform high-voltage (hi-pot) tests to determine insulation integrity

12.5.5 In-Circuit Analyzer

An in-circuit analyzer (ICA), also known as a *manufacturing defect analyzer* (MDA), tests for shorts, opens, resistance and capacitance values, and some semiconductor junctions on an unpowered circuit board. It will not test any semiconductor functions. As the Fig. 12.6 shows, the ICA/MDA conducts simple, unguarded measurements.

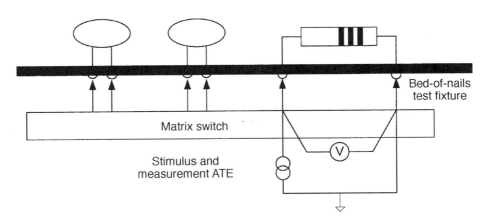

FIGURE 12.6 Simplified ICA/MDA measurement system.

The system of Fig. 12.6 shows a four-terminal Kelvin resistance measuring system, designed to improve on a two-terminal measuring system (a.k.a. an ohmmeter measurement) by not including the resistance of the measuring leads. Capacitors and inductors can be measured by this system if an ac source is used instead of a dc source. Devices in parallel can usually be measured by using a guarding system, described in the next section on in-circuit testing.

An ICA is fast and, like the shorts tester, will produce location data for rework. If the covered faults are the most common production faults, then an ICA is an appropriate test device. If performance failures are predominant, an ICA will not find them, since it does not power up the board. Typically priced under $100,000, an ICA is a cost-effective tool.

An ICA/MDA has a fault coverage of about 50 to 80%; that is, it will find up to 80% of the faults on a given board. Due to its relative simplicity, it also has high up-time.

12.5.6 In-Circuit Testing

An in-circuit tester (ICT) will detect shorts and test passive components (like the ICA), typically using a bed-of-nails fixture. It will then power up the board and test active components using a guarding technique to measure the performance of individual components by electrically isolating them from the surrounding circuitry. Typically, an ICT system will identify IC orientation and make basic digital logic tests. The ICT system has broad fault coverage with more complete fault data for rework than the ICA or shorts tester. An ICT system is cost-effective if it covers the fault spectrum needed. It makes an excellent screening system prior to functional testing. A basic ICT test circuit with guarding is shown in Fig. 12.7. Guarding is used to isolate the DUT from the surrounding circuitry.

The original circuit in the figure is Z_x in parallel with the series combination of R_1 and R_2. If $R_1 = R_2 = 47k$, and $Z_x = 100k$, a basic two or four-terminal measurement would read:

$$Z_{meas} = 100k \; // \; (47k + 47k) = 48.5k$$

The guarding system shown will work in the following way:

- The application of current source I_s will create a corresponding source voltage V_s at the left end of Z_x.

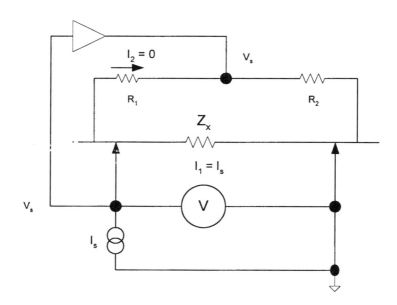

FIGURE 12.7 ICT test circuit with guarding.

- Through the buffer, the same voltage V_s will be applied to the junction of R_1 and R_2.
- The same voltage V_s will be present at each end of R_1; resulting in current $I_2 = 0$ through R_1.
- The voltage drop across Z_x will now be a result of only the current through Z_x.
- R_1 and R_2 have been "guarded" out of the measured portion of the circuit.
- Z_x will be correctly measured as 100k.

There are limitations to the guarding technique. Without using slower, more sophisticated measuring techniques, the basic guarding system will be accurate for impedances between 100 Ω and 1 MΩ. Below 100 Ω, accuracy drops off, because basic guarding measurements do not employ Kelvin measurement techniques, and lead resistances become a greater percentage of the value of the component being measured. Above 1 MΩ, the inaccuracy is due to the difficulty in blocking all parallel current paths.

At prices ranging from $100,000 to $500,000 or more, an ICT system may represent a significant capital expenditure. Software generation may take 3 to 10 weeks, and these systems typically come with automatic program generation software that assists the test engineers in development. ICT systems are therefore best suited for higher-volume board testing applications. An ICT system typically has about 85 to 95% fault coverage.

12.5.7 Functional Testing

A functional board tester simulates inputs to the board and output loading the board is expected to drive, then measures outputs to determine proper overall functional operation of the UUT. It has the broadest fault coverage, since it is intended to test all operating modes of the board. As an example, a functional tester for an automotive engine control module may provide simulated inputs such as throttle position, oil pressure, engine rpm, vehicle speed, water temperature, and manifold absolute pressure (MAP). It will provide output loads for electronic fuel injectors and the ignition system. It then powers up the board, changes the simulated inputs, and determines if the outputs respond properly. It can also test for the ability to drive outputs at each end of their specification range. For example, if it is expected that the fuel injectors may have impedances ranging from 100 to 125 Ω, the tester can test for proper output at each of those load impedances.

A properly programmed functional tester can provide rapid go/no-go information about the UUT and will have a typical fault coverage of about 85 to 98%. That is, when properly set up and programmed, a functional tester should perform a complete checkout of a product and allow very few faults to pass undetected. However, it has poor abilities for supporting fault diagnosis and rework. While it will detect, when properly programmed, most or all faults an assembled UUT might display, it will rarely be able to identify faults at the component level. This in one reason other types of testers, particularly ICA and ICT, precede functional test.

The functional tester typically requires long setup and software generation times, even when automatic generation software is included. This time is typically measured in months, and a functional tester for a complex UUT may require 4 to 8 months of development time, as well as costing a lot of money.

12.5.8 Substitution Testing

For certain types of systems consisting of several circuit boards, substitution testing may be a cost-effective test technique. As an example of this test technique, consider a laptop computer consisting of three boards, along with the display, keyboard, and mouse. The three boards may be the main board, the peripheral (display, keyboard, and mouse) driver, and the modem board. To use substitution testing on the peripheral board, all other components of the laptop system are present, including internal display, external monitor, and internal and external keyboards. The finished production board is plugged into this system, and all affected operating aspects of the laptop are verified, such as external monitor drive and internal display drive.

12.5.9 Dedicated Testers

Dedicated testers typically are in-house built test systems designed to test one specific circuit board. If ten different boards need to be tested, then ten different dedicated testers will be designed and built. Complex test fixturing requirements are the typical drivers for a dedicated test system. While any type of board, e.g. analog, digital and/or RF, can be tested with a dedicated tester, considerable development time must be allowed on both hardware and software compared to the development time which would be required by the typical purchased ATE system. If the testing to be done is on a high-volume product and the needed testing will require only a relatively simple set of tests, a dedicated tester can be an inexpensive alternative. Design of a dedicated tester will require considerable expertise on the part of the in-house test staff. It is imperative that, if a dedicated tester is the system of choice, the documentation of both design and operation be as good as the documentation that would be expected of a commercial test system. Since the designers and debuggers (typically test technicians) of the dedicated tester may leave their current assignment, personnel new to the system must have the same level of documentation that would be expected of a purchased system.

12.5.10 Fault Coverage

The faults that each of the previously discussed testers will find varies. Table 12.3 and Fig. 12.8 recap typical fault coverages and show how the coverage affects the final yield. Each type of tester must be evaluated to determine coverage for a particular product.

TABLE 12.3 Yields of Different Fault Coverages

First pass		Tester with 66% coverage (e.g., ICA)		Tester with 80% coverage (e.g., functional)	
d	yield (%)	d	yield (%)	d	yield (%)
0.5	61	0.17	84	0.1	90.55
0.3	74	0.10	90	0.06	
0.1	90	0.03	96.7	0.02	

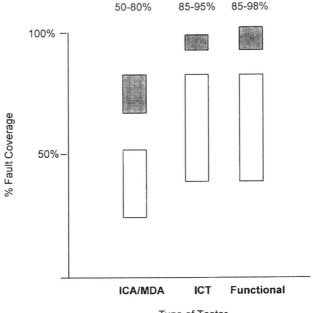

FIGURE 12.8 Typical fault coverage.

12.6 Test Fixtures

For any CA/MDA or ICT tester using a bed-of-nails (BON) fixture, the following criteria are useful in the design of the boards to be tested:

- One test node per net.
- Node access from one side only, with the understanding that two-side node access means more than double the expense and complexity of the BON.
- Probe spacing of 0.100 in (2.5 mm) nominal is ideal, less spacing (0.080 in = 2.0 mm or 0.050 in = 1.27 mm) is possible with the understanding that the probes are smaller and less robust, particularly for the 0.050-in probes.
- Contact areas for probes should be on designated test pads or trace areas, with a minimum of 0.030 in (0.9 mm) square.
- On the assumption that the BON fixture will have to serve more than one board from time to time, each board should have a unique tool hole pattern. This is normally two or more precision tooling holes positioned asymmetrically near a corners of the PCB. This ensures that the correct test head is employed in the testing of each PCB. On both the PCB and the head the position tolerance of these holes from their CAD value should be no more than 0.002 in.
- From the fixture vendor, determine the unpopulated area that must exist around the periphery of the board to allow for a good vacuum seal.
- To avoid gasketed hole plugs or vacuum covers, either fill all through holes and vias with solder or tent them with the solder mask.

12.6.1 BON Probes

BON test probes are spring-loaded, gold-plated probes with sharp points designed to be able to puncture test-pad oxides. There are as many as 50+ types of probe styles. A probe style should be custom selected to best fit your manufacturing and test process. The probe supplier should be aware of the following issues to provide proper guidance in the selection of probes:

- Presence of no-clean flux residue, as standard probes either will not reliably puncture the residue or will ultimately fail to function properly due to residue buildup on the probe and its spring
- Test pad spacing, as spacings less than 0.100 in may require special probes
- Landing area of flat test pad vs. via hole vs. through hole
- High components on the board, which may require special milling of the test fixture

12.6.2 BON Test Points

Shown in Fig. 12.9 is an example of a board with tolerances identified that must be taken into account during test pad layout. Note that the test point/pad/via location must accommodate the combined tolerances that may exist in the PCB manufacturing process as well as in the BON fixture itself. PCB tolerances may come from artwork shrink or stretch as well, from etch issues, and from tooling hole tolerances. The sum of these various tolerances is the reason most fixture and probe manufacturers recommend a minimum of 0.030-in (0.9-mm) test pad/via sizes. The tolerances noted in the figure are primarily driven by the need to accurately locate the test probes onto the test pads when the BON test fixture is activated.

The use of microvias may make these locations unsuitable for probing unless the surrounding pad is made a suitable size. Test point locations must also allow for sufficient clearance from the part body for the probe contact point so that the probe cannot land on any part of a soldered joint. Any unevenness of the contact point can force the probe sideways, with the resulting risk of damage.

The test probe contact location I.D. should correspond with the net designation in the PCB layout design. In this way, the X-Y location of each test node can be furnished to the fixture vendor for fixture

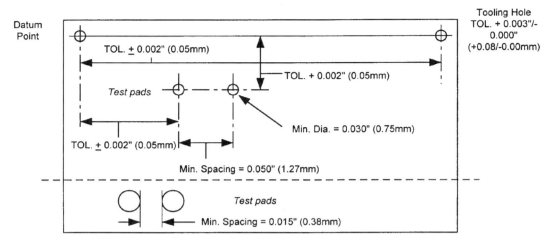

FIGURE 12.9 Examples of test probe contact locations.

development. The test development engineer will also need the node information for test software development.

As noted earlier, one of the criteria for test probe/pad locations is that they must be on an acceptable grid. Since a 0.100-in grid allows for more robust probes, it is frequently advantageous to maintain this probe/pad spacing, even if the lead pitch of an active device is less. Figure 12.10 shows a 0.050-in pitch device with 0.100-in test pad spacing:

12.7 Environmental Stress Screening

Environmental stress screening (ESS) is a technique used to expose defects that would not normally be detected with test techniques such as manufacturing defect analysis (MDA), in-circuit testing (ICT), functional testing, etc. ESS, properly applied, is intended to expose defects that are contributors to the "infant mortality" failures described in Section 12.3. As such, these defects may be the result of component weaknesses, process weaknesses, and/or substrate weaknesses.

Whereas test such as MDA or ICT are intended to increase the quality of the product by uncovering defects that are the result of component, process or substrate failures, ESS is intended to increase overall

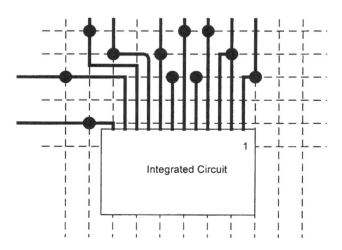

FIGURE 12.10 Examples of test probe contact locations.

product reliability by *inducing* the infant mortality failures. It is important to note that, in ESS, failures are expected. Another way to justify ESS is that it is intended to move failures that would otherwise occur during the warranty period into the production period of the product.

ESS is different from *burn-in*. Burn-in typically involves *soaking* a product at an elevated (or lowered) temperature, whereas ESS stresses a product by continuously cycling it over a range of temperature and perhaps humidity. It has been shown that an ESS thermal profile with a temperature ramp of 5° C/min will cause more failures than a burn-in at a steady 120° C. Since ESS is intended to compress the time-to-failure, it is an accelerated life test.

A major design parameter for an ESS program is to design it so that the failures that will occur as a result of the ESS cycles are those that would normally occur. That is, the "stress" is designed so it will not decrease the useful life of products that pass the screening. As such, the ESS program must initially determine the amount of stress the product can handle without affecting that useful life.

ESS is intended to accelerate failure mechanisms that occur over time. For example, dendritic growth will occur if contaminants remain on the surface of a substrate, and this growth can cause a low-impedance short between two traces over time. A properly designed ESS profile can cause rapid dendritic growth, allowing the failure mechanism to occur during the production test phase of the product. Another example is that temperature variations caused by expected environmental shifts and/or by power on/off cycling will stress solder joints through expansion and contraction due to different thermal coefficients of expansion between, e.g., the part leads and the FR-4 substrate. An appropriate stress test must consider the temperature extremes to which the product will actually be exposed, then choose a set of temperatures for the thermal cycling that will exaggerate both the low- and high-temperature ends of the spectrum, while at the same time not be the root cause of the induced failure.

Shown in Fig. 12.11 is an example of a stress test for an audio product that is expected to be used indoors. Expected shipping/storage temperature extremes are −30° to +70° C, expected operating envi-

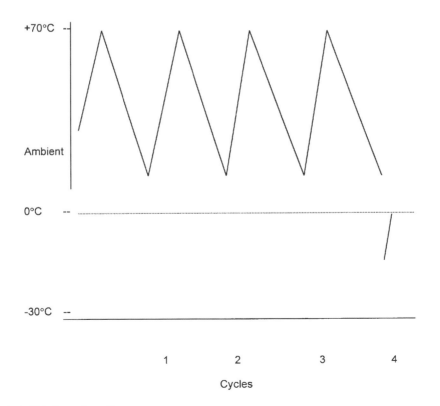

FIGURE 12.11 ESS thermal profile.

ronmental temperature extremes are +15° to +40° C, and the operating temperature range is expected to be +15° C with power off to +55° C with power on. Test temperatures must consider all of these factors, so the stress test will operate over the range of –30° to +70° C. Wider ΔTs are not desirable, since the ESS test is meant to mimic realistic temperature extremes while accelerating the time sequence.

The solder joints, and any other expected failures, would be examined in two ways. In initial developmental testing, they would be examined after a specific number of cycles. If, e.g., the overall test is expected to last 100 cycles, the joints would be examined every 10 cycles to determine if the ESS is overstressing the joints. Once the testing has been proven, the joints would only be examined at the end of the testing cycles.

As another example of a preproduction stress test, a change to a no-clean solder paste may result in a concern about long-term surface insulation resistance (SIR) changes on the substrate. To determine if the product will remain operational throughout its expected life, both temperature and humidity must be used in the test, along with a bias voltage applied across the highest-risk areas of the circuits. Humidity can be added at atmospheric pressure or, in a technique known as a *highly accelerated stress test* (HAST), humidity can be added under pressure as a stress factor. The insulation resistance is read at time zero and at ambient conditions, then in situ at intervals throughout the duration of the test. Figure 12.12 shows an example of a resistance graph for a test between conductors, for which a precipitous drop in insulation resistance can be seen around 250 hr. This is characteristic for an SIR test, as the bias voltage forces an insulation breakdown due to the dendritic growth caused by the flux remains, the bias voltage, and the increased temperature and humidity. This dendritic growth results in a relatively low-impedance path forming on the substrate.

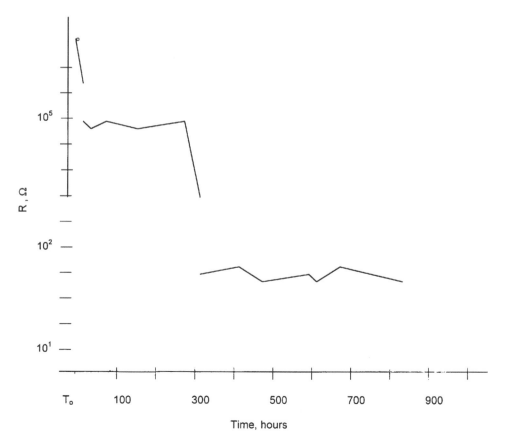

FIGURE 12.12 SIR trace insulation resistance vs. time.

If the product is expected to survive in a hostile environment, such as an oil refinery, it would be appropriate to introduce the corrosive and conductive contaminants that will be present into the test chamber. For a refinery, this could include sulfur. For other types of processing plants, such as a grain mill, dusts may be introduced. In this event, the test group must be certain that, if flammable or explosive agents are introduced, the test chamber is designed to protect against fires and explosions.

In this example of an SIR test, it must be noted that conditions will vary depending on the type of product. For example, testing conditions for a stereo receiver that will spend its life on a shelf would be different from the conditions for an refinery control module, which will spend its life in the already described environment. A first-level product may require tests at 85° C, 80% RH and 30 Vdc, while a second-level product may require 50° C, 80 RH, and 15 Vdc applied bias voltage. The humidity will support any growth due to flux residue contamination, while the bias voltage both enhances growth and allows the resulting impedance measurement to be taken.

In projecting product reliability based on stress test, it is necessary to know the number of failure mechanisms that exist in the product and the number of these failure types that the ESS will uncover. Blindly "doing" ESS may or may not have the desired results. It is also important to note the temperature limitations of all components and the substrate. Exceeding any of these may risk damage to the components.

12.8 Test Software

This section will discuss some of the issues to be considered in the selection of software to drive hardware testing systems. Systems specifically to test software programs will be discussed in section 8.

The selection of test software, like any other decision in the overall design, manufacturing and testing of an electronic product, is a decision based on all the criteria the test team considers appropriate. These criteria may include

- Platform: PC, Unix, proprietary
- Compatibility with existing system software
- Quality of operator interface
- Ability to customize the operator interface
- I/O compatible with existing systems/hardware
- Vendor support
- Ease of test development by in-house engineers/programmers
- Time for development of test code
- Quality of debugging aids
- Speed of test execution
- Documentation
- Data logging capability
- Networking capability

Like all complex decisions, the test software decision is an interactive one. No matter which issue the test team considers and specifies first, it almost assuredly will be modified as other considerations are taken into account. Each facility must make its own decisions based on its own needs. The first step is to determine what tests and test steps are necessary to verify that the product is being produced at the desired quality level. This is a description of the necessary functions identifying the desired tests, the necessary test results, the users (who has to read the report?), and the level of operator interaction necessary and/or desirable. It is also a realistic assessment of what tests are necessary to maintain the desired product quality level. Both of these impact the economic viability of the tests. It is very important at this stage to also consider future needs.

The next two interactive steps in choosing a test software package are to determine what level of platform is necessary to support the previously defined test needs of the product and then select software packages based on that platform. While the facility may have PC, Unix boxes, and name-brand test systems (e.g., Teradyne, Hewlett-Packard, etc.), typically the test for the particular product under consideration will fit best on one of the platforms. Unless an operation is truly starting with a blank check and a new facility, it is rare that the test team can afford all-new hardware.

It should be recognized that all software will require some customization to meet any product's unique test needs. The operator interface will need to be developed, as will the data storage and network needs. Data file formats must be compatible. The programming language must be considered. While powerful languages such as the several variations of C can solve most problems, the time to develop programs in C is not trivial, especially if the user is not already well versed in it. This must be considered as the test team discusses whether to buy a software package or develop its own from scratch. Rarely is the time necessary to develop a custom software package justifiable, although Romanchik, 1998, discusses decision-making among Basic, Visual Basic, C, C++, and various graphical-oriented programming environments. He discusses both development of test programs using these languages as well as maintenance of the software.

Programming flexibility needs to be considered. Most test program development requires multiple test sequences to be chained, nested, and imbedded within loops. Kolts points out that a 64-channel relay multiplexer with 4 tests on each channel will require that a test sequence of 4×64 (256) tests to be written if no loops are available. If a looping construct is available, only 4 tests would need to be written and then wrapped in a loop for the 64 channels. Obviously, the fewer lines of code that need to be written, the fewer chances for errors and the easier the code is to check, and minimal code also results in shorter development time.

Using a feature list such as the one at the beginning of this section, the test team should develop a comparison chart among vendors to be considered and narrow the list of potential vendors before a full-scale set of tests is run. Ideally, there should be no more than three software packages to undergo full testing, given the amount of time that a full-scale evaluation of each package will take.

Once the number of potential vendors has been narrowed down, the test team should conduct site visits to existing user. The supervisor of the site to be visited should understand that this is not a sales call and that the team would like to be able not only to observe the system in operation and see the output but also to talk to each member of the site development team to determine the various problems each faced in the development of a working system. The feature list that has been developed should be carried to the site by each member of the team, and each should then do an individual analysis of the features based on the demonstrations on-site. After the visit, the test team should meet to compare the results of the separate evaluations, and work out any discrepancies in real or perceived results.

Once the results of the paper evaluation and of the site evaluation are compared, it is time to perform an in-house evaluation. Sufficient time should be set aside for the team to become familiar with each feature that has been deemed to be important to the testing of the product. It is much cheaper to spend evaluation time at this point than to spend time solving unforeseen problems after the software package has been purchased.

12.9 Testing Software Programs

Glenn Blackwell, Garry Grzelak

This section discusses some of the issues to be considered in the testing of the software component of the product. The selection of software to drive hardware test systems is discussed in Section 12.7.

The software fault spectrum is different from the process and or hardware fault spectrum but a methodical and integrated software test strategy still apply in order guarantee software quality. As test engineers are faced with shrinking electronic assemblies, reduced physical access and more inherent software content we will see increased pressures on the software testing.

Much hardware is being designed today with built-in software configurability to extend the hardware's flexibility. Many products today are being designed to work in conjunction with other electronic systems, which adds new complexities around the interoperability testing with other system components.

12.9.1 Software Testing Compared to Hardware Testing

To better understand the software testing approach, let us look at Fig. 12.13, which compares the software and hardware development and test paths. These are drawn to show analogies between the hardware and software test paths, not to show the actual development process.

In the software path, we simplify the process to include the high-level design (HLD), low-level design (LLD), coding, unit test, component integration test, and system integration test and then delivered to an independent system verification test process. The arrows under each box describe the main testing activity to get to the next piece of the process. Reference points A, B, and C are used to identify transition points as the software testing progresses. In a typical software testing strategy, we will see the primary software testing ownership responsibilities shift from the development team to a test team.

Each of the stages of the software test process will be discussed briefly as follows. It is not in the scope of this book to discuss each of these stages in detail but rather to make you aware of the essential elements of each stage. For each section, the objective will be stated along with the entrance and exit criteria. A bullet form of the activities in each stage and the quality metrics and control documents for each stage is also included.

In Fig. 12.14, we can see the relationship between the various test stages from unit test through system verification test. You will notice the hierarchical integration techniques of software integration. Unit test is usually reserved to an individual developer's test of the developed code. Several unit tests are gathered into a component integration stage, several component integration stages are gathered into a system integration stage, and the system integrated software product is then delivered to the system verification test team.

Testing the software on large products or systems involves a number of issues that start at the beginning of the design phase. In any project that involves a programmable device or devices, both hardware and software must be developed by teams in a concurrent atmosphere. While the low-level designs will take place independently (the software team really doesn't care how many decoupling capacitors are required on the circuit board), the high-level design work must be integrated. The amount of RAM and ROM (whether PROM, EPROM, E^{2}PROM, etc.), e.g., will be dependent on the amount of code, the amount of intermediate data storage, etc.

The reader must recognize that each facility will need to make its own decisions about the relative importance of each of the criteria. Additionally, each facility will have to decide whether it can perform its own verifications or whether third-party verifications are either desirable or are required by the customer(s).

In the following descriptions, *severity* of problems will be noted. Generally,

- *Severity 1* defects are those that completely render the software unusable and prevent it from successfully performing any major tasks.
- *Severity 2* defects are those that render a subsystem of the software unusable while allowing the allowing other subsystems of the software to operate to some degree.
- *Severity 3* defects are present in operating software and render some function of the software unusable, but they are frequently accompanied by a workaround that allows operations to continue. These defects do not prevent the test from progressing.
- *Severity 4* defects are those minor defects that the customer is notified of and are of a nuisance nature. These include documentation errors or user interface format error, and correction can be postponed until the next software upgrade. These defects do not prevent the test from progressing.

12.9.2 Software Code and Unit Test

The objective of this process step is to refine the software low-level design into source code and to perform unit test. Source code is written in accordance with the project coding guidelines and is validated by code reviews and inspections and unit tested. Coding parts include modules, macros, messages, and screens. Unit test is the software engineer's verification that the module functions correctly and is ready for

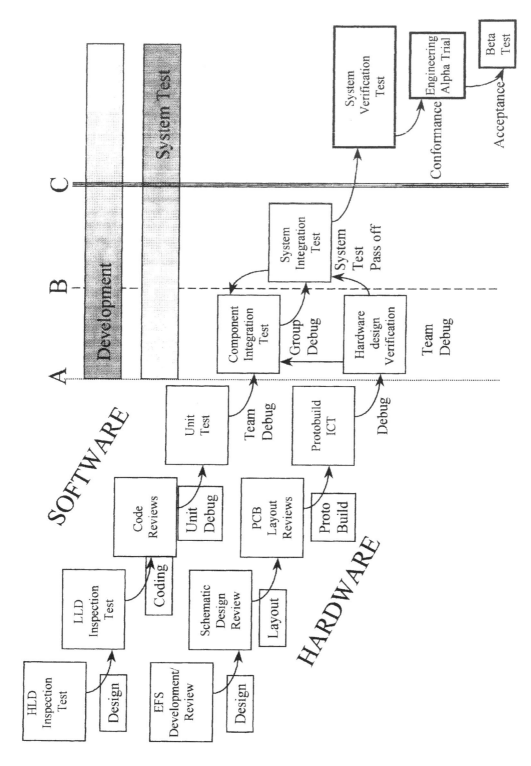

FIGURE 12.13 Hardware and software test flowchart comparison.

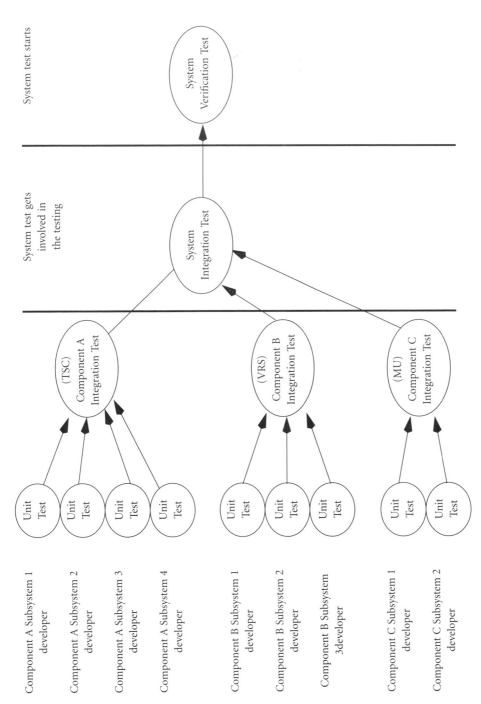

FIGURE 12.14 Unit test, component test, integration test, verification test relationship example.

integration and build. It is often done with stubbing our the code or tools that do not use or need a fully operational function build. A major goal of unit test is to test 100% of the code branches (see Table 12.4).

TABLE 12.4 Software Code and Unit Test Chart

Primary suppliers	Entry criteria	Activities	Exit criteria	Primary customers
SW engineering	Low-level design for each module is complete and documented in the software design Design review is complete Code review is complete	Write code per guidelines, including text messages, dialogs and screens Prepare for code inspections and reviews Hold code inspections and reviews Create unit test harness Unit test each module Each line of code and logic is verified Error logs and boundaries are tested	Code has been written and reviewed Unit tests are complete Unit test cases are documented Code coverage metric is available Code is under configuration management control	SW engineering Component integration team
Process control and quality records				

Metrics:
  Effort (hours) for coding
  Actual coding process against schedule (no. of code modules complete vs. date)
  Actual for software size (source lines of code)
  Inspection results form
  Defects found in unit test
Controlled documents:
  Unit test plan and results				

12.9.3 Software Component Integration Test

The objective of this process step involves integrating components into a major system component and verifying that the features still work correctly after integration. This process may be iterative, as each new feature comes out of the *code and unit* test. This component integration test process ends when all new components for the project have been integrated. Test cases are executed for each component (e.g., TSC, VRS, M are component name examples) to verify that requirements are met. When the component features have all been added and verified for correct functionality, the component is ready for the next phase of *system integration* (see Table 12.5).

12.9.4 System Integration Test

The objective of this process step involves integrating the major system components that make up the product and verifying that the features work correctly when the product is put together as a full system. This system integration test process may be iterative, as each component comes out of *component integration test*. This process ends when development work for all components of the system has been completed, the components have been integrated, and the system test integration plan verifies that the product as a whole contains all the features stated in the requirements. When this phase is finished, the product is ready for *software verification testing*. Personnel in charge of software engineering of customer applications should be responsible for coordinating this test (see Table 12.6).

12.9.5 System Verification Test—Main Phase

The objective of this system integration test process step involves verifying that the product meets customer requirements by executing the product in a system environment representing complex high-

TABLE 12.5 Component Integration Test

Primary suppliers	Entry criteria	Activities	Exit criteria	Primary customers
SW engineering Hardware engineering	Unit testing of the component to be integrated is completed. A component build is available for test. Prototype hardware is available as needed.	Unit tested modules are integrated. The component is built. Create (design and review) development test plan and test cases. Execute initial development tests on partial and full builds of the integrated software. Execute regression tests as necessary. Place product software under configuration management control. Make lab equipment available for component integration testing. Error recovery is tested.	Development test cases executed. All severity 1 and 2 defects fixed or verified. Fully integrated components placed under configuration management control.	Engineering Test Engineering

Process control and quality records

Metrics:
    Track actual integration and test process against schedule and plan
    Track actual code size against estimated (e.g., KLOC, thousand lines of code)
    Track SPRs generated (defects/day)
Controlled items:
    Component integration test plan
    Code under configuration management control

TABLE 12.6 System Integration Test

Primary suppliers	Entry criteria	Activities	Exit criteria	Primary customers
SW engineering Customer engineering Hardware engineering Documentation	Component integration testing is completed. HW design verification is completed. Market-specific requirements are identified. System integration test plan is written. Appropriate switch simulators are available. Draft customer documentation is available.	System components are put together. Create (design and review) system integration test plan and test cases. Execute system integration tests on partial and full builds of the integrated system per plan. Execute regression tests as necessary. Fix the identified integration defects.	Development in all the components is completed. All features have passed test. All components are available. All planned tests have been executed. Defect metrics are available. All severity 1 and 2 defects fixed and verified.	Test (SVT) engineering

Process control and quality records

Metrics:
    Track software problem reports (SPRs) generated (defects/day)
    Track actual effort (hours) against estimated for system integration test
Controlled items:
    System integration test plan

demand customer usage including market specific features. This step is the final in-house verification of product quality and includes functional new feature tests, regression tests, and field trial test plan validation. Completion of this SVT Main Phase allows the *engineering field trial* to begin (see Table 12.7).

TABLE 12.7 System Verification Test—Main Phase

Primary suppliers	Entry criteria	Activities	Exit criteria	Primary customers
SW engineering Customer apps. engineering Hardware engineering Documentation	System integration test has passed. Product media delivered in final customer form. Draft customer documentation w/release notes. System integration test results. Finalized requirement documentation. All product functions available Product equivalent HW available	Smoke test Execute test cases per plan Verify defect fixes through execution of regression tests Track test defects and results Review documentation SPR prioritization meetings	All tests completed (executed successfully or deferred as a result of a deferred SPR) All severity 1 and 2 defects fixed and verified, all severity 3 and 4 answered Tapes/PROMs and draft customer documentation available for engineering field trial Code is "frozen" Test results are available	Engineering field trial coordination SVT extended test

Process control and quality records

Metrics:
   Track SPRs generated (defects/day)
   Track actual effort (hours) against estimated SVT—main phase
   Track test case execution—planned, attempted, and completed
   Track number of defects deferred to future releases
Controlled items:
   Test strategy
   SVT test plan
   SVT test cases

12.9.6 System Verification Test—Extended Tests

The objective of this process step involves stress, load, throughput, and performance testing of the systems in environments that demonstrate worst case conditions the system may see in the field. This test will also test backward compatibility between various parts of the system. Installability, maintainability and serviceability will be tested by the Customer Simulation Group (CSG). Reliability testing is done as part of these tests (see Table 12.8).

12.9.7 Engineering Field Trial (Alpha Test)

The engineering field trial (see Fig. 12.15) is described in the following paragraphs. For complex systems, it is often advised that an engineering field trial be performed. The purpose of the engineering field trial is to make sure the system operates correctly in the field environment in which it will normally exist. It is highly unlikely that you could simulate in a lab environment all the complexities that a complex software system may encounter in the real world. A joint customer/supplier team is created to perform an engineering field trial. The supplier is usually in charge of driving the system through the engineering field trial while the customer rides along in the passenger seat.

Engineering Field Trial—Selection of Site

The objective of this stage of the process is to bring together marketing, engineering, customer service, and the customer to define the requirements for, and locate the site(s) for, an engineering field trial. When this step is done, installation planning for the trial can begin (see Table 12.9).

Testing

TABLE 12.8 System Verification Test—Extended Tests

Primary suppliers	Entry criteria	Activities	Exit criteria	Primary customers
SW engineering Customer apps. engineering Hardware engineering Documentation	Successful completion of system test, main testing. Extended SVT test is completed. Worst-case environments for test cases identified. Lab equipment is sufficient to perform the extended tests.	Execute test case per plan. Verify defect fixes from main throughput testing Track defects from engineering trial and monitor for test case updating. Track test case execution and progress against the test plan. Update SVT test plan, test cases are required. Fault injection and detection. SPR prioritization review.	All tests completed (executed successfully or deferred by plan). All severity 1 and 2 defects fixed, all severity 3 and 4 answered. Test results are available.	System verification test

Process control and quality records

Metrics:
  Track SPRs generated (defects/day)
  Track actual effort (hours) against estimated SVT—extended
  Track test case execution—planned, attempted, and completed during test period
  Track number of defects deferred to future releases
  MTBF reliability
Controlled items:
  Test strategy
  SVT test plan
  SVT test cases (initial)

FIGURE 12.15 Field engineering trial flowchart.

Engineering Field Trial—Installation

The objective of this stage of the process is to make the identified site ready for the field trial. Equipment needs are identified; people are identified; schedules are identified; all equipment is built, shipped, and installed in the field; and software is installed, whether by tape, CD, or PROM. When this step is complete, the system is ready to be brought up and made live (see Table 12.10).

Engineering Field Trial—Bring Up the System and Dry Run

The objective of this stage of the process is to make the engineering field trial live and take the system through a dry run of the customer acceptance procedure to confirm that all components are working as expected in the field. When this step is complete, the engineering field trial and customer acceptance trial can begin (Table 12.11).

TABLE 12.9 Engineering Field Trial—Selection of Site

Primary suppliers	Entry criteria	Activities	Exit criteria	Primary customers
Marketing Engineering Customer service Customer	List of customer environment specifics that must be present to allow a comprehensive engineering trial to be done (i.e., the field trial site requirements) Preliminary schedules for both vendor deliverables and customer availability	By reviewing customer site(s) availability, determine a configuration that will allow the least disruption to the customer's operations while providing the vendor with the best possible trial of the equipment in a realistic deployment environment.	Sites for the field trial are identified. Lead people for the vendor and the customer are identified. Equipment needs for the field trial are identified and ordered. Clear commitments are articulated for each side on which deliverables are who's responsibility.	Marketing Engineering Customer Installation group

Process control and quality records

Metrics:
 Schedules and how far (±) they were off from the plan
Controlled items:
 Field trial site requirements document

TABLE 12.10 Engineering Field Trial—Installation

Primary suppliers	Entry criteria	Activities	Exit criteria	Primary customers
Marketing Engineering Customer Installation group EF&I	Specify site(s) where the field trial will be conducted The trial plan is complete The draft customer installation documentation is complete	Identify additional equipment needed at the site(s) Identify any customer test interfaces that need to be installed Identify the installation personnel and schedules Build, ship, and install the equipment Install the software to be tested Confirm that all customer interface equipment is installed Finalize the field trial test plan	All components of the product needed for the field trial are present at the sits(s), installed and ready to be brought live A finalized field test plan exists	Marketing Engineering Customer

Process control and quality records

Metrics:
 Actual schedule (±) from the plan
 Hours involved in preparing the site and installing equipment
Controlled items:
 SPRs found during the installation process

Engineering Field Trial—Execute Field Trial Plan

The objective of this stage of the process is to take the system, in the customer environment, through a filed trial test plan designed to verify the product in environments and loads that cannot be re-created in-house. When this step is complete, engineering can sign off on the product's stability in the customer environment (Table 12.12).

Testing

TABLE 12.11 Engineering Field Trial—Bring Up the System and Dry Run

Primary suppliers	Entry criteria	Activities	Exit criteria	Primary customers
Marketing Engineering Installation group	All components of the system are fully installed in the field. Customer site support is available to assist with issues. Engineering support is available to assist with issues.	Bring up all components of the system. Bring up user terminals or user workstations. Begin using the system. Run the customer acceptance procedure. Document any defects found.	The system is stable in the customer environment and can pass the customer acceptance plan.	Marketing Engineering Customer

Process control and quality records

Metrics:
   Actual schedule vs. from the plant
Controlled items:
   SPRs found during execution of the customer acceptance plan in a customer environment

TABLE 12.12 Engineering Field Trial—Execute Field Plan

Primary suppliers	Entry criteria	Activities	Exit criteria	Primary customers
Marketing Engineering Customer	The system is fully installed and stable in the customer site(s) chosen for the engineering field trial. The engineering field trial test plan is complete.	Execute the field trial test plan. Document defects found. Verify fixes sent to the field by engineering.	The system passes the engineering field trial test plan. All bug fixes that will be made are present and stable in the field. All feature changes/modifications/ patches are present and stable in the customer environment.	Marketing Engineering Customer Production

Process control and quality records

Metrics:
   Field trial actual schedule vs. the plan
   Number of defects found during the field trial process
Controlled items:
   Defects (SPRs) found during the installation process

12.9.8 Customer Acceptance Process (Beta Test)

The customer acceptance process is similar to the engineering field trial except that, now, the customer is driving the system, and the supplier is in the passenger seat. The customer acceptance process (beta test) is shown in Fig. 12.16 and outlined in the following paragraphs.

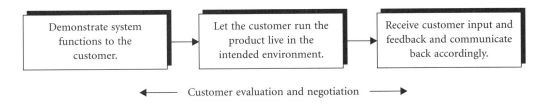

FIGURE 12.16 Customer beta test.

Customer Acceptance (Beta Test)—Demonstration to Customer

The objective of this stage is to demonstrate to the customer the new features, components and capabilities of the system in a controlled, orchestrated way so that the benefits can be explained for maximum effect. This phase begins with the training required to make the customer comfortable with the features' operations, and begins the process of integrating the features into the customers present standards and procedures (Table 12.13).

TABLE 12.13 Beta Test Demonstration to Customer

Primary suppliers	Entry criteria	Activities	Exit criteria	Primary customers
Marketing Engineering Customer	The system is installed in the engineering field trial site and has been demonstrated to pass the customer acceptance plan. Marketing has identified who are the primary decision makers and objectives for the customer acceptance trial. Customer documentation is complete.	Schedule, plan, and conduct a controlled demonstration of the system's new features, components, and capabilities to those who should be convinced of the need for the system. Identify, train, and support the customer technical lead for the customer acceptance trial. Document any issues or concerns raised by the customer as the system is presented in its entirety. Coordinate the introduction of the new features and capabilities into the present standards and procedures.	Customer has seen the new features in as favorable a light as can be demonstrated. The customer lead for the customer acceptance trial has been identified and trained.	Marketing Engineering Customer

Process control and quality records

Metrics:
   Actual schedule vs. plan
   Number of issues/concerns raised by the customer during presentation of the system(s)
Controlled items:
   Documentation changes the customer requests, and defects found

Customer Acceptance (Beta Test)—Customer Evaluation/Negotiation

The objective of this stage is to turn the product/system over to the customer for evaluation. The customer will use the product/system in the live environment to qualify the product as acceptable or unacceptable for deployment and use. Issues during this phase may be addressed in the release or postponed to a future release. The end of this phase is customer buy-off on the value of and need for the system (Table 12.14).

System Verification Test—Final Test

The objective of this process step is to verify that the final product meets the customer requirements. It is performed by executing a subset of tests to verify that all defect fixes to be included in the final product, identified in the SVT main phase, engineering field trial, SVT extended tests, and beta trial, have been made correctly. Successful completion of this phase is a requirement of the product release to production (Table 12.15).

TABLE 12.14 Beta Test Customer Evaluation

Primary suppliers	Entry criteria	Activities	Exit criteria	Primary customers
Marketing Engineering Customer	The system is fully deployed and stable in the customer environment. The customer has been introduced to the new features/capabilities and trained on how to use them. Vendor support for beta trial has been identified. A customer acceptance plan has been created.	Support the customer technical lead in the trial Provide training and issue resolution. The customer acceptance test plan is executed. Document any issues/defects found by the customer. Negotiate issues that will be addressed in this release vs. future releases. Provide verified fixes to the field to address customer concerns.	Customer has evaluated the system and has provided an accepted judgment to the supplier.	Marketing Engineering Customer Production

Process control and quality records

Metrics:
   Actual schedule vs. plan
   Number of defects documenting issues/defects found by the customer
Controlled items:
   Document issues/defects found by the customer

TABLE 12.15 System Verification Final Test

Primary suppliers	Entry criteria	Activities	Exit criteria	Primary customers
Marketing SW engineering HW engineering Documentation Customer service group	SVT main test, engineering field trial, beta, customer acceptance tests are all successfully completed. Regression test cases identified. Final software build available with all planned defect fixes included. Final customer documentation complete. SVT final test plan complete.	Execute SVT final tests per plan. Verify defect fixes through execution of regression tests. Review final customer documentation. Prepare final test report. Train operations engineer with operational features and functions of the new system. SPR prioritization review.	All tests complete per plan. All defects negotiated with customer fixed and verified. Severity 1 and 2 defects fixed, and severity 3 and 4 defects answered. Media in final customer form. Test results available. Test review meeting scheduled.	Production Market distribution

Process control and quality records

Metrics:
   Track test case execution—planned, attempted, and completed during the test period
   Track actual effort (hours) against that estimated for SVT regression phase
   Track number of defects deferred to future releases
   Track SPRs generated (defects/day)
Controlled items:
   Test strategy
   SVT test plan
   SVT test cases

Defining Terms

Backdriving. An digital-logic in-circuit test technique that applies defined logic levels to the output pins of digital devices. This application will also be applied to the next device's inputs.

Bed-of-nails (BON) test fixture. A test fixture consisting of spring-loaded test probes that are aligned in a test holder to make contact with test pads on the circuit assembly. The circuit board/assembly is typically pulled down onto the probes using vacuum applied by the fixture. See also *clamshell fixture*.

Clamshell fixture. A type of BON that mechanically clamps the UUT between two sets of tests rather than using a vacuum pull-down. Commonly used when both sides of the UUT must be probed.

Design for test (DFT). A part of concurrent engineering, DFT involves the test engineering function at the design phase of a project to develop test criteria and be certain that the necessary hardware and software are available to allow all test goals to be met during the test phase of the project.

Failure. The end of the ability of a part or assembly to perform its specified function.

Fault. Generally, a device, assembly, component, or software that fails to perform in a specified manner. May be in the components used in an assembly, as a result of the assembly process, as a result of a software problem, or as a result of a poor design in which interaction problems occur between components that individually meet specifications.

Fault coverage. The percentage of overall faults/failures that a given test will correctly detect and identify.

First-pass yield. The number of assemblies that pass all tests without any rework.

Functional test. Using simulated or real inputs and outputs to test the overall condition of a circuit assembly.

Guard, guarding, guarded circuit. A test technique in which a component is isolated from parallel components or interactive components. In analog circuits, guarding places equal voltage potentials at each end of parallel components to the DUT to prevent current flowing in the parallel component. In digital circuits, guarding disables the output of a device connected to the DUT, commonly through activating a tristate output.

In-circuit analysis. See *manufacturing defects analyzer*.

In-circuit test. Conducts both unpowered and powered tests to test both correctness of assembly and functionality of individual components. Will identify bad components.

Infant failures, infant mortality. Synonymous terms used to refer to failures that occur early in a the life product/assembly.

Manufacturing defect analyzer. A system that tests for shorts, opens, passive component values, and semiconductor junctions on an unpowered board.

Abbreviations

ATE. Automatic test equipment: a system that may conduct any of a number of tests on a circuit assembly. A generic term that may include in-circuit tests, functional tests, or other test formats, under software control.

ATPG. Automatic test program generation: computer generation of test programs based on the circuit topology.

BGA. Ball grid array, type of IC package.

BON. Bed-of-nails test fixture.

CND. Cannot duplicate, see *NFF*.

COTS. Commercial off-the-shelf, a term used by the U.S. military services.

DFT. Design for test.

DUT. Device under test.

ESS. Environmental stress screening.

NFF. No fault found, a term used when a field failure cannot be replicated.

NTF. No trouble found, see *NFF*.

ODAS. Open data acquisition standard.

QFP. Quad flat pack, type of IC package.

RTOK. Retest OK, see *NFF*.

SPR.   System problem report.
UUT.   Unit under test.

World Wide Web Related Sites

See also Chapter 9, "Design for Test."

Evaluation Engineering magazine, http://www.evaluationengineering.com
Open Data Acquisition Association, http://www.opendaq.org
Test & Measurement World magazine, http://www.tmworld.com

References

"Just Enough Test." John Fluke Mfg. Co., Inc. Everett, WA, 1991
Antoy, S, Gannon, J; "Using term rewriting to verify software." *IEEE Transactions on Software Engr.*, vol. 20, no. 4, April 1994.
Arthur, Lowell Jay; *Improving Software Quality: An Insider's Guide.* John Wiley and Sons., 1993.
Bateson, John; *In-Circuit Testing.* Van Nostrand Reinhold Co., New York, 1985.
Black, P.E., Windley, P.J.; "Verifying Resilient Software." *Advanced Technology Proceedings of the Hawaii Intl. Conf. on System Sciences,* vol. 5, IEEE Computer Society, 1997.
Dai, H., Scott, C.K.; "AVAT, a CASE tool for software verification." *Proceedings of the IEEE Intl. Workshop on Computer-Aided Software Engr.,* 1995.
Desai, S.; "How electrical test requirements will affect tomorrow's designs." *Electronic Engineering,* vol. 70, no. 861, Oct. 1998.
Devanbu, P, Stubblebine, S.; "Research Directions for automated software verification." *Proceedings of the IEEE Intl. Automated Software Engr. Conf., ASE.* IEEE Computer Soc., 1997.
Grady, Robert B.; *Software Metrics: Establishing a Company-Wide Program,* Prentice-Hall Inc., 1987.
Hamilton H.E., Morris C.H.; "Control Chip Temperature During VLSI Device Burn-In." *Test & Measurement World,* vol. 19, no. 5, April 1999, http://www.tmworld.com.
Holcombe, M; "Integrated methodology for the specification, verification and testing of systems." *Software Testing Verification and Reliability,* vol. 3, no. 3–4, Sep–Dec 1993.
Humphrey, Watts S.; *Managing the Software Process.* Addison-Wesley, 1990.
Karinthi, R., et al.; "Collaborative environment for independent verification and validation of software." *Proceedings of the 3rd Workshop on Enabling Technologies: Infrastructure for Collaborative Enterprises.* 1994.
Kolts, B.S.; "Matching Test Software with Test Applications." *Electronic Packaging and Production,* Feb. 1998.
Molnar, J.A.; "Navy initiatives to steer COTS instrument technology developments." *Proceedings of AUTOTESTCON 1998.* IEEE, 1998.
Pierce, P.; "Software verification and validation." *Northcon-Conference Record.* IEEE, 1996.
Purcell, J.; "Simulation software cuts testing time and expense." *Control,* vol. 9, no. 10, October 1996.
Pynn, Craig T.; *Strategies for Electronic Test.* McGraw-Hill, New York, 1986.
Robinson, G.D.; "Why 1149.1 (JTAG) really works." *Electro International, Conference Proceedings,* 1994. IEEE, 1994.
Romanchik, D; "Simplify electronic test development." *Quality,* vol. 37, no. 11, Nov. 1998.
Schiavo, A.; "Building an acquisition matrix for better test equipment management." *Evaluation Engineering,* vol. 35, no. 2, Feb. 1996.
Simpson, W.R.; "Cutting the cost of test; the value-added way." *IEEE International Test Conference,* 1995.
Thames, T; "Using commercial off-the shelf (COTS) equipment to meet joint service requirements." *Proceedings of AUTOTESTCON* 1998. IEEE, 1998
Turino, J; *Design to Test.* Van Nostrand Reinhold, New York, 1990.

Welzel, D, Hausen, H, Boegh, J; "Metric-based software evaluation method." *Software Testing Verification and Reliability,* vol. 3, no. 3–4, Sep–Dec 1993.

Williams, R, et. al.; "Investigation of 'cannot duplicate' failures." *Quality and Reliability Engr.*, vol. 14, no. 5, Sep–Oct 1998.

13
Inspection

Glenn R. Blackwell
Purdue University

13.1 Introduction 13.1
13.2 General Inspection Criteria 13.2
13.3 Solder Paste Deposition Volume 13.4
13.4 Solder Joint Inspection Criteria 13.5
13.5 Visual Inspection 13.8
13.6 Automated Optical Inspection 13.10
13.7 Laser Inspection 13.10
13.8 X-Ray Inspection 13.10

13.1 Introduction

In a perfect assembly process, inspection (or test, for that matter) would not be necessary. The processes would be in such good control that the product would come out right each time. In reality, we know this is not possible. Process characteristics shift with time, incoming materials vary, power gets interrupted, and humans err. These and other variations make inspection a necessity.

Having accepted inspection, the most important question to answer is, "where in the processes is inspection the most cost-effective?" Like test, inspection may take place at various points in the assembly process. As examples, inspection points in the surface mount assembly process may include

- Bare boards
- Parts
- After solder paste deposition
- After parts placement
- After reflow
- After rework

It is also important to note that the inspection criteria and the pass-fail criteria will depend on the type of product being inspected. In IPC-A-610, "Acceptability of Electronic Assemblies," the IPC defines the following three classes of electronic assemblies:

- Class 1: General Consumer Products, such as audio and video products
- Class 2: Dedicated Service Electronic Products, such as communications equipment where failures are undesirable but not critical.
- Class 3: High Performance Electronic Products, such as pacemakers and weapons systems, where downtime is unacceptable and/or the operating environment is harsh.

The most cost-effective place to inspect is at the operation in the user's process that has resulted in the largest number of assignable faults. Each user and each process/line must be examined to make this determination.

Bare board inspections are also discussed in Chapter 5, and parts inspections are discussed in Chapter 3. This chapter will concentrate on inspection during the assembly process and criteria for deciding where and how often to inspect.

It is important to note that the primary distinction between inspection and test is that inspection is intended to find faults that present themselves visually or by X-ray, whereas test is intended to find faults electronically. In most manufacturing operations, both inspection and test will be necessary.

Definitions important in the understanding of inspection include:

Bridge: two or more traces or lands that should not be electrically connected but are shorted by excess or misapplied solder.

Cold solder joint: a solder joint that did not reach a high enough temperature during the reflow or wave solder process. It is characterized by poor wetting and a dull gray, often porous, appearance.

Control chart: used in SPC, a control chart tracks rejects or their inverse, acceptable parts/assemblies, over time. Trends and sudden changes in the chart may be used to correct process problems.

Fiducial: a vision target on a circuit board/substrate. It is used by automatic inspection systems to index the board in their X-Y plane.

Solder balls: spherical solder residue that is present on the board after a soldering operation. The balls are not well adhered to the mask, trace, or laminate on which they sit, and they may short conductors.

Solder bridge: see *Bridge*.

13.2 General Inspection Criteria

There are two main goals of any inspection process. The first is to find, weed out, and allow rework of any faults/rejects, and the second is to provide feedback to the manufacturing system to allow each fault to be assigned a probable cause. Once cause is established, the system can be corrected or tuned to minimize the occurrence of that fault.

Any inspection must address the most common manufacturing defects. From the list in Section 13.1, the defects may vary, depending on where in the process the inspection is being performed. Bare-board inspection is obviously different than post-placement inspection.

Bare-board inspections may be done visually, with an automated optical inspection (AOI) system, and with automated testing equipment. A bare-board inspector or inspection system should check for

- Uniformity and thickness of solder plating on pads
- Trace widths and spaces
- Quality of solder mask
- Solder mask print registration relative to open pads and vias
- Finished via hole presence and size
- Finished through-hole presence and size
- Warp
- Twist

These items are discussed in detail in Chapter 5, "Circuit Boards."

Assembly-related problems may include solder joint shorts, opens, and insufficients, as well as component problems. The vast majority of assembly problems are solder-joint related. Oresjo (1998) reported the fault spectrum from a Hewlett-Packard study of 74 million solder joints as shown in Table 13.1.

The criteria for each fault must be defined for the inspectors or the inspection system. While some faults (e.g., opens and missing components) are easy to define, others (e.g., insufficients and excess solder) may be more difficult. Nevertheless, all inspectors and inspection systems must use the same set of criteria. Visual inspection criteria are available from the Surface Mount Technology Association, and many large companies have their own defined inspection criteria.

TABLE 13.1 Solder Joint-Related Problems

Fault	Approximate % of Overall Faults
Opens	34
Insufficients	20
Shorts	15
Missing components	12
Misaligned components	10
Defective component	8
Wrong component	5
Excess solder	3

The volume of solder paste deposited on the board has a direct relation to faults such as opens, insufficients, shorts, and excess solder. The Oresjo study shows, and industry generally accepts, the figure that 70% of assembly defects result from problems with solder paste. These include solder quality problems, solder deposition and volume problems, and reflow problems. Solder quality issues are discussed in Chapter 2.

Solder deposition, whether by pin, stencil, or syringe, is expected to result in a specific volume of solder deposited in specific locations. Since industry experts agree that solder volume problems lead directly to insufficients, solder bridges, and other problems, inspection at this point can prevent many problems farther down the assembly line. A trained inspector can make a reasonable estimate of coverage area and volume, but only an estimate. Visual inspection also provides no numerical feedback for keeping the process in control. Solder volume inspection/measurement is covered in Section 13.3.

As in test, many of the modern board design criteria work against inspection. Limited visual access due to fine and ultrafine pitch components, under-part connections such as those on flip chips and BGAs, and mechanical devices such as RF shields all make both visual and automated optical inspection difficult.

Post-placement inspection will determine missing parts and off-location parts. Post-placement inspection is difficult to do visually, since the board cannot be picked up and moved without risk of part movement. As discussed in Section 13.3, vision inspectors find that their work is most accurate when they can tilt the board to accommodate the different reflectivities of different surfaces. AOI systems also work well here. Selection criteria include

- Operator evaluation for ease of operation
- Speed of measurements
- Outputs desired: SPC data, alarms for missing and off-location parts

Post-reflow inspection is done with a variety of techniques. Vision inspection, AOI, X-ray, and laser systems are all used for post-reflow inspection. The user must first decide what inspection criteria are of most value, then consider which type of inspection system is most appropriate, perhaps by comparing demonstrations of each type, then do a comparative evaluation of manufacturers of the selected type of system.

Post-reflow inspection concentrates on solder joints and placement faults. Since some parts can move during the reflow process, it may be difficult to assign cause to some placement faults. Solder joint fault criteria have been developed by a number of entities, including the IPC. Generally, solder joint inspections will note three classifications of faults:

- *Reject faults,* which cause the board to be rejected and scrapped or sent to rework. They are likely to cause a failure in the operation of the circuits on the board. Feedback to operations is required. An example would be a solder bridge.
- *Correction faults,* which are unlikely to cause circuit failures. They are acceptable but are fed back to operations for a correction in the process. Rework is not necessary except for extreme final uses such as life support or aerospace applications. An example would be a part whose leads are 25%

off their corresponding lands, or a solder bridge between two lands that are also connected by circuit traces.
- *Cosmetic faults*, which will not cause a circuit failure but should have their cause determined. Rework is not necessary. An example would be excess solder on the wave side of a through-hole lead.

13.3 Solder Paste Deposition Volume

The need to determine solder paste volume has become more critical with the use of components whose joints are either difficult to inspect, such as BGAs, or fine-pitch components, with their greater demand for smaller paste volumes with less margin for volume error. Failure to deposit the proper amount of solder can lead to a number of problems/ faults with the associated solder joints:

- Insufficients
- Solder bridges
- Excess solder wicking, leading to brittle leads
- Rework of joints after reflow, leading to joints with higher stress levels and increased probability of later failure

Measurement and verification of actual solder volume is typically by laser-based systems that calculate true volume. Camera-based systems are less expensive but measure only the area of solder coverage and imply volume by assuming some average thickness of paste print. While less expensive and faster than laser-based systems, camera-based systems do not provide true volume information. If most solder deposition faults in a given line are detectable by area information, the camera-based system is appropriate. Selection of a volume/area measurement system should only be done after trial of selected vendors' systems in the actual line in which they would be expected to operate. Criteria for selection should include

- Operator evaluation for ease of operation
- Measurement accuracy: area, volume
- Determination of whether measurement should be continuous and in-line or sampling
- Speed of measurement vs. line speed
- Outputs desired: volume, SPC data, alarms for over- and under-volume faults

Improvements in quality at this point in the line have a very favorable impact on first-pass yield, since reject boards pulled at this point have no components mounted on them, making this a very cost-effective point for reducing scrap. Reduction of scrap and improvement of first-pass yield will have a positive effect on every manufacturer's bottom line The cost-per-defect at this point is substantially lower than any defects detected further down the line after components are placed on the board. At this point, rework consists only of assigning cause for the fault, and if the fault is not bare-board related, cleaning the board and reusing it. Ries (1998) reported that rework after the post-print inspection cost an estimated $0.45, whereas rework after in-circuit test costs closer to $30.

Additionally, it is accepted that rework rarely, if ever, produces the same quality solder joints as first-pass through a reflow oven. Engelmaier (1991) notes that any thermal shock, as is experienced in reworked joints, causes transient differential warpages in the solder joint, leading to an increased risk of joint failures.

Off-Line Inspection

Several manufacturers offer manual noncontact, laser-based systems that provide the capability to make X-Y-Z measurements on solder prints. These systems can measure paste height and X and Y dimensions of the print. While volume calculation may need to be done manually, sample boards can have their solder print volumes verified. Some of these systems will allow manual input of the Z-axis measurement, and from the X and Y dimensions they will then calculate volume. These systems obviously are designed for inspection checks and are not designed to verify every board and provide constant control feedback, although they are very useful for prototype and low production volumes.

On-Line Inspection

3-D laser-based systems will automatically scan the printed board and provide volume calculations for all or selected paste deposits. Due to their relatively slow speed, the only way some 3-D systems can keep up with the line is to examine a sample of deposits on each board, changing the sample measured from board to board, so that after scanning, e.g., ten boards, all paste deposit locations have had their volumes calculated. These systems will provide SPC data for process control feedback.

Printer-Based Inspection

Some automated solder stencil print machines have their own built-in inspection system. This eliminates the need to have a separate inspection system with its associated manufacturing floor footprint.

13.4 Solder Joint Inspection Criteria

Before considering the criteria for solder joint inspection, the user must review the solder joint design requirements in Chapter 2 and the circuit board land requirements in Chapter 5. They are partially reviewed here for chip components and for SOICs, as examples for joint inspection. The reader must refer to the most recent publication of IPC-A-610, "Acceptability of Electronic Assemblies," for the most up-to-date information regarding soldering and acceptable criteria joint. This chapter will not attempt to re-create all the joint criteria, since any update to IPC-A-610 will render this information out of date.

Chip components such as resistors and capacitors are shown in Fig. 13.1. Some chip components, such as electrolytic capacitors, may only have solderable terminations on the bottom or on the end and bottom rather than on all four sides and the end.

Sample joint criteria for chip components include:

Criteria Location	Joint Criteria
Maximum fillet height	Not over the top of the component
Minimum fillet height	1/4 of the termination height
Minimum solder thickness	Visible solder fillet under component, 0.2 mm/8 mil for AOI/X-ray

Solder joint criteria for SOICs assume that the end of each lead is not a solderable termination, since the gull-wing leads of these packages are normally cut after the leads are tinned in the manufacturing process (Fig. 13.2).

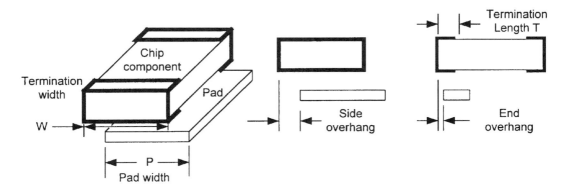

FIGURE 13.1 Joint terminations for chip components.

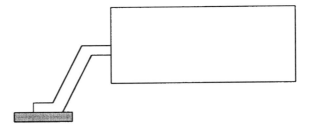

FIGURE 13.2 Joint terminations for SOICs.

Sample joint criteria for SOICs include:

Criteria Location	Joint Criteria
Maximum heel fillet height	1/2 lead height
Minimum heel fillet height	Solder thickness + lead thickness
Minimum solder thickness	Acceptable solder fillet at heel, visible fillet at toes and sides

The fillet criteria for both chip components and SOICs include proper wetting of the fillets.

Insufficient Solder

Insufficient solder generally leads to an unacceptably small fillet(s). There may also be some joints for which there is a gap between the land and the component lead/termination.

Causes of insufficient solder joints include

- Insufficient solder paste volume
- Wrong component orientation when passing through wave

Excess Solder

Excess solder can create stiff joints, which lose the ability to flex during thermal excursions. Excess solder violates the maximum fillet height criteria in the joint design guidelines.

Excess solder can be caused by

- Excess solder paste volume
- Wave solder conveyor speed too low
- HASL air volume/velocity too low
- Wrong orientation of components in wave

Solder Bridges

Bridges are the unwanted formations of an electrical connection (short) between two conductive points by means of solder extending (bridging) over nonconductive material. When a bridge causes a short circuit between two points of a circuit at different potentials, or when it forms a ground loop, it must be removed. A bridge between two points of the same potential, which does not cause a ground loop, need not be removed.

Causes of solder bridges include

- Insufficient spacing of conductors/traces, solder lands, and components
- Incorrect placement/orientation of components
- Excess solder deposition in reflow soldering
- Wave too high in wave soldering
- Warped substrate, resulting in flooding of upper surface by wave

- Insufficient flux
- Excess slump in solder paste
- Misregistration of solder mask
- Insufficient oil intermix in wave
- Insufficient pressure/volume of air in HASL process

Solder Icicles

Icicles consist of excess solder on leads after the wave solder process. They are formed when excess solder remains on a through-hole lead as the substrate breaks contact with the solder wave. They are usually regarded as cosmetic defects, but they can result in decreased air-insulation distance in high-voltage circuits.

Possible causes include

- Contamination of the solder bath
- Insufficient oil intermix
- Uneven distribution of flux

Drawbridging

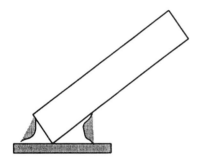

FIGURE 13.3 Drawbridge effect on chip components.

Drawbridging (a.k.a. the Manhattan effect) is when chip components leave the reflow or vapor phase oven with one end lifted off its corresponding land. The lifted end is not soldered, and the long axis of the component may be at an angle up to 90° relative to the board. It is caused by one end of the component heating and reflowing before the solder at the other end of the component becomes molten, or by a significant mismatch of molten solder tension between ends caused by missing or insufficient solder paste at one end. The molten solder at one end of the component has sufficient surface tension to rotate the other end component up off the board. A drawbridge (Fig. 13.3) is always regarded to be a major reject defect.

Causes of drawbridges include

- Missing or grossly insufficient solder paste deposit at one end of the component
- Missing or defective solder mask on trace, allowing molten solder to wick away from land
- Wrong land size
- Long axis of component parallel to direction of travel through the oven, allowing one end to reach the reflow zone before the other end

Solder Balls

Solder balls are spherical remains of solder paste that have separated from the solder deposits attached to leads or lands and exist separately on board surfaces. They are always a reject defect, as they are likely to move during the life of the board, potentially causing short circuits.

Causes of solder balls include

- Insufficient drying of solder paste during the reflow profile, resulting in sudden expansion of remaining liquids and expulsion of small quantities of solder during the reflow bump
- Excess drying of solder paste during the reflow profile, resulting in solder oxide formation and poor solderability
- Solder paste/flux with insufficient activity to reduce oxides
- Deposition of solder paste on solder mask

- Solder paste with an excess amount of oxides; the flux cannot reduce all the oxides; caused by out-of-date or defective paste
- Inherent propensity of the paste to form solder balls during reflow

As discussed in Chapter 2, solder paste can be tested for its tendency to form solder balls.

Solder on Mask

Excess solder that has reflowed on the mask exists as a web, as a skin, or as a splatter. It cannot be allowed, since it will break loose and, if not already causing a short, will cause shorts later in the lifetime of the product.

Causes of solder-on-mask include

- Incorrect flux
- Solder contamination
- Incorrect preheat time or temperature
- Defective solder paste

Component Mislocation

Component mislocation may occur as a result of an incorrect placement during that operation, or as a result of the component shifting during the molten solder phase of reflow. In the presence of mislocation, and prior to assignment of cause, it is important to determine whether the component is in fact in the wrong place or, if many components are "wrong," if the board artwork is incorrect. This may require examination of bare boards and optical comparison with the original CAD artwork data.

Adhesive Contamination

Adhesive has no solderability. Therefore, if adhesive has contaminated a land, a component termination, or both, the solder joint is rejected.

Causes of adhesive contamination include

- Adhesive deposit on land
- Adhesive volume too high, resulting in squeeze-out to land during placement
- Adhesive viscosity too low, causing stringing during deposition
- X or Y movement of component during or after placement

Solder Voids

Voids are caused by gas or liquid being trapped in the solder joint. Typically, they can be observed only by X-ray or by cross-sectioning the joint. If the entrapped gas or liquid bubble blows up as the solder solidifies, a visible blow hole may result. Whether this is a reject or a correction fault depends on the size of the blow hole. If the volume of solder in the joint is significantly reduced by the hole, the joint must be rejected.

Causes of solder voids and blow holes include

- Insufficient drying of solder paste during the preheat phase
- Defective solder paste
- Moisture in the substrate, if voids show in through holes

13.5 Visual Inspection

Visual inspection is still the most common inspection technique. As noted in Section 13.2, it is of limited value in newer technologies that use dense, fine-pitch ICs, and in flip chips and BGAs. Commonly done under magnification of 2× to 10×, visual inspection is relatively easy to implement, since operators exist at many points along the process line. However, it is not easy for inspectors to provide reliable reject information. Inspectors are commonly expected to be able to detect board surface damage or warpage, misaligned components, missing components, component damage, coplanarity/opens problems, solder bridges, solder insufficients, and solder balls. As with automatic optical inspection (AOI) equipment, any point to be inspected must be visible to the inspector and therefore the board design must allow for this.

It is also apparent that newer component technologies like flip chips and BGAs are not inspectable by this technique. Therefore, the visual inspection is limited in its scope when these components are used.

Problems that must be addressed with visual inspection include these:

- Operators are pressed into service as inspectors without specific training in inspection.
- "Anyone" can be an inspector, and inspectors are frequently expected to be able to function at many locations in the line, while operators are trained in a specific operation and perform that operation repetitively, becoming very competent at that job.
- If the same bare board is used for different products, different populations on the board are difficult to keep track of.
- Human nature prevents inspectors from performing their job at the same level of effectiveness at all times.
- Human vision varies among adults.
- Human interpretation of what is seen is subjective.

As an example of the subjectiveness of human inspection, Prasad and Simmons (1984) studied variations among inspectors. The inspectors would report different quality levels even if looking at the same set of assemblies. In analyzing data from six inspectors taken over three months, they found that there was a 6:1 ratio among the inspectors in defects reported. When they expanded the study to track ten inspectors over nine months, there was still a 5:1 ratio of defect rates among the inspectors.

For inspectors to perform to the best of their ability and to provide the most useful feedback to the assembly line, the inspection station must be set up to minimize distractions and maximize aid to the inspector. The inspector should have

- No other functions to perform
- Comfortable seating
- Excellent lighting
- UV lighting if conformal coatings or plating voids are to be inspected
- Magnification from 2× to 10×, depending on the finest pitch to be inspected
- Magnification to 100× if wire bonding is to be inspected
- Routine breaks
- An automated method of providing feedback to operators
- An automated method of providing reject information to rework operators

If the inspector is expected to consult with others about a reject, or if the inspector is expected to train others, a camera-based magnification system may be appropriate. Although CCD camera-based systems with monitors have less resolution than the eye would see directly through a magnifier, the ability to display the image allows these other functions. The use of a camera and monitor, coupled with a computer, allows digitized images to be stored and sent to others electronically for training or for rework.

Most magnified inspection lights have their magnification levels rated in *diopter*. This is a unit of measurement of the refractive power of lenses. Diopter relates the refractive power with the focal length. Higher-diopter lenses have a shorter focal length and have greater magnification. A 3-diopter lens has a magnification of 1.75×, while a 5-diopter lens of the same diameter has a 2.25× magnification. As with any other equipment, selection of new magnifiers must include a trial period with the inspector using each unit under consideration for actual inspections.

13.6 Automated Optical Inspection

Automated optical inspection (AOI) uses camera-based vision systems to inspect the board optically. Lighting systems for AOI may include fluorescent, LEDs, lasers, or UV. Like visual inspectors, AOI can be expected to find board warpage, missing components, misaligned components, component damage, coplanarity/opens problems, solder bridges, solder insufficients, and solder balls. AOI systems are set up to err on the side of showing rejects when none exist, resulting in common false-positive readings.

AOI can only inspect one side of a board. To inspect the other side, the board must either be reloaded with the other side up or run through a second AOI system after inverting. High-throughput lines may find that the AOI system cannot inspect all components and maintain line speed, requiring spot inspections.

13.7 Laser Inspection

There are two distinct types of laser inspection systems. The first uses a scanning laser as the light source for a camera-based inspection system. The inspection is based solely on the visual appearance of the joint.

The second laser technique, known as the Vanzetti system, uses a laser to heat each solder joint individually, then uses an IR detector to monitor how rapidly the joint loses heat. The heat loss rate is an indication of both proper solder volume and proper connection to the substrate. This system will detect voids as well as solder insufficients, since the volume of a joint with large voids in it will inhibit heat transfer from the joint to the board.

Relatively low in speed and high in cost, laser-based camera systems can be used for post-deposition solder-paste volume measurements, as well as post-reflow solder joint inspections. Like visual inspectors, laser systems must have direct visual access to the joints they are to inspect.

13.8 X-Ray Inspection

X-ray inspection systems are categorized into three types: 2-D X-ray, 3-D X-ray, and X-ray laminography. They all use transmission of an X-ray beam through the subject board but differ both in the analysis techniques, speed, and capabilities.

2-D X-Ray

Basic 2-D transmission X-ray systems are typically used to analyze solder joints on single-sided boards. Like basic medical X-rays, they transmit a beam through the subject circuit board and analyze the relative intensities of the transmitted beam. The image generated is a result of attenuation of the beam by all structures along the path of the beam and results in a gray-scale picture. There is no way to differentiate among the various structures along the path, since attenuation is the total result of the tin/lead in solder, trace tinning, copper, and internal metal layers. Therefore, the system must be taught the characteristics of each board location, and it must be taught from both acceptable board characteristics and unacceptable board characteristics.

2-D X-ray cannot be used on double-sided boards unless each solder joint is in a unique location relative to the opposite side of the board. X-ray systems cannot check that the correct component has been placed or that polarized components are properly oriented. They do a good job of detecting insufficient solder and voids in joints, since these mean there is less tin/lead in the path of the beam and therefore less attenuation. They cannot detect a nonwetted joint if the correct volume of solder does exist in the path of the beam, even if there is a gap between the solder on the lead and the solder on the land. They also do poorly on the joints with noncollapsible solder balls used with ceramic BGAs. Since the 10/90 Sn/Pb balls used are of a higher density than the 63/37 Sn/Pb eutectic solder used for attachment, most of the beam is attenuated by the 10/90 balls, and the transmission X-ray systems cannot differentiate between good and poor eutectic joints.

3-D X-Ray

3-D X-ray can analyze solder joints on both sides of a double-sided board. It can also see joints hidden under components such as a BGA or a flip chip, and under RF shields. Like 2-D systems, 3-D X-ray systems cannot check that the correct component has been placed or that polarized components are properly oriented.

X-Ray Laminography

X-ray laminography systems are similar to medical tomography scanners (CAT or CT scanners). They use a scanning beam to cut through the subject board at different angles, analyze the resulting transmitted beam strengths, and then use a sophisticated computer algorithm to determine the cross-sectional makeup of the board.

Like a human CT scan, the cross-sectional images created by the laminography system can be taken at different heights in the total Z-axis of the scan. In this manner, a cross section can be taken at board/trace level, and at different heights above the board, allowing the system to determine solder joint volume, and the presence of voids, as well as up the lead to determine if solder wicking has occurred. The scan cross section can also be moved through the thickness of the board to image solder joints on the opposite side of the board.

Like any X-ray system, denser material such as solder shows up as darker pixels, while less-dense material like copper or a void will show up lighter. Some laminography systems offer false-color images to allow easier identification of density differences.

Not surprisingly, laminography systems are very expensive. It may be appropriate to rent a system while getting the process under control.

References

Adams, RM, Glovatsky, A, Lindley, T, Evans, JL, Mawer, A; "PBGA reliability study for automotive applications." SAE Special Publications, vol. 1345, Feb. 1998, SAE, Warrendale, PA.

Bolliger, B, Stewart, M; X-ray laminography improves functional-test yield." *Evaluation Engineering*, vol. 35, no. 11, Nov. 1996.

Hundt, M; "High power surface mount packages." SAE Special Publications, vol. 1345, Feb 1998, SAE, Warrendale, PA.

Jalkio, J; "Choosing solder paste inspection equipment." *Surface Mount Technology*, vol. 9, no. 9, Sept. 1995.

Lecklider, T; "X-ray inspection underpins assembly process quality." *Evaluation Engineering*, vol. 37, no. 11, Nov 1998.

Ludlow, JE; "Automated die and wire bond inspection using machine vision." *Electrecon, Proceedings*, 1994. EMPF, Philadelphia, PA.

Pye, R; "Vision inspection: meeting the promise?" *Proceedings, IEEE International Test Conference, 1997*. IEEE, Piscataway, NJ

Reed, JM, Hutchinson, S; "Image fusion and subpixel parameter estimation for automated optical inspection of electronic components." *IEEE Transactions on Industrial Electronics*, vol. 43, no. 3, June 1996.

Wu, WY, Wang, MJ, Liu, CM; "Automated inspection of printed circuit boards through machine vision." *Computers in Industry*, vol. 28, no. 2, May 1996.

14
Package/Enclosure

Glenn R. Blackwell
Purdue University

14.1 Introduction 14.1
14.2 Ergonomic Considerations... 14.1
14.3 User Interfaces 14.2
14.4 Environmental Issues.. 14.42
14.5 Maintenance 14.43
14.6 Safety... 14.44

14.1 Introduction

As discussed in quality function deployment (QFD) publications, the customer is the driving force in product development, and nowhere is that more true than in the design and implementation of the external enclosure of the product. Regardless of the specifications and performance of the product as a whole, the customer's interface with the product is the enclosure. If the enclosure and associated controls, signals, etc., do not live up to expectations, the customer will always use the word *but* when describing the product, as in, "It works well, but...."

QFD describes the first of three levels of customer responses as *dissatisfiers,* features that, if missing, would cause significant displeasure and are therefore expected to be present. An example might be an on/off switch, forcing the user to unplug the product after each use. The second level is *satisfiers,* features that when done poorly cause displeasure, but if done well cause satisfaction. An example might be the quality of the display on a personal computer. The third level is *delighters,* sometimes called *wows,* that the customer did not consider but would view with unexpected pleasure if present. An simple example commonly given is the trunk net in the Ford Taurus.

Where Else?
Other chapters that include information about external enclosures include Chapter 1, "Fundamentals of the Design Process," and particularly the section on "Quality"; Chapter 6, "EMC and PCB Design," which also includes EMC control issues regarding the enclosure; and Chapter 8, "Connector Technologies." Chapter 16, on "Product Safety and Third-Party Certification," covers certain criteria necessary to meet the requirements of safety regulatory agencies such as UL, CSA, and the European Union.

14.2 Ergonomic Considerations

Ergonomics is the study and practice of designing anything a user comes in contact with in a manner that will allow the most efficient and "friendly" interface, including but not limited to items such as size, color, display size and shape, switch placement and actions, and labeling.

If the true "voice of the customer" is to be heard, ergonomic evaluations must depend not solely on the engineers' evaluations of potential use, but frequently on actual studies. Logan et al. report on a 10-week study in users' home to gather data for Thompson Consumer Electronics on user issues that should be considered in television design.

Obviously, ergonomics has as much effect on industrial and manufacturing users as it does on consumers. Helander and Burri report on ergonomics studies in IBM manufacturing facilities. The studies included analyses of equipment, processes, ambient factors, and job procedures. It is reasonable to consider all of these in any development of an industrial product. For example, suppose a new modular PLC is being designed specifically for use in the petroleum refining. The design team should consider not only the immediate operator interface of the front panel, but also expected configurations of the product modules and installation environments. Ease of routine maintenance procedures would also be considered, all in an attempt to make the new PLC user friendly. Even though the industrial user is frequently not the person writing the purchase request, that user/operator may have a profound effect on future orders.

For those readers involved in the design of office and consumer equipment that involves input and output devices such as keyboards and displays, Yamada reports on a project to develop a standardized set of guidelines for the ergonomics of those devices. While the guidelines are voluntary, they do provide good information for designers.

Ergonomics can cover simple logical situations, such as vertically mounted on/off switches always being "off" when moved in the down direction. Other examples would be an "on" light that is mounted in a visible location. Generally speaking, ergonomic design of an enclosure and the controls and indicators mounted on it should begin by considering the total package from the standpoint of the user or users. Some devices, such as a personal computer, are designed to be operated by one user at a time. Other devices, such as some process equipment, may be operated by more than one person at a time. For devices that require routine maintenance, the maintenance worker(s) must be considered in the design. This is covered further in Section 13.5.

Even devices that may fall into QFD's *wow* category may fail the ergonomic evaluation by the user. Laurie et al. report on a voice-activate dialing system (VADS) for an automotive cellular phone that was rejected by users due to use difficulties.

14.3 User Interfaces: Selection, Design, and Arrangement of Controls and Displays

Any control or part that is expected to be routinely adjusted by the user, or that requires periodic maintenance by the user, should be accessible without tools. The following material from Kroemer, Kroemer, and Kromer-Elbert[*] discusses both panel interfaces and testing done to justify certain criteria for interface controls.

14.3.1 Overview

Many traditional "knob and dial" controls were researched in the l940s and 1950s; thus, well proven design recommendations are available. Unfortunately, they rely more on design and use experience than on known psychological rules and principles.

Advances in technology, as well as voiced user complaints and preferences, have brought up new issues, related particularly to computer data entry and information display. Also, labeling and warnings have become of ergonomic importance.

14.3.2 Introduction

Controls, called *activators* in ISO standards, transmit inputs to a piece of equipment. They are usually operated by hand or foot. The results of the control inputs are shown to the operator either in terms of *displays* or *indicators* or by the ensuing actions of the machine.

---

[*]Reprinted by permission from Kroemer, K, Kroemer, H, Kroemer-Elbert, K; *Ergonomics: How to Design for Ease and Efficiency.* Chapter 11, "Selection, Design, and Arrangement of Controls and Displays." Prentice-Hall, New York, 1994.

The 1940s and 1950s are often called the *knobs and dials era* in human factors engineering because much research was performed on controls and displays. Thus, this topic is rather well researched, and summaries of the earlier findings were compiled, e.g., by Van Cott and Kinkade (1972), Woodson (1981), and McCormick and Sanders (1982). Military and industry standards (e.g., MIL-STD-1472 and HDBK-759; SAE J 1138, 1139, and 1048; HFS/ANSI 100) have established detailed design guidelines. These are matter-of-fact practical rules concerning already existent kinds of controls and displays used in well established designs and western stereotypes. Yet, overriding general *laws* based on human motion or energy principles, or on perception and sensory processes, are not usually known or stated. Thus, the current rules for selection and design are likely to change with new kinds of controls and displays, and new circumstances and applications.

The Need

Most recommendations for selection and arrangements of controls and displays are purely empirical and apply to existing devices and western stereotypes. Hardly any *general laws* are known that reflect human perceptual, decision-making, or motoric principles that would guide the ergonomics of control and display engineering.

Application 14.1 Controls

Sanders and McCormick (1987) distinguish among the following control actions.

Activate or shut down equipment, such as with an *on-off* key lock

Make a discrete setting, such as making a separate or distinct adjustment like selecting a TV channel

Make a *quantitative setting,* such as selecting a temperature on a thermostat (this is a special case of a *discrete* setting)

Apply *continuous control,* such as steering an automobile

Enter data, as on a computer keyboard

A1.1 Control Selection

Controls shall be selected for their *functional usefulness.* This includes:

- The control type shall be compatible with stereotypical or common expectations (e.g., use a pushbutton or a toggle switch to turn on a light, not a rotary knob).
- Control size and motion characteristics shall be compatible with stereotypical experience and past practice (e.g., have a fairly large steering wheel for two-handed operation in an automobile, not a small rotary control).
- Direction of operation shall be compatible with stereo-typical or common expectations (e.g., *on* control is pushed or pulled, not turned to the left).
- Operations requiring fine control and small force shall be done with the hands, while gross adjustments and large forces are usually exerted with the feet.
- The control shall be *safe* in that it will not be operated inadvertently or operated in false or excessive ways.

There are few "natural rules" for selection and design of controls. One is that hand-operated controls are expected to be used for fine control movements; in contrast, foot-operated controls are usually reserved for large-force inputs and gross control. Yet, consider the pedal arrangements in modern automobiles, where vital and finely controlled operations, nowadays requiring fairly little force, are performed with the feet on the gas and brake pedals. Furthermore, the movement of the foot from the accelerator to the brake is very complex. Probably no topic has received more treatment in course projects and master theses than this against-all-rules but commonly used arrangement.

Compatibility of Control-Machine Movement

Controls shall be selected so that the direction of the control movement is compatible with the response movement of the controlled machine, be it a vehicle, equipment, component, or accessory. Table 14.1 lists such compatible movements.

TABLE 14.1 Control Movements and Expected Effects

	Direction of control movement											
Effect	Up	Right	Forward	Clockwise	Press*, Squeeze	Down	Left	Rearward	Back	Counter-clockwise	Pull**	Push**
On	1	1	1	1	2	—	—	—	—	—	1	—
Off	—	—	—	—	—	1	2	2	—	1	—	2
Right	—	1	—	2	—	—	—	—	—	—	—	—
Left	—	—	—	—	—	—	1	—	2	—	—	—
Raise	1	—	—	—	—	—	—	2	—	—	—	—
Lower	—	—	2	—	—	1	—	—	—	—	—	—
Retract	2	—	—	—	—	—	—	1	—	—	2	—
Extend	—	—	1	—	—	2	—	—	—	—	—	2
Increase	2	2	1	2	—	—	—	—	—	—	—	—
Decrease	—	—	—	—	—	2	2	1	—	2	—	—
Open Valve	—	—	—	—	—	—	—	—	—	1	—	—
Close Valve	—	—	—	1	—	—	—	—	—	—	—	—

Note: 1 = most preferred; 2 = less preferred.
*With trigger-type control.
**With push-pull switch.
Source: Modified from Kroemer, 1988d.

The term *compatibility* usually refers to the context (Sanders and McCormick, 1987) or situation where an association appears manifest or intrinsic, e.g., locating a control next to its related display. Yet, other relationships depend on what one has learned, what is commonly used in one's civilization. In the "western world," red is conceived to mean danger or stop, and green to mean safe and go. Such a relationship is called a population stereotype, probably learned during early childhood (Ballantine, 1983), which may differ from one user group to another. For example, in Europe, a switch is often toggled downward to turn on a light, while in the United States the switch is pushed up (Kroemer, 1982). Leaving the western world, one may encounter quite different stereotypical expectations and uses. For example, in China, some durable conventions exist that are different from those in the United States (Courtney, 1988; Chapanis, 1975).

A1.1.1 Control Actuation Force or Torque

The force or torque applied by the operator for the actuation of the control shall be kept as low as feasible, particularly if the control must be operated often. If there are jerks and vibrations, it is usually better to stabilize the operator than to increase the control resistance to prevent uncontrolled or inadvertent activation.

Note that, in the following recommendations, often values for (tangential) *force* at the point of application instead of *torque* are given, even for some rotational operations. This is usually the most practical information.

A1.1.2 Control-Effect Relationships

The relationships between the control action and the resulting effect shall be made apparent through common sense, habitual use, similarity, proximity and grouping, coding, labeling, and other suitable techniques, discussed later. Certain control types are preferred for specific applications. Table 14.2 helps (along with Tables 14.1 and 14.3) in the selection of hand controls.

TABLE 14.2 Control-Effect Relations of Common Hand Controls

Effect	Keylock	Toggle Switch	Push-button	Bar knob	Round knob	Thumbwheel Discrete	Thumbwheel Continuous	Crank	Rocker switch	Lever	Joystick or ball	Legend switch	Slide*
Select ON/OFF	1	1	1	3	—	—	—	—	1	—	—	1	1
Select ON/STANDBY/OFF	—	2	1	1	—	—	—	—	—	1	—	1	1
Select OFF/MODE 1/ MODE 2	—	3	2	1	—	—	—	—	—	1	—	1	1
Select one function of several related functions	—	2	1	—	—	—	—	—	2	—	—	—	3
Select one of three or more discrete alternatives	—	—	—	1	—	—	—	—	—	—	—	—	1
Select operating condition	—	1	1	2	—	—	—	—	1	1	—	1	2
Engage or disengage	—	—	—	—	—	—	—	—	—	1	—	—	—
Select one of mutually exclusive functions	—	—	1	—	—	—	—	—	—	—	—	1	—
Set value on scale	—	—	—	—	1	—	2	3	—	3	3	—	1
Select value in discrete steps	—	—	1	1	—	1	—	—	—	—	—	—	1

Note: 1 = most preferred; 3 = least preferred.
*Estimated, no experiments known.
Source: Modified from Kroemer, 1988d.

Continuous versus Detent Controls

Continuous controls shall he selected if the control operation is anywhere within the adjustment range of the control; it does not need to be set in any given position. Yet, if the control operation is in discrete steps, these shall be marked and secured by *detents* or *stops* in which the control comes to rest.

A1.1.3 Standard Practices

Unless other solutions can be demonstrated to be better, the following rules apply to common equipment:

1. One-dimensional steering is by a steering wheel.
2. Two-dimensional steering is by a joystick or by combining levers, wheel, and pedals.
3. Primary vehicle braking is by pedal.
4. Primary vehicle acceleration is by a pedal or lever.
5. Transmission gear selection is by lever or by legend switch.
6. Valves are operated by round knobs or T-handles.
7. Selection of one (of two or more) operating modes can be by toggle switch, pushbutton, bar knob, rocker switch, lever, or legend switch.

Table 14.3 provides an overview of controls suitable for various operational requirements.

TABLE 14.3   Selection of Controls

Small operating force	
Two discrete positions	Keylock, hand-operated
	Toggle switch, hand-operated
	Pushbutton, hand-operated
	Rocker switch, hand-operated
	Legend switch, hand-operated
	Bar knob, hand-operated
	slide, hand-operated
	Push-pull switch, hand-operated
Three discrete positions	Toggle switch, hand-operated
	Bar knob, hand-operated
	Legend switch, hand-operated
	Slide, hand-operated
Four to 24 discrete positions, or continuous operation	Bar knob, hand-operated
	Round knob, hand-operated
	Joystick, hand-operated
	Continuous thumbwheel, hand-operated
	Crank, hand-operated
	Lever, hand-operated
	Slide, hand-operated
	Track ball, hand-operated
	Light pen, hand-operated
Continuous slewing, fine adjustments	Crank, hand-operated
	Round knob, hand-operated
	Track ball, hand-operated
Large operating force	
Two discrete positions	Push button, foot-operated
	Push button, hand-operated
	Detent lever, hand-operated
Three to 24 discrete positions	Detent lever, hand-operated
	Bar knob, hand-operated
Continuous operation	Hand wheel, hand-operated
	Lever, hand-operated
	Joystick, hand-operated
	Crank, hand-operated
	Pedal, foot-operated

Source: Modified from Kroemer, 1988.

A1.2 Arrangement and Grouping of Controls

Several "operational rules" govern the arrangement and grouping of controls.

"Locate for the Ease of Operation." Controls shall be oriented with respect to the operator. If the operator has different positions (such as in driving and operating a backhoe), the controls and control panels shall "move with the operator" so that in each position their arrangement and operation is the same for the operator.

"Primary Controls First." The most important and most frequently used controls shall have the best positions with respect to ease of operation and reaching.

"Group Related Controls Together." Controls that have sequential relations, that are related to a particular function, or that are operated together, shall be arranged in "functional groups" (along with their associated displays). Within each functional group, controls and displays shall be arranged according to operational importance and sequence.

"Arrange for Sequential Operation." If operation of controls follows a given pattern, controls shall be arranged to facilitate that sequence. The common arrangements are left-to-right (preferred) or top-to-bottom, as in print in the western world.

"Be Consistent." The arrangement of functionally identical or similar controls shall be the same from panel to panel.

"Dead-Man Control." If the operator becomes incapacitated and either lets go of a control, or Continues to hold onto it, a "dead-man control" design shall be utilized which either returns the system to a noncritical operation state or shuts it down.

"Guard Against Accidental Activation." Numerous ways to guard controls against inadvertent activation may be applied, such as putting mechanical shields at or around them, or requiring minimal forces or torques (see later). Note that most will reduce the speed of operation.

"Pack Tightly But Do Not Crowd." Often it is necessary to place a large number of controls into a limited space. Table 14.4 indicates the minimal separation distances for various types of controls.

TABLE 14.4 Minimal Separation Distances (in Millimeters) for Hand Controls*

	Keylock	Bar knob	Detent thumbwheel	Push-button	Legend switch	Toggle switch	Rocker switch	Knob	Slide switch
Keylock	25	19	13	13	25	19	19	19	19
Bar Knob	19	25	19	13	50	19	13	19	13
Detent thumbwheel	13	19	13	13	38	13	13	19	13
Pushbutton	13	13	13	13	50	13	13	13	13
Legend switch	25	50	38	50	50	38	38	50	38
Toggle switch	19	19	13	13	38	19	19	19	19
Rocker switch	19	13	13	13	38	19	13	13	13
Knob	19	25	19	13	50	19	13	25	13
Slide switch	19	13	13	13	38	19	13	13	13

*The given values are measured edge-to-edge with single controls in their closest positions. For arrays of controls, in some cases larger distances are recommended—see Boff and Lincoln (1988).

A1.2.1 Control Design

The following descriptions, with Tables 14.5 through 14.19 and Figs. 14.1 through 14.4, present design guidance for various detent and continuous controls. The first set (*keylocks* through *alphanumeric keyboards*) consists of *detent* controls.

Keylock. *Application*–Keylocks (also called *key-operated* switches) are used to prevent unauthorized machine operation. Keylocks usually set into *on* and *off* positions; they are always distinct.

Design Recommendations—Design recommendations are given in Fig. 14.1 and Table 14.5. Other recommendations are as follows:

1. Keys with teeth on both edges (preferred) should fit the lock with either side up. Keys with a single row of teeth should be inserted into the lock with the teeth pointing up.

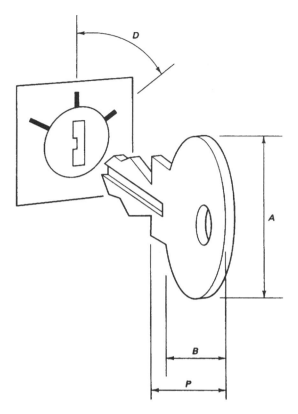

FIGURE 14.1 Keylock (modified from MIL-HDBK-759).

TABLE 14.5 Dimensions of a Keylock

	A Height (mm)	B Width (mm)	P Protrusion (mm)	D Displacement (degrees)	S* Separation (mm)	R† Resistance (Nm)
Minimum	13	13	20	45	25	0.1
Preferred	—	—	—	—	—	—
Maximum	75	38	—	90	25	0.7

Note: Letters correspond to measurements illustrated in Figure 14.1.
*Between closest edges of two adjacent keys.
†Control should "snap" into detent position and not be able to stop between detents.
Source: Modified from MIL-HDBK-759.

2. Operators should normally not be able to remove the key from the lock unless turned OFF.
3. The *on* and *off* positions should be labeled.

Bar knob. *Application*—Detent bar knobs (also called *rotary selectors*) should be used for discrete functions when two or more detented positions are required.

Shape—Knobs shall be bar shaped with parallel sides, and the index end shall be tapered to a point.

Design Recommendations—Design recommendations are illustrated in Fig. 14.2 and listed in Table 14.6.

Detent thumbwheel. *Application*—Detent (or discrete) thumbwheels for discrete settings may be used if the function requires a compact input device for discrete steps.

Design Recommendations—Design recommendations are illustrated in Fig. 14.3 and listed in Table 14.7.

Package/Enclosure

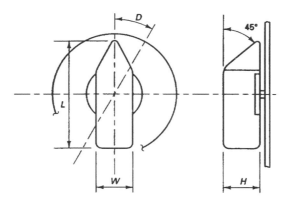

FIGURE 14.2 Bar knob (modified from MIL-HDBK-759).

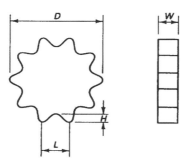

FIGURE 14.3 Discrete thumbwheel (modified from MIL-HDBK-759).

TABLE 14.6 Dimensions of a Bar Knob

	L Length (mm)	W Width (mm)	H Height (mm)	R† Resistance (Nm)	D Displacement (degrees)	S, Separation	
						One hand, Random operation (mm)	Two hands, Simultaneous operation (mm)
Minimum	25	13	26	0.1	15	25	75
	38*	—	—	—	30**	38*	100*
Preferred	—	—	—	—	—	—	125
Maximum	100	25	75	0.7	90	50	150
	—	—	—	—	—	63*	175*

Note: Letters correspond to measurements illustrated in Fig. 14.2.
*If operator wears gloves.
†High resistance with large bar knob only. Control should snap into detent position and not be able to stop between detents.
**For blind positioning.
Source: Modified from MIL-HDBK-759.

Pushbutton. *Application*—Pushbuttons should be used for single switching between two conditions, for entry of a discrete control order, or for release of a locking system (e.g., of a parking brake). Pushbuttons can be used for momentary contact or for sustained contact.

Design Recommendations—Design recommendations are given in Fig. 14.4 and listed in Table 14.8. Other recommendations are as follows:

Shape—The pushbutton surface should normally be concave (indented) to fit the finger. When this is impractical, the surface shall provide a high degree of frictional resistance to prevent slipping. Yet, the surface may be convex for operation with the palm of the hand.

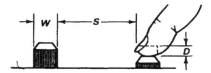

FIGURE 14.4 Pushbutton (modified from MIL-HDBK-759).

TABLE 14.7 Dimensions of a Discrete Thumbwheel

	D Diameter (mm)	W Width (mm)	L Through distance (Nm)	H Through depth (Nm)	S Separation side-by-side (degrees)	R* Resistance (N)
Minimum	38	3	11	3	10	0.2
Preferred	—	—	—	—	—	—
Maximum	65	—	19	13	—	0.6

Note: Letters correspond to measurements illustrated in Fig. 14.3.
*Control should snap into detent position and not be able to stop between detents.
Source: Modified from MIL-HDBK-759.

Positive Indication—A positive indication of control activation shall be provided (e.g., snap feel audible click, or integral light).

Push-pull switch. Push-pull controls have been used for discrete settings, commonly *on* and *off*; immediate positions have been occasionally employed, e.g., for the air-gasoline mixture of a combustion engine. They generally have a round flange under which to "hook" the fingers. Their diameter should be not less than 19 mm, protruding at least 25 cm from the mounting surface the separation between adjacent controls should be at least 38 mm. There should be at least 13 mm displacement between the settings (MIL-STD-1472D).

Legend switch. *Application*—Detent legend switches are particularly suited to display qualitative information on equipment status that requires the operator's attention and action.

Design Recommendations—Design recommendations are given in Fig. 14.5 and listed in Table 14.9.

Legend switches should be located within a 30°-cone along the operator's line of sight.

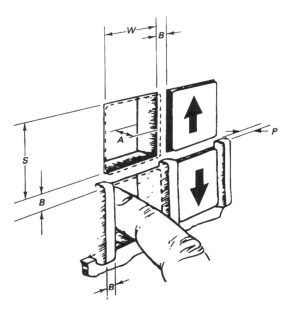

FIGURE 14.5 Legend switch (modified from MIL-HDBK-759).

TABLE 14.8 Dimensions of a Pushbutton

| | W, Width of square or diameter | | | R, Resistance | | | | D** | S, Separation | | | |
| | | | | | | | | | Single finger | | | |
Operation	Fingertip (mm)	Thumb (mm)	Palm of hand (mm)	Little finger (N)	Other finger (N)	Thumb (N)	Palm of hand (N)	Displacement (mm)	Single operation (mm)	Sequential operation (mm)	Different fingers (mm)	Palm or thumb (mm)
Minimum	10 13*	19	25	0.25	0.25	1.1	1.7	3.2 16	13 25*	6	6	25
Preferred	—	—	—	—	—	—	—	—	50	13	13	150
Maximum	19	—	—	1.5	11.1	16.7	22.2	6.5 20*	—	—	—	—

Note: Letters correspond to measurements illustrated in Fig. 14.4.
*If operator wears gloves.
**Depressed button shall stick out at least 2.5 mm.
Source: Modified from MIL-HDBK-759.

TABLE 14.9 Dimensions of a Legend Switch

	W, Width of square or diameter (mm)	D Displacement (mm)	B Barrier width (mm)	P Barrier protrusion (mm)	R Resistance (N)
Minimum	19	3		5	0.28
Preferred	—	—	—	—	—
Maximum	38	6	6	6	11

Note: Letters, W, A, B_w, and B_d correspond to measurements illustrated in Fig. 14.5.
Source: Modified from MIL-HDBK-759.

Toggle switch. *Application*—Detent toggle switches may be used if two discrete positions are required. Toggle switches with three positions shall be used only where the use of a bar knob, legend switch, array of push buttons, etc., is not feasible.

Design Recommendation—Design recommendations are given in Fig. 14.6 and listed in Table 14.10, and are as follows:

Orientation—Toggle switches should be so oriented that the handle moves in a vertical plane, with *off* in the down position. Horizontal actuation shall be employed only if compatibility with the controlled function or equipment location is desired.

Rocker switch. *Application*—Detent rocker switches may be used if two discrete positions are required. Rocker switches protrude less from the panel than do toggle switches.

Design Recommendation—Design recommendations are given in Fig. 14.7 and listed in Table 14.11.

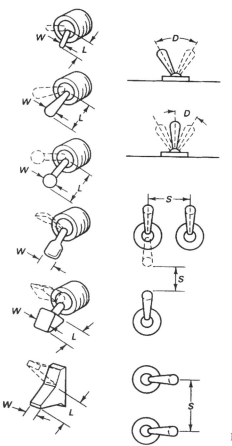

FIGURE 14.6 Toggle switch (modified from MIL-HDBK-759).

TABLE 14.10 Dimensions of a Toggle Switch

	L Arm length (mm)	W, Tip width or diameter (mm)	R† Resistance (N)	D, Displacement			S, Separation			
				2 Positions (degrees)	3 Positions (degrees)		Horizontal array, vertical operation			Vertical array horizontal operation (mm)	Toward each other tip-to-tip other (mm)
							Single finger		Several fingers simultaneous operation (mm)		
							Random operation (mm)	Sequential operation (mm)			
Minimum	9.5 38*	3	2.8	25	18		19	13	19 32*	25 38*	25
Preferred	—	—	4.5	—	25		50	25	—	—	—
Maximum	50	25	11	120	60		—	—	—	—	—

Note: Letters correspond to measurements illustrated in Fig. 14.6.
*If operator wears gloves.
†Control should snap into detent position and not be able to stop between detents.
Source: Modified from MIL-HDBK-759.

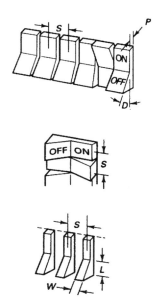

FIGURE 14.7 Rocker switch. Narrow switch (bottom) is especially desirable for tactile definition when gloves are worn (modified from MIL-HDBK-759).

TABLE 14.11 Dimensions of a Rocker Switch

	W Width (mm)	L Length (mm)	D Displacement (degrees)	P Protrusion depressed (mm)	S Separation center to center (mm)	R Resistance (N)
Minimum	6.5	13	30	2.5	19 32*	2.8
Preferred	—	—	—	—	—	—
Maximum	—	—	—	—	—	11.1

Note: Letters D, H, L, S, and W correspond to measurements illustrated in Fig. 14.7.
*If operator wears gloves.
Source: Modified from MIL-HDBK-759.

Orientation—Rocker switches should be so oriented that the handle moves in a vertical plane with *off* in the down position. Horizontal actuation shall be employed only if compatibility with the controlled function or equipment location is desired.

A1.3 Sets of Numerical Keys

Two different kinds of numerical key sets are widely used. One, usually associated with telephones, is arranged thus:

1 2 3
4 5 6
7 8 9
  0

The other arrangement is called the calculator key set:

7 8 9
4 5 6
1 2 3
  0

Keying performance may be more accurate and slightly faster with the telephone key set than with the calculator arrangement (Conrad and Hull, 1968; Lutz and Chapanis, 1955). ❏

14.3.3 Alphanumeric Keyboards

On the original typewriter keyboard, developed in the 1860s, the keys were arranged in alphabetic sequence in two rows. The QWERTY layout, patented in 1879 was adopted after many modifications as an international standard in 1966. It has since been universally used on typewriters, then on computers and on many other input devices.

The letter-to-key allocation on the QWERTY has many, often obscure, reasons. Letters that frequently follow each other in English text (such as q and u) were spaced apart so that the mechanical type bars might not entangle if struck in rapid sequence. The "columns" of keys assigned to separate fingers run diagonally, across the keyboard. This was also done originally due to mechanical constraints of the type bars. The keys are arranged in straight rows, which the fingertips are not. Obviously, the original QWERTY keyboard was designed considering mostly mechanics, not *human factors*. Many attempts were made to improve typing performance by changing the keyboard layout. They include relocating keys within the standard keyset layout or changing the keyboard layout—for example, by breaking up the columns and rows of keys, by dividing the keyboard into separate sections, by adding sets of keys, etc. (For reviews, see, for example, Alden, Daniels, and Kanarick, 1972; Gilad and Pollatsheck, 1986; Gopher and Raij, 1988; Hirsch, 1970; Kroemer, 1972, 1992; Lithrick, 1981; Martin, 1949; Michaels, 1971; Norman and Fisher, 1982; Noyes, 1983a; Seibel, 1972.) A terminology for describing major design features of keyboards is shown in Fig. 14.8 with some descriptors taken from ANSI 100 (Human Factors Society, 1988).

Application 14.2   Keys

The dynamics of force-displacement characteristics of keys are important for the user, but they are difficult to measure, because the procedures commonly used rely on static force applications. This is not indicative of the actual dynamic operation, which varies from operator to operator (Sind, 1989). Furthermore, the technology of key switches is rapidly developing. Yet, there appears to be consensus that for *full-time keying*, keys with some displacement (about 2 mm) and a "snap back resistance" or an audible "click" signal (where key activation is felt by reduced key resistance or heard) are preferable (Monty, Snyder, and Birdwell, 1983; Sind, 1989; Cushman and Rosenberg, 1991). Instead of having separate single keys, some keyboards consist of a membrane that, when pressed down in the correct location, generates the desired letter with virtually no displacement of the keyboard. A major advantage of the membrane is that dust or fluids cannot penetrate it; however, many typists dislike it. In experiments, experienced users showed better performance with convention keys than with the membrane keyboard, but the differences were not large and diminished with practice (Loeb, 1983; Sind, 1989) For *part-time* data input, keys with little or no displacement, such as membrane keyboards, may be acceptable.

Instead of simply improving conventional key and keypad designs, more radical proposals have recently appeared. These include the use of chording instead of single key activations for generating letters, words, or symbols. This has been used, for example, for mail sorting or on-the-spot recording of verbal discussions (Gopher and Raij, 1988; Keller, Becker, and Strasser, 1991; Noyes, 1983b). While traditional keys are *tapped down* for activation and have only two states, *on* and *off*, other proposals use ternary keys that are *rocked* forth and back with the fingertips, where fingers are inserted into well-like switches that can be tapped, rocked, and moved sideways; or where finger movements are registered by means of instrumented gloves (Kroemer, 1992). ❏

Traditionally, computer entries have been made by mechanical interaction between the operator's fingers and such devices as keyboard, mouse, trackball, or light pen. Yet, there are many other means to generate inputs. Voice recognition is one fairly well known method, but others can be employed that utilize, for example,

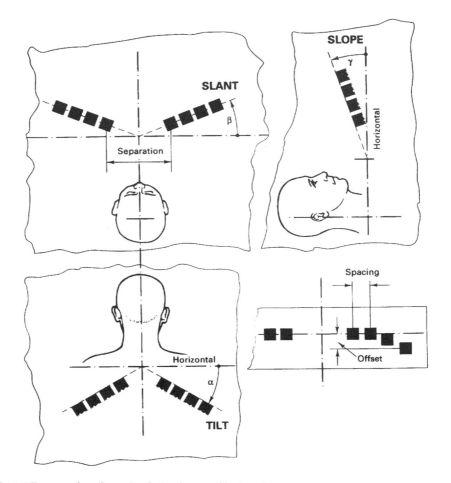

FIGURE 14.8 Terms to describe major design factors of keyboards.

- Hands and fingers for pointing, gestures, sign language, tapping, etc.
- Arms for gestures, making signs, moving or pressing, control devices
- Feet for motions and gestures, for moving and pressing of devices
- The legs, also for gestures, moving and pressing of devices
- The torso, including the shoulders, for positioning and pressing
- The head, also for positioning and pressing
- The mouth for lip movements, use of the tongue, or breathing such as through a blow/suck tube
- The face for making facial expressions
- The eyes for tracking
- Combinations and interactions of these different inputs could be used, such as in *ballet positions* (Jenkins, 1991).

Of course, the method selected must be clearly distinguishable from "environmental clutter," which may be defined as "loose energy" that interferes with sensor pickup. The ability of a sensor to detect the input signals depends on the type and intensity of the signal generated. For example, it may be quite difficult to distinguish between different facial expressions, while it is much easier to register displacement of, or pressure on, a sensor. Thus, the use of other than conventional input methods depends on the state of technology, which includes the tolerance of the system to either missed or misinterpreted input signals.

Pointing with the finger or hand, for example, is an attractive solution. Sensors are able to track position the direction of pointing, and possibly movements. Sensing could be done with either a relative or an absolute reference. The sensing could be continuous or discrete in one, two, or three dimensions. Pointing, for example, is a common means to convey information; a large number of characters could be generated by it. Pointing is a dynamic activity; motion (or position) could be sensed either on the finger or hand directly or by means of "landmarks" such as reflective surfaces, for example, attached to rings worn. Thus, pointing is a natural, easily learned and controlled activity, which could be observed fairly easily. However, there is the possibility of missed inputs, misinterpreted inputs, and of fatigue or overuse syndromes when many pointing actions are executed quickly and repeatedly (Jenkins, 1991).

While these ideas may be far-fetched, even current keyboards impose quite different motoric task on the hands of the user than did previous mechanical designs, which had much key displacement and large keying forces. Use of a smaller set of keys (such as with chording) would reduce the required band travel and allow resting the wrists on suitable pads. This might alleviate some of the problems associated with cumulative trauma disorders by virtue of reduced energy requirements and improved operator posture (Kroemer, 1992; Rose, 1991).

Application 14.3   Control Descriptions

The following set of control descriptions concerns continuous operations.

Knob.   *Application*—Continuous knobs (also called round knobs or rotary controls) should be used when little force is required and when precise adjustments of a continuous variable are required. If positions must be distinguished, an index line on the knob should point to markers on the panel.

Design Recommendations—Design recommendations are given in Fig. 14.9 and Table 14.12. Within the range specified in the table, knob size is relatively unimportant—provided the resistance is low and the knob can be easily grasped and manipulated. When panel space is extremely limited, knobs should be small and should have resistances as low as possible without permitting the setting to be changed by vibration or by inadvertent touching.

Knob Style—Unless otherwise specified, control knobs shall conform to the guidelines established in military standards, e.g., MIL-STD-1472 or MIL-HDBK-759.

Crank.   *Application*—Continuous cranks should be used primarily if the control must be rotated many times For tasks involving large slewing movements or small, fine adjustments, a crank handle may be mounted on a knob or handwheel.

Grip Handle—The crank handle shall be designed so that it turns freely around its shaft, especially if the whole hand grasps the handle.

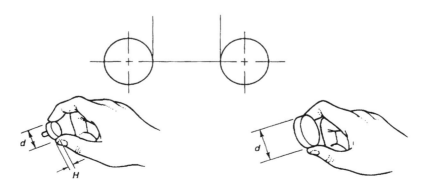

FIGURE 14.9 Knob (modified from MIL-HDBK-759).

TABLE 14.12 Dimensions of a Knob

	H Height (mm)	D, Diameter		T, Torque		S, Separation	
		Fingertip grip (mm)	Thumb and finger grasp (mm)	Up to 25 mm in diameter (N m)	Over 25 mm in diameter (N m)	One hand (mm)	Two hands simultaneously (mm)
Minimum	13	10	25	—	—	25	50
Preferred	—	—	—	—	—	50	125
Maximum	25	100	75	0.03	0.04	—	—

Note: Letters correspond to measurement illustrated in Fig. 14.9.
Source: Modified from MIL-HDBK 759.

Design Recommendations—Design recommendations are given in Fig. 14.10 and Table 14.13.

Handwheel. *Application*—Continuous handwheels that are designed for nominal two-handed operation should be used when the breakout or rotation forces are too large to be overcome with a one-hand control—provided that two hands are available for this task.

Knurling—Knurling or indentation shall be built into a handwheel to facilitate operator grasp.

Spinning Handle—When large displacements must be made rapidly, a spinner (crank) handle may be attached to the handwheel when this is not overruled by safety considerations.

Design Recommendations—Design recommendations are given in Fig. 14.11 and listed in Table 14.14.

Lever. *Application*—Continuous levers may be used when large force or displacement is required at the control and/or when multidimensional movements are required There are two kinds of levers, often called *joysticks* or simply *sticks*:

- A *force joystick,* which does not move (is *isometric*) but transmits control inputs according to the force applied to it.

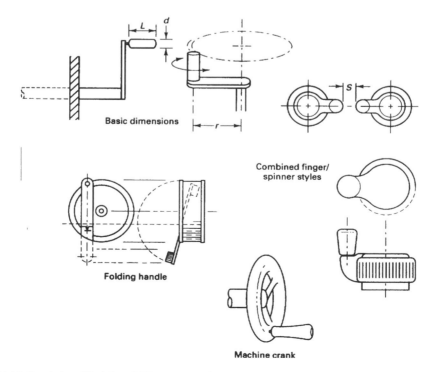

FIGURE 14.10 Crank (modified from MIL-HDBK-759).

TABLE 14.13 Dimensions of A Crank

	Operated by finger and wrist movement (Resistance below 22 N)					Operated by arm movement (Resistance below 22 N)				
	L Length (mm)	d Diameter (mm)	r, Turning radius		S Separation (mm)	L Length (mm)	d Diameter (mm)	r, Turning radius		S Separation (mm)
			Below 100 RPM (mm)	Above 100 RPM (mm)				Below 100 RPM (mm)	Above 100 RPM (mm)	
Minimum	25	9.5	38	13	75	75	25	190	125	75
Preferred	38	13	75	58	—	95	25	—	—	—
Maximum	75	16	125	115	—	—	38	510	230	—

Note: Letters correspond to measurements illustrated in Fig. 14.10.
Source: Modified from MIL-HDBK-759.

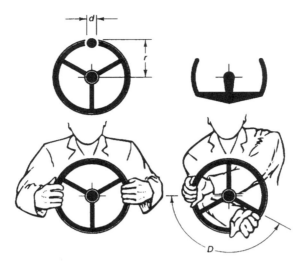

FIGURE 14.11 Handwheel (modified from MIL-HDBK-759).

TABLE 14.14 Dimensions of a Handwheel

| | r, Wheel radius | | d | | | D |
	With power steering (mm)	Without power steering (mm)	Rim diameter (mm)	Tilt from vertical (degrees)	R Resistance (N)	Displacement both hands on wheel (degrees)
Minimum	175	200	19	30 Light vehicle	20	—
Preferred	—	—	—	—	—	—
Maximum	200	255	32	45 Heavy vehicle	220	120

Note: Letters correspond to measurements illustrated in Fig. 14.11.
Source: Modified from MIL-HDBK-759.

- A *displacement joystick* (often falsely called *isotonic*), which transmits control inputs according to its spatial position, or movement direction, or speed.

The force-controlled lever is especially suitable when "return-to-center" of the system after each control input is necessary. The displacement lever is appropriate when control in two or three dimensions is needed, particularly when accuracy is more important than speed.

Design Recommendations—Design recommendations are given in Fig. 14.12 and listed in Table 14.15. Other recommendations are as follows:

Limb Support—When levers are used to make fine or continuous adjustments, support shall be provided for the appropriate limb segment:

- for large hand movements: elbow support
- for small hand movements: forearm support
- for finger movements: wrist support

Coding—when several levers are grouped in proximity to each other, the lever handles shall be coded (see below).

Labeling—When practicable, all levers shall be labeled (see below) with regard to function and direction of motion.

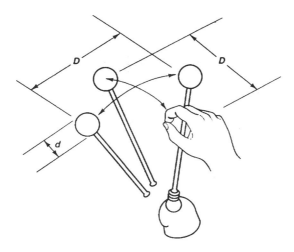

FIGURE 14.12 Lever (modified from MIL-HDBK-759).

TABLE 14.15 Dimensions of a Lever

| | d, Diameter | | R, Resistance | | | | D, Displacement | | S, Separation | |
| | | | Fore-aft | | Left-right | | | | | |
	Finger grip (mm)	Hand grip (mm)	One hand (N)	Two hands (N)	One hand (N)	Two hands (N)	Fore-aft (mm)	Left-right (mm)	One hand (mm)	Two hands (mm)
Minimum	13	32	9	9	9	9	—	—	50*	75
Preferred	—	—	—	—	—	—	—	—	—	—
Maximum	75	75	135	220	90	135	360	970	100*	125

Note: Letters A and D correspond to measurements illustrated in Fig. 14.12
*about 25 mm if one hand usually operates two adjacent levers simultaneously.
Source: Modified from MIL-HDBK-759.

Elastic Resistance—For joystick controls, elastic resistance that increases with displacement may be used to improve "stick feel."

High-Force Levers—For occasional or emergency use, high-force levers may be used. They shall be designed to be either pulled up or pulled back toward the shoulder, with an elbow angle of 150° (± 30°). The force required for operation shall not exceed 190 N. The handle diameter shall be from 25 to 38 mm, and its length shall be at least 100 mm. Displacement should not exceed 125 mm. Clearance behind the handle and along the sides of the path of the handle shall be at least 65 mm. The lever may have a thumb-button or a clip-type release.

Continuous thumbwheel. *Applications*—Thumbwheels for continuous adjustments may be used as an alternative to round knobs if a compact thumbwheel is beneficial.

Design Recommendations—Design recommendations are given in Fig. 14.13 and Table 14.16.

Slide. Slide switches are used to make continuous settings, for example, in music mix-and-control stations. Design recommendations are from Cushman and Rosenberg (1991) and MIL-STD-1472D. They are compiled in Fig. 14.14 and Table 14.17.

A3.1 Computer Input Devices

In addition to the traditional *key* (discussed above), a variety of other input devices have been used, such as the trackball, mouse, light pen, graphic tablet, touch screen, or stick (see above). These are used to move a cursor, manipulate the screen content, point (insert or retrieve information), digitize information, etc. Many tasks change with new technology and software. Therefore, any compilations current at the

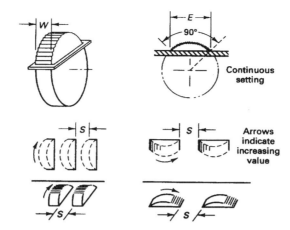

FIGURE 14.13 Continuous thumbwheel (modified from MIL-HDBK-759).

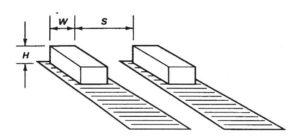

FIGURE 14.14 Slide (modified from MIL-HDBK-759).

TABLE 14.16 Design Recommendations for a Continuous Thumbwheel

	E Rim exposure (mm)	W Width (mm)	S, Separation		R Resistance (N)
			Side-by-side (mm)	Head-to-foot (mm)	
Minimum	—	—	25 38*	50 75*	—
Preferred	25	3.2	—	—	—
Maximum	100	23	—	—	3.3†

Note: Letters correspond to measurements illustrated in Fig. 14.13.
*If operator wears gloves.
†to minimize danger of inadvertent operation.
Source: Modified from MIL-STD-759.

TABLE 14.17 Dimensions of a Slide Operated by Bare Finger

	W Width (mm)	H Height (mm)	S Separation (mm)	R Resistance (N)
Minimum	6	6	19* 13† 16**	3
Maximum	25	—	50* 25† 19**	11

Note: Letters correspond to measurements illustrated in Fig. 14.14.
*Single-finger operation.
†Single-finger sequential operation.
**—simultaneous operation by different fingers.
Source: Modified from MIL-STD-1472D.

time (such as by Boff and Lincoln, 1988; Cushman and Rosenberg, 1991) become outdated quickly, and the designer must follow the current literature closely (see, for example, Chase and Casali, 1991; Han, Jorna, Miller, and Tan, 1990; Epps, 1987).

A3.2 Foot-Operated Controls

Foot-operated switches may be used if only two discrete conditions-such as *on* and *off* need to be set. Recommended design parameters are compiled in Table 14.18.

In vehicles and cranes (seldom in other applications), pedals are used for continuous adjustments, such as in automobile speed control. Recommended design dimensions are listed in Table 14.19.

A3.3 Remote Control Units

Remote control units, usually simply called *remotes* are small hand-held control panels that are manipulated at some distance from a computer, with which they communicate by cable or radiation. ❏

> Remote control units to be used with TV and radio equipment in the home got so difficult to understand and operate in the 1980s that they became almost proverbial for useless gadgetry. In 1991, a simple and inexpensive remote control unit "to replace all remotes" came on the market and was an immediate commercial success.

In the robotic industry, remote control units (often called *teach pendants*) are used by a technician to specify the point in space to which the robot effector (its tool or gripper) must be moved to operate on a workpiece. This can be done in two ways. One is to take the three-dimensional coordinates of that point from a drawing or a computer-aided design program, and simply enter these into the remote unit. The other is to use the robot as its own measuring device by positioning the effector in the desired place, with all joint angles properly selected, and then simply to command the robot computer to record these positions internally.

Many human engineering issues are pertinent. One is the need of the human operator to see, at least initially, the exact location of the robot effector in relation to the workpiece. This may be difficult to do—for instance, because of lacking illumination—and it may be dangerous for the operator who must step into the operating area of the machine. A second ergonomic concern is proper design and operation of the controls. Joysticks are preferable in principle because they allow easy commands for movement in a plane; pushbuttons, the more common solutions, permit control only in linear or angular direction. Yet, pushbuttons are easier to protect from inadvertent operation and from damage when the unit is dropped (Parsons, 1988, 1991). The proper arrangements of sticks or buttons in arrays on the manipulation surface of the remote is similar to the design aspects discussed earlier in this chapter. Particular attention must be paid to the ability to immediately stop the robot in emergencies and to move it away safely from a given position, even by an inexperienced operator. Reports about injuries to the robot operator or bystanders, or damage to property by incorrect remote operation of the robot are frequent (Etherton and Collins, 1990).

Application 14.4

A4.1 Coding

There are numerous ways to help identify hand-operated controls, to indicate the effects of their operation, and to show their status (see Example 14.1). Throughout the system. the coding principles shall be uniform. The major coding means are

- *Location.*   Controls associated with similar functions shall be in the same relative location from panel to panel.
- *Shape.*   Shaping controls to distinguish them can appeal to both the visual and tactile senses. Sharp edges shall be avoided. Various shapes and surface textures have been investigated for diverse uses. Figures 14.15 through 14.20 show examples.

TABLE 14.18 Dimensions of a Foot-Operated Switch for Two Discrete Positions

	W Width or diameter (mm)		D Displacement (mm)			F Resistance (N)	
	Operation by regular shoe	Operation by heavy boot	Operation by ankle flexion only	Operation by whole leg movement		Foot does not rest on control	Foot rests on control
Minimum	13	13	25	25		18	45
Maximum	—	65	65	100		90	90

Source: modified from MIL-STD-1472D.

TABLE 14.19 Dimensions of a Pedal for Continuous Adjustments (e.g., Accelerator or Brake)

	H Height or depth (mm)	W Width (mm)	D, Displacement (mm)			R, Resistance (N)			Edge-to-edge separation, S	
			Operation with regular shoe	Operation with heavy boot	Operation by ankle flexion only	Operation by whole leg movement	Foot does not rest on pedal	Foot rests on pedal	Operation by ankle flexion	Operation by whole leg movement	Random operation by one foot	Sequential operation by one foot
Minimum	25	75	13	25	25	25	18	45	—	45	100	50
Maximum	—	—	65	65	65	180	90	90	45	800	150	100

Source: Modified from MIL-STD 1472D.

Package/Enclosure   **14**-25

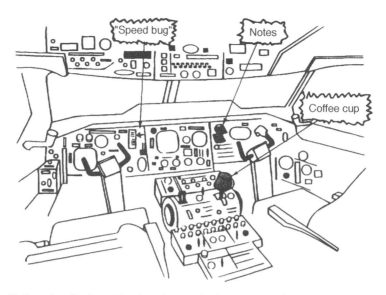

EXAMPLE 14.1 "Informal coding" used by aircraft crews (with permission from Norman, 1991).

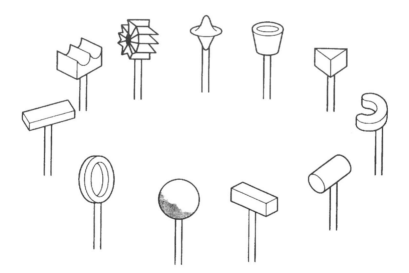

FIGURE 14.15 Examples of shape-coded aircraft controls (Jenkins, 1953).

- *Size.* Up to three different sizes of controls can be used for discrimination by size. Controls that have the same function on different items or equipments shall have the same size (and shape).
- *Mode of operation.* One can distinguish controls by different manners of operation, such as push, turn, and slide. If the operator is not familiar with the control, a false manner of operation may be tried first, which is likely to increase operation time.
- *Labeling.* While proper labeling (see below) is a secure means to identify controls, this works, only if the labels are in fact read and understood by the operator. The label must be placed so that it can be read easily, is well illuminated, and is not covered. Yet, labels take time to read. Trans-illuminated ("back-lighted") labels, possibly incorporated into the control, are often advantageous.
- *Color.* Most controls are either black (number 17038 in FED-STD 595) or gray (26231). For other colors, the following may be selected: red (11105, 21105, 31105, 14187); orange-yellow

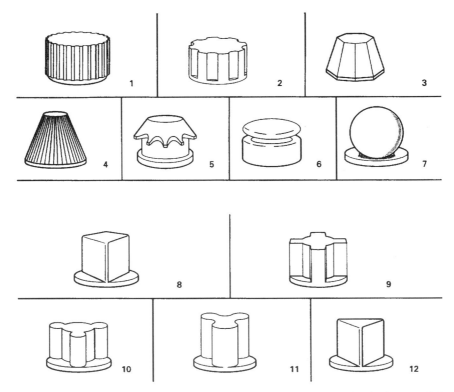

FIGURE 14.16 Examples of shape-coded knobs of approximately 2.5 cm dia., 2 cm height (Hunt and Craig, 1954). Numbers 1 through 7 are suitable for full rotation (but do not combine 1 with 2, 3 with 4, 6 with 7). Numbers 8 through 12 are suitable for partial rotation. Recommended combinations are 8 with 9, 10 with 11, 9 with 12 (but do not combine 8 with 12, 9 with 10 or 11, 11 with 3 or 4).

(13538, 23538, 33538); or white (17875, 27875, 37875). Use blue (1523, 25123) only if an additional color is absolutely necessary. Note that the use of color requires sufficient luminance of the surface.
- *Redundancy.* Often, coding methods can be combined, such as location, size, and shape, or color and labeling. This provides several advantages. One is that the combination of codes can generate a new set of codings. Second, it provides multiple ways to achieve the same kind of feedback, called redundancy: for example, if there is no chance to look at a control, one can still feel it; one knows that on a traffic light, the top signal is red, the bottom green.

The various types of codings have certain advantages and disadvantages, as listed in Table 14.20. Table 14.21 indicates the largest number of coding stimuli that can be used together.

A4.2 Preventing Accidental Activation

Often, it is necessary to prevent accidental activation of controls, particularly if this might cause injury to persons, damage to the system, or degradation of important system functions.

There are various means to prevent accidental activation, some of which may be combined:

- Locate and orient the control so that the operator is unlikely to strike it or move it accidentally in the normal sequence of control operations.
- Recess, shield, or surround the control by physical barriers.
- Cover or guard the control.
- Provide interlocks between controls so that either the prior operation of a related control is required, or an extra movement is necessary to operate the control.

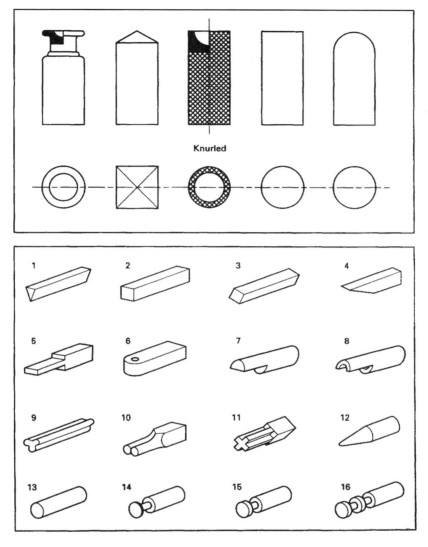

FIGURE 14.17 Examples of shape-coded toggle switches (top, Stockbridge, 1957; bottom, Green and Anderson, 1955) of approximately 1 cm dia., 2.2 cm height.

- Provide extra resistance (viscous or coulomb friction, spring-loading, or inertia) so that an unusual effort is required for actuation.
- Provide a "locking" means so that the control cannot pass through a critical position without delay.

Note that these means usually slow down the operation, which may be detrimental in an emergency case. Coding of foot-operated controls is difficult, and no comprehensive rules are known. Operation errors may have grave consequences—for example, in automobile driving (Rogers and Wierwille, 1988).

A4.3 Displays

Displays provide the operator with necessary information about the status of the equipment. Displays are either visual (lights, scales, counters; CRTs, flat panels), auditory (bells, horns, recorded voice), or tactile (shaped knobs, Braille writing). Labels and instructions/warnings are special kinds of displays.

The "four cardinal rules" for displays are:

FIGURE 14.18 Examples of pushbuttons shape-coded for tactile discrimination (with permission from Moore, 1974). Shapes 1, 4, 21, 22, 23, and 24 are best discriminable by bare-handed touch alone, but all shapes were on occasion confused.

1. Display only that information that is essential for adequate job performance.
2. Display information only as accurately as is required for the operator's decisions and control actions.
3. Present information in the most direct, simple, understandable, and usable form possible.
4. Present information in such a way that failure or malfunction of the display itself will be immediately obvious.

Selecting either an auditory or visual display depends its conditions and purpose. The objective may be to provide

Status information—the current state of the system, the text input into a word processor, etc.
Historical—information about the past state of the stem, such as the course run by a ship
Predictive—such as the future position of a ship, given certain steering settings
Instructional—telling the operator what to do, and how to do something
Commanding—giving directions or orders for a required action

Package/Enclosure 14-29

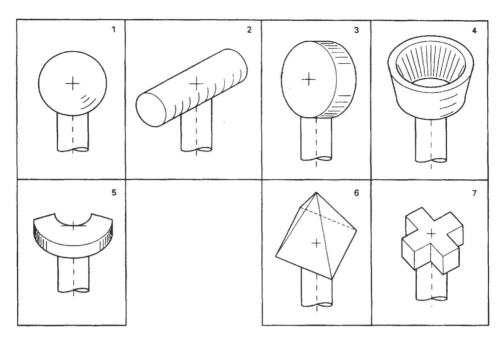

FIGURE 14.19 Shapes proposed in 1980 for use on finger-operated controls of mining equipment. Recommended for concurrent use: *Two handles:* 1 and 2, 1 and 5, 1 and 6, 1 and 7, 2 and 3, 2 and 4, 2 and 5, 2 and 6, 2 and 7, 3 and 5, 4 and 6, 3 and 7, 4 and 5, 4 and 6, 4 and 7. *Three handles:* 1, 2, 6; 1, 2, 7; 2, 3, 6; 2, 3, 7; 3, 5, 6. *Four handles:* 1, 2, 3, 6; 1, 2, 3, 7; 2, 3, 4, 6; 2, 3, 4, 7. *Five handles:* 1, 2, 4, 5, 6. Avoid combinations: 1 and 3, 3 and 4, 5 and 7.

An auditory display is appropriate if the environment must be kept dark; the operator moves around; the message is short, is simple, requires immediate attention, deals with events in time.

A visual display is appropriate if the environment is noisy; the operator stays in place; the message is long, is complex, will be referred to later, deals with spatial location.

A4.3.1 Types of Visual Displays

There are three basic types of visual displays. The *check* display indicates whether a given condition exists (*example:* a green light to indicate normal functioning). The *qualitative* display indicates the status of a changing variable or the approximate value or its trend of change (*example:* a pointer within a normal range). The *quantitative* display shows exact information that must be ascertained (*examples:* find your location on a map; read text or a drawing on a CRT) or indicates an exact numerical value that must be read (*example:* a clock). Overall guidelines are as follows:

- Arrange displays so that the operator can locate and identify them easily without unnecessary searching.
- Group displays functionally or sequentially so that the operator can use them easily.
- Make sure that all displays are properly illuminated or luminant, coded, and labeled according to their function.

Example

In some existing designs, it is advisable to reduce the information content. For example, a quantitative indicator may be reduced to a qualitative one by changing a numerical display of temperature to indicate simply *too cold, acceptable,* or *too hot.*

Light signals. Signals by light (color) are often used to indicate the status if a system (such as *on* or *off*) or to alert the operator that the system, or a portion thereof, is inoperative and that special action must be taken. Common light (color) coding systems are as follows (see Table 14.22):

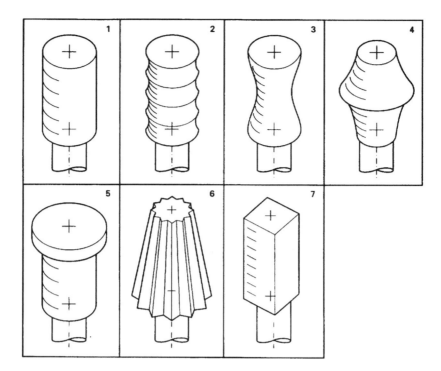

FIGURE 14.20 Shapes proposed in 1980 for use on lever handles on mining equipment. Recommended for concurrent use: *Two handles:* 1 and 5, 1 and 6, 1 and 7, 2 and 6, 2 and 7, 2 and 4, 3 and 6, 3 and 7, 4 and 5, 4 and 6, 4 and 7, 5 and 6, 5 and 7. *Three handles:* 1, 2, 6; 1, 2, 7. *Four handles:* 1, 2, 4, 5; 1, 2, 4, 6; 1, 2, 4, 7. *Five handles:* 1, 2, 3, 4, 6; 1, 2, 3, 4, 7. Avoid combinations: 2 and 5, 3 and 5, 6 and 7.

- A white signal has no correct/wrong implications but may indicate that certain functions are *on*.
- A green signal indicates that the monitored equipment is in satisfactory condition and that it is all right to proceed. For example, a green display may provide such information as *go ahead, in tolerance, ready, power on,* etc.
- A yellow signal advises that a marginal condition exists and that alertness is needed, that caution be exercised, that checking is necessary, or that an unexpected delay exists.
- A red signal alerts the operator that the system or a portion thereof is inoperative and that a successful operation is not possible until appropriate correcting or overriding action has been taken. Examples for a red light signal are to provide information about *malfunction, failure, error,* and so on.
- A flashing red signal denotes an emergency condition that requires immediate action to avert impending personal injury, equipment damage, and the like.

Emergency signals. An an emergency alert is best by auditory warning signal (see below) accompanied by a flashing light. (The operator may acknowledge the emergency by turning off one signal.)

Visual warning indicators. The warning light should be within 30° of the operator's normal line of sight. The emergency indicator should be larger than general status indicators. The luminance contrast C with the immediate background should be at least 3 to 1. Flash rate should be between thee and five pulses per second, with *on* time about equal to *off* time. If the flashing device should fail, the light shall remain on steadily; warning indicators must never turn off merely because a flasher fails. A "word" warning (such as DANGER-STOP) should be used, if feasible.

Complex displays. More involved displays provide information that is either of a qualitative kind or that actually provides exact quantitative information. Some displays provide information about special

TABLE 14.20 Advantages and Disadvantages of Coding Techniques

Advantages	Location	Shape	Size	Mode of operation	Labeling	Color
Improves visual identification	X	X	X		X	X
Improves nonvisual identification (tactual and kinesthetic)	X	X	X	X		
Helps standardization	X	X	X	X	X	X
Aids identification under low levels of illumination and colored lighting	X	X	X	X	(When transilluminated)	(When transilluminated)
May aid in identifying control position (setting)		X		X	X	
Requires little (if any) training: is not subject to forgetting					X	
Disadvantages						
May require extra space	X	X	X	X		X
Affects manipulation of the control (ease of use)	X	X	X	X		
Limited in number of available coding categories	X	X	X	X		X
May be less effective if operator wears gloves		X	X	X		
Controls must be viewed (i.e., must be with visual areas and with adequate illumination present)					X	X

Source: modified from Kroemer, 1988d.

TABLE 14.21 Maximal Number of Stimuli for Coding

Visual stimuli	
Light intensity ("brightness")	2
Color of surfaces	9
Color of lights (lamps)	3
Flash rates of lights	2
Size	3
Shape	5
Auditory stimuli	
Frequency	4
Intensity ("loudness")	3
Duration	2

Source: adapted from information compiled by Cushman and Rosenberg, 1991.

settings or conditions, or indicate the difference between an expected and the actual condition. For these purposes, usually four different kinds of displays are used: moving pointer (fixed scale), moving scale (fixed pointer) counters, or *pictorial* displays. Table 14.23 lists these four kinds of displays and their relative advantages and disadvantages.

For a quantitative display, it is usually preferable to use a moving pointer over a fixed scale. The scale may be straight (either horizontally or vertically), curved, or circular.

Scales should be simple and uncluttered; graduation and numbering of scales should be done such that correct readings can be taken quickly. Numerals should be located outside the scale markings so

TABLE 14.22 Coding of Indicator Lights

Size/type	Color			
	Red	Yellow	Green	White
13 mm diameter or smaller/steady	Malfunction; action stopped; failure; stop action	Delay; check; recheck	go ahead; in tolerance; acceptable; ready	Function or physical position; action in progress
25 mm diameter or larger/steady	Master summation (system or subsystem)	Extreme caution (impending danger)	Master summation (system or subsystem)	
25 mm diameter or larger/flashing	Emergency condition (impending personnel or equipment disaster)			

Source: modified from MIL-HDBK-759.

TABLE 14.23 Characteristics of Displays

Use	Moving Pointer	Moving Scale	Counters	Pictorial Displays
	Good	*Fair*	*Good*	*Fair*
Quantitative information	Difficult to read while pointer is in motion	Difficult to read while scale is in motion	Minimum time and error for exact numerical value, but difficult to read when moving	Direction of motion/scale relations sometimes conflict, causing ambiguity in interpretation
	Good	*Poor*	*Poor*	*Good*
Qualitative information	Location of pointer easy; numbers and scale need not be read; position changes easily detected	Difficult to judge direction and magnitude of deviation without reading numbers and scale	Numbers must be read; position changes not easily detected	Easily associated with real-world situation
	Good	*Fair*	*Good*	*Good*
Setting	Simple and direct relation of motion of pointer to motion of setting knob; position change aids monitoring	Relation to motion of setting knob may be ambiguous; no pointer position change to aid monitoring; not readable during rapid setting	Most accurate monitoring of numerical setting; relation to motion of setting knob less direct than for moving pointer; not readable during rapid setting	Control-display relationship easy to observe
	Good	*Fair*	*Poor*	*Good*
Tracking	Pointer position readily controlled and monitored; simplest relation to manual control motion	No position changes to aid monitoring; relations to control motion somewhat ambiguous	No gross position changes to aid monitoring	Same as above
	Good	*Fair*	*Poor*	*Good*
Difference estimation	Easy to calculate positively or negatively by scanning scale	Subject to reversal errors	Requires mental calculation	Easy to calculate either quantitatively or qualitatively by visual inspection
General	Requires largest exposed and illuminated area on panel; scale length limited unless multiple pointers used	Saves panel space; only small section of scale need be exposed and illuminated; use of tape allows long scale	Most economical of space and illumination; scale length limited only by available number of digit positions	Picture/symbols need to be carefully designed and pretested

Source: modified from MIL-HDBK-759.

Package/Enclosure

that they are not obscured by the pointer. The pointer, on the other side of the scale, should end with its tip directly at the markings. Figure 14.21 provides related information.

The scale mark should show only such fine divisions as the operator must read. All major marks shall be numbered. Progressions of either 1, or 5, or 10 units between major marks are best. The largest admissible number of unlabeled minor graduations between major marks is nine, but only with a minor tick at 5. Numbers should increase as follows: left-to-right; bottom-to-top; clockwise. Recommended dimensions for the scale markers are presented in Fig. 14.22. The dimensions shown there are suitable even for low illumination. ❏

14.3.4 Electronic Displays

In the 1980s and 1990s, *mechanical* displays (such as actual pointers moving over printed scales) were increasingly replaced by *electronic* displays with computer-generated images, or solid state devices such as light-emitting diodes. Whether a mechanical or electronic display, the displayed information may be coded by the following means:

Shape—straight, circular, etc
Shades—black and white or gray
Lines and crosshatched patterns
Figures, pictures, or pictorials—such as symbols or various levels of abstractions, for example, the outline of an airplane against the horizon

A – Fixed scale–moving pointer preferred: three-level marking, numbered at each major mark. Pointer adjacent to graduation marks to preclude obscuration of either marks or numbers.

B – For short, finite scale, every 5th graduation is marked; using only two-level marking.

C – When scale crowding makes pointer-mark association difficult, scale may be graduated in units of two, with two-level scale marking and numbering at each major marking.

D – When dial face is deeply inset within instrument case and visibility of numbers is more important than scale mark-pointer association, pointer may be located inside the graduations along with numbers at major markings. Pointer width should be narrowed at point at which it passes numbers.

E – Moving scale against an index mark or pointer may be used when scale length precludes the fixed scale format (i.e., graduation marks would be too close together). Open window configuration helps operator focus on significant scale area.

F – When open window configuration is oriented in vertical position, numbers should appear upright as each number passes the index mark or pointer. Total scale exposure is desirable when operator needs to refer to other portions of the scale.

FIGURE 14.21 Scale graduation, pointer position, and scale numbering alternatives (modified from MIL-HDBK-759).

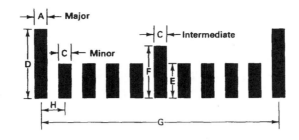

Dimension (in mm)	Viewing distance (in mm)		
	710	910	1525
A (Major index width)	0.89	1.14	1.90
B (Minor index width)	0.64	0.81	1.37
C (Intermediate index width)	0.76	0.99	1.63
D (Major index height)	5.59	7.19	12.00
E (Minor index height)	2.54	3.28	5.44
F (Intermediate index height)	4.06	5.23	8.71
G (Major index separation between midpoints)	17.80	22.90	38.00
H (Minor index separation between midpoints)	1.78	2.29	.381

Minimum scale dimensions suitable even for low illumination

FIGURE 14.22 Scale marks (reprinted from MIL-HDBK-759).

Colors—see below
Alphanumerics (letters, numbers, words, abbreviations)

In many cases, electronically generated displays have been fuzzy, overly complex and colorful, hard to read, and requiring exact focusing and close attention which may distract from the main task, for example, when driving a car (Wierwille, 1992; Wierwille, Antiln, Dingus, and Hulse, 1988). In these cases, the first three of the "four cardinal rules" listed earlier were often violated. Furthermore, many electronically generated pointers markings and alphanumerics did not comply with established human engineering guidelines especially when generated by line segments, scan lines, or dot matrices. Although some designs are acceptable according to current technology and user tolerance (Cushman and Rosenberg, 1991), requirements are changing and innovations in display techniques rapidly develop. Hence, printed statements (even if current and comprehensive when they appear, such as by Cushman and Rosenberg, 1991; Kinney, and Huey 1990; Krebs, Wolf, and Sandrig, 1978; Silverstein and Merrifield, 1985; Woodson, 1981) are becoming obsolete quickly, and none is given in this text. One must closely follow the latest information appearing in technical publications to stay up to date in this field.

Example

Consider the digital watch and digital speedometer. Although both are more accurate and easier to read than their analog counterparts, many people…prefer products with analog displays. One reason is that analog displays also provide information about deviations from reference values. For example, a watch with an analog display also shows the user how many minutes remain before a specific time. Similarly, a speedometer with an analog display shows the driver of a car both the car's speed and how far it is above (or below) the posted speed limit (Cushman and Rosenberg, 1991, pp. 92–93).

Electronic displays are widely used with computers, in cockpits of airplanes and automobiles, and in cameras. Critical physical aspects are size, luminance, contrast, viewing angle and distance, and color.

Package/Enclosure

The overall quality of electronic displays has often been found wanting; viewers usually state that it is more difficult to read the same text from a CRT than from print on paper. However, the image quality of the compared text carriers must be comparable; a suitable metric is the *modulation transfer function*, MTF (Snyder, 1985). It describes the resolution of the display using a spatial sine-wave test signal. For the same MTF area (MTFA), i.e., for comparable image quality, one can expect the same reading times whether hard copy (photograph, printed material) or soft copy (electronic display) is used. Reading time and subjective impression of quality vary with image quality, largely independent of the display technology used (Jorna and Snyder, 1991). Yet, readers have many other criteria than just "reading time" to make use and preference statements (Dillon, 1992).

Monochrome displays have only one color, preferably near the middle of the color spectrum, i.e., white (achromatic), green, yellow, amber, orange rather than blue or red. While measured performance with each of the colors appears similar, personal preferences exist that make it advisable to provide a set with switchable colors (Cushman and Rosenberg, 1991).

If several colors appear on the (chromatic) display, they should be easily discriminated. It is best to display simultaneously not more than three or four colors, including red, green, yellow or orange, cyan or purple, all strongly contrasting with the background; more may be used if necessary, if the user is experienced, and if the stimuli subtend at least 45° of visual angle (Cushman and Rosenberg, 1991; Kinney and Huey 1990).

There is a set of phenomena that have relevance to displaying visual information:

Albney effect	Desaturating a colored light (by adding white light) may introduce a hue shift.
Assimilation	A background color may appear to be blended with the color of an overlying structure (e.g., alphanumeric characters). The effect is the opposite of color contrast.
Bezold-Brucke effect	Changing the luminance of a colored light is usually accompanied by a change in perceived hue.
Chromostereopsis	Highly saturated reds and blues may appear to be located in different depth planes, i.e, in front of or behind the display plane). This may induce visual fatigue and feeling of nausea or dizziness.
Color adaptation	Prolonged viewing of a given color reduces an observer's subsequent sensitivity to that color. As a consequence, the hues of other stimuli may appear to be shifted.
Color "blindness"	About 10 percent of all western males (1 percent of females) have hereditary color deficiencies, mostly such that they see colors (especially reds and greens) differently or less vividly.
Color contract	The hue of an object (e.g., a displayed symbol) is shifted toward the complementary of the surround. The effect is the opposite of assimilation.
Desaturation	Desaturating highly saturated colors may reduce the effects of adaptation, assimilation, chromostereopsis, and color contrast but increase the Albney effect.
Liebmann effect	Removing all luminance contrast from a color image may produce subjectively fuzzy edges.
Receptor distribution	The distribution of cones (color-sensitive receptors) on the retina is uneven. Near the fovea, there is most sensitivity to red and green (particularly under high illumination); at the periphery, there is more sensitivity to blue and yellow.

Application 14.5

In spite of the many variables that (singly and interacting with each other) affect the use of complex color display, Cushman and Rosenberg (1991) complied the following guidelines for use of color in displays:

- Limit the number of colors in one display to four if users are inexperienced or if use of the display is infrequent. Not more than seven colors should ever be used.

- The particular colors chosen should be widely separated from one another in wavelength to maximize discriminability. Colors that differ from one another only with respect to the amount of one primary color (e.g., different oranges) should not be used.
- Suggested combinations include:
    - green, yellow, orange, red, white
    - blue, cyan, yellow, orange, white

  Avoid:
    - reds with blues
    - reds with cyans
    - magentas with blues

In general, avoid displaying several highly saturated, spectrally extreme colors at the same time.

- Red and green should not be used for small symbols and small shapes in peripheral areas of large displays.
- Blue (preferably desaturated) is a good color for backgrounds and large shapes. However, blue should not be used for text, thin lines, or small shapes.
- Using opponent colors (red and green, yellow and blue) adjacent to one another or in an object/background relationship is sometimes beneficial and sometimes detrimental. No general guidelines may be given.
- The color of alphanumeric characters should contrast with that of the background.
- When using color, use shape or brightness as a redundant cue (e.g., all yellow symbols are triangle, all green symbols are circles, all red symbols are squares, etc.) Redundant coding makes the display much more acceptable for users who have color deficiencies.
- As the number of colors is increased, the sizes of color-coded objects should also be increased.

A5.1 Location of Displays

Displays should be oriented within the normal viewing area of the operator with their surfaces perpendicular to the line of sight. The more critical the display, the more centered it should be within the operator's central cone of sight. Avoid glare.

A group of pointer instruments can be so arranged that all pointers are aligned under normal conditions. If one of the pointers deviates from that normal case, its displacement from the aligned configuration is particular obvious. Figure 14.23 shows examples of such arrangements.

A5.2 Control-Display Assignments

In many cases, instruments are set by controls, which should be located in suitable positions close to each other so that the control setting can be done without error, quickly, and conveniently. Proper operation is facilitated by selecting suitable control/display placements. Popular expectancies of relationships exist, but they are often not strong and may depend on the user's background and culture. The assignment is clearest when the control is directly below or to the right of the display. Figure 14.24 shows examples of suitable arrangements.

A frequent course assignment for students of ergonomics is to design a stove top in which the controls for four burners are located so that their relations are unambiguous. No solution has ever been found that satisfactorily combines good assignment, safety or operation, and easy cleaning.

Expected movement relationships are influenced by the types of controls and displays. When both are congruous, i.e., if both are linear or rotary, the stereotype is that they move in corresponding directions, e.g., both up or both clockwise. When the movements are incongruous, their preferred movement relationship can be taken from Fig. 14.25. The following rules apply generally:

Package/Enclosure

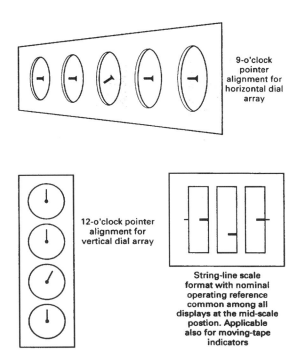

FIGURE 14.23 Aligned pointers for rapid check-reading (modified from MIL-HDBK-759).

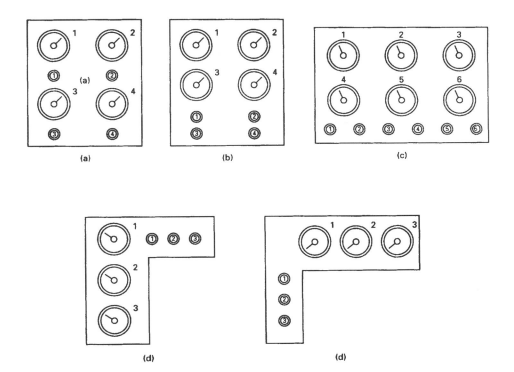

FIGURE 14.24 Control-display relationships (adapted from MIL-HDBK-759).

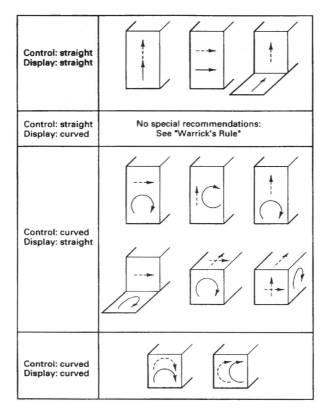

FIGURE 14.25 Compatible control-display directions (modified from Loveless, 1962).

Gear-slide ("Warrick's rule"): a display (pointer) is expected to move in the same direction as does the slide of the control close to ("geared with") the display.

Clockwise-for-increase: turning the control clockwise should cause an increase in the displayed value.

A5.3 Control/Display Ratio

The *control/display (C/D)* ratio (or *CD* gain) describes how much a control must be moved to adjust a display. The *C/D* ratio is like a gear ratio. If much control movement produces only a small display motion, one speaks of a high *C/D* ratio, and of the control as having low "sensitivity." The opposite ratio, *D/C,* is also used; this expression resembles a transfer function.

Usually, two distinct movements are involved in making a setting: first a fast primary (or slewing) motion to an approximate location, then a fine adjustment to the exact setting. The optimal *C/D* ratio is that which minimizes the sum of the two movements. For continuous rotary controls, the *C/D* ratio is usually 0.08 to 0.3, for joysticks 2.5 to 4. However, the most suitable ratios depend much on the given circumstance and must be determined for each application (Boff and Lincoln, 1988; Arnaut and Greenstein, 1990).

A5.4 Auditory Signals

As indicated earlier, auditory signals are better suited than visual displays when the message must attract attention. Therefore, auditory displays are predominantly used as warning devices, especially when the message is short or simple. often together with a flashing light.

Auditory signals may be single tones sounds (mixture of tones), or spoken messages. Tones and sounds may be continuous, periodic, or at uneven timings. They may come from horns, bells, sirens, whistles, buzzers, or loudspeakers.

Package/Enclosure

Use of tonal signals is recommended for qualitative informations, such as for indications of status, or for warnings (while speech may be appropriate for all types of messages). Tonal signals should be at least 10 dB louder than the ambient noise and in the frequency range of 400 to 1,500 Hz; if the signal undulates or warbles, 500 to 1,000 Hz. Buzzers may have frequencies as low as 150 Hz and horns as high as 4,000 Hz. The tonal signal can be made more conspicuous by increasing its intensity, by interrupting it repeatedly, or by changing its frequency. For example: increase from 700 to 1700 Hz in 0.85 s, be silent for 0. 5 s (cycle time one second), then start over, etc. Where specific warning sounds are needed, people can identify the following as different:

$1,600 \pm 50$ Hz tone, interrupted at a rate of 1 to 10 Hz

900 ± 50 Hz steady tone, plus $1,600 \pm 50$ Hz tone interrupted at a rate of 0 to 1 Hz

900 ± 50 Hz steady tone

900 ± 50 Hz steady tone, plus 400 ± 50 Hz tone interrupted at a rate of 0 to 1 Hz

400 ± 50 Hz tone, interrupted at a rate of 1 to10 Hz

Word messages may be prerecorded, digitized, or synthesized speech. The first two techniques are often used (as for telephone answering or "talking products") and are characterized by good intelligibility and natural sound. Synthesized speech uses compositions of phonemes; the result does not sound natural but is effective in converting written text to speech, and it may sound startling, thus attracting attention to the message.

A5.5 Labels and Warnings

A5.5.1 Labels

Ideally, no label should be required on equipment tor control to explain its use. Often, however, it is necessary to use labels so that one may locate, identify, read, or manipulate controls, displays, or other equipment items. Labeling must be done so that the information is provided accurately and rapidly. for this, the following guidelines apply:

- *Orientation.* A label and the information printed on it shall be oriented horizontally so that it can be read quickly and easily. (Note that this applies if the operator is used to reading horizontally, as in western countries.)
- *Location.* A label shall be placed on or very near the item that it identifies.
- *Standardization.* Placement of all labels shall be consistent throughout the equipment and system.
- *Equipment Functions.* A label shall primarily describe the function ("what does it do") of the labeled item.
- *Abbreviations.* Common abbreviations may be used. If a new abbreviation is necessary, its meaning shall be obvious to the reader. The same abbreviation shall be used for all tenses and for the singular and plural forms of a word. Capital letters shall be used, periods normally omitted.
- *Brevity.* The label inscription shall be as concise as possible without distorting the intended meaning or information. The texts shall be unambiguous, with redundancy minimized.
- *Familiarity.* Words shall be chosen, if possible, that are familiar to the operator.
- *Visibility and Legibility.* The operator shall be able to be read easily and accurately it the anticipated actual reading distances, at the anticipated worst illumination level, and within the anticipated vibration and motion environment. Important factors are contrast between the lettering and its back ground; the height, width, stroke width, spacing, and style of letters; and the specular reflection of the background, cover, or other components.
- *Font and Size.* Typography determines the legibility of written information; it refers to style, font, arrangement, and appearance.

The font (typeface) should be simple, bold, and vertical, such as Futura, Helvetica, Namel, Tempo, and Vega. Note that most electronically generated fonts (such as be LED, LCD, matrix) are generally inferior to printed fonts; thus, special attention must be paid to make these as legible as possible.

Recommended *height of characters* depends on the viewing distance, e.g.,

Viewing distance 35 cm, suggested height 22 mm
Viewing distance 70 cm, suggested height 50 mm
Viewing distance 1 m, suggested height 70 mm
Viewing distance 1.5 m, suggested height at least 1 cm

The *ratio of stroke width to character height* should be between 1:8 to 1:6 for black letters on white background, and 1:10 to 1:8 for white letters on black background.

The *ratio of character width to character height* should be about 3:5.

The *space between letters* should be at least one stroke width.

The *space between words* should be at least one character width.

For continuous text, mix upper- and lower-case letters. (For labels, use upper-case letters only.) For VDTs, see ANSI 100 (Human Factors Society), 1988, or newer edition.

A5.5.2 Warnings

Ideally, all devices should be safe to use. In reality, this often cannot be achieved through design. In this case, one must warn users of danger associated with product use and provide instructions for safe use to prevent injury or damage.

It is preferable to have an *active* warning, usually consisting of a sensor that notices inappropriate use and of an alerting device that warns the human of an impending danger. Yet, in most cases, *passive* warnings are used, usually consisting of a label attached to the product and of instructions for safe use in the user manual. Such passive warnings rely completely on the human to recognize an existing or potential dangerous situation, to remember the warning, and to behave prudently.

Factors that influence the effectiveness of product warning information have been compiled by Cushman and Rosenberg (1991). They are listed in Table 14.24.

TABLE 14.24 Factors that Influence the Effectiveness of Product Warning Information

Situation	Warning Effectiveness	
	Low[1]	High[2]
User is familiar with product	✔	
User has never used product before		✔
High accident rate associated with product		✔
Probability of an accident is low	✔	
Consequences of an accident are likely to be severe		✔
User is in a hurry	✔	
User is poorly motivated	✔	
User is fatigued or intoxicated	✔	
User has previously been injured by product		✔
User knows good safety practices		✔
Warning label is very legible and easy to understand		✔
Active warnings alert user only when some action is necessary		✔
Active warnings frequently give false alarms	✔	
Product is covered with warning labels that seem inappropriate	✔	
Warning contains only essential information		✔
Source of warning information is credible		✔

Source: reprinted with permission from Cushman and Rosenberg, 1991.
[1]Low probability of behavioral change.
[2]High probability of behavioral change.

Labels and signs for passive warnings must be carefully designed by following the most recent government laws and regulations, national and international standards, and the best applicable human engineering information. Warning labels and placards may contain text, graphics, and pictures—often graphics with redundant text. Graphics, particularly pictures and pictograms, can be used by persons with different cultural and language backgrounds, if these depictions are selected carefully. (For recent overviews, see, e.g., Cushman and Rosenberg, 1991; Lehto and Clark, 1991; Lefito and Miller, 1986. Miller and Lefito, 1986; Ryan, 1991.) One must remember, however, that the design of a *safe* product is much preferable to applying warnings to an inferior product. Furthermore, users of different ages and experiences, and users of different national and educational backgrounds, may have rather different perceptions of dangers and warnings (Chapanis, 1975; Tajima, Asao, Hill, and Leonard, 1991).

A5.5.3 Symbols

Symbols (icons) are simplified drawings of objects or abstract signs, meant to identify an object, warn of a hazard, or indicate an action. They are common in public spaces, automobiles, computer displays, and maintenance manuals. The Society of Automotive Engineers (SAE) and the International Standardization Organization (ISO), for example, have developed extensive sets of symbols and established guidelines for developing new ones. Some of the symbols for use in vehicles, construction machinery, cranes, and airport handling equipment are reproduced in Fig. 14.26. Note that both abstract symbols (e.g., *on*) and simplified pictorials may require learning or the viewer's familiarity with the object (hourglass for *elapsed hours*). If one must develop new symbols, the cultural and educational background of the viewer must be carefully considered; many symbols have ancient roots and many invoke unexpected and unwanted reactions (Frutiger, 1989).

The following guidelines have been adapted from those of ISO Technical Committee 145 (dated October 29. 1987):

- Symbols should be graphically simple, clear distinct, and logical to enhance recognition and reproduction.
- Symbols should incorporate basic symbol elements that can be used alone or combined as necessary into a logical symbolic language that, if not immediately obvious, is at least readily learned.
- Graphical clarity should prevail in disputes with logical consistency, because no symbol is recognizable, no matter how logical, if it cannot be distinguished from other symbols.
- A minimum of detail should be included; only details that enhance recognition should be allowed, even if other details are accurate renditions of the machine or equipment.   ❏

14.3.5 Summary

Although psychological rules of *how the human best perceives displayed information and operates controls* are generally missing, there are many well proven design recommendations for traditional controls and dial-type displays. These are presented, in much detail, in this chapter.

With the widespread use of the computer, new input devices are emerging, such as mouse, trackball, or touch screen. The classic typing keyboard has mushroomed: a redesign is urgently needed. Probably other key designs, possibly new input methods such as by voice, should replace the monstrous 100+ key arrangement that apparently overloads many users.

Display technology has also changed. Electronic displays, many of which caused eye strain and headaches, can be made in a quality similar to hard copy. Use of color, if properly done, makes visual information transition interesting and easy.

Warning signals, instructions on how to use equipment, and warnings to avoid misuse and danger are important, both ergonomically and juristically.

14.3.6 Challenges

- What are the implications for future control and display developments, if so few general rules exist that explain current usages?

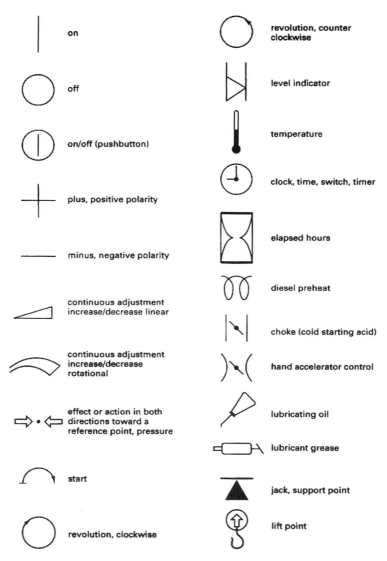

FIGURE 14.26 Sample of ISO symbols.

- Should one try to generate the same "population stereotypes" throughout the world?
- Are there better means to control the direction of an automobile than by the conventional steering wheel?
- Is it worthwhile to look for different keyboard and key arrangements if voice activation might become generally used in a decade, or two decades?
- Why is it difficult to compare the "usability" of CRT displays with that of information displayed on paper?

14.4 Environmental Issues

Environmental issues with regard to enclosure and product design include the following:

- Emission of any fluids, whether liquid or gas, during operation

- Safety operators and personnel in the immediate area (cannot be compromised by any discharge from the unit)
- Provision for safe storage or disposal of any waste materials generated by the product
- Protection of any moving equipment from incidental contact
- Reduction of any emitted EMI/RFI to an allowable level, generally described as a "not harmful" level
- For most devices, acceptance of any EMI/RFI received
- Ability to withstand anticipated shipping and storage conditions while in an unpowered state, without any damage
- Ability to withstand expected operating environments while powered, and not allow damage to the operating parts of the unit
- Ability to withstand "reasonable" abuse and continue to operate (While a personal computer would not be expected to operate properly after being dropped off a desk, a calculator would be expected to operate properly.)
- The need for all materials used in the construction of the enclosure to resistant to, or be protected from, all corrosive elements in the environment in which the device is expected to be used

As can be seen in the listed example of the PC and the calculator, many of these decisions and resulting designs are a result of marketing realities and the expectations of a reasonable customer.

The environmental issues also include the impact of the product on humans in the immediate area of the product. Klein and Marzi report on the design of a noise-dampening enclosure for a router used to automatically shape aircraft parts. The design of the enclosure was a concurrent effort of noise engineering, manufacturing, and R & D.

Any enclosure that will include power and control wiring will have to use a National Electrical Manufacturers' Association (NEMA) standard enclosure, or one conforming to NEMA standards.

EMC/EMI are discussed in anther chapter, but it is worth noting an example how design criteria can have an impact on several areas. It is common to include a fan in many electronic products, because it is an easy and inexpensive way keep the product from overheating. However, both the construction and associated electrical noise and the openings in the enclosure required by a fan are contrary to many EMC/EMI control recommendations. Newberger reports on a study done regarding the thermal performance of a RF enclosure that was totally enclosed. Other performance criteria require considering conflicting criteria and bringing forth a design that addresses all the criteria. In a similar way, Correira reports on the need for network equipment to meet specifications in areas such as enhanced security, small enclosure volume, and climate control.

14.5 Maintenance

Some products are designed to maximize efficiency of assembly. Some products are designed to allow routine maintenance to be performed efficiently. Design efforts should make these two goals compatible. Examples of routine maintenance items that most manufacturers have designed well include the oil dipstick on most newer cars. Many older cars hid the dipstick under the exhaust manifolds. Newer cars that fit the "user-friendly" category have the dipsticks in plain view, labeled in some fashion. This change in no way negatively affects the operation or the assembly of the engine.

The level of maintenance accessibility will depend on the expected use and the expected lifetime of the product, as well as whether routine maintenance is required. Home stereo systems are not expected to need routine maintenance during their expected lifetime, so they are not designed for it. Personal computers do need certain maintenance, e.g., the CMOS battery, but they are not designed to make this an easy task. The designers considered how often a CMOS battery needs to be replaced (5 to 10 years), then ranked assembly and operational criteria ahead of ease of maintenance. In most programmable logic controllers (PLCs) used in process control, the ability to replace the memory support battery and bring the unit back up to control operation is considered very important. Even though replacement times are again long, the ability to be able replace the battery rapidly means accessibility is much easier than in a PC.

14.6 Safety

Safety issues are required considerations in enclosure design. Some issues were mentioned in Section 14.4, "Environmental Issues." It should be noted that safety issues do not have concomitant user-performance issues. That is, a safety requirement may or may not consider whether the device will continue to operate properly. For example, any inspection agency will expect the frame and enclosure panels to have sufficient strength to both support the product adequately and to resist any reasonable abuse to which the device might be subjected. No one expects an electronic product to survive a collision with a tractor trailer, but "reasonable" abuse may not cause a risk of fire, electric shock, or personal injury as a result of partial collapse that may lead to, e.g., reduction in safe spacing between energized parts. In a similar manner, desktop electrical and electronic devices are routinely tested with a three-foot/one-meter drop onto a hard surface. The testing agency doesn't care if the device continues to operate properly, but the device may not catch on fire or otherwise present a hazard as a result of the drop.

Perhaps the best example of "it doesn't have to work" is in the testing of products that are expected to be used in explosive environments, be they gas, dust, etc. If a product is characterized by its manufacturer as "explosion proof," the testing agency will typically install a spark plug inside the device. They will place the device inside a chamber, which will then be filled with the most explosive mixture of the atmosphere (gas or dust + oxygen) in which the product is to be tested. The product cover will be closed, and the mixture inside the product will be ignited with the spark plug. It is expected that the device internals will be destroyed. The product will be considered explosion-proof if its enclosure does not release enough energy to ignite the atmosphere in the test chamber.

Some tests are necessary, even though they may not be considered "reasonable" abuse, if the resulting danger is severe. For example, even though hitting a cathode ray tube (CRT) face with a sharp object is not a reasonable thing to do, all safety and test agencies will subject a CRT to an impact test, because of the chance for extremely serious injury in the event of an implosion of a tube as a result of a blow.

The following paragraphs note common safety issues which must be considered, and which will be inspected for by inspection and regulatory agencies. The reader should note that these are examples of the most common tests and/or design requirements, They are no means all-inclusive. If the design team expects the product to be submitted to UL, CSA, or the like, the team must obtain copies of the relevant standards and requirements before beginning the overall design. These agencies even have safety standards that circuit boards and circuit components must meet. The general issues include:

- Accessibility of any electrically "live" part by a user. Each agency has its own definitions/measurements for a probe that is to be used to test any opening in the enclosure for accessibility.
- Containment of an internal fire or molten metal, plastic, and the like, to prevent any particles from falling on any surface on which the device rests.
- If maintenance requires the replenishment of any liquid, powder or other material, all parts that could present a safety hazard to the person performing the replenishment must be covered, insulated, or in some way protected.
- Any cover on an opening that is routinely opened, or is opened to work with a protective device such as a fuse or circuit breaker, should be hinged or similarly attached so that it cannot be inadvertently left off the product.
- Devices such as switches, fuseholders, etc. that are expected to be handled by the user may not rotate in their mounting panels.
- Any printed circuit board used in primary or hazardous secondary circuits must be acceptable for that application.

- Any exposed metal must be grounded with the third prong of the power cord, or the product must be constructed to meet double-insulation standards. Exceptions to this rule exist in certain consumer products.
- The "leakage" current available to users must be typically under 500 µA. For large, permanently connected products, the allowable leakage may be higher.
- Any moving parts must be protected/covered to reduce the risk injury.
- Any motor-operated product will have a number of safety, electrical overload, and other requirements to which is must adhere.

The enclosure and its associate controls are the user's primary interaction with the product. It therefore deserves as in-depth consideration as the functional parts of the product.

References

Collucci, D; Enclosures adapt to changing technologies." *Design News* (Boston), vol. 50, no. 7, April 10, 1995.

Correira, G; "Approaching packaging as a system." *Electronic Design,* vol. 44, no. 12, June 10 1996.

Corlett, EN; Clark, TS; *The Ergonomics of Workspaces and Machines: A Design Manual.* Taylor & Francis, London, 1995.

Helander, MG; Burri, GJ; "Cost effectiveness of ergonomics and quality improvements in electronics manufacturing." *International Journal of Industrial Ergonomics,* vol. 15, no. 2, Feb. 1995.

Klein, R; Marzi, E; "Noise enclosure for a router." *Proceedings, National Conference on Noise Control Engineering,* vol. 1. Institute of Noise Control Engineering, Poughkeepsie, NY, 1996.

Kroemer, KHE; Kroemer, HB; Kroemer-Elbert, KE; *Ergonomics: How to Design for Ease and Efficiency.* Prentice-Hall, Englewood Cliffs, NJ, 1994.

Laurie, NE, Kulatilaka, I; Arent, D; Andres, RO; Fisher, DL; "Usability evaluation of a commercial voice activated dialing system using a driving simulator." *Proceedings of the Human Factors and Ergonomics Society,* v.1, 1997. Human Factors and Ergonomics Society, Santa Monica, CA, 1997.

Logan, FJ; Augaitis, S; Miller, FH; Wehmeyer, K; "Living room culture–an anthropological study of television usage behaviors." *Proceedings of the Human Factors and Ergonomics Society,* vol. 1, 1995. Human Factors and Ergonomics Society, Santa Monica, CA, 1995.

Newberger, J; "Totally enclosed naturally cooled electronic enclosures." Thermal Phenomena in Electronic Systems, *Proceedings of the Intersociety Conference,* IEEE, 1996.

O'Shea, P; "Cooperation among designers yields quality shielded enclosures." *Evaluation Engineering,* vol. 32, no. 11, Nov. 11, 1993.

Yamada, T; "Human-machine interface for electronic equipment." *Proceedings of the TRON Project Symposium.* IEEE Computer Society Press, Los Alamitas, CA 1993.

15
Electronics Package Reliability and Failure Analysis: A Micromechanics-Based Approach

Peter M. Stipan
Bruce C. Beihoff
Rockwell Automation
Allen-Bradley

Michael C. Shaw
Design and Reliability Department
Rockwell Science Center

15.1 Introduction 15.1
15.2 Reliability 15.3
15.3 Micromechanisms of Failure in Electronic
      Packaging Materials... 15.9
15.4 Package Components... 15.17
15.5 Failure Analyses of Electronic Packages 15.25
15.6 Thermal Management... 15.27
15.7 Concluding Remarks... 15.32

15.1 Introduction

The goal of this chapter is to provide a unified framework for understanding the factors that govern the reliability of electronic packages. This is achieved by relating the reliability both to the fundamental mechanisms of failure that operate within the constituents as well as their primary driving forces. Indeed, extensive research over the past several decades has provided the electronics hardware design community with an unparalleled degree of understanding of these effects; increasingly, however, relentless miniaturization and increasing power density have required packaging approaches with new materials and architectures with unknown reliability. A predictive capability is therefore vital to minimize the costly and time-consuming procedures of prototyping and qualification testing. In addition, the commodity nature of modern electronic products dictates that they be designed for a specific operational life, and design for longer life represents unnecessary costs. For example, it is generally accepted that there is no need for reliability beyond, say, a few years for cellular telephones or pagers, whereas for military electronics, exceptionally high reliability may be required for well over two or more decades. Furthermore, the industry trend is toward ever shortening design cycles, such that the need for accurate accelerated testing is more acute than ever.

At least three primary electronic packaging technologies are relevant and are distinguished by the following characteristics (Figs. 15.1a through c).

a. High power packages (~1–150 kW) for power electronics (Fig. 15.1a) that may operate at voltages greater than 1000 V, currents greater than 1000 A, and moderate frequencies (<1 MHz).

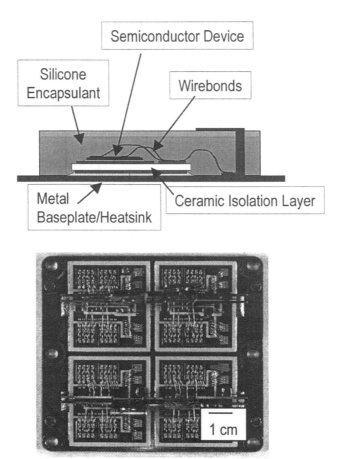

FIGURE 15.1a IGBT power package; high-power package from a motor drive application.

b. RF power packages for portable wireless communications (Fig. 15.1b) that operate at moderate power levels (~25 W), low voltages and currents, but very high frequencies (beyond 1 GHz).
c. Commodity packages (Fig. 15.1c) that operate at very low power (<<20 W) and moderate frequencies but must be extremely low in cost.

Each of these technologies is covered by the following discussion; where possible, specific reference will be made to the special requirements of each.

Several aspects to the design for reliability process must be understood at the outset. One is that failure typically occurs in a properly operated, well designed system either through a gradual *wearout* phenomenon or a random event. As will be detailed below, this allows the *mean*, or *average*, time to failure to be estimated will some degree of confidence, although the *exact* moment of failure is subject to an inherent degree of uncertainty. (Note that the uncertainty still can be quantified separately though the models described below.) Knowledge of the mechanisms of degradation thus is required for lifetime prediction. This is often obtained by accelerated testing to failure under severe conditions. By linking the conditions of the accelerated test with those encountered during service, it then becomes possible to predict the mean time to failure under service conditions. Finally, the *distribution* of lifetimes is vital to predicting the likelihood of premature failure; ideally, one would prefer a nearly deterministic life for optimum reliability prior to the anticipated failure time. These concepts will be elaborated below.

Therefore, this chapter provides the practicing engineer with a comprehensive summary and reference of the state of the art of electronics package reliability, and it is intended to be used within the context

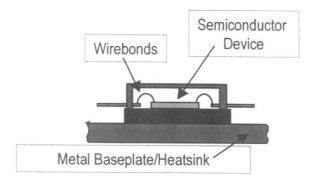

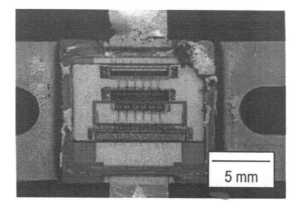

FIGURE 15.1b Power transistor package; RF package for a wireless communications application.

of actual day-to-day electronics design. Several example cases are presented to relate the concepts presented in the text to common problems routinely encountered. The emphasis is on the hardware external to the semiconductor devices themselves; semiconductor device reliability is beyond the scope of the present chapter.

The topics are organized in the following manner. First, basic reliability concepts are defined. Then, known mechanisms of failure related to electronic hardware are reviewed. Third, these concepts are unified through a discussion of the reliability of typical electronic package constituents, such as solder joints and wirebonds. Next, essential failure analysis steps in the investigation of failed components are reviewed. Finally, the thermal management of electronics packages is briefly described specifically within the context of electronic package reliability.

15.2 Reliability

Reliability refers to an estimate of the ability of a system to function properly over a specified time interval. The estimate can be one of total confidence (100%), such as the reliability that an on-board passenger assigns to the aircraft. Alternatively, the reliability can be less than perfect, i.e., <100%, with familiar examples including human life expectancies, automobile failure rates, and the reliability of electronic packages. For these cases, there are two distinctly different approaches to estimating the reliability. The first is to collect and analyze actual lifetime data and, from the analysis, establish a failure/time relationship that assumes that the characteristics of the failure rates will remain more-or-less the same in the future. For example, in some cases, during the qualification of new products, a series of life tests under normal operating conditions are performed and analyzed to assess the reliability *under these conditions*. In this approach, it is not required to determine the root cause of the failure; instead, the sum of all failures is

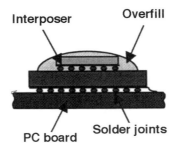

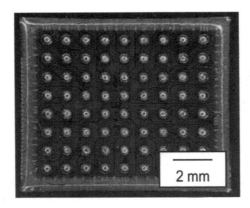

FIGURE 15.1c Area array package, for low-cost portable consumer electronics.

utilized to assess the overall failure rate. If one were to follow this approach in determining the reliability of a specific automobile design, then historical failure data would be collected and analyzed with no distinction made between engine, water pump, or chassis failures; they would all be lumped into an overall rate of failure.

An alternate approach is to study the *mechanism*, or cause, of failure—particularly the rate at which damage accumulates prior to failure. The point at which failure is reached can then be ascertained exactly, provided one has knowledge of the initial state of damage, the rate of damage accumulation, and the criterion for when the level of damage has reached a terminal state. For example, in the case of the automobile, it is well known that the seals on the valves within the engine slowly degrade with time and service. Provided there were no intermediate failures, then the reliability of the automobile would be determined by the rate of seal degradation in conjunction with the level of seal damage that constitutes failure, as well as the level of operation of the automobile. Both approaches are described in the following sections.

15.2.1 Basic Reliability Concepts

The reliability of a system can be expressed as a number between 0 and 1, or between 0 and 100%, where the system can be considered either an individual or an assembly of entities. For example, if there is a fifty-fifty chance of survival in a given case, then we might say that the reliability is 0.5 or 50%. In the following, all reliabilities will be expressed as numbers between 0 and 1. Also, reliability generally decreases with operation or time such that it refers to the probability that a system, or some fraction of an assembly, that will function properly at some point in the future. In view of this definition, we can define the time-dependent reliability, $R(t)$ as

$$R(t) = \frac{n_{\text{survivors}}(t)}{n_{\text{initial}}} \tag{15.1}$$

where $0 < R(t) < 1$, and n_{initial} represents the total number of specimens within the assembly. Although the reliability may depend on the effects of temperature, stress, etc., the following discussion will focus on the effects of time. *The essential challenge is to develop a theoretical understanding of the physical basis of the factors that influence the parameters governing $R(t)$.*

The reliability can be thus determined from experimental data representing the distribution of failures as a function of time. This is accomplished by ranking sequentially the experimentally determined failures and then assigning each failure a *reliability*, R, calculated as

$$R = \frac{i}{n_{\text{total}} + 1} \tag{15.2}$$

where i is the ranking of the individual failure, and n_{total} is the number of cases. Equation (15.2) is recognized as a variant of Eq. (15.1). These experimentally determined reliabilities are then plotted as a function of time. The resultant distributions may then be fitted according to any of the reliability models described later (Table 15.1).

TABLE 15.1 Characteristics of Typical Reliability Models[1]

Hazard Function	$h(t)$	$f(t)$	$R(t)$	Fitting Parameters
Constant	λ	$\lambda e^{\lambda t}$	$\lambda e^{\lambda t}$	λ
Linearly Increasing	λt	$(\lambda t)e^{-\frac{\lambda t^2}{2}}$	$e^{-\frac{\lambda t^2}{2}}$	λ
Weibull	$\frac{\gamma}{\theta} t^{(\gamma-1)}$	$\frac{\gamma}{\theta} t^{(\gamma-1)} \left(e^{-\frac{t^\gamma}{\theta}} \right)$	$e^{-\frac{t^\lambda}{\theta}}$	γ, θ
Exponential	$(b)e^{-\alpha t}$	$(b)e^{-\alpha t} e^{-\frac{b}{\alpha}(e^{\alpha t}-1)}$	$e^{-\frac{b}{\alpha}(e^{\alpha t}-1)}$	α, b
Arrhenius	$(A)e^{-\frac{E_a}{kT}}$	$(A)e^{-\frac{E_a}{kT}} e^{\frac{AE_a}{k}\left(e^{-\frac{E_a}{kT}}-1\right)}$	$e^{\frac{AE_a}{k}\left(e^{-\frac{E_a}{kT}}-1\right)}$	E_a, A

Analysis of the data also provides insight into several other key parameters, including the *probability density function*, $f(t)$, which is calculated as

$$f(t) = \frac{dR(t)}{dt} \tag{15.3}$$

as well as the *cumulative probability of failure*, $F(t) = 1 - R(t)$, or

$$F(t) = \int_0^t f(z)dz \tag{15.4}$$

Finally, the remaining key quantity is the hazard function, $h(t)$, which represents the *instantaneous rate of failures*:

$$h(t) = \frac{f(t)}{R(t)} \qquad (15.5)$$

The hazard rate is often construed as the *bathtub curve* within the electronics hardware industry. This is derived from the observation that the hazard rate can be high initially, as defective parts fail in early times (infant mortality), then level off to a lower, more or less constant failure rate as parts fail randomly, and finally increase in later times until parts wear out. An interesting distinction thus can be made, namely, that during the period of a constant hazard rate, failures are dominated by mechanisms that are intrinsically *random*, such as integrated circuit damage by cosmic ray interaction. However, during the wearout period, failures are dominated by mechanisms that may be well defined by physics-based constitutive behavior. The significance of this distinction will be discussed later in more detail.

These relationships are schematically illustrated in Fig. 15.2. The essential features are as follows:

1. $R(t)$ begins at 1 and monotonically approaches 0 as $t \to \infty$. That is, the reliability initially is 100% but decreases with time until, eventually, failure is assured.
2. $F(t)$ begins at 0 and monotonically approaches 1 as $t \to \infty$. That is, the number of failed units, or the probability of failure, initially is zero but increases with time until, as above, failure is assured.
3. $f(t)$ begins at 0, increases with time until a maximum is reached, and then decreases to 0 as $t \to \infty$. That is, the instantaneous rate of failure initially is zero, the rate of failures, or the likelihood of failure increases with time until a maximum is reached, and it then decreases to zero with increasing time.

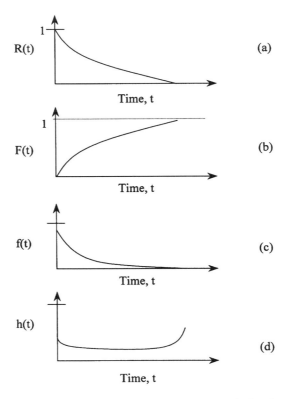

FIGURE 15.2 Schematic examples of the principal reliability functions described in the text: (a) reliability, $R(t)$; (b) cumulative failure probability, $F(t)$; (c) probability density, $f(t)$; and (d) hazard function, $h(t)$.

4. $h(t)$ initially is high, subsequently decreases to a period where the hazard rate is more or less constant, and then rises rapidly with time as the rate of failure increases.

Several key conclusions should be noted. First, the point at which f(t) reaches a maximum is of interest, since this defines the point at which the highest numbers of failures will occur. A second quantity relates to the mean time to failure, MTTF, or the time at which $R(t) = 0.5$. Note the distinction between the MTTF and the mean time between failures, MTBF, which refers to the MTTF plus the average amount of time needed to make a repair. A third aspect relates to the *scatter* of lifetimes around the MTTF, since this will dictate the relative confidence level assigned to the system at times other than the median life. For systems with very high levels of scatter, *the reliability is much lower substantially earlier* than for systems with very small scatter around the median life. Therefore, it is generally desirable to minimize the scatter of lifetimes, since in this way, the lifetime becomes more deterministic. I.e., provided the design engineer can then control the parameters that govern the median life, then the lifetime of the system can be specified with minimal premature failure.

Several analytical models for $R(t)$, $f(t)$ and $h(t)$ are appropriate for different applications. These are summarized in Table 15.1. Generally, the models may be parametrically assessed for the quality of fit in each instance. In all cases, however, at least two quantities are of primary concern: the *mean time to failure* and the *distribution of lifetimes* about the mean value.

In Table 15.1, E_a, A, α, b, λ, γ, θ, are fitting parameters that must be deduced from analysis of the data. Once these parameters are known, then the appropriate model can be used to predict the reliability under a range of operating conditions.

Example 15.1

An engineer has been assigned the task of assessing the reliability of ceramic baseplates from different commercial vendors. The individual has been given two sets of ten baseplates from each vendor (set 1 and set 2) and has been asked to develop a reliability model for the baseplates, specifically with respect to the effects of *mechanical stress*, σ. Consequently, the engineer performs a standard series of mechanical experiments with each of the ten baseplates to determine their strength, then ranks the strength of each of the baseplates in decreasing order in accordance with Eq. (15.2), and calculates the cumulative failure probability, $F(\sigma)$, of the baseplates in accordance with Eq. (15.4). These results are presented in Table 15.2. Note that, in this case, the driving force for failure was stress, σ, instead of time, t. This illustrates a key point, namely, that the failure models can be applied to a wide range of driving forces for failure: time, temperature, stress, etc.

TABLE 15.2 Plotting the Resultant Reliability Data As a Function of Failure Stress

	Strength, ksi	
$F(t)$	Set 1	Set 2
0.05	39.9	41.02
0.15	40.4	41.03
0.25	40.6	41.04
0.35	40.87	41.05
0.45	41.33	41.2
0.55	41.5	41.25
0.65	41.9	41.35
0.75	42.2	41.4
0.85	42.5	41.5
0.95	42.8	41.55

Plotting the resultant reliability data as a function of failure stress yields the results indicated in Fig. 15.3. Clearly, baseplate set 2 exhibits a much smaller scatter around the mean life than data set 1, despite the observation that the mean strength is actually slightly higher for data set 1. *In critical applications,*

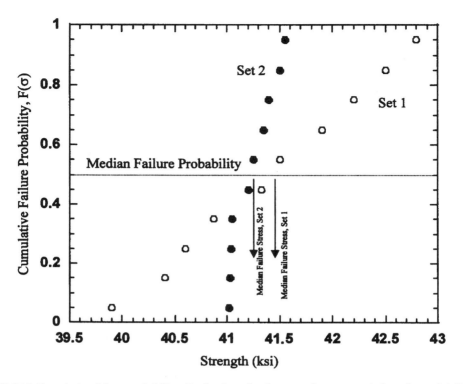

FIGURE 15.3 Cumulative failure probability distributions for the example on ceramic baseplate reliability. Two different distributions are shown for the two cases described in the text.

therefore, it may be desirable to select baseplate set 2, since the reliability is much greater at moderate stress levels than for baseplate set 1.

15.2.2 Accelerated Testing

The essential challenges in analyzing the data obtained through accelerated testing are twofold: (1) identification of the appropriate model to fit to the accelerated life data and (2) estimation of the parameters within the model. An excellent review of this subject may be found in Ref. 1. In the present case, it is sufficient to recognize that once the model has been established and fitted, the acceleration factor, A_T, may be determined as the ratio between the life predicted under normal operating conditions and the life measured under the accelerated condition.

For example, suppose that an Arrhenius model has been found to accurately represent the failure rate of a specific electronic component. In this model, we know that the lifetime, $L(t)$ is given as the inverse of the hazard rate, $h(t)$ from Table 15.1.

$$L(t) = A\exp\left(+\frac{E_a}{kT}\right) \tag{15.6}$$

where A = a fitting parameter
E_a = activation energy of the dominant reaction
k = Boltzmann's constant (8.6×10^{-5} eV/K)
T = temperature in Kelvins

If we make the simplifying assumption that both A and the activation energy remain constant with temperature, then it is straightforward to define the acceleration factor, A_T, as

$$A_T = \exp\left[\frac{E_a}{K}\left(\frac{1}{T_o} - \frac{1}{T_s}\right)\right] \qquad (15.7)$$

where T_s = temperature of the accelerated test
T_o = temperature during service

Example 15.2

An engineer performs an accelerated aging test of FR4 circuit boards at a temperature of 85° C with a constant environment and finds that the mean time to failure was 3200 hr. It is necessary to predict both the acceleration factor, A_T, and the lifetime, L_o, under a service environment of 45° C. Separately, the engineer fitted the Arrhenius equation to accelerated life data under two temperatures and determined an activation energy of 0.19 eV for the dominant failure mode. This was accomplished by determining failure times of 2702, 2325, and 1983 hr at test temperatures of 95°, 105°, and 115° C (368, 378, and 388 K, respectively). In turn, these correspond to hazard rates of 3.7×10^{-4}, 4.3×10^{-4}, and 5.0×10^{-4} hr^{-1}. Then, by fitting Eq. (15.6) to these data, plotted on semilog paper vs. $1/T$, the activation energy (0.91 eV) was ascertained. Then:

$$L_o = 3200 \exp\left[\frac{0.25}{8.6 \times 10^{-5}}\left(\frac{1}{318} - \frac{1}{473}\right)\right] = 64000 \text{ hr} \qquad (15.8)$$

and

$$A_T = \exp\left[\frac{0.25}{8.6 \times 10^{-5}}\left(\frac{1}{318} - \frac{1}{473}\right)\right] \qquad (15.9)$$

This example illustrates the basic approach to determining the acceleration factor and service life from accelerated life data. ❑

15.3 Micromechanisms of Failure in Electronic Packaging Materials

The preceding discussion provided the framework for reliability model assessment and accelerated life investigations. The next step is to develop a deeper understanding of the parameters that govern the characteristics of the life models themselves by examining the physical models for failure and degradation in materials.

The failure of an electronic system due to a hardware problem can usually be categorized with known failure mechanisms. This is the outcome of decades of basic and applied research within the electronics, materials science, mechanics, and related fields. Provided the system is operated within specified boundaries, its reliability is governed by one of a number of mechanisms that lead to the gradual accumulation of damage within the system. For example, metal fatigue mechanisms are known to operate within the packages, where the life is dictated by the environmental and operational profile. Similarly, moisture-induced chemical reactions have been extensively documented as factors that limit the operational reliability of electronics. Alternatively, if the system is suddenly exposed to conditions beyond the specified limits of operation, such as excessive heat or mechanical stress, the reliability of the system may be related to the degree of overstress. Therefore, in both cases, the reliability of the system may be thought of as a quantity that depends *both* on the nature of the environment as well as the properties of the constituent materials. *The goal of this section is to introduce to the reader the most prevalent failure mechanisms within electronics systems.*

The physics-of-failure concepts pioneered by Pecht et al.[2] have provided the contemporary electronics community with perhaps the most unified approach in dealing with electronics reliability. The key philosophy is to develop a fundamental understanding of the underlying mechanisms that govern the degradation of electronics hardware (Fig. 15.4). Then, by assigning rate constants or maximum limits to these various failure mechanisms, insight is obtained with regard to the anticipated lifetime of the electronics. *The essential challenge, therefore, is to identify the primary physical mechanism for failure, and to determine the relevant rate constants accurately.*

Example 15.3

As an illustrative example, imagine that we have a package containing metal interconnections subject to a metal fatigue damage mechanism. Additionally, preexisting defects with a size of 5 microns are known to exist in the as-fabricated condition. Microscopic cracks are known to initiate and grow from these defects immediately upon operation. Specifically, at a critical package location, a crack growth rate of 0.075 microns per operational cycle is known to occur for defects of this size. Therefore, by relating this crack growth rate to the 275-micron long zone within the interconnection required to maintain reliable operation (and assuming a constant rate of damage), the number of cycles to failure is readily ascertained as

$$L_o = \frac{275 \times 10^{-6} \text{ m} - 5 \times 10^{-6} \text{ m}}{0.075 \times 10^{-6} \text{ m/cycle}} = 3650 \text{ cycles} \quad (15.10)$$

This example illustrates the necessity for an understanding of the degradation mechanisms, the degradation rates, and the critical level defining failure. ❑

In the following section, the key micromechanisms for degradation and failure in electronic packaging materials are discussed. This will lay the foundation for the subsequent section on assessing the reliability of specific package constituents, where application-specific degradation rates and critical levels for failure are shown.

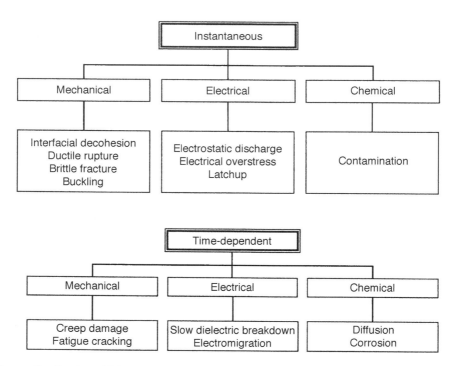

FIGURE 15.4 Classification of failure mechanisms according to their driving force and time dependence.[2]

15.3.1 Instantaneous Overload Fracture

Ductile Overload

In ductile materials, such as lead, copper, aluminum and solder, the mechanics-based design criterion is given as a critical *stress*, σ_f.

$$\sigma_f = \frac{F}{A} \tag{15.11}$$

where F = applied *force*
       A = the cross sectional, i.e., load bearing, *area*

Stress is a tensorial quantity that relates to the three-dimensional nature of most architectures; in the present case, consideration of the uniaxial tensile stress, σ, is sufficient.

The critical value for σ_f can be defined in terms of one of two ways.

1. The *ultimate stress-carrying capability of the material*, σ_{UTS}, i.e., the stress beyond which the material fractures
2. The *yield stress* σ_y, i.e., the stress beyond which a permanent deformation is imparted to the material

The yield and fracture stress of relevant materials is provided in Table 15.3.[3-9] It should be noted that the yield stress for most metals is much lower than the fracture stress; the phenomenon of increased stress-carrying capability beyond the yield stress is defined as *work hardening*. However, the yield stress typically is the more relevant design target, for several reasons. First, repeated cyclic loading at levels of stress well below the yield stress delays metal fatigue (described below) until a substantially higher numbers of cycles than if plastic yielding occurs. Also, plastic yielding typically results in deleterious residual stress within the electronic devices. *Therefore, an important aspect of any electronics design is to ensure that the maximum level of stress within the metal interconnections, and the like, remains below the yield stress.* This is often accomplished by theoretical or numerical stress analysis.

A second point is that the yield stress of metals generally decreases with increasing temperature. The relative decrease in yield stress is conveniently interpreted through the homologous temperature, T_H,

$$T_H = \frac{T}{T_m} \tag{15.12}$$

where T = instantaneous temperature
       T_m = liquidus temperature of the metal or alloy

The liquidus temperatures, T_m, of many metals of interest is provided in Table 15.3. Typically, the yield stress has reached ~10% of its room temperature level by the time the homologous temperature has reached ~0.4. The temperature dependence of yield stress for several metals is provided in Ref. 9. The homologous temperature also governs the rate of creep deformation of the metals used to form the packaging interconnections below. *In summary, the package must be designed to maintain the mechanical stress within the metallic constituents below the yield or ultimate failure strength, σ_y, or σ_{UTS}.*

Brittle Overload

A fundamentally different situation applies for brittle materials, such as glass or ceramics, where the fracture criterion is not solely governed by the applied stress. *Instead, an additional parameter is required, namely the physical dimensions of the flaw, or defect, that causes fracture to occur.* The requirement of knowledge of this length scale over which the stress is applied represents a fundamental distinction compared to ductile materials where only the stress is needed, and has its origin in the energetics of brittle fracture.[6]

Specifically, for brittle fracture, the fracture strength, σ_f, is given as

TABLE 15.3 Material Properties[3-9]

Material	Thermal Conductivity, k W/mK	Specific Heat, C_p J/gk	Density, p g/cm3	Coefficient of Thermal Expansion, CTE (ppm3)	Melting temp., K °K	Melting temp., C °C	Young's Modulus, E GPa	Yield Stress, σ_Y MPa	Ultimate Tensile Stress, σ_{uts} MPa	Fracture Toughness, K_{Ic} MPa·m$^{0.5}$	Fracture Energy, G_{Ic} J/m2
Silicon	140	0.714	2.32	3	1683	1410	107	—	—	0.7	3
Silicon carbide	200	0.69	3.2	4	3110	2837	450	—	—	3-3.5	23
Diamond–Type IIA	994	0.527	3.51	1.5	4000	3727	1000	—	—	—	—
Aluminum Oxide	40	0.82	3.9	8	2323	2050	380	—	—	3	20
Beryllium Oxide	240	1.1	3	8	2700	2427	380	—	—	—	—
Aluminum Nitride	180	—	—	6	—	—	200	—	—	—	—
Silica glass	5–10	—	2.5	0.7	1100	827	70	—	—	0.5	—
Fused quartz	4	—	2.2	0.5	1935	1665	94	—	—	0.5	—
Nickel	91	0.44	8.9	13	1726	1453	214	70	400	—	—
Copper	400	0.391	8.92	16.7	1356	1083	124	60	400	—	—
Aluminum	237	0.903	2.7	23	933	660	70–80	40	200	—	—
Lead	28–35	0.17	11	29	600	327	14	11	14	—	—
63Sn–37Pb solder	50	—	—	25	183	–90	15	~10	~20	—	—
Gold	—	—	19.3	14.2	1336	1063	82	40	220	—	—
Molybdenum	140	—	—	5	—	—	—	—	—	—	—
Conductive Epoxy	~0.5–1	—	1.1–1.4	30	340–380	70–110	3	30–100	30–120	0.3–0.5	0.1–0.3
Polyimide	0.35	1.6	1.4	50	580–630	310–360	3–5	—	50°90	—	—
RTV	0.31	—	0.9	260	—	—	0.1	—	30	—	—
Teflon®	0.25	1	2.3	12	510–530	240–260	0.37–0.7	—	~50	—	—

$$\sigma_f = Y\left(\frac{K_{Ic}}{\sqrt{a}}\right) \quad (15.13)$$

where a = relevant defect size (Fig. 15.5)
Y = a geometry dependent parameter,
K_{Ic} = the intrinsic material or interfacial property designated as *fracture toughness*

The values for Y for embedded, surface, and corner cracks configurations (Fig. 15.5) are 1.56, 1.4, and 1.3, respectively, and typical toughness values are presented in Table 15.3 for common materials. Thus, for a given combination of applied stress and defect size, catastrophic failure ensues. Furthermore, as the sizes of the defects increase, the strength decreases. This is consistent with experience—e.g., why we scribe glass to reduce the strength sufficiently to crack it by hand along the scribed line. Note that, in contrast to the case for ductile materials described above, small surface scratches can compromise the strength of brittle materials.

Example 15.4

The same engineer who studied the distribution of strengths of the ceramic baseplates is interested in calculating the surface flaw size, a, that dictated the median strength. Recalling that the median strength of set 2 was 284 MPa, (41.2 ksi) and that the fracture toughness of aluminum oxide is 3.5 MPa $\sqrt{m}$, then from Eq. (15.13),

$$a = \left(\frac{YK_c}{\sigma_f}\right)^2 = \left(1.4\frac{3.5 \times 10^6 Nm^{\frac{3}{2}}}{284 \times 10^6 Nm^{2}}\right) = 300 \text{ microns} \quad (15.14)$$

❑

These concepts also apply directly to interfacial fracture or delamination, where the interfacial fracture resistance is often a fraction of that of either of the materials that are being joined. Reliability requires that the size of the interfacial defects be decreased below the critical level for the design stress being targeted. For example, interfacial delamination or peeling often occurs (undesirably) at levels of stress

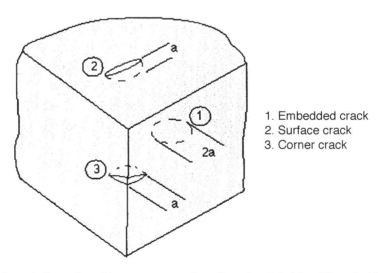

FIGURE 15.5 Schematic illustration of three common crack configurations in brittle solids: embedded cracks, surface cracks, and edge cracks. The stress intensity prefactor, Y, for each geometry is 1.3, 1.4, and 1.56, respectively [a represents the radius of the presumed circular defect, per Eq. 15.13)].

or temperature well below levels needed to induce fracture or yielding of the materials on either side of the interface. This can result *either* from the formulation of an adhesive joint with insufficient fracture toughness, K_c, or the presence of a defect population with an unacceptably large dimension, a. Furthermore, this phenomenon is qualitatively similar to the notion of an adhesive joint, such as a solder layer, that may exhibit a crack-growth resistance to fatigue crack growth that is a small fraction of the materials on either side of the joint. This results in premature failure of the solder joint under cyclic fatigue conditions, as described in more detail later.

In other words, Eq. (15.13) reveals that the critical stress required to propagate a crack along an interface is lessened by the presence of a large defect such as a void, or an unbonded region owing to a local patch of contamination. Alternatively, the same decrease in joint strength may result from an adhesive with a low intrinsic fracture toughness. Indeed, one of the primary goals of interfacial engineering might be interpreted as being *both* to raise the interfacial fracture resistance *and* to lower the interfacial defect population below those of the materials being joined.

A number of *interfacial fracture energy* experimental test specimens recently have been devised and calibrated.[10] Note especially that these measure a different quantity from engineering tests for interfacial *strength* (such as the *adhesive-tape* test) that fail to capture the length scale over which to apply the stress. In each case, it is important to introduce a defect along the interfacial region of interest under well controlled conditions. From the analysis of the results, the interfacial fracture energy, G_{Ic}, is obtained, which can be interpreted as a fracture toughness-like quantity through the relationship

$$G_{Ic} = \frac{K_{Ic}^2(1-\nu^2)}{E} \tag{15.15}$$

where ν = Poisson's ratio of the material on either side of the interface

Typically, $\nu \sim 0.25$. Specifically, G_{Ic} is related to the quantity of energy required to create fractures.

The physical origin of these differences in brittle and ductile fracture lies in the significantly different microscopic deformation mechanisms, which operate near the tips of cracks. In turn, these arise from the differences in atomic bonding between metals and ceramics. In summary terms, in metals, as cracks are loaded by tensile stress, plastic deformation occurs near their tips through dislocation emission and motion, leading to a blunting of the cracks. In brittle materials, however, the high strength or highly directional ionic/covalent bonds preclude dislocation generation, with the result that cracks remain essentially atomically sharp. In this case, the stresses become concentrated near the crack tip, leading to a locally high stress. *Because the degree of stress concentration depends on the size of the crack, we have the condition described in Eq. (15.13).*

Therefore, during the design process, it is vital first to ascertain σ_f and K_{Ic} of the materials and interfaces of interest. This is accomplished through test procedures that are now well established and can be found under, for example, a range of ASTM standard test procedures. Once these quantities are known, the next stage is to evaluate the distribution of stress, σ, within the structure. This can be accomplished either through analytical or detailed numerical analysis. Finally, the maximum tolerable defect is ascertained through Eq. (15.14). *As a consequence, the design/process specification/inspection procedure must reflect this maximum defect size.* If this defect size is impossibly small, then steps must be taken to either increase K_{Ic} or decrease the magnitudes of the stress to acceptable levels. Once this design cycle has been completed, the design is safe against simple overload fracture, provided the specified bounds of operation are maintained.

15.3.2 Time-Dependent, Progressive Failure

Several types of time-dependent degradation mechanisms are known to occur: fatigue, creep, and corrosion. These will now be reviewed briefly.

Fatigue Crack Growth

Cyclic fatigue crack growth occurs when materials are subjected to repeated cycles of stress (Fig. 15.6). Such crack growth can occur at loads that result in stresses that are well below the yield or failure stresses of the material. Thus, fatigue cracking is a particularly insidious form of damage and failure. There are three main types of fatigue, depending on whether the material contains initial defects of appreciable dimension, and whether the range of imposed stress exceeds the yield stress.

1. Fatigue of uncracked components; high cycle fatigue.

    The basic equation that has been found empirically to relate the lifetime, L_o, of uncracked materials *in the absence of stresses that cause yielding* is Basquin's Law.[4]

$$L_o = B(\Delta\sigma)^{-\theta_b} \quad (15.16)$$

    where $\Delta\sigma$ is the range of applied stress (Fig. 15.6), and θ_b and B are both material-dependent constants. For most materials, $8 < \theta_b < 15$. Thus, once θ_b and B are known for a given material, and the range of stress has been determined from analysis, Eq. (15.16) provides a simple, approximate method for predicting the number of cycles to failure.

2. Fatigue of uncracked components: low cycle fatigue.

    If the range of *stress exceeds the yields stress of the metal*, Eq. (15.16) fails to adequately represent the lifetime; instead, the Coffin-Manson Law applies.[4] In this case, the lifetime is related to the range of plastic strain, $\Delta\varepsilon_{pl}$, imposed on the structure.

$$L_o = C(\Delta\varepsilon_{pl})^{-\theta_c} \quad (15.17)$$

    where θ_c and C are both material-dependent constants. For most materials, $\theta_b \sim 2$.

3. Fatigue of cracked structures → growth controlled fracture.

    The previous two instances related the life of the structure to the point of fatigue crack *initiation*. In these cases, since the gestation period for crack initiation is so much longer than for subsequent growth and failure, the approximation is valid. However, when crack initiation sites already exist, then the fatigue life is dominated by the *growth* of cracks. Specifically, the crack growth rate is governed by the Paris Law relationship.[4]

$$\frac{da}{dN} = P(\Delta K)^{\theta_p} \quad (15.18)$$

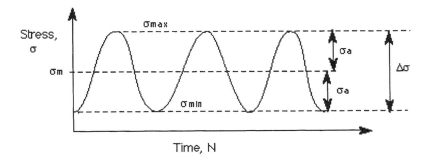

FIGURE 15.6 Schematic of the nomenclature used to describe the nature of the time-dependence of repeated, cyclic stress. σm is the mean stress, σmin and σmax are the minimum and maximum stress levels, σa is the stress amplitude, and Δs is the range of applied stress. For example, Δs is used with Eq. (15.16) to predict the lifetime of wirebonds to high cycle fatigue.

where ΔK = range of applied stress intensity factor [Eq. (15.13)]
θ_p and P = material-dependent constants

Consequently, once the initial defect size, a_i, and final acceptable crack size, a_f, are known, then the lifetime, L_o, is given as

$$L_o = \frac{1}{P} \int_{a_i}^{a_f} \frac{da}{(\Delta K)^{\theta_p}} \quad (15.19)$$

An interesting area of ongoing research that is derived from this relationship involves investigating the effects of the statistical distribution of defect sizes to the distribution of lifetimes.

Creep Deformation

Time-dependent deformation of materials occurs through creep deformation at elevated temperatures ($T_H > 0.3$–0.4). Although creep itself may lead eventually to failure through the development of internal voids and porosity, the more frequently observed effect of creep deformation is on the increase in plastic strain, ε_p, at elevated temperatures. From Eq. (15.17), it is found that an increase in plastic strain will result in a consequent decrease in cycles to failure. There are two major conditions encountered during service.

1. Constant stress—*strain, ε, increases* with time but *stress, σ, remains the same.*
2. Constant strain—*stress, σ, decreases* with time but *strain, ε, remains the same.*

An example of creep deformation that may occur under constant strain conditions within the context of electronics is the deformation of the edges of a semiconductor device that is joined to a substrate through a continuous solder joint. This is the case for power modules and other RF power devices used in cellular telephones. Experimental measurements of the stress within the center of the devices confirm that the stress remains constant, although the strain near the edges increases owing to creep relation.[13] An example of creep subject to constant strain conditions is the stress relaxation that occurs within the solder joints encapsulated within a ball grid array package; in this case, the global deformation is driven by the CTE mismatch of the ball grid array package and substrate. That is, the solder balls are effectively "frozen" in place upon encapsulation; subsequent stress relaxation occurs within the solder balls through mass diffusion. Determination of the appropriate condition therefore is important with respect to assessing the effects on the reliability of the system.

Creep deformation in materials occurs through three stages. The initial, or *primary creep* phase, and the final, or *tertiary creep* phases, typically happen relatively quickly. The intermediate stage, or *secondary creep* stage, is governed by a power law relationship as follows:

$$\dot{\varepsilon}_{ss} = L(\sigma)^{\theta_R} \quad (15.20)$$

where θ_R and L are both material-dependent constants. Typically, $3 < \theta_R < 8$. Understanding creep deformation behavior is the subject of ongoing research activity, particularly the effects on the surrounding structure on the effective deformation rates.

Corrosion and Oxidation

Chemical reactions occur between the packaging constituents themselves and the environment. In all cases, the degree of reaction is the net result of the thermodynamic driving force for reaction and the actual kinetics of the reaction, which may depend wholly on the specifics of the application.

Example 15.5

As an example, consider the driving force for the oxidation of pure aluminum metal: 1045 kJ/mol of O_2. This is well over twice the corresponding driving force for pure iron (508 kJ/mol of O_2), yet rust (iron oxide) is far more familiar than the corresponding oxide on the surfaces of aluminum. The primary reason for this is that the kinetics of the reaction, i.e., the *reaction rate,* for oxidation of aluminum is dramatically slower than for iron. This is a result of the microstructure of the aluminum oxide scale—it is very impervious to the passage of oxygen between the environment (air) and the underlying metal (aluminum). In contrast, iron oxides crack and spall, constantly exposing fresh metal for further oxidation. This is of crucial importance in assessing the reliability of aluminum/aluminum wirebonds, for example, as a function of the chemical environment. Provided that there are no chemical species such as chlorine, the aluminum oxide scale remains intact. Small levels of chlorine, coupled with moisture, however, can penetrate the aluminum oxide scale and lead to degradation of the wirebond reliability. Therefore, to relate the effect of the environment to the reliability of the electronic packaging constituents, *it is vital to first understand the relative roles of the thermodynamic driving forces and the reaction rate kinetics,* as addressed in the next section. ❑

15.4 Package Components

The intent of the present section is to relate the basic failure modes described in the previous section to the typical constituents of an electronics package.

15.4.1 Solder Joint Reliability

Solder joints in electronic packages are subject to progressive failure by metal fatigue during operation. This results from the dual phenomena of an inherently low yield stress, σ_f, coupled with typical operating temperatures (>100° Celsius, in some cases) that are well within the temperature regime where significant creep deformation may occur; $T_H > 0.6$ [Eq. (15.12)]. The driving forces for such creep deformation originate from differences between the coefficients of thermal expansion of the materials used on both sides of the soldered joints, as well as the solder material itself. This mismatch can be considerable and lead to large strains, especially near free surfaces and edges. As described above, imposition of stress or strain cycles under such conditions will lead to metal fatigue (e.g., Fig. 15.7). Assessment of the reliability is thus contingent upon knowledge of the stress and strain evolution within the solder materials.

Much work has been devoted to the micromechanics of solder fatigue; for more detailed discussions of this rich subject the reader is referred to Refs. 11 through 44. Although debate continues regarding the optimum approach for reliability prediction, for the present discussion, it is sufficient to note that reliable designs may achieved for relevant solder materials with minimal iteration through the application of the Coffin-Manson relationship defined by Eq. (15.17). For example, for eutectic lead-tin solder, 63Sn-37Pb, subject to very low-frequency mechanical stress cycles near room temperature (Lau, 1997), the lifetime, L_o, is

$$L_o = 3.013(\Delta\varepsilon_{pl})^{-0.924} \qquad (15.21)$$

The reader is cautioned, however, that in most cases, because the solder material is operated at varying temperatures and dwell times, this simple approach is of limited utility.

The most accurate predictions for deformation and stress can be achieved only through detailed numerical analysis with full account given to the temperature dependent and anisotropic material properties of the entire assembly. As a recent example, He et al.[12,13] demonstrated the effects of silicon anisotropy on the stress and strain fields within solder joints with high spatial resolution. Furthermore, it was shown that a critical die size exists beyond which the stress within the device is independent of die size. *One crucial ramification of this observation is that the reliability of the device may become independent of device size.*

To illustrate these points, measurements and analysis of stress within silicon devices attached to a thick copper substrate are shown in Fig. 15.8. Note that the thermal stress distribution can be divided into two regions; a transient region near the die edge, defined as the *shear-lag zone,* where the in-plane stress

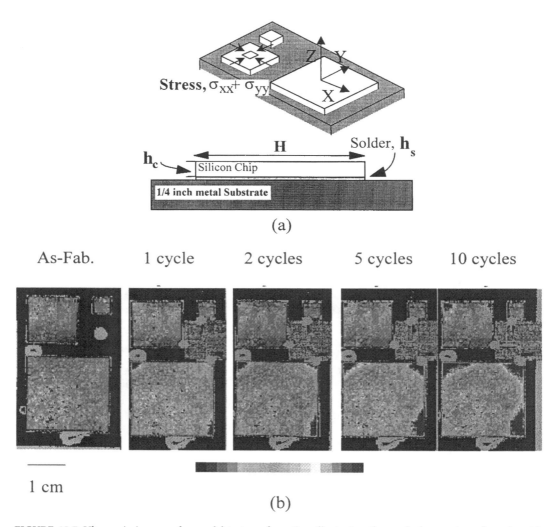

FIGURE 15.7 Ultrasonic images of a model test configuration illustrating the gradual extension of cracks with repeated thermal cycling. Three different silicon test articles are soldered to a single copper baseplate (a) and subject to thermal cycling.[12,13] Crack extension is visible as the red and orange colored regions extending inward from the corners of the devices (b). Note that no fatigue crack growth has occurred in the smaller devices.

increases with position from the edges; and a central region of constant, spatially invariant stress defined as the *plateau region*. The magnitude and distribution of these thermal stresses are governed by the CTE mismatch, the elastic and geometric properties of the device, substrates, and most importantly by the solder shear yield strength σ_y and creep rate [Eq. (15.20)]. The elastic stress distribution within the shear lag zone has been calculated by Suhir,[45] under the plane strain approximation where the dimension of the y-direction is assumed to be infinite. When the length and width of a device are comparable to each other, this approximation breaks down, and the modified stress distribution at the mid-plane of the device is given under the thin film limit, as[45]

$$\sigma_{xx}(x) + \sigma_{yy}(y) = \sigma_0 \left[2 - (1 + \nu_c) \frac{\cosh(kx)}{\cosh\left(k\frac{H}{2}\right)} \right] \quad (15.22)$$

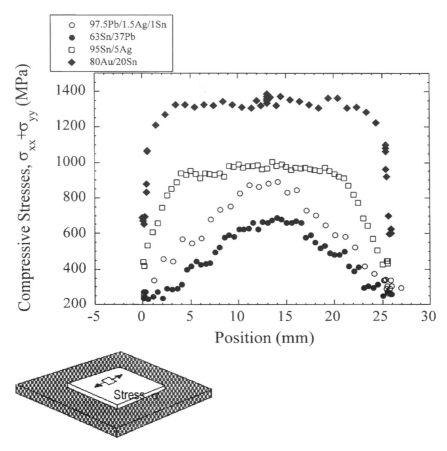

FIGURE 15.8 Ultrasonic images of a model test configuration illustrating the gradual extension of cracks with repeated thermal cycling. Three different silicon test articles are soldered to a single copper baseplate (a) and subject to thermal cycling.[12,13] Crack extension is visible as the red and orange colored regions extending inward from the corners of the devices (b). Note that no fatigue crack growth has occurred in the smaller devices.

and $k = [G_s(1 - v_c^2)/(E_c h_c h_s)]^{1/2}$. E_c and v_c are the Young's modulus and Poisson's ratio of the device, respectively. σ_0 is the maximum in-plane thermal stress that can be reached in the die. Equation (15.22) can be extended to general cases by simply replacing σ_0 and k by Suhir's definition.[45]

For most solder alloys, however, the *shear* yield strength, τ_y, is far less than the interfacial shear stress imposed by the rapid stress buildup near the edge. In these cases, the in-plane stress within the silicon increases linearly with the distance from the edge before reaching its maximum value as given by Ref. 12,

$$\sigma_{xx}(x) = \begin{cases} \dfrac{\tau_y(H/2 - |x|)}{h_c} & \text{for } (H/2 - |x|) \le L \\ \sigma_0 & \text{for } (H/2 - |x|) > L \end{cases} \quad (15.23)$$

where the plastic shear lag zone length $L = \sigma_0 h_c / \tau_y$. Results in Fig. 15.8 clearly demonstrate the plastic response of three solder alloys with the values of τ_y determined by fitting the data using Eq. (15.23).

In addition to the shear lag zone size, the properties of the solder also determine the value of σ_0. For example, it is well known that the maximum thermal stress is proportional to $\Delta\alpha\Delta T$, where $\Delta\alpha$ is the difference in CTE between the substrate and the device, and ΔT is the temperature range over which the stresses develop. *Note that ΔT does not correspond to the difference between the processing temperature and room temperature, T_{room}; instead, ΔT is*

$$\Delta T = T^* - T_{room} \quad (15.24)$$

where T^* *is designated as the "stress onset temperature,"* below which the creep of solder alloy is so slow in comparison to cooling rate that there is no notable stress relaxation during subsequent cooling.[13] Because of the complex nature of solder creep processes and their dependence on the actual processing conditions, the value of T^* can be determined only through experiments, as described by He et al.[12,13]

Finally, the degree of uniformity of the solder joint is known to influence the lifetime of the package. In particular, voids within the solder are known to lead to local heating, and consequent increase in creep relaxation of the solder, as well as potential thermal runaway. An example of voids within a solder joint as observed through transmission X-ray microscopy is shown in Fig. 15.9, where the voids within the solder joints are visible as the light, irregularly shaped regions. Clearly, extreme caution must be exercised during fabrication to avoid the development of such voids.

15.4.2 Wirebond Reliability

Wirebonding is a mature, well established method of forming electrical interconnections between semiconductor devices and the external environment and is ubiquitous within the electronics industry (e.g., Fig. 15.10). The success of wirebonding as compared to other techniques such as tape automated bonding (TAB), area array solder joining, or conductive epoxies lies in the significant flexibility of interconnection

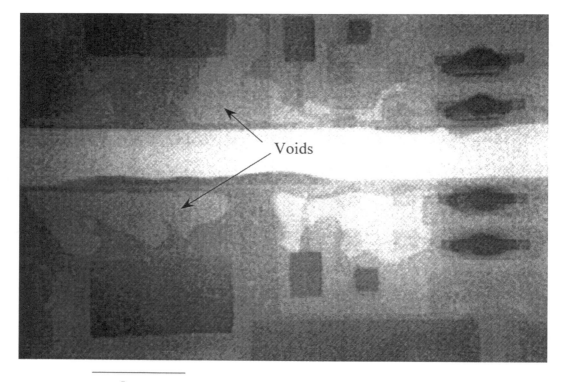

FIGURE 15.9 Transmission X-ray micrograph of an electronic package exhibiting voids within the solder joints.

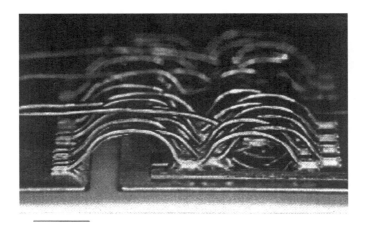

FIGURE 15.10 Optical micrograph of wirebonds used within a high-power package.

layout, low cost, mature equipment supply infrastructure and well known reliability of the wirebonding technology. Both aluminum and gold wire material are used in contemporary wirebonding applications.

The basic process involves making electrical connections through a 0.7 to 20 mil dia. wire, whereby the wire is physically attached at both ends to a metallized surface through a combination of either diffusion bonding and/or cold welding. To accomplish the attachment, an end of the wire is physically held against the surface to which it is to be joined. Then, ultrasonic power is applied, with or without the simultaneous application of heat (thermosonic). The strengths of wirebonded joints can approach the strength of the monolithic metal. Achieving this strength, however, and establishing reliable joint integrity over time, is possible only by careful design. Critical ultrasonic parameters include the wirebonding pressure, ultrasonic power, ultrasonic frequency, duration of power application and, for thermosonic bonding, the temperature. Work by Harmon[46] has demonstrated that the pull strength of the wirebond joint varies with ultrasonic power—specifically, that there exists a critical ultrasonic power for achieving the maximum bond strength. An example wherein incomplete bonding has occurred is illustrated in Fig. 15.11.

The primary material properties that are important include the chemical composition of the bond wire and of the pad/terminal or device, as well as the geometric parameters such as the wire diameter, loop height, and length. Pull strength or bond shear tests serve as methods for characterizing the quality of wirebonded interconnections and are well established, e.g., ASTM Standard Specification, F 458, *Standard Methods for Measuring Pull Strength of Microelectronic Wire Bonds*.

The wirebonded structure is subject to failure at several locations (Fig. 15.12)[47–59]; however, the reliability of the wirebonds is now known to be dictated primarily by the mechanical stress that is induced within the wirebonds and associated structure. This stress arises as a result of the differences in thermal expansion mismatch between the materials composing the wire and the underlying device/substrate assembly. Of special concern is that this stress is repeatedly generated and removed during typical service, since a number of progressive damage mechanisms described earlier are known to occur under these conditions. Specifically, this critical stress is known to lead to the development of cracks in both the heel of the wire itself as well as within the interfacial region of the wire and underlying electronics (Fig. 15.12).[46]

Calculation of the stress is best achieved with high resolution through numerical analysis of the application of interest. Then, application of Eq. (15.16), with the appropriate parameters for θ_b and B, enables the number of cycles to failure to be calculated. Alternatively, Pecht[2] has provided an analytical approach to estimating the maximum stress within the heel of a wirebond based on its structure and the properties of the underlying material:

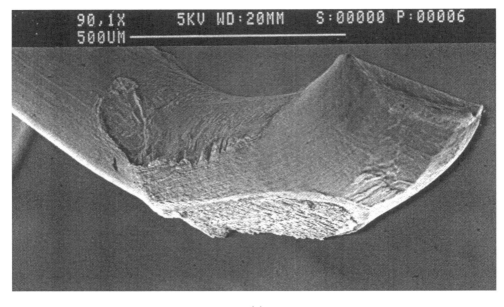

(a)

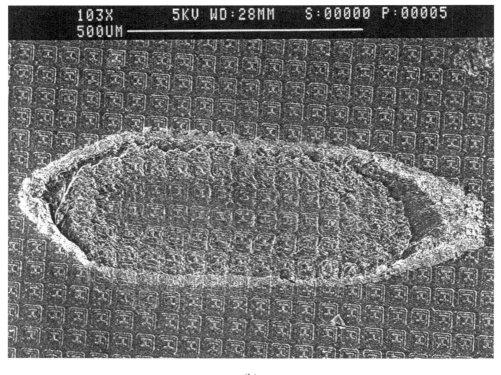

(b)

FIGURE 15.11 High-magnification electron micrographs of a wirebonded interconnection that failed because of poor bonding. In (a), the underside of the debonded wire exhibited evidence of contact with the emitter metallization (b) on the power device.

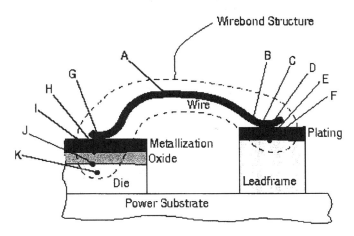

FIGURE 15.12 Schematic of the known failure locations in wirebonded interconnections.[47]

$$\sigma_b = 6E_{\text{wire}}(\Delta T)\left(\frac{r}{d_h}\right)\left(\frac{l_h}{d_h} - 1\right)^{0.5}\left[2a_s + \frac{\alpha_s - \alpha_w}{\left(1 - \frac{d_h}{l_h}\right)}\right] \quad (15.25)$$

where r = radius of the wire
E_{wire} = Young's modulus of the wire
d_h and l_h = height and length of the wirebond
α_s and α_w = coefficients of thermal expansion of the substrate assemble and wire
ΔT = imposed temperature range of fluctuation

Finally, it should be noted that, in some instances, melting of the wirebond/metallization site has been observed (Fig. 15.13). Typically, this is the result of an electrical anomaly such as a power transient or device failure.

Example 15.6

A new engineer has been assigned the task of increasing the reliability of the wirebonded interconnections within a power electronics package that have been known to fail through metal fatigue. It has been observed that fatigue failures occur within the heel of the wire, with a typical life of 125,000 cycles and $\theta_b = 10$ under a specific environmental and power cycling set of conditions. A new design decreases the cyclic stress within the heel of the wire from

$$\Delta\sigma = 45 \text{ MPa to } 32 \text{ MPa} \quad (15.26)$$

The task is to determine the effect of the new design on the fatigue lifetime.
From Eq. (15.16),

$$L_o = B(\Delta\sigma)^{\theta_b}$$

thus,

$$L_o^2 = L_o^1\left(\frac{\Delta\sigma_2}{\Delta\sigma_1}\right)^{\theta_b} = (125000)\left(\frac{32}{45}\right)^{10} = 3.8\times10^6$$

This represents nearly a 30× increase in reliability. ❏

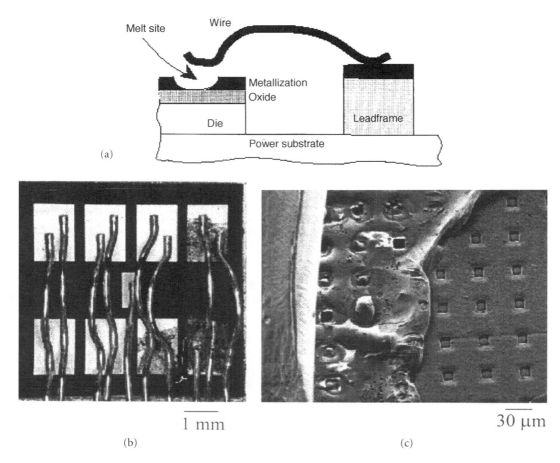

FIGURE 15.13 Example of a wirebonded interconnection that failed as a result of melting of the bond pad: (a) schematic of the melt site, (b) low-magnification micrograph of the failed region on the device, and (c) high-magnification micrograph of the region on the emitter bond pad that exhibited melting.

15.4.3 Chip Capacitors and Ceramic Baseplates

Brittle surface mount components such as chip capacitors, resistors, and inductors may fail in a catastrophic manner from the presence of stress originating from a number of sources. Examples of cracking from the thermal stress imposed as a result of the difference in thermal expansion mismatch between the component and the solder are well known.[2] In these cases, care must be taken to ensure that the fracture toughness of the material is not exceeded by the applied stress intensity factor, as described by Eq. (15.13).

A similar phenomenon has already been exemplified for ceramic baseplates. Here, the presence of minute surface flaws, as well as internal voids, can lead to the generation of relatively high stress intensity factors at relatively modest loads. Thus, care again must be taken to maintain the stress intensity factor within the baseplate material below the critical fracture toughness.

15.4.4 Semiconductor Devices

Semiconductor devices can fail as a result of applied electric fields in excess of their dielectric breakdown potential. Such failure is typically associated with a local "filament" of current, similar to a lightning bolt, that forms along a locally low-resistance path within the material. This filament generates extremely high

Electronics Package Reliability and Failure Analysis

temperatures within the silicon. Furthermore, because of the negative coefficient of resistance of silicon (i.e., its resistivity decreases with increasing temperature), a positive feedback loop is established, which leads to thermal runaway that may vaporize the silicon. This, in turn, leads to high internal pressure, causing cracking and further failure of the silicon. Examples of high-power devices that have failed in this manner are shown in Figs. 15.14 and 15.15.

15.5 Failure Analyses of Electronic Packages

The failure analysis of electronics hardware is a challenging undertaking that requires knowledge of electronics, materials science, mechanics, chemistry, and related disciplines. The overall objective is to

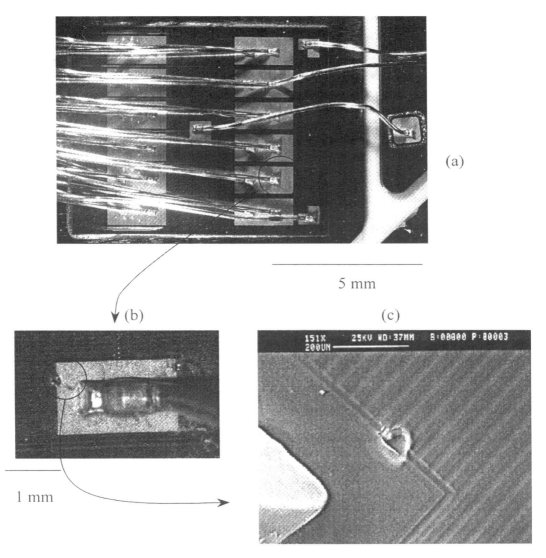

FIGURE 15.14 Example of a high-power device that exhibited electrical device failure: (a) low magnification photograph of the power module illustrating the general layout, (b) high-magnification micrograph illustrating the edge of the bond pad where failure occurred, and (c) high-magnification electron micrograph of the edge of the bond pad where failure occurred. The site of electrical arcing is visible in (c).

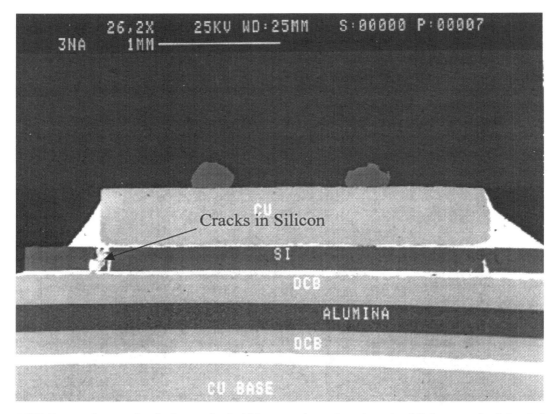

FIGURE 15.15 Cross-sectional micrograph of a high-power electronic package that failed as a result of electrical overload to the device. Cracking of the silicon device is visible as indicated by the arrows; penetration of the cracked silicon by the melted solder is visible as well, confirming the transient high temperatures associated with the high-power failure.

develop a clear understanding of the root cause of failure and to relate the root cause to the normal operating conditions of the system. Several descriptions of the specific chemical and mechanics analysis that may be performed are provided in Ref. 9; this chapter aims to introduce the reader only to the basic concepts of electronics hardware failure analysis.

All failure analyses must begin with a thorough understanding of the context of the failure—whether it was encountered under nominally normal conditions, what the track record of the failed system was, etc. The utility of the information collected during this stage of the investigation cannot be overemphasized. This stage of the investigation also involves full documentation of the remains of the system by optical photography and microscopy (e.g., Figs. 15.1, 15.10, 15.11, and 15.13 through 15.15). Analysis by transmission X-ray microscopy and reflected ultrasonic examinations may also be performed (e.g., Figs. 15.7 and 15.9). At this point, a preliminary fault tree may be constructed to guide subsequent analyses.

The second stage of failure analysis involves destructive tests. These range from sampling of material specimens to confirm conformance to chemical composition and mechanical property specifications, to cleaning of surfaces, etc. An important subset of investigations at this stage may be the fabrication of model test specimens that may be utilized to simulate the postulated failure scenario under controlled conditions.

Finally, the principal causes of failure are established and ranked in accordance to likelihood. These conclusions are then related back to the predicted lifetime of the system (if available) to assess whether revisions should be made to the predictions or operating guidelines.

15.6 Thermal Management

Because of the crucial role of temperature in controlling the reliability of electronic packages, basic thermal management concepts now will be reviewed; a more comprehensive treatment is provided in Chapter 11. The primary goal of the thermal design of an electronics package is to maintain the device junction temperature, T_j, below a critical level (Fig. 15.16).[3] This temperature is a design invariant, and it applies to steady-state power dissipation as well as to transient power pulses or surges. This limitation arises primarily from the physics of the semiconductor material, namely, that the concentration of electrons and holes that naturally exist within the material increases exponentially with temperature. Above a certain temperature, this natural level, or *intrinsic carrier concentration*, is so great that the semiconducting properties are lost, and the device becomes inoperable. Furthermore, this process is driven by the high power density, measured in W/m³, from the device junctions, through the various interfaces associated with the package until the power is finally dissipated into the ambient environment. This process is depicted in Fig. 15.16 for the case of a power package for a motor drive application.

A secondary goal is to minimize the temperatures throughout the assembly during operation. This is because elevated temperatures and the associated increased range of temperature cycling that occurs during operation generally accelerate the failure mechanisms described earlier. Consequently, decreasing the temperature of the system usually has a noticeable effect on reliability.

Therefore, the principal application-specific design invariant is the chip power dissipation, P_{chip}, such that the temperature of the device junction, T_j, required to transfer the heat into the surrounding medium (at reference temperature, T_o) through a package with a total thermal resistance, Θ_{th}, is,

$$T_j = T_o + \Theta_{th} P_{chip} \tag{15.27}$$

or, alternatively,

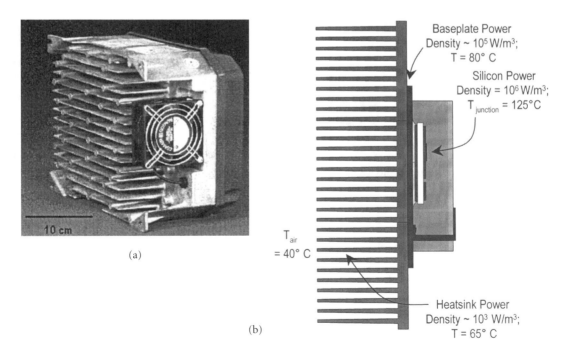

FIGURE 15.16 Examples of system-level packaging considerations. In (a), a commercial motor drive is indicated, with the heatsink/fan assembly shown. In (b), this architecture is shown schematically, with typical temperatures and power densities indicated at the various interfaces.

$$\Delta T = T_j - T_o = \Theta_{th} P_{chip} \qquad (15.28)$$

where Θ_{jc}, Θ_{cs}, Θ_{sa} = junction-to-case, case-to-sink, and sink-to-ambient thermal resistances
Θ_{th} = sum of Θ_{jc}, Θ_{cs}, and Θ_{sa}

Clearly, the lower Θ_{th} is for a given power level, P_{chip}, the lower the junction temperature, T_j, of the device. This relationship is graphically represented in Fig. 15.17, which represents a system-level design guide for the total package thermal resistance.

In turn, prediction and determination of Θ_{th} is a critical design challenge, since it represents one of the most important system parameters that the design engineer must minimize. The highest level of confidence in predicting the thermal resistance of a system is performed today through sophisticated conjugate thermal analysis, namely, where the coupled effects of fluid flow, temperature dependent material properties, and specific architectural arrangements may be related. However, insight may be gained into the primary effects of the parameters that govern the thermal resistance of the package through the analysis of simple, physics-based models for heat flow subject to Fourier conduction and convective heat transfer into air or liquids.

The typical approach during the design of electronic packages is to partition the thermal resistance into three separate quantities, namely, the junction-to-case, the case-to-sink, and the sink-to-ambient thermal resistance. In the former, the thermal resistance is dominated by conductive heat transfer, whereas for the latter, substantial heat transfer occurs through convection. This partitioning is convenient for the electronics designer in that the specification of the package is usually performed separately from that of the associated mounting and thermal management hardware of the surrounding system. For example, the thermal resistance of a given ball grid array package is separately quoted as compared to the thermal resistance of the heatsink and fan assembly.

The estimate of the thermal resistance of the package itself is typically straightforward. Specifically, in the simplest case of one-dimensional heat transfer by conduction, the thermal resistance, Θ_{th}, of a layer of material is simply:[3]

$$\Theta_{th} = \frac{t}{kA} \qquad (6.29)$$

where t and A = thickness and cross-sectional area of the layer
k = thermal conductivity of the material

Values of the thermal conductivity of many relevant materials are provided in Table 15.3.

For the case where heat conduction occurs from layers of different sizes, e.g., for the case where the device is smaller than the substrate (Fig. 15.18), then the decrease in thermal resistance of the larger layer can be estimated according to the Kennedy modification,[3] namely,

$$\Theta_{th} = \frac{H}{k\pi A} \qquad (15.30)$$

where the parameter H has the values listed in Table 15.4 for typical package dimensions and the specific boundary condition of heat flowing from one layer and out the opposite face of the underlying layer (Fig. 15.18). In this approach, the cross sectional areas are approximated by discs, where the radius of the device and underlying layers are a and b, respectively, and the thickness of the underlying layer is w (Fig. 15.18).

Example 15.7

An RF device that is 2 × 2 mm dissipates 5 W and is attached to a metallized aluminum nitride layer of dimensions 10 × 10 × 1 mm; the aluminum nitride material in turn is attached to a copper flange of dimensions 20 × 20 × 2 mm. The thermal conductivity of the AlN and copper are 190 and 400 W/mK. All of the heat that is dissipated in the device must pass through the bottom of the flange. This config-

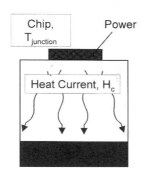

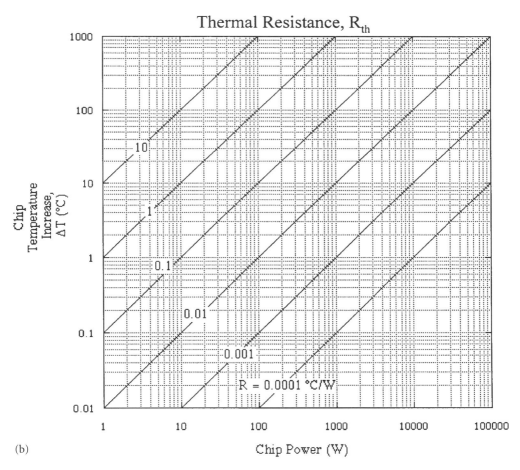

FIGURE 15.17 Basic relationships between the device power, package thermal resistance, and temperature difference required for heat transmission. In (b), a generic design guide is shown that relates the increase in chip temperature, ΔT, as compared to ambient temperature, to the chip power, P, through the package thermal resistance, R_{TH}.

uration is shown schematically in Fig. 15.1b. The total thermal resistance of this stack is required. The thermal resistance of the device is neglected in this example.

First, appropriate radii, a, must be calculated for the device, AlN, and Cu layers; these are 1.13, 5.64, and 11.3 mm, respectively. Next the Kennedy factors, H, must be obtained. The ratios a/b and w/b for the AlN and Cu layers are

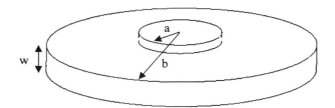

FIGURE 15.18 Schematic of the geometry used to determine the Kennedy factor for calculating the thermal resistance of layers of dissimilar sizes. The radius of the disc with the same area as the upper layer is a, the radius of the lower layer is b, and the thickness of the lower layer is w.

TABLE 15.4 Kennedy Factors for Heat Flux from Opposite Face (see Fig. 15.18 for nomenclature)[3]

	Kennedy factor, H				
w/b	a/b = 1	a/b = 0.3	a/b = 0.1	a/b = 0.05	a/b = 0.01
0.01	0.01	0.035	0.1	0.2	0.7
0.02	0.02	0.07	0.2	0.36	0.8
0.05	0.05	0.16	0.45	0.65	0.95
0.10	0.10	0.35	0.65	0.8	0.97
0.20	0.20	0.55	0.8	0.9	0.99
0.50	0.50	0.8	0.95	0.99	1
1	1	1	1	1	1
2	2	1.2	1.05	1.01	1
5	5	2.1	1.4	1.1	1
10	10	3.2	1.9	1.45	1.1

$$\left(\frac{a}{b}\right)^{AlN} = \frac{1.13}{5.64} = 0.2; \quad \left(\frac{w}{b}\right)^{AlN} = \frac{1}{5.64} = 0.17; \quad H^{AlN} = 0.45$$

$$\left(\frac{a}{b}\right)^{Cu} = \frac{5.64}{11.3} = 0.5; \quad \left(\frac{w}{b}\right)^{Cu} = \frac{2}{11.3} = 0.17; \quad H^{Cu} = 0.40 \quad (15.31)$$

Next, the thermal resistance of the two layers is obtained as

$$\Theta_{th}^{AlN} = \frac{0.45}{190(\pi)1\times10^{-4}} = 7.5° \text{ C/W}$$

$$\Theta_{th}^{Cu} = \frac{0.40}{400(\pi)4\times10^{-4}} = 0.8° \text{ C/W}$$

$$\Theta_{th}^{total} = \Theta_{th}^{AlN} + \Theta_{th}^{Cu} = 8.3° \text{ C/W}$$

Finally, the increase in temperature, ΔT, is

$$\Delta T = P\Theta_{th} = (5 \text{ W})(8.3° \text{ C/W}) = 41.5° \text{ C}$$

This result may then be used in conjunction with Eqs. (15.25, 15.16, and 15.17) to determine, for example, the stress and hence fatigue life for the wirebonds or solder joints. ❏

The next step is to consider is the thermal resistance between the case and the ambient environment. Because of the many variables involved, this is the subject of intense ongoing research and development. However, the details of heat transfer from surfaces into either air or a liquid environment can be conveniently summarized by the relationships shown in Fig. 15.19 for several modes of heat transfer.[3] This information then can be used in conjunction with the properties of heatsinks to determine their thermal resistance; this process has been greatly simplified by commercial heatsink manufacturers who routinely publish detailed tables of the thermal resistance of their heatsinks.

The above discussion has been for steady-state power dissipation conditions; other thermal management schemes are forced to cope with thermal transients. Although this is described elsewhere in this book, it is worth calculating the effective distance, or pulse penetration depth, d, reached by a transient thermal pulse of duration t_p. This is given by

$$d = \left(\frac{t_p k}{\rho C_p}\right)^{\frac{1}{2}} \tag{15.32}$$

where ρ = density
          C_p = specific heat of the material

This relationship is plotted in Fig. 15.20 for a range of material systems, with basic material data provided in Table 15.3. Thus, by knowing the nature of transient power surges that may occur within a specific application, the design of the package, e.g., insertion of high-conductivity layers, may be performed with good confidence.

Finally, a key quantity that can govern the thermal resistance of electronic packages is the thermal resistance of the interfaces between the various layers. This interfacial thermal resistance can be extremely difficult to predict in advance, yet it can exert a substantial influence over the overall thermal resistance.

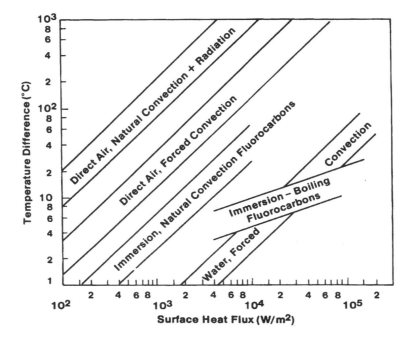

FIGURE 15.19 Relationships for the increase in temperature required to drive various heat fluxes from surfaces subject to different cooling technologies. The ultimate goal of any packaging approach is to transfer the heat into the ambient environment.

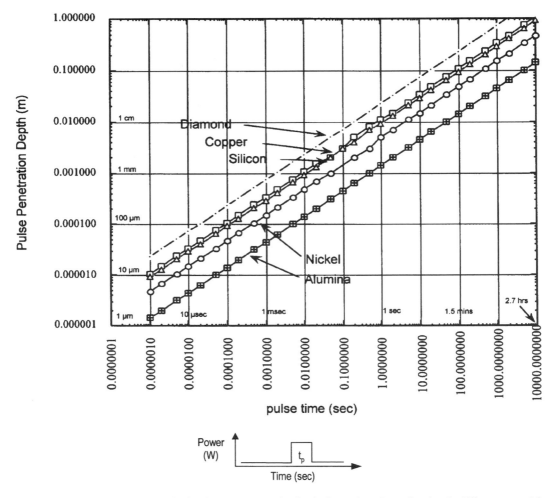

FIGURE 15.20 Approximate results for the penetration depth, d, of transient thermal pulses in different materials as given by the relationship shown in Eq. (15.32).

In the best case, the thermal resistance of the interface is negligible with respect to the thermal resistance of the remainder of the package. Unfortunately, there are several reasons that this is often not the case. Among them are voids or other low-conductivity defects that may be present along the interfaces (e.g., Fig. 15.9). Also, for interfaces that rely on mechanical contact to achieve high thermal conductivity, the presence of poor or nonuniform contact can lead to degraded interfacial thermal resistance. In addition, chemical constituents and voids can interact to promote the growth of oxides and intermetallic compounds. These compounds typically exhibit poor thermal conductivity. As a result, a suite of interfacial compounds, ranging from greases to compliant polymer layers, has been developed to combat this phenomenon. Indeed, conventional assembly instructions typically provided detailed descriptions of the thickness and mounting torque required to produce interfacial thermal resistance within predictable levels.

15.7 Concluding Remarks

The goal of this chapter has been to provide a mechanism-based approach for understanding the factors that dictate the reliability of electronic packages. This has been accomplished by relating the reliability both to the fundamental mechanisms of failure that operate within the constituents, as well as their primary driving forces. Also, basic reliability principles have been reviewed and utilized within the context of the approach, with a description given to the role of accelerated testing. Essential thermal management concepts have been discussed as they relate especially to the driving forces and reaction kinetics of the degradation mechanisms. Finally, the reliability of key packaging constituents has been discussed within the context of the primary failure mechanisms that are known to operate.

Acknowledgments

The authors wish to acknowledge several stimulating discussions with M. Ashby, B. Cox, M. Dadkhah, J. He, M. James, P. Lee, J. Mather, V. Mehrotra, H. Marcy, W. L. Morris, L. Radosevich, and S. Schroeder in preparation of this work. Also, the financial support of Rockwell International is gratefully acknowledged.

References

1. Elsayed, A., *Reliability Engineering*, Addison Wesley Longman, Inc., Reading, MA, 1996.
2. Pecht, Michael, *Integrated Circuit, Hybrid, and Multichip Module Package Design Guidelines: A Focus on Reliability*, John Wiley & Sons, Inc. New York, 1994.
3. Tummala, Rao R. and Rymaszewski, Eugene J., *Microelectronics Packaging Handbook,* Van Nostrand Reinhold, New York, 1989.
4. Ashby, Michael F., & Jones, David R.H., *Engineering Materials: An Introduction to Their Properties and Applications,* International Series on Materials Science and Technology, Vol. 34. Pergamon Press, Oxford, 1980.
5. McCluskey, F. Patrick, Grzybowski, Richard, & Podlesak, E., eds., *High Temperature Electronics*, CRC Press, New York, 1997.
6. Hertzberg, Richard W., *Deformation and Fracture Mechanics of Engineering Materials,* second ed., John Wiley & Sons, New York, 1983.
7. Barrett, Craig R., Nix, William D., & Tetelman, Slan S., *The Principles of Engineering Materials,* Prentice-Hall, Inc., Englewood Cliffs, NJ, 1973.
8. Ashby, M.F., *Materials Selection in Mechanical Design,* Pergamon Press, Oxford, 1992.
9. ASM International, *Metals Handbook Ninth Edition, Vol. 11, Failure Analysis and Prevention,* American Society for Metals, Metals Park, OH, 1986.
10. Hutchinson, J.W., & Suo, Z., "Mixed Mode Cracking in Layered Materials," in *Advances in Applied Mechanics,* Vol. 29, pp.63-191, 1992.
11. Schroeder, S.A., Morris, W.L., Mitchell, M.R., & James, M.R., "A Model for Primary Creep of 63Sn-37Pb Solder," in *Standard Technical Publication 1153,* American Society for Testing and Materials, Philadelphia, PA, 1995.
12. He, Jun, Shaw, M.C., Sridhar, N., Cox, B.N., & Clarke, D.R., "Direct Measurements of Thermal Stress Distributions in Large Die Bonds for Power Electronics," in *Proceedings of the MRS Spring Meeting,* San Francisco, CA, Vol. 515, April 1998.
13. He, Jun, Shaw, M.C., Mather, J.C., & Addison, R.C. Jr., "Direct Measurement and Analysis of the Time-Dependent Evolution of Stress in Silicon Devices and Solder Interconnections in Power Assemblies," in *Proceedings of IEEE Industry Application Society Conference,* St. Louis, MO, October 1998.
14. Ling, S., & Dasgupta, A., "A Nonlinear Multi-Domain Stress Analysis Method for Surface-Mount Solder Joints," in *Transactions of the ASME,* Vol. 118, June 1996.
15. Engelmaier, W., "Design for Reliability for Surface Mount Solder Attachments: Physics of Failure and Statistical Failure Distributions, in *Proceedings of the Technical Conference of the International Electronics Packaging Society,* Vol. 1, 1992.

16. Engelmaier, W., "Design for Reliability of Surface Mount Solder Attachments: The Physics of Failure and Statistical Failure Distribution," in *Interconnection Technology,* April 1993.
17. Wen, Liang-Chi, & Ross, Ronald G., Jr., "Comparison of LCC Solder Joint Life Predictions With Experimental Data," in *Journal of Electronic Packaging,* Vol. 117, June 1995.
18. Engelmaier, W., "Generic Reliability Figures of Merit Design Tools for Surface Mount Solder Attachments," in *IEEE Transactions on Components, Hybrids, and Manufacturing Technology,* Vol. 16, No.1, February 1993.
19. Lau, John H., & Pao, Yi-Hsin, *Solder Joint Reliability of BGA, CSP, Flip Chip, and Fine Pitch SMAT Assemblies,* McGraw-Hill, New York, 1997.
20. Wu, Wuchen, Held, Jacob, Peter, Scacco, Birolini, Allesandro, "Investigation on the Long Term Reliability of Power IGBT Modules," in *Proceedings of 1995 International Symposium on Power Semiconductor Devices and ICs,* Yokohama, Japan.
21. Lau, John H., "Creep of Solder Interconnects under Combined Loads," in *IEEE Transactions on Components, Hybrids, and Manufacturing Technology,* Vol. 16, No. 8, December 1993.
22. Adachi, Mitsuri, Ohuchi, Shinji, and Totsuka, Norio, "New Mode Crack of LSI Package in the Solder Reflow Process," in *IEEE Transactions on Components, Hybrids, and Manufacturing Technology,* Vol. 16, NO. 5, August 1993.
23. Puttlitz, Karl J., & Shutler, William F., "C-4/CBGA Comparison with Other MLC Single Chip Package Alternatives," in *IEEE Transactions on Components, Packaging, and Manufacturing Technology—Part B,* Vol. 18, No. 2, May 1995.
24. Busso, E.P., Kitano, M., & Kumazawa, T., "Modeling Complex Inelastic Deformation Processes in IC Packages' Solder Joints," in *Transactions of the ASME,* Vol. 116, March 1994.
25. Busso, E.P., Kitano, M., & Kumazawa, T., "A Visco-Plastic Constitutive Model for 60/40 Tin-Lead Solder Used in IC Package Joints," in *Journal of Engineering Materials and Technology,* Vol. 114, July 1992.
26. Busso, E.P., Kitano, M., & Kumazawa, T., "A Forward Gradient Time Integration Procedure for An Internal Variable Constitutive Model of Sn-Pb Solder, in *International Journal for Numerical Methods in Engineering,* Vol. 37, 1994.
27. Tong, Ho-Ming, Mok, Lawrence S., Grebe, K.R., Yeh, Helen L., Srivastava, Kamalesh K., & Coffin, J.T., "Effects of Parylene Coating on the Thermal Fatigue Life of Solder Joints in Ceramic Packages, in *IEEE Transactions on Components, Hybrids, and Manufacturing Technology,* Vol. 16, August 1993.
28. Heinrich, S.M., Wang, Y., Shakya, S., Schroeder, S., & Lee, P.S., "Selection of Design and Process Parameters for Non-Uniform Ball-Grid Arrays," in *Interpack '95,* Lahina, HW.
29. Heinrich, Stephen M., Shakya, Shilak, Wang, Yanhua, Lee, Ping S., & Schroeder, Scott A., "Improved Yield and Performance of Ball-Grid Array Packages: Design and Processing Guidelines for Uniform and Nonuniform Arrays," in *IEEE Transactions on Components, Packaging, and Manufacturing Technology—Part B.,* Vol. 19, No.2, May 1996.
30. Hatsuda, T., Doi, H., Sakata, S., & Hayashida, T., "Creep Behavior of Flip-Chip Solder Joints," at The American Society of Mechanical Engineers Winter Annual Meeting, November, 1992, Anaheim, CA.
31. Desaulnier, Wilfred E., Jr., DeSantis, Charles V., & Binder, Martin C., "A Framework for Comparing Life Prediction Models for SnPb Solder: Part II—Application," in *EEP, Vol. 10-2, Advances in Electronic Packaging,* ASME, 1995.
32. Gupta, Vineet K., Barker, D.B., & Dasgupta, A., "Modeling Solder Joint Fatigue Life for Gullwing Leaded Packages: Part II—Creep Model and Life Calculation," *EEP, Vol. 10-2, Advances in Electronic Packaging,* ASME, 1995.
33. Ju, S.H., & Sandor, B.L., "The Reversibility of Creep-Fatigue Damage in 63SN-37PB Solder," *EEP, Vol. 10-2, Advances in Electronic Packaging,* ASME, 1995.
34. Guo, Zhenfeng, & Conrad, Hans, "Fatigue Cracking Kinetics and Lifetime Prediction of Electronic Solder Joints," in *EEP, Vol. 2, Advances In Electronic Packaging,* ASME 1995.

35. Uegai, Tani, Inoue, Yoshioka, & Tamua, Keiichi, "A Method of Fatigue Life Prediction for Surface-mount Solder Joints of Electronic Devices by Mechanical Fatigue Test," in *EEP, Vol. 4-1, Advances in Electronic Packaging*, ASME, 1993.
36. Clech, Jean-Paul, Manock, John C., Noctor, Donna M., Bader, Frank E., & Augis, Jacques A., "A Comprehensive Surface Mount Reliability Model Covering Several Generations of Packaging and Assembly Technology," in *IEEE Transactions on Components, Hybrids, and Manufacturing Technology*, Vol. 16, No. 8, December 1993.
37. Syed, Ahmer R., "Creep Crack Growth Prediction of Solder Joints During Temperature Cycling—An Engineering Approach," in *Transactions of the ASME*, Vol. 117, June 1995.
38. Solomon, H.D., & Tolksdorf, E.D., "Energy Approach to the Fatigue of 60/40 Solder: Part I—Influence of Temperature and Cycle Frequency," in *Transactions of the ASME*, Vol. 117, June 1995.
39. Hu, Jimmy M., "An Empirical Crack Propagation Model and its Applications for Solder Joints," in *Transactions of the ASME*, Vol. 118, June 1996.
40. Engelmaier, W., "Surface Mount Solder Joint Long-term Reliability: Designing, Testing, Prediction," in *Soldering & Surface Mount Technology*, No. 1, February 1989.
41. Guo, Zhenfeng, & Conrad, Hans, "Effect of Microstructure Size on Deformation Kinetics and Thermo-Mechanical Fatigue of 63Sn37Pb Solder Joints," in *Journal of Electronic Packaging*, Vol. 118, June 1996.
42. Solomon, H.D., & Tolksdorf, E.D., "Energy Approach to the Fatigue of 60/40 Solder: Part II—Influence of Hold Time and Asymmetric Loading," in *Journal of Electronic Packaging*, Vol. 118, June 1996.
43. Banks, Donald R., Burnette, Terry E., Gerke, R. David, Mammo, Ephraim, & Mattay, Shyam, "Reliability Comparison of Two Metallurgies for Ceramic Ball Grid Array," in *IEEE Transactions on Components, Packaging, and Manufacturing Technology—Part B*, Vol. 18, No. 1, February 1995.
44. Hwang, Jennie S. *Ball Grid Array & Fine Pitch Peripheral Interconnections*, Electrochemical Publications LTD, Port Erin, Isle of Man, 1995.
45. Suhir, E., "Stresses in Adhesively Bonded Bi-material Assemblies Used in Electronic Packaging," in *Proceedings of the MRS Spring Meeting*, Vol. 72, Palo Alto, CA, 1986.
46. Harman, George, *Reliability and Yield Problems in Wirebonding in Microelectronics*, ISHM, Reston, VA, 1991.
47. Schafft, H.A., *Testing and Fabrication of Wirebond Electrical Connections: A Comprehensive Survey.* National Bureau of Standards, Technical Note 726, 1972.
48. Trigwell, Steve, "Failure Mechanisms of Wire and Die Bonding," in *Solid State Technology*, May, 1993.
49. Pitt, Veronica A., & Needes, Christopher R.S., "Thermosonic Gold Wire Bonding to Copper Conductors," in *IEEE Transactions on Components, Hybrids, and Manufacturing Technology*, Vol. CHMT-5, No. 4, December, 1982.
50. Olsen, Dennis R., and James, Kristi L., "Effects of Ambient Atmosphere on Aluminum-Copper Wirebond Reliability," in *IEEE Transactions on Components, Hybrids, and Manufacturing Technology*, Vol. CHMT-7, No. 4, December 1984.
51. Pecht, M., Dasgupta, and Lali, P., "A Failure Prediction Model for Wire Bonds," in *ISHM 1989 Proceedings*, Baltimore, MD.
52. Nesheim, Joel K., "The Effects of Ionic and Organic Contamination on Wirebond Reliability," in *Proceedings of the 1984 International Symposium on Microelectronics (ISSHM)*, Dallas TX, 1984.
53. James, Kristi, "Reliability Study of Wire Bonds to Silver Plated Surfaces," in *IEEE Transactions of Parts, Hybrids, and Packaging*, PHP-13, 1977.
54. Harman, George, G., and Wilson, Charles L., "Materials Problems Affecting Reliability and Yield of Wire Bonding in VLSI Devices," in *Materials Research Society Symposium Proceedings*, Vol. 154, 1989.
55. Pitt, V.A., & Needes, C.R.S., "Ultrasonic Aluminum Wire Bonding to Copper Conductors," in *IEEE Transactions on Components, Hybrids, and Manufacturing Technology*, CHMT-10, 1987.

56. Harman, G.G., "Metallurgical Failure Modes of Wirebonds," at 12th International Reliability Physics Symposium, 1974.
57. Ravi, K.V., & Philosky, E.M., "Reliability Improvement of Wirebonds Subjected to Fatigue Stresses," in *10th Annual Proceedings of the International Reliability Physics Symposium,* 1972.
58. Onuki, Jin, & Koizumi, Masahiro, "Reliability of Thick AL Wire Bonds in IGBT Modules for Traction Motor Drives," in *Proceedings of 1995 International Symposium on Power Semiconductor Devices and ICs,* Yokohama, Japan.
59. Wu, Wuchen, Held, Marcel, Jacob, Peter, Scacco, Paolo, & Birolini, Alessandro, "Thermal Stress Related Packaging Failure in Power IGBT Modules," in *1995 Proceedings of International Symposium on Power Semiconductor Devices & ICs,* Yokohama, Japan.
60. Mahalingham, Mali, "Thermal Management in Semiconductor Device Packaging," in *Proceedings of the IEEE,* Vol. 73, No. 9, September 1995.
61. Beck, J.V., Osman, A.M., & Lu, G., "Maximum Temperatures in Diamond Heat Spreaders Using the Surface Element Method," in *Journal of Heat Transfer,* Vol. 115, February 1993.

16
Product Safety and Third-Party Certification

Steli Loznen
The Standards Institution of Israel

Constantin Bolintineanu
Digital Security Controls Ltd.

16.1 Essential Requirements for Safety.................................... 16.1
16.2 Compliance Assessment Procedures............................... 16.9
16.3 Design for Compliance.. 16.18
16.4 Product Certification 16.31
16.5 Appendix.. 16.38

16.1 Essential Requirements for Safety

Constantin Bolintineanu

16.1.1 Safety Philosophy of Electronic Equipment

Safety is a term commonly applied to a wide range of issues. Safety, as related to equipment, refers to all the features built into a piece of equipment to prevent an accident. The accident can be a result of one or more *hazards* that can appear during the installation or use in normal or abnormal conditions of equipment operation. The basic philosophy of safety is that the safety features must be incorporated into the equipment from the design stage, continuing with the manufacturing, shipping, installation, and operation of the equipment.

All potential hazards should be identified within the design stage, which is the stage at which all the problems can be corrected and/or eliminated with no impact on the performance of the equipment and with a minimal financial efforts.

From the point of view of safety, the design stage has four major steps:

1. Identification of all the potential hazards in normal conditions and in all foreseeable fault conditions
2. Design of safety features built into the equipment
3. Identifying safety features that cannot be built into equipment, which leads to the process of designing safeguards
4. Design of safety instructions, installation instructions, marking, warnings, etc. to cover all the hazards identified during steps 1 and 3

In all of the above-mentioned steps, the responsibility for the achieved safety level of the equipment belongs to the manufacturer. It is important to understand that, throughout the life of the equipment, the manufacturer is solely liable for the equipment's behavior.

The *hazards* associated with the *safety* of the electrical equipment are designated according to their characteristics and are discussed in this section as follows:

Electric Shock and Energy Hazards
Mechanical Hazards
Heat, Fire, and Tracking Hazards
Moisture, Liquids, and Corrosion Hazards
Radiation, Toxicity, and Similar Hazards
Sonic and Ultrasonic Pressure Hazards
Explosion and Implosion Hazards
Abnormal Operation Hazards
Human Factors Hazards
Ergonomic Hazards

To provide a generally acceptable level of equipment safety, appropriate national rules and regulations and/or applicable requirements of one or more industrial standards are required to be incorporated as features of the equipment during the design stage. It is important for a manufacturer to be aware, from the design stage, of the standards applicable to the equipment intended to be designed, manufactured, and marketed.

At the same time, the manufacturer must be aware of the specific requirements desirable by the intended market. When the equipment is transported, installed, in standby conditions, in normal operating conditions, or during foreseeable abnormal operating conditions, the equipment must be *reasonably safe*. A safe equipment means equipment incapable of producing harmful effects during its life at each stage mentioned above.

Product liability cases are filed mainly based on design defects, manufacturing defects, or misrepresentation of the equipment.

Electric Shock and Energy Hazard

Electric shock is a hazard that is manifested as an electric current passing through the human body. Currents of a milliampere level can produce reactions in a person and may cause an indirect hazard due to the involuntary reactions stimulated by the electric shock. Physiological reactions of the current passing can range from a simple sensation to death.

Electric shock and energy hazards can be produced by

- Unsafe access to hazardous voltages
- Inadequate insulation
- Inadequate grounding

Generally, in normal conditions the accessible voltages shall not exceed 30 Vrms or 60 Vdc. Accessible parts of the equipment shall not be hazardous live nor a source of discharge energy at a level that can produce electric shock.

Several factors that influence the body's susceptibility to electric shock are as follows:

- Impedance of the human body (wetness of the skin, internal resistance of human body, etc.)
- The pathway of current through the human body (The heart, the neck, and the head are the most dangerous pathways for the current.)
- The duration and intensity of the current
- The frequency of the current

For frequencies nominally from dc to 1000 Hz, threshold values of the effects of current flow through the body are as follows:

1 mA	threshold of feeling
10 mA	threshold of "let go" (involuntary peripheral muscle contractions)

100 mA   threshold of ventricular fibrillation
1 A   threshold of burns and sustained large muscle contractions

All of these effects depend on the actual path of current through the body.

Tests such as the dielectric strength test, the leakage current test, the capacitor discharge test, etc. are used to evaluate the risk of electric shock hazards.

Mechanical Hazards

A mechanical hazard can be produced by factors that include

- Inadequate stability
- Inadequate mechanical strength
- Inadequate instructions or safeguards
- Inadequate finishing of the accessible parts of the enclosure
- Inadequate lifting and carrying means
- Expelled parts

The equipment shall not become physically unstable to the degree that it could become a hazard to the operator or to the service personnel. The equipment shall not cause a hazard when it is subjected to shock, vibration, impact, drop, and handling likely to occur in normal use. The equipment shall have adequate mechanical strength, meaning that the components must be reliably secured, as must the electrical connections. Handling in normal use must not lead to a hazard. Protection against hazards created by expelled parts shall be provided, generally when single-fault conditions are considered. Sufficient stability is required for all the equipment except equipment that, in normal use, is secured to the building structure.

Moving parts shall not be able to crush, cut, or pierce parts of the body of an operator likely to contact them, nor severely pinch the operator's skin. When hazardous moving parts involved in the process cannot be made completely inaccessible during operation, and where the hazard associated with parts is necessary to the proper operation and is obvious to the operator, provision of a warning will be considered as an adequate protection.

If the equipment is portable and has a weight higher than a certain limit (e.g., 18 kg, ~35 lb), a means for handling and lifting, with appropriate mechanical strength, must be provided. The means shall be capable of withstanding a force several times higher (safety factor) than the mass of the equipment.

All easily touched edges, corners, openings, guards, and handles must be rounded and smoothed so as not to cause injury to the operator or service personnel.

Tests such as the mechanical strength test, the stability test, the drop test, the vibration test, etc. are tests used to evaluate the risk of mechanical hazards.

Heat, Fire, and Tracking Hazards

The risk of fire is a very complex phenomenon. Fire can produce, in a *chain reaction*, other types of hazards such as electric shock, mechanical hazards, toxicity, etc. When we consider the risk of fire, we must take into account additional factors of power dissipation, maximum admissible temperature, the necessary energy to start the ignition, and so on.

The selection and application of the components and materials must be made to minimize the possibility of ignition and spread of flame. The flammability of materials and the tracking properties must be taken into consideration according to each type of equipment. Flammability is defined as a *class* and denotes the capacity of ignition and flame production of a material. Any excess heating can cause a hazard of fire within the equipment in normal and in single-fault conditions, and can cause fire to spread outside the equipment.

Electrical components shall be used in such a way that their maximum working temperature under the maximum normal load condition will be less than the necessary temperature to cause ignition of the materials with which they are likely to come into contact. Components working at high temperatures shall be effectively shielded or separated to prevent overheating of their surroundings. Adequate clearances must be provided and such components must be mounted on materials with the proper flammability class as per requirements of the applicable standard.

Tracking is a progressive formation of conducting paths that are produced on the surface of a solid insulating material due to the combined effects of electric stress and electrolytic contamination of surface. The *comparative tracking index* (CTI) represents the numerical value of the maximum voltage (in volts) at which a material withstands 50 drops of a defined test liquid without tracking. The tracking properties are related to the pollution degree of the environment in which the equipment will work. Tests such as the heating test, input test, etc. are used to evaluate the risk of fire.

Moisture, Liquids, and Corrosion Hazards

Electrical equipment that contains liquids, or is intended to be used in measurement of processes that involve liquids, must be designed to give adequate protection to the operator and surrounding area against hazards arising from moisture and liquids in normal use. The equipment that must be considered under the above requirement can include equipment that uses:

- Liquids with continuous contact (e.g., liquid pumps, containers, etc.)
- Liquids with occasional contact (e.g., cleaning fluids, liquids having accidental contact, etc.)
- Liquids in the area surrounding the equipment

Where cleaning or decontamination is specified by the manufacturer, this shall not cause a direct safety hazard, nor an electrical hazard, nor a hazard resulting from corrosion. The cleaning method shall be described in detail in the user's manual. Special attention must be given to spillage, overflow, equipment containing liquids, etc. Tests such as spillage, overflow, etc. are tests used to evaluate the risk of liquid hazards.

Radiation, Toxicity, and Similar Hazards

Hazardous radiation can emanate from X-ray generators, ultraviolet sources, lasers, microwaves, etc. The hazard presented by radiation and similar effects depends on a number of factors including distance from the body, level of radiation, time of exposure, etc.

Radiation and/or toxic substances can damage the living tissues by interaction. Harmful consequences can result from constant exposure, even if the human body is subjected to low-level radiation. It is extremely important to control all types of radiation to which operators, service personnel, patients, and others may be exposed. Each type of equipment that presents a radiation hazard shall carry warning label(s) in accordance with the requirements of the applicable standard. If the size or design of the product makes labeling impractical, an adequate warning should be included within the user information or on the package. Generally, the labels shall be affixed on protective covers, safety-interlocked panels, etc.

Special attention must be paid to high-power radiation sources that have the potential ability to change the shape of the materials by heating and/or to penetrate the walls of protective housing by melting or vaporizing the materials of the enclosure. These effects depend on the type of radiation, exposure time, environmental factors, and the thermomechanical properties of the material irradiated. A protective housing may be made of passive and/or active guards. A *passive* guard relies on the intrinsic ability of the material to resist penetration by radiation. An *active* guard uses sensors or equivalent devices to limit the time during which hazardous levels of radiation can persist within the protective housing. Generally, in the case of equipment that presents a radiation hazard, safety interlocks shall be provided for access panels of protective housings if those panels can be removed or displaced during maintenance or operation and thus give human access to a radiation level in excess of the maximum acceptable level.

Measurements of the emitting radiation and the evaluation of the capacity to generate radiation are methods used to evaluate the risk of radiation, toxicity, and similar hazards.

Sonic and Ultrasonic Pressure Hazards

At certain levels, sound pressure and ultrasonic pressure can become a hazard to the operator and the service personnel. Moreover, health and safety regulations impose certain maximum levels of sound pressure, ultrasonic pressure, and so on within the industrial environment. Measurements of the sound or ultrasonic pressure are used to evaluate the risks of sonic or ultrasonic hazards.

Explosion and Implosion Hazards

The equipment that incorporates components under pressure or under high vacuum can produce the risk of explosion or implosion in normal or single-fault conditions. Appropriate safety features, enclosures, protective screens, and so forth shall be provided for such situations so that, when devices will release pressure (positive or negative), that release will not cause danger to the operator and/or surroundings.

Other types of equipment can liberate an amount of poisonous or injurious gases. The amount shall be kept within acceptable limits, and references should be made to each specific case and condition.

Special attention must be paid to equipment that incorporates batteries. Batteries shall not cause explosions or produce a fire hazard as a result of leaking or ventilation, excessive charge or discharge, or installation with an incorrect polarity. When a current passes through a cell of a rechargeable battery, gases are nearly always produced as a result of the electrochemical reactions involved. When these gases are not reabsorbed or vented to the atmosphere, the pressure inside the cell rises to a level that can cause an explosion. The higher-capacity batteries require special ventilation to release the gases to the atmosphere. Special attention must be paid to lithium batteries from the point of view of accessibility and disposal. Generally, tests in single-fault condition are used to evaluate the risks of explosion and implosion hazards.

Abnormal Operation Hazards

The analysis of the equipment behavior, the design of the circuits, and the mechanical design shall give the necessary information regarding the hazards that can rise during abnormal (single-fault conditions) operation of the equipment.

Tests such as the locked rotor test or those designed to detect short circuits on the secondary of the transformers or on the capacitors, overload, impairment of cooling, bypassing of the protective devices or interlocks, etc. will give the necessary information regarding the behavior of the equipment when it is operated in an abnormal condition.

Human Factor Hazards

The failure to design for reasonable human factors can lead to hazards. Contributing factors are as follows:

- Unavailable or inadequate operating instructions or overcomplicated instructions
- Inadequate accessories
- Inadequate warning of side effects
- Erroneous software
- Incorrect diagnosis
- Erroneous data transfer
- Misinterpretation of decision

Proper instructions and training can eliminate many human error hazards.

Ergonomics Hazards

It is important to minimize, by use of ergonomic principles, the possibility of errors and thus of hazards and operator risk. The following examples illustrate the ways in which ergonomic principles may be used to lessen hazards:

- It is important to provide the status indicators for the power supplies.
- All displays should be unambiguous and easily readable.
- Controls and commands that must be activated and operated by the operator's hands should be arranged and positioned in a logical manner (e.g., a group of keys with the same classification

should be put together, in the same area) and in such a way that they will not obstruct the operator's view and access to emergency switches or emergency stop commands.
- Spacing must be provided between keys so as not to allow the chance for errors caused by touching the wrong key.
- Colors of indicators and controls shall give the meaning of the function as per applicable standard requirements.
- Auditory signals shall accompany critical situations.
- The outer surfaces of the equipment present in the field of vision from a normal operating position should not be reflective.
- There should be no sharp edges, corners, or other physical hazards that could harm the operator.
- Safe distances and adequate clearances must be maintained for fingers, hands, arms, legs, and the body where these parts can be exposed to hazards during the operation of the equipment, and they must be provided in such a manner to minimize any risk. Where the design cannot provide safe distances, then covers, rails, etc. shall be provided to prevent access to dangerous areas. These devices must not give rise to any additional hazards.

16.1.2 Specific Concepts

To understand the requirements of safety standards, it is important to be familiar with all the definitions and terminology as they are described within the applicable standard. This paragraph will present a few definitions and terms referring to electrical safety.

Generally, there are two categories of persons normally concerned with electrical equipment: operators and service personnel.

Operator is a term applicable to all persons other than service personnel.
Service personnel refers to persons having appropriate technical training and experience to be aware of hazards to which they are exposed in performing a task and of measures to minimize the danger to themselves and other persons.
Operator access area refers to an area to which, under normal operating conditions, one of the following applies:

- Access can be gained without the use of a tool.
- The means of access is deliberately provided to the operator.
- The operator is instructed to enter regardless of whether a tool is needed to gain access.

Rated voltage is the primary voltage as declared by the manufacturer. It also can be expressed as *rated voltage range* (if applicable).
Rated current is the input current of the equipment as declared by the manufacturer.
Rated frequency is the primary power frequency as declared by the manufacturer.
Primary circuit is an internal circuit directly connected to the external supply mains or other equivalent source to supply electric power.
Secondary circuit is a circuit that has no direct connection to primary power and derives its power from a transformer, converter, or equivalent isolation device, or from a battery.
Normal load is the mode of operation of the equipment that approximates as closely as possible the most severe conditions of normal use in accordance with the manufacturer's operating instructions.

16.1.3 International Regulations Mandating Product Safety

Around the world, organizations have been created and given a mandate for product safety. For example, in 1970, the U.S. Congress created the OSHA (Occupational Safety and Health Administration) to ensure safe and healthful working conditions. The electrical safety of occupancies is based on the National

Electrical Code written by the NFPA (National Fire Protection Association). The NEC requires that all products used in the marketplace be listed. The listing must be with a recognized agency, such as UL, CSA, etc. The Canadian Electrical Code specifies that "electrical equipment used in electrical installations …shall be approved and shall be of a kind or type and rating approved for the specific purpose for which it is to be employed. Approved, as applied to electrical equipment, means that the equipment has been submitted for examination and testing to an acceptable certification agency…." To prove that a product conforms to local rules and regulations (as spelled out by, e.g., the National Electrical Code, Canadian Electrical Code, or other standards or guidelines), the manufacturer of the electrical equipment must mark it with an indication of the approval status required for the intended market. Several forms of approval include certification, special inspection, special acceptance certification, field engineering services, component recognition, classification, and follow-up services.

Almost every country imposes general and/or specific requirements related to the safety level of equipment. The states belonging to the European Union have reached an agreement regarding a "new approach to harmonization and standardization" to ensure the free movement of products throughout the community, presuming that the products manufactured in conformity with the harmonized standards meet the essential requirements. Public authorities in each country are responsible for the protection requirements of their territory. Safety clauses require the member states to take all appropriate measures to withdraw unsafe products from the market. All the rules and regulations pertaining to product safety have been included in well known EU directives (e.g., machinery directive, medical device directive, low-voltage directive, toys directive, etc.). In Europe, the standards are made by one of three European Standard Organization: CENELEC, CEN, and ETSI. In 1993, the member states of the European Union reached an agreement on affixing and use of the CE conformity marking, which is intended to be used in the technical harmonization directives. The CE marking symbolizes conformity with all the obligations incumbent on the manufacturer of the product by virtue of the community directives provided for its affixation. The CE marking affixed to industrial products assures users that the product conforms to all the community harmonization provisions that apply to it, and that it has been the subject of the conformity evaluation procedures.

16.1.4 Standards

Standards have been issued to set a minimum acceptable safety levels for electrical equipment. Standards for safety may be issued by a recognized testing laboratory or agency such as UL, CSA, VDE or other specialized organizations, ANSI, IEC, etc.

A standard for safety is a written document that comprises a collection of requirements for a particular product category, requirements to be used as a basis for determining the eligibility of a product to be evaluated, tested, and accepted.

A Standard for safety comprises the following main parts:

1. Introduction (foreword)
2. Scope and objects
3. General requirements
4. Tests
5. Marking and instructions

Introduction

The introduction gives information regarding the applicability of the standard (latest date of publication, latest date of withdrawal of conflicting standards). For each evaluation, one must use the latest edition of the standard to avoid having to repeat the process of evaluation and/or certification after a short period of time.

The introduction enumerates the fields for which the standard is applied (e.g. electric shock, energy hazards, fire, etc.). Within a safety standard, the following terminology is encountered:

- *Shall* means *mandatory.*
- *Should* means *recommended but not mandatory.*
- *May* means *is permissible.*

Scope and Objects

The scope is one of the most important parts of the standard. The manufacturer must be able to identify the applicability of the standard to manufactured equipment, including ratings, application, and restrictions.

General Requirements

The general requirements give all the information applicable to the construction of the equipment.

Tests

Generally, the tests are separated into two distinctive categories: type tests and routine tests.

A type test is carried out on one or more pieces of equipment, representative of a type, to determine whether the construction and manufacturing methods comply with the requirements according to the applicable standard.

Routine tests (manufacturing production line tests) are tests applied during or after manufacture to detect manufacturing failures and unacceptable tolerances in manufacturing and materials.

Marking and Instructions

This part of the standard gives all the requirements regarding the marking that must be affixed on the equipment and instructions that must be provided to the end user.

For a particular piece of equipment, more than one standard can be applicable. The requirements of the standards are based on engineering principles, research, test records, and field experience. All aspects of the manufacture, installation, and operation of the equipment included in a standard are based on extended consultation with manufacturers, users, inspection authorities, and others having specialized experience. Generally, a standard passes several steps before it becomes mandatory. Different categories of specialists are involved in the process of elaborating a standard, including government representatives, manufacturers, consumers, importers, insurance specialists, installers, and retailers.

When the evaluated equipment performs a safety function, the applicable performance standards shall be used in conjunction with the safety standards for evaluation.

16.1.5 Criteria for Compliance, Required Documentation

To be allowed to introduce a piece of equipment on the market, as per the requirements of local rules and regulations, the equipment must bear an acceptable approval status. To obtain a certification mark or a special acceptance label or equivalent, the equipment must be submitted for evaluation to an accredited test house. A test house must have the capability to perform all the applicable tests and to examine the equipment to verify conformance with all appropriate standards. Different organizations are mandated to issue the accreditation for a testing house to confer to that house the status of *Nationally Recognized Testing Laboratory* (NRTL).

When a product is submitted to an NRTL, the manufacturer must complete a submittal process. The required documentation that must be forwarded to an NRTL comprises the following:

1. One or more product samples, as required by the standard, and a detailed description of the product and its intended use. A complete list of all models, types, or product variations that the manufacturer intends to cover shall be submitted along with the description. The description shall describe in detail similarities and differences among models, types, variations. It is mandatory that the rating of each model, type, or product variation be submitted along with the product.
2. A complete list of all components and materials used in the equipment. For each component, the following information must be submitted: manufacturer, ratings, approval/listing status. For each

material, the name of the material, its manufacturer, type designation, technical characteristics (e.g., flammability rating, maximum admissible temperature) must be submitted.
3. A block diagram and a detailed wiring diagram illustrating the function of the equipment and electrical or electronic circuits; design drawings and/or photographs of the product.
4. All the installation manuals, instruction manuals, safety instructions, and, if available, the marking of the product as per the standard requirements and any other markings intended to appear on the equipment and/or on the package; associated software to allow continuous operation at maximum load.
5. A list with all intended alternate components, materials, or alternate arrangements of parts intended to be used in the future (if applicable).
6. Name and address of each factory where the product is intended to be manufactured; the name and address of authorized representatives who will receive all communications and will provide supplementary information to the testing house.
7. Information regarding previous listings or certification marks obtained for the same product or category.
8. Additional information that the manufacturer deems necessary (conductor sizes used, insulation class, overload protections, printed circuit board layout etc.).
9. Participant's name and address—name of company submitting the product and who is financially responsible.
10. The system configuration representing the realistic worst-case load in normal condition and all the characteristics regarding the type of connection, type of operation of the equipment, and so on.

When the evaluation is finished, and the product has been found to be in compliance with all requirements, the NRTL will issue the necessary documentation and will provide for the period of the certification. It will also establish the required follow-up services to ensure the necessary level of performance as established during the initial examination.

16.2 Compliance Assessment Procedures

Steli Loznen

16.2.1 Risk Management

The increasing complexity of modern electric and electronic equipment, and the resulting systemic failures, often elude traditional testing and assessment. It is necessary to know how well particular devices perform in relieving certain conditions and what characteristics are associated with better and worse performance. The user must know how often devices fail and why.

It is impractical to expect absolute safety in the use of electric and electronic equipment. Generally it is accepted that no system can be completely fail-safe, and any associated risk should be reduced to a level that is "as low as reasonably practicable" (ALARP).

To reach this objective during the design phase of an electronic equipment, it is necessary to analyze the *harms* (physical injury and/or damage to health or property), *hazards* (potential sources of harm), and *risks* (the probable rate of occurrence of a hazard causing harm and the degree of severity of the harm) associated with the use of the equipment. The realistic expectation must be that risks are kept as low as possible, taking into account the cost that would be incurred in further reducing risk and the benefits resulting from use of the product.

If drawn up and carried out correctly, a risk analysis should detect any risk associated with an equipment. The overall process for the analysis and control of risk is referred to as *risk management*.

The first step in the risk management is the identification of qualitative and quantitative characteristics of the equipment or accessory under consideration. The following list includes some of the factors that could affect the safety of the evaluated product:

- What is the intended use?
- Which are the safety related materials and/or components used?
- Are equipment measurements made?
- Is the equipment interpretative?
- Is the equipment intended to control other equipment?
- Are there unwanted outputs of energy?
- Is the equipment susceptible to environmental influences?
- Is the equipment provided with accessories?
- Are maintenance and calibration required?
- Does the equipment contain software?
- Can the equipment be affected by long-time use or by retard effects?
- To what mechanical forces will the equipment be subjected?
- What determines the lifetime of the equipment?

The answers to all these questions, and a supplementary functional description, provide a general view of the evaluated product. At this, one can consider the next steps of the risk management process: *risk analysis* and *risk control*.

The second step of risk management process is the *risk analysis*, which deals with the identification of hazards and estimation of risks.

Start with *identifying hazards*, which on electrical and electronic equipment can be divided into categories of electrical, mechanical, energy, radiation, fire, environmental, functional, and aging.

Hazards should be identified for all reasonably foreseeable circumstances, including the normal use, incorrect use, and fault conditions, and should include appropriate hazards that affect the operator, service personnel, bystanders, or the environment. A list of the identified hazards can be compiled as shown in Table 16.1.[*]

TABLE 16.1 Hazard Identification Table

No.	Hazard	Conditions NU–Normal use SFC–Single fault IU–Incorrect use	Effect Operator Service personnel Bystanders Environment	Hazard Rating Number (HRN)	Countermeasures SD–Safe design PM–Protection measures IU–Informing user	New Hazard Rating Number (NHRN)

Associated with identification of hazards should be the *initiating causes*, which can be operator error, component failure, software error, integration errors, or environmental impact.[†] Next, one must estimate the risk for each hazard.

---

[*]Table 16.1 refers also to other aspects (hazard rating number, countermeasures) of risk management, which will be presented later.

[†]For the analysis of risks, different techniques may include AEA (action error analysis), ETA (event tree analysis), FTA (fault tree analysis), FMEA (failure mode and effects analysis), hazard and operability (HAZOP) studies, etc. The selection of and use of such techniques depends on the nature of the equipment (system, subsystem, independent unit, etc.).

The risk is regarded as the probable rate of occurrence of a hazard causing harm, and the degree of severity of the harm. We can consider *direct* risks as resulting from inadequate design, equipment failure, or misuse resulting in physical harm to the operator, the equipment, or the environment, or *indirect* risks such as inadequate equipment performance that can adversely affect the operator, the equipment, or the environment.

To better analyze risks, their components (consequences and probability) should be analyzed if they occur in absence of a failure, in a failure mode, or only in a multiple failure condition.

To characterize the risks, evaluate the *likelihood* of a hazard occurring as:

- *Frequent* (likely to occur frequently)
- *Probable* (occurs several times in life of an item)
- *Occasional* (likely to occur some time in the life of the item)
- *Remote* (unlikely, but can reasonably be expected to occur in the life of the item)
- *Improbable* (very unlikely to occur, but possible)
- *Incredible* (it can be assumed that occurrence may not be experienced)

Also evaluate the potential *severity* (qualitative measure of the possible consequences of a hazardous event) should it occur as:

- *Catastrophic* (has the potential of resulting in multiple deaths or serious injuries)
- *Critical* or *major* (has the potential of resulting in death or serious injury)
- *Marginal* or *moderate* (has the potential of resulting in injury)
- *Negligible* or *minor* (has little or no potential to result in injury)

At this point, one must attribute to each hazard the *HNR (hazard rating number)* (refer to Table 16.1), which represents the relationship between the likelihood of a hazard occurring and the potential severity should it occur.[*]

$$HRN = likelihood \times severity$$

According to the HNR, one comes to the specific *risk level* as

- Acceptable
- Tolerable
- Undesirable
- Intolerable

One example of the relationship between the risk level, the likelihood of a hazard occurring, and the severity should it occur is shown in Table 16.2.

A risk is acceptable if the risk is less than the maximum tolerable risk and the risk is made as low as reasonably practicable (ALARP). Establishing the level of safety required and obtaining agreement on acceptable risk levels it is a political and social activity involving a wide range of interests.

Practically, the intolerable and undesirable risks will be reduced by reducing the severity and/or the likelihood of the hazard, and the HRN should indicate *acceptable* or *tolerable* risks.

The reduction of the severity and/or the likelihood of the hazards will bring us to the third step of risk management process: the *risk control*. The risk shall be controlled so that the estimated risk of each identified hazard is made acceptable.

---

[*] Along with the hazard identification, the attribution of the hazard rating number is the most difficult stage on risk management process. This action requires a great deal of experience and special skills from assessor. Additional data can be obtained from scientific data, field data from similar equipment, reported incidents, relevant standards, etc.

TABLE 16.2 Risk Level, Likelihood of Occurrence, and Severity

	Minor	Moderate	Major	Catastrophic
Frequent	Undesirable	Intolerable	Intolerable	Intolerable
Probable	Tolerable	Undesirable	Intolerable	Intolerable
Occasional	Tolerable	Tolerable	Undesirable	Intolerable
Remote	Acceptable	Tolerable	Tolerable	Undesirable
Improbable	Acceptable	Acceptable	Tolerable	Tolerable
Incredible	Acceptable	Acceptable	Acceptable	Acceptable

The risk control serves to outline the procedures applied to reduce any possible risk, and it refers to solution for reduction of the likelihood or potential severity of a hazard, or both. These solutions should be directed at the cause of the hazard or introduce protective measures that operate when the cause of the hazard is present, specifying safety requirements to eliminate or minimize hazards and achieve an acceptable risk level.

To achieve the above aim, the following *countermeasures* (refer to Table 16.1) can be useful:

- Reduction of the hazard by "redesign" of the equipment (inherent safe design)
- Additional measures such as warnings, alarms, installation requirements, formal commissioning, routine maintenance, safety testing
- Measures requiring the user to be aware of the residual risks (need for particular precautions in the use of certain types of equipment or in certain procedures)

A summary of risk analysis (see Table 16.3) should be provided by each manufacturer. This summary must refer to result of risk management and especially to residual risks and to all countermeasures applied to reduce the identified risks. This document is very useful for engineering departments and can be considered as consumer report which can differentiate between the same kind of equipment manufactured by various manufacturers.

TABLE 16.3 Risk Analysis Summary

Manufacturer..

Type of equipment ....   Model ....

Hazard identified...

Techniques used for identification ....

Causes..

Severity of hazard: ❏ catastrophic   ❏ critical   ❏ marginal   ❏ negligible

Estimated likelihood: ❏ frequent   ❏ probable   ❏ occasional   ❏ remote   ❏ improbable   ❏ incredible

Does the hazard occur in: ❏ the absence of a failure   ❏ failure mode only   ❏ multiple-failure mode only

Techniques employed to reduce the likelihood of the hazard ....

Can a failure be detected before a hazard occurs?   ❏ yes   ❏ no

During what time interval? ....

Estimated residual risk...

16.2.2 Safety and EMC Tests

During the assessment of an equipment, a series of tests is conducted to prove compliance with the standard's requirements. Parts of these tests are performed by inspection: marking, symbols, warnings, damages after heating test, protective earth connection, arrangement of accessible terminals, connection of external cords, etc.

Other tests are conducted on the *inoperative apparatus* (i.e., ground continuity, dielectric strength, mechanical tests, accessibility to hazardous parts, etc.) and in *normal work condition* or *abnormal condition*.

Ground Continuity

The impedance between the earth terminal of the appliance inlet and any accessible metal parts protectively earthed that could become energized in case of single fault shall not exceed 0.1Ω (for equipment with detachable power supply cord) or 0.2 Ω (for equipment with nondetachable power supply cord). This test applies a current of 25 Arms or dc (depending on the standard) from a low-voltage source (less than 6 V) and measures the voltage drop between any two points tested. The resistance of the ground connection is calculated and must not exceed the above mentioned values. Conductors to be used for measuring circuit shall be copper wires with at least 3.5 mm^2 in cross-sectional area and shall be wired as short as possible.

Dielectric Strength

The purpose of dielectric tests is to evaluate characteristics of insulation under electrical stress. These tests are performed by applying a specified test voltage for a period of one minute. During the test, no breakdown or flashover shall occur. Common waveforms used in dielectric strength tests include ac voltage, dc voltage, and the 1.2 × 50 s impulse voltage. The test voltage (specific for each standard) is dependent on the working voltage (U) through the insulation tested (i.e., for information technology equipment with a working voltage up to 250 Vac, with basic insulation between line and neutral connected together and the earthed metal enclosure, the test voltage is 1500 Vac or 2121 Vdc; the test voltage for double insulation is 3000 Vac or 4242 Vdc). For insulation between two isolated parts or between an isolated part and an earthed part, the working (reference) voltage is equal to the arithmetic sum of the highest voltages between any two points within both parts plus a fixed voltage (i.e., in a medical application for an insulation transformer 1:1 at 230 Vac, the working voltage becomes 460 Vac, and the test voltage for basic insulation is 2U + 1000 Vac and, for double or reinforced insulation, 4U + 3000 Vac).

When performing these tests, power-dissipating component parts, electronic devices, and capacitors located between the circuits under test may be removed or disconnected so that the spacings and insulation rather than the component parts are subjected to the test voltage. All switches and controls will be adjusted to ensure that all conductors intended to be tested are indeed connected to the circuit under test.

Note: The dielectric and current leakage tests are carried out after humidity treatment. Parts sensitive to humidity, normally used in equipment and having no influence on safety, need not be subjected to a humidity test (i.e., disk drives).

Mechanical Tests

Enclosure Rigidity Test
The rigidity of the enclosure is tested by applying a specified force of anywhere on the surface. This action must not reduce the air clearances and creepage distances to values below those allowed.

Drop Test
The equipment shall comply with the standard after being dropped from a specified height on a hardwood board.

Loading Test
Carrying handles or grips on portable equipment shall withstand a force equal to four times the weight of the equipment. If more than one handle is provided, the force shall be distributed between the handles. After the test, no adverse effects (breaking loose from equipment, distortion, cracking, etc.) may be present.

Impact Test
The strength of the enclosure is tested by application of blows with a specified impact energy by means of the spring-operated impact test apparatus. The blows are applied also to handles, levers, knobs, displays, and to signal lamps with area exceeding 4 cm^2. Any damage sustained shall produce no safety hazard.

Equipment not likely to be dropped, such as stationary, fixed, or floor-supported or counter-topped supported appliances, are subject to a steel *ball impact test.* In the horizontal test, a steel ball is dropped from a specified height. Vertical surfaces of the enclosure are subjected to a ball impact by allowing the ball to swing as a pendulum and strike the vertical surface. The polymeric enclosures are required to survive the impact test without breaking, cracking, or otherwise opening to the extent that hazardous, live, or moving parts are exposed.

Stability

The equipment shall not overbalance when tilted at an angle of 10°.

Actuating parts

All actuating parts shall be secured so they cannot be pulled off or work loose in normal use. For rotating controls, a torque (which depends on the diameter of control knob) is applied between the control knob and the shaft in each direction alternately. The knob shall not rotate with respect to the shaft.

Power Input

The steady-state current or power input of the equipment shall not exceed the rated current or power by more than a specified percentage (generally ±10%) under normal load, according of the type of equipment and the rated input power. The measurement is to be performed with an rms instrument.

Limitation of Voltage and Energy (Capacitor Discharge)

The voltage between the pins of a supply mains plug and between either pin and the enclosure of an electrical equipment is limited to 37% of the voltage in the moment after disconnection or 60 V (for medical electrical equipment). The measurement should be conducted 1s after disconnection of the plug.

Leakage Currents

The leakage currents on electrical equipment are as follows:

- *Earth leakage current* is the leakage through the earth conductor.
- *Enclosure leakage current* is from the enclosure (accessible parts nonprotectively earthed) to earth. The enclosure leakage current measured with the earth disconnected should be the same as that found for earth leakage under normal conditions.

For a medical application add also

- *Patient leakage current* is current from those parts in contact with the patient to earth.
- *Patient auxiliary current* is the current that flows through the patient, between various elements of applied parts.

The assessment of leakage current for electrical equipment is made for both the normal and single-fault condition, on maximum load, with on/off switch in both position, with normal supply polarity, and then repeated with the polarity of the supply reversed, with a supply equal to 110% of the highest rated main voltage and with the highest rated supply frequency. Possible single-fault conditions are interruption of the supply (phase or neutral conductors open) and interruption of protective earth conductor (not applicable for earth leakage current). The enclosure leakage current should be measure also for 110% main voltage applied between earth and any signal input or output parts.

Practical considerations for reduction of leakage current include:

- Use a low-leakage power supply cord.
- Use only three-wire power cords and proper grounds.
- Use layouts and insulating materials that minimize the capacitance between all hot conductors and the chassis.
- Use devices that have stray a low capacitance (RFI filters, power transformers, power wires, etc.).

It is recommended that MOVs not be used between primary circuits and earth ground. Their leakage characteristics can be unpredictable and change with time.

Product Safety and Third-Party Certification

Normal Heating

Any part of equipment that has a safety function shall not exceed specified values of temperature during normal conditions. The test is conducted for the supply voltage, which requires the highest power input. The thermocouple method is used for measurement except for windings, in which case the resistance method is recommended.

Abnormal Operation and Fault Condition

Equipment shall be tested such that, even in single-fault condition (only one fault at a time), no safety hazard shall exist. The following single-fault conditions are considered:

- Interruption of protective earth conductor or of one supply conductor
- Failure of an electrical component or of a mechanical part that might cause a safety hazard
- Failure of temperature-limiting device
- Failure of the software
- Impairment of cooling
- Locking of moving parts
- Overloading of mains supply transformers
- Short circuit of either constituent part of a double insulation
- Interruption and short-circuiting of motor capacitors
- Etc.

During the abnormal operation the temperature of safety parts is monitored and shall not exceed specified values. Also the leakage current shall remain at allowable values. After abnormal operation and fault condition, the insulation between the mains part and the enclosure, when cooled down to approximately room temperature, shall withstand relevant dielectric strength tests.

Equipment Used for Safety Tests

The following equipment may be necessary for safety tests:

- direct current and rms voltmeters
- power meter
- ac and dc ammeters
- temperature data logger, oscilloscope
- variable ac and dc power supplies
- electronic loads
- rheostats
- isolation transformer
- dielectric tester
- leakage current tester
- ground continuity tester
- winding resistance tester
- socket outlet torque balance
- humidity chamber
- ball pressure test apparatus
- sharp edge tester kit
- creepage and clearance gauge set, 1–8 mm,
- micrometer
- caliper

- mechanical test equipment such as
  - test hook
  - force gauges
  - test finger
  - test pin
  - impact test ball
  - impact hammer
  - torque screwdriver and wrench

Radio Frequency Interference and Immunity

The emission requirements for equipment are those of CISPR 11 or CISPR 22. The equipment is categorized in two classes according to where it will be used.

- Class A, suited for "heavy" industrial areas
- Class B, suited for residential areas

Each class is divided in two groups.

- Group 1, equipment that uses RF energy internally
- Group 2, equipment that uses the RF energy in the form of electromagnetic radiation for treatments

The allowable limits for emission differ between class and groups. The equipment is tested for conducted emission in the range 0.15 to 30 MHz, and for radiated emission in the range 150 kHz to 1 GHz.

In the U.S.A., the emission requirements for equipment are those of FCC, and the proper standard should be consulted. These are discussed further in Chapter 6.

Radiated RF field immunity is tested according to IEC 61000-4-3. An immunity level in volts per meter (V/m) is required using a signal for amplitude modulation over a specified frequency range. All standards do not allow any performance degradation of the equipment during or after the test.

Test Procedure for Conducted Emission on AC Mains Lines (0.15 to 30 MHz)
The conducted emission test is performed inside a shielded room (to minimize background noise interference), with the equipment placed on a 0.8-m high wooden table, 0.4 m from the room's vertical wall.

The equipment is powered from ac line via a 50 Ω/50 µH line impedance stabilization network (LISN) on the phase and neutral lines. The LISNs are grounded to the shielded room ground plane (floor) and kept at least 0.8 m from the nearest boundary of the equipment. The emission voltages at the LISN's outputs are measured using a receiver, which complies with CISPR 16 requirements. A frequency scan between 0.15 and 30 MHz is performed at 9 kHz intermediate frequency (IF) bandwidth, and using peak detection. The spectral components having the highest level on each line are measured using a quasi-peak and average detector.

Test Procedure for Radiated Emissions (30 to 1000 MHz)
A preliminary measurement to characterize the equipment is performed inside a shielded room at a distance of 3 m, using peak detection mode and broadband antennas. The preliminary measurement produced a list of the highest emissions and their antenna polarization. The equipment is then transferred to the open site and placed on a remotely controlled turntable. The equipment is placed on a nonmetallic table, 0.8 m above the ground. The effect of varying the position of the cables shall be investigated to find the configuration that produces maximum emission. The frequency range 30 to 1000 MHz is scanned, and the list of the highest emissions is verified and updated accordingly.

The emissions are measured using a EMI receiver complying to CISPR 16 requirements. The readings are maximized by adjusting the antenna height (between 1 and 4 m), the turntable azimuth (between 0 and 360°), and antenna polarization.

Verification of the equipment emissions is conducted within the following conditions:

- Turning the equipment on and off
- Using frequency span less than 10 MHz
- Observation of the signal level during the turntable rotation

Background noise shall not be affected by the rotation of the equipment; the emissions are measured at a distance of 10 or 30 m.

Test Procedure for Immunity to Radiated Field
The equipment is placed in the shielded room and subjected to a field strength of 10 V/m, nonmodulated, at the frequency range of 27 to 500 MHz. The radiated field is monitored by an isotropic E-field sensor, connected via a fiber optic link to a remote readout unit, out of the shielded room. The radiated field is applied in vertical and horizontal polarization, using biconical and log periodical antennas. The performance in normal use of the equipment is verified during the test.

Immunity to Electrostatic Discharge

The electrostatic discharge (ESD) event may couple into a tested device by conduction (common-mode signals) or radiation. ESD is most likely to occur through cracks in an equipment's nonconductive (plastic) enclosure. Immunity to ESD is tested according to the method given in IEC 61000-4-2. A test voltage is indicated for contact discharge (applied to accessible conductive parts) and another for air discharge, which can result in peak discharge currents. After the test, the unit shall continue to function or, if it fails, it shall not cause a safety hazard. During the test degradation of performance is allowed, except the actual operating state or stored data.

Test Procedure for Immunity to Electrostatic Discharge
The equipment is set up on an insulating support 0.1 m thick, which is placed on a wooden table 0.80 m high, which is covered by an earth reference plane.

Electrostatic discharges are applied to the test points, which are normally accessible to the operator. The amplitude is gradually increased from 2 kV to 8 kV, in increments of 2 kV. Discharges to objects placed near the equipment are simulated by applying the discharges of the ESD generator to the earth reference plane. All points are tested by a minimum of ten single discharges.

Fast Transient and Surge Immunity

Fast transients bursts are tested using the methods of IEC 61000-4-4 with a rise time and low-energy bursts of pulses. Alternating current lines, interconnecting cables, and dc lines must withstand specified pulses.

Surge immunity is tested according with IEC 61000-4-5. Differential and common mode pulses are applied on ac lines. After the test, the unit shall continue to function or, if it fails, shall not cause a safety hazard.

Test Procedure for Immunity to Electrical Fast Transient/Burst
The equipment is placed on a 0.8-m high wooden table, which is placed on the shielded room floor. The EFT/B generator is placed on, and grounded to, the shielded room floor as well.

A test signal having the standard wave is applied to the phase neutral and ground lines of the equipment mains input, at a distance of 1 m from the equipment. The test signal voltage is 1 kV, and it is applied for 1 min to each line, in negative and positive polarities.

The same test signal (with a voltage level of 0.5 kV in this case) is applied to the signal lines, control and dc lines (as applicable) that are connected to the equipment.

Other EMC Tests

Other common EMC tests include the following:

- Conducted RF immunity in the range 10 kHz to 100 MHz

- Magnetic field immunity in the range 30 Hz to 100 kHz
- Immunity to a quasistatic field of 0.5 Hz with an intensity of 2000 V/m
- Immunity to dips, 120 to 90 V for 500 ms every 10 s
- Immunity to dropouts, 95 to 0 V for 10 ms every 10 s
- Immunity to voltage variation ±20 V from 115 V

These requirements are described in IEC 61000-4-X Series standards.

EMC Test Site Description

- A shielded room should consists of a main room (7.4 × 4.35 × 3.75 m) and a control room (3.12 × 2.5 × 2.5 m) with following shielding performance:
    - magnetic field (60 dB at 10 kHz rising linearly to 100 dB at 100 kHz)
    - electric field (better than 110 dB between 50 MHz and 1 GHz)
    - plane wave (110 dB between 50 MHz and 1 GHz)

  All the power lines entering both shielded rooms are filtered.
- An open test site should consists of 3-m and 10-m ranges, using a 7 × 14-m solid metal ground plane, a remote-controlled turntable, and an antenna master. The turntable and the tested equipment placed on it are environmentally protected. All power, control, and signal lines are routed under the ground plane. Antenna master position and polarization shall be controlled. Also, the antenna position shall be adjustable, and the position of the turntable should be controlled. The turntable shall be mounted in a pit, and its surface is flush with the open site ground plane.
- Recommended test equipment includes
    - EMI receiver, fully compliant with CISPR 16 requirements
    - oscilloscope with probes
    - transient waveform monitor
    - phase control amplifier
    - single-phase isolated backfilter
    - transient generator
    - E-field meter
    - signal generator
    - spectrum analyzer
    - RF amplifier
    - close-field probe
    - ESD simulator
    - power amplifier
    - antenna, biconical
    - antenna, log-periodic

16.3 Design for Compliance

16.3.1 Selection of Components

Steli Loznen

All safety-related components, as listed below, shall be approved according to the specific standard. For other components, there may be no need for approval of every single component. The flowchart shown in Fig. 16.1 summarizes the compliance options.

Product Safety and Third-Party Certification

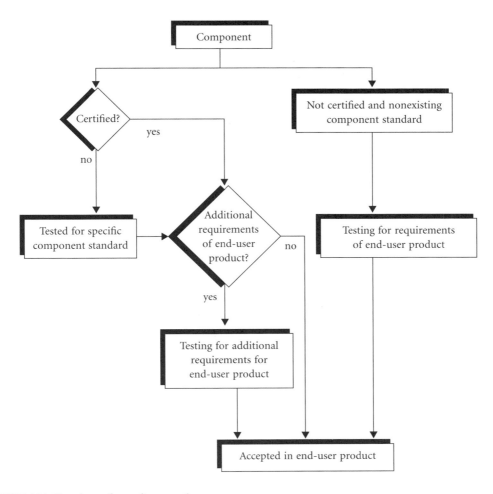

FIGURE 16.1 Flowchart of compliance options.

If it can be determined that a hazard does not exist when a component is used outside of its electrical rating, the component may be accepted in the end-use product (e.g., a component rated 230 V in an end-use product rated 220 to 230 V).

Safety-related components include the following:

- Power supply
- Transformer
- Power supply cord
- Fuse and fuseholder
- Batteries
- Switches
- Appliance inlet
- Line filter
- Power entry module
- Motor and ac or dc fans
- Thermoplastic materials
- Wiring

Power Supplies

A power supply is an example of a device that may fall under different standards, depending on its use.

- For use in an ITE device, it requires standard IEC 60950 approval.
- For use in medical equipment, IEC 60601-1 approval is needed.
- For use in laboratory, measurement, or control equipment, which need to comply with IEC 61010-1, a power supply approved to IEC 60950 is acceptable.

If the power supplies are approved to other standards, they must powered via a safety isolation transformer or tested supplementarily. As an example, for a medical application, a power supply approved to IEC 60950 must be tested supplementarily for:

- Limitation of voltage/energy
- Primary-to-secondary separation, leakage currents
- Dielectric strength
- EMC
- Connection to mains supply
- Overload transformer test
- Construction of transformer

The normal heating test must be conducted only if a discrepancy exists between IEC 60601-1 requirement and IEC 60950 test results.

Transformers

Mains supply transformers shall comply with the following requirements:

- Overheating of insulation in the event of short-circuit, or overload on any output winding
- Dielectric strength (after the humidity conditioning treatment) of the electrical insulation
- Creepage distances and air clearance distances (must meet at least the values required by the type of insulation in the transformer)

Construction Aspects

The most common method of providing suitable protection against overheating due to overloading and short circuits is the use of the proper wire gauges in the transformer primary and secondary, and fuse values with respect to the ampacity of the transformer. If the fuse alone is not sufficient for overload tests, a suitably rated thermal cutoff may be used.

Plastic materials such as used for the bobbin and outer cover must be flame rated to at least 94V-2. At least one electrostatic shield should be incorporated in the transformer between the primary and secondary coils, and this should be connected to the protective earthing. This shield will improve the isolation by reducing the capacitive coupling between the primary and secondary windings.

If the equipment is supplied through an isolation transformer, install the transformer close to the powered equipment to keep the ground capacity small.

Leads connecting the primary to ac power source shall be sized to carry at least the maximum input current, and a minimum of 300 V, 105° C.

Select a load wire size that, at an absolute minimum, is heavy enough to carry the output current that would flow if the load terminals were short circuited.

If an isolation transformer is connected between the ac power source and the power supply input terminals, it should be rated for at least 200% of the maximum rms current required by the power supply.

Power Supply Cord Set

All equipment shall be supplied with a power supply cord set suitable for the rating of the unit and marked according to the Harmonization Wire Coding System or U.S. requirements. Alternatively, the equipment can be provided with an appliance inlet and, in each country, a detachable cord set will be supplied. (Example: For a unit rated up to and including 6 A, the cross-sectional area of the conductors in the power supply cord must be minimum 0.75 mm². For products rated between 6 A and 10 A, the minimum cross-sectional area required is 1.0 mm².)

Construction Aspects

Nondetachable power supply cords must be provided with a bushing and a strain relief. To make sure that the bushing and cord anchorage work properly, the cord is subject to 25 pulls at up to 100 N. It should no displace more than 2 mm. The bushing should not be removable without the aid of a tool.

In the U.S., the power supply cord shall have conductors not smaller than No. 18 AWG (American Wire Gauge) and shall be for transportable equipment type SJ: SJT (trade name "Junior Hard Service"), SJE, SJO, or SJTO PVC-jacketed coded cords. For other equipment, it must be at least that of type SV.

In the U.S.A., for medical applications, *Hospital Only* or *Hospital Grade* plugs are to be used. The unit shall be provided with instructions to indicate that grounding reliability can only be achieved when connected to an equivalent socket marked *Hospital Only* or *Hospital Grade*.

Permanently connected equipment in the U.S.A. shall have provision for the connection of the wiring systems according to the National Electrical Code (NEC), ANSI/NFPA 70. The free length of a lead inside an outlet box or field-wiring compartment shall be minimum of 152 mm. When preparing power supply cords for permanent attachment to equipment, the protective-grounding conductor should be made slightly longer than the circuit conductors so that the protective-grounding connection will be the last to break in the event of failure of the strain relief mechanism.

Fuses/Fuseholders

In all mains-connected equipment, at least the "hot" supply lead shall be fused. For a Class I (earthed) medical equipment, each supply lead (live and neutral) must be fused or protected against overcurrent. For permanently installed equipment, the neutral conductor shall not be fused. Standards requirements for fuses are IEC 60127, UL 198G, and CSA 22.2 No. 59-1972. IEC 60127 requires fuses to withstand currents of 120% of rated current. UL and CSA specify that the fuse must not open with load conditions less than 110% of rated current. The time-current characteristics of fuse are known by the following abbreviations and color codes:

- FF, super-quick acting (black)
- F, quick acting (red)
- M, medium time lag (yellow)
- T, time lag (blue)
- TT, long time lag (gray)

UL and CSA define *Non-Time Delay* (fast-acting) and *Time-Delay* classes of fuses. A fast-acting (F) fuse must to be sized from 1.5 to 3.0 times the full load current; a time-lag (T) fuse must to be sized from 1.25 to 1.5 times the full load current.

It is usual to refer to low (L) breaking capacity in the order of magnitude of 30 to 200 A, and high (H) breaking capacity from 1500 to 15000 A. (In the event of a fault, an inadequate breaking capacity can result in a persistent arc, causing serious damage.)

The fuse current rating must be selected on the basis of

- Fault-mode current, to prevent any overcurrent that would result in unacceptable overheating
- Normal current mode, which must be measured with an rms instrument
- Voltage rating equal or exceeding the line voltage
- Fast-acting fuses in equipment with surge currents that are 10 times the full load current during the first 10 ms of operation

Additional design considerations are as follows:

- Place the fuseholder and fuse in the equipment at a point where there is little temperature rise.
- The fuseholder shall be mounted so as to prevent rotation around its mounting axis.
- Fuseholder terminals must be protected from inadvertent contact by means of shrink tubing or other acceptable means.

PTC resistors should not be considered overcurrent protective devices, because they are not designed to meet the same characteristic curves as fuses.

Batteries

When selecting a battery, analyze the following factors:

- Backup run time required, in amp-hours
- Maximum discharge current expected
- Minimum acceptable recharge time
- Power level
- Memory effect allowed
- Charge retention with time (how long a charged battery can remain off the charger before its charge drops below an acceptable level)
- Service life, in time or number of recharges
- Low-temperature use

The selection of a battery should be based on the highest discharge rate and the lowest operating temperature expected. Select the battery voltage according to the power level of the equipment: 12 V for up to 100 to 150 W, 24 V for up to 300 to 500 W, and 48 V for greater than 500 W.

Construction Aspects

- Equipment that uses batteries shall be provided with a means of preventing incorrect polarity connection and an appropriately rated device (fuse, diodes, resistors, etc.) for protection against fire hazards caused by a short circuit.
- When rechargeable batteries are used, provide a means to indicate when the battery is in the charging stage and instruct the user as to which battery charger shall be used for safety compliance.
- Provide a means of ventilation on the batteries compartments where gases can escape during charging or discharging.
- Materials used for the housing compartment must be resistant to corrosion by acids, alkalis, and their vapors.
- Use a separate compartment from the electronics for wet cell batteries, and the compartment must contain the spillage of electrolyte in the event of rupture of cells.
- Provide instructions for removing the primary (nonrechargeable) batteries from an equipment that is not likely to be used for some time.
- Provide instructions for safety use of lithium batteries. Provide recommended disposal information for all batteries.

Switches

Switches must be rated to equal or exceed the load they control and must disconnect all ungrounded (hot) connectors (switch-type double-pole, single-throw, DPST). Spacings for switches shall comply with IEC 60328 requirements. In a case where a product includes a one-third or greater horsepower motor, a separate switch is required to control it.

Construction Aspects
Switches need to be mounted securely so they do not rotate in normal use. The indicator lights (if provided) shall comply with IEC60073 with regard to coloring.

Interlock Switches

Usually provided on the doors and covers of equipment to prevent risk of electrical shock, energy, and/or mechanical injury hazards to the operator or to the service personnel, these must have certain construction features. They must be of two-pole construction and must incorporate two springs inside. Two-pole construction is applicable only for interlocks provided to prevent risk of electrical shock or energy. For stationary equipment, a one-spring switch may be used if device is located in such a manner that the law of gravity applies to disconnect the switch upon failure of the one spring. The interlock system must pass a 10,000-cycle endurance test, making and breaking the intended load without failure other than in a safe mode. This requirement does not apply to the switch only but also to components in the interlock system, such as a relay connected to the switch mechanism.

Construction Aspects

Interlock switches must be designed or installed so that they are unlikely to be operated when the doors/covers/panels are in the open position. If the interlock can be operated by means of the IEC test finger, then it is considered to be likely to cause unintentional operation and is unacceptable. Any bypass means provided for service personnel must be such that the safety interlock function (a) is self-restoring when the equipment is returned to intended operation and (b) requires a tool for operation when in operator access areas. The switch must not be capable of being operated by means of the IEC test finger.

The interlock must function prior to the openings being large enough to provide access to hazardous moving parts or parts that involve risk of electric shock or energy. Such determination is made by using the IEC test finger.

Appliance Inlet, Line Filters, Capacitors, Resistors, Transient Voltage Surge Suppressors, and Power Entry Modules

If the inlet used is without additional functions such as integral fuseholders, line filters, or selector switches, then it shall comply with IEC 60320.

An appliance inlet with integral filter or a separate filter unit in the equipment requires the voltage (in the event of a capacitor discharge) between the pins of the inlet and between any pin and the earthed enclosure of the equipment not to exceed a standard specified value, one second after disconnection of the supply.

A PTC (positive temperature coefficient) resistor is considered to be a fixed (current-limiting) impedance and could be used in electrical circuits as current limits.

Transient voltage suppressors can be connected across the mains. A TVSS device used in the primary circuit could contribute to a hazard (e.g., exploding) if not suitably investigated. If a transient suppressor is separately approved, it can be used without a protective device, but if is not separately approved, a protective device against the short circuit is required.

Construction Aspects

- Since the inlet terminals are live after switching the unit on, they must be protected by means of shrink tubing or insulated quick-disconnect connectors. If shrink tubing and insulated quick-disconnect connectors are used, they must be rated to a minimum of 300 V and 105° C.
- If in a symmetrical (differential mode) line filter, the capacitor placed between line and neutral (nominated with the suffix X) is insufficient, and additional inductance is necessary.
- A varistor may be used across an X-capacitor to absorb and dissipate high voltage spikes from the power supply.
- The value of the Y-capacitors used to reduce the asymmetrical (common mode) interference and connected line to ground and neutral to ground (acting as HF-shunts) is determined by the maximum permissible earth leakage current of the equipment.
- If the across-the-line (X) capacitors in the filter are too large to comply with the capacitor discharge requirement, it is necessary to add a discharge resistor.

- Capacitors shall not be connected between live parts and nonearthed accessible parts. Any capacitors rated greater than 100 nF must be marked with the voltage rating and capacitance. Capacitors should not be connected across thermal cutouts. Capacitors connected between mains parts and earthed accessible metal parts shall comply with IEC 60384-14.
- Approved Y1 capacitors (line-to-line) are accepted across reinforced insulation.
- Approved Y1 and Y2 capacitors are accepted across basic and supplementary insulation.
- Approved X1 capacitors (line-to-ground) should be used in a primary circuit before a rectifier.
- One resistor can bridge basic or supplementary insulation.
- Two resistors in series are acceptable across double or reinforced insulation.
- Resistors that are connected across the mains or between the mains and the earth do not need to comply with component standards.

Motors and Fans

Direct current motors will be tested to ensure that they do not cause a fire hazard. Manufacturers should avoid motors approved for construction only, since these motors have not been investigated for locked rotor and running overload protection. Only motors that are recognized as having impedance or thermal protection should be used.

Motors are tested in single-fault condition for locked rotor for seven hours (a cheesecloth placed over the motor under test should not ignite), interruption and short-circuiting of motor capacitors, and running overload protection. Motor winding temperatures are determined for steady-state condition.

Construction Aspects

The cooling dc or ac fans should provide the air flow rate necessary for transfer of the excessive heat from the equipment. The air flow rate is expressed in cubic feet per minute (CFM), cubic meters per second (m^3/s), cubic meters per minute (m^3/m), liters per second (L/s), or cubic meters per hour (m^3/hr).

Critical systems that incorporate fan cooling shall be provided with thermal warning or shutdown when loss of air flow is detected.

Thermoplastic Materials

The level of burn performance of polymeric materials is characterized by relative thermal index and classified by the flame rating. The relative thermal index is assigned in various thickness for both electrical and mechanical properties. All polymeric parts and printed wiring (circuit) board must have acceptable flame rating. Common flame ratings are 94-5V, 94V-0, V-1, V-2, HB. The least flammable is 5V material. The minimum flame ratings required for the common used materials are as follows:

```
Printed circuit board ..........................................................94V-1
Bobbin and outer cover for transformers.........................94V-2
Internal polymeric parts, including connectors...............94V-2
Enclosure for transportable equipment............................94V-1
Enclosure for fixed and stationary equipment................94V-0 and 94-5V
Air filter externally mounted and decorative parts.........94HB
```

The terminal blocks intended for making the internal connection between the power cord and the primary circuitry of the equipment shall be selected so that creepage and clearance distances between lines as well as line and ground are maintained.

For a terminal block that supports uninsulated parts of the mains, a ball pressure test to a temperature of 125° C is required.

Evaluating the suitability of polymeric material for an application, according to flammability, requires the following steps:

1. Identify the polymeric materials that will be used in the product.
2. Determine the function that each material played in the product.

3. Obtain information concerning material properties.
4. Evaluate the suitability of the material in the end product application by checking the end product standard.

An alternative for polymeric material flammability identification is to conduct the combustibility test.

Note: When wood is used as an enclosure or as a decorative or internal part, no flammability requirements apply except for an exposed surface area greater than 0.93 m^2 or a single dimension larger than 1.83 m, in which cases the minimum flame-spread requirements apply.

Wiring

All wiring used within the equipment must be suitable for the voltage, current, and temperatures to which the wires are subjected. If the insulation of wiring from different circuits come in contact, the insulation for each wire must be rated for the highest voltage in either of the wires. The wire must be insulated with PVC, PFE, PTFE, FEP, or neoprene.

Construction Aspects

- Wiring must be protected from coming into contact with sharp edges.
- When passing through metal guides, smooth surfaces or bushings are required.
- Any wire that is subject to motion requires some form of auxiliary mechanical protection, such as helical wraps.
- Splices are permitted but must be mechanically secured prior to soldering, and wires connecting to screw terminal should include a solder lug with upturned ends.
- All wiring involved in earthing must be provided with green/yellow colored insulation.
- All wiring between the input and the fuses must have conductors with the same cross-sectional area as that of the supply cord.
- All electrical connections other than soldering shall be provided with positive-detent, crimp-type insulated connectors suitable for the voltage and temperatures involved. They shall be sized for the wire and mounting terminations. Where hazardous voltage or energy is involved, all wire connections to connectors shall employ a method of double securement (crimp both the conductor and the insulator). Where fork-type lugs are used, they shall be snap-on or upturned lug type.
- All soldered connections shall be made mechanically secure before soldering. Tack soldering is not acceptable. Prior to soldering, the lead must be inserted through an eyelet or opening of a terminal block, or into a U or V shaped slot in the terminal, or wrapped around a terminal post. Another form of mechanical securement is to tie off the lead to adjacent lead with wire tie-wrap near termination point.
- All primary wiring should be routed away from low-voltage circuits and, similarly, low-voltage wiring should be routed away from primary circuits. Separation should be such that, even after a 10-N force on the wires, there is at least 4.0 mm of separation. Alternatively, the wires may be separated by an additional level of insulation such as 0.4-mm tubing.
- The low-voltage leads from the secondary circuitry should be additionally fixed by tie-wrap or similar fixing to prevent the wires from contacting hazardous voltage primary circuits, if they were to become free. Similarly, wires at hazardous voltages must be secured so that they cannot touch secondary circuitry if they were to break free from a connector.
- Wiring with only basic insulation shall be protected by additional fixed sleeving.

16.3.2 Construction Details

Constantin Bolintineanu

It is important to understand that, to engineer safe equipment, the designer must take in consideration all the causes and effects of the possible hazards able to rise during the transportation, installation, and operation of the equipment. The ability of the equipment to acquire by design the required level of safety depends on many mechanical requirements and requirements related to the material characteristics.

Each equipment must be designed by taking into consideration the applicable standard's requirements as well as specific rules and regulations. It is important to underline that each equipment can present special characteristics, and thus more than one standard will be used for design and evaluation.

While it is not possible to cover all situations, generally, the safety level must be reached in each of the following situations:

- *Protection against electric shock.* The design will consider in this situation the insulation requirements, conditions related to the integrity of protective earth connection, segregation of circuits, etc.
- *Protection against energy hazards.* The design will consider the requirements regarding the separation of circuits and components able to expose energy hazards, shielding of these components, and the use of safety interlocks by removing the hazard condition when access is gained and in the energy hazard area.
- *Protection against the risk of fire.* The design will consider these requirements:
  - selection of the materials to be sure that they will perform the required function without developing any risk of hazard
  - requirements related to the connections
  - protections in overload situations
  - limiting the quantity of combustible materials used
  - designing the positioning of the components
  - proper heatsinks
  - the design of the enclosure (Characteristics as flammability, maximum accessible temperature, and temperature limits for materials must be taken in consideration.)
- *Protection against mechanical hazards.* The design will consider the requirements regarding the following: the adequate selection of guards and interlocks intended to be provided to limit the access to the moving parts; emergency switches; ergonomic principles; etc.

16.3.3 Practical Construction Recommendations

Steli Loznen

Enclosures

Enclosures, barriers, and guards must be made of materials that are not highly combustible. They include steel; aluminum (must be corrosion protected); or heat-resistant, tempered, wired, or laminated glass. All easily corroded metals shall be provided with a means (painting, plating, galvanizing) to protect them from corrosion. Dissimilar metals shall not be employed where reliable continuity is required.

Plastics can be also used, but flame ratings must be observed, and they should be capable of withstanding a thermal conditioning test-ball pressure test at 75°C for 1 hr. When a thermoplastic material is molded into a desired shape of the finished product, internal molding stresses may be present in the final part. With time and elevated temperature, the internal stresses may cause gradual distortion and possible exposure of internal parts. After placement in an oven for seven hours at a specified temperature, the sample of the complete product is removed and examined for possible mold stress relief distortion. Any distortion or opening of the enclosure is considered unacceptable if it results in exposure to hazardous, live, or moving parts of tested equipment.

In addition,

- To determine if hazardous parts (live, moving, shaped, hot, etc.) are accessible to the operator, openings are tested by using the IEC standard test finger, test hook, and test pin.

- Conductive coatings applied to nonmetallic surfaces shall comply with UL 746C.
- Avoid sharp corners and edges, with special attention to flange or frame edges and the removal of burrs.
- Enclosures designed to give a degree of protection against ingress of water shall provide this protection in accordance with the IEC 529 (IP classification).
- Enclosures protecting against contact with hazardous parts shall be removable only with the aid of a tool, or an automatic device (interlock) shall make these parts not hazardous when the enclosure is opened.
- Special requirements refer to flame spread (rating of 75 or less) on the external surfaces of combustible material. Selection of the test method (UL E84 Steiner tunnel, the radiant panel test ASTM E 162, or ASTM E 84 Steiner tunnel) it is according with the dimensions of the surface tested.

Circuit Separation and Spacings

Protective spacing is a system for separating uninsulated, or unacceptably insulated, conductive parts from other such parts, or from earthed parts, or from both, in such way that allowable leakage currents are not exceeded in normal and in single-fault condition. The assessment is performed by making physical measurements of spacings (creepage distance and air clearance) and testing of the dielectric strength to verify compliance with the separation criteria. The values for creepage distance (the shortest distance between two conducting parts measured along the surface) and air clearance (the shortest distance between two conducting parts measured through air) are specified for each specific equipment in relevant standard. Usually, clearance is less than creepage.

Air clearance alone is acceptable only for isolation between hazardous parts and unearthed accessible parts if the parts are rigidly positioned (i.e., molding, etc.). Part secured by screws must have lock washers or, for the purpose of evaluating spacing, they are considered likely to come loose.

Generally, in electronic equipment, following separation must be maintained:

- Between an earthed, accessible metal part and hazardous parts (i.e., mains), by basic insulation
- Between a non-earthed accessible metal part and hazardous parts, by double or reinforced insulation or by an intermediate earthed circuit
- Between primary mains part and secondary parts, by double or reinforced insulation

Earthing (Grounding)

Provide the equipment with a protective earthing system for connecting nonconductive parts of equipment to earth. The earth conductor allows a low-resistance path to earth for any fault currents that could be produced by leakage, improper wiring, or misuse.

Functional earth terminals shall not be used to provide protective earthing.

Connect to earth all accessible conductive parts that are likely to render an electrical shock in the event of a fault condition. Faults to be considered include loosening or accidental disconnection of wiring connections, deterioration or breaking of insulation, etc.

A protective-earthing conductor must be at least the size of the largest circuit conductor for which it affords protection.

The protective-earthing system shall not be interrupted by switches, overcurrent devices, or the like.

Secure mechanically and solder the earthing lead to the appliance inlet earthing terminal. The other end of the earthing lead must be terminated by means of closed-loop connector (or the equivalent), which is secured to a threaded stud on the chassis by means of rivets, bolts, nut, or star lock washer. (If connection is made on a painted or coated surface, the use of a star-toothed washer ensures good continuity.) Other earths can share the same stud but must be placed over the nut holding the earth conductor in place, and each wire must have its own closed-loop connector and washer.

Protective earth terminals and screw and nuts which clamp earth conductors shall not serve to fix any other components not related to earthing, or for mechanical securements between parts of equipment.

Where plugs and sockets are used to supply power to internal components, the plug/socket combination must be such that the earth connection is made first and broken last. The standard plug/socket that complies with this requirement has the earth pin longer than the line and neutral pins.

16.3.4 Marking Requirements
Constantin Bolintineanu

Marking represents the information provided on or in connection with a product to identify all the characteristics of that product related to electrical safety.

The electrical equipment shall bear *markings* as necessary to identify the equipment and to ensure that the equipment is suitable for the intended application.

Generally, the marking consists of the following categories of information:

1. The manufacturer's name, trademark, or other recognized symbol of identification
2. Catalog number and serial number or type
3. Ratings: the applicable characteristics as follows:
    - voltage
    - current consumption
    - power consumption
    - frequency
    - number of phases
    - rated load
    - etc.
4. Type of power supply (ac, dc, or both)
5. Evidence of approval: certification mark, listing mark, inspection label, etc.
6. Other marking as may be required to ensure the *safe* and *proper* installation, operation and maintenance of the equipment (e.g., short-circuit interrupting capacity of the overcurrent protective devices, diagram number, adjustment of the input voltage, mode of operation, etc.)

Where appropriate, the method for adjusting the input voltage shall be fully described in the installation instructions or service manual. Unless the means of adjustment is a simple control near the rating marking, and the setting is obvious by inspection, a marking with the following instructions or similar ones must appear near the rating marking:

See installation instructions before connecting to the supply.

Additional markings are permitted, provided that they do not give rise to misunderstanding.

The marking shall be legibly and durably fixed to the enclosure of the equipment and plainly visible to persons where the equipment is installed and operated. In considering the durability of the marking, the effect of normal use shall be taken into account (corrosive agents, environmental conditions, etc.)

Control devices, visual indicators, and displays (particularly those related to safety functions) shall be clearly marked with regard to their functions. All control devices and components shall be plainly identified with the same designation as it is shown in the technical documentation.

Marking related to fuses shall be located on or adjacent to each fuseholder or in another location, provided that it is obvious to which fuse holder the marking applies, and giving the following information:

- Fuse rated current
- Fuse rated voltage (where fuses of different rated voltage value could be fitted)
- Special fusing characteristics (if required)

For fuses located in an operator area, and for soldered-in fuses, it is permitted to provide an unambiguous cross-reference to service documentation, which shall include the necessary information.

Protective earthing terminals, terminals for external primary power supply conductors, shall be marked with the applicable symbols. These symbols shall not be placed on screws or other parts that might be removed when conductors are being connected. The marking shall not be placed on any removable parts that can be replaced.

For the regulatory devices intended to be adjusted during installation or in normal use, an indication shall be provided for the direction of adjustment to increase or decrease the value of the characteristic being adjusted.

The symbols and identification of components and devices shall be consistent throughout all documentation submitted with the equipment. Symbols shall conform to the standards where the appropriate symbol exists (e.g., ISO 7000, IEC 417, IEC 878, etc.).

Each particular standard has specific requirements for different *warning labels* that must appear on the equipment, related to the specific designation of the equipment (e.g., telephone equipment, spas, bathtubs, medical equipment, electric signs, etc.). The *warnings* shall be placed close to the subject or in the operator's manual (e.g., replaceable batteries, radiation hazards, etc.).

For each type of equipment, it is the responsibility of the manufacturer to consult the applicable standards to cover all the requirements from the point of view of the marking.

16.3.5 Documentation Requirements

Constantin Bolintineanu

With each supplied equipment, it is the responsibility of the manufacturer to provide the necessary information for transport, installation, operation, and maintenance of the equipment.

The required information shall be supplied in the form of instructions, schematics, diagrams, charts, tables, etc. All the above documentation shall be in the language acceptable by the end user, or in a language agreed upon by the end user and supplier. When the equipment is designed to be installed by service personnel, some standards accept English as a language to be used for Instructions.

The complexity of the equipment defines the characteristics of the instructions. The provided instructions shall give the information regarding the following issues:

1. Normal operating conditions of the equipment
2. Description of the equipment
3. Installation instructions: location, environmental conditions, methods of installation, guidance for mounting, spacing, etc.
4. Electrical supply requirements and the method of connection to the electrical supply (adjustments of the power supply, if applicable, protective earthing requirements)
5. Information on the physical environment (operating temperature, humidity, vibration, noise, type of location, etc.)
6. Operating instructions as follows:
    - Block diagram, circuit diagram, schematics
    - Programming instructions
    - Sequence of operation
    - List of the critical components
    - Specific instructions for the connection and disconnection of detachable parts and accessories and for the replacement of material that is consumed during operation
    - Precautionary instructions if a cover is removed and this condition can affect safety as related to access to moving parts, to hazardous voltages, etc.
    - Instructions for cleaning
    - Instructions for the use and maintenance of the rechargeable batteries, protective devices, etc.
    - Instructions for replacement of fuses and other parts

– Inappropriate uses

– Other instructions considered essential for a safe operation

7. Instructions for transportation and storage

– Instructions regarding the suitable precautions that shall be taken to protect against hazards during transportation, conditions of storage for long and short periods, etc.

8. Warning statements and the explanation of warning symbols (marked on the equipment)
9. An address to which the end-user can refer

The accompanying documentation shall be regarded as a component part of the supplied equipment.

For each equipment, depending on the type of equipment, the applicable standard can ask for specific details. For example, in the case of laser equipment, the instructions must explain in detail the warnings and must state the precautions needed to avoid possible exposure to hazardous laser radiation, the pulse duration and the maximum output, the locations of laser apertures, the maximum permissible exposure, the conditions for ventilation, the relevant national regulations, the safety classification of each laser product, etc.

The block diagram shall represent with symbols the electrical equipment along with its functional characteristics (waveforms, test prints, etc.) without necessarily showing all the interconnections. Circuits shall be shown in a manner that facilitates the understanding of their function as well as maintenance and fault location. Characteristics related to the function of the control devices and components that are not evident from their symbolic representation shall be included on the adjacent diagrams or referenced by a footnote or by some equivalent notation.

The operating instructions shall detail proper procedures for the installation and operation of the equipment. All necessary details must be provided as follows:

- The methods of programming
- The program verification
- The equipment required for testing and servicing
- Safety instructions
- Safety procedures

Regarding maintenance and service of the equipment, a list of controls, adjustments and procedures for operation and maintenance, and service procedures shall be provided, along with a list of those controls and procedures that could be used by persons other than the manufacturer or his agents. Protective procedures and precautions, maintenance and service requirements for service personnel, and the procedure and schedule for calibration of the measurement systems (where such a system is involved), shall be also provided.

Because of potential hazards, *only* persons who have received appropriate training can operate some equipment. Information regarding the requisites must be mentioned within the manuals. The manufacturer or the supplier (dealer) of the equipment or a recommended external organization can provide the required training by offering specific information as follows:

- Familiarization with the equipment and operating procedures
- Description of proper use of the control procedures, interlocks, etc.
- Basic requirements for personal protection-identification of the potential hazards
- Accident reporting procedures and first aid

It is recommended that all amendments to the instructions be given in writing. The manufacturer must keep accurate records of all the revisions of the instructions to ensure that each end user has received the updated version.

It is recommended that, as far as possible, all the documents be included in only one book; where this is not feasible, it is recommended that a numeration of the volumes be provided. The end user shall be aware that there is more than one volume to the user's manual (the instructions). Each document shall carry a cross-reference number to all the rest of the documents belonging to the equipment.

The documentation related to the safety must be provided in writing.

16.4 Product Certification

Steli Loznen

Designing electronic equipment for product safety compliance is a complex task, especially when attempting to satisfy multiple sets of requirements for diverse markets. The certification is a technical procedure that ensures satisfactory minimum quality control for products according to a predefined standard, norm, or document. Product certification increases user confidence in the safety of the product, helps to establish a firm reputation for producing quality products, and can represent protection against product liability lawsuits.

Third-party certification is a form of certification in which the producer's claim of conformity is validated, as part of a third-party certification program, by a technically and otherwise competent independent body, in no way controlled by the producer or buyer. The third party may be responsible for collecting the required data, generating test results, or conducting inspections, in addition to reviewing the results of such activities and making a final determination on the product's conformance or lack of conformance.

Self-certification is the process by which a manufacturer or supplier declares that a product meets one or more standards based on confidence in the internal quality control system or the results of testing or inspection undertaken in house or performed by others.

A quality control or quality assurance (QA) system is used by product manufacturers to control all the variables to produce a product of consistent quality and to meet defined specifications. The International Organization for Standardization (ISO) has published a series of international standards (ISO 9000, 9001, 9002, 9003, and 9004) on the subject of quality assurance (see also Chapter 16.1) which, along with the terminology and definitions contained in ISO Standard 8402, provide a practical and generally applicable set of principles for QA systems. ISO 9000 provides guidance on the selection of the specific quality management program most likely to be appropriate for a manufacturer's operations. ISO 9001, 9002, and 9003 describe distinct quality assurance model (general quality system, production quality system, product quality system), and ISO 9004 examines each of the quality system elements in greater detail. The ISO 9000 series standards provides guidance and information on basic requirements for quality management in manufacturing industries.

16.4.1 CE Marking

To ensure safety, health, environmental conservation, and consumer protection, the European Union (EU) drafted *New Approach Directives* for industrial products. These directives establish the requirements which electronic equipment manufacturers must meet to gain CE (Conformity European) marking for their products. What are the directives? Basically, the directives are a series of rules that member states and manufacturers must follow to prove that products are safe to use. By following the directives, the manufacturer can earn the right to affix the CE mark to his product. At the moment, there are several hundred directives, but only ≈20 of them are CE marking directives.

The CE marking is the means by which the EU is regulating and harmonizing technical and regulatory standards with respect to safety, health, and the environment. This harmonization means that manufacturers need obtain only one approval for all countries in the EU and EFTA (European Free Trade Association), except Switzerland.

The person responsible for placing the device on the market must be established in a member state of the EU.

All stand-alone electrical and electronic equipment shall be CE marked to qualify for sale in the EU. Prototypes shall not be CE marked but must be safe. Electronic components do not need marking. The CE mark is the key to entry to the European market and the key to free movement of goods within the EU. Non-European manufacturers do not have an automatic right to put their products on the European market.

The CE marking is not a safety, quality, or certification mark; it is a manufacturer's statement that all of the applicable requirements in the directives that apply to his product have been met. As such, it is a legal statement. It is the manufacturer's responsibility to ensure compliance.

A CE mark is awarded by manufacturer's self-declaration based on the tests conducted for compliance with a specific directive, or based on the certification provided by a European notified or certified body, which is specific for the appropriate product category (medical, telecommunication, EMC, etc.). Many of the EU directives require manufacturers to have a certified quality system conforming to ISO 9000. A documented quality system is a important route to CE marking. *Notified bodies* and *competent bodies* are appointed by the government of the EU country in which they are located and included in the official listing of bodies who are competent to conduct the tests in question.

The conformity assessment of a product or family of products may or may not require the certification by a notified body as regulated in the respective directive. Lists of notified bodies, the tasks and responsibilities which have been assigned to them, and their unique four digit identification number is published and updated in the *Official Journal (OJ) of the European Communities*. If certification is a requirement as part of the conformity assessment procedure, the manufacturer has the option to choose any of the notified bodies in any of the member states of the European Union. Notified bodies are and must remain third parties, independent of their clients and other interested parties.

4.1.1 Procedure for CE Marking

1. The procedure of CE marking begins with the question as to which directives apply to the product in question. For electronic equipment, some or all of the following directives may apply:

- Low Voltage Directive 73/23/EEC, amended by Directive 93/465/EEC
- EMC Directive 89/336/EEC, amended by Directive 92/31/EEC
- Active Implantable Medical Devices 90/385/EEC
- Telecommunication Terminal Equipment Directive 91/263/EEC
- Medical Devices Directive 93/42/EEC
- Satellite Earth Station Equipment Directive 93/97/EEC
- Explosive Atmospheres Directive 94/9/EEC
- In Vitro Medical Devices Directive 98/79/EEC

    The following directives apply to all types of equipment, whether they fit into one of the above directives or not:

- Liability of Defective Products Directive 85/374/EEC
- General Product Safety Directive 92/59/EEC
- CE Marking Directive 93/68/EEC and Decision 93/465/EEC

2. After establishing which directives apply, the second step in the process is to fulfill the essential requirements (general requirements and requirements regarding design and construction) of each directive. For this reason, the product should be compliant with a European harmonized standard to which conformity is declared.

    As an example, a power supply would fit into different directives, depending on its use:
    – If used in ITE (Information Technology Equipment), it should comply with EN60950 standard (without clause 6).

- If used in equipment connected to the telecommunication network, the TTE (Telecommunication Terminal Equipment) Directive applies, and it should comply with EN60950, including clause 6.
- If used in a medical equipment covered by the MDD (Medical Devices Directive), it should comply with EN60601-1 standard.
- If used in equipment for measurement, control, and/or laboratory applications, compliance with EN61010-1 standard is required.

The manufacturer may either carry out the necessary safety tests itself (if the in-house testing laboratory is certified) or use an independent testing house that has been certified by a *Competent European Authority*. If a manufacturer puts a product on the market without testing it as required, that manufacturer may be guilty of a criminal offense.

3. As soon as the test is completed, a declaration of conformity to the essential requirements must be prepared. The declaration of conformity includes:
   - Manufacturer's name and address
   - European representative's name and address
   - Identification of device: name, type, and model
   - Name and number of directives complied with
   - Number and year of harmonized standard to which conformity is declared
   - The statement, "We, the undersigned, hereby declare that the equipment specified above conforms to the above Directive and Standards."
   - Manufacturer's and representative's signatures

   The declaration of conformity shall be provided for each product. If the manufacturer is not established in the EU, or if there is no EU representative, the importer of the product is responsible according to law.

4. After signing the declaration of conformity, the manufacturer affixes a "CE" mark to the product. The CE marking must have a vertical dimension not less than 5 mm and must be affixed visibly, legibly, and indelibly.

5. A technical file must be maintained that contains approved documentation, showing the manufacturer's reasoning as to how the product satisfies EU directives. This must be kept for at least 10 years after the manufacture of the last unit of the product. This documentation will be made available at the request of an enforcement authority. The technical file will comprise:
   - A copy of the declaration of conformity
   - A general description of the unit, including all variants, functional description, interconnection (for systems), step-by-step operating and use instructions, photographs spotlighting the product, and the usage
   - Product specifications including assembly drawings, circuit diagrams, packaging specs, and labeling
   - Product verification: list of standards applied, all test reports and certificates, safety-related component licenses, risk analysis, validation of packaging, and aging studies

Low Voltage Directive

The Low Voltage Directive (LVD) applies for equipment with a specified ac supply voltage of 50 V to 1000 V or dc supply voltage from 75 to 1500 V. The LVD applies to all electrical equipment, including domestic, professional, and industrial products. Certain specialized equipment, such as medical equipment, telecommunication terminal equipment, and equipment for use in explosive atmospheres are the subject of separate directives. The most reliable method of demonstrating compliance with the LVD is to comply with a "harmonized" European safety standard. Harmonized standards are those that have been ratified and published by CENELEC (European Committee for Electrotechnical Standardization) after having been approved by a majority of EU states. These standards are listed in the *Official Journal of the European Union*. If no EN standard exists for a specific product, look for other international (IEC, ISO) or national standards. Adequate manufacturing control and testing should be carried out as part of an organization's manufacturing quality system. The technical file must be available from the manufacturer. The LVD covers only complete equipment, but sometimes a component of complex equipment can be considered a piece of equipment in its own right. A packaged resistor placed on the market separately does not fall within the LVD definition of electrical equipment, but a PCB with components attached to it (i.e., a CPU, an A/D converter, etc.) may be within the scope of LVD when placed on the market separately. Electrical components that fall outside the scope of the EMC Directive are usually exempt from the LVD, too.

EMC Directive

Compliance with the EMC Directive can be demonstrated in one of the following ways:

- Self certification, which may be accomplished by having the tests done by at a specialist test house or by acceptable in-house testing.
- Technical construction file (TCF), which requires detailing the design, specifications, drawings, and all information that will convince the authorities that, if the product was tested independently, it would comply. These documents are submitted to a "competent body" for approval. The TCF is used when tests cannot be performed in accordance with relevant standards. This approach is used when standard test procedures are not possible or when alternative test standards and/or methods are used because of the product's specific constraints.
- Type examination by a relevant body, which is a compliance process for products designed for the transmission of radio communications as defined by the ITU (International Telecommunications Union).

For all approvals required under the EMC directive, the manufacturer must use a *competent body* (there is no functional difference between a *competent body* and a *notified body*), which can appoint an affiliated test laboratory. As such, the tests done by this laboratory are accepted by the competent body.

Upon successful completion of testing and documentation of the results, the manufacturer issues a *declaration of conformity* and affixes the CE Mark to the product. As an alternative to affixing the CE marking to product, the LVD and EMC Directives allow the marking to be placed on the packaging or documentation accompanying the equipment (the instructions for use, or the guarantee certificate).

Medical Devices Directive

Devices covered by Medical Device Directive are assigned to one of four classes (I, IIa, IIb, and III) according to the risks that exist in the use of the product. Software associated with a medical device falls under the same classification as its host device. Procedures for product assessment vary according to the device's classification.

In addition to compliance with the essential requirements of MDD, an assessment of quality assurance procedures is required for all specified medical devices. The most important regulations for quality assurance in the production of medical devices are the EN 29000 Series and EN 46000 Series standards, and for technical safety the EN 60601 Series standards.

For medical devices in Class I, the manufacturer declares compliance of the product with the relevant requirements (EN 60601 Series). Sterile products and measuring devices are expected to be evaluated by a notified body.

Product Safety and Third-Party Certification

For medical devices covered by Class IIa, the conformity declaration of the manufacturer must be accompanied by a conformity assessment from a notified body. This assessment may consist of an audit for quality assurance system in conformity with EN29001/EN46001, an audit for production quality assurance in conformity with EN29002/EN46002, an audit for product quality assurance (end-production checks: inspection and testing) in conformity with EN 29003, or by examination and testing of sample products (EN 60601 Series). The assessment method is the manufacturer's choice.

For medical devices included in Class IIb, a notified body will carry out either a full audit for quality assurance system according to EN29001/EN46001 or *type examination* combined with audit for production quality assurance in conformity with EN 29002/EN46002 or audit for product quality assurance (end-production checks: inspection and testing) in conformity with EN 29003 or examination and testing of sample product (EN60601 Series).

For medical devices from Class III, a notified body will carry out either a full audit for quality assurance system according to EN29001/EN46001 with product dossier examination or a type examination combined with an audit of production quality assurance according to EN 29002/EN46002 or an examination and testing of sample product (EN60601 Series).

Steps for application of CE marking on a medical devices are as follows:

1. The manufacturer classifies the product as a Class I, II, or III (including subsets) product.
2. The manufacturer selects a notified body, if required, with the ability to carry out the appropriate conformity assessment procedure.
3. The manufacturer ensures that the product complies with the essential requirements of the directive.
4. The manufacturer chooses the conformity assessment procedure in accord with the notified body.
5. The manufacturer elaborates the product dossier.
6. The manufacturer provides the notified body with the product dossier and the samples required for evaluation (if applicable).
7. The notified body will conduct the assessment of the product and quality management auditing of the production.
8. The notified body will decide about the conformity with the directive and supply the certification.
9. The manufacturer writes the *declaration of conformity*.
10. The manufacturer affixes the CE mark to the product.

Note: For medical devices, the notified bodies are designated by the Ministry of Health in each country and are charged with the task of ensuring products meet essential safety requirements and perform according to the manufacturer's specifications.

16.4.2 UL Mark

Underwriters Laboratories Inc. (UL) is the leading third-party product safety certification organization in the United States. The UL mark is awarded for listing, classification, and recognition.

UL's *listing* means that UL has tested and evaluated representative samples of the product and determined that they meet UL's standards for safety. (Products that bear the UL mark for Canada have been tested and evaluated to the appropriate Canadian standard for safety.)

UL's *classification* means that UL has tested and evaluated a representative sample of the product with respect to certain properties of the product, a limited range of hazards, or suitability of use under limited or special conditions. UL classifies products to applicable UL requirements and standards for safety, as well as to the standards of other national and international organizations such as the NFPA, ASTM, ISO and IEC.

UL's *recognition* covers the evaluation of components or materials (i.e., plastics, etc.) that will be used in a complete product or system. These components are products that are incomplete in construction, restricted in performance capabilities, or otherwise intended only for incorporation into other end-use products that may be eligible for UL's listing or classification.

For all components, the "conditions of acceptability" shall be considered for the components' inclusion in end-use product. UL's *Recognized Component Directory* (the *Yellow Book*) contains recognition for different components. This directory is a resource for product designers, end-product manufacturers and other to select (early in the product development cycle) suitable components that fit their specific needs.

Note: Products certified by other organizations, even if they are tested to UL standards, cannot bear any UL mark.

The follow-up services protect and control the use of the UL mark. Each year, UL field representatives visit factories to observe, inspect, and select samples to countercheck that products continue to meet UL requirements.

16.4.3 Reciprocal Certification Scheme

A reciprocal certification scheme may be established whereby a certificate of conformity (or license to label the product with a certification mark) is granted in one participating country and recognized in other participating country. These schemes can be for a regional level or for an international level. Examples of regional schemes include

- The CENELEC HAR Agreement, an agreement on the use of a marking scheme for electrical cables and cords complying with harmonized specifications
- The CENELEC Electronic Components Committee System (CECC) under which participants must accept certified electronic components produced under this scheme without further testing
- The CENELEC ENEC (European Norms Electrical Certification) mark, a voluntary safety mark issued by 20 European signatory bodies and used for information technology equipment and electrical business equipment, transformers and luminaries products

- The *Keymark*, which is the CEN (European Committee for Standardization)/ CENELEC European Mark, a voluntary registered third-party product certification mark to show conformity of products to ENs European standards

Although the CE marking was introduced by the EU to facilitate free trade throughout European countries, it is largely self-declared by manufacturers, often without a basis in safety testing and conformance to recognized standards, and often without a basis in independent, third-party certification.

Product Safety and Third-Party Certification

The *Keymark* gives appliance manufacturers a true safety mark, based on European Norms from a third-party certifier. The Keymark also requires confirmation of the manufacturer's quality management system. Both the manufacturer and the product will be subjected to regular assessment. The Keymark may replace the numerous national certification marks for products that are sold in Europe. The Keymark does not replace the CE Mark which is a mandatory mark.

CB Scheme

The CB Scheme is an international network of product certification organizations, called National Certification Bodies (NCB) in CB-Scheme language, established by the IEC (International Electrotechnical Committee) for Conformity Testing to Standards for Electrical Equipment (ICEE). It provides a means for the mutual acceptance of test reports for product certification at the national level. Then, when a National Certification Body issues a CB Test Certificate (CBTC) and CB Test Report (CBTR), the manufacturer can use this information to obtain national certifications in 36 participating countries: Austria, Australia, Belgium, Canada, China, Czech Republic, Germany, Denmark, Finland, France, Greece, Hungary, Iceland, India, Ireland, Israel, Italy, Japan, Luxembourg, Netherlands, Norway, Poland, Portugal, Russian Federation, Singapore, Slovakia, Slovenia, South Africa, South Korea, Spain, Sweden, Switzerland, Ukraine, United Kingdom, U.S.A., and Yugoslavia.

For each NCB, specific IEC standards are assigned. Difference requirements for other countries can be fulfilled by a supplementary test report. Some NCBs require inspection of the manufacturer's factory and may use a local agency for that purpose.

The following categories of products are covered by the CB scheme:

- CABL, Cables and cords
- CAP, Capacitors as components
- CONT, Switches for appliances and automatic controls for electrical household equipment
- HOUS, Household and similar equipment
- INST, Installation accessories and connection devices
- LITE, Lighting
- MEAS, Measuring instruments
- MED, Electrical equipment for medical use
- OFF, Information Technology and office equipment
- POW, Low voltage, high power switching equipment
- PROT, Installation protective equipment
- SAFE, Safety transformers and similar equipment
- TOOL, Portable tools
- TRON, Electronics, entertainment

Mutual Recognition Agreements

Mutual recognition agreements (MRAs) for conformity assessment are considered as a means to reduce barriers to trade. This is one of several important instruments available that can help to achieve the single market.

In 1997, an MRA was started between United States and the European Union by creating a mechanism for each to recognize the other's conformity assessment bodies (CABs). The MRA covers the following areas:

- Telecommunications
- Electromagnetic compatibility (EMC)
- Electrical safety
- Pharmaceuticals
- Recreational craft
- Medical devices.

The MRA is intended to give both U.S. and EU certifiers the opportunity to evaluate products for each other's home territory, but it does not necessarily affect the marketplace acceptance of those certifications. The MRA does not create a direct equivalence between European and U.S. regulations in the above areas. Rather, it allows for mutual recognition of test results and other conformity assessment documentation. Components recognized by the UL or by the Canadian Standards Association (CSA) as complying with North American component requirements do not meet the European requirements unless CSA or North American standards are equivalent to or more demanding than the European standards. Likewise, European standards may or may not automatically be accepted by UL or CSA.

Other MRAs are signed between EU and Canada, Australia, and New Zealand or are in progress with Japan, Eastern European countries, and others.

16.4.4 Methods Used in Third-Party Certification

- Type testing (conformity to a particular standard)
- Surveillance of the manufacturing process (assessment of materials, components, production and control processes)
- Audit testing (test samples are random selected from the market)
- Field investigation (determination of the cause of failures of products during the use)
- Batch testing (testing for conformance to the standard of a sample selected from current production)
- 100% product testing (each individual product is tested)

The choice of methods depends on the needs of both the manufacturer and the user and on the nature of the product. To be accredited, a third-party test house shall meet the requirements of ISO Guide 25 and of EN45001 standard.

16.5 Appendix

Table 16.4 is a sample list of various Underwriters Laboratories' standards. Note that this is only a representative list and contains *less than one-third of UL's standards*. A complete listing can be found from UL's web site, www.ul.com. Note also that similar listings are available from agencies such as the Canadian Standards Association (CSA), the International Electrotechnical Commission (IEC) (www.iec.ch), the European Committee For Electrotechnical Standardization (CENELEC) (www.cenelec.be), and TÜV Rheinland Japan Ltd. (www.jpn.tuv.com). The reader must carefully determine the appropriate agency and applicable standard(s) in the evaluation of a product.

TABLE 16.4 Sample UL Standards

UL Std. No.	Title	Latest Ed. No.	Year
22	Amusement and Gaming Machines	fifth	1999
50	Enclosures for Electrical Equipment	eleventh	1995
65	Wired Cabinets	sixth	1997
67	Panelboards	eleventh	1993
73	Motor-Operated Appliances	eighth	1993
79	Power-Operated Pumps for Petroleum Product Dispensing Systems	eighth	1996
187	X-Ray Equipment	seventh	1998
197	Commercial Electric Cooking Appliances	eighth	1993
198B	Class H Fuses	fifth	1995
198C	High-Interrupting-Capacity Fuses, Current-Limiting Types	fifth	1986
198D	Class K Fuses	fifth	1995
198E	Class R Fuses	fourth	1988
248-16	Low-Voltage Fuses—Part 16: Test Limiters	first	1996
250	Household Refrigerators and Freezers	tenth	1993
268	Smoke Detectors for Fire Protective Signaling Systems	fourth	1996
268A	Smoke Detectors for Duct Application	third	1998
275	Automotive Glass-Tube Fuses	ninth	1993
291	Automated Teller Systems	fourth	1995
294	Access Control System Units	fourth	1994
346	Waterflow Indicators for Fire Protective Signaling Systems	fourth	1994
347	High Voltage Industrial Control Equipment	fourth	1993
351	Rosettes	seventh	1995
352	Constant-Level Oil Valves	seventh	1997
353	Limit Controls	fifth	1994
416	Refrigerated Medical Equipment	fourth	1993
427	Refrigerating Units	third	1994
429	Electrically Operated Valves	fifth	1999
464	Audible Signal Appliances	seventh	1996
466	Electric Scales	eighth	1999
467	Grounding and Bonding Equipment	seventh	1993
469	Musical Instruments and Accessories	third	1997
471	Commercial Refrigerators and Freezers	eighth	1995
474	Dehumidifiers	seventh	1993
482	Portable Sun/Heat Lamps	seventh	1994
484	Room Air Conditioners	seventh	1993
486A	Wire Connectors and Soldering Lugs for Use with Copper Conductors	ninth	1997
499	Electric Heating Appliances	twelfth	1997
506	Specialty Transformers	eleventh	1994
507	Electric Fans	eighth	1994
508	Industrial Control Equipment	sixteenth	1993
508C	Power Conversion Equipment	second	1996
510	Polyvinyl Chloride, Polyethylene, and Rubber Insulating Tape	seventh	1994
511	Porcelain Cleats, Knobs, and Tubes	sixth	1994
512	Fuseholders	tenth	1993
514A	Metallic Outlet Boxes	ninth	1996
514B	Fittings for Cable and Conduit	forth	1997
514C	Nonmetallic Outlet Boxes, Flush-Device Boxes, and Covers	third	1996
521	Heat Detectors for Fire Protective Signaling Systems	seventh	1999
525	Flame Arresters	sixth	1994
536	Flexible Metallic Hose	ninth	1997
539	Single and Multiple Station Heat Detectors	fourth	1995
541	Refrigerated Vending Machines	sixth	1995
542	Lampholders, Starters, and Starter Holders for Fluorescent Lamps	seventh	1994
544	Medical and Dental Equipment	fourth	1998
551	Transformer-Type Arc-Welding Machines	seventh	1994

A
Definitions

Additive Process.   A process for obtaining conductive patterns by the selective deposition of conductive material on clad or unclad base material. (See also "Semi-additive Process" and "Fully Additive Process.")

Adhesive.   A substance such as glue or cement used to fasten objects together. In surface mounting, an epoxy adhesive is used to adhere SMDs to the substrate.

Adhesion Failure.   The rupture of an adhesive bond such that the separation appears to be at the adhesive-adhered interface.

Alloy.   Two or more elements, at least one of them being a metal, combined.

Ambient.   The contacting environment surrounding a system, assembly, or component.

Ambient Temperature.   The average or mean temperature of the surrounding air that comes in contact with the equipment under test.

Annular Ring.   That portion of conductive material completely surrounding a hole.

ANSI.   American National Standards Institute

Aqueous Flux.   An organic soldering flux that is soluble in distilled water.

Aramid.   Fibers made from polyamide, or amide polymers that have at least 85% of the linkages directly attached to two benzene rings.

ASIC.   Application specific integrated circuit. Custom semiconductors, designed for very specific functions and are used in electronic products such as camcorders, automobile air bag systems, and printers.

Aspect Ratio (Hole).   The ratio of the length or depth of a hole to its preplated diameter.

Attenuation.   The reduction in the amplitude of a signal due to losses in the media through which it is transmitted. The unit of measure is decibels (dB).

BGA or Ball Grid Array.   A surface mount package attached to a substrate using an array of solder balls as the interconnect technology.

B-Stage.   An intermediate stage in the reaction of a thermosetting resin in which the material softens when heated and swells, but does not entirely fuse or dissolve, when it is in contact with certain liquids. (See also "C-Stage Resin.")

B-Staged Material.   See "Prepreg."

Backward Crosstalk.   Noise induced into a quiet line, as seen at the end of the quiet line that is closest to the signal source, because the quiet line has been placed next to an active line. (See also "Forward Crosstalk.")

Balanced Transmission Line.   A transmission line that has distributed inductance, capacitance, resistance, and conductance elements that are equally distributed between its conductors.

Ball Bond.   The thermocompression termination of the ball-shaped end of an interconnecting wire to a land. (See also "Wedge bond.")

Bare Board.   An unassembled (unpopulated) printed board.

Base Material.   The insulating material upon which a conductive pattern may be formed. (The base material may be rigid or flexible, or both. It may be a dielectric or insulated metal sheet).

Basis Material. Material upon which coatings are deposited.

Bathtub Curve. A plot of typical failures versus time.

Bed-of-Nails Fixture. A test fixture consisting of a frame and a holder containing a field of spring-loaded pins that make electrical contact with a planar test object.

Benchmark, Testing. A standard measure of the performance of testers relative to each other, including setup time, test program generation, and fixturing.

Bifurcated Contact. A type of connector contact that usually consists of a flat spring that has been slotted lengthwise to provide independent contact points with the mating part.

Blind Via. A via extending from an interior layer to only one surface of a printed board.

Blister. Delamination in the form of a localized swelling and separation between any of the layers of a lamination base material, or between base material and conductive foil or protective coating.

Bond. An interconnection that performs a permanent electrical and/or mechanical function.

Bond-to-Bond Distance. The distance from the bonding site on a die to the corresponding bonding site on a lead frame, interconnecting base material, etc.

Bonding Wire. Fine gold or aluminum wire used for making electrical connections between lands, lead frames, and terminals.

Bump (Die). A raised metal feature on a die land or tape carrier tape that facilitates inner-lead bonding.

Bumped Die. A semiconductor die with raised metal features that facilitate inner-lead bonding.

Bumped Wafer. A semiconductor wafer with raised metal feature on its die lands that facilitate inner-lead bonding.

Buried Via. A via that connects two or more interior layers of a printed board and does not extend to either surface of that printed board.

Burn-In. The process of electrically stressing a device at an elevated temperature, for a sufficient amount of time to cause the failure of marginal devices (a.k.a. *Infant Mortality*).

C-Stage Resin. A resin that has reached its final cure stage. See also "B-stage resin."

C4 or Controlled Collapse Chip Connection. A methodology of solder joint flip chip connection to a substrate where the surface tension forces of the molten solder controls the height of the joint and supports the weight of the chip.

CBGA. Ceramic Ball Grid Array.

CD. Compact Disc.

Capacitive Coupling. The electrical interaction between two conductors that is caused by the capacitance between them.

Castellation. A recessed metallized feature on the edge of a leadless chip carrier that is used to interconnect conducting surface or plans within or on the chip carrier.

Center-to-Center Spacing. The nominal distance between the centers of adjacent features on any single layer of a printed board. (See also "Pitch.")

Ceramic substrates. Inorganic, nonmetallic materials used primarily because of their long life, low loss, and ability to withstand high operating temperatures. Examples include aluminum oxide, beryllium oxide, glass, and ceramics.

Characteristic Impedance. The resistance of a parallel conductor structure to the flow of alternating current (ac), usually applied to high-speed circuits, and normally consisting of a constant value over a wide range of frequencies.

Chip. See "Die."

Chip Carrier. A low-profile, usually square, surface-mount component semiconductor package whose die cavity or die mounting area is as large fraction of the package size and whose external connections are usually on all four sides of the package. (It may be leaded or leadless.)

Chip scale package, chip size package or CSP. An active, multi-I/O package that is typically no more than 20% larger than the IC.

Chip-and-Wire. An assembly method that uses discrete wires to interconnect back-bonding die to lands, lead frames, etc.

Chip-on-Board (COB). A printed board assembly technology that places unpackaged semiconductor dice and interconnects them by wire bonding or similar attachment techniques. Silicon area density is usually less than of the printed board.

Circuit. A number of electrical elements and devices that have been interconnected to perform a desired electrical function.

Clad (adj.). A condition of the base material to which a relatively thin layer or sheet of metal foil has been bonded to one or both of its sides, e.g., "metal-clad base material."

Clean Room. A manufacturing area that contains air specially filtered to remove dust particles.

Clearance Hole. A hold in a conductive pattern that is larger than, and coaxial with, a hole in the base material of a printed board.

Clinched Lead. A component lead that is inserted through a hole in a printed board and is then formed to retain the component in place and in order to make metal-to-metal contact with a land prior to soldering. (See also "Partially-Clinched Lead.")

Closed-Entry Contact. A type of female connector contact that prevents the entry of an oversized mating part.

CMOS or Complementary Metal Oxide Semiconductor. CMOS is a widely used type of semiconductor that uses both NMOS (negative polarity) and PMOS (positive polarity) circuits. CMOS chips require less power than chips using just one type of transistor. They are used in many battery-powered devices, such as portable computers and PCs, which also contain a small amount of battery-powered CMOS memory to hold the date, time, and system setup parameters.

Co-Firing. The simultaneous processing of thick-film circuit elements during one firing cycle.

Coaxial Cable. A cable in the form of a central wire surrounded by a conductor tubing or sheathing that serves as a shield and return.

Coefficient of Thermal Expansion (CTE). The linear dimensional change of a material per unit change in temperature. (See also "Thermal Expansion Mismatch.")

Cohesion Failure. The rupture of an adhesive bond such that the separation appears to be within the adhesive.

Coined Lead. The end of a round lead that has been formed to have parallel surfaces that approximate the shape of a ribbon lead.

Cold Solder Connection. A solder connection that exhibits poor wetting and is characterized by a greyish porous appearance. (This is due to excessive impurities in the solder, inadequate cleaning prior to soldering, and/or the insufficient application of heat during the soldering process.) (See also "Rosin Solder Connection.")

Common Cause. A source of variation that affects all the individual values of the output of a process.

Component Density. The quantity of components on a unit area of printed board.

Component Lead. The solid or stranded wire or formed conductor that extends from a component to serve as a mechanical or electrical connector, or both.

Component Side. See "Primary Side."

Computer-Aided Design (CAD). The interactive use of computer systems, programs, and procedures in the design process wherein the decision-making activity rests with the human operator, and a computer provides the data manipulation function.

Computer-Aided Manufacturing (CAM). The interactive use of computer systems, programs, and procedures in various phases of a manufacturing process wherein the decision-making activity rests with the human operator and a computer provides the data manipulation functions.

Condensation Soldering. See "Vapor Phase Soldering."

Conductive Pattern. The configuration or design of the conductive material on a base material. (This includes conductors, lands, vias, heatsinks, and passive components when these are integral part of the printed board manufacturing process.)

Conductivity (Electrical). The ability of a substance or material to conduct electricity.

Conductor. A single conductive path in a conductive pattern.

Conductor Layer No. 1. The first layer of a printed board that has a conductive pattern on or adjacent to its primary side.

Conductor Spacing. The observable distance between adjacent edges (not center-to-center spacing) of isolated conductive patterns in a conductor layer. (See also "Center-to-Center Spacing.")

Conductor Width. The observable width of a conductor at any point chosen at random on a printed board as viewed from directly above unless otherwise specified.

Conformal Coating. An insulating protective covering that conforms to the configuration of the objects coated (e.g., printed boards, printed board assembly) providing a protective barrier against deleterious effects from environmental conditions.

Confounding. A situation whereby certain effects cannot be separated from other effects.

Connector. A device used to provide mechanical connect/disconnect service for electrical terminations.

Connector, One-Part. See "Edge-Board Connector."

Connector, Two-Part. A connector containing two sets of discretely formed mating metal contacts.

Constraining Core. A supporting plane that is internal to a packaging and interconnecting structure.

Contact Angle (Soldering). The angle of a solder fillet that is enclosed between a plane that is tangent to the solder/basis-metal surface and a plane that is tangent to the solder/air interface.

Contact Retention Force. The minimum axial load in either direction that a contact withstands while it is in its normal position in a connector insert.

Contact Spacing. See "Pitch."

Coplanarity. When measured on a flat surface, the distance between the lowest and highest points on a wafer, substrate, or package. In flip chip, *coplanarity* typically refers to the delta between the lowest and highest bump heights.

Cpk Index (Cpk). A measure of the relationship between the scaled distance between the process mean value and the closest specification limit.

Creep. Time-dependent strain occurring under stress.

Critical Defect. Any anomaly specified as being unacceptable.

Crosstalk. The undesirable interference caused by the coupling of energy between signal paths. (See also "Backward Crosstalk" and "Forward Crosstalk.")

CSP. See chip scale package.

DCA or Direct Chip Attach. Process by which the silicon die is interconnected directly to the PCB; includes flip chip and COB.

Decoupling. The absorbing of noise pulses in power supply lines, which were generated by switching logic, so as to prevent the lines from disturbing other logic in the same power-supply circuit.

Defect. Any nonconformance to specified requirements by a unit or product.

Delamination. A separation between plies within a base material, between a base material and a conductive foil, or any other planar separation with a printed board. (See also "Blister.")

Dendritic Growth. Metallic filaments that grow between conductors in the presence of condensed moisture and an electric bias.

Design Rule. Guidelines that determine automatic conductor routing behavior with respect to specified design parameters.

Design-Rule Checking. The use of a computer-aided design program to perform continuity verification of all conductor routing in accordance with appropriate design rules.

Desmear. The removal of friction-melted resin and drilling debris from a hole wall.

Dewetting. A condition that results when molten solder coats a surface and then recedes to leave irregularly shaped mounds of solder that are separated by areas that are covered with a thin film of solder and with the basis metal not exposed.

Dice. Two or more die.

Die. The uncased and normally leadless form of an electronic component that is either active or passive, discrete or integrated. (See also "Dice.")

Die Bonding. The attachment of a die to base material.

DIPs. See Dual Inline Package.

Dip Soldering. The making of soldered terminations simultaneously by bringing the solder side of a printed board with through-hole mounted components into contact with the surface of a static pool of molten solder.

Discrete Component. A separate part of a printed board assembly that performs a circuit function, e.g., a resistor, a capacitor, a transistor, etc.

Don't Care Area. See "Exclusion Area."

Doping. The addition of an impurity to alter the conductivity of a semiconductor die.

Double-Sided Printed Board. A printed board with a conductive pattern on both of its sides.

Drawbridged Component. See "Tombstoned Component."

Dross. Oxide and other contaminants that form on the surface of molten solder.

DSP or Digital Signal Processing. DSP devices process signals, such as image and sound signals or radar pulses, by converting analog signals to digital and enhancing, filtering, modulating, and otherwise manipulating the signals at high speeds.

Dual-Inline Package (DIP). A basically rectangular component package that has a row of leads extending from each of the longer side of its body that are formed at right angles to a plane that is parallel to the base of its body. DIPs are intended for through-hole mounting in a board.

Edge Rate. The rate of change in voltage with time of a logic signal transition.

Edge-Board Connector. A connector that is used specifically for making nonpermanent interconnections with the edge-board contacts on a printed board.

Edge-Board Contact(s). Printed contact(s) on or near any edge of a printed board that are used specifically for mating with edge-board connectors.

EIA. Electronic Industries Alliance (formerly Electronic Industries Association).

EIAJ. Electronic Industries Association of Japan.

Electrodeposition. The deposition of a conductive material from a plating solution by the application of electrical current.

Electroless Deposition. The deposition of conductive material from an autocatalytic plating solution without the application of electrical current.

Electroless Plating. See "Electroless Deposition."

Electromagnetic Interference (EMI). Unwanted radiated electromagnetic energy that couples into electrical conductors.

Embedded Component. A discrete component that is fabricated as an integral part of a printed board.

Epoxy Smear. See "Resin Smear."

Escapes. Critical defects that are missed by an inspection system.

Etchback. The controlled removal by a chemical process, to a specific depth, of nonmetallic materials from the sidewalls of holes to remove resin smear and to expose additional internal conductor surfaces.

Etching. The removal of excess metallic substance applied to a base via chemical or chemical and electrolytic means.

Eutectic. An alloy of two or more metals where the liquidus and solidus points are equal.

Eutectic Die Attach. The mounting of a semiconductor die to a base material with a perform of a eutectic metal alloy that is brought to its eutectic melting temperature.

Fault. Any condition that causes a device or circuit to fail to operate in a proper manner.

Fault Dictionary. A list of elements in which each element consists of a fault signature that can be used to detect a fault.

Fault Isolation. The identification process used to determine the location of a fault to within a small number of replaceable components.

Fault Modes. The various ways faults may occur.

Fault Resolution. A measure of the capability of a test process to perform failure isolation.

Fault Signature. The characteristic, unique erroneous response produced by a specific fault.

Fault Simulation. A process that allows for the prediction or observation of a system's behavior in the presence of a specific fault with actually having that fault occur.

Fault Tolerance. The ability to execute tasks regardless of the failure of strategic components.
FCB. Flip chip bump.
FCOB. Flip chip on board.
Feature. The general term that is applied to a physical portion of a part, such as a surface, hole, or slot.
FEOL. Front-end of line.
Fiducial Mark. A printed board artwork feature (or features) that is created in the same process as the conductive pattern and that provides a common measurable point for component mounting with respect to a land pattern or land patterns.
Fine-Pitch Technology (FPT). A surface mount assembly technology with component terminations on less than 0.625-mm (0.025-in) centers.
Fire (v.). To heat a circuit so that its thick-film components are transformed into their final form.
Fixture, Test. A device that interfaces between test equipment and the unit under test.
Flat Cable. Two or more parallel, round or flat, conductors that are contained in the same plane of a flat insulating base material.
Flat Pack. A rectangular component package that has a row of leads extending from each of the longer sides of its body that are parallel to the base of its body.
Flexible Printed Board. A printed board using a flexible base material only. May be partially provided with electrically nonfunctional stiffeners and/or cover lay.
Flexible Printed Circuit. Printed wiring and components utilizing flexible base materials with or without flexible cover layers.
Flip Chip. A leadless, monolithic circuit element structure that electrically and mechanically interconnects to a base material through the use of conductive bumps.
Flip Chip Mounting. The mounting and interconnecting of a flip chip component to a base material.
Flux. A chemically and physically active compound that, when heated, promotes the wetting of a base metal surface by molten solder by removing minor surface oxidation and other surface films and by protecting the surfaces from reoxidation during a soldering operation.
Flux Activity. The degree or efficiency with which a flux promotes wetting of a surface with molten solder.
Flux Residue. A flux-related contaminant that is present on or near the surface of a solder connection.
Footprint. See "Land Pattern."
Forward Crosstalk. Noise induced into a quiet line, as seen at the end of the quiet line that is the farthest from the signal source, because the quiet line has been placed next to an active line. (See also "Backward Crosstalk.")
Fully Additive Process. An additive process wherein the entire thickness of electrically isolated conductors is obtained by the use of electroless deposition. (See also "Semi-Additive Process.")
Functional Tester. Equipment that analyzes the unit under test as a complete functional entity by applying inputs and sensing outputs.
Glass Transition Temperature. The temperature at which an amorphous polymer, or the amorphous regions in a partially-crystalline polymer, changes from being in a hard and relatively brittle condition to being in a viscous or rubbery condition.
Golden Board. See "Known Good Board."
Ground Plane. A conductive layer that serves as the ground, or common, reference for electrical power, signals, and shielding.
Guarding. The in-circuit testing process of ensuring that a shunt path does not interfere with the testing of a device.
Gull Wing Leads. An SMT lead form. Leads extending horizontally from the component body centerline, bent downward immediately past the body, and then bend outward just below the bottom of the body, thus forming the shape of a gull's wing.
HDI. High density interconnect.
Header (Connector). A pin field that is positioned in a three- or four-sided plastic housing that mounts directly onto a printed board.

Heatsink. A mechanical device that is made of a high-thermal-conductivity and low-specific-heat material that dissipates heat generated by a component or assembly.

Hi rel. A contraction of high reliability.

Hole Breakout. A condition in which a hole is not completely surrounded by the land.

Hole Density. The quantity of holes in a unit area of printed board.

HTOL. High temperature operating life (accelerated stress test).

Hybrid Circuit. An insulating base material with various combinations of interconnected film conductors, film components, semiconductor dice, passive components, and bonding wire that form an electronic circuit.

Hybrid Microcircuit. An insulating base material with various combinations of interconnected film conductors, film components, semiconductor dice, passive components, and bonding wire that form an electronic microcircuit.

IEEE. Institute of Electrical and Electronics Engineers.

IMAPS. International Microelectronics and Packaging Society.

IMC. Intermetallic compound.

Impedance. The resistance to the flow of current, represented by an electrical network of combined resistance, capacitance, and inductance reaction, in a conductor as seen by an ac source of varying time voltage. The unit of measure is ohms.

In-Circuit Testing. The application of test signals directly to a device's input terminals and sensing the results directly from the device's output terminals.

Infrared Reflow (IR). Remelting of solder using infrared heating as the primary source of energy.

Insufficient Solder Connection. A solder connection that is characterized by the incomplete coverage of one or more of the surfaces of the connected metals and/or by the presence of incomplete solder fillets.

Integrated Circuit. A combination of inseparable associated circuit elements that are formed in place and interconnected on or within a single base material to perform a microcircuit function.

Interconnect. A highly conductive material that carries electrical signals to different parts of a die.

Interposer. The electrical interconnection structure between an IC and its package.

Intermetallic Compound, Solder. An intermediate layer in a wetted solder connection between the wetted surface and the solder, consisting of the solution of at least one constituent of the wetted surface and at least one constituent of the solder.

I/O. Input/output.

IPC. Institute for Interconnecting and Packaging Electronic Circuits.

ISO 9002. An international quality standard that includes development, production, installation and service.

JEDEC. Joint Electronic Devices Engineering Council of the EIA.

J-Leads. The preferred surface mount lead form used on PLCCs, so named because the lead departs the package body near its Z-axis centerline, is formed down then rolled under the package. Leads so formed are shaped like the letter "J."

Junction Temperature. The temperature of the region of a transition between the p-type and n-type semiconductor material in a transistor or diode element.

Known Good Board (KGB). A correctly fabricated printed board that serves as a standard unit by which others can be compared.

Known Good Die (KGD). An unpackaged integrated circuit chip known to meet all performance and electrical specifications established for the life of an application.

Laminate. The raw material for printed circuits made of a sheet of plastic with copper foil adhered to one or both sides.

Land. A portion of a conductive pattern usually, but not exclusively, used for the connection and/or attachment of components.

Laser Trimming. The modification of a film component's value by the removal of film by applying heat from a focused laser source.

Leaching, (v.) Metallization. The loss or removal of a basis metal or coating during a soldering operation.

Lead. A length of insulated or uninsulated metallic conductor that is used for electrical interconnections.

Lead Frame. The metallic portion of a component package that is used to interconnect with semiconductor die by wire bonding and to provide output terminal leads.

Leaded Chip Carrier. A chip carrier whose external connections consist of leads that are around and down the side of the package. (See also "Leadless Chip Carrier.")

Leaded Surface-Mount Component. A surface-mount component for which external connections consist of leads that are around and down the side of the package.

Leadless Chip Carrier. A chip carrier whose external connections consist of metallized terminations that are integral part of the component body. (See also "Leaded Chip Carrier.")

Leakage current. The undesired flow of electrical current over or through an insulator.

LED. Light emitting diode.

Lithography. A process in which a masked patter in projected onto a photoresist coating that covers a substrate.

Local Fiducial. A fiducial mark (or marks) used to locate the position of a land pattern for an individual component on a printed board.

LSI. Large-scale integration (integrated circuit).

Mask. See "Resist."

MCM-C. Multichip module ceramic.

MCM-L. Multichip module laminate.

MCM. Multichip module. A collection of more than one bare die or micropackage on a common substrate. Typically enclosed.

MCP. Multichip package. A small enclosed module with an external form factor that matches a single chip package and typically contains two to five bare dice.

MES. Manufacturing execution systems. A term defined by Advanced manufacturing Research (AMR, Cambridge, MA) to describe a system that centers on the product itself as it moves through the plant and out to the customer rather than focusing on measurements of material usage or process control.

Metallization (n.). A deposited or plated thin metallic film that is used for its protective and/or electrical properties.

Micron or Micrometer (μm). A metric unit of linear measure that equal 1/1,000,000 m, 0.0010 mm, 0.0004 in, or approximately 10,152 Å. The diameter of a human hair is approximately 75 microns.

Microstrip. A transmission-line configuration that consists of a conductor that is positioned over, and parallel to, a ground plane with a dielectric between them.

mil. A unit of length equal to 0.001 in.

mHz. Millihertz.

MHz. Megahertz.

Minimum Electrical Spacing. The minimum allowable distance between adjacent conductors, at a given voltage and altitude, that is sufficient to prevent dielectric breakdown, corona, or both, from occurring between the conductors.

MIS. Management information system. Computerized network used in effectively structuring critical information in a form usable for identification of inefficiencies.

MLB. Multilayer board.

Motherboard. The principal board that has connectors for attaching devices to the bus. Typically, the motherboard contains the CPU, memory, and basic controllers for the system. On PCs, the motherboard is often called the system board or mainboard.

Montreal Protocol. An agreement by industrialized nations, at a meeting held in Montreal, Canada, to eliminate chlorofluorocarbons from all processes by 1995.

Multichip Integrated Circuit. See "Multichip Module."

Definitions A-9

Multichip Module (MCM). A microcircuit module consisting primarily of closely spaced, unpackaged semiconductor dice or chip-scale packages that have a silicon area density of usually more than of the module area.

Multilayer Printed Board. The general term for a printed board that consists of rigid or flexible insulation materials and three or more alternate printed wiring and/or printed circuit layers that have been bonded together and electrically interconnected.

Negative Etchback. Etchback in which the inner conductor layer material is recessed relative to the surrounding base material.

Negative (n.). An artwork, artwork master, or production master in which the pattern being fabricated is transparent to light and the other areas are opaque.

NEMI. National Electronics Manufacturing Initiative.

Net. An entire string of electrical connections from the first source point to the last target point, including lands and vias.

Net List. A list of alphanumeric representations, each of which is used to describe a group of two or more points that are electrically common.

NIST. National Institute of Standards and Technology.

OEM. Original equipment manufacturer.

Oxidation. Addition of oxygen to a metal.

Package. A container for die (often plastic or ceramic) that provides protection and connection to the next level of integration.

Pad. See "Land."

Partially Clinched Lead. A component lead that is inserted through a hole in a printed board and is then formed to retain the component in place, and but not necessarily to make metal-to-metal contact with a land prior to soldering. (See also "Clinched Lead.")

Passivation. The formation of an insulating layer over the semiconductor surface that acts as a barrier to further oxidation or corrosion.

Paste Flux. A flux formulated in the form of a paste to facilitate its application. (See also "Solder Paste" and "Solder-Paste Flux.")

PBGA. Plastic ball grid array.

PGA. Pin grid array; pad grid array. Pin grid arrays are square chips in which the pins are arranged in concentric squares.

Photoresist. A material that is sensitive to portions of the light spectrum and that, when properly exposed, can mask portions of a base metal with a high degree of integrity.

Pitch. The distance, from center to center, of the leads on a component.

Plasma. Ionized gas used to remove resist, to etch, or to deposit various layers onto a wafer.

Porosity (Solder). A solder coating with an uneven surface and a spongy appearance that may contain a concentration of small pin-holes and pits.

ppm. Parts per million.

Preheating (v.). The raising of the temperature of a material(s) above the ambient temperature to reduce the thermal shock and to influence the dwell time during subsequent elevated-temperature processing.

Prepreg. A sheet of material that has been impregnated with a resin cured to an intermediate stage, i.e., B-staged resin.

Primary Side. The side of a packaging and interconnecting structure that is so defined on the master drawing. (It is usually the side that contains the most complex or the most number of components.)

Printed Board (PB). The general term for completely processed printed circuit and printed wiring configurations. (This includes single-sided, double-sided, and multilayer boards with rigid, flexible, and rigid-flex base materials.)

Printed Board Assembly. The generic term for an assembly that uses a printed board for component mounting and interconnecting purposes.

Printed Circuit. A conductive pattern composed of printed components, printed wiring, discrete wiring, or a combination thereof, which is formed in a predetermined arrangement on a common base. (This is also a generic term used to describe a printed board produced by any of a number of techniques.)

Printed Circuit Board. Printed board that provides both point-to-point connections and printed components in a predetermined arrangement on a common base. (See also "Printed Wiring Board.")

Printed Circuit Board Assembly. An assembly that uses a printed circuit board for component mounting and interconnecting purposes.

Printed Wiring Board. A printed board that provides point-to-point connections but not printed components in a predetermined arrangement on a common base. (See also "Printed Circuit Board.")

Propagation Delay. The time required for an electronic signal to travel along a transmission line, or the time required for a logic device to perform its function and to present a signal at its output.

QFP. Quad flat package.

Reflow Soldering. The joining of surfaces that have been tinned and/or have solder between them, placing them together, heating them until the solder flows, and allowing the surface and the solder to cool in the joined position.

Reflow Spike. The portion of the reflow soldering process during which the temperature of the solder is raised to a value that is sufficient to cause the solder to melt.

Registration. The degree of conformity of the position of a pattern (or portion thereof), a hole, or other feature to its intended position on a product.

Resin. A natural or synthetic resinous material. (See also "Rosin.")

Resin Smear. Base material resin that covers the exposed edge of conductive material in the wall of a drilled hole. (This resin transfer is usually caused by the drilling operation.)

RF. Radio frequency.

RH. Relative humidity.

Rosin. A hard, natural resin consisting of abietic and primaric acids and their isomers, some fatty acids, and terpene hydrocarbons, which is extracted from pine trees and subsequently refined.

Saponifer. An aqueous organic- or inorganic-base solution with additives that promote the removal of rosin and/or water-soluble flux.

SBB. Solder ball bumping or solder bump bonding, also stud bump bonding.

Secondary Side. That side of a packaging and interconnecting structure that is opposite the primary side. (It is the same as the "solder side" on through-hole mounting technology.)

SEM. Scanning electron microscope.

SEMATECH. SEmiconductor Manufacutirng TECHnology consortium.

SEMI. Semiconductor Equipment and Materials International.

Semi-additive Process. An additive process wherein the entire thickness of electrically isolated conductors is obtained by the combined use of electroless metal deposition and electroplating, etching, or both.

Semiconductor. A solid material, such as silicon, that has a resistivity midway between that of a conductor and of a resistor.

Si. Silicon.

SIA. Semiconductor Industry Association.

Signal Plane. A conductor layer that carries electrical signals. (See also "Ground Plane" and "Voltage Plane.")

Single-Inline Package (SIP). A component package with one straight row of pins or wire leads.

SIPs or Single Inline Packages. Chips that have just one row of legs in a straight line, like a comb.

Skin Effect. The increase in resistance of a conductor at microwave frequencies that is caused by the tendency of electric current to concentrate at the conductor's surface.

SMD. Surface mount device.

Smear Removal. See "Desmear."

Socket Contact. A female connector contact.

Solder. A metal alloy used to bond other metals together. Tin and lead are used in "soft" solders, which melt rather easily. The more tin, the harder the solder, and the lower the temperature required to reflow. Copper and zinc are used in "hard" soldering, or brazing, which requires considerably more heat to melt the alloy.

Solder Ball. A small sphere of solder adhering to a laminate, resist, or conductor surface. (This generally occurs after wave solder or reflow soldering.)

Solder Bridging. The unwanted formation of a conductive path of solder between conductors.

Solder Bump. A round ball of solder used to make interconnections between a flip-chip component and a base material during controlled-collapse soldering.

Solder Cream. See "Solder Paste."

Solder Mask. An insulating pattern applied to a printed circuit board that exposes only the areas to be soldered.

Solder Paste. Finely divided particles of solder, with additives to promote wetting and to control viscosity, tackiness, slumping, drying rate, etc. that are suspended in a cream flux.

Solder Resist. A heat-resisting coating material applied to selected areas to prevent the deposition of solder upon those areas during subsequent soldering.

Solder Side. The secondary side of a single-sided assembly.

Solder Terminal. An electrical/mechanical connection device that is used to terminate a discrete wire or wires by soldering.

Solder Wicking. The capillary movement of solder between metals surfaces, such as strands of wire.

Solderability. The ability of a metal to be wetted by molten solder.

Soldering. The joining of metallic surfaces with solder and without the melting of the base material.

Soldering Iron. Common name for a tool used to heat components and to reflow solder.

Solid state. Technology using solid semiconductors in place of vacuum tubes.

SPA. Silicon-package architecture.

SPC. Statistical process control.

Specification Limits. The requirements for judging acceptability of a particular characteristic.

Statistical Control. The condition of describing a process from which all special causes of variation have been eliminated and, as a result, only common causes remain.

Stripline. A transmission-line configuration that consists of a conductor that is positioned equidistant between, and parallel to, ground planes with a dielectric among them.

Substrate. The base material that is the support structure of an IC or PCB. A base with interconnecting electrical conductors acting as a support for the attachment of electronic components. Substrates can be made from a variety of materials, from polymers to ceramic, and exhibit certain specific thermal conductivity and strength characteristics.

Subtractive Process. The fabricating of a conductive pattern by the selective removal of unwanted portions of a conductive foil.

Surface Insulation Resistance (SIR). The electrical resistance of an insulating material between a pair of contacts, conductors, or grounding devices in various combinations, determined under specified environmental and electrical conditions.

Surface Mount Device (SMD). See "Surface Mount Component (SMC)."

Surface-Mount Component (SMC). A leaded or leadless device (part) that is capable of being attached to a printed board by surface mounting.

TAB. See "Tape Automated Bonding."

TAP. Test, Assembly, and Packaging, or BEOL (the back-end of semiconductor manufacturing).

Tape and Reel. A continuous strip of tape cavities that hold components. It can be wound around a reel to be presented to a automated component placement machine.

Tape Automated Bonding. A fine-pitch technology that provides interconnections between die and base materials with conductors that are on a carrier tape.

TBGA. Tape ball grid array.

TCP. Tape carrier package.

T_g. Glass transition temperature.

Thermal Conductivity. The property of a material that describes the rate at which heat will be conducted through a unit area of the material for a given driving force.

Thermocompression Bonding. The joining together of two materials without an intermediate material by the application of pressure and heat in the absence of electrical current.

Thixotrophy. A property of a substance, e.g., an adhesive system, that allows it get thinner upon agitation and thicker upon subsequent rest.

Through-Hole Mounting. The electrical connection of components to a conductive pattern by the use of component holes.

Throughput. The number of wafers processed on a given production machine during a measured period of time.

Tombstoned Component. A defect condition whereby a leadless device has only one of its metallized terminations soldered to a land and has the other metallized termination elevated above and not soldered to its land.

Tooling Hole. A tooling feature in the form of a hole in a printed board or fabrication panel.

Transmission Line. A signal-carrying circuit with controlled conductor and dielectric material physical relationship such that its electrical characteristics are suitable for the transmission of high-frequency or narrow-pulse electrical signals. (See also "Balanced Transmission Line," "Microstrip" "Stripline," and "Unbalanced Transmission Line.")

TSOP. Thin small-outline package.

UBM. Under bump metallurgy, also known as pad limiting metallurgy (PLM).

Ultrasonic Bonding. A termination process that uses ultrasonic-frequency vibration energy and pressure to make the joint.

Unbalanced Transmission Line. A transmission line that has elements (such as inductance, capacitance, resistance, and conductance) that are not equally distributed between its conductors.

Underfill. Encapsulant material typically deposited between a flip chip device and substrate used to reduce a mismatch in CTE between the device and the substrate.

Vapor-Phase Soldering. A reflow soldering method that is based on the exposure of the parts to be soldered to hot vapors of a liquid that has a boiling point that is sufficiently high to melt the solder being used.

Via. A plated-through hole that is used as an interlayer connection, but in which there is no intention to insert a component lead or other reinforcing material. (See also "Blind Via" and "Buried Via.")

Voltage Plane. A conductive layer at other-than-ground potential that serves as a voltage source for the circuitry on a multilayer board.

VLSI. Very large scale integration.

Wafer. A thin slice of semiconductor crystal ingot that is used as a base material.

W. Watt.

Wave Soldering. A process wherein an assembled printed board is brought in contact with the surface of a continuously flowing and circulating mass of solder.

Waveguide. A tube used to transmit microwave-frequency electromagnetic wave energy.

Wedge Bond. A wire bond made with a wedge tool. (See also "Ball Bond.")

Wetting, Solder. The formation of a relatively uniform, smooth, unbroken, and adherent film of solder to a basis metal.

Wire Bonding. Microbonding between a die and base material, lead frame, etc.

Yield. A measure of manufacturing efficiency; the percentage of acceptable production obtained from a specific manufacturing process.

Z. Impedance.

μ. μm or micron. A metric unit of linear measure that equals 1/1,000,000 m, 0.0010 mm, 0.0004 in, or approximately 10,152 Å.

Index

Symbols
µBGA, see MicroBGA

Numerics
2-D X-ray test 9.15
3-D X-ray test 9.15
3-W rule 6.50–6.53

A
Absorption 8.19, 8.48
Abstraction 1.22
Accidental activation, controls 14.26
Acrylic adhesives 10.5
Activators, ISO standard 14.2
Activity, flux 2.28
Adhesion 7.2–7.5
Adhesives 2.12, 10.1
    acrylic 10.5
    anisotropic 2.23
    application 2.15
    application methods 10.8
    bond strength 10.2
    cohesive failure 10.3
    conductive 4.7
    contamination 13.8
    cure profile 10.17
    cure properties 10.2
    cure strength 10.15
    curing 10.13
    cyanoacrylates 10.5
    deposition 2.15
    dielectric strength 2.21
    differential scanning calorimetry 10.19
    dots 2.17, 10.10
    elastomeric 2.14, 10.4
    electrically conductive 2.23, 10.5
    epoxy 10.4
    failure 2.14, 10.3
    failure characteristics 2.14
    general classification 10.3
    green strength 10.2
    metal electrode leadless face 10.1
    one-part 2.14
    physical characteristics 2.13
    pin-transfer technique 2.16, 10.9
    postcure properties 10.3
    pot life 2.13, 10.2
    precure 10.2
    process 2.13
    process problems 2.19
    repair 2.22
    rework 2.14
    shelf life 10.2
    small outline IC 10.1
    small outline transistors 10.1
    stencil printing 2.17, 10.9
    syringe deposition 2.17
    thermal conduction 2.20
    thermal cure 10.13
    thermal resistance 2.21
    thermally conductive 10.7, 11.24
    thermoplastic 2.14, 10.4
    thermosetting 2.14, 10.4
    two-part 2.14
    ultraviolet cure 10.12
    voids 10.16, 10.20
    volume 2.17
    z-axis 10.7
    z-axis conductive 2.23
Aerospace applications 7.19
Airflow 11.22
Alpha particles 7.18
Alumina, see Aluminum oxide
Aluminum 7.3, 7.7, 7.18
Aluminum nitride 7.1, 7.9
Aluminum oxide 7.2, 7.9
Aluminum wedge bonding 7.11
Ambient temperature 11.18
American National Standards Institute 1.45, 5.29
American Society for Testing and Materials 1.45, 5.29
Ampere's law 6.7, 6.9–6.10
Analog design, circuit boards 5.24
Analysis of variance 1.29
Anisotropic adhesives 2.23
ANOVA, see Analysis of variance
Antiresonant frequency 6.70
Application-specific integrated circuits 7.20
Area array packaging 4.8

Arrhenius equation 11.1
Aspect ratio 6.22
Assembly types 2.5
ASTM, see American Society for Testing and Materials
Asymmetric stripline, see Dual stripline topology
ATE, see Automatic test equipment
ATPG, see Automatic test program generation
Attenuation 8.48, 8.56
    fiber optics 8.19
Automated optical inspection, see Inspection, automated optical
Automatic test equipment 9.24
Automatic test program generation 9.4, 9.24
Autorouter 5.8
Axiomatic theory of design 1.35

B

B vs. C method 1.30
Back-bonded chips 4.3, 4.5
Backdriving, defined 9.23, 12.31
Backward crosstalk 8.2
Baking, IC packages 2.34
Ball grid array 3.7–3.8, 5.21, 7.9–7.10
Bandwidth, fiber-optic 8.21
Bare board test 9.2
Bare die 3.14, 4.2, 7.11–7.13, 7.17–7.20, 7.22–7.23
Base materials, propagation delay 6.40
Baseline shift 6.82
Bed-of-nails fixtures
    defined 9.23
    design for 9.18
    requirements 9.20
    wireless 9.22
Beryllia, see Beryllium oxide
Beryllium oxide 7.1
BGA, see Ball grid array
$Bi_2Ru_2O_5$ 7.3
Bifurcated trace 6.45
Binder 7.2–7.5
Bismuth borosilicate glass 7.4
Board
    defined 2.2
    double-sided 6.79
    flex during test 9.19
    single-sided 6.79
Bode plot 6.71
Bond pads
    die 7.11
    die bond 7.3, 7.7
    solder 7.7, 7.11
    termination pads 7.4
    terminations 7.7
    wire 7.7
    wire bond 7.3

Bonding
    aluminum wedge 7.11
    die bonding 7.3
    epoxy 7.3
    eutectic 7.3, 7.12
    flip chip 7.10, 7.18, 7.23
    gang 7.12
    gold ball 7.11
    inner lead 7.12
    mass 7.12
    outer lead 7.12
    single-point 7.12
    tape automated 7.11, 7.17–7.18
    thermocompression 7.12
    ultrasonic wedge 7.11
    wire 7.3, 7.9, 7.11, 7.17–7.18, 7.22–7.23
Boundary register 9.6
Boundary scan 9.4
Boundary scan descriptive language 9.14–9.15
Boundary-register cell 9.6
Boundary-scan register cells 9.12
BQFP, see Bumpered quad flat pack
Brazing 7.9
Bridge, defined 13.2
BSDL, see Boundary scan descriptive language
B-stage 5.3
Bulk 6.61
Bulk capacitor selection 6.83
Bump 4.5
Bumpered quad flat pack 2.8
Bumping 4.5
Bumping die 4.4
Buried capacitance 6.74
Buried passive 7.10–7.11
Burn-in 7.11, 7.19
Bypass capacitor 6.26
Bypassing 6.61

C

C4, see Controlled collapse chip connect
Cable, fiber-optic 8.29
Canadian Standards Association 1.45, 16.38
Capacitance
    calculation 6.73, 6.75
    distributed 6.43
    intrinsic 6.43
    power and ground plane 6.71, 6.73
Capacitive input values 6.42
Capacitive loading 6.42–6.43
    trace impedance 6.42
Capacitor equations 6.81
Capacitors
    behavioral characteristics 6.5
    bulk, selection 6.83
    bypass 6.26
    calculating values 6.80

capacitive effects 6.80
ceramic 7.8
charge equation 6.81
decoupling 6.76, 6.86, 7.21, 7.23
decoupling, placement 6.80
decoupling, selection 6.80
dielectric material 6.75, 6.78
discharge equation 6.81
energy storage 6.65
flat-pack 6.77
impedance 6.64–6.65
inductance 6.67
lead-length inductance 6.67
loop area 6.79
multilayer ceramic 7.8
parallel 6.68–6.69, 6.79
physical characteristic impedance 6.63
placement 6.78
resonance 6.66
retrofit 6.77–6.78
self-resonance 6.66
self-resonant frequency 6.67
self-resonant frequency range 6.68
surface-mount 6.67–6.68, 6.86
through-hole 6.67–6.68, 6.77
tolerance rating 6.67
trimmed 7.8
voltage rating 6.83
CBGA, see Ceramic ball grid array
CCGA, see Ceramic column grid array
CE, see Concurrent engineering
Ceramic 7.17
  aluminum nitride 7.1, 7.9
  aluminum oxide 7.1–7.2, 7.9
  beryllium oxide 7.1
  cofire 7.10
  glass 7.9
  low-temperature cofired 7.9
  multilayer 7.10
  substrate 7.1, 7.10
Ceramic ball grid array 3.8
Ceramic column grid array 3.8
Ceramic dual in-line package 3.5, 3.15
Ceramic packages (see also Packages) 7.9
CerDIP, see Ceramic dual in-line package
Characteristic impedance 6.43, 6.45, 6.55, 8.57, 8.66
Chassis grounds 6.21–6.22
Chemical vapor deposition 7.7
Chip carrier 3.9
Chip on board 3.1, 3.14, 4.1, 4.3
Chip on flex 3.14
Chip resistors 7.8
Chip-scale package 3.13–3.14, 4.1, 4.7, 4.25, 7.9–7.10, 7.23
Chromium 7.7

Circuit boards
  additive 5.3
  analog design 5.24
  cleaning 2.40
  coefficient of thermal expansion 5.14
  design for manufacturability 5.18–5.19
  design for testability 5.18–5.19
  digital design 5.25
  electromagnetic compatibility 5.6
  electromagnetic interference 5.6
  introduction 5.1
  panellizing 5.19
  partitioning 5.24
  propagation constants 5.9
  propagation delay 5.9
  RF design 5.26
  subtractive 5.3
Cladding 8.48
Classes of electronic products 2.2
Cleaning
  circuit board 2.40
  sockets 8.14
Closed-loop circuit 6.9
Coaxial cable 8.49
  connectors 8.54
  designations 8.53
  long interconnects 8.55
  power handling 8.51
  short interconnects 8.53
  theory 8.50
COB, see Chip on board
Coding, controls 14.23
Coefficient of thermal expansion 2.41, 5.4, 7.1–7.2, 7.4, 7.9, 7.18
  adhesives 2.22
  circuit board 5.14
COF, see Chip on flex
Cofired 7.1
Cofiring 7.3–7.4, 7.8–7.11, 7.14
Cohesive failure 2.14
Cold solder joint, defined 13.2
Commercial off-the-shelf 9.24, 12.10
Common-mode current 6.11–6.12
Compatibility, ergonomic 14.4
Complementary metal-oxide semiconductor 7.21
Components
  buried passive 7.10
  faults 9.1
  integrated passive 7.10
  mislocation 13.8
  packaging 6.85
  search 1.25–1.26
  sockets 8.11
  surface mount 7.9
  through-hole 7.9

Computational fluid dynamics 2.12, 11.21
Computer input devices 14.21
Computer-aided design 7.20
Concurrent engineering 1.1–1.2
    and SMT 2.2
Conductive adhesive 4.7
Conductor ink 7.3–7.4
Connectors 8.5
    coaxial 8.54
    contact normal force 8.7
    contact types 8.9
    contact wipe length 8.7
    electrical considerations 8.5
    environmental considerations 8.9
    fiber-optic 8.16, 8.46
    mating force 8.7
    mechanical considerations 8.6
    normal force 8.7
    zero-insertion-force 8.7
Constriction resistance 11.19
Construction
    circuit separation 16.27
    enclosures 16.26
    grounding 16.27
Contacts
    retention force 8.14
    types 8.9, 8.13
Containment 6.22
Continuous quality improvement 1.3
Control chart, defined 13.2
Control/display ratio 14.38
Controlled collapse chip connection 4.6, 4.8
Controlling dimension 2.6
Controls
    accidental activation 14.26
    bar knob 14.8
    dead-man 14.7
    descriptions 14.17
    detent thumbwheel 14.8
    grouping 14.7
    keylock 14.7
    legend switch 14.10
    pushbutton 14.9
    push-pull switch 14.10
    rocker switch 14.12
    selection 14.3
    toggle switch 14.12
Convection ovens 2.37
Coplanarity 2.32–2.34, 2.41
Copper 7.3, 7.7, 7.9, 7.15, 7.17
Copper wire, dc resistance 6.17
Corrective action program, ISO9000 1.41
CQI, see Continuous quality improvement
Crosstalk 5.12, 6.16, 6.51, 8.2
    adhesives 2.24

CSA, see Canadian Standards Association
CSP, see Chip scale package
CTE, see Coefficient of thermal expansion
Customer attribute 1.5

D

DAQ, see Data acquisition
Data acquisition 9.17
Data rate limitations 8.20
dc resistance, copper wire 6.17
dc sputtering 7.7
DCA, see Direct chip attach
Dead-bug prototype 5.18
Dead-man control 14.7
Decomposition 1.8, 1.22
Decoupling 6.61
    capacitor selection 6.80
    capacitors 6.76
    parallel 6.70
Decoupling capacitors 6.86, 7.21, 7.23
    placement 6.80
Defects 1.23
    design 9.1
    manufacturing 9.1
    solder joint 2.29
Department of Defense 5.28
Department of Defense Standardization Documents
    1.45
Deposition
    chemical vapor deposition 7.7
    electroless plating 7.7
    electroplating 7.7
    evaporation 7.6–7.7
    laser assisted chemical vapor deposition 7.7
    metalorganic decomposition 7.7
    pulsed laser deposition (PLD) 7.7
    reactive sputtering 7.7
    RF sputtering 7.7
    sol-gel 7.7
    sputtering 7.6
Design 7.20, 7.22
    defects 9.1
Design for compliance 16.18
    batteries 16.22
    fuses and fuseholders 16.21
    misc. components 16.23
    motors and fans 16.24
    power supplies 16.20
    power supply and cord set 16.20
    switches 16.22–16.23
    thermoplastic materials 16.24
    transformers 16.20
    wiring 16.25
Design for manufacturability 1.35, 2.2, 5.18
Design for testability 2.2, 5.18, 9.1, 9.23
    defined 12.32

Index

Design of experiments 1.4, 1.24
Design rule check 5.8, 5.17
DFM, see Design for manufacturability
DFT, see Design for testability
Die
    bond pads 7.3, 7.7, 7.11
    bonding 7.3
    shrink 6.85
Dielectric constant 5.10, 6.39–6.40
Dielectric ink 7.2, 7.4
Dielectric material 6.75
    permittivity 6.33
Dielectric strength, adhesives 2.21
Differential microstrip topology 6.38
Differential pair routing 6.54
Differential scanning calorimetry 10.22, 10.24
    with adhesives 10.19
Differential stripline topology 6.38
Differential-mode current 6.11–6.12
Diffusion patterning 7.4
Diode network 6.59
DIP, see Dual in-line package
Direct chip attach 4.1, 4.7
Discontinuities, image plane 6.50
Dispersion 8.48
Displays 14.2, 14.27
    characteristics 14.32
    control/display ratio 14.38
    electronic 14.33
    location 14.36
    rules 14.28
    visual 14.29
Divergence theorem 6.7
Doctor blade 7.1–7.2
Document control program, ISO9000 1.41
Documentation 1.32
    requirements 16.29
DODISS, see Department of Defense Index of Specifications and Standards
DOE, see Design of experiments
Double-sided PCB 6.79
Drawbridging, solder 13.7
DRC, see Design rule check
Dual in-line memory module 3.16
Dual in-line package 3.2, 3.15
Dual stripline topology 6.36
    characteristic impedance 6.36
    propagation delay 6.37
Dynamic logic 7.18
Dynamic random access memory 7.18
Dynamics 1.22

E

ECO, see Engineering change order
EDA, see Electronic design automation
Effective relative permittivity 6.39
EIA, see Electronic Industries Alliance
EIAJ, see Electronic Industries Alliance of Japan
Elastomeric adhesives 2.14
Electric field 6.8
Electrically conductive adhesives 2.23, 10.5
Electrically long traces 6.8, 6.44, 6.54
    calculating length 6.47
Electroless plating 7.7
Electromagnetic compatibility 5.6, 5.15, 6.1
Electromagnetic field 6.39
Electromagnetic interference 5.6, 5.15
Electronic assemblies, classes 13.1
Electronic CAD (ECAD) 5.1, 5.16
Electronic design automation 2.9
Electronic Industries Alliance 1.46, 5.29
Electronic products, classes 2.2
Electroplated solder bump 4.13
Electroplating 7.7
Electrostatic shield 6.7
Embedded microstrip 6.34
    characteristic impedance 6.34
    propagation delay 6.35
EMC, see Electromagnetic compatibility
Emergency signals 14.30
Emergent properties 1.22
EMI suppression
    fundamental concepts 6.29
    fundamental principles 6.28
EMI, see Electromagnetic interference
Emulsion 7.5
Encapsulant 7.12
Encapsulated 7.12
Enclosures
    environmental issues 14.42
    maintenance 14.43
    safety 14.44
End termination 6.56
Engineering change order 1.21
Environmental issues, enclosures 14.42
Environmental stress screening 9.24, 12.16
Epoxy
    adhesives 10.4
    bonding 7.3
Equivalent series inductance 6.64, 6.75
Equivalent series resistance 6.64, 6.75
Ergonomics 14.1
Escape routing 7.9
ESL, see Equivalent series inductance
ESR, see Equivalent series resistance
Etching 5.6, 7.5, 7.8
European Committee For Electrotechnical Standardization 16.38
Eutectic bond 7.11
Eutectic bonding 7.3, 7.12

Eutectic solder 2.29
Evaporation 7.6–7.7
External inductance 6.17

F

Face-bonded chips 4.5
Failure
    adhesive 2.14
    analysis 15.25
    brittle overload 15.11
    cohesive 2.14
    creep deformation 15.16
    defined 12.32
    ductile overload 15.11
    fatigue crack growth 15.15
    micromechanisms in materials 15.9
    temperature rise effects 11.1
Failure characteristics of adhesives 2.14
Fanout 7.9, 7.12
Faraday cage 6.7
Faraday's law of induction 6.7
Fault
    coverage 9.23, 12.14
    coverage, defined 12.32
    criteria 13.2
Faults 7.22, 9.23
    component 9.1, 12.5
    distribution 12.8
    functional 9.1
    manufacturing 12.7
    performance 12.7
    software 12.8
    sources 12.4
FCA, see Flip chip attachment
FCC, see Federal Communications Commission
Federal Communications Commission 1.46
Ferrite beads 6.5
Ferrite material, plot of behavioral characteristics 6.6
Ferrites 8.65
Fiber
    connectors 8.46
    manufacturing 8.27
    modulation 8.26
    multimode 8.22
    nonlinearities 8.24
    polarization maintaining 8.24
    radiation-hardened 8.24
    scattering 8.25
    single-mode 8.22
    special use 8.24
    termination 8.44
    tests 8.47
    wavelength 8.18
Fiber-optic interconnects 8.16
Fiducial marks 7.6

Fiducial, defined 13.2
Fiducials 5.20–5.21
Filters, microwave 8.64
First-pass yield 9.23
Flat pack capacitor 6.77
Flexible laminates 7.13
Flip chip 2.25, 4.1, 4.4, 4.7, 7.10, 7.12, 7.18, 7.21, 7.23
    attachment 4.7
    electroless nickel under-bump metallurgy 4.10, 4.19
    electroplated solder bump 4.10, 4.12
    evaporated solder bump 4.10
    printed solder bump 4.10
    printed solder paste bump 4.16
    solder ball bumping 4.10, 4.18
    stud bump bonding 4.10, 4.18
Floor life 2.34
Flux 2.27
    activity 2.28
    application 2.28
    cancellation 6.10
    cancellation or minimization 6.14
    flip chip assembly 4.21
    minimization 6.10
    no-clean 2.27
    RMA 2.27
    water-soluble 2.27
Flying probe testers 9.22
Foam fluxing 2.28
Footprint 2.9–2.10
Forced convection cooling 11.11, 11.13
Forward crosstalk 8.2
FR-4 5.21
Frequency domain 6.10
Full factorial analysis 1.29
Functional faults 9.1
Functional material 7.2, 7.4
Functional test 9.15

G

Gallium arsenide 7.21
Gang bonding 7.12
Gauss' law 6.7
Gaussian structure 6.7
GenCAM 5.17
Gerber 5.17
Glass ceramic 7.9
Glass sealing 7.9
Glass transition temperature 2.11, 2.42, 5.2
    substrate 2.12
Gold 7.3, 7.7, 7.15
Gold ball bonding 7.11
Gold-palladium 7.3
Gold-platinum 7.3
Green sheet 7.9

Green tape 7.2, 7.8, 7.13
Ground
  bounce 6.69
  chassis 6.21–6.22
  loops 6.21
  loops, control of 6.22, 6.25
  loops, RF 6.22
  shift 6.13
  stitch 6.21–6.22, 6.26
  trace 6.51
  vias 6.50–6.51
Ground plane 5.12–5.13, 5.18, 5.23, 6.14, 6.20
  discontinuities 6.25
Grounding 16.27
  hybrid 6.19
  methodology 6.18
  multipoint 6.19–6.20
  single-point 6.19
Ground-noise voltage 6.13
Grounds 5.23
Grouping of controls 14.7
Guarding 9.23
Gull-wing leads 2.42, 7.9

H

HASL, see Hot-air solder leveling
Hazards
  abnormal operation 16.5
  defined 16.1
  electric shock 16.2–16.3
  ergonomic 16.5–16.6
  explosion and implosion 16.5
  heat, fire, and tracking 16.3–16.4
  human factor 16.5
  mechanical 16.3–16.4
  moisture, liquids, and corrosion 16.3–16.4
  radiation 16.4
  toxicity 16.4
  types of 16.2
Heat
  ambient temperature 11.18
  capacity 11.3
  fundamentals 11.3
  removal 11.18
  removal techniques 11.17
  slugs 11.17
  specific 11.4
  spreaders 11.17
  spreading resistance 11.18
  temperature 11.3
  thermal conductivity 11.4
  thermal expansion 11.4
Heat sink 7.22, 11.18–11.19
  compounds 11.20
  material 2.11

Heat transfer
  ambient temperature influence 11.16
  conductive 11.8
  convective 11.10
  forced convective 11.21
  fundamentals 11.8
  radiant 11.14
  to substrate 11.24
Hermetic 7.9
Hermetic DIP 3.16
Hidden RF characteristic of passive components 6.2
Hidden schematic 6.3
Hot-air solder leveling 2.26, 5.22
House of quality 1.5
Hybrid 7.20, 7.23
  assemblies 7.1, 7.9
  circuits 7.1
  grounding 6.19

I

I leads 7.9
I/O Buffer Information Specification 5.27
IBIS, see I/O Buffer Information Specification
IC, thermal model 11.16
IEC, see International Electrotechnical Commission
IEEE standard 1149.1 9.5
IEEE standard 1149.1, see also Test access port 9.10
Image plane 6.13, 6.24, 6.26, 6.49, 6.76
  discontinuities 6.50
  violation 6.24
IMAPS, see International Microelectronics and Packaging Society
Impedance
  control 6.31
  copper plane 6.20
  equation 6.64
  matching 6.44
  power and ground plane 6.73
  trace 6.48
  transformers, microwave 8.60
In-circuit analysis 9.24
In-circuit test 9.15
Independence axiom 1.36
Indicators 14.2
Inductance
  capacitors 6.67
  conductor 6.18
  external 6.17
  lead-length 6.65, 6.75
  trace 6.8, 6.20, 6.48
Inductive reactance 6.8
Inductors, behavioral characteristics 6.5
Information axiom 1.36
Injection molding 7.9
Ink 7.13

Inner lead bonding 7.12
Input devices, computer 14.21
Insertion loss 8.48, 8.66
Inside vapor disposition 8.27
Inspection 13.1
    2-D X-ray 13.10
    3-D X-ray 13.11
    automated optical 13.10
    criteria 13.2
    criteria, solder joint 13.5
    flip chips and BGAs 13.9
    human 13.9
    laser 13.4, 13.10
    points 13.1
    post-placement 13.3
    post-reflow 2.39, 13.3
    solder paste volume 13.4
    visual 13.2, 13.8–13.9
    X-ray laminography 13.11
Institute for Interconnecting and Packaging Electronic Circuits 1.46, 2.2, 2.6–2.7, 5.27
Integrated passive 7.10–7.11
Interconnect devices 8.1
Interconnects
    board 8.10
    fiber-optic 8.16
    levels of 8.3
    single-point 8.4
    terminals 8.4
    wires 8.3
Interface, user 14.2
Internal quality auditing program, ISO9000 1.41
Internal reflection 8.49
International Electrotechnical Commission 1.46, 16.38
International Organization for Standardization 1.46
IPC, see Institute for Interconnecting and Packaging Electronic Circuits
IR ovens 2.37
IrO_2 7.3
ISHM, see International Microelectronics and Packaging Society
ISO 9000 1.37–1.38
ISO 9001 1.38
ISO 9001 process flowchart 1.42
ISO 9002 1.38
ISO 9003 1.38
ISO 9004 1.38
ISO, see International Standards Organization
Isolation test 9.2
Isomorphism 1.22

J

J leads 7.9
JEDEC, see Joint Electronic Devices Engineering Council
JLCC, see J-lead chip carrier
J-lead chip carrier 3.4
Joint Electron Device Engineering Council 1.46, 3.2
Joint Test Automation Group 9.4
Joystick 14.20
Junction temperature 2.11

K

Keyboards 14.15
KGD, see Known good die
Kirchhoff's law 6.10
Kirchhoff's voltage law 6.9
Known good die 3.1, 4.2, 7.11, 7.19, 7.23

L

LaB_6 7.3
Labels and warnings 14.39
Laminar flow 11.10
Laminate 7.15
Laminate multichip modules 7.13
Laminated packages 7.2
Lamination 7.8–7.9, 7.13
Land 2.42
Land grid array 3.10
Land-to-land clearance 2.10
Laser assisted chemical vapor deposition 7.7
Laser inspection, see Inspection, laser
Laser trimming 7.4
Laser welding 7.9
Layer jumping 6.50–6.51
LCCC, see Leadless ceramic chip carrier
Lead borosilicate glass 7.4
Lead coplanarity 2.32
Lead frame 2.11
Lead inductance 6.16
Leaded multichip module 3.16
Leadframe 7.9, 7.11–7.12
Lead-length inductance 6.65, 6.71, 6.75
Leadless ceramic chip carrier 2.11, 3.10
Leadless chip carrier 3.9
Leads, parallel 6.70
LGA, see Land grid array
Lift-off patterning 7.8
Loaded board test 9.2
Logic families 6.27
Loop area 6.15
    magnetic 6.22
Low-temperature cofired ceramic 7.9, 7.11

M

Macrobending 8.20, 8.48
Magnetic field 6.8
Magnetic lines of flux 6.8, 6.10
Magnetic loop area 6.22
Manganese oxide 7.7

Index

Manhattan effect 13.7
Manufacturing defect analyzer 9.1, 9.24, 12.32
Manufacturing defects 9.1
Marking 16.28–16.29, 16.31
Mask 7.4–7.5, 7.8
Mass bonding 7.12
Materials 7.7
Mating force, connectors 8.7
Maxwell's equations 6.6
    made simple 6.7
mBGA, see Mini-BGA
MCM, see Multichip module
MELF, see Adhesives, metal electrode leadless face
Memory cube 3.16
Mesh count 7.5
Mesh number 7.5
Metal packages 7.9
Metal quad flatpack 3.10
Metallization dissolution 2.25
Metallorganic decomposition 7.7
Metallurgical bond 7.11
Microbending 8.20, 8.48
MicroBGA 3.13
Micro-Q devices 6.78
Microstrip 5.11–5.12, 5.14, 5.23
    characteristic impedance 6.32
    coated 6.34
    embedded 6.34
    propagation delay 6.33
Microstrip topology
    dielectric constant 6.41
    differential routing 6.38
Microvias
    as test points 9.20
Microwave
    filters 8.64
    guides 8.57
    impedance transformers 8.60
    monolithic integrated circuit 7.21
    transmission lines 8.59
    tuning elements 8.62
Migration resistance 7.3
Military 7.3
Mini-BGA 3.10
Minibump 4.9
Mixed-signal devices 7.21, 7.23
Mixed-technology components 2.9
Modeling 1.9, 1.22
Module assemblies 3.16
Moisture absorption, substrate 2.12
Moisture sensitivity 2.34
    IC packages 2.32
Molybdenum 7.3, 7.9
MQFP, see Metal quad flatpack
MQuad 3.10–3.11

Multichip modules 3.1, 3.16, 6.86, 7.9, 7.11–7.12, 7.19–7.20, 7.22
    MCM-C 7.13–7.15, 7.17, 7.20–7.21, 7.23
    MCM-D 7.14–7.15, 7.17, 7.20–7.21, 7.23
    MCM-L 7.13–7.15, 7.17, 7.20–7.21, 7.23
    packaging 6.87
Multilayer ceramic packages 7.9
Multilayer substrate 7.10
Multipoint grounding 6.19–6.20
Multivari chart 1.25

N

National Electrical Code, and fiber cables 8.38
National Electrical Manufacturers' Association 14.43
National Fire Protection Association 1.46
National Institute of Standards and Technology 1.39
Nationally Recognized Testing Laboratory, required documentation 16.8–16.9, 16.29–16.31
Natural convection cooling 11.13
NEMA, see National Electrical Manufacturers Association
NEPCON 2.6
Netlist 5.8, 5.15–5.16
NFPA, see National Fire Protection Association
Nichrome 7.7
Nickel 7.3, 7.7
NIST, see National Institute of Standards and Technology
Nitrogen firing 7.3
Nonrecurring engineering cost 7.22
Nonregulated products 1.45
NRTL, see Nationally Recognized Testing Laboratory
Nucleate boiling 11.11

O

Ohm's law 6.7
Omega layer 6.57
OMPAC, see Overmolded pad-array carrier
One-part adhesives 2.14
Open Data Acquisition Association 12.33
Operator
    defined 16.6
    instructions 16.1, 16.3, 16.5, 16.7–16.9
Optical fiber, see Fiber
Optical time domain reflectometer 8.49
Optimization 1.10, 1.22
Organic solderability preservative 2.27, 5.22
Orthogonal arrays 1.24
Orthogonally routed traces 6.37
OSP, see Organic solderability preservative
Outer lead bonding 7.12
Outside vapor disposition 8.27
Ovens
    convection 2.37
    IR 2.37
    vapor phase 2.37

Over the wall 1.2
Overmolded pad-array carrier 3.7
Overshoot 6.44

P

Packages
    ball grid array 7.9
    ceramic 7.9
    chip-scale 7.9–7.10
    hermetic 7.9
    injection molded 7.9
    MCM 7.21
    metal 7.9
    multichip modules 7.12
    multilayer ceramic 7.9
    pin grid array 7.9–7.10
    plastic 7.9
    thermal enhanced 11.17
Packaging
    levels 11.2
    major functions 11.2
Packaging materials, physical and thermal parameters 11.11
Paired comparisons 1.25, 1.28
Palladium 7.3
Panellizing, circuit boards 5.19
Parallel capacitors 6.68–6.69
    usage 6.79
Parallel C–series RL resonance 6.63
Parallel decoupling 6.70
Parallel leads 6.70
Parallel resonance 6.63, 6.74
Parallel termination 6.56–6.57
Parallel wires 6.14
Parasitics, IC package 4.5, 4.8
Part placement 2.34
Partitioning 6.26–6.28
Passivated bond pads 4.5
Paste-in-hole technology 2.31
Patterning 7.5–7.6, 7.8
PBGA, see Plastic encapsulated ball grid array
PC, see Process control
PCB traces, behavioral characteristics 6.4
Peak surge current 6.83–6.84
Perfboard 5.18
PGA, see Pin grid array
Photodefinable 7.4
Photolithography 7.4, 7.8, 7.14–7.15
Photoresist 7.5, 7.8, 7.15, 7.17
Physical characteristics of wire 6.17
Pin grid array 3.5, 3.7, 3.15, 7.9–7.10, 7.13, 7.20
Pinholes 7.4
Pin-transfer, adhesives 2.16
Pitch, lead center to adjacent lead center 2.7

Placement
    accuracy 2.36
    equipment 2.35–2.36
    power and ground planes 6.76
Planarity 2.42
Plastic encapsulated ball grid array 3.11
Plastic J-lead chip carrier 3.3
Plastic leaded chip carrier 2.8, 3.2, 3.11
Plastic packages 7.9
Plastic quad flatpack 3.11
Plated-through holes 2.11
Plating
    electroless 5.22
    electroplate 5.22
Platinum 7.3
PLCC, see Plastic leaded chip carrier
PMD, see Polarization mode dispersion
Pointing accuracy, test probe 9.20
Polarization maintaining fiber 8.24
Polarization mode dispersion 8.22
Polyimide 7.14
Popcorning 2.33
Positrol 1.31
Post-placement inspection 13.3
Post-reflow inspection 2.39, 13.3
Pot life, adhesives 2.13
Power and ground plane capacitance 6.73
Power and ground planes 6.68–6.69
    capacitance 6.71, 6.75
    impedance 6.73
    placement 6.76
    self-resonant frequency 6.68
Power density 15.27
Power dissipation 2.11, 11.18
Power handling, coaxial cable 8.51
Precontrol 1.31
Prepreg 5.3, 5.5
Printed circuit boards
    mixed-technology 2.4
    thermal design 2.10
    types 2.3, 2.5
Printed wiring boards 5.2
    design 5.6
    interconnects 5.8
Process control 1.4, 1.31
Product life cycle 1.2, 1.4
Profile, adhesive cure 10.17
Project manager 1.19
Propagation
    constants, circuit board 5.9
    delay, circuit boards 5.9
    time 6.44
    velocity 6.41, 6.47
Propagation delay 6.39–6.40, 6.43, 6.45
    base materials 6.40

Prototype systems 2.40
Prototyping 5.17
PTH, see Plated-through holes
Pulsed laser deposition 7.7
PWB, see Printed wiring boards

Q

Q9000 1.38
QFD, see Quality function deployment
QFP, see Quad flat pack
QS13000 1.38
QS9000 1.38
Quad flatpack 3.2, 3.7, 3.11
Quality 1.22, 2.2
Quality function deployment 1.4–1.5
    delighters 14.1
    dissatisfiers 14.1
    satisfiers 14.1
QWERTY layout 14.15

R

RAB, see Registrar Accreditation Board
Radiation-hardened fiber 8.24
Rat's nest 5.16
Rayleigh scattering 8.19
RC network 6.58–6.59
Reactive sputtering 7.7
Realistic tolerances parallelogram plot 1.30
Redesign 1.2
Reflected wave 6.46
Reflection equation 6.46
Reflection, optical 8.49
Reflections 6.44, 6.46
Reflow
    convection 2.37
    IR 2.37
    profile 2.37–2.38
    soldering 2.37
    vapor phase 2.37
Reflow soldering, simultaneous 10.1
Refraction, optics 8.17
Refractive index, optical 8.49
Registrar Accreditation Board 1.39
Registration, ISO9000 1.38
Regulated products 1.45
Relative dielectric constant 6.39
Relative permittivity 6.39
Reliability 2.2
    basic concepts 15.4
    chip capacitors and ceramic baseplates 15.24
    electronic packages 15.3
    semiconductor devices 15.24
    wirebond 15.20
Remote control units 14.23
Repair 7.18, 7.22
Resist 2.9

Resistor ink 7.3–7.4
Resistors
    behavioral characteristics 6.4
    chip 7.8
Resonance 6.61
    capacitors 6.66
    parallel 6.63
    parallel C-series RL 6.63
    series 6.62
Resonant angular frequency 6.62
Retention force 8.14
Return on investment 1.18
Rework 7.12, 13.4
Rework, adhesives 2.14
RF current
    common-mode 6.11
    density distribution 6.15
    differential-mode 6.11
    loop 6.8
    return 6.14
RF ground loops 6.22, 6.26
RF return path 6.13, 6.24
RF spectral energy 6.27
RF sputtering 7.7
Right-hand rule 6.8
Ringback 6.46
Ringing 6.44, 6.55
Risk management 16.9
Robot control 14.23
ROI, see Return on investment
Routing 6.44
    layers 6.25, 6.49
RuO_2 7.3
Ruthenium 7.7

S

Safety 16.1
    defined 16.1
    design stage 16.1
    enclosures 14.44
    philosophy 16.1
SBB, see Solder bump bonding
Schottky diodes 6.60
Screen 7.5–7.6
Screen printing 7.5–7.6
Sealing 7.9
Sealing glass 7.3
Self-resonant frequency, power and ground planes
    6.68
Semiconductor die 4.1
Series resonance 6.62, 6.74
Series termination 6.55
Serpentine 7.3–7.4
Service personnel, defined 16.6
Shanin, Dorian 1.24
Sheet resistance 7.3, 7.7

Shrink quad flatpack 3.12
Shrink small outline package 3.12
Shrinkage 7.2, 7.9
Signal integrity 6.44, 8.1
Signal return loop control 6.21
Silicon 7.17–7.18
Silicon monoxide 7.7
Silver 7.3, 7.9
SIM, see Single in-line module
Single in-line memory module 3.17
Single in-line module 3.15
Single in-line package 3.15
Single stripline topology 6.35
    characteristic impedance 6.35
    propagation delay 6.36
Single-point bonding 7.12
Single-point grounding 6.19
Single-sided PCB 6.79
Sintering 7.2, 7.5, 7.9, 8.27
SIP, see Single in-line package
Skin effect 6.7, 6.16, 6.26
    losses 6.18
SLICC, see Slightly larger than IC carrier
Slightly larger than IC carrier 3.13
Small outline 3.12
    IC 3.3
    package 2.7, 3.12
    package, with J leads 3.12
    transistor 3.12
SMD, see Surface mount devices
SMI, see Surface Mount International
SMOBC, see Solder mask, on bare copper
SMT, see Surface mount technology
SnO_2 7.3
SO, see Small outline
Sockets 8.11–8.12
    area array 8.15
    cleaning 8.14
    DIP 8.16
    PGA 8.15
    PLCC 8.15
    selection 8.13
    zero-insertion-force 8.11, 8.13
Software
    code and unit test 12.21
    component integration test 12.24
    system integration test 12.24
    system verification test 12.24
    testing programs 12.20
    testing, vs. hardware 12.21
SOIC, see Small outline IC
SOJ, see Small outline package with J leads
Solder 7.3, 7.15, 7.18, 7.21
    balls 2.30, 2.39, 4.9, 13.7
    balls, defined 13.2

    balls, in reflow 2.39
    bond pads 7.7, 7.11
    bridge 2.39, 13.4, 13.6
    bridge, defined 13.2
    bumping 4.7
    bumps 4.13, 7.22
    drawbridging 13.7
    eutectic 2.29
    excess 13.6
    fillet 2.10
    icicles 13.7
    insufficients 13.4, 13.6
    joint defects 2.29
    joint fatigue 4.9
    joint inspection 2.39
    joint inspection criteria 13.5
    joint reliability 15.17
    joint-related problems 13.3
    joints 2.25
    leach resistance 7.3
    mask 5.3
    mask, on bare copper 2.42
    paste 2.29
    paste deposition volume 13.4
    resist 2.9
    sphere 2.30
    sphere, formation 2.30
    voids 13.8
Solder paste 2.30
    deposit 2.34
    printing 2.29
    viscosity 2.30
Solderability 2.25–2.26
Soldering, simultaneous reflow 10.1
Sol-gel 7.7
Source impedances 6.32
Source termination 6.55
Spectral width 8.20
Split planes 6.24
Spray fluxing 2.28
Spreading resistance 11.18
Sputtering 7.6–7.7
SQFP, see Shrink quad flatpack
Squeegee blade 7.6
$SrRuO_3$ 7.3
SSOP, Shrink small outline package
Standards organizations 1.45
Static pressure drop 11.22
Static random access memory 7.21
Stencil printing 7.5–7.6
    adhesives 2.17
    solder paste 2.29
Stencils 7.5–7.6
    solder paste 2.29
    stepped 2.29

Index

Stratified experiment 1.26
Stripline 5.11–5.12, 5.14, 5.23
Stripline topology
    dielectric constant 6.41
    differential routing 6.38
Substrate 2.2, 7.5, 7.7, 7.12
    ceramic 7.1, 7.10, 7.17
    cofire 7.10
    cofired 7.4, 7.8
    cofiring 7.1
    constraining core materials 5.4
    design 2.8
    low-temperture cofired ceramic 7.11
    multilayer 7.10
Surface mount devices 7.9
    capacitors 6.67–6.68, 6.80
    definition 2.6
Surface Mount International 2.6
Surface mount technology 1.23, 2.1
Surface Mount Technology Association 1.46, 2.6
Surface tension, cleaning 2.28
Surge suppression 7.3
Susceptibility 5.15
Swiss cheese syndrome 6.25
Synthesis 1.9, 1.22
Syringe deposition, adhesives 2.17
Systems
    engineer 1.20
    engineering 1.7
    engineering, defined 1.19
    hard 1.22
    soft 1.22

T

Taguchi 1.23–1.24
TaN 7.3
Tantalum 7.7
Tantalum nitride 7.7
Tantalum oxynitride 7.7
Tantalum pentoxide 7.7
TAP, see Test access port
Tape automated bonding 3.14, 7.11, 7.17
Tape ball grid array 3.12
Tape carrier package 3.14
Tape carrier ring 3.14
Tape casting, see Doctor blade
Targeting
    ballistic 2.36
    guidance 2.36
    tracer 2.36
TBGA, see Tape ball grid array
TCR, see Tape carrier ring
Teach pendants 14.23
Telecommunications 7.3
Temperature coefficient of capacitance 11.6
Temperature coefficient of resistance 11.6
Temperature sensitive parameter 2.11
Termination 7.3
    methods 6.54
    pads 7.4
    types 6.55
Terminations 6.44, 7.7
    optical fiber 8.44
    trace 6.52
    wraparound 7.9
Test 7.18–7.19, 7.22
    2-D X-ray 9.15
    3-D X-ray 9.15
    accelerated 15.8
    automatic test equipment 9.24
    automatic test program generation 9.4
    automation 12.9
    backdriving 9.23
    bare board 9.2
    bed-of-nails fixtures 9.23
    board 12.2
    board flex during 9.19–9.20
    burn-in 12.17
    cable and backplane 12.11
    data acquisition 9.17
    dedicated test systems 12.14
    design for 9.1
    dielectric strength 16.13
    EMC 16.12, 16.17
    equipment for safety test 16.15
    ESD 16.17
    fast transient 16.17
    fault coverage 9.23
    first-pass yield 9.23
    fixtures 9.18, 12.15
    flying probe 9.22
    functional 9.3, 9.15, 12.2, 12.13
    ground continuity 16.13
    guarding 9.23
    in-circuit 9.3, 9.15, 9.24, 12.2, 12.12
    IPC requirements 9.2
    isolation 9.2
    leakage currents 16.14
    loaded board 9.2
    manual 12.10
    manufacturing defect 12.11
    manufacturing defects analyzer 9.3, 9.15, 9.24
    mechanical 16.13
    PC-based 12.10
    performance 12.2
    philosophies 12.1
    power input 16.14
    probe layout 9.20–9.21
    probe pointing accuracy 9.20
    RFI 16.16
    safety 16.12

shorts 12.11
shorts/opens 9.3
site description 16.18
software 12.19
software component integration 12.24
software vs. hardware 12.21
strategies 12.4
stress screening 12.2
stress testing 12.2
substitution 12.13
surge 16.17
system integration 12.24
system verification 12.24
techniques 9.15
testing software programs 12.20
vector 9.4
vectorless 9.15
vision 9.15
wireless fixtures 9.22
x-y probe 9.22
Test access port 9.5, 9.7
hardware 9.11
modes 9.8
states 9.7
Test points 2.10
Test probe layout 9.20–9.21
Testability 9.2
T_g 2.11, 2.42
Thermal
adhesive process 11.24
conductivity 11.4
conductivity, substrate 2.12
demand 2.25
design, circuit boards 2.10
expansion 11.4
management 2.10, 11.1, 15.27
profile, SMT reflow 2.37
resistance 11.16
resistance, adhesive 2.21
Thermal coefficient of X-Y expansion, substrate 2.12
Thermal conduction
adhesives 2.20
Thermal expansion coefficient, see Coefficient of thermal expansion
Thermally conductive adhesives 10.7
Thermistors 11.7
Thermocompression 7.11–7.12
Thermoplastic adhesives 2.14
Thermosetting adhesives 2.14
Thermosonic 7.11
Thevenin
equivalent circuit 6.80
impedance 6.82
network 6.57
termination 6.58

Thick film 7.2, 7.8–7.9
ink 7.2
Thick film paste, see Thick film ink
Thick-film copper 7.3
Thin film 7.6–7.7
Thin small outline package 3.2, 3.12
Thin-film multichip modules 7.14
Through holes, defined 2.42
Through-hole
capacitors 6.68
components 6.25, 7.9
packages 3.15
technology 2.2
Time domain 6.10
Time domain reflectometer 6.46
Time to market 1.3
$TiSi_2$ 7.3
Titanium 7.7
Tolerance rating, capacitors 6.67
Top-hat 7.4
Total quality management 1.3, 1.5
TP-101 Guidelines 5.19
TQM, see Total quality management
Trace impedance 6.42, 6.48
capacitive loading 6.42
Trace inductance 6.8, 6.20
Trace separation 6.51
Trace width
base dimension 6.32
crest dimension 6.32
Traces
daisy-chained 6.53
electrically long 6.8, 6.54
orthogonally routed 6.37
Traces, defined 5.15
Trade studies 1.10
Trade study tree 1.11
Trade table 1.12
Transformers
behavioral characteristics 6.5
Transmission line 6.31, 6.39
characteristic impedance 6.31–6.32, 6.42
effects 6.31
electromagnetic field component 6.31
microwave 8.59
Trimming 7.4, 7.11
TSOP, see Thin small outline package
Tungsten 7.3, 7.9
Turbulent flow 11.10
TÜV Rheinland Japan Ltd. 16.38
Two-part adhesives 2.14
Types, printed circuit board 2.3

U

UBM evaporation 4.11

UL, see Underwriters Laboratories
Ultrasonic wedge bonding 7.11
Under-bump metallurgy 4.9
Underfill 2.25, 4.6, 4.21, 7.12
    adhesives for 2.24
    volume 4.6
Undershoot 6.44
Underwriters Laboratories 1.46, 5.29, 16.38
User interface 14.2

V

Value added 1.2
Vapor axial deposition 8.27
Vapor phase ovens 2.37
Variables search 1.28
VDIP, see Vertically mounted DIP module
Vector tests 9.4
Vectorless test 9.15
Vehicle 7.2, 7.4, 7.6
Velocity of propagation 6.39, 6.41, 6.47
Vertical inline package 3.15
Vertical mount package 3.12
Vertically mounted DIP module 3.17
Very small outline package 3.12
Vias 6.50, 6.79, 7.9, 7.14–7.15
    defined 5.15
    effects in power and ground planes 6.87
    fill 7.3–7.4, 7.9
    ground pin 6.51
VIL, see Vertical in-line package
Viscosity 7.4, 7.6
    solder paste 2.30
Vision test 9.15
Visual displays 14.29
Visual inspection station 13.9

Visual inspection, see Inspection, visual
Voice of the customer 1.4
Voids, adhesive 10.16, 10.20
Voltage standing wave ratio 8.58
    defined 8.66
VPAK, see Vertical mount package
VSOP, see Very small outline package

W

Wafer bumping 4.7
Warning labels 14.40
Warnings and labels 14.39
Wave fluxing 2.28
Wave velocity 6.40
Wettability 2.25
Wire
    ampacity 8.4
    behavioral characteristics 6.4
    bond 7.9, 7.18
    bond pads 7.3, 7.7
    bonding 4.3, 7.3, 7.11, 7.17, 7.22–7.23
    physical characteristics 6.17
Wraparound terminations 7.8–7.9

X

X-ray inspection, see Inspection, X-ray
x-y probe testers 9.22

Z

z-axis adhesives 10.7
z-axis conductive adhesives 2.23
z-axis expansion 2.11
Zero insertion force 8.7
ZIF, see Zero insertion force